A CITIZEN'S DISCLOSURE ON UFOS AND ETI

VOLUME THREE

MILITARY INTELLIGENCE INDUSTRIAL COMPLEX, USAPS AND COVERT BLACK PROJECTS

TERENCE M. TIBANDO

Copyright Page

In writing this book, I sought out the best possible evidence available on this subject whether that was from numerous UFO and ETI related books, networking with other UFO authors, researchers and first-hand witnesses to UFO sightings, from films and TV documentaries, from internet searches, or just from my personal sightings and contact experiences.

When material is quoted in this book full acknowledgement is given to the author or source of that material as indicated by the extensive bibliography, webliography and videography at the back of the book.

When photographic images are used in this book that are obtained from the internet, usually from Google Images, a full search was made to determine copyright information, the author's name, or address or email address or phone number or copyright mark in order to asked permission to use their photographs. In almost all cases where such images are posted to the internet, there was no satisfactory way to identify the owner of the image even through Google because they left no identification of themselves to be found. When an author's name does appear on an image, written permission was sought or it was not used at all; more often than not, there usually was no reply or response back from the owner.

This lack of due diligence to place a copyright mark or the owner's name is all it would take for that person to claim ownership of a picture, yet the lack of it creates major problems for many people, especially for other authors who lawfully seek their permission to use their photo images.

Because this book is one of six volumes in a series created as public educational material and is of a transitional nature, and I have quoted or referenced the websites from where I obtained the photo images and therefore, I am invoking the Fair Use Doctrine also known as Fair Usage Clause to publish these images in my book.

I will of course give full acknowledgement and credit to the author's and owners of such images in all my future book publications in recognition of their work if they come forward to be identified.

With this stated, this book and all future books in this series are copyrighted and cannot be reproduced in whole or part, in any form or by any means electronic or mechanical, including photocopying, recording, or by any information storage and retrieval system now known or hereafter invented, without the expressed permission of its author.

A CITIZEN'S DISCLOSURE ON UFOS AND ETI

VOLUME THREE

MILITARY INTELLIGENCE INDUSTRIAL COMPLEX, USAPS AND COVERT BLACK PROJECTS

A CITIZEN'S DISCLOSURE ON UFOS AND ETI

VOLUME THREE

MILITARY INTELLIGENCE INDUSTRIAL COMPLEX, USAPS AND COVERT BLACK PROJECTS

TERENCE M. TIBANDO

"Hggna"

A Cosmic Cousin
Publication

Other Publications by the Author Page

Although, this book is the author's first publication it forms a part of six smaller books or volumes that was originally written as one massive tome of UFO and ETI information entitled: **"A Citizen's Disclosure On UFOs And ETI Visiting The Earth"** which began in March 2009 and was completed in August 2016.

Other books/volumes by the author in this series:

1. Book One (Volume One): **"Global Evidence of the UFO and ETI Presence"**

2. Book Two (Volume Two): **"UFO Disclosure and Covert Programs of Deception"**

3. Book Three (Volume Three): **"Military Intelligence Industrial Complex, USAPs and Covert Black Projects"**

4. Book Four (Volume four): **"In Search of Extraterrestrial Intelligence"**

5. Book Five (Volume Five): **"Evidence of a Type Two ET Civilization in Our Solar System"**

6. Book Six (Volume Six): **"The Rosetta Stone of ETI Contact and Communications"**

Introduction

This book is the third volume of "A Citizen's Disclosure on UFOs and ETI" which originally formed part of a 3500 page encyclopedic tome. It provides the latest and best evidence on this subject matter that comes from my 65years of experience dating back to 1953 when I and my mother witnessed flying saucer type craft hover over a Canadian Air Force Base in St. Jeans, Quebec. This was followed by an ET visitation by three ghost-like beings that came into my bedroom a couple of days later.

Over my lifetime, there followed many more UFO sightings and ETI encounters and interactions many in group witness settings as a part of a CSETI field expedition to initiate human contact and communications with these diverse extraterrestrial beings visiting our planet.

This phenomenon is real, regardless of what you have read or have been told by the mainstream media or the science community. There is no transparency in their public disclosure but rather deception, lies and cover-ups designed to betray the public's trust in officialdom. Increasingly, the distrust toward the official position that dispenses their version of knowledge and truth has become suspect which has led to a public awaking to start investigating truth for themselves.

This third volume in a series of six volumes hopes to add to the public's resource for truth and honesty which will lead the reader to do further research and investigation on this subject. The unfettered self-investigation of truth is a god-given right of everyone to understand the world in which they live in.

As you read through this volume and the other volumes in this series, there will be much that is familiar to the senior UFO researcher and much that is new, especially with some alternative and perhaps, controversial perspectives that re-evaluate some long, widely-held traditional beliefs held within the UFO Community.

This book and the other volumes of "A Citizen's Disclosure on UFOs and ETI" in this series connects the dots and assembles the pieces of the UFO and ETI puzzle to form a much larger picture of this phenomenon. The implications and potentialities for humanity derived from understanding this interstellar phenomenon are vast which could positively revolutionize our future for generations to come. The task at hand is for the mainstream science community and our respective governments and its officials is to be honest and straight forward with the general public in what it discloses on this subject.

When writing a book of this complexity and length, it is the hope that errors in content are minimal or non-existent, although the reality is that no matter how careful the research and entry of evidence, inevitably errors will creep in because new evidence comes to light to prove an alternative reality. This author has endeavoured to minimize such errors as the UFO database is already corrupted and such errors seem to be perpetuated due to poor research or the deliberate ignoring or marginalizing of real facts. With this in mind, corrections have been made where necessary while still maintaining the integrity of the evidence as it was originally reported.

With such material gleaned from my predecessors in understanding the "Big Picture" of this phenomenon, full credit and acknowledgement is given to these Ufologists in the footnotes (author's rants), web links and bibliography of this book.

Please note the colour format when quoting from a source is used to distinguish the web links from video links and bibliography which is indicated in the Bibliography section of this book.

In essence, this book is an interactive book, perhaps the first of its kind, where web links and video links are used to enhance the research value for the student and the seasoned Ufologist and thus, it extends a 3500-page textbook into a 10,000-page tome!

The reader will be able to follow the web link references to see where I have collected the information for this book. Unfortunately, at the time of publication of this book, some of these web links may no longer be active, it is the nature of the internet and it ephemeral web links!

My apologies for any inconvenience this may cause, however, try "googling" a specific word or phrase; this may reveal the same or a similar website!

This book should be considered a reference manual only, leading to further research and investigation for the serious UFO researcher.

As the reader goes through this third volume, they will realize the extent to which there has been a suppression and cover -up of this phenomenon by many sectors of society, especially by the military, the intelligence and science communities, as well as the private industrial sector and by our elected government officials. The cover up of the existence UFOs and ETI is interlaced with deceptive and misleading research programs from the above mentioned communities of officialdom.

Back-engineering of the alien spacecraft by WWII Nazi Germany, and by postwar Britain, Canada, and especially the USA has been in full over-drive by those covert arms of the military to discover the science of the alien propulsion system. This reverse engineering also includes the understanding and manipulation of alien DNA and its various hybridization PLF programs. This also includes military engagements and shoot-down programs as well as tracking, monitoring and crash retrieval programs. All the while the news media has been coerced and threatened to no longer function as the public's "Fifth Estate" having been usurped by the military industrial intelligence community. Marginalization and ridicule hold the public's interest in this phenomenon is at bay while the M.I.I.C. leap generations ahead of mainstream science jealously guarding its secrets with threats, even death to those who get to close to the truth.

The depths and convolutions of going down the rabbit hole to discover the truth behind the UFO and ETI phenomenon include deep black projects, SAPs, USAPs, military super weapons, even a secret black space program and a false flag alien invasion agenda! The corrupt and evil machinations of the M.I.I.C. know no bounds and are being played out as a lob-side chess game against an unsuspecting public.

This book can help in changing the outcome of that 'chess game' in favour of the common people and the laying of the foundation of a good hopeful future! **Enjoy the ride in discovery!**

Terry Tibando - September 2015

CHAPTER 50
THE AGENCY OF UFO SECRECY - MAJIC 12(AKA. MAJESTIC 12, MJ12....... 490

CHAPTER 51
WHEN IT COMES TO NATIONAL SECURITY LIE, LIE. LIE,
DENY, DENY, DENY!..526

CHAPTER 52

DEEP BLACK PROJECTS, SAPS, USAPS, MILITARY SUPER WEAPONS.........560

CHAPTER 53

**ARE THE LUCRATIVE BLACK PROJECTS AND PROGRAMS THE REAL
CAUSE FOR THE COLLAPSE OF THE AMERICAN ECONOMY?.......................608**

CHAPTER 57
WHAT GOES ON DEEP DOWN IN A FEW D.U.M.B.S..712

CHAPTER 38
THE NEWS MEDIA: THE PEOPLE'S "WATCHDOG" OR
A PUPPET OF THE INTELLIGENCE COMMUNITY?

In this section, we are going to unequivocally identify to the best of our ability just who are the power players that control the information on the reality and truth of the **UFO** and **ETI** phenomenon. We will show the importance of keeping the lid of suppression and cover-up tightly on the potentially explosive barrel that is the knowledge of the UFO question. We will see how this UFO cover-up has lead to the corruption of those who are entrusted to protect us and look out for our best interests. We will discover the hidden agenda to control and manipulate the general public through a methodical campaign of ridicule, lies, disinformation, misinformation, denials and outright threats and even murder to keep us away from knowing the truth and reality of the existence of Extraterrestrial Intelligences that are currently visiting the planet.

In order to get to the place where we can point an accusatory finger at the perpetrators of the truth embargo on the subject of UFOs and ETI, we need to build a second case similar to the first section of this book in which the hardcore facts and evidence presented establish beyond a doubt, that the UFO/ETI phenomenon is real. The evidence for this is probably more elusive than a proverbial flying saucer but, not impossible to uncover or expose to the light of truth and scrutiny.

The evidence is multi-layered; it is transnational in its scope and global in its influence. It is comprised of a small wealthy corporate elite group, who members come from some of the most powerful families and corporations in the industrialized world. This **cabal** wields almost unimaginable power through control of natural resources, control of governments and corporate finances, influence and peddling of politicians and governments, all with the ability to back up its hidden agenda with the use and deployment of one's of the world's largest military forces, the U.S. Military, as well as strategic coercing of other governments and their militaries to do its bidding!

As almost absurd and incredible as this may seem, it is, in fact, a cold harsh reality that is slowly distilling down into the minds of a catatonic America while simultaneously placing a death grip on the hearts of her people, the effects of which are being felt by all people around the world.

A Bygone Age of Honest Unbiased Media Reporting and Journalism

People will tell you that in the good old days of the mid19[th] to the mid 20th centuries, newspapers were, for the most part, the voice and conscious expression of the common people in society. Newspapers were more than a source of daily information about local and national events that shaped the country's identity; it opened the many faces of multi-national consciousness to a greater concept of global awareness and to the reality of a shared commonwealth of human oneness with everyone else who inhabited the planet.

Newspapers expressed our hopes, our dreams, our goals and our achievements. It also informed us of our miseries, our crimes, our wars and our injustices against our fellow man. It was a mirror of clarity that we held up to ourselves to gauge our humanity and our growth toward a better, more advanced civilization.

The news press opens to us, stories of human interest, of strange faraway and exotic places, foreign countries in the midst of political intrigue and turmoil, events, and circumstances to mystify and awe us which have "imprinted" (pardon the pun) upon our collective consciousness, a world of increasing daily familiarity. There were always the many editorial comments of high-minded intelligence expressed in eloquence to both educate and to shake us from our everyday life of complacency and apathy. The mass media forms our image of the world and then tell us what to think about that image. Essentially everything we know -- or think we know -- about events outside our own neighborhood or circle of acquaintances comes to us via our daily newspaper, our weekly news magazine, our radio, or our television.

The news press was the benchmark for unadulterated truth and honesty in a society! The power of the printed word was a force to be reckoned with and many news magnets of the late 19th and early 20th centuries knew its power and the influence that it had over the masses. There is no greater power in the world today than that wielded by the manipulators of public opinion in America. No king or pope of old, no conquering general or philosopher ever exercised a power even remotely approaching that of the few dozen men who control America's mass news media and entertainment industry. In the 19th to the 21st centuries, information is king and its power and control is wielded like a mighty weapon!

Owners of the news press became powerful through the increased circulation of their newspapers and through the increase of advertising as a primary source of revenue. They were some of the most powerful and influential "giants of industry and capitalism and more importantly of thought". This last aspect was not lost later on the military and the intelligence communities in many countries after the First World War.

The many global news presses were the forerunners of tabloids and magazines of today, and radio, television and the internet became the digital and electrical versions of the news press. They were all part of the family known as the news media. Society had once upon a time viewed the news media as a reliable, dependable source of unbiased and honest journalistic reporting of the world news. It was the *"Free Press"* and the news media family is, after all, the *"watchdogs"* of truth and order in society.

So, what the hell happened to these *"watchdogs"* of public trust in the last 60 plus years?

It is not just the heavy-handed suppression of certain news stories from our newspapers or the blatant propagandizing of history-distorting TV "docudramas" that characterizes the opinion-manipulating techniques of the media masters. They exercise both subtlety and thoroughness in their management of the news and the entertainment that they present to us.

It has become clearer over the decades that misplaced interests and the pursuit of money, power, manipulation and coercion in the name of national security and protectionism is what transpired and overtaken the news media. These materialistic pursuits became the all-consuming focal points of the news media but, more importantly, it became the focus for the military and the intelligence communities who viewed all media venues as useful tools for national security and intelligence gathering.

It has been stated that anyone journalist who was on the payroll of one of the agencies in the intelligence community (most notably the **CIA**) was worth 20 spies out in the field of intelligence gathering. Journalists are by profession excellent in tracking down a news story and gathering information sometimes, going into difficult or dangerous situations to get their story.

"The CIA's secret activities, covert missions, and connections of control are all done under the pretense and protection of national security with no accountability whatsoever, at least in their minds. Considering the public is held accountable for everything we think, say, and do there is something seriously wrong with this picture. The CIA is the President's secret army, who have been and continue to be conveniently above the law with unlimited power and authority, to conduct a reign of terror around the globe. The "old boy network" of socializing, talking shop, and tapping each other for favors outside the halls of government made it inevitable that the CIA and Corporate America would become allies, thus the systematic infiltration and takeover of the media." **"Operation Mockingbird: CIA Media Manipulation" Mary Louise, 03 August, 2007, The National Expositor** http://www.bibliotecapleyades.net/sociopolitica/sociopol_mediacontrol05.htm

Under the guise of 'American' objectives and lack of congressional oversight, the CIA accomplish their exploits by using every trick in the book (*and they know quite a few*) that they actually teach in the notorious "School of the Americas", nicknamed the "School of Dictators" and "School of Assassins" by critics. The **Association for Responsible Dissent** estimates that 6 million people had died by 1987 as a result of CIA covert operations, called an "American Holocaust" by former State Department official **William Blum**. http://www.bibliotecapleyades.net/sociopolitica/sociopol_mediacontrol05.htm

"Journalists **Carl Bernstein** and **Bob Woodward**, who broke the stories on Watergate (late 70's) in the Washington Post having gained access to what the CIA was trying to keep from congress about its program of using journalists at home and abroad, in deliberate propaganda campaigns. It was later revealed that Woodward was a naval intelligence briefer to the White House and knew many insiders including **General Alexander Haig**. A high-level source told Bernstein, "One journalist is worth twenty agents." CFR/Trilateralist **Katharine Graham**, in a 1988 speech given to senior CIA employees at Agency headquarters said, *"We live in a dirty and dangerous world. There are some things the general public does not need to know and shouldn't. I believe democracy flourishes when the government can take legitimate steps to keep its secrets and when the press can decide whether to print what it knows."* http://www.bibliotecapleyades.net/sociopolitica/sociopol_mediacontrol05.htm

On the surface this statement by Katherine Graham seems to make a lot of sense and may have been true in the late 80s but, this is the Twenty-First century and this attitude really wasn't relevant then, nor is it relevant now or ever has or will be relevant. The world becomes **"a dirty** *and dangerous"* place to live in because governments and their intelligence agencies create situations in the world that are filled with suspicion and self-serving foreign political agendas.

Saying that *"there are some things the general public does not need to know",* keeps the public at arms left with a total disregard to the fact that governments which are elected into office are ultimately answerable to any and all public inquiry without suppression or cover-up of information and this applies to all governmental agencies, commissions, and departments, no matter what the reason may be, whether in the name of national security or any other supposed legitimate reason.

Governments are the **"public servants"** to the people who elected them into power in the first place! That means that it is the *general public* who are in essence the bosses, the CEOs, the executive officers or whatever terms that may be applicable to which all governments and their departments, agencies, etc. are ultimately answerable to and not the other way around! **People should not be afraid of their governments, rather it Governments that should always be afraid of their people!** Governments are not independent entities unto themselves or beyond public reproach believing that given a vote of confidence means that they can do whatever they want and thus, mistakenly think that they only answerable to themselves.

They cannot be allowed to become some **"Frankenstein monster"** that comes to life by the common vote of the people only to run amok with a will of its own, without regard for who it was that bestowed power upon them or without a healthy respect for its creator, the general public.

Democracy does not flourish *"when the government takes legitimate steps to keep its secrets or when the press can decide whether to print what it knows".* Such a belief creates a cancerous rot from within that not only threatens to destroy that government eventually making it impotent but also, it becomes globally dangerous to other world governments. It will throw the country and its people into anarchy, violence and possible political upheaval resulting in revolution. World history is replete with many disastrous examples of countries that have experienced such revolutions forcing old governmental political systems to be overhauled and replaced with new political systems thus, ultimately placing people back in the "driver's seat!"

Many countries, governments and their political leaders for the most part trust most foreign journalists to report and print only truthful and honest accounts of any story in their country. But cold war tensions after the second World War created suspicion in many countries towards foreign journalists because it was discovered that a few and then increasingly more journalists were doing more than just journalism. They were spying for their respective countries gathering intelligence information from other governments or influencing political thought in those other governments.

In an article by Carl Bernstein originally published in Rolling Stone, October 20, 1977 *"How Americas Most Powerful News Media Worked Hand in Glove with the Central Intelligence Agency and Why the Church Committee Covered It Up",* under the subtitle: *"The CIA And The Media",* he describes how more than 400 American journalists (during the late 70s) were for the past twenty-five years secretly carrying out assignments for the Central Intelligence Agency, according to documents on file at CIA headquarters.

"Starting in the early days of the **Cold War** (late 40's), the CIA began a secret project called **Operation Mockingbird**, with the intent of buying influence behind the scenes at major media outlets and putting reporters on the CIA payroll, which has proven to be a stunning ongoing success. The CIA effort to recruit American news organizations and journalists to become spies and disseminators of propaganda was headed up by **Frank Wisner, Allen Dulles, Richard Helms, and Philip Graham** (publisher of The Washington Post).

Wisner had taken Graham under his wing to direct the program code-named Operation Mockingbird and both have presumably committed suicide."
http://www.bibliotecapleyades.net/sociopolitica/sociopol_mediacontrol05.htm

"Some of these journalists' relationships with the Agency were tacit; some were explicit. There were cooperation, accommodation, and overlap. Journalists provided a full range of clandestine services—from simple intelligence gathering to serving as go-betweens with spies in Communist countries. Reporters shared their notebooks with the CIA. Editors shared their staffs. Some of the journalists were Pulitzer Prize winners, distinguished reporters who considered themselves ambassadors-without-portfolio for their country. Most were less exalted: foreign correspondents who found that their association with the Agency helped their work; stringers and freelancers who were as interested in the derring-do of the spy business as in filing articles; and, the smallest category, full-time CIA employees masquerading as journalists abroad. In many instances, CIA documents show, journalists were engaged to perform tasks for the CIA with the consent of the managements of America's leading news organizations."
http://tmh.floonet.net/articles/cia_press.html

"The CIA had infiltrated the nation's businesses, media, and universities with tens of thousands of on-call operatives by the 1950's. **CIA Director Dulles** had staffed the CIA almost exclusively with Ivy League graduates, especially from Yale with figures like **George Herbert Walker Bush** from the **"Skull and Crossbones" Society.**
http://www.bibliotecapleyades.net/sociopolitica/sociopol_mediacontrol05.htm

According to Bernstein, there has been a long history of CIA involvement with the American press which continues to this day to be palled by an official policy of obfuscation and deception. The primary reasons for this have been that the use of journalists are among the most productive means of intelligence-gathering employed by the CIA. The Agency has claimed, however, that they have cut back sharply on the use of reporters since 1973 primarily as a result of pressure from the media, itself, although, some journalist-operatives are still posted abroad. The other reason is that investigation into CIA files and statements reveal that between the 1950s and 1960s there were embarrassing relationships with some of the most powerful organizations and individuals in American journalism.

Most notable of these special cooperative relations with the CIA have been with **American Broadcasting Company**, the **National Broadcasting Company**, the **Associated Press, United Press International, Reuters, Hearst Newspapers, Scripps-Howard**, *Newsweek* **magazine,** the **Mutual Broadcasting System**, the *Miami Herald* and the old *Saturday Evening Post* and *New York Herald-Tribune*. But, by far the most valuable of these associations have been with the *New York Times*, **CBS** and **Time Inc.** http://tmh.floonet.net/articles/cia_press.html

The **Columbia Broadcasting System (CBS)** was unquestionably the CIAs most valuable broadcasting asset. **CBS President William Paley** and **Allen Dulles** enjoyed an easy working and social relationship. Over the years, the network provided cover for CIA employees, including at least one well-known foreign correspondent and several stringers; it supplied outtakes of news film to the CIA; established a formal channel of communication between the Washington bureau chief and the Agency; gave the Agency access to the CBS news film library; and allowed reports by CBS correspondents to the Washington and New York newsrooms to be routinely monitored by the CIA. Once a year during the 1950s and early 1960s, CBS correspondents joined the CIA hierarchy for private dinners and briefings." http://tmh.floonet.net/articles/cia_press.html

Many Americans under the illusion in believing that we still have a free press, yet their main source of news comes from state-controlled television, while under the misconception that reporters are meant to serve the public. Reporters are paid employees and serve the media owners, who usually cower when challenged by advertisers or major government figures.

It is not unusual to find that groundbreaking stories like the **Iran-Contra scandal** which were covered by trustworthy top-notch reporters for such news media as Associated Press were largely ignored by the press and congress. Even, retractions of true stories for political reasons is a common practice in magazines like Newsweek as reported by Robert Parry. In "Fooling America", Parry stated in his talk, *The people who succeeded and did well were those who didn't stand up, who didn't write the big stories, who looked the other way when history was happening in front of them, and went along either consciously or just by cowardice with the deception of the American people."*

Bernstein explains further that "in the field, journalists were used to help recruit and handle foreigners as agents; to acquire and evaluate information, and to plant false information with officials of foreign governments. Many signed secrecy agreements, pledging never to divulge anything about their dealings with the Agency; some signed employment contracts, some were assigned case officers and treated with unusual deference. Others had less structured relationships with the Agency, even though they performed similar tasks: they were briefed by CIA personnel before trips abroad, debriefed afterward, and used as intermediaries with foreign agents." http://tmh.floonet.net/articles/cia_press.html

"During the 1976 investigation of the CIA by the **Senate Intelligence Committee**, chaired by **Senator Frank Church**, the dimensions of the Agency's involvement with the press became apparent to several members of the panel, as well as to two or three investigators on the staff. But top officials of the CIA, including former **CIA directors William Colby and George Bush**, persuaded the committee to restrict its inquiry into the matter and to deliberately misrepresent the actual scope of the activities in its final report." http://tmh.floonet.net/articles/cia_press.html

"CIA officials almost always refuse to divulge the names of journalists who have cooperated with the Agency. They say it would be unfair to judge these individuals in a context different from the one that spawned the relationships in the first place. "There was a time when it wasn't considered a crime to serve your government," said one high-level CIA official who makes no secret of his bitterness. "This all has to be considered in the context of the morality of the times, rather than against latter-day standards—and hypocritical standards at that."

The CIA even ran a formal training program in the 1950s to teach its agents to be journalists. Intelligence officers were "taught to make noises like reporters," explained a high CIA official, and were then placed in major news organizations with help from management. "These were the guys who went through the ranks and were told, "You're going to be a journalist," the CIA official said. Relatively few of the 400-some relationships described in Agency files followed that pattern, however; most involved persons who were already bona fide journalists when they began undertaking tasks for the Agency.

"The CIA's use of journalists continued virtually unabated until 1973 when, in response to public disclosure that the Agency had secretly employed American reporters, William Colby began scaling down the program. In his public statements, Colby conveyed the impression that the use of journalists had been minimal and of limited importance to the Agency."

"Meanwhile, important CIA operatives who had been placed on the staffs of some major newspaper and broadcast outlets were told to resign and become stringers or freelancers, thus enabling Colby to assure concerned editors that members of their staffs were not CIA employees. Colby also feared that some valuable stringer operatives might find their covers blown if scrutiny of the Agency's ties with journalists continued. Some of these individuals were reassigned to jobs on so-called proprietary publications—foreign periodicals and broadcast outlets secretly funded and staffed by the CIA. Other journalists who had signed formal contracts with the CIA—making them employees of the Agency—were released from their contracts, and asked to continue working under less formal arrangements."
http://tmh.floonet.net/articles/cia_press.html

"Yet even while the Senate Intelligence Committee was holding its hearings in 1976, according to high-level CIA sources, the CIA continued to maintain ties with seventy-five to ninety journalists of every description—executives, reporters, stringers, photographers, columnists, bureau clerks and members of broadcast technical crews. More than half of these had been moved off CIA contracts and payrolls but they were still bound by other secret agreements with the Agency. According to an unpublished report by the **House Select Committee on Intelligence**, at least fifteen news organizations were still providing cover for CIA operatives as of 1976." http://tmh.floonet.net/articles/cia_press.html

"Major networks are primarily controlled by giant corporations that are obligated by law, to put the profits of their investors ahead of all other considerations which are often in conflict with the practice of responsible journalism. There were around 50 corporations a couple of decades ago, which was considered monopolistic by many and yet today, these companies have become larger and fewer in number as the biggest ones absorb their rivals. Such aggressive maneuvers and hostile take-overs clearly demonstrate that the profit motive of free enterprise within large

24

corporations is alive and flourishing in America. This concentration of ownership and power reduces the diversity of media voices, as news falls into the hands of large conglomerates with holdings in many industries that interfere in newsgathering, because of conflicts of interest. **"Mockingbird"** was an immense financial undertaking with funds flowing from the CIA largely through the **Congress for Cultural Freedom (CCF)**." **"Operation Mockingbird: CIA Media Manipulation" Mary Louise, 03 August, 2007, The National Expositor** http://www.bibliotecapleyades.net/sociopolitica/sociopol_mediacontrol05.htm

"Media corporations share members of the board of directors with a variety of other large corporations including banks, investment companies, oil companies, health care, pharmaceutical, and technology companies. Until the 1980's, media systems were generally domestically owned, regulated, and national in scope. However, pressure from the IMF, World Bank, and US government to deregulate and privatize, the media, communication, and new technology resulted in a global commercial media system dominated by a small number of super-powerful transnational media corporations (mostly US based), working to advance the cause of global markets and the CIA agenda."
http://www.bibliotecapleyades.net/sociopolitica/sociopol_mediacontrol05.htm

"The **first tier of the nine giant Media firms that dominate the world** are **Time Warner/AOL, Disney/ABC, Bertelsmann, Viacom/CBS, Rupert Murdoch's News Corporation/Fox, General Electric/NBC, Sony, Universal/Seagram, Tele-Communications, Inc. or TCI and AT&T.** This is just the head of the octopus which has its second and third tier Media 'tentacles" working together in unison or feigned division. This would include The **Washington Post/Newsweek, The New York Times/Weekly Standard, Tribune Co., US News, Gannett/USA Today, Dow Jones/Wall Street Journal, Washington Times, Knight-Ridder**, and etcetera. Media propaganda tactics include blackouts, misdirections, expert opinions to echo the Establishment line, smears, defining popular opinions, mass entertainment distractions, and Hobson's Choice (the media presents the so-called conservative and liberal positions)." http://www.bibliotecapleyades.net/sociopolitica/sociopol_mediacontrol05.htm

"The Agency's unwillingness to end its use of journalists and its continued relationships with some news executives is largely the product of two basic facts of the intelligence game: journalistic cover is ideal because of the inquisitive nature of a reporter's job; and many other sources of institutional cover have been denied the CIA in recent years by businesses, foundations and educational institutions that once cooperated with the Agency."

The CIA is not now, nor has it ever been a central intelligence agency but rather the covert action arm of the President's foreign policy advisors.
http://www.bibliotecapleyades.net/sociopolitica/sociopol_mediacontrol05.htm

The CIA is by far America's most widely known intelligence agency and its infamy and political intrigue into the foreign affairs of other governments overseas is legendary. Many people domestically and globally see the CIA as at the core of American corruption like an octopus with its many pervasive tentacles reaching out and invading into all aspects of private and public life both nationally and internationally. The CIA is not now, nor has it ever been a central intelligence agency but rather the covert action arm of the President's foreign policy advisors, of which disinformation is a large part of its responsibility and the American people are the primary target of its lies.

"One of the *many* [bold italic added] primary reasons John F. Kennedy was assassinated had to do with the fact he dared to interfere in the framework of power. Kennedy was intent on exercising his ELECTED powers and not allowing them to be usurped by power-crazed individuals in the intelligence community, threatening to "splinter the CIA into a thousand pieces and scatter it to the wind." There were four things that filled the CIA with rage and sealed his fate; JFK fired Allen Dulles, was in the process of founding a panel to investigate the CIA's numerous crimes, put a damper on the breadth and scope of the CIA, and limited their ability to act under National Security Memoranda 55."

Mary Louise sums up her article on Operation Mockingbird with the sentiments and the outrage that is echoed by hundreds of millions of America who have been unwittingly betrayed by their

26

elected government, its politicians and agencies. "Personally, I have come to the conclusion that the media is not only influenced by the CIA.....the media is the CIA. Many Americans think of their supposedly free press as a watchdog on government, mainly because *the press itself shamelessly promotes that myth.* One of the first tenets for the control of a population is to control all sources of information the population receives and mostly because of the pervasive CIA and Operation Mockingbird, the mainstream American Press is a controlled multi-national corporate/government megaphone. They are up to their eyeballs in dirty deeds and there will never be an end to the corruption that prevails unless the CIA is abolished. Otherwise, the CIA will just keep on using their tricks of propaganda, stuffed ballot boxes, purchased elections, extortion, blackmail, drug trafficking, sexual intrigue, kidnapping, beating, torture, intimidation, economic sabotage, false stories about opponents in the local media, infiltration and disruption of opposing political parties, demolition and evacuation procedures, death squads, and politically motivated assassinations. The CIA is the epitome of organized crime run amuck!" **"Operation Mockingbird: CIA Media Manipulation" Mary Louise, 03 August, 2007, The National Expositor** and http://www.bibliotecapleyades.net/sociopolitica/sociopol_mediacontrol05.htm and https://www.youtube.com/watch?v=nWQHAOUKgJk

The question at this point is how does the CIA's control and manipulation of the news media affect the reality of the UFO and ETI phenomenon via the daily news releases to the general public?

In the intelligence community and the military industrial complex, information control is a powerful weapon to keep order within one's country and to provide cover-up of one's strengths, weaknesses, and military assets while at the same time distilling propaganda and disinformation to other governments.

What has been illustrated up to this point is that the basic cornerstone of suppression, cover-up, "spin doctoring" and sanitizing of the news has been firmly laid upon a foundation of lies, denials, disinformation, misinformation and obfuscation with the objective to control what people think and know, and thus, ultimately to control their lives.

UFOs and ETI invade the comfortable paradigms of our reality on this planet not by physical threat or war but, by disrupting and altering the status quo of the power structure of our governments, our financial institutions, religions and science and most of all, the mysterious hidden agenda of the wealthy corporate power-elite.

UFOS and in particular their ET occupants represents a new view of the universe. A view that says we are not alone in the universe and their presence here in our skies and upon our Earth with their interaction with some of Earth's population is confirmation of their reality.

Their visitation to our planet suggests very strongly that there is a new way to generate clean infinite power rather, than the pillaging and raping of mother Earth of her resources while leaving only environmental damage and pollution behind.

A prime example is the recent (April 2010) destruction of the **BP (British Petroleum)** floating oil rig in the Gulf of Mexico and its resultant drill-hole on the ocean's floor gushing millions of gallons of oil per day into the Gulf and up to the ocean's surface. It is so out of control that the black leaking ooze is killing all manner of sea life and ocean mammals like dolphins, seals, and birds but, it is spreading onto the American southern shorelines destroying animal and bird habitats as well as various plant life. This floating "black death" is being carried by the natural ocean currents all the way around the shores of Florida and is threatening the Eastern seaboard with the same lethality it has inflicted onto the southern coastline and from that point ocean currents can carry it eastward toward European shorelines.

After many failed attempts to effectively cap the oil drill-hole on the bottom of the Gulf, BP has finally (as of the writing of this book) been able to cap the drill-hole, however, the clean-up of the massive miles wide oil slick will take years and the destructive environmental side effects upon future generations to come will be incalculable.

Author's Rant: If this is not a wake-up call for action to everyone on the planet and particularly to all oil corporations, governments and their leaders to step up to the plate, halt all ocean oil drilling immediately, and place a moratorium on all other oil drilling as well as to implement without further hesitation cheap, clean and non-polluting alternative sources of energy then, I don't know what else is. If we continue with this fetish of destroying the environment for natural resources, just to satisfy a few deep pockets of the wealthy corporate elite, so that they can wield absolute power and control over the planet and her people then, we are a doomed race and are deserving of the consequences of our actions or in this case, inactions.

ETI visitation represents an established fact that interplanetary and interstellar travel between stars is not just mere conjecture or a hypothetical possibility but a routine reality. It represents an opportunity to explore our vast universe and expand the knowledge of humanity. It also represents the independence away from the many grid systems of energy in which we are forced into accepting. It represents the public's ability to control once and for all, the strong arm tactics of their governments and agencies that have run amok beyond oversight and control. It represents the ability for each individual to choose lifestyles that improve the human spirit, the way we learn, they way we live together, the removal of crushing and oppressive poverty and the ability to heal the planet and ourselves.

This is the ultimate in paradigm shifts which is in direct opposition to those few who wish to control the Earth and its people. Thus, this secretive cabal has found necessary to control the information sources on this subject matter and remove any challenge or opposition to its power and influence. Any hope for a brighter future for mankind has been in the interim of the current time period been crushed.

There is a hidden agenda under way that has been carefully crafted and meticulously orchestrated and unless, and until people awaken from their beds of heedlessness, they may eventually awaken to a world not to their choosing or to their liking in which they are enslaved to its ruthless system. They will awaken to a world order based on materialism and not a new world

28

order based on spiritual values. By the time the populace figures it all out, it may already be too late!

The "Giggle Factor" in Journalism, "Sanitizing "and "Spin Doctoring" the News

How is it possible, particularly in the last 60 plus years that the power players in the arena of information control via the news media have been able to control and influence what we see, what we know and what we believe and in particular how the subject of UFOs and ETI should be perceived by the public?

Most American people have no idea of the serious moral dilemma in which their country is sinking into nor understand that murder and war crimes have been conducted in their name under the guise of national security and protectionism of American foreign interests around the planet. **Peter Phillips and Mickey Huff** in **Project Censored**'s *Media Freedom Int'l* their news article, *"Inside the Military Media Industrial Complex: Impacts on Movements for Peace and Social Justice"* state that "corporate mainstream media are in large part to blame. The question then becomes how can this mass ignorance and corporate media deception exist in the United States and what impact does this have on peace and social justice movements in the country?"

The answer is not one that most people want to hear or accept! We are extremely gullible because we have allowed ourselves to be dumb down constantly by our political leaders, by the trust we place in our scientists, and in our clergy and by the trust we place in public servants in positions of power and authority, even by our educational institutions, but mostly through a dumb down news media. We believe that the news media as the fourth estate acts as a free press in an open and free society when in reality it has abandon its public duty of checks and balances on hidden government projects that are illegal, secretive and abusive. It is a delusional fantasy, a lie that has been perpetrated upon the public under the guise of an air of factual legitimacy. It propounds truth without substance or mettle, all the while expelling the written word in a daily dosage of diarrheal inconsequence. We are made to believe that this is unadulterated and honest truth and thus, our perception of the free media must also, be honest and truthful while in reality, it is anything but!

Peter Phillips and Mickey Huff have reported further in their news article that "In the United States today, the rift between reality and reporting has peaked. There is no longer a mere credibility gap, but rather a literal *"Truth Emergency"* in which the most important information affecting people is concealed from view. Many Americans, relying on the mainstream corporate media, have serious difficulty accessing the truth while still believing that the information they receive is the reality. A Truth Emergency reflects cumulative failures of the fourth estate to act as a truly free press. This truth emergency is seen in inadequate coverage of fraudulent elections, pseudo 9/11 investigations, illegal preemptive wars, torture camps, doctored intelligence, and domestic surveillance. Reliable information on these issues is systematically missing in corporate media outlets, where the vast majority of the American people continue to turn for news and information."

In a survey conducted by Project Censored among 376 randomly selected **National Conference for Media Reform (NCMR)** attendees at a Free Press or Media Reform Movement conference in Minneapolis, it "sought to determine the level of belief in a *truth emergency*, a systematic hiding of critical information in the US. Not surprisingly, for a sample of independent media reform activists, majorities in the 90% plus range agreed on most criticisms of mainstream media, that corporate media failed to keep the American people informed on important issues facing the nation and that a *truth emergency* does indeed exist in the US. Regarding the reasons, 87% of the participants believed that a military-industrial-media complex exists in the US for the promotion of the US military domination of the world and most agreed with research conclusions by Project Censored, and others, that a continuing powerful ***global dominance group*** inside the US government, the US media, and the national policy structure is responsible."

"While most progressive media activists do not believe in some omnipotent conspiracy, an overwhelming portion of NCMR participants do believe the leadership class in the US is *dominated* by a neo-conservative group of some several hundred people who share a goal of asserting US military power worldwide. This **Global Dominance Group (GDM)** continues under both Republican and Democratic rule. In cooperation with major military contractors, the corporate media, and conservative foundations, the GDM has become a powerful long-term force in military unilateralism and US political processes."
http://www.mediafreedominternational.org/2009/12/21/inside-the-military-media-industrial-complex-impacts-on-movements-for-peace-and-social-justice/

The Global Dominance Group and Information Control

"A long thread of sociological research documents the existence of a dominant ruling class in the US, which sets policy and determines national political priorities. C. Wright Mills, in his 1956 book ***The Power Elite***, documented how World War II solidified a trinity of power in the US that comprised corporate, military and government elites in a centralized power structure working in unison through "higher circles" of contact and agreement. This power has grown through the Cold War and, after 9/11, the Global War on Terror.

At present, the global dominance agenda includes penetration into the boardrooms of the corporate media in the US. Only 118 people comprise the membership on the boards of director of the ten big media giants. These 118 individuals, in turn, sit on the corporate boards of 288 national and international corporations. Four of the top 10 media corporations share board director positions with the major defense contractors including:

William Kennard: New York Times, Carlyle Group
Douglas Warner III, GE (NBC), Bechtel
John Bryson: Disney (ABC), Boeing
Alwyn Lewis: Disney (ABC), Halliburton
Douglas McCorkindale: Gannett, Lockheed-Martin.

Given an interlocked media network of connections with defense and other economic sectors, big media in the United States effectively represent the interests of corporate America. Media critic and historian Norman Solomon described the close financial and social links between the boards of large media-related corporations and Washington's foreign-policy establishment: ***"One way or another, a military-industrial complex now extends to much of corporate media."*** The

30

Homeland Security Act Title II Section 201(d)(5) provides an example of the interlocked military-industrial-media complex. This Act specifically asks the directorate to "develop a comprehensive plan for securing the key resources and critical infrastructure of the United States including information technology and telecommunications systems (including satellites) emergency preparedness communications systems."'
http://www.mediafreedominternational.org/2009/12/21/inside-the-military-media-industrial-complex-impacts-on-movements-for-peace-and-social-justice/

In **Dr. Steven Greer's** paper on "*Media Play*", he states that "*Truth... as long as it is inconsequential" and "is exercised within a certain sphere of influence that does not reach a critical mass of significance*" is the type of news that we have become accustomed to reading. Real honest and truthful news based on facts can never be read by the public, it is not what the corporations of **Big Media** or the government want in its readership. Ask any insider or frustrated journalist who writes about truth but, cannot get it out in public circulation. **"MEDIA PLAY" By Steven M. Greer MD; 29 April 2004**

If truth and facts are what you expect to read or hear daily then, you may find it in the much smaller news presses found in the smaller communities of the land, because their sphere of influence and significance is limited. **Big Media** in America is utterly corrupted, you would find more significance of truth from the media of many of the third world countries, whether in Cuba, Iraq, China or India than you will from any American media.

We know that Big Media is in bed with the CIA and no doubt with other intelligence agencies and "without the cooperation and compliance and corruption" of corporate media the covert programs that keep the world in a chaotic state could not endure. Every UFO investigator has heard the story of how **Mike Wallace** of CBS' 60 Minutes was given government documents on UFOs and wanted to do an expose on the subject but was pressured from someone, from somewhere within the media to drop the story.

 "The Big Media is vertically and horizontally integrated into a large corporate and quasi-governmental matrix of shadowy interests and corrupting influences. In no way is the major media in America, free nor has it been for decades." **"MEDIA PLAY" By Steven M. Greer MD; 29 April 2004**

Ask any news media outlet about the UFO matter and you will get one of two responses, that the subject is closed, that the Air Force determined through Project Bluebook that UFOs are not real and that the **Condon Committee** proved that there was no threat to national security and that there was no benefit to the advancement of science, in other words, there was nothing to it. Officialdom backed with credible scientists and scientific investigation is the way the US military have given their definitive death-knell pronouncement that there is nothing more to be gain with any further research or investigation. End of subject! **The Big Lie** is then perpetuated even when there is voluminous evidence to the contrary.

The second response to the UFO matter is to debunk or ridicule the story by exploiting weaknesses of the sighting such as a possible misidentification of an everyday common occurrence like the planet Venus or a meteor or a balloon or Chinese paper lanterns.

When the story can't be debunked effectively then, attack the witness by ridiculing his credibility, by finding a flaw in his character like social awkwardness or some source of embarrassment like his physical appearance. If his appearance is less than one of the "beautiful people" like some toothless, bespectacled wonder from some small hick town and his intelligence is slightly above vegetable level then, as a reporter, your job his made easy.

Another Big Media ploy to discredit and debunk the witness and story is to play up the good old boys from the bar angle having a few laughs at the gullible public's expense by pulling a hoax on them and then, blame it all on the silly season time of year. Stories about crop circles fit into this category and thus, this phenomenon becomes nothing more than a publicity-seeking hoaxed story.

This type of journalism is referred to in the Big Media as the **"Giggle Factor"** even when some other news media tries to give it legitimacy, it becomes obligatory by the other big corporate news media to debunk it.

Failing to discredit the evidence or the witness, the media can always interview someone else, who is preferably a professional skeptic to give an opposing point of view. This creates doubt in the minds of the readers or TV viewers about the whole subject matter.

It is all news propaganda, a psychological weapon used in wartime thousands of years ago but, developed and refined in the last century into a science. It is much like the movie: ***"Wag the Dog"*** starring Robert De Niro and Dustin Hoffman who try to save the political career of a president who has had a sexual affair before his re-election for a second term by creating a false war scenario.

Any integrity in real factual news has long since disappeared after the mid to late 60s. An example of this is the news coverage of the Vietnam War, where news correspondents competed with each other to cover the grizzly details of the war. One could sit down with their family for supper during which the six o'clock national newscast would serve up generous blood splattering insights into the graphic reality of war. This news was factual, it was real and unbiased and it wasn't manipulated by the news media or the intelligence community.

The **"supper time war"** was too horrific for most peace-loving Americans to digest daily and seeing loved ones carried into a MASH unit missing an arm or a leg or with an open abdominal wound or even worse dead was too much for anyone to handle on a daily basis. American citizens and much of the world's democratic populace were protesting weekly, if not almost daily America's involvement in Vietnam. Politicians were on the hot seat of public opinion to end the war immediately as their popularity was doing a nose dive with the prolongation of the Vietnam War. The US Military were fairing no better by public protests that were now global. The sixties coined the adage, "don't trust the establishment" for good reason because people were quickly realizing that this was a false war to get American influence and muscle into the East as a challenge to Communism and to further its national interests and foreign policies.

For that reason the government and the intelligence community decided that they had to dig their heels in, in order to carry on with their agenda and from that lesson, any future wars had to be sanitized for public consumption. This would be perceived by the public as doing the right thing, however, the agenda by the government was one of dumbing-down and making the public insensitive to the reasons for going to war. It needed to focus less on the carnage and more of the human aspects like soldiers going off to war, like leaving their families behind or their return home from a tour of duty or highlighting war as a fight for democracy and freedom, the overcoming of tyranny or finding **"weapons of mass destruction" (WMDs).**

Thus, the Iraq War was the **"new sanitized war"** which depicted **"clean air strikes",** no war casualties with missing limbs, no massive property destruction. This was a precise surgical-like war where the actual number of dead was never revealed but was only an approximation. It was a war to find the "elusive weapons of mass destruction" and to bring democratic rule back into the country. It was, in reality, a war of lies and propaganda supported by Big Media whose scripted newscasts came right from the Pentagon.

In the last decade particularly in the United States, we have seen government-produced "news" which according to a report by Pulitzer-prize winning journalist **David Barstow** in *The New York Times*, the Bush administration has been increasingly criticized for its aggressive use of a public relations type news that has been previously prepared, a sort of ready-to-serve news or an "instant news" that big corporations regularly distribute to TV stations in order to sell products or services. This is propaganda journalism that is usually distributed through the use of a [Video News Release](http://en.wikipedia.org/wiki/News_propaganda) **(VNR).** http://en.wikipedia.org/wiki/News_propaganda

A *New York Times* editorial (March 16, 2005) entitled *"And now, the counterfeit news"* affirms that at least 20 U.S. federal agencies, like the **Department of Defense** and the **U.S. Census Bureau**, as well as the **U.S. Department of Agriculture** produced and distributed hundreds of TV news reports since 2001 that were aired as if they were produced by the media. The same report says that this practice was also utilized by the Clinton Administration. http://en.wikipedia.org/wiki/News_propaganda

'**The Media Elite**, a key component of the **Higher Circle Policy Elite** in the US, are the watchdogs of acceptable ideological messages, the controllers of news and information content, and the decision makers regarding media resources. Their goal is to create symbiotic global news distribution in a deliberate attempt to control the news and information available to society. The two most prominent methods used to accomplish this task are censorship and propaganda".

"A broader definition of contemporary censorship needs to include any interference, deliberate or not, with the free flow of vital news information to the public. Modern censorship can be seen as the subtle yet constant and sophisticated manipulation of reality in our mass media outlets. On a daily basis, censorship refers to the intentional non-inclusion of a news story – or piece of a news story – based on anything other than a desire, to tell the truth. Such manipulation can take the form of political pressure (from government officials and powerful individuals), economic pressure (from advertisers and funders), and legal pressure (the threat of lawsuits from deep-pocket individuals, corporations, and institutions) or threats to reduce future access to governmental and corporate sources of news. Following are a few examples of **censorship and propaganda**.

1. Omitted or Undercovered Stories- The failure of the corporate media to cover human consequences, like one million, mostly civilian deaths of Iraqis, reduces public response to the wars being conducted by the US. Even when activists do mobilize, the media coverage of anti-war demonstrations has been negligible and denigrating from the start. When journalists of the so-called free press ignore the anti-war movement, they serve the interests of their masters in the military media industrial complex.

2. Repetition of Slogans and Sound Bites- The corporate media in the US present themselves as unbiased and accurate. *The New York Times* motto of "all the news that's fit to print" is a clear example, as is CNN's authoritative "most trusted name in news" and Fox's mantra of "fair and balanced." Through constant repetition, the metaphors and symbols that pervade our media turn into unquestioned beliefs. Terms like "liberal media," "welfare cheaters," "war on terror," illegal aliens," "tax burden," "support our troops," are all distorted images serving to conceal a transfer of wealth from people needing a safety net to corporations seeking profitable markets and military expansion.

3. Embedded Journalism- The media are increasingly dependent on governmental and corporate sources of news. Maintenance of continuous news shows requires a constant feed and an ever-entertaining supply of stimulating events and breaking news bites. The 24-hour news shows on MSNBC, Fox and CNN maintain constant contact with the **White House**, **Pentagon**, and public relations companies representing both government and private corporations."

Investigations into other related militarization of news showed Pentagon propaganda penetration on mainstream corporate news in the guise of retired Generals as "experts" or pundits who turned out to be nothing more than paid shills for government war policy.

The problem then becomes more complex. What happens to a society that begins to believe such lies as truth? The run up to the 2003 war in Iraq concerning **weapons of mass destruction (WMDs)** is a case in point. It illustrates the power of propaganda in creating not only public support for an ill-begotten war but also reduces the possibility of a peace movement, even when fueled by the truth, to stop a war based on falsehoods. The current war in Iraq was the most globally protested war in recorded history. This did nothing to stop it and has done little to end it even under a Democratic president who promised such on the campaign trail."
http://www.mediafreedominternational.org/2009/12/21/inside-the-military-media-industrial-complex-impacts-on-movements-for-peace-and-social-justice/

The paint on the canvass of journalism is still wet with the brush strokes of carefully worded scripts, reporting lies, half truths and spin doctoring facts to produce a forgery that is touted as truth and integrity in news, all the while trying to convince the public that it is a masterpiece of journalism.

The point being made here is that Big Media and the intelligence community are in bed together (perhaps, initially Big Media may have been an unwilling bed partner) and rarely, if ever, has it had the best interests of the general public in mind, especially when they have an agenda to fulfill and where the news media gives full support to that covert agenda, no matter how deceptive or secretive. Secrecy and deception go hand in hand when it comes to the UFO question.

In matters of advanced technologies, new energy generation systems, propulsion systems and the reality of UFOs – these are considered the **"crown jewels of secrecy"...** "that once disclosed, would end the need for oil, gas, coal or nuclear power. The corruption and secrecy surrounding this issue is like none other – it is in a class of its own. The media can only cover the subject either in a cavalier or dismissive way – or through direct disinformation and ridicule." **"MEDIA PLAY" By Steven M. Greer MD; 29 April 2004**

When a story is investigated and is found to be potentially explosive in its implications particularly when the story is about the hardcore evidence that UFOs and ETI exist and there has been 60 years or more of government and military cover-up to suppress its reality then, it becomes absolutely mandatory to not to print it. Shills abound throughout all levels of the news media and are the main reason why the suppression of truth about many stories of great importance never get told.

"Of course, many people in the media never look into these issues since they have blindly accepted the party line and bought into the ridicule and disinformation surrounding the subject. A lack of independent investigation, and a prevailing prejudice prevents most journalists from even giving a cursory look into these controversies." **"MEDIA PLAY" By Steven M. Greer MD; 29 April 2004**

There had been some hope within the UFO community that the subject of UFOs would be given fair treatment when ABC ran the special **Peter Jennings** Reporting: **"UFOs - Seeing Is Believing"** on February 24, 2005. However, the program had little to do with objectively reporting on the UFO phenomenon and instead dealt more with debunking the extraterrestrial hypothesis, treating the subject as if this was the only explanation for UFOs. For those who have been active over the years in UFO research, the ABC show was just more of the same when it comes to the media's mindset towards the phenomena.

Any journalist who does investigate the UFO matter or any story that Big Media didn't want published faces professional suicide and discovers how ruthless media suppression can be. Fear among journalists can be palpable and loss of one's career a certainty, while bodily threats to one's person or even "accidental death" can be an outcome that is all too real and has been some of the ways to control and intimidate those journalists who do not play ball on the news media team. However, if the news story is the impeachment of a US President via the **Watergate Scandal** or the sexual peccadilloes of another President with one of his female interns, then the news media is perceived by the public to have done its job as the public watchdog. Truth has been ferreted out and the public interest has been serviced. If only this was true in all cases of journalism but, sadly it is not reality. You would have to be completely naive to still cling to this notion of honesty and integrity in the news media!

Who Controls the Media?

Below is a chart of major news media networks and who really owns and controls the media, it may surprise you that a lot of them are owned by the military or by military contractors!

General Electric $100.5 billion 1998 revenues	Time Warner $26.8 billion 1998 revenues		The Walt Disney Co. $23 billion 1998 revenues	Viacom $18.9 billion 1998 revenues		News Corporation $13 billion 1998 revenues
GE/NBC's ranks No. 1 on the Forbes 500. Prior to its merger with NBC and an alliance with Microsoft, GE specialized in electronics. The peacock owns many New York sports team. It also owns or has equity stakes in many popular websites, including Snap.com and iVillage.	The largest media corporation in the world, Time Warner owns film and music production companies, theme parks, sports teams, magazines, websites and book publishers as well as Turner Broadcasting		With its 1995 merger with Capital Cities/ABC, Disney has become a fully-integrated media giant. In addition to its theme parks, the company profits from retail outlets, magazines, book publishers, websites, motion pictures, sports teams, TV, cable, radio, music, and newspapers.	Viacom's purchase of Paramount, CBS, and Blockbuster Video enables them to use cable, television, movies, comic books, theme parks, music publishing and book publishing to cross-market their products. Broadcasting alone brings in over $6 billion in revenues.		CEO Rupert Murdoch's style has inspired respect and fear, and it has also made his multinational ventures in publishing, television and satellite services very successful. The company owns 20th Century Fox, the New York Post, the London Times, TV Guide, many stadiums, the LA Dodgers and five New York sports teams.
NBC includes programming, news and more than 13 TV and radio stations	**TURNER BROADCASTING** includes sports teams, programming, production, retail, book publishing and multimedia	**WB** Television Network	**ABC** includes ABC Radio, ABC Video, and ABC Network News	**CBS** includes stations, CBS Radio, CBS Telenoticias and CBS Network News	**UPN** includes programming and TV stations (50%)	**FOX** includes programming and stations
Owns 25-50% of the following: • A & E (with Disney	• HBO (75%) • Cinemax • HBO Direct Broadcast • Court TV (33% with GE) • TBS Superstation		• Disney Channel • Disney Television (58 hours/week	• Nickelodeon • MTV • M2: Music Television • VH1 • Showtime		• Fox Family Channel (50%) • Fox News Channel • fx (50% with TCI's

and Hearst) • American Movie Classics (25%) • Biography Channel (with Disney and Hearst) • Bravo (50%) • Bravo International • CNBC • Court TV (with Time Warner) • Fox Sports Net • History Channel (with Disney and Hearst) • Independent Film Channel • MSG Network • MSNBC (50%) • National Geographic Worldwide • News Sport • Prime • Prism (with Rainbow, a subsidiary of Cablevision, and Liberty Media, a subsidiary of TCI) • Romance Classics • Sports Channel Cincinnati, Chicago, Florida, New England, Pacific,	• Turner Classic Movies • TNT • Cartoon Network • Comedy Central (37.5% with Viacom) • Sega Channel • OVATION (50%) • Women's Information Television (WIN) (partial) • TVKO (75%) • 4 regional all-news channels • CNN • CNN/SI (with *Sports Illustrated*) • CNNfn (financial network) • CNNRadio • Headline News • Sportsouth • CNN International • CNN Airport Network	syndicated programming) • Toon Disney • Touchstone Television • A&E (37.5% with Hearst and GE) • Lifetime Network (50%) • ESPN (80% with Hearst) • ESPN2 (80% with Hearst) • ESPN Classic (80% with Hearst) • ESPN West (80% with Hearst) • ESPNews (80% with Hearst) • Buena Vista Television • Biography Channel (with GE and Hearst) • History Channel (37.5% with Hearst and GE) • Classic Sports Network • E! (35%)	• Nick at Nite's TVLand • Paramount Networks Comedy Central (50% with Time Warner) • TNN: The Nashville Network • Movie Channel • FLIX • All News Channel (50%) • Sundance Channel (45%) • Midwest Sports Channel • CBS Telenoticias (30%) • Home Team Sports (66% with News Corporation)	Liberty Media) • fxM (50% with TCI's Liberty Media) • Fox Sports Net (25% with TCI, GE and Cablevision) • The National Geographic Channel (50%) • FIT TV Partnership • Regional networks, including TV Guide Channel and Fox Sports New York

Ohio, Philadelphia				

Other Major Players

AT&T (TCI) - Recently acquired by AT&T, TCI's hold on cable, internet, and local phone services contributed to $7.6 billion in 1997 revenues. TCI is the second-largest US cable television system provider, and it has 10% ownership of Time-Warner/Turner. The company owns all or part of USA Network, Sci-Fi Network, E!. Court TV, Starz! and Starz! 2, Black Entertainment Television, BET on Jazz, BET Movies/Starz! 3, CNN, TNT, Headline News, Prime Sports Channel, The Learning Channel, Discovery Channel, QVC, Q2, Fox Sports Net, The Travel Channel, Prevue Channel, Animal Planet, The Box, Telemundo, International Channel, Encore, MSG Network, Action Pay-per-view, and the Home Shopping Network.

Sony - Sony's main media interests, earning $9 billion in 1997 sales, are in film and television production, movie theaters and music.

Universal (Seagram) - In addition to Universal Studios, with its production facilities and theme parks, the company owns the USA and Sci-Fi cable networks.

The above "Who Owns the Media" graph is taken from the National Organization For Women Foundation website.
http://www.nowfoundation.org/issues/communications/tv/mediacontrol.html and
https://www.youtube.com/watch?v=9ona0jYWa6s

Author's Rant: Personally, I distrust newspapers in reporting any accurate and factual journalism and rarely read from cover to cover. Sitting down in front of the television watching a whole hour of primetime national news, whether it be American or Canadian news more often than not is a waste of my time. Watching an additional hour of local and regional news content has proven to be no better as it is really just "fluff" news of 15 to 30 second sound bites that butcher any serious news story!

In my teenage years, I had a keen interest in the UFO subject matter because of personal experiences with the phenomenon. I would clip news articles on the subject out of the local newspapers, it became for me like a barometer to gauge the level of journalistic interest in the phenomenon and the accuracy of their reporting on this and other matters of interest. I quickly became disillusioned after the Condon Report was released and with the drop in the number of UFO sightings that used to be reported in the local press. The barometric gauge seems to be at an all time low and "bad weather" seems to be on the horizon for UFO investigation and reporting!

As previously stated, when the Condon Committee Report on UFOs was made public and went worldwide but, particularly so in North America, the news media "coincidently" seem no longer interested in such matters. Like most newspapers, they believed in the final conclusions of the Report, that there was no substance at all in the UFO reports, that there was no threat to national security and nothing new could be added to science with continued research into the phenomenon. People it seems were either mistaken or were delusional in what they saw. This was the proverbial slap in the face and the kick in the pants for the public to stop wasting the time of the military and the scientists. The news media supported this decision with the "ridicule factor" in reporting any UFO accounts that came across their news desks. This was not a red letter day for the advancement of science and the complete abandonment of any and all public issues and concerns the UFO matter.

In many ways, we live in an **Orwellian world** that has not moved beyond **"1984"** in its control of the **"Public Speak"** and **"Public Think"** of the news media**.** If you want truth, honesty and integrity in the news media then, do not look toward Big Media to provide it. You must search long and hard and if you are lucky enough, you may find it in some small local "news rag" but its influence and impact on the public will be small and inconsequential. You may need to read newspapers or watch television from other countries to get an honest appraisal of what's really going on in the world because, in the Western world, the new is, for the most part, a propaganda exercise in tabloid journalism.

Terry Hansen, an independent journalist with an interest in scientific controversies and the politics of mass media, states in his book "The Missing Times" that big new media have consistently ignored any in-depth investigation into the UFO phenomenon since the late '60s including up to the current age. Case in point is the overflights of the Minuteman Missile sites and nuclear bases in Montana back in the late "70s by strange unidentified flying objects which received little attention outside of Montana in the national press or big news media networks with the exception by the "**National Inquirer**" a light-hearted, ridicule type tabloid newspaper operated or influenced by the CIA and by Montana's much small local newspapers, the story didn't make national headlines until two years after the event!

Hansen re-asserts in his book much of the tactics of doctored journalism used by news media and the intelligence agencies as already mentioned earlier, with regard to sensitive news stories like the presence of UFOs that have alleged national security issues.

Similar anomalous type stories like the Hudson Valley, New York sightings of large triangle UFO were also ignore by major new media as were similar UFO sightings in Belgium in the early '90s or the Cash-Lundrum Incident or the Phoenix Lights sightings and the Disclosure press conference which received about a couple of days of media attention and this too was quickly blown off into obscurity or ridicule by Big Media in the USA. Most countries globally will see such news stories as worthy of investigative time and reporting to the general public, but not in the US.

"The Missing Times, News Media Complicity in the UFO Cover-up" by Terry Hansen; 2000; USA; published by Xlibris Corporation; ISBN 0-7388-3611-7 (Hard) and ISBN 0-7388-3612-5 (soft) and https://www.youtube.com/watch?v=VD7ewtykM98 also https://www.youtube.com/watch?v=1qAYxmFsxOM Terry Hansen

The subject of UFOs, its associated technology and power generation will never get a serious treatment by Big media at this time until, the day comes when the American government and military come clean on what its knows through official disclosure or when ETs decide to show themselves in full sight of the world's governments and populace. If and when that momentous day arrives, you can bet the family farm and the kids that the Big Media will be the first in line and on the UFO and ETI gravy train of publicity stating how they have always felt that this was a real phenomenon and the newscasts on the subject will run non-stop 24/7.

United States Intelligence Community

The **United States Intelligence Community** (**IC**) is a cooperative federation of 16 separate United States government agencies that work separately and together to conduct intelligence activities considered necessary for the conduct of foreign relations and the protection of the national security of the United States. Member organizations of the IC include intelligence agencies, military intelligence, and civilian intelligence and analysis offices within federal executive departments. The IC is led by the **Director of National Intelligence (DNI),** who reports to the President of the United States.

The Intelligence Community seals represent the 16 major US intelligence agencies all promoting the anthem of "National Security" and the power play for information.

https://en.wikipedia.org/wiki/United_States_Intelligence_Community

Among their varied responsibilities, the members of the Community collect and produce foreign and domestic intelligence, contribute to military planning, and perform espionage. The IC was established by **Executive Order 12333**, signed on December 4, 1981, by **President Ronald Reagan**. The sixteen seals of the major intelligence agencies are well known but, how many other covert and deep black intelligence agencies and their sub-agencies exist and operate in the US that we know nothing about?

United States Intelligence Community seal

Supporting the work of the 16 main agencies, The Washington Post has reported that there are 1,271 government organizations and 1,931 private companies in 10,000 locations in the United States that are working on counterterrorism, homeland security and intelligence; and that the intelligence community as a whole includes 854,000 people who hold top-secret clearances.

Intelligence is information that agencies collect, analyze and distribute in response to government leaders' questions and requirements. Intelligence is a broad term that entails:

- Collection, analysis, and production of sensitive information to support national security leaders, including policymakers, military commanders and Members of Congress.
- Safeguarding these processes and this information through counterintelligence activities.
- Execution of covert operations approved by the President. **The IC** strives to provide valuable insight on important issues by gathering raw intelligence, analyzing that data in context, and producing timely and relevant products for customers at all levels of national security—from the war-fighter on the ground to the President in Washington.

Executive Order 12333 charged the IC with six primary objectives:

- Collection of information needed by the President, the National Security Council, the Secretary of State, the Secretary of Defense, and other executive branch officials for the performance of their duties and responsibilities;
- Production and dissemination of intelligence;
- Collection of information concerning, and the conduct of activities to protect against, intelligence activities directed against the U.S., international terrorist and/or narcotics activities, and other hostile activities directed against the U.S. by foreign powers, organizations, persons and their agents;
- Special activities (defined as activities conducted in support of U.S. foreign policy objectives abroad which are planned and executed so that the "role of the United States Government is not apparent or acknowledged publicly," and functions in support of such

42

activities, but which are not intended to influence United States political processes, public opinion, policies, or media and do not include diplomatic activities or the collection and production of intelligence or related support functions);
- Administrative and support activities within the U.S. and abroad necessary for the performance of authorized activities; and
- Such other intelligence activities as the President may direct from time to time.

The IC consists of 16 members (also called *elements*). The Central Intelligence Agency is an independent agency of the United States government. The other 15 elements are offices or bureaus within federal executive departments. The IC is led by the **Director of National Intelligence**, whose office, the **Office of the Director of National Intelligence (ODNI)**, is not listed as a member of the IC.

- **Independent agencies**
  - Central Intelligence Agency **(CIA)**
- **United States Department of Defense**
  - Air Force Intelligence, Surveillance and Reconnaissance Agency **(AFISRA)**
  - Army Military Intelligence **(MI)**
  - Defense Intelligence Agency **(DIA)**
  - Marine Corps Intelligence Activity **(MCIA)**
  - National Geospatial-Intelligence Agency **(NGA)**
  - National Reconnaissance Office **(NRO)**
  - National Security Agency **(NSA)**
  - Office of Naval Intelligence **(ONI)**
- **United States Department of Energy**
  - Office of Intelligence and Counterintelligence **(OICI)**
- **United States Department of Homeland Security**
  - Office of Intelligence and Analysis **(I&A)**
  - Coast Guard Intelligence **(CGI)**
- **United States Department of Justice**
  - Federal Bureau of Investigation **(FBI)**
  - Drug Enforcement Administration **(DEA)**
- **United States Department of State**
  - Bureau of Intelligence and Research **(INR)**
- **United States Department of the Treasury**
  - Office of Terrorism and Financial Intelligence **(TFI)**

http://en.wikipedia.org/wiki/US_intelligence_community

Hollywood Does its Part to Educate, Amuse, and Scare the Crap Out of Us!

The movie moguls of Hollywood euphemistically referred to as the **Movie Mafia** or the **Tinsel Town Mafia** wield a double-edged sword when it comes to the subject of UFOs, flying saucers, and aliens. Flying saucers are either piloted by peace loving ETs visiting the Earth who are ready to help mankind in solving their problems and thus, advance mankind to a higher level of civilization, or they are an evil invading race of terrifying ugly beings ready to destroy the earth and wipe out all of humanity or enslave and perform cruel medical experiments upon any survivors or perhaps, they here to eat us for lunch!

In Hollywood, the "Sword of Damocles" hangs over the heads of the movie-going public, ready to fall at any moment without any prior warning and we have no idea of the outcome to humanity when it comes to the sci-fi genre until, near the very end of the movie. Now, that's entertainment, Hollywood style!

Movies were usually the best form of entertainment in any town and a little scared every once in a while was enough to keep us on the edge of our seats and when the movies were over, we laughed and felt good knowing that none of it was real and that what we sat through for nearly two hours was only make believe that came from some imaginative mind in Hollywood. And this is perhaps, the way it should be, movies for entertainment value only with some positive message to keep us inspired, filled with hope or with an object lesson to learn from and with a view toward a rewarding future.

If we wax nostalgic a little, we see that movies always ended happily, the good guys got the girl in the end and the bad guys received their just deserts. It was a formula that was familiar and it worked for a long time. Warner Brothers, Paramount, Twentieth Century Fox, Walt Disney, etc all vied with each other to produce cinematic epics, musicals, mysteries, adventures, westerns, war and horror flicks and the audiences ate it up. As audiences matured and new generations came along, they wanted more than the easy familiarity that their parents grew up with. They wanted movies that edgy with more excitement, more action, more of things that got blown up and not always a happy ending but one filled with uncertainty and perhaps, even sadness or tragedy.

It seems that some of the **Hollywood Mafia** were listening! We are all familiar with the war movies that were prevalent during the Second World War and then later, the smaller conflict wars that the U.S. found itself involved in like the Korean War, the Vietnam War, and the Iraqi War. Here we saw the US as the good guys who fought against the forces of tyranny, Nazis or Communists or radical Islamic fundamentalists. Americans in a war movie were never seen as the bad guys and their politics never came into the movie plot, other than the good old fashion values of the American way of life, Mom, apple pie and the girl next door!

The Hollywood Mafia, like all free enterprising businesses, kept their eyes on the money because in Hollywood it all about the money. For the movie mafia, bigger pictures meant bigger profits most of the time. Infusions of big money into the movie industry came from the most unlikely sources. They found that the US military were willing participants, particularly when a war movie was being made and producers needed more realism be brought to the silver screen. The US military would supply the planes, tanks and warships for a re-enactment of a particular war event as well as the consultation of military strategy and even the soldiers, airmen or seamen to move all that hardware around. The only stipulation initially was to portray the US military in a favourable light.

This worked well for everyone involved and it seems that Hollywood was on the roll toward bigger epics and more profit but as time went on, The military and the intelligence community's view toward Hollywood was inevitably and militaristically predictable. Movies were to become another weapon in the US military arsenal.

Movies were the ideal psychological weapon for the intelligence community as well as the military industrial complex to condition the American public and the public audiences of other countries who imported American films. Agendas and perceptions could be easily manipulated in dialogue or action to illicit a visceral or emotional or intellectual reaction by the audience. If you don't believe that this is possible, the next time you go to a movie regardless of the movie type, whether it be a romance, an action or horror flick, science fiction or war movie, it doesn't matter what type of movie genre, watch and listen to the audience's reaction when the screen displays something that is an intense emotional scene and you will find the audience will react to it in unison with oohs, aahs, screams, hisses, boos, or plain revulsion and disgust, and of course with laughter. Movie directors and producers do this to illicit a particular response whether for entertainment value or as a deliberate ploy to engender a particular audience conditioning.

When the military industrial complex got involved in Hollywood movies and saw the potential it represented as a psychological weapon, they decided to buy into movie companies but remained the silent partners with the majority of the controlling interest of the business. Now they could influence the direction of how plots, dialogues, and action were to be presented on the silver screen.

Wars movies were no longer the prescribed movie genre necessary to convince the public of a particular viewpoint or a certain perspective. Subliminal messages in the movies were initially one way to present an idea or solicit an emotional response but, audiences became wise and smart enough to figure out this deceptive Hollywood ploy. An example of this is *"The Exorcist"* where archaic Christian fundamentalism was viewed as an acceptable method to deal with an emotionally vexed child possessed of demons.

It was later found that if the audience were slowly conditioned to a particular perception or mindset by viewing a number of different movie genres with similar plots or subtext, then subliminal messages weren't necessary because the whole movie itself was the subliminal message. Classic examples of manipulation of the audience's perception and emotional response as it relates to our subject of UFOs and aliens were movies like the Travis Walton's UFO/ETI experience, *"Fire in the Sky"*, or *"Independence Day"*, and *"Aliens"* and all its sequels. These terrible movies are considered entertaining yet, frightening by many were designed to portray Extraterrestrials as malevolent creatures who invade Earth to abduct us, or to harm us, or to destroy us, our cities and our way of life. They were meant to be perceived as bad and terrible creatures who could not be trusted, who were physically different from us. We were slowly being conditioned to become xenophobic toward anything not human!
Hollywood's movie mafia believe that it's "what the public wants". If you believe and accept that this is the way it is then, you are one of the hundreds of millions of people that have bought into the Hollywood lie as entertainment.

Dr. Steven Greer tells of meeting with Tracey Torme who directed the movie *"Fire in the Sky"* after having interviewed **Travis Walton** about his UFO and ETI experience. Dr. Greer found that the movie was an inaccurate recounting of Walton ET experience and asked Torme why this was. Travis had reported his five days of missing time away from his comrades as a frightening experience initially but, certainly not one of abusive torture by alien beings. In reality, the ETs only wanted to help him recover from his ordeal of being accidently hit by an electrical discharge from their spacecraft. They even permitted Travis the opportunity to pilot their craft. Greer asked Torme why this was not portrayed or stated in the movie to which Torme replied that he was well aware of the true account of Walton's experience but, had been pressured by the "**Tinsel Town Mafia"** to change the storyline to portray a more sinister perspective of the ETs as one of abducting humans and performing horrible medical experiments. It was according to his financial backers, the Hollywood Mafia aka. the **Military Industrial Complex (M.I.C.)** "what the public wanted!"

The story line of the movie, *"Independence Day"* is premised on the age-old question of whether or not we are alone in the universe. The answer comes with Earth's first contact with a race of hostile Extraterrestrials who have arrived to wipe out mankind, destroy our cities and plunder the natural resources of the Earth. This carefully crafted movie is full of special effects for the hard-core action junkie, it illustrates a growing xenophobic disgust and revulsion for anything not human while declaring that Independence Day be remember not just as an American holiday but as a holiday for all mankind under the notion that the American way of life should prevail!

Did I read too much into this movie? I don't think so! Reality, as they say, is stranger than fiction or in this case science fiction. What we have here is a movie that unabashedly tests the public waters of susceptibility and their reaction to something alien to this planet while also judging their patriotism to their country in a crisis situation.

Norio Hayakawa, a UFO researcher, and investigator sees this manipulation of public perception and response as part of the "Grand Deception" in which a false flag Extraterrestrial invasion will be perpetrated on the world to forcefully bring about a New World Order with a one world government. According to, to Norio Hayakawa, this is a distinct possibility and will likely happen at a time of great world distress to scare the world into a new global order. Hayakawa postulates that UFOs may be from another dimension. What we are witnessing may not be of our physical reality but that of another intelligent race of another dimension. Based on Hayakawa's personal Christian belief that this **Grand Deception** is biblically referenced in the **New Testament** as a pre-emptive strategy heralding the return of **Christ**. Many **Christians** who are also believers in UFOs and aliens as minions of the son of Satan and who are in league with the corporate wealthy elite follow this type of thinking as supporters of Hayakawa and his efforts to bring this deception to the public's attention.

"My own feeling is that the Illuminati is driving the world to World War 3. They will send the world into a period of great distress, destruction, deaths… Societies will collapse, cultures, languages, religions... will be destroyed. The main religions of Christianity, Islam, and Judaism are unlikely to fall easily via world war. At a certain time during this period, the Illuminists will stage a Grand Deception, an actual made up invasion of extraterrestrials. These will involve the Nephilim, fallen angels and black ops, Illuminist intelligence agencies and their hidden technology (Project Bluebeam)."

"Before this Grand Deception is staged, the world will be primed to look for a saviour, someone who will give them world peace. The appearance of these extraterrestrials 'gods' will cause **Christianity**, **Islam,** and **Judaism** to fall. Why believe in an invisible God when you can see inter-dimensional light beings with magical powers telling you they created you in the past? This will be the great falling away mentioned by apostle Paul."
http://socioecohistory.wordpress.com/2010/07/29/coast-to-coast-am-norio-hayak

Hayakawa in true form goes on to give a Christian interpretation and extrapolation of the evidence as he knows it and foresees the following scenario unfolding based on biblical prophecy.

"This will be a lie so big, a delusion so strong; the Christian church will likely see a 90% falling away. After a certain period, perhaps a year or 2, the man of sin will be revealed as the leader of these 'gods'.
The UFO lie is 2 fold. The first is that they are our creators and they are here to help us. They will masquerade as benevolent beings. The 2nd is the lie that there is another group who is malevolent, who is coming and will be out to destroy the human race. This is the lie these Nephilim and fallen angels will sell to prepare the world to go to war against Jesus Christ on his 2nd coming!"

I do not believe, Christians will avoid World War 3 via the rapture. The most likely time of the rapture of the church is after the Grand Deception, strong delusion but before the unveiling of the **Anti-Christ**. This Grand Deception plan is multi-decades in the making. The seeds of this deception are being sown since the 1940s via the UFO and aliens phenomenon. A new generation is being raised believing in ETs, light ships, inter-dimensional space travel, time travel, transhumanism….etc. This is simply a repeat of the **Days of Noah.** A period when fallen angels and their demigod children, the **Nephilim**, ruled the world."
http://socioecohistory.wordpress.com/2010/07/29/coast-to-coast-am-norio-hayak

Dr. Greer's view on this false flag Extraterrestrial invasion, an ala "Independence Day" scenario is founded on a far less biblical apocalyptic scenario but is partially rooted in an eschatological belief system based on radical **Christian fundamentalism**. Greer's spiritual belief in case anyone is wondering is founded on the **Baha'i Faith** as he states in his book, *"Hidden Truth – Forbidden Knowledge".* There are similarities to Hayakawa's perception of a materialistic concept of a new world order supported by a world government controlled by the corporate wealthy elite where large mega-corporations become the new national governments.

In this scenario, the US military as the first part of the **Military Industrial Complex** acts much like obedient puppets to the second part of the Complex namely, the large multi-transnational corporations. These transnational corporations are quickly becoming the new governments in the world and are for the most part in charge of the world's financial resources and the affect the affairs of many governments. They have orchestrated a slow but deliberate chain of events to create chaos and disruption in the world starting within the USA and then, spreading out into other western nations then, into third world nations and ultimately into the nations that are considered enemies of America. Much of this planned global upheaval and chaos is reminiscence in the same way that Nazi Germany prior to WWII spread its pernicious influence throughout its own country then, Europe and eventually around the rest of the world. Most of the Earth's population of 6 to 7 billion people are wiped out except for about 500 million people and most of these are to become the worker-slave class to the corporate wealthy elite, much in the same way that was depicted in the original movie, *"Rollerball"* starring **James Caan.**

The difference here, between Dr. Greer and Hayakawa, is that Dr. Steven Greer's scenario is supported with documentation and first-hand witnesses involved in the actual planning of this alien, ala "Independence Day" invasion scenario.

One of Greer's witnesses who is a part of the intelligence community was ready to come out and testify at the **National Press Club** as part of the **Disclosure Project** campaign to bring top level witnesses before Congress and swear to their involvement in the UFO/ETI secrecy and cover-up. He had been given clearance by his superiors to testify but at the last hour, before he going to the Press Club venue, he was recalled back to headquarters and thus, missed the Disclosure Event. He was going to disclose his part and the agency's involvement in a false flag alien invasion ala "Independence Day" scenario upon the public which would be hoaxed by the Military Industrial Complex. Dr. Greer found a note by him indicating his inability to be present at the Press Club. When the event was over, Greer and company were returning back to their hotel. This witness reappeared and explained to Dr. Greer that he was ordered to hold off with his testimony because everything was "in free fall" within the agency because of this Disclosure Event. According to this witness, half of his organization wants disclosure on the subject matter but, the old guard half within the organization wanted the UFO/ETI cover-up to remain in place.

Greer and Hayakawa have presented similar information, though through different sources like biblical references, the Disclosure Project and with personal radio interviews, they are alerting the public to a covert agenda by the MIC to hoax a false flag alien invasion upon humanity. Because of their initial research, other Ufologists are now coming forward with similar reports.

It should be realized by anyone reading this book or taking the course derived from this text that this is more than just a single witness's testimony with regard to this sinister agenda and with coming from more than just one informational source. There is a real potential threat to mankind that is being perpetrated by our fellow man and not by Extraterrestrials. This information needs to be openly discussed and monitored by everyone, in case this rogue cabal decides to unleash their covert agenda upon an unsuspecting public. We need to be ready to speak up against any hoaxing of an alien invasion. This is a terrestrial based agenda not Extraterrestrial based invasion plot!

As I have been stating and alluding to all along throughout this book that there is a sinister agenda under way that will be unleashed upon the public in the near future. President Eisenhower in the 50s was aware of this usurping of power from within the American political system and has warned us in his last address to the public that we need to be guarded against the rise of "unwarranted influence" and "misplaced power" by the Military Industrial Complex.

President Eisenhower's farewell address to the nation, January 1961 (in part): *"In the counsels of Government, we must guard against the acquisition of unwarranted influence, whether sought or unsought, by the Military Industrial Complex. The potential for the disastrous rise of misplaced power exists, and will persist. We must never let the weight of this combination endanger our* *liberties or democratic processes. We should take nothing for granted. Only an alert and knowledgeable citizenry can compel the proper meshing of the huge industrial and military machinery of defense with our peaceful methods and goals so that security and liberty may prosper together."*

The Military Industrial Complex is controlled by the private Corporate Wealthy Elite **(CWE)** and these corporations are transnational and multi-level in structure and are in turn controlled by a cabal of wealthy industrialists and their families! We are making a powerful accusatory statement here, to be sure but, one that can be proven through documentation and logical deductive reasoning!

The balance of this section of the book will show this power structure and something of the covert and hidden agenda of the MIC, the various intelligence communities, and the transnational multi-level industrial private sector. The proof will be so outrageous in its conception, that it will boggle the minds of most peace-loving, rational and sane people that the idea of UFOs and ETI visiting the planet will seem like a normal everyday common occurrence! The real action appears to be what's taking place down here on terra firma in the way of a silent coup d'état for global power and domination. Reversed engineered alien technology is being utilized like pawns in a grand chess game to achieve US global dominance. But, like any chess game, the end game is where the game is won or lost and pawns sometimes have a way of becoming powerful queens or rooks and thus, reversing the fortunes in favour of the other player!

If you have been a science fiction fan over the long years when movie industry has been cranking out this type of genre, you'll have realized that the plot lines have decidedly become more malevolent and horrific toward the poor unsuspecting human victims in the movie. This increasing malevolence of this particular genre is more than just a fetish to scare the shit out of us, because it's supposedly what we want, in reality, it's what the Movie Mafia say we need. Hollywood's negative ET theme is hardly audience driven but, part of a deep psychological conditioning. Supposedly, the audience craves more excitement and according to **Paola Harris**, Italian-American journalist and UFO researcher in answer to a question to **Michael Salla**, Honolulu Exopolitics Examiner UFO reporter, "Hollywood gives us what we want even if it distorts the reality of alien visitation and the UFO Phenomenon." The negative mindset of horror in science fiction and the negative stereotyping of any aliens invading Earth is part of the hidden covert agenda of the intelligence community to keep us off balance to the truth of the UFO/ETI matter. http://www.examiner.com/exopolitics-in-honolulu/hollywood-ufos-and-extraterrestrial-disclosure

We are slowly being psychologically conditioned to becoming xenophobic toward any alien beings that might be visiting the Earth, no matter, however, unusual their appearance may be, and this is particularly true of the ET stereotypes that we call the **"Grays"** and the **"Reptoids".** These two particular xeno-types that are currently flying around the skies are not extraterrestrial in the true sense of the word but are in fact man-made genetic engineered beings created here on Earth by a few rogue scientists working for the Military Industrial Complex.

Recall the cloning projects in the previous section of this text, the genetic manipulation of alien DNA, the mind control experiments and then the breakthrough technologies of micro or nano computerization and you will understand that these two types of ET beings are **PLFs, Programmable Life Forms**. They are **"sentient cybernetic life forms"** developed and created here on this planet by Earthmen! Not all is as it appears to be!

However, many Ufologists have bought into the **"Big Lie"** of abduction scenarios claiming these particular alien beings are real Extraterrestrial life forms from which they have built their reputations on reporting cases of these type of ET beings. Grays and Reptoids have been held responsible for the case reports of people being abducted, of being medically probed and prodded, having their DNA or sperm and ova samples extracted from their bodies with the resultant hybrid alien/human baby flying around in space, somewhere! Or they have come to a few people peeking through their windows or offering off-planet knowledge and mathematical formulae about the secret of their spacecraft's propulsion system or revealing what star system they originated from via the dream state of sleeping people, a case example of this is the **Stan Romanek** ET visitation case.

The events are exactly as he describes them but his nocturnal star hopping visitors are more of the human manufactured kind! The publicity generated from this one case has taken on a life of its own. Keep in mind just who owns and manages the Big Media as it is the same people who want this story to come out and into the public's awareness. These types of reports get more air time than is necessary because they are supporting a particular viewpoint and a hidden agenda. There are far better documented UFO and ETI cases than the Romanek case (with all due respect to Stan Romanek) but, these accounts never see the light of day in the Big Media press or if they do make it into the newspapers or on the 6 o'clock news, they are presented with a ridicule factor usually where a counter opposing position from a professional debunker is given equal or greater credibility in the news story. The person who UFO or ETI encounter comes off looking foolish or his experience was a case of misunderstanding of some commonplace circumstance, thus discrediting his testimony and evidence.

Ufologists, however, will defend this type of research like pit bulls guarding an illegal grow-op farm or house full of marijuana in a blind rage of illogic and poor research. But that's exactly what the **MIC** wants and comes to expect from Ufologists. UFO investigators and alien abduction researchers are playing right into the hands of the covert agenda of the MIC, although, they will deny that this is happening at all. Most abduction researchers know nothing about PLFs and disbelieve that mankind has the technical ability to create cyborgs that it's too much like science fiction! But the old adage, "Science is stranger than fiction!" is most apropos in this situation. In fact the more unbelievable the facts and evidence, the harder it is to accept the reality of its existence! The present day science community could never accept this situation, nor tolerate it as essentially, such technology has left them behind coughing in the dust of the covert **black world of science** trying to play catch-up if at all possible to a rogue science community that's at least a 100 years ahead of them.

If there was ever a need for a psychologist to wade in on this subject matter, now would be the time because, in Ufology, we are up against the varsity team of deception and mind control and manipulation. They have created technological realities to boggle the minds of most people who research this subject matter and ensured that it would be difficult for anyone to accept it as anything but extraterrestrial in nature.

Enter the **Movie Mafia** who are in the hip pocket of the Military Industrial Complex. Their job now, with the UFO and ETI research community having already been convinced of the deception perpetrated by the MIC, is to get the word out to as many people as possible and to thoroughly instill a strong sense of fear and foreboding about any contact, or abduction, or possible invasion from Extraterrestrial beings. Case in point is that horrible movie that came out recently called *"The Fourth Kind"* about alien abductions of hapless and unsuspecting humans by Gray ETs. This was a deliberate ploy by the Movie Mafia and the owners of the movie company, namely the MIC to gauge the reaction and response of the public to alien abductions and in a sense to slap down the other more positive movie, *"Close encounters of the Third Kind"* by Steven Spielberg claiming this new movie release about alien abductions was based on facts of actual reported cases. The truth is that the facts were contrived over a sixty year period of an unofficial disclosure campaign of lies, misinformation and disinformation of UFOs and ETI by the news media, Hollywood and the MIC to convince UFO researchers and the public that it was all real while not officially giving public disclosure to its reality.

With the movie industry on board with the hidden covert agenda and supported unwittingly by the Ufologists already involved in abduction research and to a lesser degree by other UFO researchers, The mass conversion of the public is under way until, the time is nearly ripe to spring the false flag hoaxed alien invasion then, all hell will break loose.

We have already and continue to steadily give up our human rights and privacy after the 9/11 tragedy which was an inside job by the rogue groups within the Military Industrial Complex and the carefully orchestrated agenda is slowly being unfurled with greater intensity, chaos, and destruction to which we will give up more of our freedoms, rights and privacy to the point that very little of any of it will remain. Does this not bring the horrific memories of pre-war Nazi Germany to mind that lead up to most of Europe and the rest of the world becoming engulfed in global conflict and world war?

Yet, there are movie producers and directors who have decided not to play ball with the rest of the Tinsel Town Mafia and go it alone in their own creative licensing that is not at all negative but, is instead positive, uplifting and redemptive with a spin on a more hopeful future for mankind. **Steven Spielberg** is most notably the foremost director of this science fiction genre and his movie *"Close encounters of the Third Kind"* was hopeful for a good future for mankind. This went in direct opposition to the hidden covert agenda by the cabal of the wealthy corporate elite. Much of what is in the movie, is very accurate from the sightings and behaviour of the ETI spacecraft, the high strangeness and psychic impressions portrayed by the lead actors of the movie, the close encounters and interactions with ETs which on the surface appeared to be abductions but, were in reality a relationship between humans and ETI, leading up to the final conclusive moments in the movie of open contact between ETI and humans that was mutual and peaceful. The audience left their theatres in awe, inspired and hopeful toward the day of eventual first contact with ETs.

In **Linda Moulton Howe's** website **Earthfiles**, she reports that "two British authors, Matthew Alford and Robbie Graham were investigating the concept that the CIA and other American intelligence agencies manipulate Hollywood to produce television and movies that can alter public perceptions and be used to monitor public reactions. While researching Hollywood history and the background of Steven Spielberg specifically, Matt Alford discovered a 1978 interview with film director Spielberg about his 1977 blockbuster *Close Encounters of the Third Kind.* The interview was published in a 1978 edition of the now out-of-print magazine called *Cinema Papers.* Below in the section highlighted by yellow lines, Spielberg answers **the interviewer's question:** *At any point during the setting up of the film were you more in doubt than not?"*

Spielberg's reply*: "I really found my faith when I heard that the government was opposed to the film. If NASA took the time to write me a 20-page letter, then I knew there must be something happening."*

"I had wanted co-operation from them, but when they read the script, they got very angry and felt that it was a film that would be dangerous. I think they mainly wrote the letter because (Spielberg's previous) Jaws convinced so many people around the world that there were sharks in toilets and bathtubs, not just in the oceans and rivers. They were afraid the same kind of epidemic would happen with UFOs."
http://www.earthfiles.com/news.php?ID=1760&category=Environment

It was said that former **President Jimmy Carter** was extremely interested in the UFO subject having had a personal UFO sighting of his own in 1969, before being elected President. Carter was asked during his presidential running that he would release all the UFO information on the subject to the public. This, however, he was unable to do, even though some UFO documents were released, he wondered how much had not been released. Later in 1976 when the now President-elect asked outgoing **Director of Central Intelligence George Bush Sr.** for, *"the information that we have on UFOs and Extraterrestrial intelligence,"* Bush denied Carter's request, stating that, *"simple curiosity on the part of the President"* was inadequate, as the information existed, **"on a need to know basis only."**
http://www.sightings.com/1.reports2010/spielsaucersecrets.html; UFO/ET Headlines Spielberg's Saucer Secrets By Robbie Graham & Matthew Alford; ©2010 Robbie Graham & Matthew Alford *Reprinted By Permission*

[This is an intriguing response from the then **CIA Director George Bush Sr.** indicating that he must have known something more about the UFO matter but, was unwilling to discuss it with President Carter. If the President of the United States, who is the highest chief Executive Officer in the land doesn't have a need to know then, who does? This question would continue to plague all the presidents in the US particularly, who was extremely avid to know what was going on and why he couldn't get access to the information on the UFO cover-up.]

There were some rumours circulating, though, not officially acknowledged by White House staff, in which it was claimed that Carter had a meeting with Spielberg to let him know how much he enjoyed his movie, "Close Encounters". There is a photographed of the two of them together having a casual conversation and is signed: "To **Steven Spielberg**, [from] **President**

52

Jimmy Carter", indicating that a meeting between the two did, in fact, take place. The President's continued interest in getting the truth out on UFOs triggered a red flag response from "the military-intelligence community – and even among Carter's staff – to keep the Administration publicly from being further associated with flying saucers. A UFO-spotting President viewing the ultimate UFO movie at the **White House** and having get-togethers with its alien-obsessed director would have been a PR nightmare. Not only that, but it would likely have exacerbated the White House's existing UFO-related problems.
http://www.sightings.com/1.reports2010/spielsaucersecrets.html

"Spielberg's most successful alien film, *E.T.*, was also an Oval Office favourite. Incumbent, this time, was **President Ronald Reagan**, who, like both Carter and Spielberg, was also known to be keenly interested in UFOs. During a private screening of *E.T.* at the White House in 1982 in which Reagan and Spielberg were seated together, the President is reported to have leaned over to his guest and whispered: *"You know, there aren't six people in this room who know how true this really is."* Spielberg related this story to Hollywood television producer Jamie Shandera shortly after the screening." http://www.sightings.com/1.reports2010/spielsaucersecrets.html

It is not hard to imagine Reagan having made the statement. Like President Carter, Reagan had also seen a UFO, not once but twice (whilst Governor of California), and his daughter, Patti Davis, had spoken of her father as being, "fascinated with stories about unidentified flying objects and the possibility of life on other worlds." Reagan's belief in aliens even affected the content of key policy speeches. In a September 1987 address to the General Assembly of the United Nations, the President said:

"I occasionally think how quickly our differences worldwide would vanish if we were facing an alien threat from outside this world..." He then paused, before posing a cryptic question: *"And yet, I ask you, is not an alien force already among us?"*

Most assumed he was speaking metaphorically, but a discussion two years previous at the Geneva summit conference between **Reagan** and Soviet leader **Mikhail Gorbachev** raised the possibility that he may actually have meant it. Reagan had suggested that, in the event of an alien invasion, their two countries should fight back as one. Gorbachev himself acknowledged this discussion during a public speech at the Kremlin in 1987:

"The US President said that if the Earth faced an invasion by extraterrestrials, the United States and the Soviet Union would join forces to repel such an invasion," Gorbachev said, before adding, somewhat wryly, *"I shall not dispute the hypothesis, though I think it's early yet to worry about such an intrusion."*

It's no secret that Spielberg has discussed the UFO topic with at least two Presidents besides Reagan – these most likely being Democrats Carter (for the reasons cited above) and Clinton (with whom the director was known to be chummy). According to Spielberg, the Commanders in Chief in question had no special access to the government's alleged UFO secrets. He even implied they had been deliberately kept out of the loop:

"I know a couple of Presidents who certainly have not been 'clued in' – or at least they're telling me they haven't been clued in – as to the existence of any hard evidence that we've been gathering."

This would seem to gel with the allegations about President Carter's failed attempts to access sensitive UFO information. Clinton, too, publicly has voiced his frustration at being stonewalled on the issue. At a speech in Belfast in 1995, Clinton made a point of bringing up the famous **'Roswell Incident' of 1947**:

"If the United States Air Force did recover alien bodies, they didn't tell me about it, either, and I want to know."

He was even more direct in a question and answer session following a speech in Hong Kong in 2005. When asked about Roswell, the President replied:

"I did attempt to find out if there were any secret government documents that revealed things. If there were, they were concealed from me too. And if there were, well I wouldn't be the first American President that underlings have lied to, or that career bureaucrats have waited out. But there may be some career person sitting around somewhere, hiding these dark secrets, even from elected presidents. But if so, they successfully eluded me."

Spielberg has produced and directed both positive and negative sci-fi movies. His **"War of the Worlds"** (a re-make of the original 1953 **George Pal** classic) is definitely a negative "Bad Alien" movie and his made for television mini-series, **"Taken"** is considered positive by many Ufologists but, perception is everything. When many Ufologists and the general public still buy into the **"bad Gray ET"** abduction scenario, whether the reason for aliens committing the act is positive or not or whether the act is fully understood by humans, the perception is still profoundly **negative**!

This brings up the question that given **Spielberg's** knowledge about the UFO phenomenon with regard to the movies he has produced and directed, one has to ask, did he have insider information from some higher intelligent source or did he merely do his homework on the subject?

Researched work of any kind is for the most part always built upon the previous work of others. The fact this book is written upon the investigative and researched work of other Ufologists gives each succeeding researcher such as this author, a greater perspective of the UFO/ETI subject matter than our predecessors. It is the natural way by which science grows and develops. Anyone, including Spielberg, can with a little reasoning, logic and some intuitive insight and perhaps, some personal experience with the phenomenon, come up with some intriguing answers that would surprise even those who are in the know about this subject. How much information you uncover depends on *"how far down the rabbit hole you want to go"* to get to the answers!

Thus, the answer is that he probably has done his homework aided by his own personal research team but, he may also have had some limited insider information divulged to him by those who are in the loop because his own personal research work uncovered more than the intelligence agencies thought was possible for someone in his position. He is sometimes referred to as the "go to guy" in what is thought to be a secret program to acclimatize the public to the reality of extraterrestrials through entertainment media.

Author's Rant: In the opinion of this writer, the only reason why the intelligence community feels the needs to acclimate the public to the reality of Extraterrestrial beings visiting the planet is because the public, for the most part, is deeply entrench in religious dogma and the superstitious beliefs and traditions of the church-state.

This would explain his early positive ET movies but, it is suspected that his negative stereotyping of the "Bad Aliens" was probably due to pressure placed upon him from the **Movie Mafia** or from sources higher up in the intelligence community. No doubt, in the near future Ufologists and the general public will find out the real reasons and the decisions behind why he made the science fiction movies in the manner that he did and his connection to insider presidential information on the UFO question.

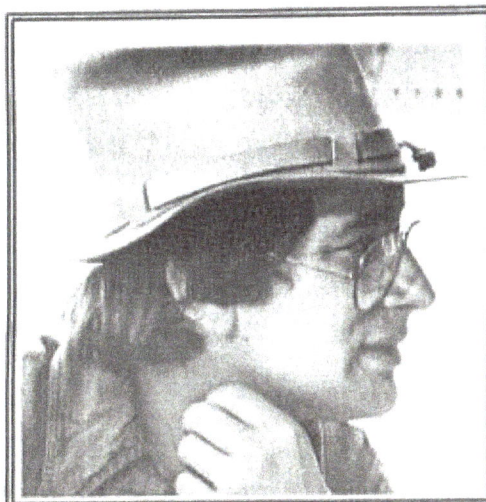

STEVEN SPIELBERG

Since you are scriptwriter and director of this film, you must have a certain attitude to the UFO phenomenon. Do you believe in close encounters?

I believe in the possibility, in the 30 years of evidence. I am not 100 per cent convinced, and I haven't had any direct experiences, my attitude has always been "Prove it". But I am naïve convinced now than I was three years ago.

Steven Spielberg's "Close Encounters of the Third Kind" is at present outpacing "Star Wars" at the box-office and may possibly become the biggest grossing film of all time. If so, Spielberg will have twice achieved that feat; the other time being with "Jaws".

Spielberg graduated from UCLA in 1970 and went straight to Universal where he directed episodes for several television series, including: "Marcus Welby, MD", "Columbo" and "Name of the Game". He

from major airlines, air traffic controllers, U.S. Air Force officers, even four security people at the Pentagon who, during the early 1950s, had worked in the intelligence corps and were around when UFOs buzzed the capital; there was a great flap in Washington. It sounds like a wonderful science fiction film, but Washington took it very seriously.

The best people I talked to, however, were the average family types who never expect anything

Out-of-print *Cinema Papers* © 1978 interview with Steven Spielberg, discovered by author Matthew Alford while researching the history of Hollywood cinema.
https://issuu.com/libuow/docs/cinemapaper1978aprno016

have been stronger if you had not shown these extra-terrestrials?

Not for most people, because they would have been frustrated at not having seen the vision completed. A lot of people think I should not have shown the shark in **Jaws**, that I should have continued the mystery of the water, so that the water itself became the threat. But that's my duality — the philosopher-filmmaker and the commercial-filmmaker-entertainer. I try to make those two things work for each other.

Did you consider not showing the creatures?

Yes, for a long time, and I personally felt a great disappointment in not knowing what piloted those things. In **2001** Stanley Kubrick considered the same thing because he shot many aliens — but he never used them in the final film. That was fine for **2001**, because from the beginning it had promised an esoteric payoff; you didn't ever expect to see an extra-terrestrial.

My film isn't so technologically intellectual, and because of this it would be wrong not to show the creatures.

Why did you choose Dr Allen Hyneck as technical advisor on the film?

I knew of Hyneck when I first began researching the film because he was famous for saying how it was all a bunch of bunk. He had been hired by the Air Force to give easy explanations to complicated phenomena and he was very good at it.

Hyneck would say a phenomena

Francois Truffaut as the French scientist, Claude Lacombe, and Bob Balaban as his interpreter. **Close Encounters of the Third Kind.**

up of the film were you more in doubt than not?

Sure, when I met a lot of kooks whose stories weren't consistent the second and third time round. I felt very disappointed, suspecting that maybe only the more intelligent people knew how to make up a good story. But fortunately it didn't happen too often.

I really found my faith when I heard that the government was opposed to the film. If NASA took the time to write me a 20-page letter, then I knew there must be something happening.

I had wanted co-operation from them, but when they read the script they got very angry and felt that it was a film that would be dangerous. I think they mainly wrote the letter because **Jaws** convinced so many people around the world that there were sharks in toilets and bathtubs, not just in the oceans and rivers. They were afraid the same kind of epidemic would happen with UFOs.

It was the same with the Air Force; they gave us no co-operation at all. So when I was shooting the scenes with the army and air force, I had to do it the old-fashioned way, and go into a

Who's directing who? The two 'directors' — Francois Truffaut and Steven Spielberg.

admitted that UFOs exist, and that they are interested.

Was it at any point a moral issue for you — that you might cause panic?

Not really. When Orson Welles did his famous "War of the Worlds" broadcast in 1938, he was not so much writing a radio program about Martians invading New Jersey as about America's fear of invasion from Europe. War

Today it's just the opposite. I knew that if this film was to be popular it wouldn't be because people were afraid of the phenomena, but because the UFOs are a seductive alternative for a lot of people who no longer have faith in anything.

Did you require your actors to have a similar degree of belief as yourself?

No. Melinda Dillon believes,

Venus. Then he began coming across reports that were too extraordinary to be discounted easily. He found he could explain away 80 per cent of reported sightings, but there was still 20 per cent he couldn't, and he became fascinated by it. Finally, he went to the Air Force and said, "Hey, I think there's something here; this isn't just public psychosis."

The Air Force got very nervous and told Hyneck to mind his own business and just do his job. He got very angry and quit. He then wrote a book attacking the department.

I met Hyneck because he was a man who had suddenly learned to believe, and that was a very uncommon thing to do. I felt he was a very valuable man to have on my team because he could give me the feeling that I wasn't just making a film about chiffon; that it wouldn't be something that couldn't stand up under a hot light.

At any point during the setting

costume store and buy the army suits and gear.

Apparently President Carter has seen the film . . .

Yes. Carter likes it very much. He has reported UFOs on two occasions, and I think he's a believer. In fact, one of his campaign promises was that he would try and find out what UFOs were all about. But the minute he took office and was asked whether he was going to follow through the promise, he side-stepped the issue.

Since then, the White House has been very quiet concerning UFOs. It seems that every president, including Gerry Ford, who is interested in UFOs, stops being interested the minute they get to the White House.

There is something going on which many governments in the world feel that people should not be made aware of yet. France and Brazil are the only two countries whose governments have

Welles' invasion was not the Stuka, it was the Martian; it preyed on the vulnerability of that time.

does Richard Dreyfuss nor Truffaut. When Truffaut was asked if he believed in UFOs, he said, "I believe in the cinema".

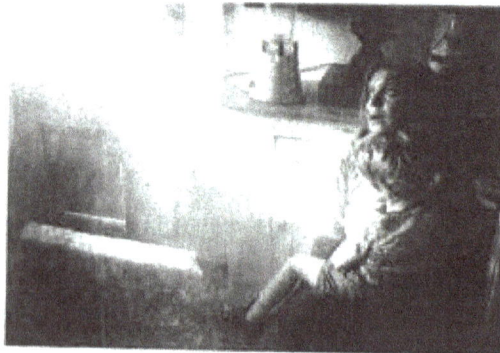

The mysterious light generated by a UFO. While a mother (Melinda Dillon) is terrified, her son (Gary Guffey) is more trusting. **Close Encounters of the Third Kind.**

Other movies with similar theme portrayals of the "Good Aliens" are "Starman", "ET – The Extraterrestrial", "The Abyss", "Mission To Mars", "The Day The Earth Stood Still" (the original 1951 version with actor Michael Rennie), etc.

A Historical Meeting: Mankind's First Open Contact with Extraterrestrials

It has been stated by some researchers that the movie, "The Day the Earth Stood Still" was, in fact, a docudrama based on real events that occurred with contactees leading up to the eventual close contact meeting with ETIs that occurred at Muroc Air Base in 1953. During an afternoon of golf near Muroc Air Base, it was reported that President Eisenhower was taken from the golf course under an escort of staff vehicles which meant that something major had taken place. Reporters were caught off guard and rumours immediately began to fly as to what had happened to the President. It was speculated that he suddenly had a heart attack and had died and thus his sudden disappearance from the golf course for several hours. The eventual cover story given to reporters was that **President Eisenhower** had developed a toothache or a chipped tooth and was immediately taken to see a dentist to get it fixed.

However, the real story was that he was taken to Muroc Air Base Somewhere he had a historical meeting with Extraterrestrial beings who had landed in a flying saucer at the air base. This historical meeting was humanities first open contact with an intelligence from the stars but, this meeting was immediately covered up by then, nascent Military Industrial Complex. The Extraterrestrial beings were concerned for humanity's warlike nature, our ability to develop, test and explode many nuclear weapons which was impacting not only the Earth's biosphere but other dimensions of reality. We had also demonstrated our ability to build rockets and to leave our own planet. Such Terrestrial technological development spelled potential danger to other galactic civilizations by a warring warrior race that was now capable of travelling to the stars armed to the teeth with devastating nuclear weapons.

The interstellar beings were willing to offer us the technology of an infinite source of energy now known in the science community as **Zero Point Energy**, *an energy derived from the quantum flux of space. This new technology would be infinite, environmentally friendly and pollution-free, cheap (next to nothing in cost) in its extraction and deployment and could be microminiaturized in its utilization.* It would literally have changed the course and direction of social development on this planet for all mankind. We could have been already in the beginning period of a golden age of civilization on Earth but, this was unfortunately not the case.

It was said that **President Eisenhower** did not accept this generous offer in return to our abandoning our nuclear capabilities and warring behaviour stating that such a new source of energy would cause an economic and social upheavals on the planet. At this time such new technology would create chaos particularly for the mega corporations like the oil, coal, and utility companies as well as the financial institutions around the planet.

**Gort and Klaatu (starring Michael Rennie) from the 1951
sci-fi thriller, "The Day The Earth Stood Still"**
https://www.britannica.com/topic/The-Day-the-Earth-Stood-Still-film-1951/images-videos

Truth is never what it appears as the real facts were that President Eisenhower wanted this information to come out into the public domain but, he was dissuaded by the military and intelligence departments around him which began the tightening of the noose around the necks of Americans and the whole world in general. Our hopeful good future had been hijacked by a rogue group who had another agenda which would benefit a select handful of the earth's population. Eisenhower at this point had lost control of the situation and eventually, as were most of the US Presidents who followed after him, was cut out of the loop of a need to know.

 Now this account or one similar to it has been around for some time with some witnesses coming forward to state that it was a real event. **Dr. Greer** has a document in his possession from one of the witnesses who was there at the time which reveals that the Presidential meeting with ET beings from another planet was, in fact, real! This is a historical document describing a

58

historical event that truly is of incredible importance to all of humanity! Greer states that President Eisenhower wanted to disclose this information to the public but was dissuaded not to by his Military Staff and thus, he lost control of the situation and it became apparent then, that his last public speech was to warn the people to be aware of the **Military Industrial Complex.**

With such an historical event as this, **mankind's first official meeting with another intelligent civilization from the stars**, its impact requires all of us to take a few steps back a bit to re-examine and re-evaluate the official position and statements presented to the public by our military, government and intelligence officials, communities, and agencies. There is a distinct boundary here between what investigators into UFO and ETI research are saying and what is the people in officialdom who supposedly represented us in all matters are saying. There is a position here that is black and white where one party is telling the truth and the other party is lying. Both can't be right or both be wrong. It's either one or the other and so far, the evidence that we have presented in this book is stacking up very high against officialdom and it does not look very good for them in the long run of this subject matter.

We also, need to re-examine the evidence and claims from the 1950s and 60s of contactee cases to which similar information was imparted to those fortunate few who stated that they had open contact and communication with ETI. Perhaps, not all those claims were hoaxed and prevarications, some of those claims may actually have been real! It appears that the warnings of nuclear danger, of polluting and raping the environment and a new energy source which powers the flying saucers as reported by the contactees seems to be in keeping with what was related to Eisenhower during his ETI meeting at Muroc Air Base in 1953!

The Internet of Truth: Believe Everything You Read

Arthur C. Clarke predicted in *Popular Science* magazine in May 1970 that satellites would one day "bring the accumulated knowledge of the world to our fingertips" using an office console that would combine the functionality of the xerox, telephone, TV, and a small computer so as to allow both data transfer and video conferencing around the globe.

The internet is a great source of information on almost anything subject that tickles your fancy or anything that inspires you to learn about the world around you or in the universe. It is like going to a library and finding that book you always want to read without being assigned to a priority waiting list for it. In fact, it can be considered a pass to every library throughout the world from the smallest to the largest, from the most insignificant to the most prestigious, from the private to the most public, from the most current sources of information to the most ancient and hidden! If you possess a computer you are merely a couple of clicks away from any subject matter that may interest in which you desire to become more familiar with.

But the internet is more than merely a source to all the world's libraries. It is also, a source to all the museums and art galleries of the world, from the awe-inspiring statues and motifs of antiquity to impressionistic sculptures of the current age, to all the paintings from the masters of bygone eras to all the latest modern day art nouveau, and digital photographic impressionists. Absolutely anything you can think of which is not restricted or suppressed to the general public can be viewed from your computer in the comfort and privacy of one's home. The computer age linked to the lightning quick access from satellites and microwave networks has literally changed and revolutionized the way we view the world, how we do business and how we see and relate to each other. It has shrunk the world where distance becomes almost meaningless and time irrelevant. Computers and the internet have quickened our intellectual and spiritual evolution on this planet and have multiple our physical capabilities in whatever we decide to undertake.

A quick little history lesson on computers is necessary here to illustrate that the military mindset views all things as a potential weapon or a tool to further the military agenda particularly in the engagement of war scenarios.

Early computers were electronic – mechanical in their operation and functioned by a binary code of ones and zeros. Concurrently a suitable computer was needed to serve the military's requirement for war and post-war time usage. **ENIAC (Electronic Numerical Integrator And Computer)**, was the first general-purpose, electronic computer. It was a digital computer capable of being reprogrammed to solve a full range of computing problems.

ENIAC was designed to calculate artillery firing tables for the United States Army's Ballistic Research Laboratory, but its first use was in calculations for the hydrogen bomb. When ENIAC was announced in 1946 it was heralded in the press as a "Giant Brain" not only because of it number crunching ability but, mainly because the computing equipment was housed in a very large room. Today, the computer memory housed in such a room could easily in fit into an everyday handheld calculator. An **iPod touch or iPhone** would be comparable to a high-rise apartment with each floor filled with these early ENIAC computers. It boasted speeds thousands of times faster than these antique electro-mechanical machines, a quantum leap in computing power that no single EINAC machine could match.

The ENIAC's design and construction was financed by the United States Army during World War II. The construction began in secret by the University of Pennsylvania's Moore School of Electrical Engineering under the code name **"Project PX".** The completed machine was unveiled and was formally accepted by the U.S. Army Ordnance Corps in July 1946.

"The concept of data communication - transmitting data between two different places, connected via some kind of electromagnetic medium, such as radio or an electrical wire - actually predates the introduction of the first computers. Such communication systems were typically limited to point to point communication between two end devices. However, the point to point communication model was limited, as it did not allow for direct communication between any two arbitrary systems; a physical link was necessary. The technology was also deemed as inherently unsafe for strategic and military use because there were no alternative paths for the communication in case of an enemy attack."
http://en.wikipedia.org/wiki/History_of_the_Internet

The room size (thirty by fifty feet) "Giant Brain", ENIAC Computer in operation being attended by its programmers in BRL building 328. (U.S. Army photo)
http://www.computerhistory.org/revolution/birth-of-the-computer/4/78/321

The USSR's launch of Sputnik spurred the United States to create the **Advanced Research Projects Agency (ARPA or DARPA)** in February 1958 to regain a technological lead.[2][3] ARPA created the **Information Processing Technology Office (IPTO)** to further the research of the **Semi Automatic Ground Environment (SAGE)** program, which had networked country-wide radar systems together for the first time. The IPTO's purpose was to find ways to address the US Military's concern about survivability of their communications networks, and as a first step interconnect their computers at the Pentagon, **Cheyenne Mountain, and SAC HQ**. In the 1983 movie, "WarGames" with Matthew Broderick, the underground SAGE Site in Cheyenne Mountain was central to the movie's theme and provided the viewing audience of the computer facility layout of the complex. Like the movie, WarGames, some areas in this underground complex are open to the general public for tours on certain days of the week.

Canada also, as a part of **NORAD (North American Aerospace Defense Command),** has its own **SAGE Site** located ten miles south outside of North Bay, Ontario and ten miles east from the **CFB North Bay Airbase**. With the formation of NORAD in the 1950s and the US's introduction of the SAGE system, CFB North Bay was started in 1959 as the Canadian counterpart to the US's Cheyenne Mountain control center. Unlike their US counterparts which were at ground level, in North Bay, the entire standard three-story installation was buried underground in what became known as "the hole". It is situated about 700feet under a granite base mountain beside Trout Lake and is linked to **Cheyenne Mountain Command Headquarters** and like Cheyenne Mtn. it is built on giant seismic tampering springs which can withstand a direct hit by a thermonuclear bomb! The **Trout Mountain SAGE Site** could house 2500 air force personnel and their families in the complex. Interestingly, back in 1964, the underground complex had one floor devoted entirely to a large room-size computers, its associated equipment, and the communication nerve centre. Like Cheyenne Mtn. the entrance into the complex could be closed off with a very large four-foot thick steel door.

The **Semi-Automatic Ground Environment (SAGE)** was a revolutionary piece of computer technology used to support the air defense for all of North America. This created one of the first large-scale computer networks and kept thousands of **SAGE** operators linked to information about possible threats. CFB North Bay was the ONLY military base in North America that held both a SAGE Combat Centre and a SAGE Direction Centre, and was the only Canadian location with equipment of this caliber". Later the base was also used as the control center for the Ontario portion of the two-site **BOMARC missile system** installed in the 1960s located about 10 miles north of the city. The BOMARC missiles were later decommissioned in 1973.
http://www.wordiq.com/definition/CFB_North_Bay

Author's Rant: As a child, my brothers and I would go to the air shows at CFB North Bay yearly to view military aircraft as well as marvel at a scale model of the NORAD SAGE Site complex located near the base and the city. My father as a Corporal instructed and trained other servicemen in nuclear defense protection and survival, his other duties often found him working in the underground SAGE Site and Chalk River Nuclear Laboratory Facilities.

No doubt, upgrades to this complex and many computers now fill the same floor that use to house one large mainframe computer. Most major military bases throughout the US and Canada also have these large underground complexes utilized by all the various other military departments including **Space Command** which is housed in Cheyenne Mountain. Most of these complexes are connected together with an underground tube-tunnel with a very high speed, electro-gravitic train to facilitate travel between military bases.

ARPANET (Advanced Research Projects Agency Network), was thus, created by a small research team at the head of the **Massachusetts Institute of Technology (MIT)** and the **Defense Advanced Research Projects Agency (DARPA)** of the United States Department of Defense. ARPANET was the world's first operational **"packet switching network",** and became the technical core of what would become the Internet and a primary tool in developing the technologies used.   http://en.wikipedia.org/wiki/ARPANET

Once again, it is the military industrial sector that developed a technology of mass information processing and communication as a tool for war which eventually has found its way into the hands of the public sector for commercial business and personal use. The computer is primarily a tool for data storage, processing, and retrieval which communicates to other computers via the internet. Think of the internet as a network of networks, much like the human brain with its billions of neurons and nerve endings all interconnected to each other coursing through the brain and throughout the body. The internet, however, is not the Web!

The **World Wide Web**, or simply the **Web**, is a way of accessing information over the medium of the Internet. It is an information-sharing model that is built on top of the Internet. The Web uses the HTTP protocol, one of the many languages spoken over the Internet, to transmit data. Think of the Web as one of the many human senses in which we access and obtain information around us.

3D Map of the World Wide Web illustrates the actual domains and connections of the world wide web. Colors have been added to represent .edu, .gov, .com, etc. domains.
http://www.vlib.us/web/worldwideweb3d.html and (courtesy: www.opte.org)

What we need to understand here is that the computer is one more form of the written word but into digital or virtual form and like many good things it is a wonderful tool for learning but when certain people have an agenda that benefits themselves at the expense of the many then it can become a tool of deception.

The two top website areas on the internet that people view the most are Sex sites and UFOs, in that order! The Sex websites are self-explanatory and are for self-indulgent prurient interest only. It is also possible that some websites are created by government and intelligence communities for their own covert agendas as a deliberate distraction away from other things on the internet. Thus, such websites become a self-generating mechanism by the addictive personalities who seek out this baser form of human sexuality and who desire to propagate such material. No more detail is required to further explain this aspect of the internet. UFOs and ETI websites, on the other hand, are a whole other story.

UFO information suppression by the Big Media and the Big News Press is an established fact which we have already discussed but, with the internet, we now have the opposite extreme in that there is an avalanche of competing UFO and ETI "news" sources.

So, whereas the traditional media provides too little information on UFOs, the internet *has* too much information, which many ufologists see as both a boon in UFO research but also, somewhat detrimental to ufology with websites filled with misinformation and disinformation. Today, it is possible for anybody to set up a site and upload fabricated stories or web share other people's fantastic conspiracy theories which spread like wildfire, never really being substantiated by the person sharing the conspiracy. To some people it just a big joke to share with their buddies or to play upon the gullibility of serious researchers., some do it as deliberate disinformation, while others such as CGI hoaxers to promote their ability and make money or in selling a book or a video. For these reasons and more, people surfing the internet for information often get the impression UFO websites and the subject, in general, are for the "nut bars" and the delusional. In such cases, they simply surf to a web suit that is more relevant to them.

People visit UFO websites for two reasons. One, out of a genuine curiosity for a need to know about something unusual in their environment which has been with mankind since recorded time but, has never been adequately explained. Two, due to popular press and media coverage in the days before the media became heavily controlled by the intelligence community, there was an ever increasing concern that government and military officials were covering up and suppressing evidence of alien visitations, crash retrievals of spacecraft and alien bodies. The public grew suspicious of officialdom's explanations and pat answers to a genuine mystery that was not being publically recognized by our governments and military or being seriously investigated by the science community. Thus, they turned to the internet to search for answers and for an exchange of dialogue with other people in their country and with others of like mind around the world.

The internet continues to be a great source of both old reports and more current UFO and ETI accounts from around the world. Youtube, Stumble Upon are excellent sources of recent UFO video but, both sites, unfortunately, are also, susceptible to CGI fakery and a lot of tomfoolery and nonsense! Like everything in life, you have to wash and sift the silt and dirt out of the pan before you discover the flakes or nuggets of gold; UFOs and ETI research is very much like that.

64

You have to remain a healthy skeptic willing to investigate the absurdity, the outright fakery and hoaxes, the physically impossible to get at the nuggets of truth and you have to it ad nauseam. Eventually, with continued investigative effort and many other pieces of corroborating evidence, you strike pay dirt!

People will often say, "You can't trust everything you read or see on the internet"! In fact, you can't trust anything you read in a book or watch on TV or even trust anything anybody tells you! The world, in general, has become so cynical and skeptical in its outlook to life that we rarely accept the value of truth first hand from anyone! Call me old fashion but, whatever happen to a man's word was his bond? Somewhere between the industrial revolution of the late 18 century and seeking world domination in the early 20th century, any honesty in a person has become an increasingly rare and valuable commodity! This negative mindset carries over to what we read and see on the internet. When the industrial world runs almost all its business via the computer and the internet, we can ill-afford to ignore whatever is virtually printed on the World Wide Web. To do so is to leave ourselves open to fraud and manipulation by the unscrupulous of the world. We have to be participants in the virtual world to keep it honest at all times, we have no choice unless we want to start living in a cave of ignorance.

So far, it appears that there is no active suppression of information on the subject of UFOs and ETI as there is a constant stream of reported sightings and accounts being put up on the internet from around the planet. Since the first reported cases that came to the public's attention in the modern era, the UFO phenomenon has never been more active. In keeping pace with this ever growing public interest, the agencies of disinformation and misinformation have been busy by placing false stories on the internet of hostile encounters or interactions with ETs, whether by the military or by the public citizenry. Such stories become more believable when they are accompanied with faked photographs or videos (usually slightly out of focus or shaky and unstable) depicting ET craft and alien beings. **CGI (Computer Generated Imagery),** artists who wish to establish a reputation in the movie industry will often claim ownership of such hoaxed filmed events after they have run their course on the "radar of public opinion" or have been thoroughly researched and proven after investigation by Ufologists as hoaxes. The hope, of course, is to catch out many Ufologists in an elaborate hoax for which they have given "thumbs up" on a case as the genuine article or event. This then sets into motion the debunkers and skeptics, who have a public field day at the expense of the "pseudo science" call Ufology with the ultimate aim of discrediting those who investigate such claims by saying, "We told you so!"

A recent UFO video has surfaced in the last year or two that showed a UFO flying over some palm trees in the early twilight evening in Haiti which seemed initially impressive and which had a lot of detail in perfect clarity to the point, one could almost see the nuts and bolts of the ET spacecraft. Careful analysis of the video indicated that it was a clever CGI hoax and since then, similar videos have also shown up on YouTube, a major source for UFO and ET videos.

Another case was the video of a UFO film in Italy in the early 2000s. Here the UFO or flying saucer takes off from a river bed, moves toward a tower structure, hovers and morphs by projecting some fin-like surface structure then, moves off down the river embankment and disappears over and beyond a bridge. This may very well have been a genuine flying craft of advanced technology but, it is possible that it was manmade and not Extraterrestrial in nature.

Since then, other similar videos have appeared on YouTube showing the morphing structure of spacecraft in remarkable clarity. These too have proven to be clever CGI hoaxes.

Although these video of CGI hoaxed UFOs may have been innocently created to reflect the talent of their creator who hopes to break into the film industry and work for the likes of **Industrial Light and Magic**, they are unwittingly or intentionally creating a false perception that all clear UFO videos are real. It is just another form of debunking the phenomenon and causing Ufologist to waste their time on chasing down false cases or to create the perception that when a real filmed event occurs that ipso facto, it too, must be a clever CGI hoax and in the mind of the internet user interested in this subject, he will dismiss a lot of what he views as not real but hoaxed and thus, he will lose any further interest in UFOs.

Another case in point, an event known in Ufology as the **Ray Santilli** "Alien Autopsy" film of an alleged Extraterrestrial being retrieved from a crashed saucer near Roswell, New Mexico in 1947.

The supposed autopsy first gained prominence during the 1990s when Ray Santilli, a London-based video entrepreneur, promoted a 17-minute black and white film purporting to be footage of the autopsy. In 1995, the film was sold to television networks internationally and broadcast with high viewer ratings in more than 32 countries. It has since been accessible on the internet.

In 2006, Santilli admitted the film was not entirely authentic, claiming that it was a reconstruction of lost footage of an actual alien autopsy film that he viewed and that a few of the frames embedded in his video were from the original. After this admission, Santilli's film is largely considered a hoax.

While dismissed as fantasy by most people, belief that alien autopsies have been carried out forms a core component of a number of UFO conspiracy theories, though the term itself is used within Ufology, fiction, and in popular culture, regardless of the factual status of the imagery that is being presented. http://www.imdb.com/title/tt0163521 **Beyond Roswell: The Alien Autopsy Film, Area 51, & the U.S. Government Coverup of UFOs by Michael Hesemann (Author), Philip Mantle (Author), Bob Shell (Author) Paperback: 303 pages publisher: Marlowe & Company (August 1998) ISBN 1-56924-709-9 ISBN 978-1-56924-709-9**

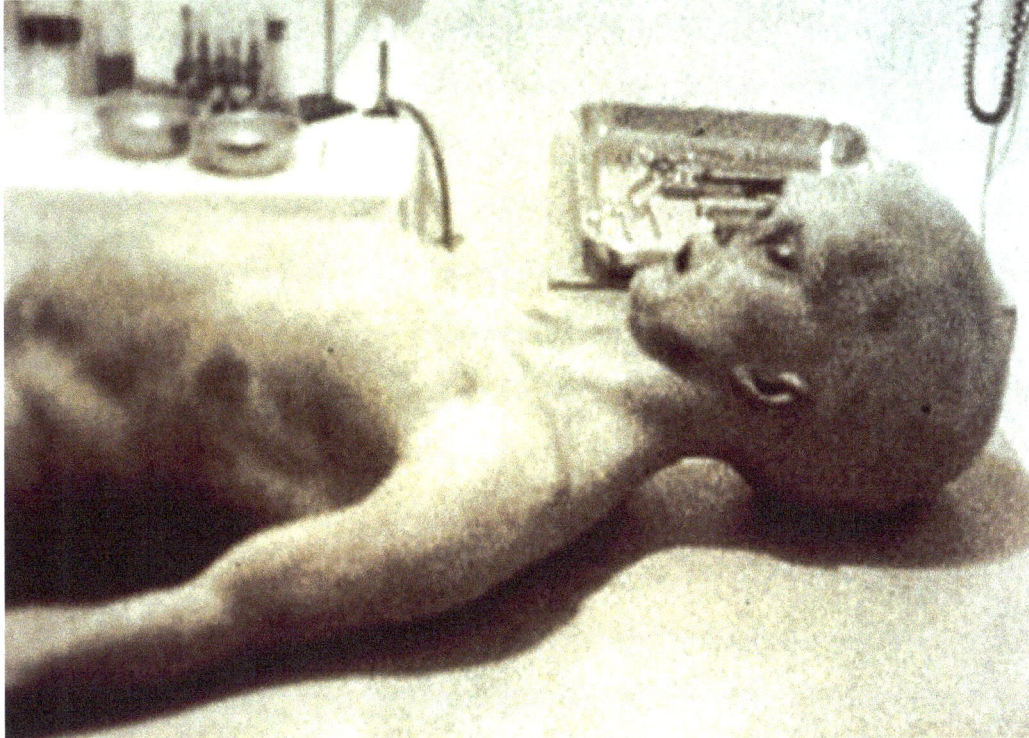

The Santilli "Alien Autopsy" film of an alleged Extraterrestrial Being retrieved from a crashed saucer near Roswell, New Mexico in 1947.
http://www.openminds.tv/the-shroud-of-ufology-part-1

In this particular case, it demonstrated that Ufology was up on its game of doing a thorough investigation into the reported claim. Score one for the "pseudo science" of Ufology! Professional debunkers both new and old came out of the woodwork to discredit the UFO researchers but, quickly scurried back to their holes of skeptic magazines and lecture tours having lost this round to Ufology.

CHAPTER 39

DISCLOSURE VS. THE MILITARY INDUSTRIAL COMPLEX

The UFO/ETI Truth Embargo

As the UFO/ETI phenomenon evolves closer to the day when there is final official disclosure, finds the subject is placed under a **"Truth Embargo"**. This Truth Embargo, a term coined by Washington lobbyist and UFO activist, **Steven Bassett** to describe the suppression and control of UFO information, has been co-opted by the covert **Majestic 12 (aka. Majic 12 or MJ 12, etc.)**, the actual controlling rogue group within the Military Industrial Complex, which is tightening its stranglehold on the release of this information.

The UFO/ETI subject, however, is like fine sand; the tighter you try to hold onto this sand of information from public disclosure, the more the sand will run through your fingers and pour out of your hand onto the ground until, it become obvious to you and everyone else, that there is a conspicuous pile of sand before your feet. This conspicuous pile of incontrovertible evidence is the nature and reality of the UFO and ETI phenomenon and it cannot be hidden forever!

Like a mighty crescendo in the musical score of an operatic play that has reached its final act, it appears that the phenomenon is about to unfold in a full blown disclosure event the likes of which may take most military and government officials and their countries by surprise. There have been massive releases of UFO documents and files from many governments of the world that seem to be indicating that one or several governments may be ready to step forward to publicly disclose the present of one or more Extraterrestrial Intelligences currently visiting the Earth.

CSETI International Director, **Dr. Steven Greer** has publicly stated that unless the USA steps forward with unwavering leadership to end the **Truth Embargo** on the subject of UFOS and ETI then, there is a **G8** nation which has approached Dr. Greer and asked that with his assistance, their government will take the lead in establishing full open ET contact and disclosure! This could in all probability occur in less than two years, barring any unforeseen aggressive military or intelligence action or any political sanctions toward Dr. Greer or towards this G8 nation.

According to Washington lobbyist Steven Bassett, such a bold move by this G8 nation could irreparably displace the US from its world leadership role. Ever since, the U.S.'s heavy involvement in the late 40s, in reverse engineering alien technology and its genetic research program into Extraterrestrial biology, compartmentalized rogue elements within the **Military Industrial Complex**, who hold ultimate control on the UFO/ETI subject, have made it nearly impossible for the US to share this knowledge with the rest of the world's science communities. For this reason, Bassett feels that the U.S. may be incapable at any time in the foreseeable future of ever providing full disclosure on this phenomenon but, he does continue to hold out hope for a positive outcome.

Now, in the last few years, many countries have opened up their "X-Files" declassifying and releasing their documents on UFO accounts that the public has reported and sent to their local police, military and government officials over the decades. These UFO reports number in the tens to hundreds of thousands of cases and are usually released over the World Wide Web either yearly, semi-annually or even quarterly. These declassified reports come from the government agencies and departments of many countries responsible for the collecting such reports. Declassification of UFO accounts have come from Britain, Canada, Australia, Sweden, Russia, Ukraine, Finland, France, Denmark, Spain, Italy, Mexico, Brazil, Peru, Ecuador, Uruguay, and many countries follow that decision during 2009 making public all the material they had in possession. The government of New Zealand has stated that it will release UFO files in 2010 after personal information is removed. Below is a growing list of countries declassifying and releasing their UFO documents.

Countries That Have Officially Disclosed the UFO Presence

Brazil -an emerging world power—disclosed its UFO documents to its people in 2005. The content of this material has now been examined and proved to be very interesting and revelatory of Brazilian official strategy to cover UFO cases, often openly regarded as "from external sources" in these new documents, meaning from outside Earth. The cases covered and the methods used to investigate them by the military are also very significant.

Australia -In 2006, Australia joined the list of countries that have been releasing government-held UFO files into the public domain.

France -March 2007 -some 100,000 documents on supposed UFOs and sightings of other unexplained phenomena that the French space agency is publishing on the Internet. France is the first country to put its entire weird sightings archive online, said **Jacques Patenet**, who heads the **French Space Agency's (CNES)** UFO cell — the **Group for Study and Information on Unidentified Aerospace Phenomena**.

United Kingdom - The release from the UK has happened in stages starting in May of 2008-2010: The release is by decade which indicates the mass quantity of evidence... They are available online through the **UK National archives**

Canada -February 2009 The Canadian Government has authorized open public access to thousands of federal government documents concerning UFOs. A total of 9500 digitized documents spanning the years 1947 to the early 1980s has been made available through the **Library and Archives Canada** website. Titled "Canada's UFOs: The Search for the Unknown" the files include correspondence, reports, memos and procedures, some of which specifically deal with UFOs. The files come from Canada's **National Defense Department**, the **Department of Transport**, the **National Research Council**, and the **Royal Canadian Mounted Police**.

Denmark -The Danish Air Force has followed France and the UK, and released UFO Files to the Public (released on the 28th January 2009)

Sweden -Sweden is releasing thousands of UFO files beginning May 9-10, 2009, according to a news story in Swedish online news magazine Expressen. Translation of the story was provided by Exopolitics Sweden and editor of Soul Travel Magazine. Sassersson's translation states that:"Besides the 18,000 Swedish cases are thousands of cases from Denmark and other Nordic countries. Thousands of US cases on microfilm are [also] included. The archive (Norrkoping) will be open to the public domain starting with May 9-10, 2009.

Uruguay - June 2009 -**Air Force Declassified Uruguay Files**, ET Hypothesis not dismissed. **Russia - Russian Navy releases UFO files** July 2009, The Russian Navy recently declassified and released UFO records for the first time. One of the more interesting details is that fifty percent of the UFO encounters reported occurred near large bodies of water, including many in the **Bermuda Triangle.** Also, **Lake Baikal** in Russia, the deepest body of fresh water on the planet, has been the location for numerous sightings of UFOs entering and leaving the water itself, as reported by a fisherman. Some of the witnesses are high-ranking Navy officials, including an admiral. One of the most fascinating encounters involved Russian military divers encountering humanoid beings in silver suits fifty meters under Lake Baikal.

Ukraine - October 2009-Ukraine releases UFO national archive. The **Ukrainian Ufology Club (UFODOS)** has released and placed (ufobua.org.ua) a national archive of UFO evidence on the Internet. The 'secret files' comprise about 500 eyewitness testimonies of UFOs in the Ukraine starting from the 17th century. According to UFODOS chief Yaroslav Sochka, the material was collected from various sources, including Hydrometeorological Center of Ukraine Air Force and public ufology organizations.

Finland - November 2009 - For decades, Finnish Defence Forces have been interested in the unusual incidences within Finnish airspace, things that people usually call UFOs. A recently published book reveals official UFO studies in Finland from 1933 to 1979. It is based on three hundred official documents, which has been released for researchers – until the last year, when public military archives were closed down and after that only a part of this material has been released under the **National Archives of Finland**. Important to note, the first cases in the book are from years 1933-1937 when hundreds of citizens in the Finland saw mysterious phantom aircraft or ghost flyers.

Italy - March 2010 Col. Marco Picciau, a security boss for the **Italian Air Force** participated in major media coverage to publicize Italy's own UFO — or OVNI — puzzle. A rough reading from the official Web site indicates the Italian military has been collecting UFO reports since 1978.

New Zealand - New Zealand announces on June 16, 2010, that it will be the next country to officially release its UFO files... In the past few months, nation after nation has released UFO files to the public in an effort to inform the masses and attempt to strengthen trust and cooperation between governments and their citizenry. Now New Zealand is joining that campaign in the hopes that the United States will finally supply a final explanation on the subject of UFOs. A spokeswoman from the **New Zealand Defense Force** said in a press conference that the files New Zealand will be processed so private information is removed, but all other information will be released regarding the UFO encounters submitted to New Zealand authorities. Of course, those following New Zealand UFO sightings will be pleased as this information means the details of the often reported Kaikoura incident will finally be released to

the public for the first time. Reportedly, for three days the US Air Force also made numerous visits to the area and even patrolled the skies...
http://blogs.myspace.com/index.cfm?fuseaction=blog.view&friendId=23406090&blogId=535986038

What's intriguing is that the one country you would expect to be in the forefront of government UFO disclosure is the one country that has the most to hide…the United States of America. The simple fact is that the US by its continued lies and denials of the existence of the UFO phenomenon and its cover-up of secret reverse engineering programs of crashed alien spacecraft has unwittingly shone a bright spotlight on its government's inaction to step forward before the general public and disclose what it knows about the UFO question.

What's interesting to note here, is that these multi-national disclosures are in no way an official government disclosure by any country but, rather it has been characterized by researchers in Ufology as "dangling the carrot on a string" in order to gauge public reaction and interest in the subject. Full official disclosure has yet to happen! The only public event to come close to a disclosure was the initiative by the **Disclosure Project** spearheaded by **Dr. Greer** in 2001 at the **National Press Club** in Washington, DC.

Disclosure and Who Controls the Internet

Dr. Greer's Disclosure Project venue in Washington had all the major media and press coverage of the country show up for the event and it was simultaneously carried over the internet. On a good day, the Press Club on its website with certain new releases will get about 50,000 "hits", however, when the UFO Disclosure event took place in May 2001, the Press Club website received 250,000 hits in the first hour and then crashed for the second hour, this was later determined to have been caused by outside jamming of the internet signal by a covert intelligence group! It has been estimated that nearly a billion people worldwide had picked up the UFO disclosure event on the internet that day and when all was said and done, over two billion people worldwide had seen, heard or read about the Disclosure Project event!

Dr. Greer later reported that according to some of his insiders in the intelligence communities and in MJ 12, the whole UFO matter from the Washington National Press Club event was in "free fall". It appears Dr. Greer and his company of first-hand disclosure witnesses have turned the whole UFO/ETI matter upside down on its head! Damage control, spin doctoring and outright suppression of new releases by the media was ordered at every level to downplay the impact of the disclosure event. UFO news stories and news documentaries downplayed and countered the reality of the phenomenon with old outdated explanations and old fashion debunking to get the public confused and disinterested in the whole subject matter.

This ploy by the Big Media had very little effect on the general public's interest as more reports of UFO sightings increased not only nationally but also, globally.

Since that time, many other UFO organizations and UFO activists and lobbyists have followed in Dr. Greer's footsteps with disclosure programs and events of their own at the National Press Club, though, with perhaps less press coverage or success in generating further public or congressional support. Dr. Steven Greer is considered by many in Ufology as the **"Father of Disclosure"** for having truly pioneered the way in the disclosure movement by helping to lift the veil of secrecy and cover-up on the UFO/ETI phenomenon.

The Disclosure Project does bring up an interesting question of who controls the internet? Dr. Greer has seen by firsthand experience that his Disclosure event in Washington was electronically jammed by an outside source as reported to him by the National Press Club staff monitoring the internet site of people logging on to their website. They told Greer that things were flowing fine and that the live broadcast was going out into the ether or over the World Wide Web and then about half way through broadcasting the signal to the outside was either cut or jammed!

What should have been a secure internet line particularly coming out of the White House from the National Press Club represents a national security problem of the highest order! If the subject of UFO Disclosure can be interfered with then, what other important national matters designed as news releases can be interrupted and jammed electronically from the public's attention? But the problem doesn't just end with jamming the White House's National Press Club internet services but, has far reaching consequences globally for many countries using the internet to do government, military, and commercial business. So again, we need to ask, who controls the internet?

According to many online news media sources like ComputerWeekly.com; Slate.com; and ForeignAffairs.com; there are many of the world's governments like China, Russia, and Europe that are expressing concern over the US dominance and leverage over the global coordination of the internet. This overwhelming balance of power lies with a private-sector nonprofit organization, the **Internet Corporation for Assigned Names and Numbers (ICANN) ,** based in Marina Del Rey, California.

ICANN is a not-for-profit organization that regulates online addresses, known as domain names, and their suffixes, such as ".com" and ".org". Since ICANN reports to the US government's Department of Commerce, the domain name process is effectively overseen by the US government. http://www.computerweekly.com/Articles/2009/04/30/235865/Who-controls-the-Internet.htm

"The Web has become just another front in the battle between the United States and the rest of the world, and Tunisia was a convenient time and place to vent strong anti-American feelings. Although the United States government has not meddled in ICANN's operations yet, our U.N. brethren fear that an America with a unilateral foreign policy will eventually become an America with a unilateral Internet policy. Other countries have every right to be suspicious. If it wanted, the U.S. government could take over ICANN and block Internet traffic to a nation that harbors terrorists. It could access the databases that house domain names and use the information to take down computers serving up anti-American rhetoric or locate state enemies."
http://www.slate.com/id/2131182/ **"Who Controls the Internet?" "Why it doesn't matter if the United States is in charge."By Adam L. Penenberg Updated Tuesday, Nov. 29, 2005, at 3:31 PM ET**

72

What other nations like China has called for the creation of a new international treaty organization. France wants an intergovernmental approach, but one fundamentally based on democratic values under the purview of the United nations This would be much like global phone system, administered by the world's oldest international treaty organization, the International Telecommunication Union, founded in 1865 and now a part of the UN family. Many governments feel that, like the phone network, the Internet should be administered under a multilateral treaty. ICANN, in their view, is an instrument of American hegemony over cyberspace: its private-sector approach favors the United States, Washington retains oversight authority, and its Governmental Advisory Committee, composed of delegates from other nations, has no real powers. http://www.foreignaffairs.com/articles/61192/kenneth-neil-cukier/who-will-control-the-internet "Who Will Control the Internet?" By Kenneth Neil Cukier November/December 2005 Published by the Council on Foreign Affairs

It would seem that the ICANN certainly has the ability to disrupt or disconnect any new release, regardless of the news content, be it about UFO disclosure or otherwise but, for what purpose would it serve this Internet Corporation? In all likelihood, they would merely follow the sanctions that the US government orders, if it truly believed that any news release may pose a threat to national security but, clearly the disclosure of UFOS and ETI currently visiting the Earth in reality poses no threat to the government. But, a public disclosure of the reality of intelligent Extraterrestrial life forms visiting the planet is, however, to the power brokers who have high-jacked our future and sent us all careening down a dark abyss of chaos that is not only both geopolitically and socially destructive, but whose ruthless behaviour has environmentally impacted our planet to the point of **"Planeticide"** (a term coined by Dr. Steven Greer to describe **"the killing of a planet"**) without any consciousness or regard for the planet or for humanity, represents a major threaten to the secretive deep black programs and the covert agenda that is being played out by the Military Industrial Complex or more precisely by the Majestic 12 cabal!

In this subsection, we have seen how the media, whether it be the press , radio, television, the Hollywood entertainment business or the internet has over the last century been infiltrated by various intelligent agencies. This is especially true for the CIA, who have usurped power and control away from the private media moguls and supplanted their own agency controlled shills and lackeys in the newly managed Big Media establishment. The control of the printed word by most of the intelligence communities of the world governments is understood and fully utilized in controlling their nation's populations and in the spread of propaganda to other nations. The USA, Britain, Canada and Australia are in the forefront of media control when it comes to information on UFOS and ETI with the US as the predominant nation on Big Media spin control. Throughout this book, it should be obvious that most of the information contained herein centers on the US handling and cover-up of the UFO matter. The United States is still viewed in the world as a leader, whether that perception by other nations is positive or negative. Most nations feel that that US is not revealing all that it knows about the subject and is heavily into the suppression and cover-up of UFO knowledge.

The adage: *"Nothing is more powerful than an idea whose time has come"* is viewed in both positive and negative concepts as it relates to the UFO/ETI question. Positive to those who embrace new ideas that bring change with the expectations of a brighter and more hopeful future. New ideas that are perceived as negative by those who cling to outworn shibbolethic standards and practices fear the loss of power and control within their own lives and when it was exercised over others.

History is replete with all kinds of examples of how populations of many societies have benefitted from new ideas that advanced civilization but, have also felt the suppression, control, and manipulation by the ideas that have benefitted only a few and not the many in society. Thoughts become ideas, ideas become words and the word is made manifest. As stated earlier in this book, Dr. Greer has posited in many of his lectures that *"Words are powerful encapsulations of thought"* and as such we need to be wise and careful in our choice of words when we communicate with each other!

Yet, with all the most powerful spiritual concepts and ideas current in society today, it has not had the profound effect on those in positions of power to bring about a more advanced civilization on this planet. In fact, quite the opposite has occurred and not because the spiritual values of the day were not relevant or powerful enough but because those in power and officialdom have hijacked our bright hopeful future from us with old outdated and impotent spiritual values that have no relevance in this modern world. This eschatological agenda to preserved old decaying spiritual values which is in reality, a last dying spasmodic attempt by those in officialdom to cling to a position of power and control over others. Remember, that those who desire to have control over others have no control within their own lives thus, they seek that which they require of themselves, from controlling society.

Once again, let's remember a little history from a previous section of this book. Power and authority which was originally bestowed upon Kings and rulers from biblical to medieval times by the process of **"divine right"** became obsolescent by the mid-1800s as power was seized from the rulers and ecclesiastics of that day and given to the common people. Failure to safeguard the rights of people and to ensure justice for all was the reason for the fall and collapse of many monarchies and kingdoms throughout Europe and the Middle East. Failure to lead people to accepting a new revelation in the Middle East during the nineteenth century and becoming the cause of indifference as well as instigating religious persecution and intolerance resulting in brutal and barbaric torture and death to tens of thousands of people was the main reason why power was ceased from the religious leaders of the time. Witness today that few monarchies still exist and that many of the major world's religions are impotent and corrupt from within having not heeded the warnings and messages of given to them in that tumultuous time. This is not only a well documented historical fact but, biblically a fulfillment of prophecy.

The problem today, as has been stated earlier in this book, is that most people on the planet do not yet recognize that they have been given the power of these former rulers and ecclesiastics. With the exception of a few, who unscrupulously though ignore of the source of that power which they wield have foolishly assumed the positions of former kings and religious leaders via the control of financial resources and information! This has become their source of power which the general public has initially placed their trust in and are only now recognizing that they are being mislead and being manipulated by a covert agenda. If you control the printed word via the Big Media then, you control the hearts and minds of the people in society. However, it seems that people are stirring from their beds of ignorance and heedlessness and are realizing the dawn of a new day of hope and promise as their gaze in now focused upon the way they are being

74

governed by those in power and they are not very happy with the situation. Perhaps, now is the time for renewed action to rectify the injustices in the world!

Author's Rant: I do not want people to think that I'm proposing anarchy against their government or any government but, what I do propose as citizens of any country is to always be obedient to our government and to the laws of the land, with one caveat, to do so not blindly or ignorantly! As responsible citizens, we need to take action to elect officials who truly represent us and our country and to hold them to the highest moral standards and ideals of society. We need to remove by the democratic electoral process those people who fail stand up for the highest good of society. It is incumbent upon every duly elected official to be the trustworthy servant of their constituents and not to view their official positions for personal self-gain or for usurping power away from the central body politic. It may even be necessary to revise the way government operates which is transparent and answerable to the grassroots of society.

The ideals and standards in society have unfortunately not evolved to these high altruistic precepts because we, the society are either apathetic, timid or complacent and possibly lazy in not performing out civic duty in electing those who represent us in government. We hope and expect that those who do their civic duty will make the wisest choice in who they for vote which will benefit everyone else. The sad reality is that those who do vote are often the minority of the population and they will elect someone who may not truly represent society's populace. This elected official may fulfill the electorate minority's own particular self-serving agenda and thus, not represent the common good of everyone. This is how corruption gains a foothold in government and in society. Fortunately for all of us, material control is ephemeral as are the ideas that motivate materialistic pursuits!

In the next section we will illustrate the black murky depths of this Machiavellian power structure, we will expose once and for all the full agenda of the Military Industrial Complex, the rogue subgroups within the MIC, the military hardware that they intend to deploy upon an unsuspecting public, and their motivation behind these covert programs. It will be a shocking and disturbing look behind the Hall of Mirrors of deception!

The Military Industrial Complex

The **Military Industrial Complex** is a phrase used to signify a comfortable relationship between parties that are charged to manage wars (the military, the presidential administration and congress) and companies that produce weapons and equipment for war (industry). To put it simply, the Military Industrial Complex is described as an all-too-friendly relationship that may develop between defense contractors and government forces, where both sides receive what they are perceivably looking for: a successful military engagement for war planners and financial profit for those manning the corporate boardrooms. It can be viewed as a "war for profit" theory.

The idea of war for profit is nothing new in the realm of human history and can be traced back centuries earlier where arms races and the power of navy ships ruled an empire's reach. The arms race between the European powers of France, Spain, and Britain could arguably be a primal version of today's modern so-called military industrial complex. The idea was that a country must build up and maintain a ready military - the largest in the world at that - to remain a world power. Centuries ago, such a military was necessitated to protect aggression from neighboring countries. These days, an invasion of the American homeland may seem ridiculous and contrary to the building of a global community founded on trust and respect. Others might argue differently. http://www.militaryindustrialcomplex.com/what-is-the-military-industrial-complex.asp

In any case, the theory of a mutually beneficial relationship may not appear to be so far-fetched. It is no secret that the defense industry profits most when a nation commits to a lengthy war overseas. As any military will spare no expense for victory, it only makes sense to tap the resources of the defense industry to accomplish the mission. A sort of pseudo-world dominance through the basic form of imperialism can be seen to be just as important to a military force as is protecting one's homeland. The bottom line: war is good business for those invested in it - manufacturing, production, servicing, etc. To the war-minded industry, a wartime economy is just as profitable as a solid growing one, where shells and ammunition take precedence over the production of peacetime light bulbs or pencils. One need only to peruse the list of manufacturers participating in production during the Second World War to see just how a wartime economy can alter a single factory. http://www.militaryindustrialcomplex.com/what-is-the-military-industrial-complex.asp

The term **Military Industrial Complex** is sometimes used more broadly to include the entire network of contracts and flows of money and resources among individuals as well as institutions of the defense contractors, The Pentagon, and the Congress and the executive branch. For this reason, the Military Industrial Complex is sometimes known as the **"Military Financial Complex"** along with its inherent forms of financial and political corruption which have also surfaced with regularity.

The phrase *Military Industrial Complex* was first utilized in an American report at the turn of the 20th Century. "Military Industrial Complex" was later immortalized by outgoing United States **President Dwight D. Eisenhower** in his January 17, 1961, farewell address to the nation. In his speech, he cites the Military Industrial Complex as a warning to the American people – to not let this establishment begin to dictate America's actions at home or abroad. The original usage appeared in the form of *Military Industrial Congressional Complex* but later removed. http://www.militaryindustrialcomplex.com/what-is-the-military-industrial-complex.asp

The first modern MICs arose in Britain, France, and Germany in the 1880s and 1890s as part of the need to defend their respective empires either on the ground or at sea. The naval rivalry between Britain and Germany and France and their revenge sentiment against German Empire that followed the Franco-Prussian war was of utmost significance in the inception, growth, and development of these MICs. Conversely, the existence of these three nations' respective MICs may have been the source of these military tensions.

Similar MICs soon followed in nations like Japan and the United States. Furthermore, the length of time necessary to build weapon systems of high complexity and massive integration required pre-planning and construction even during times of peace; thus a portion of the economies of the great powers (and, later, the superpowers), was dedicated and maintained solely for the purpose

76

of defense. This trend of coupling some industries towards military activity gave rise to the concept of a "partnership" between the military and private enterprise.

Prior to World War I, the U.S. maintained a small military (in comparison to its peers) in times of peace and instead relied on militia or, in later years, reserves, in the event of war.

Though the United States never completely demobilized following World War I, and standing forces were maintained to a greater extent in the years that followed it, World War II was the driving force that utterly changed this historical pattern of general neglect of the military. During the Second World War, the United States underwent total mobilization of all available national resources to fight and win, alongside her allies, a total war against Nazi Germany and Imperial Japan, a mobilization of resources far greater than that which took place during the entire previous history of the United States. At the end of the war, East Asia was gravely damaged, and Europe was devastated; several European states abandoned their colonial empires, faced by prohibitive costs as well as a loss of moral legitimacy, national will, and military strength; and the United States and the Soviet Union stood as the two remaining great powers left in the world, from that point, known as **superpowers**.

The United States and the Soviet Union grew suspicious and hostile to one another; faced with a threat immediately following the Second World War, the U.S. only partially demobilized and left in place a sizable apparatus of military production and large naval, air, and land forces. This period, called the Cold War, represented a 45-year period of low-intensity, unconventional conflict between the superpowers, with the ongoing potential to metastasize into a nuclear conflict that could happen with only minutes of notice, could possibly destroy both superpowers, cause a new Dark Age, and might even result in the extinction of the human species.
http://en.wikipedia.org/wiki/Military_industrial_complex

Attempts to conceptualize something similar to a modern "military-industrial complex" existed before Eisenhower's address. In 1956, sociologist C. Wright Mills had claimed in his book The Power Elite that a class of military, business, and political leaders, driven by mutual interests, were the real leaders of the state, and were effectively beyond democratic control.
http://en.wikipedia.org/wiki/Military_industrial_complex

Most Ufologists are familiar with **President Eisenhower's Farewell Address to the Nation** on January 17, 1961 however; the real reason for this speech was to signal to the public that there was more than just a rise of unchecked power by the Military Industrial Complex.

In the subsection, **A Historical Meeting: Mankind's First Open Contact with Extraterrestrials,** we determined that Eisenhower in 1954 had mankind's first contact meeting with Extraterrestrial beings visiting the Earth. He was dissuaded by his administration officials from accepting an Extraterrestrial exchange of technology for the promise that the US would disarm their nuclear weapons technology. Eisenhower was promised an infinite energy source that was pollution free and would rid our reliance on oil, coal and nuclear power but his attending officials thought that this would destabilize the world's economy and thus he lost control of the situation. Realizing that the military and the contractors of industry were financially trying to control the outcome of this whole UFO/ET situation, felt it necessary to alert the public to what was going on, this then is the real reason for his speech to the nation. Below is his speech in part as it relates to the Military Industrial Complex.

President Dwight Eisenhower famously warned the US about the "military-industrial complex" in his farewell address
https://en.wikipedia.org/wiki/Military%E2%80%93industrial_complex

"A vital element in keeping the peace is our military establishment. Our arms must be mighty, ready for instant action so that no potential aggressor may be tempted to risk his own destruction..."

*"This conjunction of an immense military establishment and a large arms industry is new in the American experience. The total influence — economic, political, even spiritual — is felt in every city, every statehouse, every office of the federal government. We recognize the imperative need for this development. Yet we must not fail to comprehend its grave implications. Our toil, resources, and livelihood are all involved; so is the very structure of our society. **In the councils of government, we must guard against the acquisition of unwarranted influence, whether sought or unsought, by the military-industrial complex**. The potential for the disastrous rise of misplaced power exists and will persist. We must never let the weight of this combination endanger our liberties or democratic processes. We should take nothing for granted. Only an alert and knowledgeable citizenry can compel the proper meshing of the huge industrial and military machinery of defense with our peaceful methods and goals so that security and liberty may prosper together."* http://www.h-net.org/~hst306/documents/indust.html and
https://www.youtube.com/watch?v=KvrlFPD1fLE

78

Recently in 2010, the British government released another volume of UFO files from their National Archives into the public domain. The 18 files released cover UFO sightings reported to the Ministry of Defence from 1995-2003 and hold copies of original correspondence from members of the public reporting close encounters. In one of these recent batch of declassified UFO files containing approximately 5000 pages of eyewitness accounts, there are statements attributed to **Winston Churchill** during the Second World War ordering a cover-up of a Second World War encounter between a UFO and an RAF bomber because he feared public *"panic"* and loss of faith in religion.

The former Prime Minister allegedly banned reporting of the "bizarre" incident, off the east coast of England, for half a century amid fears disclosures about unidentified flying objects would create public hysteria.

He is said to have made the orders during a secret war meeting with US **General Dwight Eisenhower**, the then commander of the Allied Forces, at an undisclosed location in America during the latter part of the conflict.

Eisenhower and Churchill confer at a secret location in America on the highly unusual matters of strange aerial craft seen over the European and Pacific theatres of War
https://blogofthecourtier.com/2011/06/06/tonight-on-capitol-hill-eisenhower-fireworks/

The allegations involving Churchill were made by the grandson of one his personal bodyguards, an RAF officer who overheard the discussion, who wrote to the Ministry of Defence in 1999 inquiring about the incident after his grandfather disclosed details to his family.

According to the series of letters, written by the guard's grandson who is now a physicist from Leicester, a reconnaissance plane, and its crew were returning from a mission over occupied Europe when they were involved in the war incident.

During their flight, on the English coast, possibly near Cumbria, their aircraft was approached by a metallic UFO which shadowed them.

Photographs of the object, which the crew claimed had "hovered noiselessly" near the plane, were taken by the crew.

Later, during discussions about the unexplained incident, the two men were claimed to have become so concerned by the incident that Churchill ordered it remain secret for 50 years or more and reviewed by the prime minister to stop "panic" spreading.

During the meeting, a weapons expert dismissed suggestions the object was a missile as the event was "totally beyond any imagined capabilities of the time".

"There was a general inability for either side to match a plausible account to these observations, and this caused a high degree of concern," wrote the scientist, whose details are redacted.

It is known that Churchill had an interest in UFOs, even asking for a report in 1952 on "flying saucers" and what it "amounted to".
http://www.telegraph.co.uk/news/newstopics/howaboutthat/ufo/7926037/UFO-files-Winston-Churchill-feared-panic-over-Second-World-War-RAF-incident.html
UFO files: Winston Churchill 'feared panic' over Second World War RAF incident; By Andrew Hough, and Peter Hutchison; Published: 12:01AM BST 05 Aug 2010

This recent release of UK declassified UFO files clearly supports other reports that came out at the end of WWII indicating that some of the high-level government and military officials on both sides of the Atlantic were well aware of the bizarre occurrences of strange aerial craft that were not the advanced prototype aircraft of either the allied or axis powers. These UFOs or Feu Fighters had superior aerial capabilities that could easily out maneuver in or around any aircraft that had been developed during the war. They had been reportedly seen and photographed in both the European and Pacific theatres of war.

The general conclusion reached by the US, Britain, and a few other countries before the end of the war was that these strange craft belonged to no country on Earth and were, in fact, Extraterrestrial in origin.

Mr. Churchill is reported to have made a declaration to the effect of the following: 'This event should be immediately classified since it would create mass panic among the general population and destroy one's belief in the Church'."

80

A Military Official Speaks Out on Extraterrestrial Intelligences Visiting the Earth

Contrary to the popular notion held by most Ufologists that the Military have never disclosed the present of Extraterrestrial beings visiting the Earth, one has only to check the personal history and public statements made by high ranking military officials to see that they were well aware of the UFO situation particularly during and after the cessation of the Second World War. Initially, they were just as mystified by the reports from pilots and infantry and knew little as to the nature of these flying objects. After a cursory assessment of these airmen reports it was concluded that they were Extraterrestrial craft from another star system or planet. Like Churchill and Eisenhower, those military brasses who were aware of the situation thought it best to keep this information secret from the public. Unfortunately, not all top ranking military officials were informed of this decision thus, people like General Douglas MacArthur was one of these exceptions.

General Douglas MacArthur

Famous US Army **General Douglas MacArthur** (1884-1964) was the Commanding Officer of the US armed forces in the Pacific during WWII and was a career officer in the making at an early age with a strong commitment to winning the war. He was a tall man was very presence seem to command respect and allegiance and many people thought he would even become president after the war.

In 1943, the General was seriously concerned about reports of strange lights and metallic-looking objects that shadowed or even chased Allied Aircraft.

MacArthur felt these were not Japanese or German Aircraft, reconnaissance devices or secret weapons. He based that opinion on the fact that intelligence reports indicated that Japan Pilots were experiencing the same phenomenon and had no idea what they were. As a result of that information, MacArthur ordered an investigation into the Foo Fighter encounters in the Pacific.

Since Allied Pilots were already routinely debriefed after their missions, it didn't take a great deal of time or effort to ask those that spotted **Foo Fighters** a few extra questions or have them make drawings of what they saw. That information became the foundation for **MacArthur's Report**.

After the war ended, captured Japanese war records confirmed that the Foo Fighters had also been seen by Japanese and German Pilots wherever they flew.

At 12.15 during the night of 25 February 1942, a formation of 12-15 unidentified craft was seen in the sky over Los Angeles, California. They were neither ours nor the Navy's nor the Marines'. Not having established the identity or radio contact, they decided to attack in the fear that they were unknown enemies. The coastal artillerymen discharged 1430 shots against the targets.

There was no reaction, no bombs, no plane shot down, no damage to property, no victims, and the craft disappeared. **General Marshall** related the incident to **President Franklyn Delano Roosevelt** of the United States, the very next day. This UFO incident has now become known as **"The Battle of LA",** even a movie has been made on this event.

General Douglas MacArthur

In 1943 in the Pacific, **General Douglas MacArthur** asked General Doolittle for news of an unusual object which had faced our fighters and bombers. At the end of 1943, Doolittle (the famous Air Force Pilot of World War 2 who was with General MacArthur during the Second World War) informed MacArthur that some "spectators" (as they called "them" in those days) had followed the main military action. They were not terrestrial and (maybe) they were not hostile. That's how Doolittle and MacArthur described them. We didn't know much about it All we know is that something happened in China which convinced Doolittle, and MacArthur, without a doubt, of UFO's, based on events which led to the recovery of a spacecraft. **General James Harold Doolittle** led the 1st Air Raid over Tokyo in April 1942.

General Douglas MacArthur is said to have been involved in a very early UFO research project called the **Interplanetary Phenomenon Unit (IPU),** that was allegedly formed to investigate

UFOs or study captured UFOs before the official Air Force investigation Project Sign started.

General Douglas MacArthur - Oct. 8, 1955. "You now face a new world, a world of change. We speak in strange terms, of harnessing the cosmic energy, of ultimate conflict between a united human race and the sinister forces of some other planetary galaxy." The nations of the world will have to unite, for the next war will be an interplanetary war. The nations of the earth must someday make a common front against attack by people from other planets." **"Above Top Secret" by Timothy Good**

Once again, in order to hammer home the point without revealing any detail of his UFO research involvement or his findings, **General MacArthur** implies further the need for mankind to unite against a common enemy from space. He makes a surprising statement printed in the New York Times for October 8, 1955, saying that *"because of the developments of science, all the countries on earth will have to unite to survive and to make a common front against attack by people from other planets. The politics of the future will be cosmic, or interplanetary."*

He is also, said to have made the following statement, quite unorganized, in an address to the United States Military Academy at West Point, on May 12, 1962:

"You now face a new world - a world of change. The thrust into outer space of the satellite, spheres, and missiles marked the beginning of another epoch in the long story of mankind - the chapter of the space age... We speak in strange terms: of harnessing the cosmic energy... of the primary target in war, no longer limited to the armed forces of an enemy, but instead to include his civil populations; of ultimate conflict between a united human race and the sinister forces of some other planetary galaxy..." http://www.ufologie.net/htm/m.htm

These statements are historically important in that a top ranking officer who is professional, of high caliber intelligence and ability and who is the commanding officer of the Pacific theatre of war for the US and the Allied Forces and whose word is beyond question or doubt as men fight and die by military orders. He has the ear and attention of all top military brass, departments and even, the President of America. So, when he comes out and makes a few statements about Extraterrestrial Intelligence visiting the Earth and a possible threat, this is no mere outburst from a delusional or psychotic mind but from one who is in a position of high authority and responsibility for the most powerful nation on the Earth.

Did he have all the facts of the situation? Probably not and his statements were characterized by a military mindset but, it indicates for the first time on a high level of authority that the existence of ETI is a reality that needed to be taken seriously to which **General Jimmy Doolittle** and **MacArthur** conferred on the matter even after the war. There is credence for the story that MacArthur may have retrieved a crashed saucer from China in the late 1930s or early '40s. General Doolittle made an investigative visit to Europe to find out what was the nature of Foo Fighters, the unusual glowing craft that was able to fly circles around Allied planes in Europe during the war. At the time they were believed to be a top secret weapon of the Nazis and they in

83

turn thought that these strange crafts were secret Allied aircraft. The assessment of General Doolittle was made directly to President Roosevelt upon his return to America and he told him that these flying objects were, in fact, interplanetary spacecraft which were probably investigating what humans were doing in the war.

This is strategically important to the Allied Forces and in particular to the US Forces as there was no ambiguity about what these things were. They were not unidentified flying objects a term to characterize this phenomenon which really was a carefully thought out nomenclature to misdirect and throw off any public inquiry or investigation into the matter that may have arisen from air servicemen or ground infantrymen who told family and friends either during or after the war.

After the war ended, in the minds of the high ranking military officials of both the US and the UK, there was no doubt whatsoever, of the reality and existence of alien spacecraft visiting the Earth. The overriding concern then turned to the operational performance characteristics of these foo fighter craft. Speculation as to the propulsion system of these spacecraft was thought to be based on an anti-gravity or electro-gravitic technology because, it was what Earth scientists were working on at the time, most notably the work of **T. Townsend Brown** during the '20s and '30s.

These crafts operated beyond the capability of Earth sciences at the time and no recorded account from any airmen during the WWII indicated any hostility from these alien craft, merely that we were being observed. As to who or what was piloting these flying saucers, whether it was robotic or remote controlled or by some Extraterrestrial Intelligent beings and the possible nature of these ETs, their motives or agenda was a whole other matter.

The world had emerged from a period in human history where we had once again engaged in global conflict but, it was a world where many national borders had changed and cold war tensions were increasingly developing between communistic countries of the east and democratic countries from the west. The post-war mindset under these circumstances was that a possible threat for a more devastating global war, this time with thermonuclear weapons loomed ever higher on the horizon. Compounding and exacerbating this situation was the introduction of a whole new element unforeseen in it consequences and its implications, that of an Extraterrestrial presence. It is small wonder then, that the US Military although the victors from a world war also, found themselves inadequately prepared to deal with a higher intelligence than themselves. An intelligence originating from the stars that possessed superior technology that was beyond the military's immediate comprehension.

If as MacArthur speculated that these aliens may be hostile in nature then, the US military needed to understand what they were up against and do some catch up work, so as to even out the playing field in the likelihood of a worst-case scenario, that being an aggressive action from an alien race invading our planet.

This assessment, however, did not rule out the possibility of a technological breakthrough by Nazi Germany in a radical advancement in aeronautical propulsion and aero-form design. It was a well-known fact that both sides of the war were experimenting with disc or saucer shape designs for aircraft development. Notably, **T. Townsend Brown** had been doing research and

84

experiments in the 1930s in the US on this type of design. Even, Canada, after the war ended, had made some advancements in this particular saucer shape design during the mid-'50s.

Now, in order to understand the thinking about a need for designing a saucer shape aircraft, it is necessary to take a tangential step sideways to discuss some classic examples. We will explore some of the countries that designed and created saucer shape craft also, when they built them and why these radically shaped crafts were built in the first place as well as why these saucer engineering projects went "deep black" particularly in the US.

CHAPTER 40

COUNTRIES THAT HAVE HAD FLYING SAUCER PROGRAMS

AVRO Canada, the Avro Arrow and Canada's Saucer Program

When we think of technological advancements internationally in the aerospace industry, it would never occur to most people that Canada at one time was one of the world major leaders in aeronautical engineering and design. Canada had one of the most remarkable aircraft companies in the world that had made incredible technological strides in advanced aerospace design that was literally several generations ahead of most aircraft of its time. That company was **A.V. Roe Canada,** better known as **AVRO Canada** which became internationally famous for building two of the most advanced aircraft in the world at the time, the **Avro Arrow** and the saucer shape craft, the **Avrocar** both of which flew. In 1945, the UK-based **Hawker Siddeley Group** purchased Victory Aircraft from the Canadian government, creating **A.V. Roe Canada** as the wholly owned Canadian branch of its aircraft manufacturing subsidiary, A.V. Roe and Company situated at Malton, Ontario. We'll look first at the Avro Arrow and then, examine the more radical designed flying saucers, the Avrocar and its more advanced prototypes, the **Project Y-2**. Yes, Canada actually built more than just one type of flying saucer, other than just the famous Avrocar!

The **Royal Canadian Air Force (RCAF)** had commissioned AVRO Canada to build them an advanced supersonic jet interceptor with performance requirements unheard of in the military world at that time, the aircraft became known as the Avro Arrow! This was a delta wing fighter-bomber financed by the **Liberal government** and was unlike anything Canadian aeronautical engineers had ever built before. It was built to ensure Canada's defense and sovereignty of its northern territories and borders against incursion by Russian long-range bombers. This aircraft was considered even, by today's military standards to have been 30 or more years ahead of its time in overall aeronautical performance.

The Arrow would be powered by two **Iroquois jet turbine engines** built and fitted by **Orenda Engines Ltd.**, these too, were a radical advancement in aeronautical power plants. The Iroquois engines were capable of pushing the aircraft forward at mach 2.3 (the RCAF requirement) and some speculated that it could go as fast as mach 2.5 and even above mach 3.0 with the right refinements and tweaking of the engines. One of these Iroquois engines was tested by retrofitting it onto the rear fuselage of a Boeing B47 (a four-engine bomber) on loan from the USAF. The pilots of the bomber reported that when the cut off their four main engines and turn on the Iroquois engine, the bomber suddenly flew faster than it was built or designed to fly. This was proof positive of Orenda's design concept!

The unfortunate cancellation of financial resources and the ultimate destruction of the advanced aircraft during the **Conservative government** under **Prime Minister Diefenbaker** created a political uproar in the **House of Commons** that even to this day has left a bitter taste in the mouths of many politicians. Diefenbaker after conferring with President Eisenhower felt that with the recent launch of the **USSR Sputnik 1 satellite** on the same day as the **CF-105 Arrow** was rolled out on October 4, 1957, was no mere coincidence. The RCAF so we are told, no longer needed a long range interceptor bomber but, wanted surface to air missiles instead and therefore, went ahead and cancelled the Arrow program. Sputnik 1 heralded the dawn of the space age and potentially the end of the long-range bomber, the Avro Arrow being its main target.

Avro Arrow 201 flew 25 flights between 25 March 1958 and 19 February 1959
http://www.avro-arrow.org/Arrow/picts.html

The destruction of Avro Arrow with its highly advanced Iroquois turbine engines was considered in hindsight, a military and technological travesty in Canadian aerospace history. Canada had many European aircraft companies knocking at their door to buy up any plans. Blueprints, engines and even aircraft parts from the Arrow. Canada was being recognized just before the demise of the Arrow as a major military power in the aeronautical sector of industry but, due to the short-sightedness of the Conservative government at the time, we threw that all away.
https://www.youtube.com/watch?v=MDvZu50HvLs

Today, the Avro Arrow legacy can be found in most of the world's military powers who had built similar aircraft at the time. Countries like USA., UK, France, the USSR, even Sweden continued building delta wing craft regardless, of any possible threat from the **Communist Eastern Bloc**. The Avro Arrow introduced the concepts of cantering in the delta wings of aircraft, enabling jet to fly supersonic with stability, the revolutionary design in turbine jet engines, a better guidance system for missiles, and a computerized autopilot navigation system, a quick interchangeable engine - nacelle assembly system and many other aerospace refinements.

There is even talk of re-building the Arrow instead of buying the F-35s for Canada's defense. https://www.youtube.com/watch?v=HLLWH56uULY

Contrary to these accomplishments, there were erroneously reported stories in the Canadian press stating that the aircraft was over budgeted and was never capable of the performance requirements set for it by the Canadian military. These stories were not true and in reality, they were a cover story for what was really taking place between the US and Canada. The true agenda that had been hatched by the US government and its military, more precisely by the Military Industrial Complex was a North American air defense program for both countries.

NORAD

This air defense program became known as **NORAD (North American Air Defence Command).** The US government convinced and sold the Canadian government and its military on a grand strategic program for continental defense. Essentially, the NORAD program **[which this author's father worked for while in the RCAF at that time]** went something along these lines:

Canada and the United States would be partners in controlling, monitoring, patrolling and defending the whole continent of North America.

A system of radar detection sites known as the **DEW Line (Detection and Early Warning)** would be set up in Canada's high Arctic north to monitor any aircraft or inbound missiles from Russia or elsewhere.

Radar and missile launch sites would be set up along the US-Canada border on the States side with both American and Canadian military missile servicemen manning the sites. *(In joint US-Canadian war games, it appears on several occasions that Canadian soldiers were more willing to press the red button to launch the nuclear missiles than their American counterparts!).*

Both countries would also have underground command centres to monitor the whole of North America. These being situated in Cheyenne Mountain, Colorado Springs as the head control centre and at Trout Lake just outside of North Bay, Ontario as the secondary control centre and each was linked electronically to each other. *(Both sites have been decommissioned with Canada's being decommissioned in the early '80s. Supposedly the US site is no longer operational or has been moved to another more secretive location).*

Canada would also be given the **Bomarc Missile System** and site locations constructed across Canada (one of the sites being near North Bay) so that these missiles could intercept inbound enemy aircraft and their **Intercontinental Ballistic Missiles (ICBMs).** *(These had nuclear payloads however, they were incapable of reaching their targets with any precise accuracy and the launch roofs housing the missile would freeze up during winter making launch almost impossible.)*

88

The US would also give Canada the **Voodoo jet interceptor fighter** to replace our now-defunct Avro Arrow jet interceptor fighter-bomber. *(The Voodoo jet was a substandard aircraft technology barely capable of mach 1 speed; the Avro Arrow was capable of mach 2.3 or better)!*

Both countries would participate in simulated war games to iron out the bugs and flaws in their defense system on a regular basis.

There were many other aspects to this NORAD program which naturally upset the Russians and her allies and this in turn help promote a greater arms race between the two big military powers. When all was said and done, the system was upgraded several times, cost more to both nations but, more so to Canada than would have been spent on the Avro Arrow and all of its subsidiary programs. The Canadian Army, Navy, and Air Force were amalgamated into one unified military armed force thus reducing its effectiveness, resourcefulness and centralizing our military power essentially under US military control. Our military weapons, aircraft, and armament upgrade purchases are usually not the latest technology but, more often outdated second, third or more generations old, substandard in other words to the requirements needed in an ever advancing militaristic world.

Although the Canadian public feels that we still have our own military, this is only an outward appearance to which many will hotly debate this point but, our forces are pretty much a puppet military to US control and to whatever is their current agenda. Canada and the US have been bed partners far too long now, that a divorce in relationship is unthinkable and untenable to imagine! Canada has thrown open wide its borders to allow American military aircraft to fly through anytime they desire even, when the usual courtesy of asking or informing the Canadian military or government first, they will nevertheless, just cross the border if they feel justified in doing so, because the NORAD agreement which we are partners permits them to do so.

Canada at the end of the Second World War was considered by many as one of the top ten military powers in the world, based on the fact that we as a nation were still financially, politically and militarily wholly intact, sadly, however, today we are ranked 23rd in power, thanks, largely in part to the NORAD agreement. With the defunct Avro Arrow program and the dissolution our military power, a lot of our top scientists and aerospace engineers headed south across the border to work in other more lucrative US aerospace companies including NASA or Boeing and Lockheed or they went to work in European aeronautical companies thus, the **"Brain Drain"** of the '60s in Canada! We have recouped some of that military strength in the '90s but, nothing to where we could or should have been, if we had stayed true to ourselves as a nation and followed our own path and destiny. http://www.globalfirepower.com/

Now, as I said, there were two parts to this story that involved **Avro Canada**. The first part was necessary to show how this country became a second rate military force in the world even though we are considered technically well equipped, have one of the best paying military forces and the best trained. The Avro company had proven its technical excellence and proficiency in building and test flying one of the world's superior jet interceptors to the point that foreign governments and their military were lining up to put in their orders for the plane . However, Avro's ambitions and engineering prowess was far greater and went far beyond the Arrow aircraft in which very few Canadians or the world at that time were aware of, with the exception of some of our neighbours to the south. Some US aerospace companies felt we were too good at what we were producing and felt competitively threatened by our high-spirited technical development. This would be one of the reasons why the Canadian aerospace industry went 'south" in more ways than one.

Most people are now familiar with Canada's **"flying saucer",** the **Avrocar** which demonstrated successfully the **Coanda Effect** on a lifting body which was the forerunner to the British hovercraft that came out years later (another, Canadian first in aerodynamics design)! The Avrocar required no skirt around it perimeter (as does the hovercraft) to hover off the ground or to move forward. http://en.wikipedia.org/wiki/Coand%C4%83_effect

Special Projects A.V. Roe (SPAR)

Avro Canada had developed a sector of their company for special projects that included jet propelled trucks (a jet-propelled Mack truck was actually built and tested successfully) and jet powered sports cars besides the Avrocar flying saucer. This new sector of the company became known as **Special Projects A.v. Roe** or **(SPAR)** which was comparable to the US's **"Skunk Works"**, a highly super secret sector of aeronautical research and development. *The Arrow Scrapbook – Rebuilding a Dream and a Nation by Peter Zuuring; Arrow Alliance Press; Dalkeith, Ontario; 1999 ISBN1-55056-690-3)*

SPAR had been approached by the USAF and the US Army and was financed to build them a fast battlefield **VTOL (Vertical Take Off and Landing)** all terrain craft. No doubt their engineering expertise for solving difficult and challenging problems with radical solutions had preceded them. Today, helicopters and the Harrier jet perform the same tasks that the Avrocar was incapable of doing. Was this the project that broke the camel's back or did they actually succeed in their challenge?

The Avrocar initially perform like a wobbly hovercraft in its first few test flights but, this instability was soon rectified, however, it was only able to achieve three to four feet of "altitude" off the ground as it accelerated forward. In actuality, the prototype never flew out of **ground effect,** that is it never could climb and cruise at high altitude. The Avrocar with instability problems corrected was nevertheless, considered for all intents and purposes a dismal failure even though, it successfully demonstrated the Coanda Effect, still it wasn't enough for the US Military to sink hundreds of millions of dollars more into full production for squadrons of low flying saucers. There were two prototypes of the Avrocar as proof of concept, one worked and the other never flew thus, the US Military went home with their poorly operating toys, only to put one in the **Smithsonian Museum** (the one that didn't fly) and the other one in a military warehouse, somewhere.

End of story? Not quite!

90

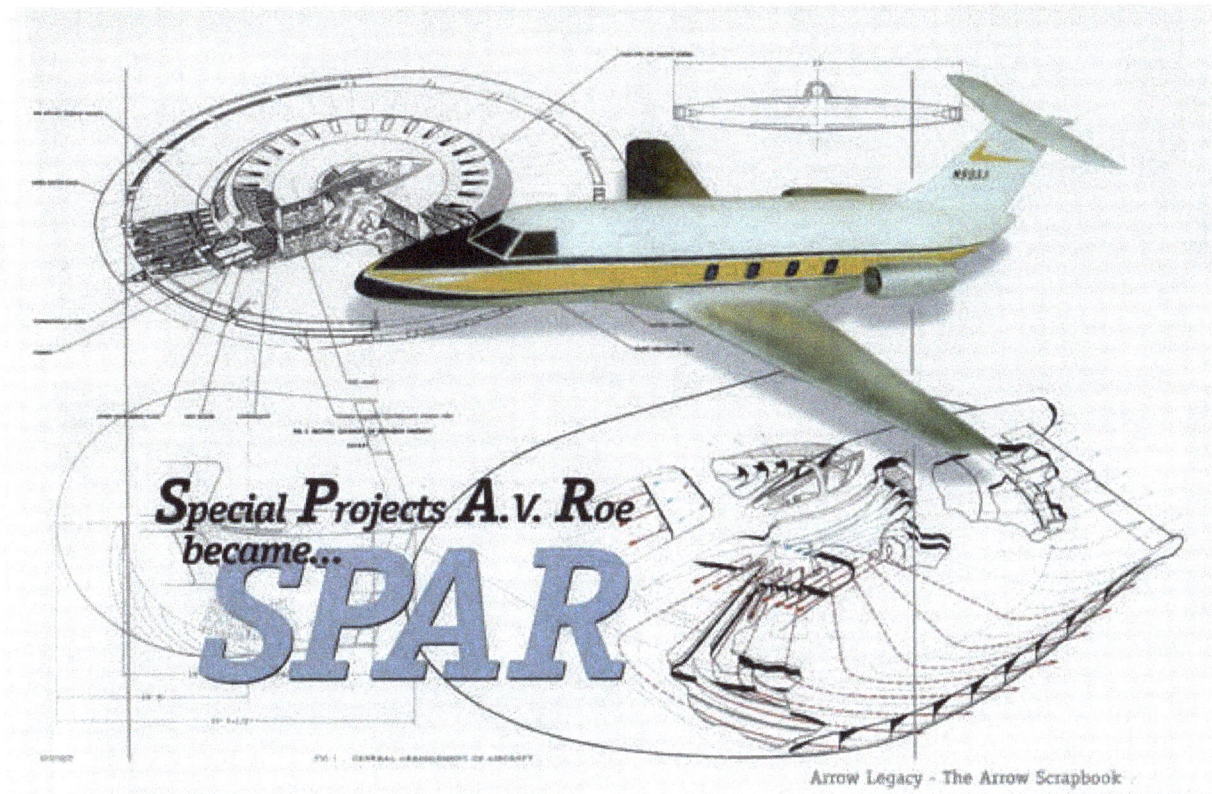

Special Projects A.v. Roe became...
SPAR

Arrow Legacy - The Arrow Scrapbook

SPAR (Special Projects A.v. Roe) Became Canada's Skunk Works. Arm Guards and Special Pass Cards Ensured a Tight Security
(From: The Arrow Scrapbook by Peter Zuuring)

John Frost, who was the chief aeronautical engine for Orenda Engines Ltd in partnership with Avro Canada made a proposal that Avro start an experimental project based on vertical takeoff and landing concepts. The idea of a saucer-like flying machine had revolutionary implications then and still does even in this current age. A conventional aircraft is very inefficient, aerodynamically. Like a bumble bee, there's no way it should fly. It only does so because of the wing which gives it lift and the engine's power to overcome the drag of the fuselage, the load, the tailplane, the stabilizers, fins and the engines."

Shortly after its formation in 1952, Frost's **Special Projects Group** started a paper study on a **"pancake" engine**, a jet turbine that had its main components arranged in a circular design. From the outset, the Special Projects Group had a cloak-and-dagger feel to it. Housed in a Second World War-era structure, across from the company headquarters, the group had all the accoutrements of a top-secret operation, including security guards, locked doors, and special pass cards. Within the confines of this technical fortress, Frost surrounded himself with a collection of like-minded dreamers and maverick engineers. There he encouraged close cooperation and, while ostensibly the boss, he was collegial and very much one of the boys.

Research undertaken by Frost on the **"Coanda effect"** confirmed that the concept of ground cushion could be the basis for a vehicle he had envisioned that could have both **Vertical Take-Off and Landing (VTOL)** capabilities and could still operate as a high-performance aircraft. As

Frost developed further studies, his ideas on revolutionary vertical takeoff systems led to the patent of **"Aircraft Propulsion and Control"**.

In 1952, **"Project Y-2"**, a **"spade-shaped" fighter** powered by Frost's revolutionary pancake engine proceeded to mock-up stage. By 1953, with the company having little more than a wooden mock-up, paper drawings and promises to show for a $4-million (Cdn) outlay, a more critical eye was cast on the project. Not surprisingly, the plug got pulled when government funding from the **Defence Research Board** dried up.
http://en.wikipedia.org/wiki/John_Carver_Meadows_Frost

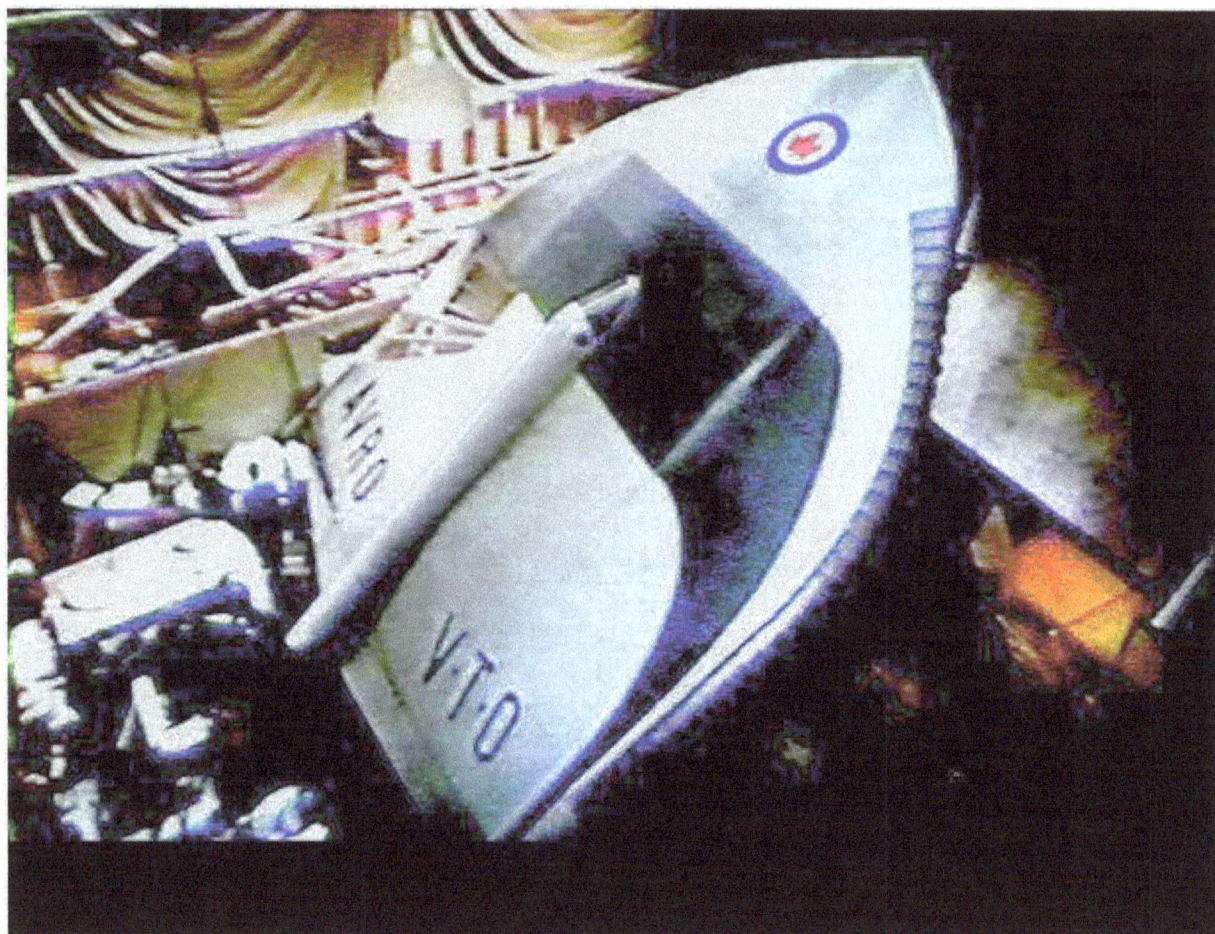

Project Y-2 a.k.a. "Jump Gyros". This appears to be a mock-up version but it rumoured that they were actually built and flown out of Georgetown, Ontario
https://en.wikipedia.org/wiki/Avro_Canada_VZ-9_Avrocar

Frost's later ideas revolved around a disk or saucer shape - a **"flying saucer"** and resulted in a number of patents in Great Britain, the United States and Canada on the unique concepts of propulsion, control and stabilization systems that were incorporated. Frost continued to lobby for the project now called the **"Project Y-2"** and achieved a remarkable breakthrough by demonstrating the project to the United States Air Force. With funding from the Americans, Frost was able to proceed with his research. From 1955 to 1959, the design team concentrated on the new VTOL supersonic studies known as **Weapon Systems 606A** which Avro Canada continued to support through an associated **private venture program**, the **PV-704** which resulted in the construction of an engine test rig in 1957.

The PV-704 supersonic test model, powered by six Armstrong-Siddeley Viper jet engines driving a central rotor, was built and housed inside a small, brick testing rig. The test model **Project Y2** was abandoned in favour of a simpler flying model lead to the only design that materialized from the **Avro Special Project Group**, a **"proof-of-concept"** vehicle, the **VZ-9-AV "Avrocar"**. Two Avrocar prototypes were constructed and completed a series of wind tunnel tests at NASA Ames in California and a 75-hour flying program at the Malton home of Avro Canada.

The results of the testing revealed a stability problem and degraded performance due to turbo-rotor tolerances. To solve problems of stability in ground effect the designers came up with a pneumatic analog control system using the huge vertical-axis lift fan as the sensing element. The obvious wobbliness of the ship in its original configuration completely disappeared thanks to this light, simple solution. Contrary to previous reports, the stability problem with the Avrocar was corrected with some modifications. The instability factor of the craft was a cover story even before modifications could be achieved, or that funding ran out with the final flight test program completed in March 1961.

The PV-704 Engine Test Rig Constructed in 1957 as a
Proof of Concept of the Y2 Flying Saucers
http://www.thelivingmoon.com/47brotherthebig/03files/Project_Silver_Bug.html

93

In light of a historical review of engineered disc shape craft of which the Avrocar and some of its predecessor saucer craft from France and particularly from Germany were all successful proofs of concepts, it was no wonder that the US Military were sufficiently convinced to secretly pursue further research and development with their own **"Skunk Works"** programs coupled with the highly secretive and compartmentalized reverse-engineering programs of alien spacecraft.

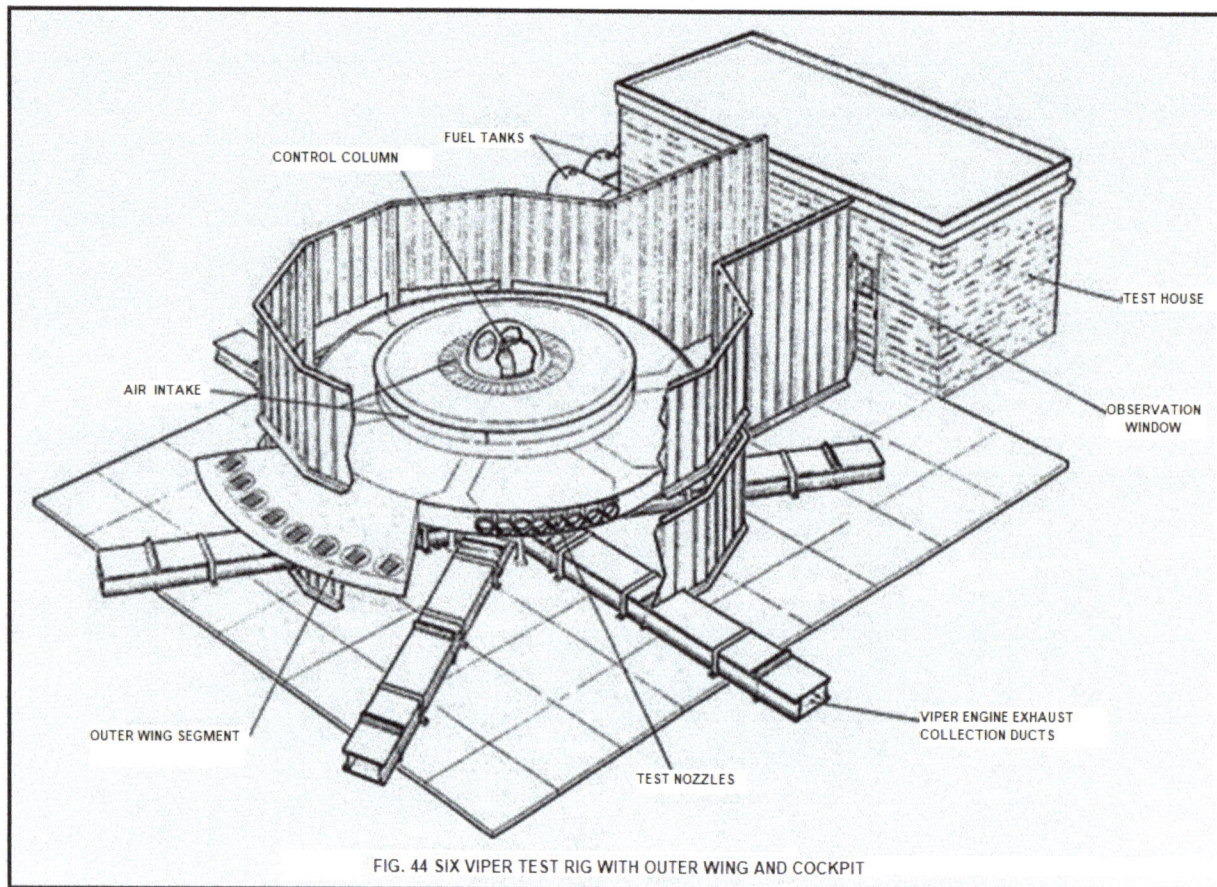

FIG. 44 SIX VIPER TEST RIG WITH OUTER WING AND COCKPIT

This diagram illustrates the Engine Test Rig used to test the aerodynamics of engine design, power, and thrust

http://ufxufo.org/avro/pv704/pv704.htm

Avro Canada's SPAR Project Y2 flying saucer a.k.a. by the US Military as Project "Silver Bug". Note the Canadian ensign emblem indicating a Canadian built project

http://lurch2.blogspot.ca/2014/03/air-force-flying-saucers.html

In fact, Avro Canada's SPAR program had many disc shape craft designs on the drawing board and in mock ups of various sizes and configurations based on the Avrocar and the **Project Y and Y2 flying saucers**. Many of these designs were further developed by the US Military and appear in later prototype construction. What appeared to be abysmal failures in jet aircraft and exotic aircraft design by talented Canadian engineers were taken very seriously by our neighbours to the south. Hundreds of schematics in the US patent office show just how serious this obsession for flying disc shape craft was for the USAF and other military departments.

As the result of his work in **(VTOL), vertical takeoff and landing** systems **John Frost** was invited to become a fellow of the **Canadian Aeronautics and Space Institute** on 25 May 1961. The citation noted that Frost had discovered and patented the air cushion effect that had been evident in his work on flying saucers and that U.S. Patent #3124323 "Aircraft Propulsion and Control" was one of a series of US, Canadian and British patents to became known as the **"Frost patents."** http://en.wikipedia.org/wiki/John_Carver_Meadows_Frost and **Campagna, Palmiro.** *The UFO Files: The Canadian Connection Exposed.* **Toronto: Stoddart Publishing, 1998. ISBN 0-7737-5973-5**

The end result from Project Y and Project Y2 was the Avrocar as a final proof of concept funded by the USAF and US Army. A USA emblem was placed on the craft by Avro Canada indicating US purchase and ownership

http://lurch2.blogspot.ca/2014/03/air-force-flying-saucers.html

The question now becomes, why was the Canadian aerospace industry financially crippled when it was on the verge of a major breakthrough in aeronautical research, development, and engineering? Why did we accept substandard aircraft and weaponry from the US Military and a North American partnership in a defense program that by the late '80s and early '90s soon became obsolete or decommissioned at least in Canada? What replaced it, if anything, to enable us to protect ourselves in case of another global conflict? What did the US Military determine to be an alternative defensive System? Was it the **"Star Wars" - Strategic Defence Initiative (SDI)?** Did the US have a major breakthrough in flying saucer development of their own and what about the deep black reverse engineering programs developed from the alien saucer crash in Roswell, New Mexico in June 1947? https://www.youtube.com/watch?v=UI0Z6qZkFYo and https://www.youtube.com/watch?v=0EtFOEkf75E

CENTRAL INTELLIGENCE AGENCY   REPORT NO. OO-V-2743

INFORMATION FROM

FOREIGN DOCUMENTS OR RADIO BROADCASTS   CD NO.

COUNTRY   Germany, USSR, French Equatorial Africa, Syria,
Iran

SUBJECT   Military - Unconventional aircraft

HOW
PUBLISHED   Daily, thrice-weekly newspapers

WHERE
PUBLISHED   Athens, Brazzaville, Tehran

DATE
PUBLISHED   11 Mar - 20 May 1953

LANGUAGE   Greek, French, Persian

DATE OF
INFORMATION   1952 -

DATE DIST. *18* Aug 195

CENTRAL INTELLIGENCE. COVERAGES   3
CLASSIFICATION
Cancelled
Changed to
BY AUTHORITY
SUPPLEMENT TO
REPORT NO.
Date   16 Mar 1955

THIS IS UNEVALUATED INFORMATION

SOURCE   Newspapers as indicated.

ENGINEER CLAIMS "SAUCER" PLANS ARE IN SOVIET HANDS;
SIGHTINGS IN AFRICA, IRAN, SYRIA

GERMAN ENGINEER STATES SOVIETS HAVE GERMAN FLYING SAUCER EXPERTS AND PLANS --
Athens, I Vradyni, 13 May 53

Vienna (Special Service) -- According to recent reports from Toronto, a
number of Canadian Air Force engineers are engaged in the construction of a
"flying saucer" to be used as a future weapon of war. The work of these engineers is being carried out in great secrecy at the A. B. Roe Company [translit-
eration from the Greek] factories.

"Flying saucers" have been known to be an actuality since the possibility
of their construction was proven in plans drawn up by German engineers toward
the end of World War II.

Georg Klein, a German engineer, stated recently that though many people
believe the "flying saucers" to be a postwar development, they were actually
in the planning stage in German aircraft factories as early as 1941.

Klein said that he was an engineer in the Ministry of Speer [probably refers to Albert Speer, who, in 1942, was Minister for Armament and Ammunition
for the Third Reich] and was present in Prague on 14 February 1945, at the
first experimental flight of a "flying saucer."

During the experiment, Klein reported, the "flying saucer" reached --
meters within   es and a --

1952 CIA document mentioning German flying saucers and the Canadian
AVRO saucer.

**This document indicates the seriousness in which US intelligence took the development of
manmade flying saucers by other countries not only from its neighbor
to the north but from its cold war enemy, the USSR (Russia)**

http://lurch2.blogspot.ca/2014/03/air-force-flying-saucers.html

Nazi Germany's Flying Saucer Program

Before we can look at the **"Holy Grail" legend of Ufology**, the famous Roswell saucer crash of 1947, it is necessary to examine the stories surrounding the Nazi development of advanced aircraft and the highly compartmentalized flying saucers projects being built in Germany during the Second World War. It has also been suggested by some UFO researchers that there may be a connection between advanced Avro's saucer concepts and construction with that of Nazi Germany's black programs of the **"Bell Project"** and engineering of other types of disc shape craft. Could manmade flying saucers be the explanation to the whole UFO and ETI phenomenon or is the evidence still heavily in support of the ET Hypothesis? Or is it possible that both theories hold the answers to the elusive mystery? Like everything stated in this book so far, more questions are raised then, answers thus; further scrutiny is required to sort out the truth and the facts from the fiction.

The study of Nazi "flying saucers" or UFOs is as enigmatic a subject as the UFO and ETI phenomenon, for there is so much nonsense and fantasy prone speculation that is mixed in with actual truth and verifiable evidence that it is in itself, a whole separate topic of discussion. Therefore this small section will not do this subject justice and will merely give the reader or student just the "bare bones" of information requiring him or her to investigate other books to get the full story. Like any controversial subject that requires serious research and in-depth investigation, party lines of opposition have arisen within the investigation of **Nazi flying saucer development programs**. So, we have skeptics on one side debunking the subject and promoters on the other side fiercely defending and promoting the reality of Nazi saucer design, engineering, and construction.

"The "German saucers" are often known also as the "**V-7 legend**": this comes from a reportedly circular aircraft named "V-7" and claimed to have flown in Prague on February 14, 1945.

As the folklore and legends grow, the latest developments include claims of German space trips to the Moon, Mars and nearby star systems, and we will also take these myths into consideration in this section. Also, of interest are all those rumours about secret Allied developments of original German projects, Nazi underground bases in Antarctica and related stories, like Hitler's escape and mysterious U-Boats, sighted after the end of WWII.

It is contended by skeptics and debunkers "that there are no original first-hand historical documents about the development of saucer-shaped aircraft by the Germans. The supporters of the saucers' reality say that most but all documents and blueprints were destroyed by the Nazis before surrender or captured by the Allied and never released, due to their extreme strategic importance. Investigation for locating possible undisputable sources is still running."
http://www.naziufos.com/#join

It will be seen that there is more than enough evidence to support that the German saucer development was not merely fanciful dream but an ambitious evolvement from prototypes to full production of tested, war-ready flying craft, some actually seeing combat action.

Germany after the First World War had suffered a humiliating defeat and the country was thrown into a political and economic downward spiral but, in the minds and hearts of many of her people and particularly in the mind of a new upstart leader, **Adolf Hitler**, such defeat was unacceptable. It would seem that Germany would have another turn to prove what it was unable to do in the first war under an imperial dictatorship. This time, it would be sudden, more devastating, technologically more advanced and involve more countries on the planet with millions of more deaths. It would be like no other war before it!

Bear in mind that the focus here is not on the Nazi Party's political agenda or the individual battles fought but, rather, the technological advancements Germany achieved that were considered outside the box in aeronautical design and engineering which would later be adopted by other nations such as Britain, Canada, the USA and Russia after the war.

After the great Battle of Britain which saw incredible "dogfights" between aircraft of Germany and the **Axis Powers** against Britain and her **Allied Forces**, Germany's air domination over Europe began to decline. This devastating aerial defeat was followed by continuous bombardment of German arms factories, munitions plants, airfields, naval yards, oil production and refining facilities. These key areas were necessary for Germany to maintain the war effort and thus Hitler ordered that these factories and plants be moved underground and be moved constantly making their detection and destruction more difficult for Allied bombers.

V-1 and V-2 rockets were built in these underground camouflaged shelters along with the construction of submarine parts and advanced jets and rocket planes with their final construction taking place near naval yards or airfields. Even synthetic petroleum refining facilities and uranium mines and processing plants went underground. Hitler also, ordered development of powerful and advanced weapons which would turn the war in favour of a Germany victory over Europe.

"Though Germany's air defense system was the best of any warring nation, it was clear that if Germany was to survive, improvement was imperative. Germany experimented with radically new types of air defense systems. Anti-aircraft rockets, guided both from the ground and by infra-red homing devices were invented. Vortex cannons, sun cannons, air-explosive turbulence bombs, rockets trailing long wire to ensnare enemy propellers, numerous electronic jamming devices, electronic devices designed to stop ignition-based engines, magnetically repulsed projectile sand long-range x-ray "death rays" were all under development as the conflict ended. Among these exotic solutions were saucer-shaped interceptor aircraft." Could some of these weapon developments be the "**Wonder Weapons**" that Hilter spoke about?

"The Germans already had jet and rocket interceptors as well as jet and rocket attack vehicles. German skies were full of these and other exotic aircraft so this new saucer shape was not considered as important then as we do today looking back upon it from a UFO perspective. To the German military and civilians alike these were just more new weapons." **Henry Stevens, Hitler's Flying Saucers: A Guide to German Flying Discs of the Second World War, 2003, Adventures Unlimited Press, Kempton, Illinois, 60946 ISBN 1-931862-13-4** http://www.bibliotecapleyades.net/ufo_aleman/rfz/index.htm#menu

Part of the development Germany's advanced and exotic weapons was also the growth and ascendancy of the **SS (Schultz Staffel)** which initially was Hitler's personal bodyguard unit but grew to become the most powerful national security entity in the **Third Reich** after Hitler himself. They oversaw seized control of almost every aspect of the military from the army, navy and air force to the underground manufacturing factories and production plants of weapons research and development. "The SS became an empire within an empire answerable only to Adolf Hitler."

"As the SS rose within Germany, so did the fortunes of Doctor of Engineering, **General Hans Kammler**. Kammler seems to come into prominence through his talent at designing and building massive underground facilities. Soon Kammler was placed, by Hitler, in charge of **V-weapons (Vergeltungswaffen).** This means Kammler was in charge of the facilities at **Peenemunde** and **Nordhausen.** He was **Dr./General Walhter Dornberger's** boss who, in turn, was **Dr. Wernher von Braun's** boss. Further, Kammler headed up an advanced research and development group, associated with the **Skoda Works**, called the **Kammler Group (16).** This group held the most advanced technical secrets of the Third Reich."
http://www.bibliotecapleyades.net/ufo_aleman/rfz/chapter1.htm

At the end of the war, General Kammler had disappeared and was never found. Being no fool, he undoubtedly took many copies of the most advanced German Technology.
"Numerous countries would have dealt with Kammler, regardless of his past. This includes the U.S.A. Couple this with the fact that no search was ever made for General Kammler in spite of the fact that he extensively employed slave labour in his projects."

"Did Kammler do a secret deal with an Allied government, exchanging information for a new identity? Or did Kammler escape Allied clutches to some safe haven such as South America? It is known that the Nazis set up shop in large, secure tracts of land between Chile and Argentina. It is also known that UFOs were seen earlier in that region than in the USA after the war. Many post-war stories involve German scientists relocating in South American countries formerly friendly to the Nazis and there building and flying German saucers."
http://www.bibliotecapleyades.net/ufo_aleman/rfz/chapter1.htm

German Nazi leaders besides being ruthless, extremely prejudiced toward most minorities, ethnic and religious groups like homosexuals, Polish people and Jews they also, leaned heavily toward occultism and mythological beliefs. These occult beliefs inspired and motivated almost everything they did or set their minds upon. They saw themselves as a superior race descended from an all-white **Aryan race** from the lost continents of **Atlantis and Lemuria**. This psychotic attitude and schizophrenic behaviour arising out of one country under the leadership of a deranged warmongering lunatic with delusions of grandeur and superiority is reminiscent to ancient Roman caesarean megalomania which inevitably crippled and brought the empire to its knees. Thus, Nazi Germany was doomed from the start, history would repeat itself as it has numerous times in the past. To say that German Nazi leaders were fanatical about the occult would be an understatement, they were obsessed with it!!!

One such occult group was the SS occult of the **"Order of the Black Sun"** who were tasked with researching alternative energies to make the Third Reich independent of scarce fuel oil for war production. Their work included developing alternative energies and fuels.

The Order of the Black sun was not the only Nazi occult society that was involved in the development of such unconventional saucer craft. The **"Vril Society"** was allegedly **"channelling"** messages from an alien civilization in the Aldebaran solar system and planned to develop a craft that could make physical contact with the civilization there. This may or may not be true; but there was certainly a high level of occult activity in mid-Europe at that time, and no doubt organizations did exist then, with unconventional beliefs just as they do today. Whatever the truth of this, by 1934 the Vril Society had apparently developed its first UFO shaped aircraft, known as the **Vril 1**, which was propelled by an anti-gravity effect. (This was the same year as **Viktor Schauberge**r discussed his flying disk ideas with Hitler.)
http://www.stevequayle.com/High.Jump/Vril.and.Andromeda.html

Many flying saucers were built by the Germans during WWII, the exact number is debatable, but is estimated between three and eight different types, along with their many prototypes and design variants. These saucer crafts were built in different places by groups of people specializing in conventional propulsion methods and in more exotic propulsion applications. In many instances, Jewish or polish prison of war slaves who had engineering and technical background were used in the construction of the disc-shaped craft because of their cheap labour and expendability. It would be an interesting project for some historian or Ufologist to locate surviving POW slaves and do "death bed interviews" before they die. Such a project would record for posterity their involvement in these secret German saucer programs adding yet, another dimension of validity to these stories about manmade saucers.

In considering the Nazi saucer programs, the best known of these projects was the Schriever-Habermohl project. Both men were engineers but, only Schriever had the distinction of also being a test pilot. The construction and development began at the Prag-Obell airport sometime between 1941 and 1943 with many German aerospace firms involved in the program significantly, the Skoda Works. It evolved as a Luffwaffe project which came under the control of Speer's Armament Ministry and eventual control was overseen by the SS with **General Hans Kammler** in command.

The first official flight of the saucer craft on February 14, 1944, some say it was in February14, 1945 but, the first unofficial test flight was witnessed by many employees and students in late August of 1943 or 1944 (this point of exactly when depends on the individual eyewitness). The saucer was really more of a disc-shaped supersonic helicopter – constructed by **Schriever and Habermohl** under the **V7 project** – that was equipped with twelve turbo-units BMW 028 was flown by the test pilot **Joachim Roehlike** at **Peenemunde.**

The craft was described as aluminum coloured, about 15 to 18 feet (5 to 6 meters) in diameter supported on four thin, long legs and about 6 feet (2 meters) in height. There was a cupola or doom on the top centre of the craft for the pilot's cockpit which was surrounded by rotating adjustable wing-vanes forming a circle and attached rim of the craft. In essence, this saucer operated much like a helicopter and auto- gyrocopter with the wing-vanes and small rockets being adjusted for pitch to enable vertical take-off and then adjusted for forward flight. The vertical rate of ascent was 800 meters per minute; it reached a height of 24,200 meters and in horizontal flight a speed of 2,200 km/h. It could also be driven with unconventional energy. But the helicopter never saw action since **Peenemunde** was bombed in 1944 and the subsequent move to Prague didn't work out either because the Americans and the Russians occupied Prague before the flying machines were ready again. Two fuzzy still pictures were taken and so no real structural detail can be discerned. Secret Societies and their Power in The 20th Century by Jan Van Helsing (pseudonym); Translation and typesetting: Urs Thoenen, Zurich. Original Title: Geheimgesellschaften und ihre Macht im 20. Jahrhundert http://www.bibliotecapleyades.net/sociopolitica/secretsoc_20century/secretsoc_20century06.htm#CHAPTER%2033

The overall flight characteristics obtained by this design were high altitudes of over 37,000 feet (12,400 meters) and a speed of mach or 750 miles per hour (1200 Kph). This was achieved during a night time flight when the saucer craft broke the sound barrier as the test pilot said he felt frighten by the vibration encountered from the shockwave at that time. Later test flights reported questionable supersonic speeds of 2200 Kph because air resistance and drag on the leading edge of the craft would mean slower speeds than claimed.

This brings up the controversial question; did the Germans break the sound barrier before America's **Chuck Yeager** did in 1947 with the **Bell X1 rocket plane**? Needless to say, many aerospace experts particularly the Americans resist the claims of the Germans doing it first and much earlier.

As the war drew to a close with the advancement of Russian Army, it was asserted that the saucer prototypes were destroyed by being pushed out onto the tarmac and burned before the Russians could get their hands on a working prototype. Although, it was stated by **Georg Klein**, another engineer in the **Speer's Armament Ministry** that there were three saucer prototypes that were destroyed on the tarmac, one of those being **Dr. Miethe's** saucer model. It is claimed that **Schriever** took the plans and models for the saucer and fled to Southern Germany but, these were later stolen by foreign agents. Schriever never claimed that his saucer ever flew which contradicts the information cited above. So, what really happened?

Now, remember **Habermohl**, the other engineer? The discrepancy may lie in the fact that the Schriever-Habermohl team built two separate lines of saucer craft as individual projects. But, according to **J. Andeas Epp**, another engineer, who consulted with both Schriever and Habermohl on their projects that it was his original design and model which formed the basis of the project and the reason Schriever's saucer prototype didn't fly, was due to a design change in the wing vane length. This was later corrected back to the original design specifications on the Habermohl prototype and his saucer actually flew.
http://www.bibliotecapleyades.net/ufo_aleman/rfz/chapter1.htm

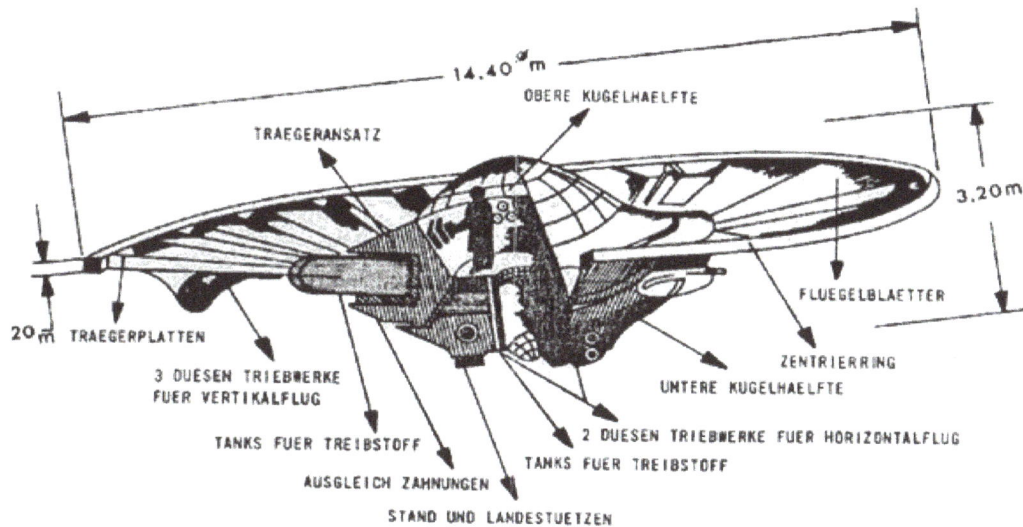

A Cross Section of the Schriever Saucer with Saucer Parts in German
https://www.thinkaboutit-ufos.com/wwii-german-flying-disk-schematic-drawing-found/

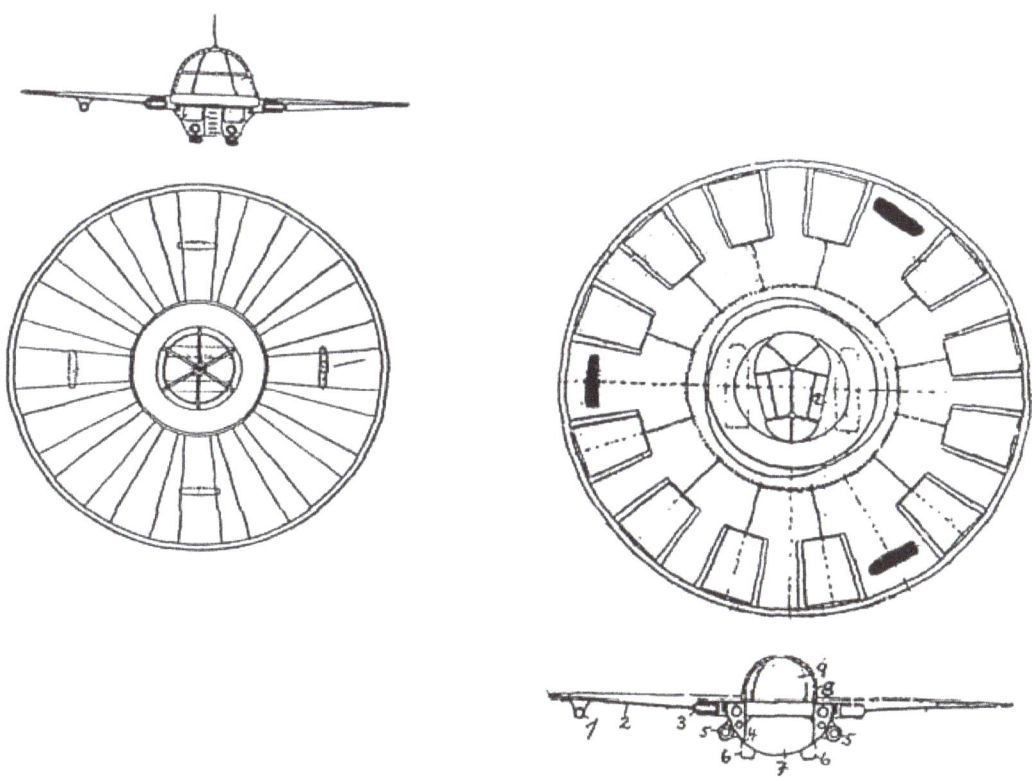

A Comparison of the Schriever and Habermohl Designs On the left is the Schriever design while on the right is the Habermohl design. Please note the differing dimensions of the vane blades.
http://www.bibliotecapleyades.net/ufo_aleman/rfz/chapter1.htm

103

The Miethe-Belluzzo Saucer Project

The next saucer project under scrutiny is the Miethe-Belluzzo Project which was headed up by **Dr. Heinrich Richard Miethe** of Germany and Italian engineer and Steam Turbine specialist, **Professor Belluzzo**. Not much is known about **Dr. Miethe** before the war. After the war, Dr. Miethe is rumoured to have worked on the Anglo-American saucer project at the firm of Avro Aircraft Limited of Canada. Unlike Miethe, Belluzzo went on record about German flying discs after the war. He is quoted on the subject in The Mirror, a major Los Angeles newspaper in 1950. This may be the first mention of the subject in the American press. In his obituary in the New York Times, his work on the German saucer program is mentioned.

Miethe is described as a "known V-weapons designer" working at Peenemuende, and its nearby test facility at Stettin, which retained and developed the Miethe design as an unmanned vehicle.

Miethe and Belluzzo worked primarily in Dresden and Breslau but for a brief time, they may have actually joined forces with Schriever and Habermohl in Prag, as evidenced by Klein's statement that three saucer models were destroyed on the Prag tarmac. Henry Stevens, Hitler's Flying Saucers: A Guide to German Flying Discs of the Second World War, 2003, Adventures Unlimited Press, Kempton, Illinois, 60946 ISBN 1-931862-13-4 http://www.bibliotecapleyades.net/ufo_aleman/rfz/chapter3b.htm

The designs envisioned by Dr. Miethe and Professor Belluzzo were quite different from those of Schriever and Habermohl. Designs of this project consisted of a discus-shaped craft whose outer periphery did not rotate. Two designs have positively been attributed to Miethe and Belluzzo, although, three designs exist as part of their legacy.

The first saucer design was not intended to take-off vertically but at an angle as does a conventional airplane. In this design, twelve jet engines are shown to be mounted "outboard" to power the craft. The cockpit was mounted at the rear of the vehicle and a periscope used to monitor directions visually impaired. Notably, a large gyroscope mounted internally at the center of the craft provided stability. This and other Miethe-Belluzzo designs were said to be 42 meters or 138 feet in diameter.

The second Miethe design version shows a cockpit above and below the center of the craft. Four jet engines lying behind the cockpits are shown as the power plants.

The third design attributed to the **Miethe-Belluzzo Project** was capable of vertical take-off. Klaas provides internal detail which has been reproduced here.

The Miethe-Belluzzo Disc--Design One

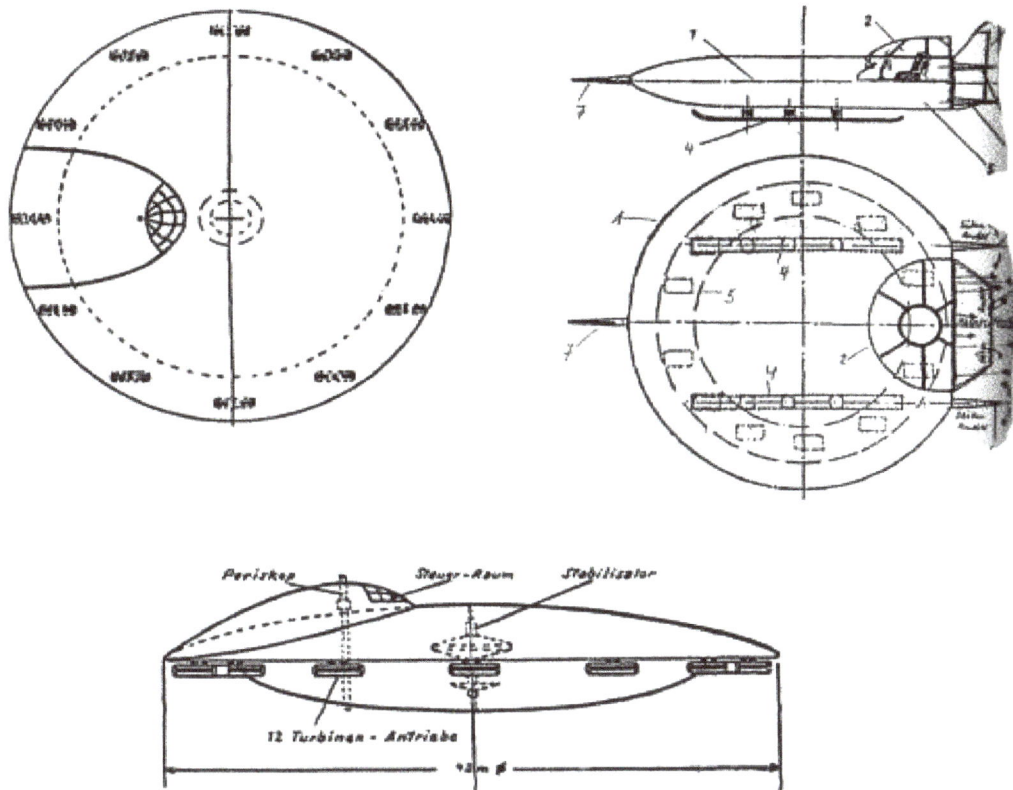

Periskop   Steuer-Raum   Stabilisator

12 Turbinen - Antriebe

On the left is a reconstruction by Georg Klein, 10/16/54, from the Swiss newspaper Tages-Anzeiger. Note the small "Stabilisator" and the outboard jet engines.
On the right is Klaus-Peter Rothkugel's more probable reconstruction incorporating fins, skids, and the inner-lying Rene Leduc engine.
http://www.bibliotecapleyades.net/ufo_aleman/rfz/chapter3b.htm

At first, this appears to be a push-pull propeller system driven by a single engine. It is not. Design one differs from design three in that the latter, with its centrally located cabin and symmetrical arrangement of twelve adjustable jet nozzles, is controlled by selectively shutting off various jets through the use of surrounding ring. This allows the saucer to make turns and to take off vertically.

It has been suggested that an engine invented by a French engineer, **Rene Leduc** and probably acquired by the Germans during their occupation of France links designs one and three, and possibly even design two while supplying the missing pieces needed to make the engine depicted air-worthy and resolves other problems.

105

The Miethe-Belluzzo Disc Designs Two and Three

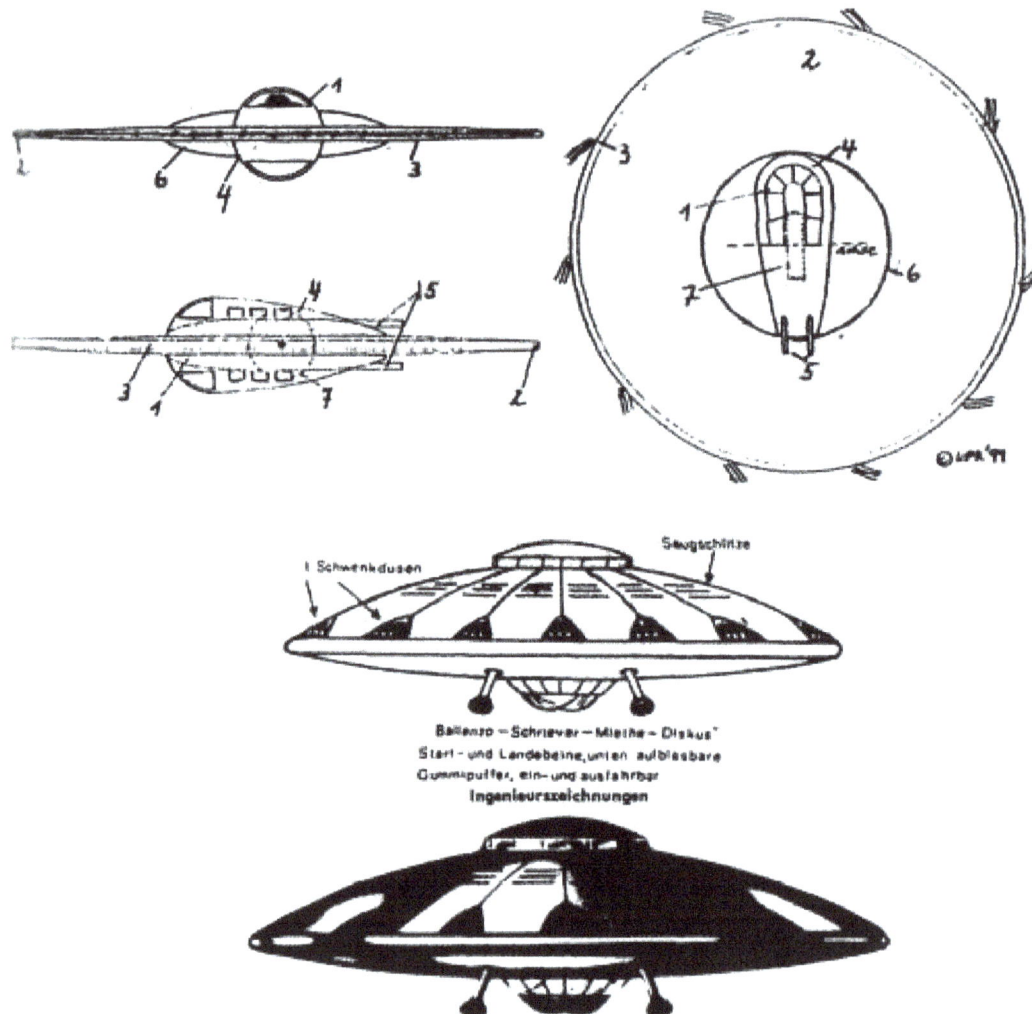

On the top is Miethe-Belluzzo design two. Note rotating disc (2) and stabilizing wheel (7) acting as a gyroscope. On the bottom is Miethe-Belluzzo design three, capable of vertical take-off.

http://www.bibliotecapleyades.net/ufo_aleman/rfz/chapter3b.htm

If a flying saucer equipped with this engine were viewed from the outside, no rotating parts would be visible. This is because the engine was totally contained within the metal skin of the saucer. It did rotate but this rotation was within the saucer itself and not visible from the outside. An air space existed all around the spinning engine, between it and the non-rotating outer skin. This engine was a type of radial flow jet engine. It was this type of engine which probably powered all of Dr. Miethe's saucer designs. It is also the prime candidate for the post-war design of John Frost, the "**Flying Manta**."

The Flying Manta actually did fly. Pictures of it during a test flight are unmistakable. They were taken on July 7, 1947, by William A. Rhodes over Phoenix, Arizona. It almost goes without saying that the time frame, July of 1947, as well as the geographical location, the American

Southwest, as well as the description of the flying object itself, beg comparisons to the saucer which crashed at Roswell, New Mexico, earlier that same month.
http://www.bibliotecapleyades.net/ufo_aleman/rfz/chapter3b.htm

If one looks at what is known of **Dr. Miethe's saucer design**, the Leduc engine, and the Frost Manta, it must be acknowledged that a connection between these three not only explains apparent inconsistencies in the existing Miethe designs but also links them to the post-war American Southwest, the precise spot where captured German World War Two technology was being tested and evaluated.

The Rene Leduc Engine

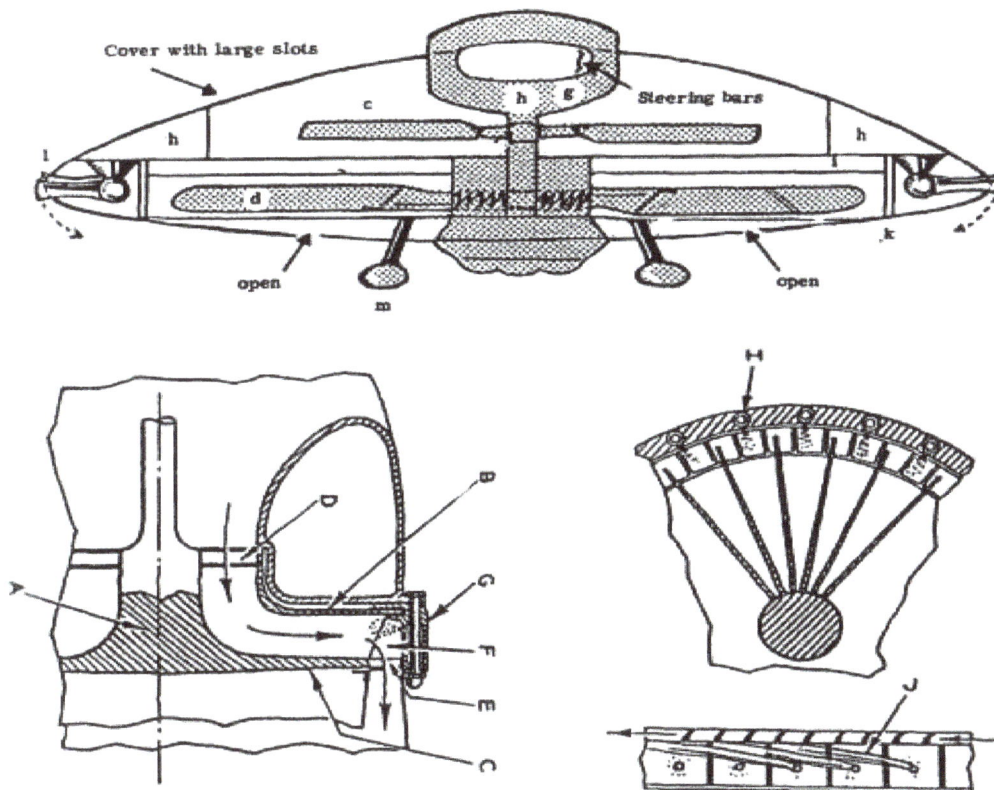

**Top: Hermann Klaas' diagram of the workings of the Miethe-Belluzzo Disc. Note: intake screw (c) Carrying wing blade (d) affixed to a piston engine, jets nozzles (e) with no apparent engines. Close but not exactly right. Bottom: Leduc design.
A-Rotor, B-Front Bulk-head, C-Rear Bulk-head,D- Intake Vane,
E-Compressor Vane,F-Combustion Chamber, G-Bulk- head,
H-Fuel Injection Jets, J-Fixed Flame Ring**
http://www.bibliotecapleyades.net/ufo_aleman/rfz/chapter3b.htm

One big difference between the Miethe-Belluzzo design and the Schriever-Habermohl designs is that the former craft was alleged to have, or be designed to have a longer flight range. This point is reinforced by the Spitzbergen flight mentioned above. Klein states that the Germans considered long range, remote-controlled attack from Germany to New York using this craft.

Politically, in 1944, Heinrich Himmler, head of the SS, replaced **Albert Speer**'s appointee, Georg Klein, with Dr. Hans Kammler as overseer of this combined saucer project but, retained Klein as his employee. Perhaps a more practical way to look at this is that Kammler, Himmler's employee, replaced Speer while Klein did what he always did. The result was that the SS took direct and absolute control over these projects from this point until the end of the war. http://www.bibliotecapleyades.net/ufo_aleman/rfz/chapter3b.htm

The Rene Leduc Engine Part Two

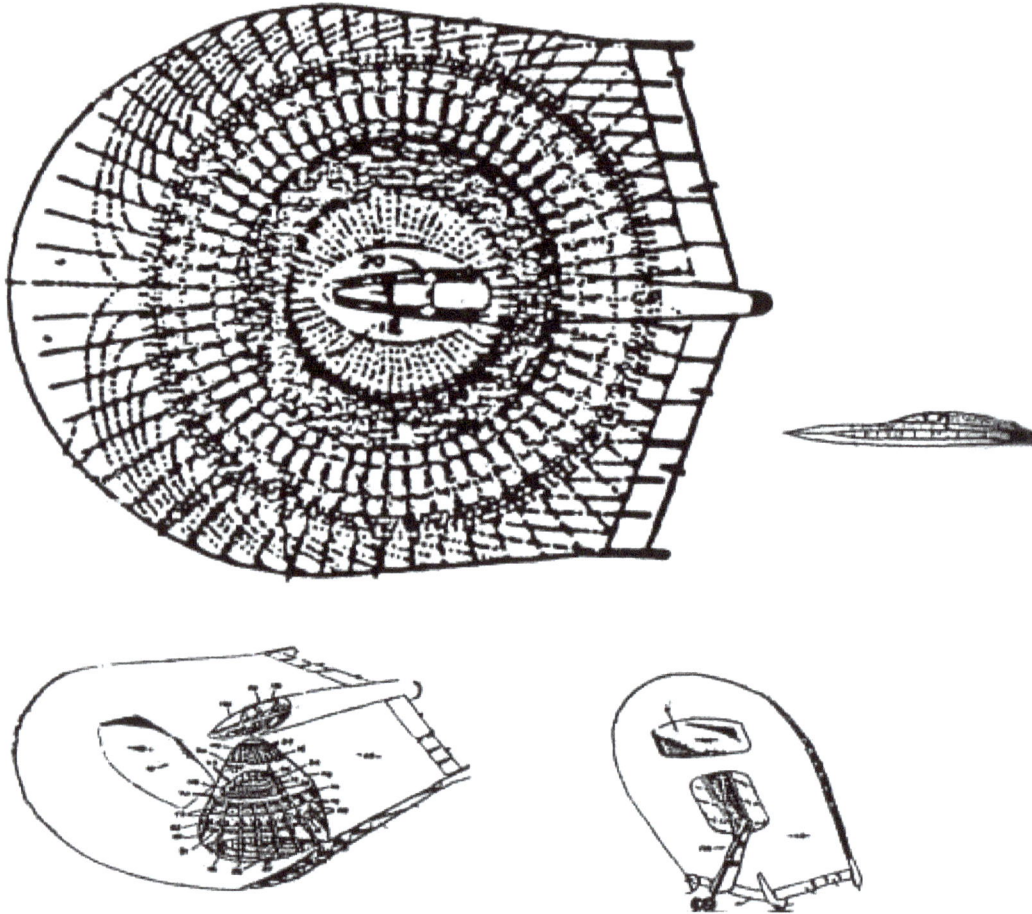

This is the mounting of the Leduc engine as illustrated by the later Avro diagram (Canada - USA). The outer hull is fixed. The inner rotating engine draws in the air from between it and the hull and exhausts through rear or sides as needed for steering. Compare this design to Miethe- Belluzzo designs, especially to the first design.

http://www.bibliotecapleyades.net/ufo_aleman/rfz/chapter3b.htm

Prior to this happening, news of these designs or application itself was made to the **German Patent Office**. All German wartime patents were carried off as booty by the Allies. This amounted to truckloads of information. Fortunately, **Rudolf Lusar**, an engineer who worked in the German Patent Office during this time period, wrote a book in the 1950s listing and describing some of the more interesting patents and processes based upon his memory of them.

They are surprisingly detailed. Included is the Schriever saucer design with detail. Also discussed is the Miethe project.

Miethe-Belluzzo Saucer in America?

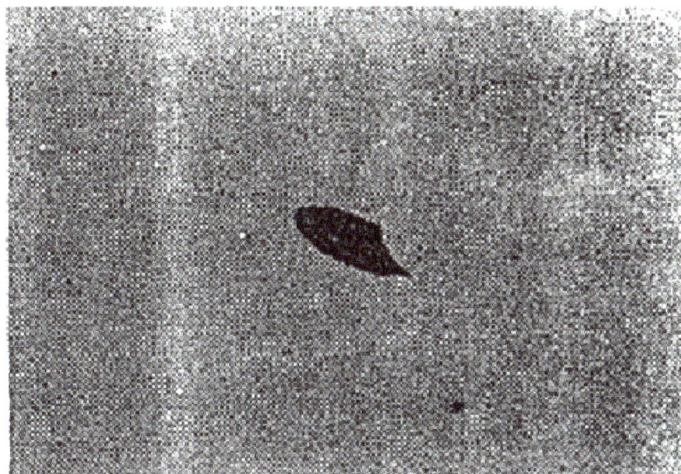

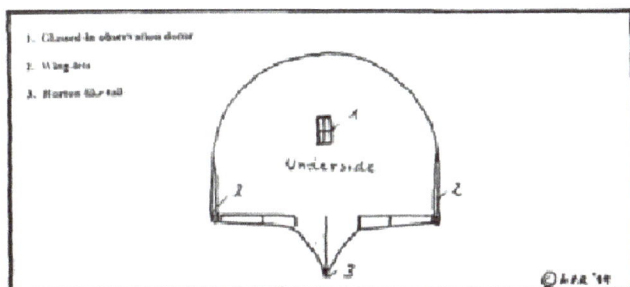

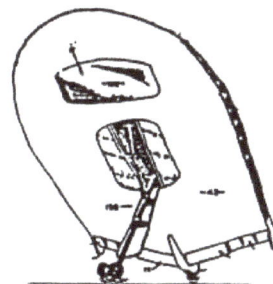

Top: a picture from the July 9, 1947, edition of the Arizona Republic taken by William A. Rhodes as it flew over his home in Phoenix. Lower Left: a drawing of the craft by Klaus-Peter Rothkugel. Lower Right: view of the Avro Frost-Manta design, predating the Silver Bug Project. Was this a captured Miethe-Belluzzo-Leduc saucer?

http://www.bibliotecapleyades.net/ufo_aleman/rfz/chapter3b.htm

The significance of these two teams cannot be minimized in the history of flying saucers or UFOs. Already in this brief discussion, the evidence, taken as a whole, is overwhelming. Please compare this to any and all extraterrestrial explanations of flying saucers. Here we have Germans who claim to have invented the idea of the flying saucer. We have Germans who claim to have designed flying saucers. We have Germans who claim to have built flying saucers. We have Germans who claim to have flown flying saucers. We have Germans who claim to be witnesses to flying saucers known beforehand to be of German construction. We have German construction details. And finally, we have a man who took pictures of a known German flying saucer in flight. The facts speak for themselves. During the Second World War, the Germans built devices we would all call today "flying saucers". No other UFO explanation can even approach this in terms of level of proof. http://www.bibliotecapleyades.net/ufo_aleman/rfz/chapter3b.htm

Nowadays, we would say, *"If it walks like a duck, quacks like a duck, craps like a duck, and flaps its wings like a duck, it must be a duck!"* Case closed!!!

Foo Fighters (A German "Wonder Weapon"?)

We have already discussed the phenomenon known as foo Fighters earlier but, another examination of these objects is required to determine if these flying fireballs were manmade by Nazi engineers or they are truly the devices of an Extraterrestrial civilization that appeared during Earth's bloodiest global conflict for surveillance purposes.

Small, round flying objects which followed Allied bombers over Germany during the latter stages of the air war were referred to as **"Foo fighter"** a name given to them by pilots because of their fiery appearance. Besides the name foo fighter, this device is sometimes called **"Feuerball"**, its German name or its English translation, fireball. Accounts of foo fighters seen in the Pacific theater of the war were also reported besides being seen in the European skies. These objects would display interesting characteristics that were unique in aerial performance outside of conventional wartime aircraft action. They would appear as small metallic globes during the day sometimes, flying singularly but more often seen flying in group formation and by night they glowed with various colors. These objects would approach Allied bombers closely frightening the bomber crews who assumed they were hostile and might explode, much like a proximity bomb. In some instances, whenever, pilots took evasive maneuvers, they found the foo fighters would keep pace with them, move for move. These objects definitely move and respond as if under intelligent control, as to whose intelligence, this needs to be determined for the arguments for both manmade and Extraterrestrial is very strong!

"In modern times, if they are mentioned at all by mainstream UFO magazines or books, an attempt is sometimes made to confuse the issue of the origin of foo fighters in one of three ways. First, they say or imply that both sides in World War Two thought foo fighters were a weapon belonging to the opposite side. They may cite as a source some German pilot obviously "out of the loop" who claims the Germans did not know their origin. Second, they attempt to advance the idea that foo fighters are still unknown and a mystery or possibly a naturally occurring phenomenon. Third, they advance an extraterrestrial origin."
http://www.bibliotecapleyades.net/ufo_aleman/rfz/foofighter.htm

There are also, many stories speaking of small, round flying balls sent to Japan by the Germans via submarine. The accounts of strange aerial objects seen in two separate theatres of war on opposite sides of the planet by pilots who flew combat missions in these areas are irrefutable fact but it is, however, not proof that foo fighter technology transfer from Germany to Japan ever occurred. There was no mad rush into Japan by the Allies in the same way that occurred in Germany by the Americans, British, French, and Russians. No scientists working on secret technology were rounded up and given safe sanction in America. What did occur was the Americans arrived in **Hiroshima** and **Nagasaki** the day after they bombed both cities with the first use of the atom bombs. They wanted to see first-hand, the effects of a nuclear weapon used

110

on people and buildings, although they had a pretty good idea of what they expected to see from prior nuclear tests at the **White Sands Proving Grounds** in New Mexico.

Author's Rant: This is an important point to keep in mind when we discuss the Roswell saucer incident and the motives of Extraterrestrial Intelligences.

Personally, I think that such stories gave Nazi Germany far more credit than is factual. If Japan acquired such technology from the Germans then, after the war their scientists and engineers would have resumed the production and development of these Foo Fighter devices for whatever purposes. However, there has never been any such technology to come out of Japan after the war in which the rest of the world and particularly the Allies would have known about, nor was any such technology ever recovered from Japan's war munitions factories or laboratories. Japan's economy after the war was devastated like much of Europe so; their main concern like their European counterparts was to rebuild their country's infrastructure.

Foo fighters or **"Feuerballs"** are described as radio-controlled missiles, built in Austria, under the control of an arm of the SS technical division. "It was armored, circular in shape, resembling the shell of a tortoise. The device was powered by special flat, circular a turbojet engine. After being guided to the proximity of the target from the ground, an automatic infra-red tracking device took over control. The circular spinning turbojet exhaust created a visual effect of a bright, fiery ball in the nighttime sky. Within the craft itself, a **klystron tube** pulsated at the frequency of Allied radar making it almost invisible to those remote eyes. A thin sheet of aluminum encircled the device immediately under the layer of protective armor but was electrically insulated from the armor. Once a bullet pierced the armor and the thin aluminum sheet, a circuit was formed which had the effect of triggering the Feuerball to climb out of danger at full speed."

Foo Fighters in Flight. The top photo is one of the most famous, taken over Europe taken at night or in dim light; the bottom was taken over the Sea of Japan between Japan and Korea in 1943. In strong light foo fighters appeared as silvery balls.
http://www.bibliotecapleyades.net/ufo_aleman/rfz/foofighter.htm

Floating Mystery Ball Is New Nazi Air Weapon

SUPREME HEADQUARTERS, Allied Expeditionary Force, Dec. 13—A new German weapon has made its appearance on the western air front, it was disclosed today.

Airmen of the American Air Force report that they are encountering silver colored spheres in the air over German territory. The spheres are encountered either singly or in clusters. Sometimes they are semi-translucent.

SUPREME HEADQUARTERS, Dec. 13 (Reuter)—The Germans have produced a "secret" weapon in keeping with the Christmas season.

The new device, apparently an air defense weapon, resembles the huge glass balls that adorn Christmas trees.

There was no information available as to what holds them up like stars in the sky, what is in them, or what their purpose is supposed to be.

**More Foo Fighters in Flight With an Accompanying Newspaper Article
Acknowledging Allied Forces' Pilots Sightings of the Strange Objects**

http://www.laesieworks.com/ifo/lib/WW2discs.html

"Once within range, special chemical additives were added to the fuel mixture which caused the air in the vicinity of the device to become ionized. This meant that electricity could be conducted directly through the air itself. Any ignition-based engine coming into range of the ionized region would become useless, misfiring, stalling and eventually crashing." **Renato Vesco; Intercept UFO; Zebra Publications, Inc. 275 Madison Ave, New York, NY. 10016; Grove Press, Inc.; 1974. ISBN: 0-8468-0010-1 (Recently reissued as Man-Made UFOs 1944-1994 by Adventures Unlimited Press)**

The U.S. military has always denied knowledge of foo fighters or any incidence with them. **Freedom Of Information Act (FOIA)** requests filed by Ufologists and other researchers asking for information on foo fighters would always receive a **"no record"** response followed. It matters not which U.S. governmental agency was contacted and queried the response was that they had never heard of foo fighters. According to Henry Steven in his book, *"Hitler's Flying Saucers: A Guide to German Flying Discs of the Second World War"* "this happened in spite of the fact that all known alternate names for foo fighters were submitted as well as a detailed description of the device itself. This was the situation until the late 1990s."

"A German researcher, Friedrich Georg, recognized a valuable entry in a microfilm roll, titled a 1944 U.S. Strategic Air Forces In Europe summary titled An Evaluation Of German Capabilities In1945, which, somehow, had eluded the censors. In that summary report, German devices called by American Intelligence "**Phoo Bombs**" are discussed. Sources for this summary were reports of pilots and testimony of prisoners of war. Phoo bombs were described as "radio-controlled, jet-propelled, still-nosed, short-range, high-performance ramming weapons for use against bombing formations". Speed was estimated at 525 miles per hour."
http://www.bibliotecapleyades.net/ufo_aleman/rfz/foofighter.htm

The Germans are credited with developing other radar jamming devices besides the Foo fighters described above and also, they experimented with silicon and germanium crystals which are used in making semiconductors and is the basis of the transistor. "The Invention of the transistor, however, is credited to **William Shockley**, for which he won the Nobel Prize, about two years after the Second World War." History, of course, is written by the victors and if truth and honesty were ever a part of history then, we need to rethink and rewrite history giving credit where credit is due even, if we dislike the person or country that we have beaten!
http://www.bibliotecapleyades.net/ufo_aleman/rfz/foofighter.htm

The argument of who invented the transistor flies in contradiction to the late Colonel   Philip J. Corso assertion in his book, *The Day After Roswell"* that transistors were based partly upon alien technology. **Corso, Phillip J., Col., 1997, page 161, The Day After Roswell. Pocket Books, a division of Simon & Schuster Inc., 1230 Avenue of the Americans, New York, NY. 10020**

Is there some deception at work here? From whom is the deception coming from, the US Military or the Nazi Germany's Military? Is it possible that both are telling the truth that the technology was developed independently and the Americans got recognition for it, before the defeated German scientists could? Could the answer be that Germany's incredible technological leap in aeronautics and weaponry has come from the retrieval of an **Extraterrestrial flying saucer** a few years ahead of the US Army retrieval of the crashed saucer in Roswell, New Mexico in 1947? Were Extraterrestrial spacecraft, unfortunately, crashing upon terra firma at different times and different locations? Remember that **Eisenhower** and **Churchill** as well as **General MacArthur**, **General Doolittle** and **President Roosevelt** had all concluded with clear evidence that these crafts were Extraterrestrial in origin and not manmade devices from Germany. So, who is telling the truth here or is the truth somewhere in the middle and more incredible than has yet been revealed?

It is no big mystery that secrets are kept by military in the name of national security and that deception, lies and propaganda of all kinds are used confuse the enemy into making bad strategic decisions and initiating reckless action. The public does not think in militaristic terms and therefore buys into whatever the military releases into the public domain. As far as the public is concerned, the military and the government hopefully, knows what they are doing and have the best interests of the public at heart. However, In this day and age, if you still believe in the "truth and honesty" mantra and that all is well with the world then, you may have been living in a cave for the last seventy years. Reality is far different now than it was back then.

114

To coin a phrase from the popular TV Sci-Fi program, the X-Files series (filmed in Vancouver, British Columbia), "The truth is out there"! **Foo Fighters** are "Phoo Bombs", a name given by the US military intelligence much like the term UFO was used as a cover name for Extraterrestrial spacecraft to confuse and make the research and access of freedom of information material more difficult by public and UFO investigators.

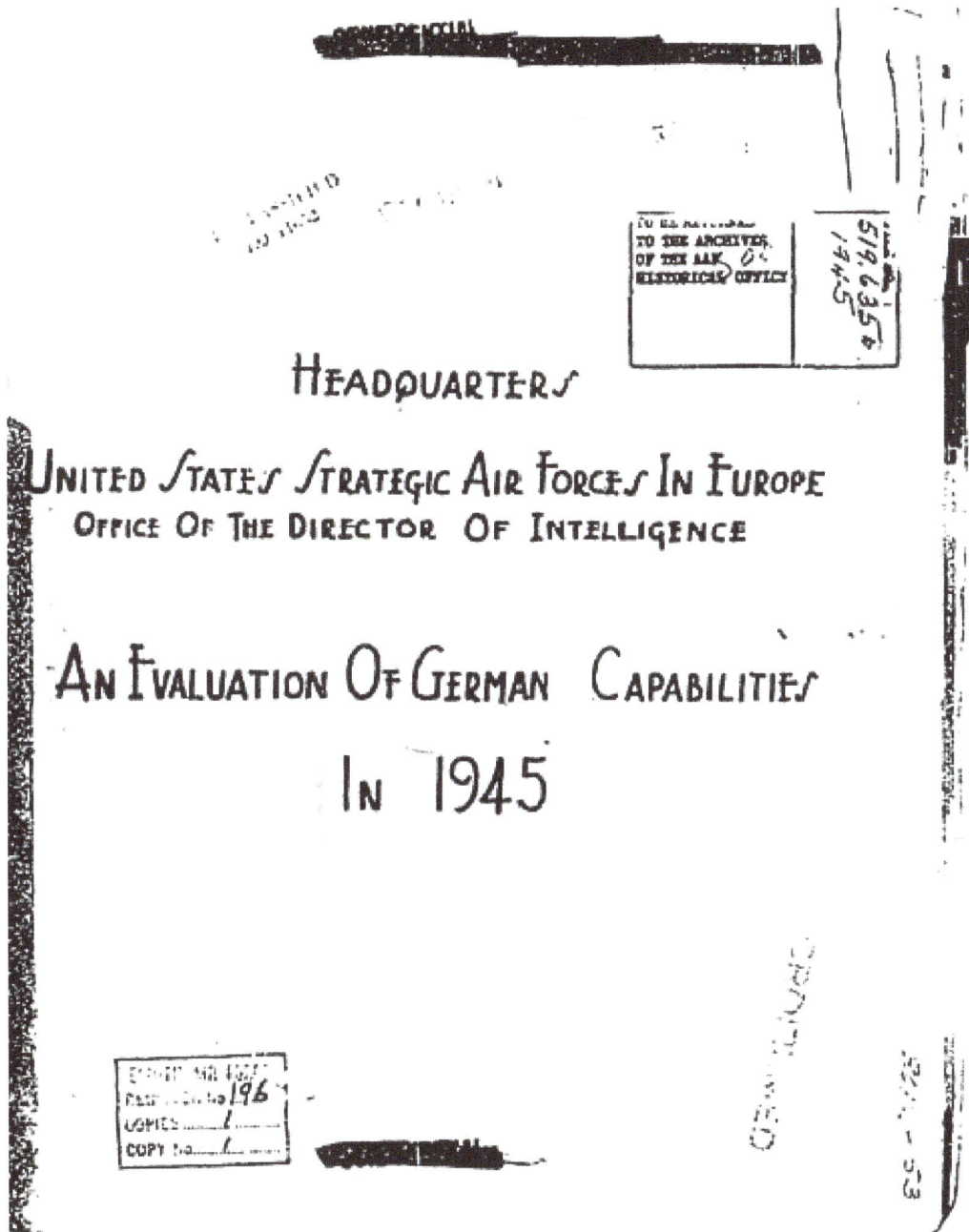

HEADQUARTERS

UNITED STATES STRATEGIC AIR FORCES IN EUROPE
OFFICE OF THE DIRECTOR OF INTELLIGENCE

AN EVALUATION OF GERMAN CAPABILITIES

IN 1945

A translation of a German document in which the US Government acknowledges that Foo Fighters and Phoo Bombs are one and the same thing.

http://www.bibliotecapleyades.net/ufo_aleman/rfz/foofighter.htm

TOP SECRET

PART SIX - OTHER WEAPONS

1. In the following paragrpahs are listed the actual or potential weapons which the Germans may use against USSTAF operations in 1945. For the most part they include the so-called V weapons. No consideration is given to those for which there is lacking evidence of possible use for some time to come. Both V-1 and V-2 are considered in the analysis because, even though they are, in effect, long-range artillery, they do possess the ability to affect our operations by hitting airfields, and supplies enroute and in concentrations.

2. V-2:

    a. Present status. The V-2, or rocket projectile, with a warhead of approximately one ton, and a current range of 225 miles, is being fired at London at the rate of 180/250 per month, and against Continental ports at the rate of approximately 300 per month.

    Against London its accuracy is currently rated at 3.2/1,000 per square mile at the main point of impact. Against Continental ports it is estimated at the least 6.1/1,000 per square mile at the main point of impact. The best record was 75 in a twenty-four hour period within a four square mile area of the Antwerp Docks.

    b. 1945 Potential: The German plan calls for an increase in monthly production from 600 to 1200. It is known, however, that any increase would be at the expense of the aircraft industry in radio equipment and certain essential components. An increase in accuracy would depend upon increased firings and increased use of already proved radio equipment, without which the majority of firings are conducted today. It is thought unlikely that range will be materially increased. Accuracy begins to fall off somewhere between 165 and 190 miles, and becomes increasingly inaccurate to the maximum of 225 miles. Whether or not V-2 becomes an increased menace in 1945 must depend upon the position of the aircraft industry and its requirements. Its potential lies in stabilization of the expanding aircraft program.

    Larger rockets (68 feet in length as against 45 feet) are known to exist, and may appear in small quantities during the year. They would have a considerably larger warhead.

3. V-1:

    a. Present Status: The so-called Flying Bomb is being fired from launching ramps against Continental targets, ports and supply concentrations, at the rate of 600 per month, and against England by airborne launchings, at the rate of 250 per month. Accuracy against Continental targets is now between 11.0/1,000 per square mile at main point of impact, and against England at 3.3/1,000 per square mile at main point of impact.

    b. 1945 Potential: Here again, the German plan calls for an expansion in production, but, as in the case of V-2, this expansion must be at the expense of other vital industries. Authoritative estimates state that airborne launchings against England may reach 450 per month, and that a very substantial increase of launchings on the Continent will take place. On the other hand, the number of He-111s available for airborne launchings is distinctly limited, and the demands of other industries are such that the expanded production may not be carried out as planned.

4. "FROG" BOMBS: Occasionally reports by pilots and the testimony of prisoners of war and escapees describe this weapon as a radio-controlled, jet-propelled, stull-nosed, short-range, high performance ramming weapon, for use against bombing formations. Its speed is estimated at 525 mph

TOP SECRET

~~SECRET~~

and it is estimated to have an endurance of 25 minutes. These bombs are launched from local airfields, and are radio-controlled, either from the ground, or possibly by aircraft. The few incidents reported by pilots indicate no success. They have passed over formations, andperformed various antics in the vicinity of formations. It is believed that in order to be effective some 100/200 would have to be launched against a formation, and it is also believed that they will not be produced in sufficient quantities to prove a real menace in 1945.

5.   MAGNETIC WAVE: The best information available is from very secret and reliable sources, and forces the conclusion that this weapon exists os a possibility. It is designed to cause failure of various electrical apparatus in aircraft. Technically it does not appear to be a possible serious threat in 1945. At most it would be effective at a few locations for preventing ground strafing. Evidence to date indicates that it could have little effect against high level attack, since the apparatus would be too cumbersome to permit its use in aircraft.

6.   GASES APPLICABLE TO AIRCRAFT: Two types of gases applicable to aircraft are known. One is designed to cause pre-ignition, blowing the heads off cylinders; and the other is designed to break down the viscosity of lubricating oils. Under laboratory conditions, free from operational considerations, these gases are a distinct possibility. It is doubtful, however, that with proper fighter escorts a sufficient concentration of either of these gases could be thrown against our formations to have any serious effect. Similarly, it is doubted whether sufficient anti-aircraft guns are available to produce an effective concentration, and it is probably that any possible concentration would be no more effective than a similar amount of well-directed flak.

7.   ATOMIC BOMB: Close check of every report, and close surveillance of the area in which tests are alleged to have taken place lead to the conclusion that such bombs are not a likelihood in 1945.

Whenever there is potential access to obtaining technological advancement in the aerospace industry along with its associated weaponry, whether originating from a friendly country or from a hostile nation or if it comes from another star system, as the military power of your country, you want to possess and secure that technology before some other foreign power does and then keep it off the public radar. The adage, *"Loose lips sink ships!"* is one of the military mantras of most countries. The public is viewed by the military as the biggest informants of top secret information to other nations! In reality, it has always been the insiders within the military that leak out secrets to other militaries, rarely is it ever the public!!!

Author's Rant: To this, I would say, well who is it then on the inside of your military department and/or intelligence agency that is leaking information in the first place because they would have to be the biggest informants of top secret and classified information! It certainly is not the public because they certainly don't have access to secret military information!

Getting back to the topic at hand, there is no doubt in this writer's mind that Germany's aeronautical technology was further ahead of the Allied forces but, they were incapable of carrying out a counter defensive late in the war against the Allies with their technological advancements.

Renaldo Vesco in his book *"Intercept UFO"* makes claims that there were other German developments in saucer designs beside the foo fighter technology resulting in a manned saucer project he called **"Kugelblitz"** (ball-lightning). As well as claiming the Kugelblitz actually flew, Vesco gave us some tantalizing details of the development of German saucer technology by the Anglo-Americans after the war, this being the **Avro Canada Saucer Program.**

There is evidence that the Americans did acquire working examples of these foo fighters after the war ended. Which were in all probability test flown out of their test facilities in the Southwestern United States along with other captured German technology. Support of this can be found in The Arizona Republic newspaper report of a sighting dated July 8, 1947, involving two flying silvery balls which may be the infamous foo fighters. (See Below)

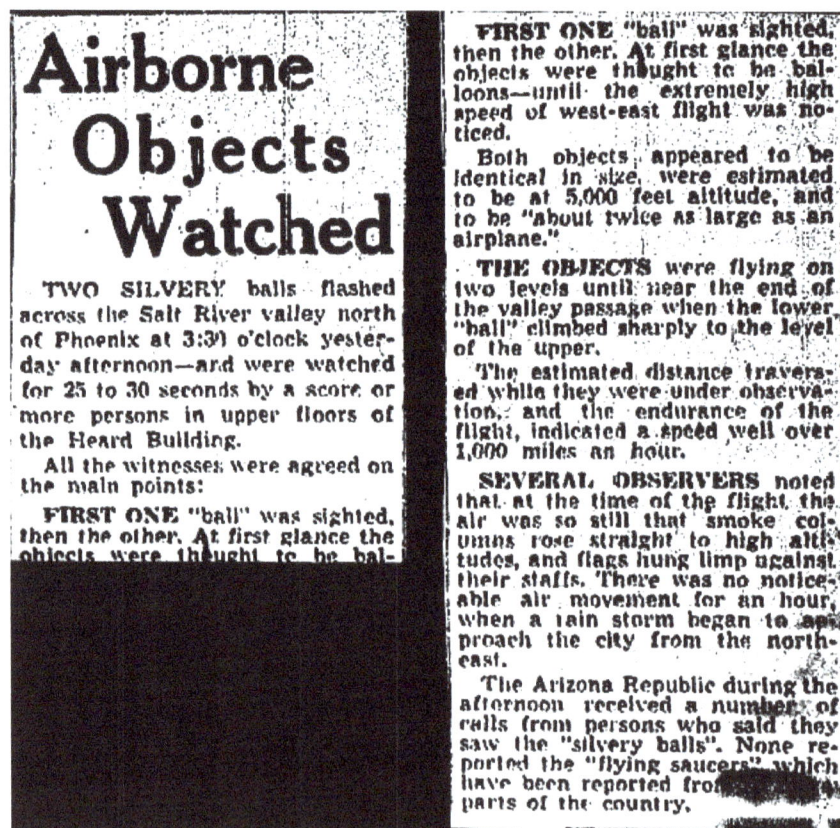

Airborne Objects Watched

TWO SILVERY balls flashed across the Salt River valley north of Phoenix at 3:30 o'clock yesterday afternoon—and were watched for 25 to 30 seconds by a score or more persons in upper floors of the Heard Building.

All the witnesses were agreed on the main points:

FIRST ONE "ball" was sighted, then the other. At first glance the objects were thought to be bal-loons—until the extremely high speed of west-east flight was noticed.

Both objects appeared to be identical in size, were estimated to be at 5,000 feet altitude, and to be "about twice as large as an airplane."

THE OBJECTS were flying on two levels until near the end of the valley passage when the lower "ball" climbed sharply to the level of the upper.

The estimated distance traversed while they were under observation, and the endurance of the flight, indicated a speed well over 1,000 miles an hour.

SEVERAL OBSERVERS noted that at the time of the flight, the air was so still that smoke columns rose straight to high altitudes, and flags hung limp against their staffs. There was no noticeable air movement for an hour, when a rain storm began to approach the city from the northeast.

The Arizona Republic during the afternoon received a number of calls from persons who said they saw the "silvery balls". None reported the "flying saucers" which have been reported fro[m] parts of the country.

Anyone fought in the Second World War in Europe or who were a postwar baby growing up with parents who live or served in Europe will tell you that that the name Peenemuende, Germany was a major target of allied aircraft bombing. This was the testing site and launch centre of the infamous V1 and V2 rockets used against the British Isles. But hardly anyone knew that it was also home to at least three different saucer programs. We already know that both the

118

Schriever-Habermohl and the Miethe-Bellonzo projects were under the control of officials from Peenemuende but, the best saucer design was the **Heinrich Fleissne saucer program**.

"Fleissner was an engineer, designer, and advisor to what he calls a **"Flugscheibe"** project based at Peenemuende during the war. It is interesting to note that Fliessner's area of expertise, fluidics, is exactly the specialty involved in investigating problems with boundary layer flow. Fleissner reports that the saucer with which he was involved would have been capable of speeds up to 3,000 kilometers per hour within the earth's atmosphere and up to 10,000 kilometers outside the earth's atmosphere."
 http://www.bibliotecapleyades.net/ufo_aleman/rfz/chapter3d.htm

Heinrich Fleissner filed a patent application with the United States Patent Office for a flying saucer (Patent Number 2,939,648) on March 28, 1955, but a patent wasn't granted until more than 5 years later. Fleissner's saucer had its engine rotating around the cabin on the outside of the saucer disc itself. Starter rockets set the engine in motion, similar to the Schriever and Habermohl saucer model but in reality, the engine was a form of ram-jet engine. "It featured slots running around the periphery of the saucer into which air was scooped. The slots continued obliquely right through the saucer disc so that jet thrust was aimed slightly downward and backward from the direction of rotation. Within the slots, fuel injectors, and a timed ignition insured a proper power curve which was in accordance with the speed and direction of the saucer much like an automobile's fuel injection is timed to match the firing of the spark plugs. Steering was accomplished by directing the airflow using internal channels containing a rudder and flaps which ran alongside of the central cabin. The cabin itself was held stationary or turned in the desired direction of flight using a system of electromagnets and servo-motors coupled with a gyroscope."

Recall the American **Silver Bug Project** which was being developed in the mid-'50s a further development of the Miethe design or a variance of its design and was simply referred to as a "radial jet engine". "But we now know this Miethe project was not the equal of the Peenemuende project in terms of speed. The Americans must have realized this sometime after the filing of Fliessner's patent in 1955. There can be little doubt that the reason for the delay of the Fleissner patent was the evaluation and possibly the pirating of his design by the Americans. At about the same moment that Fieissner's patent was granted, it was announced that the joint Canadian-American saucer project, Silver Bug, and its derivatives, had been abandoned by those governments. The only possible reason for this abandoning was that they had found something better and the better design, by far, was Fieissner's."
http://www.bibliotecapleyades.net/ufo_aleman/rfz/chapter3d.htm

Fieissner's saucer performed like a ram-jet and a chemical rocket allowing the craft to operate in the atmosphere under it ram-jet like engine and when leaving the earth's stratosphere into space, the craft could be powered by any number of fuels: "liquid, dust, powder, gas or solid." Different fuel mixtures and types could be accommodated simply by varying or adjusting the type of injectors and ignition used, for example, liquefied air and coal dust or liquid hydrogen and liquid oxygen which is a standard type of rocket fuel would be used when in space.

Fieissner's saucer could have taken off as a ram-jet, gained speed and altitude but at some point, reached a limit of diminishing returns. At this point, the saucer would have been able to slowly bleed liquid oxygen into the ram-jets for further performance enhancement. Further, it could slowly have replaced jet fuel with liquid hydrogen. This would be accompanied by a closing of the air intake apparatus. At this point there is no reason this saucer cannot become a space ship, that is, able to operate beyond the fringes of the earth's atmosphere. This performance would be more than enough to impress the U.S. Air Force and the civilian population of the late 1940s and early 1950's to drop the **Avro Silver Bug Project** in favour of the Fieissner saucer and its further developments.

Like the alleged flight of the last Avro Arrow 206 equipped with the highly advanced Iroquois engines which flew off to unknown parts of the continent before it could be de-constructed for demolition, a similar outcome was reported with Fieissner saucers.

Supposedly, an eyewitness, known by Fleissner, told him this: "Shortly before the Capitulation, on April 24, 1945 a squadron of four flying discs took off manned with two pilots whose names are unknown under heavy artillery barrage from the German and Russian sides from the Berlin-Lichterfelde Airport to a still today unknown destination." (Neue Press, 5/2/80, page 3) http://www.bibliotecapleyades.net/ufo_aleman/rfz/chapter3d.htm

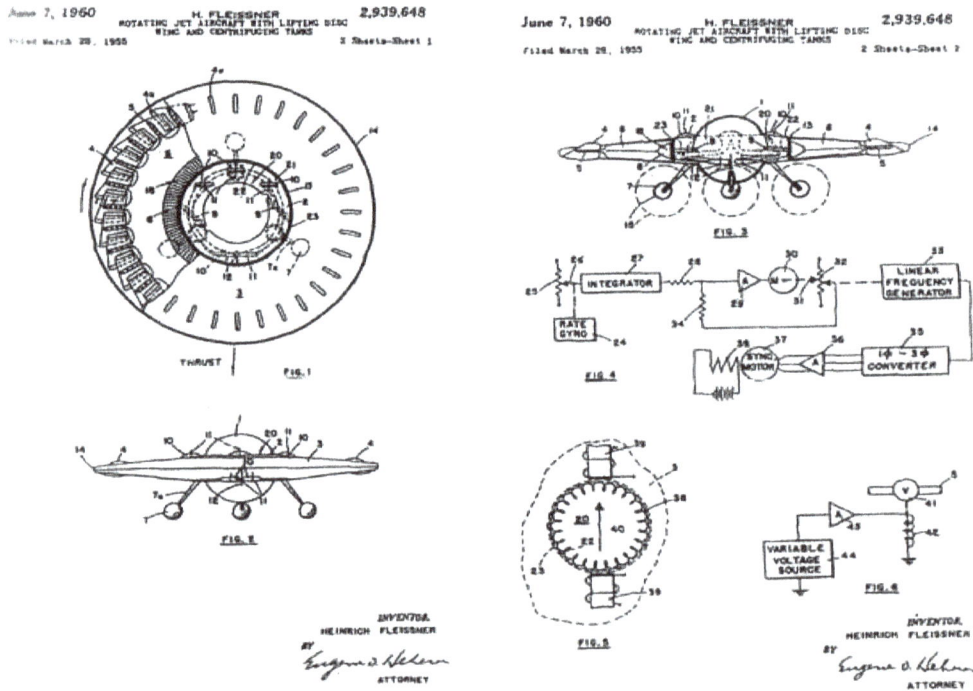

The Fleissner saucer is similar to the Avro Y2 saucer but it is superior in design and performance characteristics
http://www.bibliotecapleyades.net/ufo_aleman/rfz/chapter3d.htm

We have seen some of the more conventional and factually better known designs of disc-shaped craft along with their variants utilizing conventional propulsion methods being built in wartime Germany at different places by different groups of people. We will now examine the lesser known and less well-documented saucer projects which deal with the more controversial exotic propulsion systems.

Viktor Schauberger and the Repulsin

A pioneer in the forefront of exotic propulsion systems is Austrian born, **Viktor Schauberger** a naturalist, who observed and postulated that the movement of water as found in naturally occurring streams, rivers and lakes had energetic properties. These energetic properties were also believed to be applicable in the air as a possible means of propulsion. Even today, his thinking outside the box of conventional aerodynamics is little understood and frequently debated by scientists and aeronautical engineers. Schauberger believed that the vortex phenomenon was the natural method by which water flowed both in the earth and in streams. Viktor Schauberger believed that energy flowed naturally in a vortex but that this movement was only visible through another medium such as water or air, examples of this vortex phenomenon is the tornado, the whirlpool and the hurricane, all natural forces that exhibit the whirling spiral or **vortex effect**.

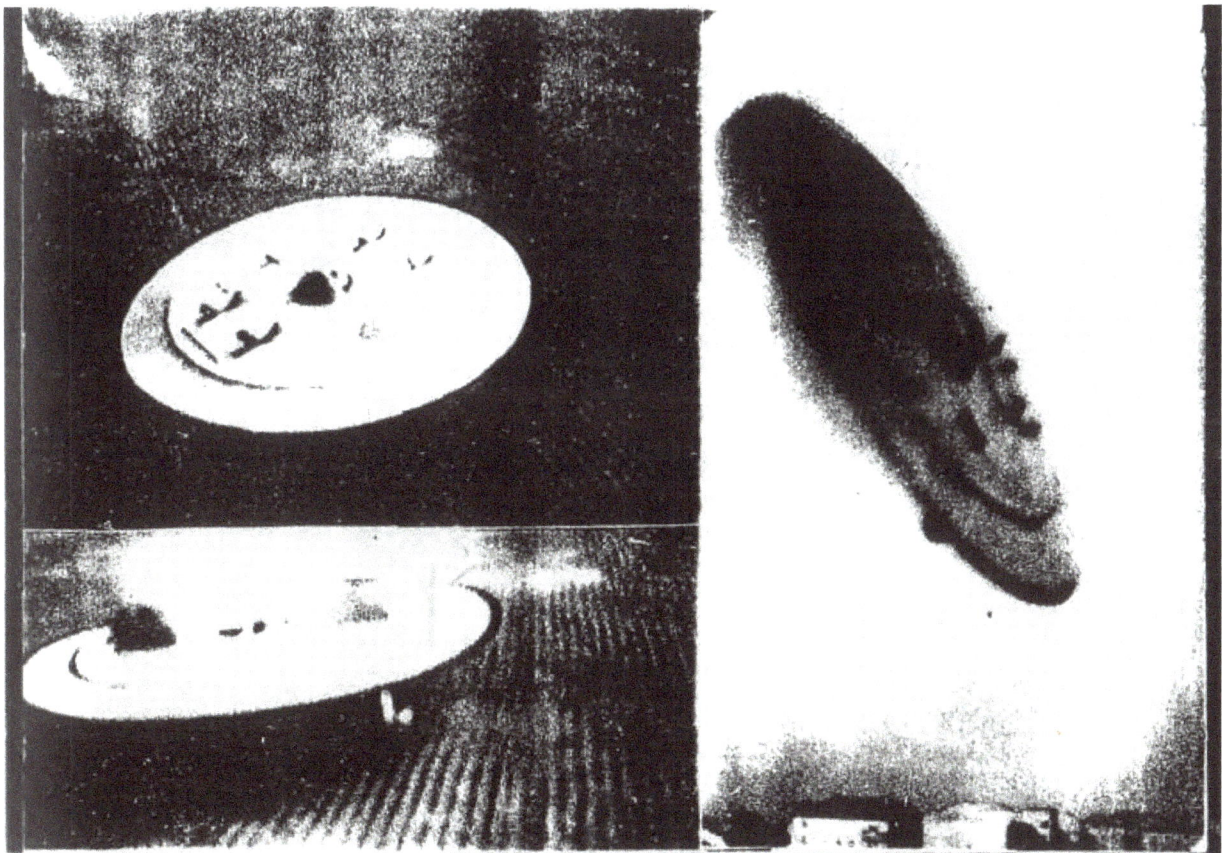

Photographs of flying discs allegedly taken in the Prag area, March 1945
http://www.bibliotecapleyades.net/ufo_aleman/rfz/chapter3d.htm

Utilizing this vortex phenomenon, Schauberger's saucer models would have an energy flow such as air entering in at the top where a small high-speed electric engine would spin the air around an axis using a paddle-like propeller. The motor would continue accelerating the rotating air until it reached the critical speed of auto-rotation at a rotational speed between10,000 and 20,000 revolutions per minute. At this point, the process was self-sustaining with air being drawn in and expelled with no additional input of energy. Air could be drawn in on a continuous basis since it was being cooled and thus made more dense in the vortex spiral. Greater density would mean a loss of volume which in turn created lower pressure at the air inlet. Air would flow centripetally inward as a vortex in a figure 8 configuration, right through to the center of the saucer with an implosive force. Greater airspeed at the point of ejection also served to lower the pressure (the same process which makes an airplane wing lift the aircraft), thus helping to lower the pressure even more at the inlet. This Implosion is always accompanied by an explosion as the fluid expands once again, in an outward centrifugal spiral. The Schauberger vortex was an open system, spiralling centripetally inward and spiralling centrifugally outward. Energy is radiated out from the center of the vortex at the narrow mid-point between the upper and lower loops. This energy, believed to be **diamagnetic energy** produces levitation causing the craft to rise.

**The Brilliant Austrian Scientist Viktor Schauberger
inventor of the Repulsine Saucer**
http://schauberger.co.uk/

Henry Stevens quotes in his own book from Callum Coats' book, *"Living Energies"* how the form of the vortex is a function of the energy flow:

"The form this vortex took is really dictated by function according to Schauberger. The "function" is the energy flow. The spiral vortex is the shape the energy flow takes in its movement. Energy flows in at the top of the vortex in the characteristic double-spiral manner. These air molecules are imploded, that is, they are made more dense and they yield heat as they progress. Air molecules are squeezed tighter and tighter together as they move down the vortex until the sub-atomic particles themselves become unglued transforming into new and unrecognized forms of energy. As the vortex itself decreases in diameter implosion and speed are increased until they reach the point within the vortex where centripetal forces stop and centrifugal forces take over."
http://www.bibliotecapleyades.net/ufo_aleman/rfz/schauberger.htm

The mechanism by which **Schauberger's saucer models** flew, were due to their construction of diamagnetic materials which repel magnetic fields. Schauberger considered copper diamagnetic and copper was used on the surfaces of the saucer models coming in contact with air.

The sub-atomic particles as described above of electrons and protons would be stripped away from the nuclei or neutrons. "Their opposing charges were free and attracted one another resulting in their mutual annihilation of one another yielding a release of energy. This would occur exactly at the point where centripetal forces ceased and centrifugal forces began, these air particles reversed their spin and altered their rotation."

An Original Surviving Repulsin Saucer Model From WWII
http://discaircraft.greyfalcon.us/Viktor%20Schauberger.htm

These neutrons were also magnetic and "were expelled from the saucer centrifugally along with water, water vapor and air which had not reacted as stated above. These "magnetic" neutrons, on the outside of the saucer body, may serve to increase the diamagnetic reaction of the copper saucer which would be both pushing away from the earth and the cloud of surrounding "magnetically charged" neutrons. In other words, the Schauberger disc may have been repelling away from the magnetically charged atmosphere which it itself had just created."
http://www.bibliotecapleyades.net/ufo_aleman/rfz/schauberger.htm

The dissolution of these sub-atomic particles which are transformed into the unrecognized energy forms may be considered in this day and age to be what is known as the "**quantum flux of space**" or "**zero point energy.**"

"The history of the Schauberger flying disc models is as follows. Viktor Schauberger began work in the early days of the Second World War producing certain parts for a "flying object" in the Kertl Works in Vienna, Austria which was the site of this work at the time. It was at the **Kertl Works** in 1940, that one of these strange contraptions had already prematurely flown in what was described as an accident where it had gone right through the roof of the Kertl plant.

Original Old Photos and Some Surprisingly Newer Photos of a Surviving Repulsin
http://www.laesieworks.com/ifo/lib/Viktor_Schauberger.html

"The purpose of this device was twofold. First, it was to investigate free energy production. This could be done by running a shaft to the rapidly rotating wheel-like component which was auto-rotating at between 10,000 and 20,000rpm. Using reduction gearing, some of that energy could be mechanically coupled to an electric generator producing electricity at no cost. The second purpose of these experiments was to test Schauberger's theories on levitation and flight."

"Two prototypes were said to have been built at Kertl. The test flight was done without Schauberger's presence or even his permission to do the test. The model flew as described above but it did considerable destruction to the Kertl Works so there were mixed feelings concerning the success of this flight. The force of levitation was so strong that it sheared six 1/4 inch diameter high-tensile steel anchor bolts on its way to the roof. According to Schauberger's calculations based upon previous tests, a 20 centimeter diameter device of this sort, with a rotational velocity of 20,000 rpm, would have lifted a weight of 228 tons!"

"In 1938, at a technical college in Dresden, Viktor Schauberger and his son Walter went to work with a Dr. Winter on a plan to extract electrical energy directly from a water flow. Some success was achieved in which a potential of 50,000 volts was achieved but, that no practical results came from this at that time."

"Probably based upon the fact that Viktor Schauberger was a veteran of the Great War, he was inducted into the Waffen-SS in 1943. This put him under the direct control of **SS Chief Heinrich Himmler.** Schauberger was ordered to castle Schoenbrunn near the Mauthausen concentration camp in Austria. There he was to select qualified prisoners, twenty to thirty craftsmen, technicians and engineers, and begin work on a new, secret weapon. Schauberger arranged improved conditions for his team and produced another flying saucer model."

"In May 1945, because of the deteriorating circumstances of the war, Schauberger was re-located to Leonstein in Upper Austria by the SS. There, just after a successful test flight of his latest device, it was seized by an American intelligence unit which appeared to be well informed about it. Schauberger was debriefed by American intelligence detained and told not to participate in further research."

"There is ample evidence that the Schauberger saucer model flew. The fact that the Schaubergers were brought to the United States to continue the work leads to the assessment that they worked on something of value to the government of the United States. The U.S. government was neither interested in his water research nor was it interested in his work on agriculture which was his main interest at the time when he was contacted by American agents. We are left to conclude that it was his work on a new form of levitation, his saucer work, which brought Viktor Schauberger and his son Walter Schauberger to the United States. After learning all they could, the government of the United States dismissed the Schaubergers rather badly, foreshadowing the treatment of the German rocket scientists two decades later."

This whole episode proved extremely unsatisfactory for both Schaubergers and they returned to Austria after about three months. As a condition for their return, Viktor was asked to sign away his knowledge to this powerful concern. Viktor was given a contract in English, which he did not read. Nevertheless, the deal was done and the Schaubergers returned to Austria. Viktor, already

in poor health, died five days later on September25, 1958. Walter continued advancing his father's research in Austria until his death in 1997."
http://www.bibliotecapleyades.net/ufo_aleman/rfz/schauberger.htm

"Sometimes the **Coanda Effect** is cited as a reason this saucer flew. Coanda effects, if present at all, are only a secondary force if Viktor Schauberger's calculations are correct. Coanda effects alone could never be powerful enough to generate the lifting force equal to 228 tons which Schauberger estimated his small model produced."

"The important factual points to keep in mind are these: First, this saucer-model, probably in more than one version, actually flew. Second, one model still exists today. Therefore this "flying saucer" is a physical reality which can be photographed, touched and studied. Third, explanations of the mysterious energetic forces causing lift in this device should not be forgotten. The German scientific leadership was interested in implosion and in what Schauberger had to teach them but they did not necessarily want to be limited by the use of air to achieve these results. Instead, they may have wanted to use Schauberger ideas but actuating these principles with electronic components."
http://www.bibliotecapleyades.net/ufo_aleman/rfz/schauberger.htm

Examination of Germany's secret weapons of war reveals that some of their scientists were working on an "electromagnetic rocket", known as the "KM-2 rocket" towards the latter stages of the war. **The KM-2 rocket** flew by means of the **Biefield-Brown Effect**, whereby the gravitational force is overcome through the use of a strong electrostatic charge, a concept developed by the work of T. Townsend Brown in the early 1920s. More on Townsend T. Brown and his work will be dealt with in a later section.

The rocket similar to a torpedo in shape flies forward with a positive charge on its leading edge created by a strong high voltage electrostatic charge while the negative charge flows off the surface of the device toward the rear. Imagine a gravity hill whose slope increases with the intensity of the ion charge. The craft being powered simply slides down the gravity gradient like a surfboard on a wave.

"This concept is a new, non-conventional form of flight. It is a non-aerodynamic method of flight. All previous conventional flight had either been through the use of lighter than air balloons, winged craft powered by propellers, jets or rockets or the brute force of rockets themselves. Using this method, Brown advocated defeating gravity by generating another form of energy over which gravity could be surfed. It is a form of field propulsion."
http://www.bibliotecapleyades.net/ufo_aleman/rfz/rocket.htm

126

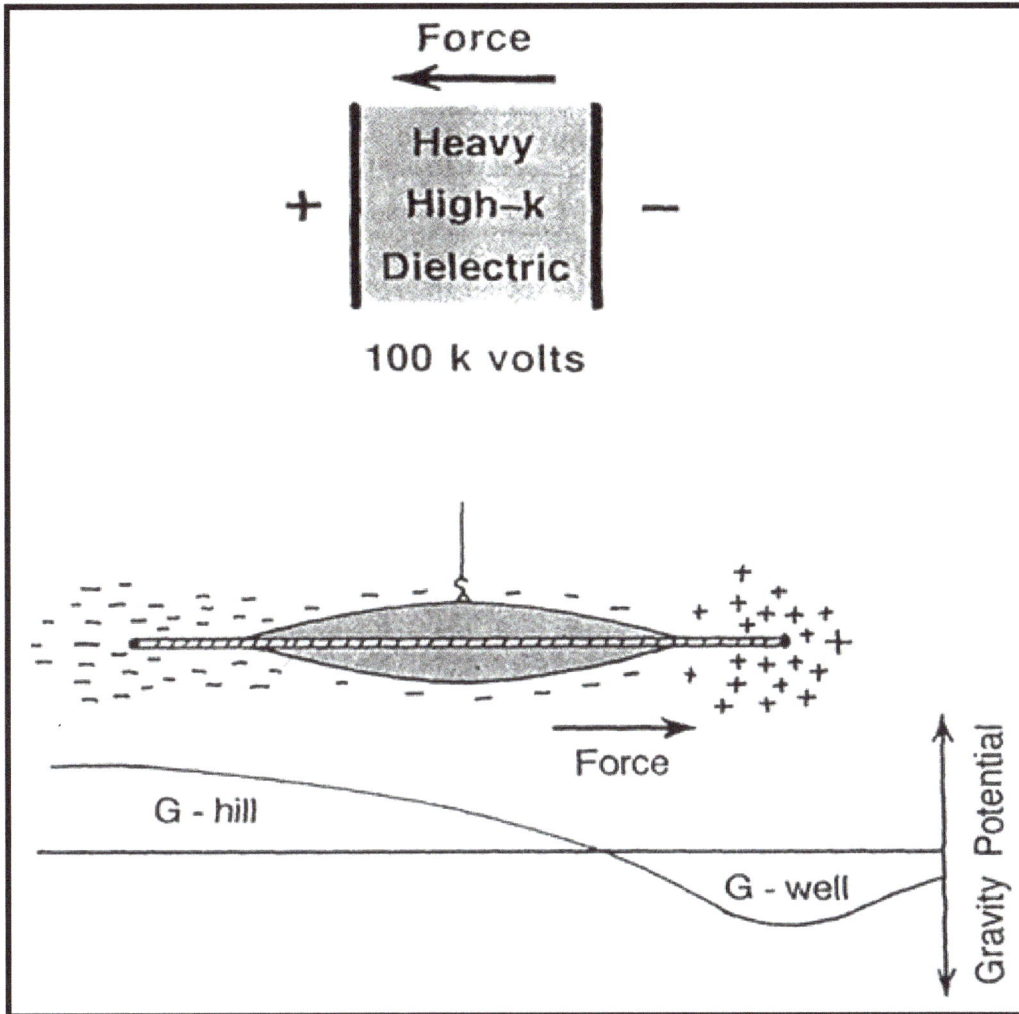

T.T. Brown found that when strong electric charges are separated by a dielectric, movement occurs towards the positive charge as if gravity were reduced on the positive side or as if the charged object were sliding down a hill.

http://www.bibliotecapleyades.net/ufo_aleman/rfz/rocket.htm

It is clear that the Germans seem to be well aware of the **Biefield – Brown Effect** as Dr. Biefield was a German speaker who was a fellow student of Albert Einstein in Switzerland and had shared his studies and experiments with other international physicists prior to the war. The Germans may simply have applied his research to their electromagnetic KM-2 rocket which was to be powered by a small atomic engine thus, producing a field propulsion device.

To accept the possibility that Nazi Germany could have developed field propulsion flying saucers is still a contestable topic which most aviation historians and Ufologists vehemently deny as a possibly. How could such an evil empire ever develop and build such advanced weapons technology that would be several decades ahead of everyone else is beyond comprehension. Yet, this attitude shows an extreme prejudice against seeking out the truth, no matter where it leads to or how awful it may be in its reality.

The reality and truth is that the Nazi Germany during the war had to think outside the box in order to survive and to carry forward their wartime agenda. The research and development for building saucers was an idea whose time had come. It wasn't just an exclusive German concept but, a technology built upon the work of many brilliant scientists in many branches of science and engineering over a period of decades and possibly centuries. Germany and the Third Reich were merely the first ones in the modern era to put all the pieces of a technology puzzle together to come up with an alternative to conventional aircraft design and an alternative propulsion method other than fossil fuel combustion. Desperate times demanded desperate measures and it came from both sides of the war, the Germans simply got a head start ahead of the rest of the planet!

It is also this writer's opinion that the Germans may have achieved a technological advance over other nations by a fortuitous recovery of a downed or crashed Extraterrestrial spacecraft and thus, were able to reverse engineer some of the alien technology to their advantage.

Field propulsion as a means to propel saucers seem to be an evolutionary process that developed prior to and then very quickly during the war. It's rudimentary beginnings began with people like Nicola Tesla, Townsend T. Brown, Paul Biefield, Vicktor Schauberger and others, all of who were visionary men ahead of their time.

Field propulsion development in Germany is a little sketchy at best in terms of strong historical records or first-hand accounts unlike the classic saucer constructions of Miethe-Belluzzo, Schriever-Habermohl, Klein and Schauberger that were fairly well documented. Apart from an eye–witness account by an unknown German pilot landing in the remote forested countryside of Germany and seeing three saucer-shaped craft in alone hangar building and then being warned off with threats by a military security guard, this seems more like hearsay evidence than anything concrete or substantial.

There does, however, seem to be some circumstantial evidence in the way of American FOIA documents from the *Combined Intelligence Objectives Sub-Committee Report number 146* regarding **Dr. Georg Otto Erb** and his work on numerous and highly classified or top secret projects, one being "rearward" impulse propulsion for aircraft and vehicles.

"Given the nature of Dr. Erb's other work, it is a safe bet that this "impulse propulsion" was not of the jet or rocket nature. In fact, there remains little doubt that this propulsion was, in fact, field propulsion. Dr. Erb was experimenting on means to apply forward motion using rearward impulse propulsion to aircraft and "vehs" (vehicles)." Henry Stevens, Hitler's Flying Saucers: A Guide to German Flying Discs of the Second World War, 2003, Adventures Unlimited Press, Kempton, Illinois, 60946   ISBN 1-931862-13-4

DR. GEORG OTTO ERB

HEAVY WIRE A 5000 MARK WEAPONS
(Target No. C1/659, C2/558)

Erb with his two assistants, Georg Buhler and Ullrich Lewitz are available at the above address and are known to Mil Gov Detachment at BORKEN.

To be investigated as early as possible. Suggest C.S.A.R. and C.E.A.D. (Fuzes) will be interested.

This man was interviewed by two members of the staff of D.D.O.S., and the attached report is compiled from their notes and a statement by Dr. ERB.

Dr. ERB was born in 1912. He is a doctor of Physics. At the outbreak of war he opened a small laboratory and had made a few inventions. In 1940 he was called up, but after a few weeks was released for research work and returned to Berlin to work for Physikalische Reichsanstalt. At this time he worked on acoustic heads for torpedoes. In 1941 he was again working in his own laboratory - mainly on fuzes. He was arrested in Berlin by the Gestapo in Nov 1944 on a charge of "favouring the enemy and sabotage". Buhler was also jailed. They escaped on 22 Apr 45 as the Russians were advancing into Berlin.

It is felt that he is reliable and likely to be of value as a source of information about the lines on which German development was proceeding in the field of experiment in which he was concerned.

Statement of the work of Dr. Otto ERB

1. Before the war, Dr. Erb developed measuring instruments of all sorts. The following are examples of his work:

(i)   Measuring apparatus for interference free determination of the hardness and temper of steel.
(ii)   Electrical measuring apparatus for automatic control of storage temperature.
(iii)   Apparatus for conversion of residual heat into electrical energy.
(iv)   Electrical medical apparatus of various sorts.
(v)   High tension apparatus.
(vi)   Warning mechanisms for excessive temperatures.
(vii)   Electric fire fighting apparatus.
(viii) Electric sources of energy of various kinds.
(ix)   Apparatus for turning the energy of the sun's rays into electrical energy.
(x)   Rearward impulse propulsion for vehs and aircraft.
(xi)   Wood gas generator for high performance.

After outbreak of war he had to devote his research to armament work.

CIC 75/139   - 1 -   Enclosure

As we delve deeper into the German secret saucer programs of the second world war, we come upon yet, another scientist whose involvement in saucer propulsion was radical and outside the box of conventional scientific thinking.

Karl Schappeller and "Reverse Thermodynamics"

Karl Schappeller (1875-1947) was born in Austria and was a self-taught lay-scientist and inventor, whose work culminated in the invention of a free-energy device attracting worldwide attention around 1930. His free-energy device had many applications developed for running homes, factories and industry but like many scientists, his work came to the attention of the SS who immediately saw its weapon potential for a possible death-ray or laser-like weapon but, it final application was as a power source in the levitation of saucers.

Most of us have heard of the two **Thermodynamic Laws**. These are laws of heat. The **First Law of Thermodynamics** states that **"energy is conserved",** meaning that the total amount of energy in the universe always remains the same. This is no surprise for most of us and it is not the real concern here.

What is of concern is the **Second Law of Thermodynamics** which discusses heat and entropy. The word entropy might be thought of as "**a state of randomness or chaos**". Negative entropy would then mean movement toward the less random or the more ordered in any particular thing. If we apply this to a system, then entropy tends to increase until the system breaks down in utter chaos. This will occur unless the system is recharged with additional outside energy.

Schappeller said there was another and unknown thermodynamic cycle which runs opposite the Second Law of Thermodynamics, a law referred to as **"Reverse Thermodynamics".** It is the reverse of the Second Law of Thermodynamics in that it leads to an increase in entropy. Not only is there an increase in order but there is an increase in cold! Building upon Tesla's work and theories, Karl Schapeller developed an aether engine based upon the opposite Second Law of Thermodynamics, the law of Reverse Thermodynamics.

A combustion engine such as used in your car or a wood burning stove operates on the **Second Law of Thermodynamics**. It is explosive and it is destructive in its operation producing heat and increasing entropy from which various by-products are produced most being pollution, which is the current world we live in.

In Reverse Thermodynamics a device would operate in reverse to the example given above but also be creative rather than destructive in performance, this is negative entropy. The machine would absorb heat and various additives into the machine. Making the process implosive instead of explosive so, there is an increase in cold. **Entropy** is increased and the by-products is oxygen using the combustion engine example or wood is created in the example of the wood burning stove.

An actual example of Reverse Thermodynamics is the **"Machine of Life"** that can be found all around us. It is in everything! A seed is such a machine where sunlight, water, various gases from the atmosphere and nutrients from the earth are combined and imputed into the seed through an "active implosive cold process" which is creative and thus, a tree is formed and comes into being. This type of cold force is a life-giving force or vital force **"Vril"**!

Each saucer program thus far discussed was building upon the success and breakthroughs of the previous conventional designs in the German saucer program which was aiming in the direction of a field propulsion operated saucer also, known in the inner circles of the SS organization as an **"ether ship"**.

It was a type of atomic engine that harnessed the "**aether**" **(free energy)** that is all around us. This energy is also known as **"space energy", infinite energy** or as the **"quantum flux of space"** or **"zero-point energy"** but, incorrectly called **"perpetual motion"** as most of, if not all these devices require input energy to kick start the operation of the device.

Early 20th Century physics regarded aether incorrectly as a type of matter according to the Michelson-Morley experiment. Particles that move at the speed of light are measured as a wave energy rather than as a particle of matter.

According to **Henry Stevens**, "Physics was hijacked early in the 20th Century by the alleged results of the **Michelson-Morley experiment**. The Michelson-Morley experiment assumed **"aether"** was matter. There is some confusion here. We know now that particles moving near the speed of light are measured as waves, that is energy, rather than as matter. Nevertheless, aether theory has been discredited among physicists who, in turn, discredit others who raise the subject. It is only through the efforts of "free energy devices" and free energy researchers that this knowledge is being returned to us. Without this aether theory, the reason these devices work cannot be explained at all. Rejection of aether theory allows these devices to be dismissed as" theoretically impossible" and so fraudulent by simple deduction. They are marginalized and dismissed as **"perpetual motion devices"**. According to established physics, perpetual motion devices violate physical laws of conservation of energy. Without an aether theory as an explanation, they do violate laws of conservation of energy and so their detractors are able to simply dismiss them out of hand. The simple fact that some of these free energy devices actually work does not seem to bother these scientists in the least. Rather than change the theory to accommodate the observed facts, the facts are ignored and substituted by dogma. Whether we like it or not, we are living in an **Energy Dark Age**." **Henry Stevens, Hitler's Flying Saucers: A Guide to German Flying Discs of the Second World War, 2003, Adventures Unlimited Press, Kempton, Illinois, 60946   ISBN 1-931862-13-4**

It was more than Physics that had been hi-jacked in the 20th Century, it was the 20th Century itself that had been hi-jacked as we will soon see further on in a later section of this book.

When we consider the advancement in science with physics as its main driving force, we see that chief among its luminaries, **Einstein** is the first to come to mind with his all-pervasive **Theory of Relativity**. Any theory about aether has been shoved to the shadows of pseudo science. Henry Stevens in his book, *"Hitler's Flying Saucers: A Guide to German Flying Discs of the Second World War"* rightfully points out that "'Two or three generations of scientists have wasted themselves on "trying to prove Einstein right". This misguided thinking has resulted in stagnation.'" Every time physicists try to move forward in a new direction using the Theory of

Relativity to explain or understand the universe in which we live (e.g. **"Dark Matter"** with all its intricacies), they find that falls short to adequately explain those mysteries thus, newer and stranger theories are propagated to come to terms with the unknown.

Theorizing is not bad science but hanging to a theory that has limited efficacy is not good science for the progress of civilization. What is truly amazing is the fact that every tear and almost every month, some new scientist or garage type inventor stumbles upon the "free energy" (aether) concept and builds the proof of concept device contrary to and "in spite of the accepted scientific theory" yet, they get no notice in the media or by the hallowed halls of academia other than some mysterious men from either the intelligence community or the military or the privatized industrial sector shows up and sequesters or prevents any further development of the free energy device. Such is the advancement of science!

"Needless to say, German scientists of the Nazi period labored under no such illusions. They never abandoned aether physics. This was the fundamental reason why field propulsion UFOs were first developed in Germany. After the Second World War, two different sciences developed called "Physics". One was the **relativism** taught in schools. The second more *esoteric* type was utilized only secretly, by the secret government, for deep black projects."
http://www.bibliotecapleyades.net/ufo_aleman/rfz/schappellerchapter4b.htm
Henry Stevens, Hitler's Flying Saucers: A Guide to German Flying Discs of the Second World War, 2003, Adventures Unlimited Press, Kempton, Illinois, 60946   ISBN 1-931862-13-4

It is believed by some scientists that "nuclear doping" of electrical components and circuits using radium chloride, like the stuff that is used on the hour and minute hands of a watch, may be the be the missing link in free energy devices and is the possible connection electromagnetic and atomic energy and thus, to **field propulsion**. It may be once again, that the Germans may have pioneered this development with the work of Karl Schappeller.

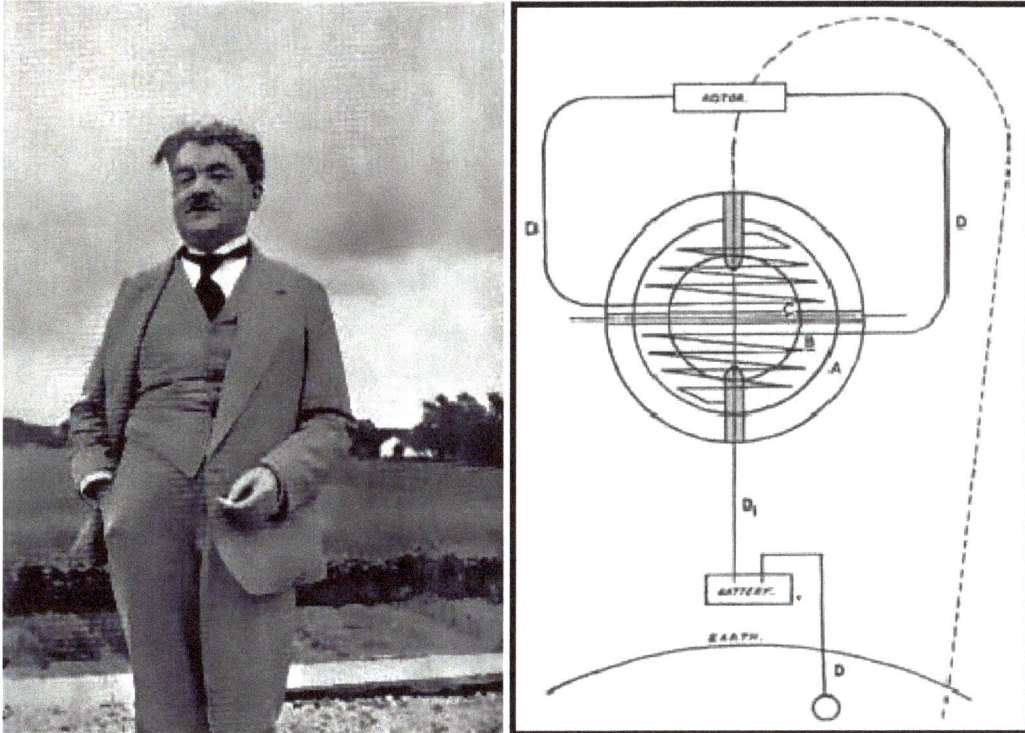

Left: Inventor Karl Schappeller Right: Karl Schappeller's Device. A. Steel outer casing. B. Special ceramic lining in which tubes are embedded. C. Hollow center, filled by glowing magnetism when in operation. D. Tubes, circuit, and earthling.
A careful examination of this sketch shows the similarity of the circuit with the functioning of the Earth's central core!

http://thirdreichoculthistory.blogspot.ca/2014/04/deutsch-occult-physik-german-occult_2.html

"In a book titled *The Physics of the Primary State of Matter*, published in the 1930s, Karl Schappeller described his Prime Mover, a 10-inch steel sphere with quarter-inch copper tubing coils. These were filled with a material not named specifically, but which is said to have hardened under the influence of direct current and a magnetic field [electro-rheological fluid]. With such polarization, it might be guessed to act like a dielectric capacitor and as a diode.

The same material is inside the rotor. The transfer of electrons upward and the magnetic field of the sphere combine to turn the rotor. No direct conversion of this energy to electricity is described, but the rotor could be attached to a generator...

According to Schappeller, the amount of this sustaining rotation of the device would be termed 'glowing magnetism' and be located at the center of the sphere. Present day terminology might be 'plasma'". http://www.rexresearch.com/schapp/schapp.htm

Without getting technical in the physics of the operation of Schappeller device which is beyond the capacity of this book at present, suffice it to say, that a **magnetic field propulsion** is generated from the device causing it to levitate.

The Schappeller device was real, it did exist, papers prior to the war were written about it and a British engineer spent a period of three years studying it and concluded that it was genuine. Eventually, it came under the control and funding of the German government of the time.

The Schappeller devices worked on the theory of implosion, similar to the Schauberger engine employing an aether-as-matter explanation for its operation which in turn complimented Nikola Tesla's evidence on radiant energy.

"Since this quest for a new science with the accompanying new machines had a relatively long history in Germany, certainly pre- dating the 3rd Reich, it is almost certain that the Schappeller device or others built along a similar understanding were further developed during the Nazi period. What became of it after the war is unknown. It can be assumed that this device did not escape the scrutiny of the numerous Allied intelligence units tasked with combing Germany for examples of German science."

Perhaps, the answer lies within the American **"Operation Paperclip"** program of ferreting German scientists and secret weapons technology out of Germany and back to the States after the war.

"The German eyewitness account of **"*Magnetscheibe*"** prompted investigation into U.S. governmental sources for corroboration. The CIOS report and the F.B.I. report provided corroboration. In the CIOS report, we find a U.S. governmental admission of experiments in field propulsion for aircraft undertaken in wartime Germany by Dr. Erb. The F.B.I. report on a field propulsion German saucer must be taken seriously because the F.B.I took it seriously. The F.B.I. carefully took the report and investigated the veracity of their subject. The F.B.I. then sent copies of this report to other intelligence agencies within the U.S. government which is indicated on the F.B.I. report itself. The Bureau saved the report all these years. The fact that this report deals with German technology but was taken by a domestic law enforcement agency, one whose "spy" activities are geographically restricted to within the USA, is noteworthy. It may indicate that the F.B.I.'s Director, J. Edgar Hoover, was kept "in the dark" about the real nature of flying saucers and may have wanted to show the other intelligence agencies that he was not so easily cut out of the information loop." Henry Stevens, Hitler's Flying Saucers: A Guide to German Flying Discs of the Second World War, 2003, Adventures Unlimited Press, Kempton, Illinois, 60946   ISBN 1-931862-13-4

The German-Canada Saucer Development Connection

Another mystery is the post-war activities of **Dr. Richard Miethe**. Most all sources state that Dr. Miethe went to Canada after the war and worked on a joint Canadian-American saucer project at an aircraft facility near Toronto, Ontario. Unfortunately, all **Freedom of Information Act** inquiries concerning Dr. Miethe run into the solid wall of "no record". Only one researcher ever claimed to have a document naming Dr. Miethe in association with this **Avro Aircraft Limited** project, (also known as **A.V. Roe Limited**), and that one researcher later admitted to being "a government asset" which throws a cloud of doubt on all his work.

There is no doubt, however, that by early 1955, work was commenced by Avro to build a mach 3 flying saucer which is reminiscent of some of the designs attributed to Dr. Miethe. Two designs were proposed, the difference being the engine used to power the saucer. One proposal was to use several axial-flow jet engines. The second and preferred proposal was to use one large radial-flow jet engine. The axial type is the type most commonly used in jet aircraft today. The radial type was similar to the first jet engine flown by the Germans in 1937. In fact, the radial engine actually under study in Canada may have had some similarities with the **Rene Leduc** engine used by Dr. Miethe.

Work continued until the early 1960s under various names including **Project Silver Bug** and **Project 1794**. Finally, a small hovercraft was unveiled by Avro as the final outcome of their saucer experimentation. This **"Avrocar"** had nothing to do with either Dr. Miethe's work or a mach three interceptor. The Avrocar was probably a cover project for something else. This "something else" was more advanced.

The Avro Aircraft, Limited experimentation with saucer-craft was always an open secret which was at times exploited by the government. Information regarding this project has been obtained via Freedom of Information Act using their American partner, the United States Air Force at Wright-Patterson Air Force Base, by this researcher as well as other researchers.

There exists a sub-story to the Canadian involvement which should be mentioned. In an article in a British UFO magazine, writer **Palmiro Campagna** revealed a previously unknown connection between the Canadian government and the history of German saucers. It seems that an SS technical liaison officer, **Count Rudolf von Meerscheidt-Huellessem**, (erroneously spelled "Hullessem" in the article), contacted the Canadian government in March of 1952, offering technical information about a German saucer which could attain speeds "limited only by the strength of the metals used and the saucer's construction". According to the article, von Meerscheidt-Huellessem wanted a large sum of money as down payment, a monthly salary and Canadian citizenship and police protection in exchange. Support for this claim comes in the form of copies of Canadian government documents describing this offer. Mr. Campagna states in the article that the Canadian government ultimately declined the offer but that the American government may have taken over negotiations and accepted.

This is a general diagram of the exterior of an Avro saucer. It is representative of how the Avro radial engine saucers worked. Within this outer hull, an inner, flat radial-type engine was situated. This is exactly the scheme designed by Dr. Richard Miethe it and has been copied from his designs. Dr. Frost and his engine designs were overblown, heavy, complicated frauds.

AVRO FLYING SAUCER: SECTIONAL DIAGRAM

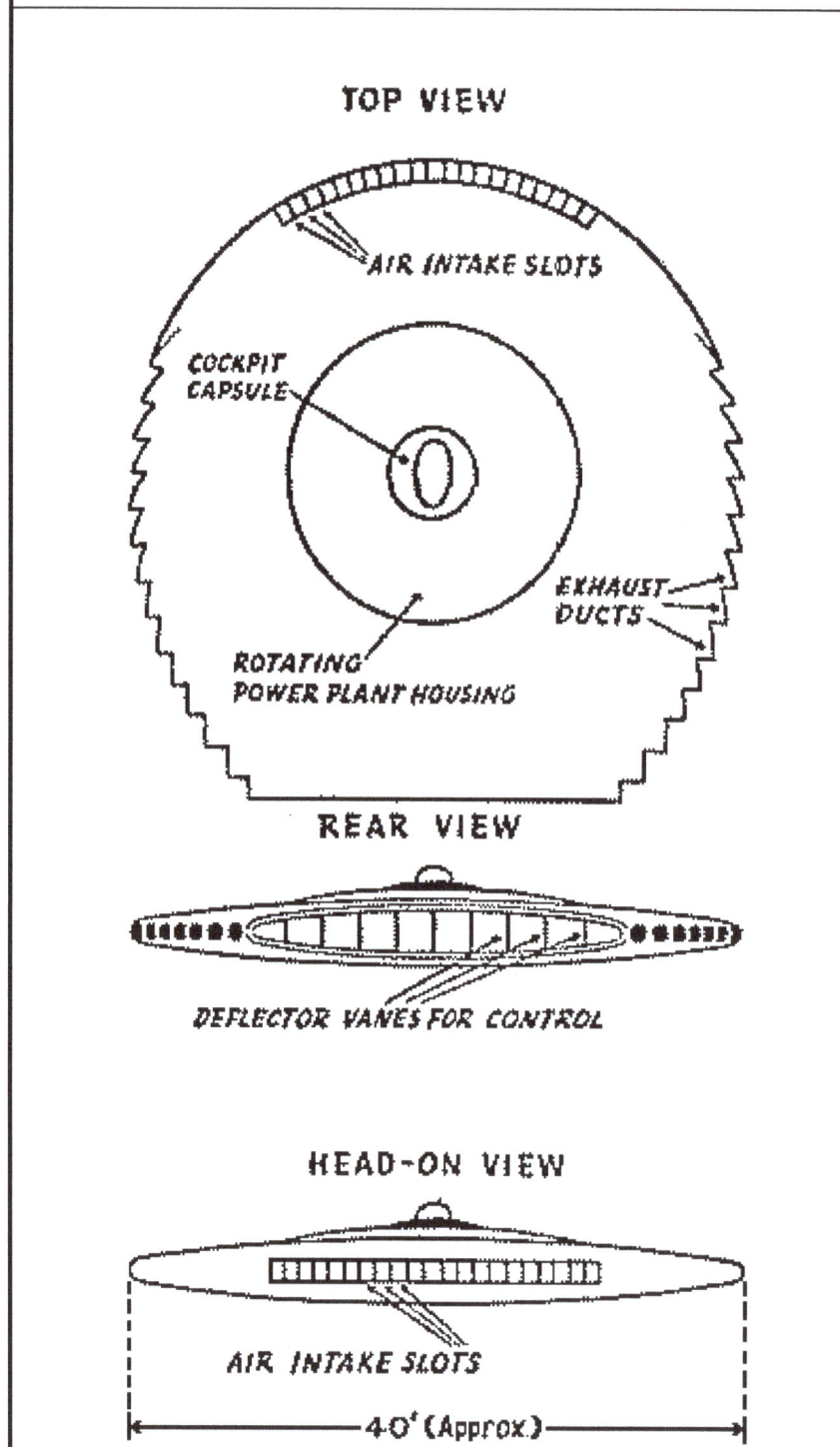

TOP VIEW

AIR INTAKE SLOTS

COCKPIT CAPSULE

EXHAUST DUCTS

ROTATING POWER PLANT HOUSING

REAR VIEW

DEFLECTOR VANES FOR CONTROL

HEAD-ON VIEW

AIR INTAKE SLOTS

40' (Approx.)

Avro Flying Saucer
http://www.bibliotecapleyades.net/ufo_aleman/rfz/chapter5.htm

Contact was made by this researcher with a daughter of Count Rudolf von Meerscheidt-Huellessem's who verified that her father was a technical liaison officer. She had little contact with her father since she was two years old since he had remarried and moved to Canada. She was able, however, to provide the address of another relative, **Countess von Huellessem,** who was Count Rudolf von Meerscheidt-Huellessem's widow.

Contact was made with Countess von Huellessem. **Count von Meerscheidt-Huellessem** died in 1988. But the Countess did know a little something about the story in question. Her late husband did discuss the flying saucer with her but only once. He told her that the "drawings" had been given to a representative of the Canadian government in 1952. After review of the drawings, the only comment from the representative was that they were "outdated". The drawings were never returned. The Canadians had succeeded in obtaining hard information concerning a real flying saucer and in paying for it with an insult. Count von Meerscheidt-Huellessem was somewhat despondent over the rejection. He never mentioned the subject to his wife again. At the time they were both making new lives for themselves in Canada and the subject never resurfaced.

In discussing these events with the Countess over the telephone, she told me that she herself had seen these drawings. She was asked if these were drawings or technical plans. She said they were technical drawings on rolls of paper. I said the word "blueprint" and she said "yes". She stated again that these drawings were given only to the Canadians and not to the Americans.

Countess von Huellessem was sent a copy of the aforementioned article. Her only comment was that her husband would not have asked for a large sum of money in exchange for this information. They already had the means. He might have asked for a position, she said, since her husband would have enjoyed working on this project.

Returning to the **Avro Aircraft Limited** - U. S. Air Force saucer project, we have to ask ourselves some questions. First, was this information, these plans, the real basis of the Avro saucer project? This would certainly explain the reason for the partnership between the two governments involved since the Americans would have needed the Canadians at that point and the Canadians would have insisted upon control of the project on their home soil. Second, did the technology brought to the Canadians by von Meerscheidt-Huellessem have anything to do with the jet technology obtained under the **Freedom of Information Act**? It certainly had nothing to do with a hovercraft which was the outcome of this project according to the government. How could a floppy hover-craft barely capable of 300 miles per hour under the best of estimates have had anything to do with a saucer whose speed was "only limited by the strength of metals used in the saucer's construction"? Could **Project Silver Bug, Project 1794**, and some of the other patents of **John Frost** attributed to this collaboration all be nothing more than an elaborate cover story?

Another point is that until recently, the only connection linking the German saucer projects to the **Avro Project** and to the Americans was the involvement of **Dr. Richard Miethe**. With the new evidence of the Peenemuende Project's connection to the American saucer projects run out of Wright Field, Ohio and the **Count von Meerscheidt-Huellessem** connections to the Canadians, the weak linkage of Dr. Miethe to these projects is superfluous. There is now more than enough evidence to make these connections with or without Dr. Miethe. Further, considering the Count

von Meerscheidt-Huellessem evidence, there is now a direct link between German saucer technology and the Canadian government's saucer project at **A.V. Roe Limited**. That link turns out to be the SS in the form of an SS technical liaison officer.
https://www.youtube.com/watch?v=KyrlFPD1fLE

Before leaving the topic of the secret German saucer program, we should take one final look at a few more controversial super secret Nazi saucer programs commonly referred in the UFO literature as the **Vril**, the **Haunebu, and the Bell saucer projects.**

One could contend that all previous saucer programs so far discussed were of more conventional proof of concept design leading up to a more exotic saucer program employing advanced engineering concepts, metallurgy, esoteric theories of propulsion, super advanced weaponry utilizing atomic bombs, lasers, electromagnetic pulse cannons and microwave emitters, computers and microelectronic circuitry coupled with the capability of tremendous speed, aerial maneuverability, long flight duration, with the ability of space flight to other stellar bodies! Fantastic and beyond belief? Certainly!

Nazi scientists were pushing the envelope of science in all directions, they were thinking outside the box and it appears that no barriers were impossible to for them to conquer. Little wonder then that the Allied forces were extremely alarmed by the information they were gathering from their spy network throughout Europe.

The eugenics of a superior, tall, strong, blue-eyed blond Aryan race was fully developed with progeny still living today from that time period. Even, genetic engineering of super soldiers was researched and experiments were carried out using prisoners and slaves of the German concentration camps!

No research areas were off limits to the diabolic minds of the leaders of the Third Reich, if it advanced the agenda and future of the German people! **Hitler**, his Chiefs of Staff and the Axis leaders often referred to these advancements as their **"Wonder Weapons"** which were so dreadful has to cause the heart to stop from fright and which would snatch victory from the jaws of defeat away from the Allies. Documents have gradually over time have leaked out or surfaced from **FOIA** searches that support these historical developments.

The problem with history is that it is often written by the victors and rarely if at all, from the viewpoint of the defeated! The Americans, Russians, British and the French certainly wrote the history of World War II according to their perception and take on the outcome of their victory over the Axis Powers. But, much of what was gleaned from German technology has not fully been disclosed to the general public as yet. We think we know all that there is from the Second War but, the reality is far different than our contemporary perceptions as taught in school. There are still secrets that are suppressed and held by the victors of the war because they are tired up in the R&D of some of the world's most powerful nations and their militaries.

An examination of the more advanced German saucer programs such as the Vril and the Bell projects initially seems either to be a flight of fancy (no pun intended) or a hard cold reality of incredible proportions, scarcely imaginable! Somewhere in between the two possibilities lies the

138

truth of the matter or is there a third possibility? The third possibility we will consider later. The **Vril**, Haunebu, and the Bell saucers operated on the principles of implosion which we have already discussed these theories with the work of Viktor Schauberger and the esoteric principles of the **Vril and Thule Societies** whose basic beliefs stem from telepathic or channeled information from an interstellar Aryan race from the star system of Aldebaran.

The Thule Society, Nazi Party founders and some of their Icons
https://richardbrenneman.wordpress.com/2012/05/08/fascism-redux-videos-from-greeces-golden-dawn/ and
https://www.pinterest.com/johnnymcmonagle/occult-of-personality/ and https://www.pinterest.com/pin/453315518720798347/

The Vril Society whose members consisted of Hitler and many high ranking Third Reich officers believed that a technology based on explosion is destructive runs against the Vril divine principle of creation **"other-side flying machine"**. "Thus they wanted to create a technology based on **Implosion**. Schauberger's theory of oscillation (principle of the overtone sequence, monochord)

takes up the knowledge of Implosion. To put it simply: **Implosion** instead of **Explosion!** Following the energy paths of the monochord and the implosion technology, one reaches the realm of antimatter and thus the cancellation of gravity." **"Secret Societies and their Power in the 20th Century" by Jan Van Helsing (Pseudonym); copyrighted 1995; ISBN 3-89478-654 – X; Translated by EWERTVERLAG S.L.; P.O. Box 35290, Playa del Ingles, Gran Canaria, Spain**
http://www.bibliotecapleyades.net/sociopolitica/secretsoc_20century/secretsoc_20century06.htm#CHAPTER%2033

Impulsion Propulsion Flying Saucers

In the summer of 1922, the first saucer-shaped flying machine was built whose drive was based on the **implosion principle**. It had a disk eight meters across with a second disk with a diameter of six and a half meters above and a third disk of seven meters diameter below. These three disks had a hole at the center of one meter eighty across in which the drive, which was two meters forty high, was mounted. At the bottom, the central body was cone-shaped, and there a pendulum reaching the cellar was hung that served for stabilization. In the activated state the top and bottom, disk revolved in opposing directions to build up an electromagnetic rotating field.

The performance of this first flying disk is not known. But experiments were carried out with it for two years before it was dismantled and probably stored in the Augsburg works of *Messerschmidt.* In the books of several German industrial companies, entries under the codename **JFM (*Jenseitsflugmachine*) (Otherworld Flight Machine),** Germany's first disc can be found that show payments towards financing this work. Certainly, the *Vril Drive* (formally called **Schumann SM-Levitator**) emerged from this machine.

In principle, the "other-side flying machine" should create an extremely strong field around itself extending somewhat into its surroundings which would render the space thus enclosed including the machine a microcosm absolutely independent of the earthbound space. At maximum strength, this field would be independent of all surrounding universal forces – like gravitation, electromagnetism, radiation and matter of any kind – and could, therefore, maneuver within the gravitational or any other field at will, without the acceleration forces being effective or perceptible.

In June 1934, Viktor Schauberger was invited by Hitler and the highest representatives of the Thule and Vril Societies and from then on worked with them. After the initial failure, the first so-called German UFO also came out in June 1934. Under the leadership of **Dr. W.O. Schumann**, the first experimental round flying machine, the **RFZ 1 (Rundflugzeug I)** was developed on the grounds of the aircraft factory Arado in Brandenburg. In its first and only flight, it rose vertically to around 60 meters, then wobbled and danced in the air for minutes. The Arado 196 guiding system was utterly useless. The pilot Lothar Waiz just managed somehow to bring it down to the ground, jump out and run away before it ripped to pieces. That was the end of the **RFZ 1**, but the beginning of the VRIL flying machines.

Before the end of 1934, the **RFZ 2** was ready, with a Vril drive and a "magnetic field impulse steering unit". It had a diameter of five meters and the following flying characteristics: With

140

rising speed the visible counters became blurred and the craft showed the colors typical of UFOs: depending on the drive setting red, orange, yellow, green, white, blue or purple. It worked – and it would meet a remarkable destiny in 1941, during the Battle of Britain, when it was used as transatlantic reconnaissance craft, because for these flights the German standard fighters ME 109 had an insufficient range.

By the end of 1941, it was photographed over the southern Atlantic on its way to the German cruiser 'Atlantis' in Antarctic waters. It could not be used as a fighter, though. The impulse steering allowed it only changes of direction at 90 degrees, 45 degrees or 22.6 degrees, but that is exactly the right-angled flying pattern associated with and typical for UFOs today!
After the success of the small RFZ 2 as a distant reconnoiter craft the **Vril-Gesselschaft** or **"Not All Good comes From Above"** got its own test area in Brandenburg.

Flugelrad-IIV2 (Flugelrad series 1943 - 1945)
http://www.laesieworks.com/ifo/lib/WW2discs.html

RFZ (RundFlugZeug Discs) of the Thule/Vril Type (1934 – 1945) RFZ Series: 1,2,3,4,6

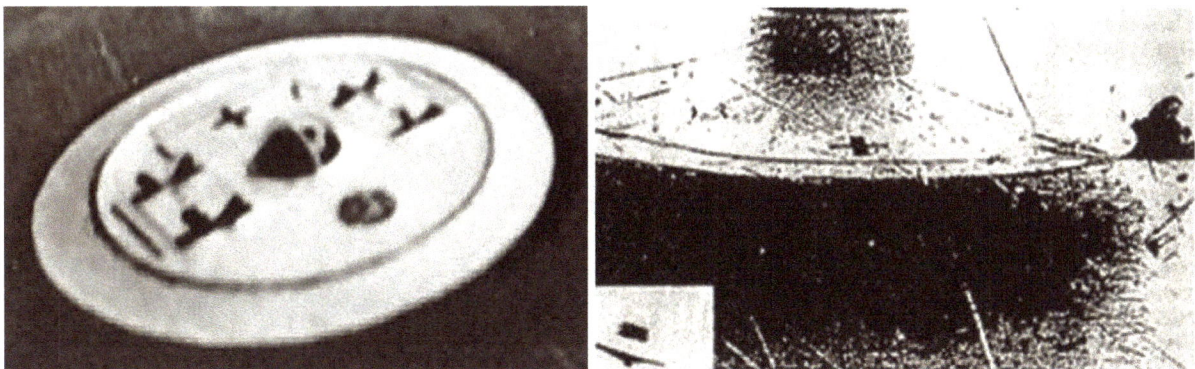

JFM - Jenseitsflugmaschine, (Other World Flight Machine) (1922 - 1924) left and right RFZ – 1 (1937)
http://www.bibliotecapleyades.net/ufo_aleman/esp_ufoaleman_5.htm

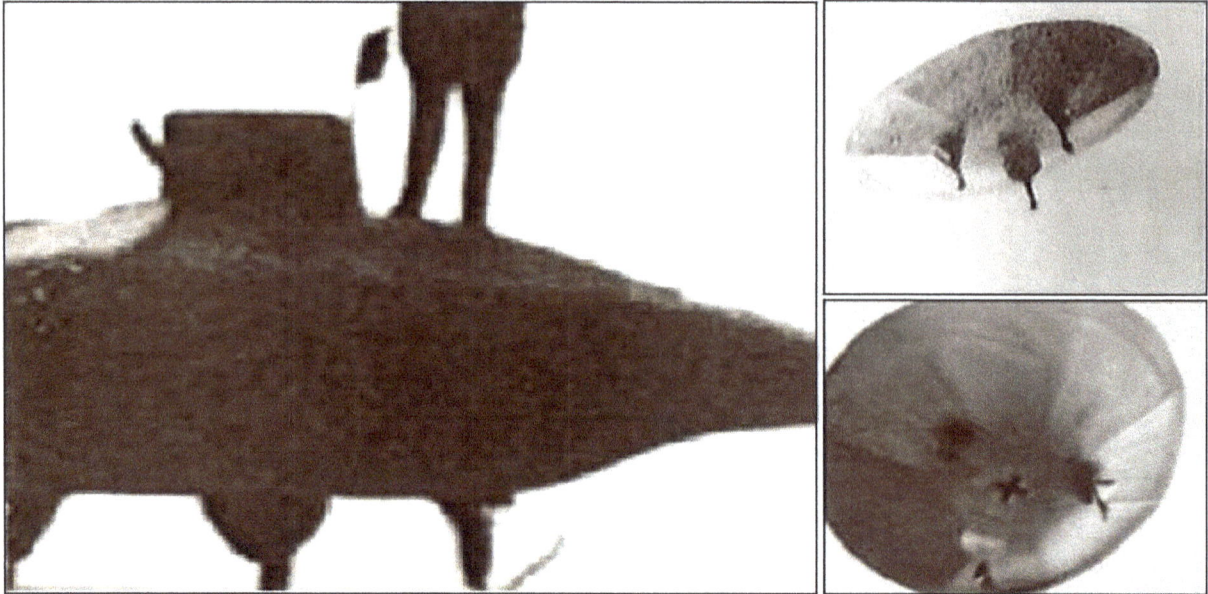

RFZ – 2 (1937)
http://www.bibliotecapleyades.net/ufo_aleman/esp_ufoaleman_5.htm

RFZ – 3 (1937)
http://www.bibliotecapleyades.net/ufo_aleman/esp_ufoaleman_5.htm

142

RFZ – 4 (1938)

RFZ – 6 (1940)

RFZ - 5 was changed to Haunebu 1 (1939)

143

VRIL 1

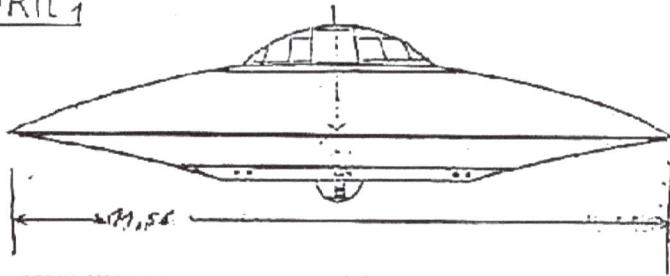

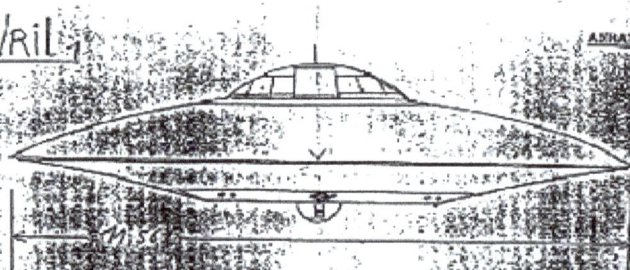

LEICHTE BEWAFFNETE FLUGSCHEIBE (JÄGER), TYPE „VRIL"
(Schumann-Gruppe)

Durchmesser: 11,50 Meter
Antrieb: Schumann-Levitator (gepanzert)
Steuerung: Mag-Feld-Impulser 5a
Geschwindigkeit: 2900 Kilometer p.Stunde (bisher), bis zu ca. 12000 mögl.
Reichweite (in Flugdauer): 5 1/2 Stunden ((Flug-Aufladung mittels K5 von
Hauneau aus wird erprobt))
Bewaffnung: 1 5cm KSK, fernsteuerbar, unten, + 2 x MG 108 u. 2 x MG 17
Außenpanzerung: Doppel-Viktalen
Besatzung: (je nach Einsatzart) 1 bis 3 Mann
Weltallfähigkeit: 100 %
Stillschwebefähigkeit: 12 Minuten
Allgemeine Flugfähigkeit: Wetterunabhängig Tag und Nacht
Grundsätzliche Einsatzreife: ca. Sept. 1944, ev. früher.

VRIL-1
VRIL-2
VRIL-3

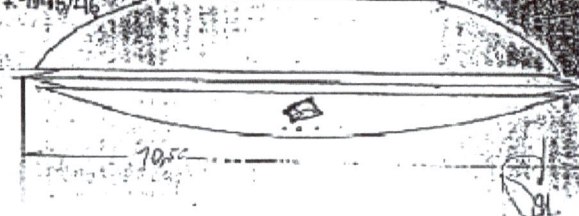

Vril -1 and the Vril-2 Zerstörer (projected for 1945/6)

144

Vril 1 - "Jäger (Hunter)"

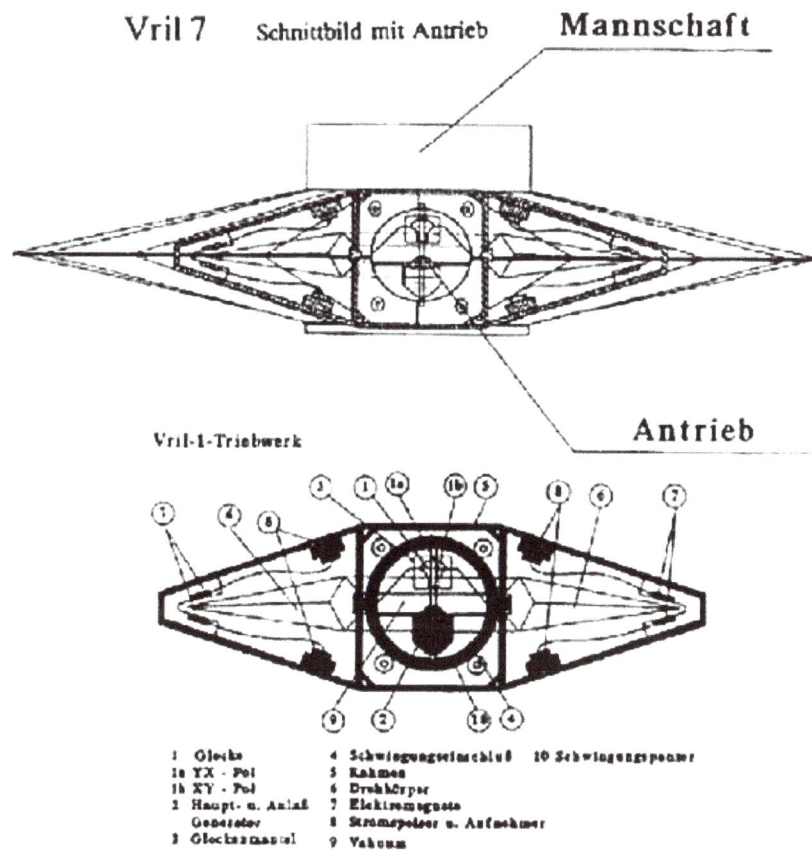

Vril 7   Schnittbild mit Antrieb   **Mannschaft**

Antrieb

Vril-1-Triebwerk

1 Gleiks	4 Schwingungseinschluß	10 Schwingungsspeaar
1a YX - Pol	5 Rahmen	
1h XY - Pol	6 Drehkörper	
2 Haupt- u. Anlaß	7 Elektromagnete	
Generator	8 Stromspeicer u. Aufnehmer	
3 Glockenmantel	9 Vakuum	

Gemeinschaft des Schwarzen Steins

Rekonstruktionsversuch   **Durchmesser des Geräts ca. 45 m**

The Electromagnetic Propulsion System of the Vri -1 Triebwerk in a VRIL 7

"VRIL-ODIN"

Vereinfachte Planskizze (Querschnitt) eines Elektrogravitationsraumschiffes nach dem Dynamoprinzip

a) Kommando- und Steuerraum des Raumschiffes
β) Mannschaftsräume
I   Vertikaler Magnetring
II   Energiespeicher
III Dreiphasenspulen
IV   Erregerspule
V   Drei Kondensatoren
VI   Ausgangsspule
VII Ferritring
VIII Barium-Strontium-Titanat-Ring

MÖGLICHKEIT B

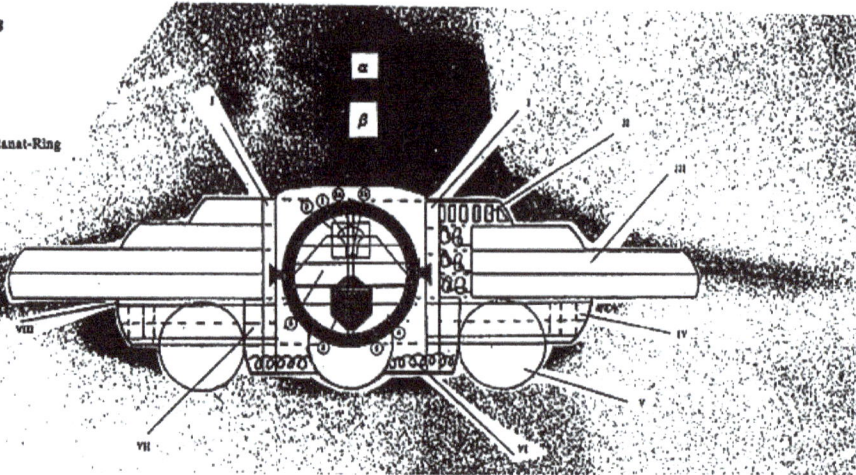

1 Glocke   4 Schwingungsanschluß
1a YX - Pol   5 Vakuum
1b XY - Pol   6 Schwingungspanzer
2 Haupt- u. Anlaß
   Generator
3 Glockenmantel

VRIL 8 Odin Propulsion System Superimposed Over a Photograph of the Craft

http://www.bibliotecapleyades.net/ufo_aleman/esp_ufoaleman_5.htm

SS-V9

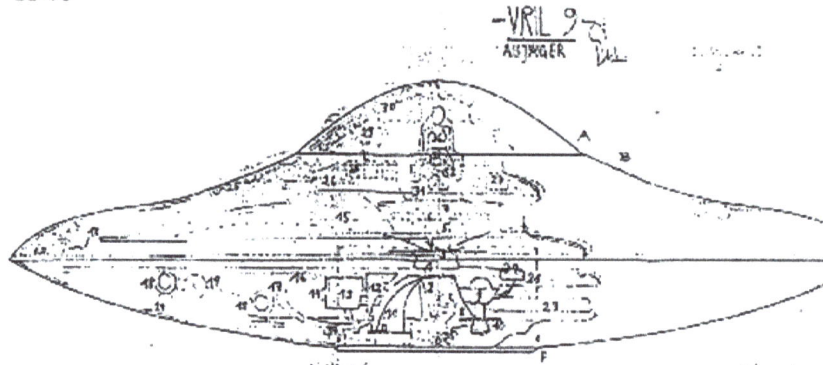

VRIL 9 Aujager, also known as the SS-V9. No known photos of it exist however, modern photos taken in various locations show a striking similarity to the above sketch

http://www.bibliotecapleyades.net/ufo_aleman/esp_ufoaleman_5.htm

146

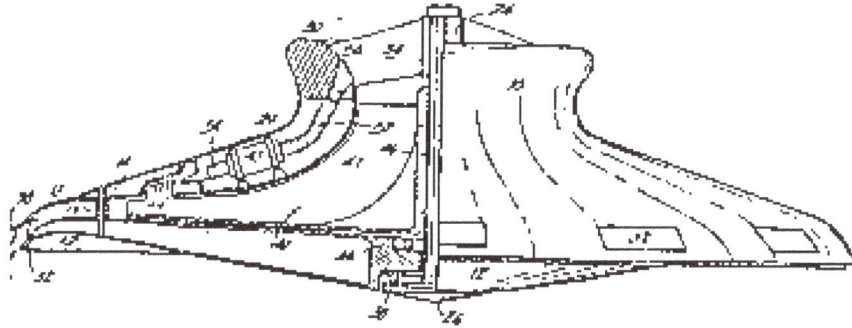

VRIL 10 Fledermaus This is Unknown Disc Diagram of Vril 10 Configuration

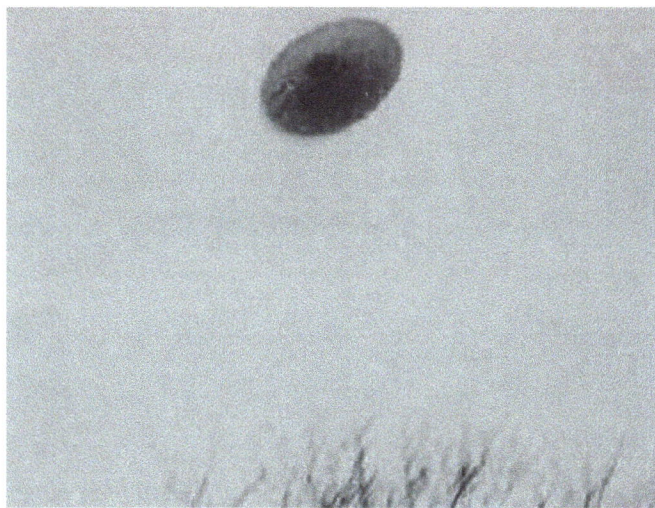

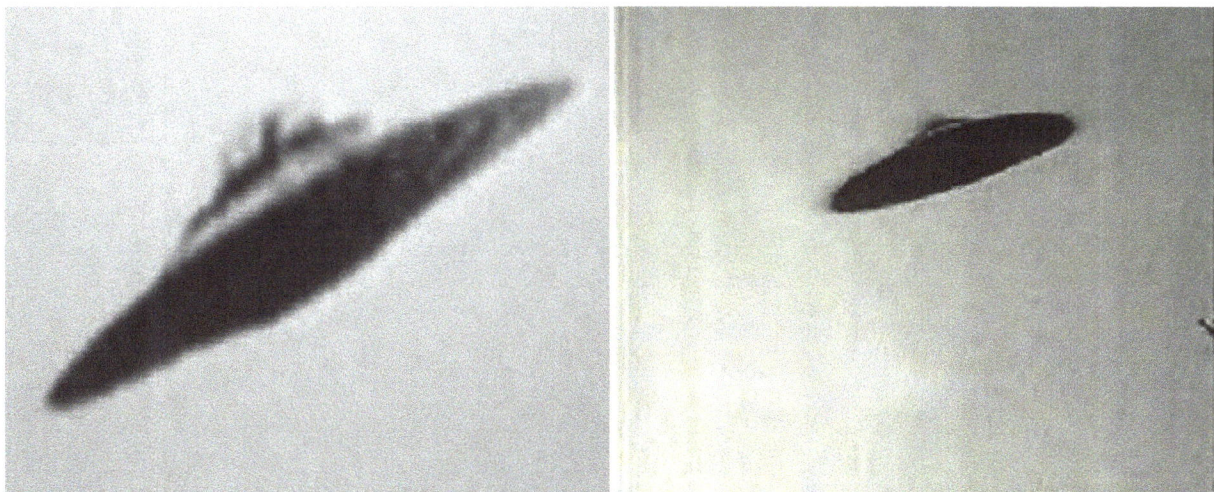

3 Photos of the VRIL 1 (Jäger) in Flight (1942)

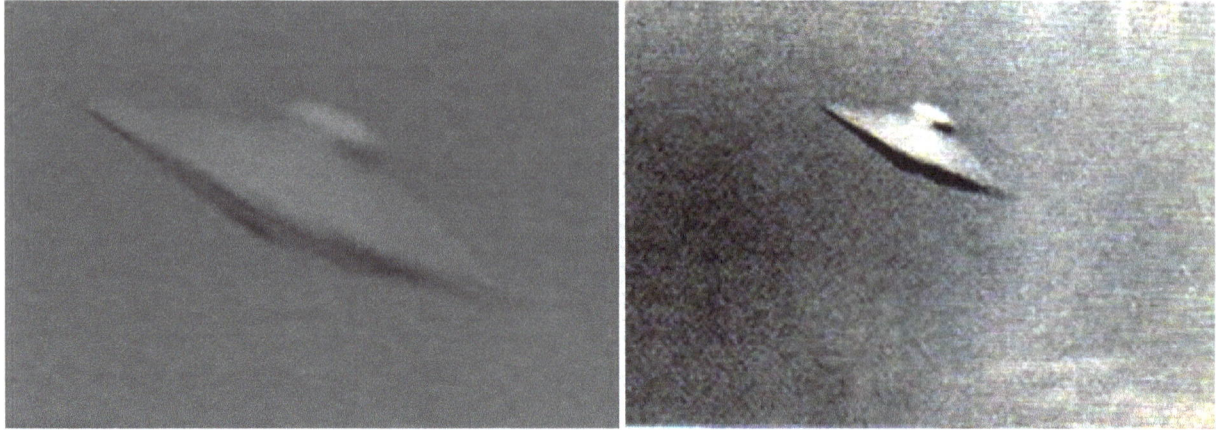

Alleged Black and White Photo and Colourized Photo of VRIL-2
(No Dates were given as to when These Photos Were Taken)

VRIL 7 Geist (1944)

148

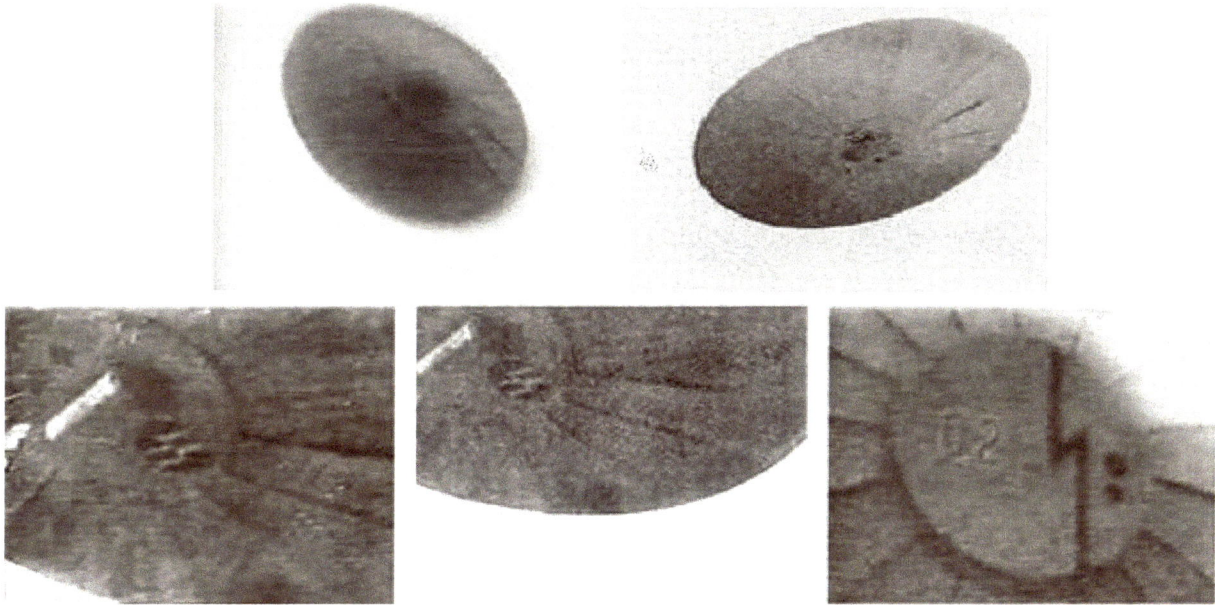

**Note the SS Insignia on the Bottom of the VRIL 7 Craft
and the Thule insignia on the Other Craft (Right)**

VRIL 7 Seen in a Probable Test flight

The VRIL 7 Craft Appears to be Mounted with Cannons in the Cupola Area

149

**VRIL 7 Testing the KSF 'Donar' Cannon (top and bottom) the top photo
has been colourized, probably based on a real photograph,**

VRIL 7 with German Cross on Cupola

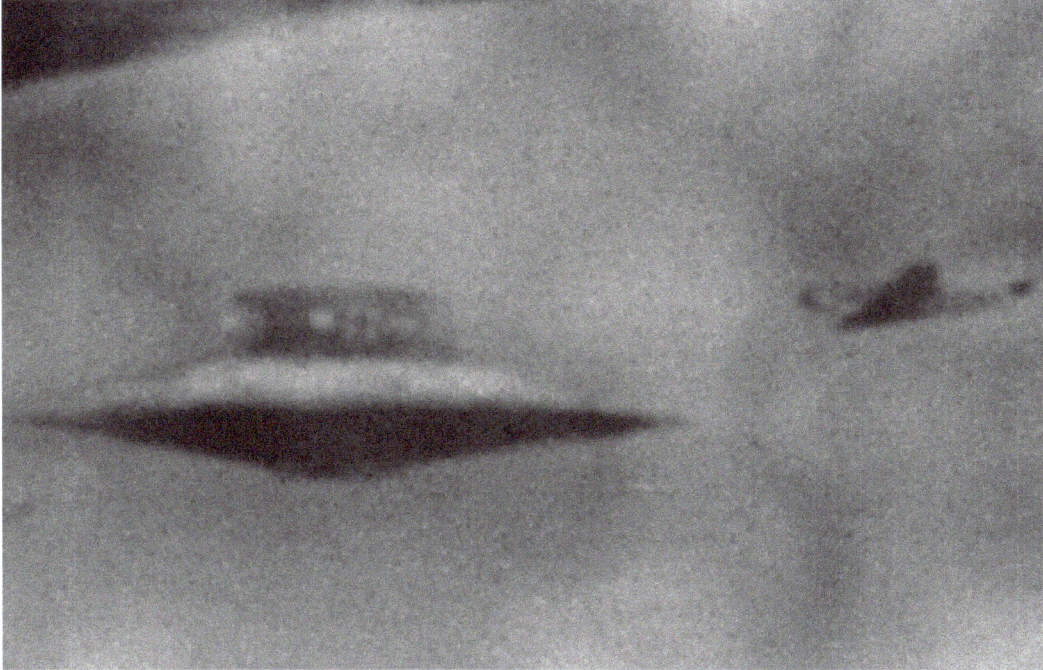

A VRIL 7 is escorted in this photo by a Me 109, not by a Ju-52 or a Focke Wulfe 190G which has usually been the case with this particular photo
http://discaircraft.greyfalcon.us/The%20Vril%20Discs.htm

VRIL 8 Odin (1945)
http://discaircraft.greyfalcon.us/The%20Vril%20Discs.htm

VRIL 8 (top photo) and the VRIL 9 Abjäger

http://discaircraft.greyfalcon.us/The%20Vril%20Discs.htm

By the end of 1941, the lightly armed **RFZ-7** was renamed the **Vril 1 Jäger (Vril Gesellschaft Arado)** which flew out of Arado-Brandenburg. It measured 11.5 meters across, carried one person, and had a Schumann-Levitator drive and a "magnetic field impulse steering unit". It reached speeds of 2,900 to 12,000 km/h. could change direction at a right angle at full speed without affecting the pilot, could fly in any weather and had a 100% space capability. Seventeen VRIL-1s were built and some versions had two seats and glass domes.

Within the SS there was a group studying alternative energy, the SS-**E-IV (Entwicklungsstelle 4), (Development Group IV – "Order of the Black Sun")** whose main task was to render

Germany independent of foreign oil. Their work included developing alternative energies and fuel sources through coal gasification, research into grain alcohol fuels, less complicated coal burning engines for vehicles and generators, as well as highly advanced liquid oxygen turbines, total reaction turbines, **AIP (Air Independent Propulsion)** motors and even **EMG (Electro-Magnetic-Gravitic) engines.**

This group developed by 1939 a revolutionary electro-magnetic-gravitic engine which improved Hans Coler's free energy machine into an energy Konverter coupled to a Van De Graaf band generator and Marconi vortex dynamo (a spherical tank of mercury) to create powerful rotating electromagnetic fields that affected gravity and reduced mass. It was designated the **Thule Triebwerk (Thrustwork, a.ka. Tachyonator-7 drive)** and was to be installed into a Thule designed disc. http://discaircraft.greyfalcon.us/HAUNEBU.htm

Vril-1-Triebwerk

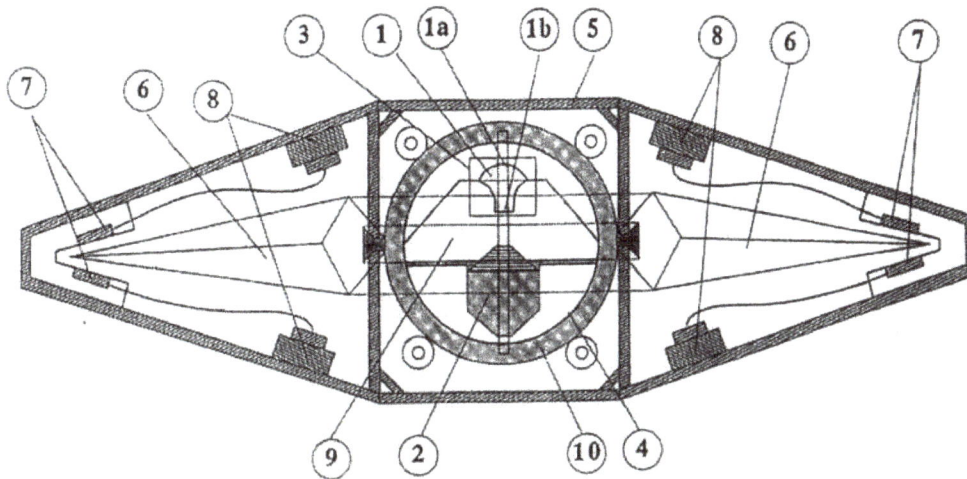

1	Glocke	4 Schwingungseinschluß	10 Schwingungspanzer
1a	YX - Pol	5 Rahmen	
1b	XY - Pol	6 Drehkörper	
2	Haupt- u. Anlaß	7 Elektromagnete	
	Generator	8 Stromspeiser u. Aufnehmer	
3	Glockenmantel	9 Vakuum	

http://discaircraft.greyfalcon.us/The%20Vril%20Discs.htm

In August 1939 the first **RFZ 5** took off. It was an armed flying gyro with the odd name **Haunebu I /SS Entwickstellung IV** and was based at Hauneburg, NW Germany. It was an all-weather craft 24.95 meters across and carried a crew of eight. Its propulsion system was a **Thule Tachyonator (Triebwerk) 7b Drive** and with **Mag Field Impulser 4 Control**. It could reach a speed of 4,800 km/h up to 17,000 km/h. It was equipped with two 80 mm **KSK (Kraftstrahlkanhonen, power ray guns)** in revolving towers and four MK-108 machine guns. It had 60% space capability. It had a double Victalen armoured hull and a Flight time of 18h and could hover for 8 mins.

By the end of 1942, the **Haunebu II** was ready. It was a Heavily Armed Flight Gyro Do-Stra craft with varied diameters from twenty-six to thirty-two meters and their height from nine to eleven meters. It had a Thule Tachyonator (Thule Triebwerk) 7c Drive with a Mag Field Impulser 4a Control. They carried between nine to twenty people at an atmospheric speed of 6,000 km/h and could fly in space at 21,000km/h theoretically. It had a flight duration of fifty-five flying hours and could hover for 19 mins. It carried an armament of six 80 mm KSK in 3 rotating turrets and one 110 mm KSK in a rotating turret. The craft had a triple Victalen hull. At this time there existed already plans for a large-capacity craft, the **VRIL 7** or the **Haunebu IV** with a diameter of 120m considered as the flying battleship of the air fleet.

A short while later the **Haunebu III**, the showpiece of all disks was ready that was seventy-one meters across. It was filmed flying in 1945. The craft came with the latest in propulsion drives, Thule Tachyonator (Thule Triebwerk) 7c plus SM-Levitators with a Mag Field Impulser 4a Control. It could transport thirty-two men and up to 70 people, and could remain airborne for eight weeks. It reached at least 7,000 km/h (according to documents in the secret SS archives up to 40,000 km/h in space). And a 25-minute quiet or hover time. It boasted an armament of 4 x 110 mm KSK in 4 rotating turrets (3 lower/1 upper), 10 x 80 mm KSK in rotating turrets plus 6 x MK 108 and 8 x 50 mm KSK all contained in a triple Victalen hull.
http://discaircraft.greyfalcon.us/HAUNEBU.htm

The large 71 meter German Haunebu III flying saucer capable of space flight

154

Vril, SS Military Technical Branch E-IV/E-V

- Vril 1 "Jäger" (Hunter) disc aircraft, 1941, 17 manufactured
- Vril 2 "Zerstörer" (Destroyer) disc aircraft project
- Vril 3 disc aircraft prototype
- Vril 4 disc aircraft prototype
- Vril 5 disc aircraft prototype
- Vril 6 disc aircraft prototypes, 2 built
- Vril 7 "Geist" (Spirit) disc aircraft, 1944, several built
- Vril 8 "Odin" (God Wotan) disc aircraft prototype, 1945
- Vril 9 "Abjäger" (Universal Hunter) disc aircraft prototype, 1945 over occupied Germany
- Vril 10 "Fledermaus" (Bat) disc project
- Vril 11 "Teufel" (Devil) disc project
- Vril Andromeda-Gerät (Andromeda Device), 139 meter cylindrical Raumschiffe, 1945 1 built, 1 under construction, built exclusively by SS E-V Unit, powered by 4 Thule Triebwerk EMG engines plus 8 SM-Levitators
- Vril Andromeda-1 Freyr (Norse God), captured by US Army 1945 partially completed
- Vril Andromeda-2 Freya (Norse Goddess), one built
- Vril DORN "Verteidiger" (Defender) unmanned delta craft weapon. "DORN" is either short for DORNier or means (Thorn), Sighted near Pescara, Italy postwar.
- Vril Gammagische Auge, "Magic Eye" recon drone, prototype only

(all discs except Andromeda-Gerät powered by Vril Triebwerk EMG engines plus Schumann SM-Levitators)

Actual Nazi German Haunebu I Flying Saucer in a test flight
Courtesy of Dr Greer from his movie "Unacknowledged" https://www.youtube.com/watch?v=ehmnGolJZq0

(SS)

Flugkreisel-Erprobung, Stand / Anzahl Erprobungsflüge:
_____

HAUNEBU I   (vorhanden 2 Stück)   52   E-IV
RAUNEBU II   (vorhanden 7 Stück)   106   E-IV
HAUNEBU III (vorhanden 1 Stück)   19   E-IV
(VRIL I)   (vorhanden 17 Stück)   84   (Schumann)

Empfehlung:
Beschleunigen von Abschlußerprobung
und Produktion „Haunebu II"
+ „VRIL I"

HAUNEBU I

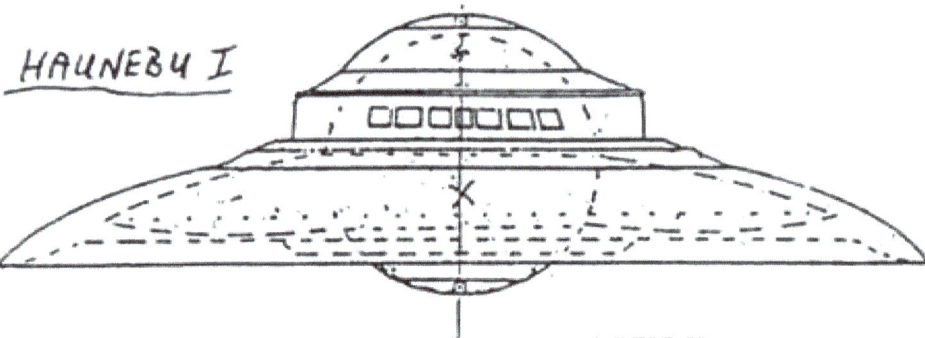

MITTELSCHWERER BEWAFFNETER FLUGKREISEL, TYPE „HAUNEBU I"

Durchmesser: 25 Meter
Antrieb: Thule-Tachyonator 7b
Steuerung: Mag-Feld-Impulser 4
Geschwindigkeit: 4800 Kilom.p.Std. (rechn. bis 17000)
Reichweite in Flugzeit: 18 Stunden
Bewaffnung: 2 x 8cm KSK in Drehtürmen und 4 x Mk 106, starr nach vorn
Außenpanzerung: Doppel-Victalen
Besatzung: 8 Mann
Weltallfähigkeit: 60 %
Stillschwebefähigkeit: 8 Minuten
Allgemeine Flugfähigkeit: Tag wie Nacht
Grundsätzliche Einsatztauglichkeit: 60 %
Frontverfügbarkeit: Nicht vor Jahresende 44

Bemerkung: Die SS-E-IV hält Konzentration auf bereits im Versuch
stehende „Haunebu II" für sinnvoller als an beiden Typen parallel
weiterzuarbeiten. „Haunebu II" verspricht entscheidende Verbesserungen
in nahezu allen Punkten. Höhere Herstellungskosten scheinen gerecht-
fertigt – besonders mit Blick auf Führer-Sonderbefehl, Flugkreisel
betreffend.

156

HAUNEBU II

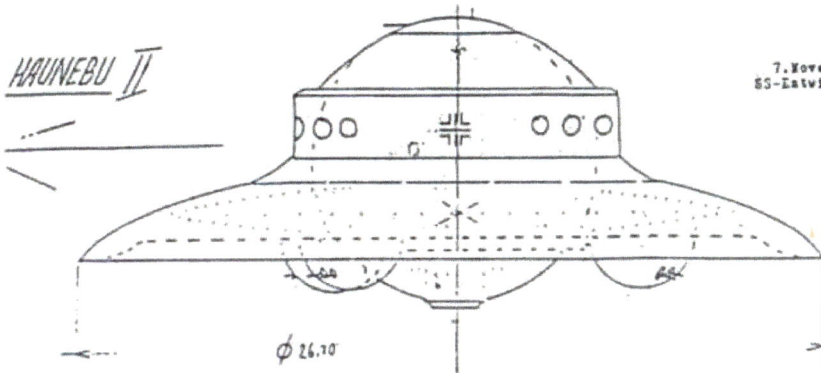

Ø 26.10

MITTELSCHWERER BEWAFFNETER FLUGKREISEL, TYPE „HAUNEBU II"

Durchmesser: 26,3 Meter
Antrieb: „Thule"-Tachyonator 7c (gepanzert; Ø TY.-Scheibe: 23,1 Meter)
Steuerung: Mag-Feld-Impulser 4a
Geschwindigkeit: 6000 Kilometer p.Stunde (rechnerisch bis ca. 21000 möglich)
Reichweite (in Flugdauer): ca. 55 Stunden
Bewaffnung: 6 8 cm KSK in drei Drehtürmen, unten, eine 11 cm KSK in einem Drehturm, oben
Außenpanzerung: Dreischott-„Victalen"
Besatzung:9 Mann (erg. Transportverm.[bis zu 20 Mann)
Weltallfähigkeit: 100 %
Stillschwebefähigkeit: 19 Minuten
Allgemeines Flugvermögen: Tag und Nacht, Wetterunabhängig
Grundsätzliche Einsatztauglichkeit (VT): 85 %

Verfügbarkeit „Haunebu II" (bei weiter gutem Erprobungsverlauf wie V7) ab Oktober.
Dann Serienherstellung ab Jahreswende 1943/44, jedoch noch ohne verbesserte Kraftstrahl-
kanones „Donar-Kan IIIV", deren Probreife nicht vor Frühsommer 1944 angenommen werden
kann.
Vom Führer verlangte hundertzehnagrotestige Einsatzreife rundum kann allerdings
nicht vor Ende nächsten Jahres erwartet werden.Erst ab etwa Serie 9.

Bemerkung zuständige SS-Entwicklungsstelle IV: Die neue deutsche Technik-und.
damit vor allem Flugkreisel und KSKs-wird wegen der noch zeitraubenden Herstel-

http://discaircraft.greyfalcon.us/HAUNEBU.htm

HAUNEBU III

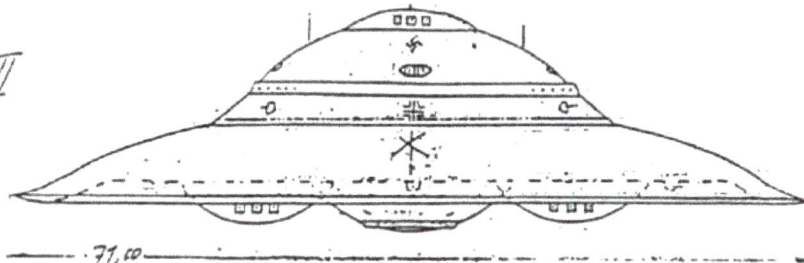

71,00

SCHWERER BEWAFFNETER FLUGKREISEL „HAUNEBU III"

Durchmesser: 71 Meter
Antrieb: Thule-Tachionator 7c plus Schumann-Levitatoren (gepanzert)
Steuerung: Mag-Feld-Impulser 4a
Geschwindigkeit: ca. 7000 Kilom.p.Stunde (rechnerisch bis zu 40000)
Reichweite (in Flugdauer): ca. 8 Wochen (bei S-L-Flug 40% mehr)
Bewaffnung: 4 x 11cm KSK in Drehtürmen (3 unten, 1 oben), 10 x 8cm KSK
in Drehringen plus 6 x Mk 108, 8 x 5cm KSK ferngesteuert
Außenpanzerung: Dreischott-Victalen
Besatzung: 32 Mann (erg. Transportverm. max. 70 Personen)
Weltallfähigkeit: 100 %.
Stillschwebefähigkeit: 25 Minuten
Allgemeines Flugvermögen: Wetterunabhängig Tag und Nacht
Grundsätzliche Einsatztauglichkeit: Etwa 1945.

Bemerkung: SS-E-IV hält den Hinweis für notwendig, daß in
„Haunebu III" ein großartiges Werk deutscher Technik im ent-
stehen ist, wegen der allgemeinen Materiallage aber alle
Kräfte auf das schneller verfügbare Haunebu II gesetzt
werden sollten.
Gemeinsam mit dem leichten Flugkreisel „Vril" der Schumann-
Gruppe könnte „Haunebu II" die von Führer aufgestellten
Forderungen sicherlich erfüllen.

http://discaircraft.greyfalcon.us/HAUNEBU.htm

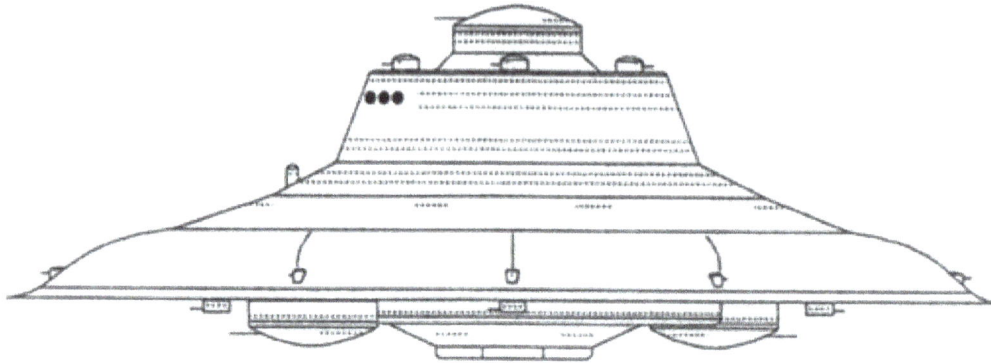

HAUNEBU IV
Heavily Armed Flight Gyro
Diameter: 120 m
Project only projected for 1946

THE ANDROMEDA MACHINE

THE 'ANDROMEDA MACHINE' PICTURED BELOW, MEASURED 300 METERS ACROSS, AND WAS CAPABLE OF CARRYING A CREW OF 200. THE FIRST TEST FLIGHT WAS MADE ON DECEMBER 8, 1942. DUE TO THE SIZE OF THE 'CIGAR SHAPED CRAFT', THE ANDROMEDA MACHINE WAS STORED IN A MODIFIED ZEPPLIN HANGER, NEAR BERLIN. THE DEVICE CARRIED ENOUGH FOOD AND WATER FOR TRIPS LASTING UP TO THREE YEARS. THE INTERIOR COMPARTMENS HELD 2 SCOUT SAUCERS, 2 ATTACK SHIPS, AND ONE 200 FT. DIAMETER BATTLE SAUCER. THE PROPULSION SYSTEM CONSISTED OF 3 ROWS OF 'COLER CONVERTERS', THAT INTERACTED WITH 50 ELECTRO-MAGNETIC INDUCERS. A STATIC ELECTRIC CHARGE WAS THEN DIRECTED TO THE ON-BOARD TESLA COILS, WHICH GENERATED A NEAR 100% ZERO-POINT ENERGY CONVERSION. THE CRAFT WAS USED TO LOCATE NEW TERRITORY FOR THE THIRD REICH, IN CASE THINGS WENT WRONG FOR THE NAZI REGIME. A TOTAL OF 11 TRIPS WERE MADE TO MARS BETWEEN 1942 AND 1945, INCLUDING 21 TRIPS TO GALAXIES OVER 3 LIGHT YEARS FROM EARTH. THE EXTRATERRESTRIAL BIOLOGICAL ENTITIES BROUGHT BACK FROM THESE MISSIONS, WERE EVENTUALLY STORED AT AREA 51 IN THE NEVADA DESERT.

①	FLUGALHAUFFEN	⑨	BRIEFING ROOM
②	SCHOONFRYDER	⑩	LIQUID MERCURY TANKS
③	GOOGALSTIEN	⑪	ELECTRO-MAGNETIC INDUCERS
④	KLIMENSCHTACH	⑫	DRIVE UNIT/TESLA COILS
⑤	SCHRAUSENHAU	⑬	RADAR ROOM
⑥	SLEEPING QUARTERS	⑭	INTER-PLANETARY SAMPLE STORAGE
⑦	WATER SUPPLY	⑮	WEAPONS SUPPLY
⑧	FOOD SUPPLY	⑯	EXTRA-TERRESTRIAL DEBRIEFING ROOM

The Andromeda Machine also known as The Andromeda Device is considered the "Mothership" (Aircraft Carrier) of the Vril and Haunebu Craft

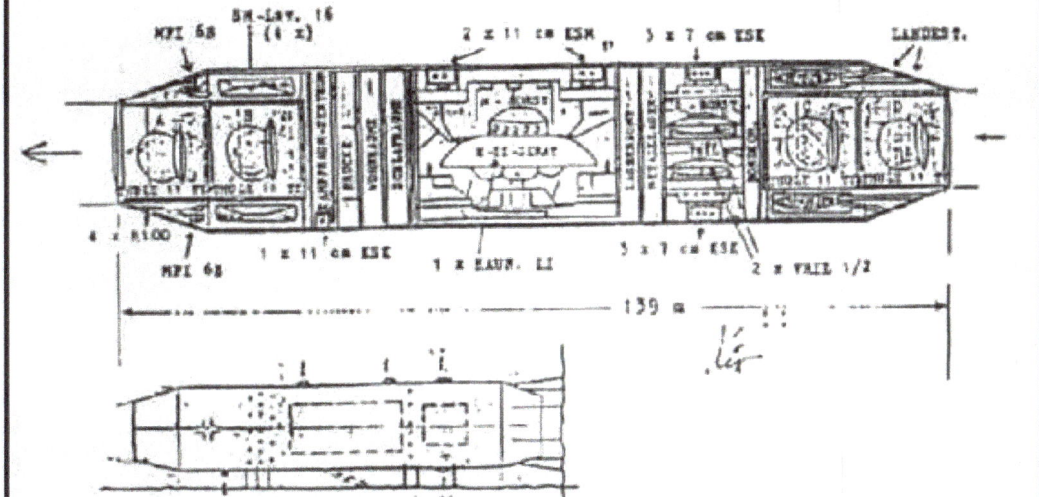

ANDROMEDA-GERÄT

A Secret Document from the German Third Reich SS Archives
Depicting the Andromeda Device

Haunebu Advanced Flying Saucer Craft and the Andromeda Device

Here is a summary of the Haunebu series of advanced saucer craft, (all discs powered by Thule Triebwerk EMG engines):

Thule, SS Military Technical Branch E-IV

Haunebu
- Thule H-Gerät Hauneburg Device, **Haunebu I disc aircraft**, 1939, 2 produced
- **Thule Haunebu II disc aircraft** 1942, 5 produced
- **Thule Haunebu II Do-Stra disc aircraft** co-produced by Dornier. Do-Stra = DOrnier STRAtosphären Flugzeug, 1944, 2 produced
- **Thule Haunebu III disc aircraft,** 1945, 1 produced
- **Thule Haunebu IV disc aircraft project**

Close-ups of a Haunebu I showing a *KraftStrahlKanaone* (KSK Power Ray Cannon)

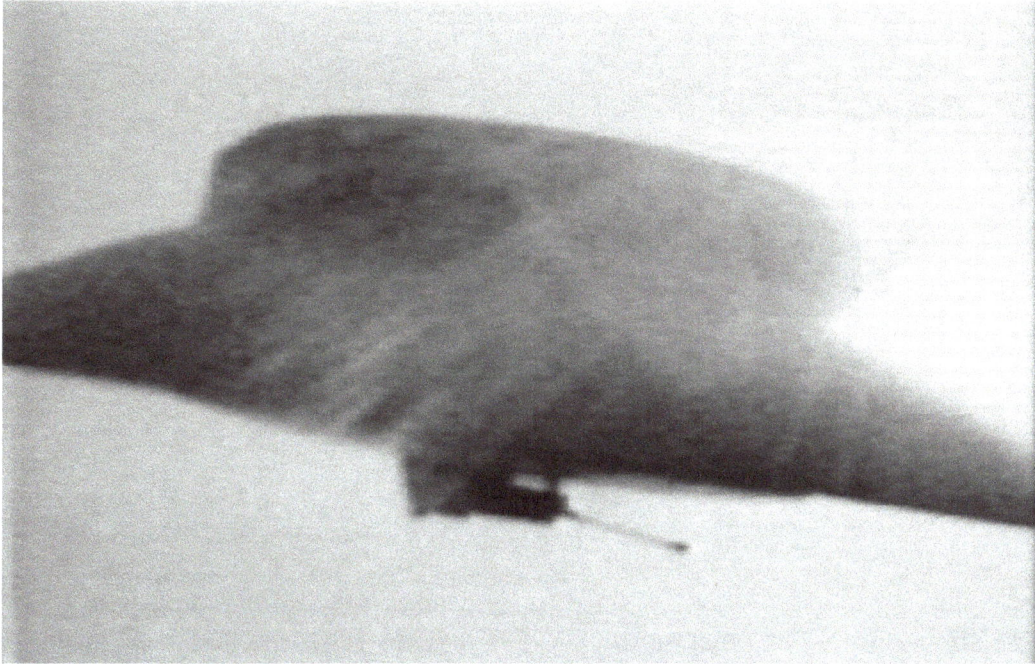

http://www.laesieworks.com/ifo/lib/WW2discs.html

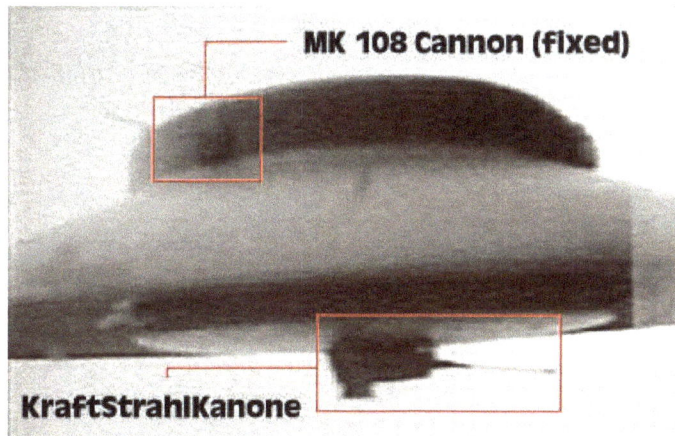

MK 108 Cannon (fixed)

KraftStrahlKanone

The MK 108 30mm cannon manufactured by Rheinmetall-Borsig for use in aircraft, was also mounted in subsequent versions of Haunebuand Vril disks. Because of its slow muzzle velocity, the cannon was difficult to aim and its range too short. However, it proved to be very effective, reliable and easy to manufacture.

http://www.laesieworks.com/ifo/lib/WW2discs.html and http://www.abovetopsecret.com/forum/thread484908/pg1

The Haunebu II. The truck in this photo is usually misidentified as a Hanomag, but Hanomags did not have such long hoods; it is a 5 ton Magirus

A Haunebu II Photographed During a Winter's Sunset Perhaps, on a Test Flight Somewhere in Northern Germany

German Haunebu II in action.

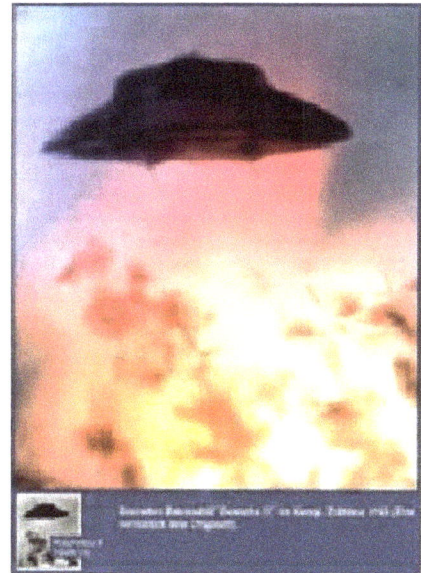

**The Haunebu II disc craft in wartime action. (Note that the picture at right is
a colourized photo From a black and white photo indicated in bottom left
hand corner and that the KSK guns appear to be shooting at
something below creating explosive devastation**

http://www.laesieworks.com/ifo/lib/WW2discs.html

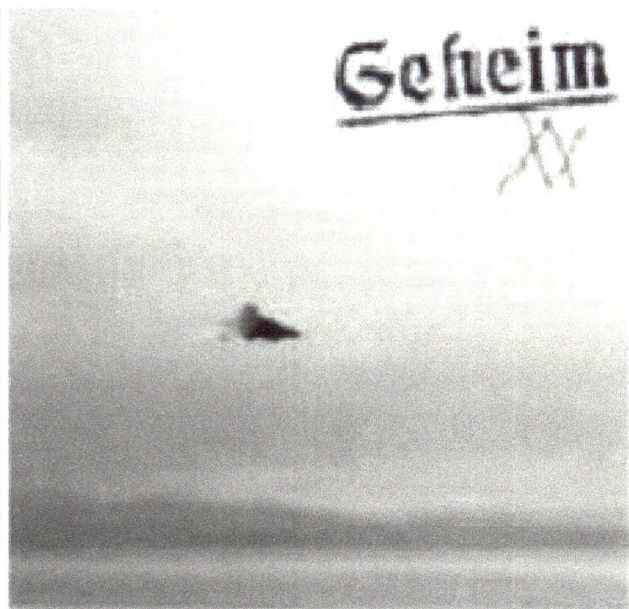

Geheim

**These photos are of the Haunebu III, the one on the left was taken two days after
Germany's surrender and is being hunted by a U.S. P-51 Mustang,
the one on the right is sometime in 1945**

http://www.abovetopsecret.com/forum/thread484908/pg1 and http://www.laesieworks.com/ifo/lib/WW2discs.html

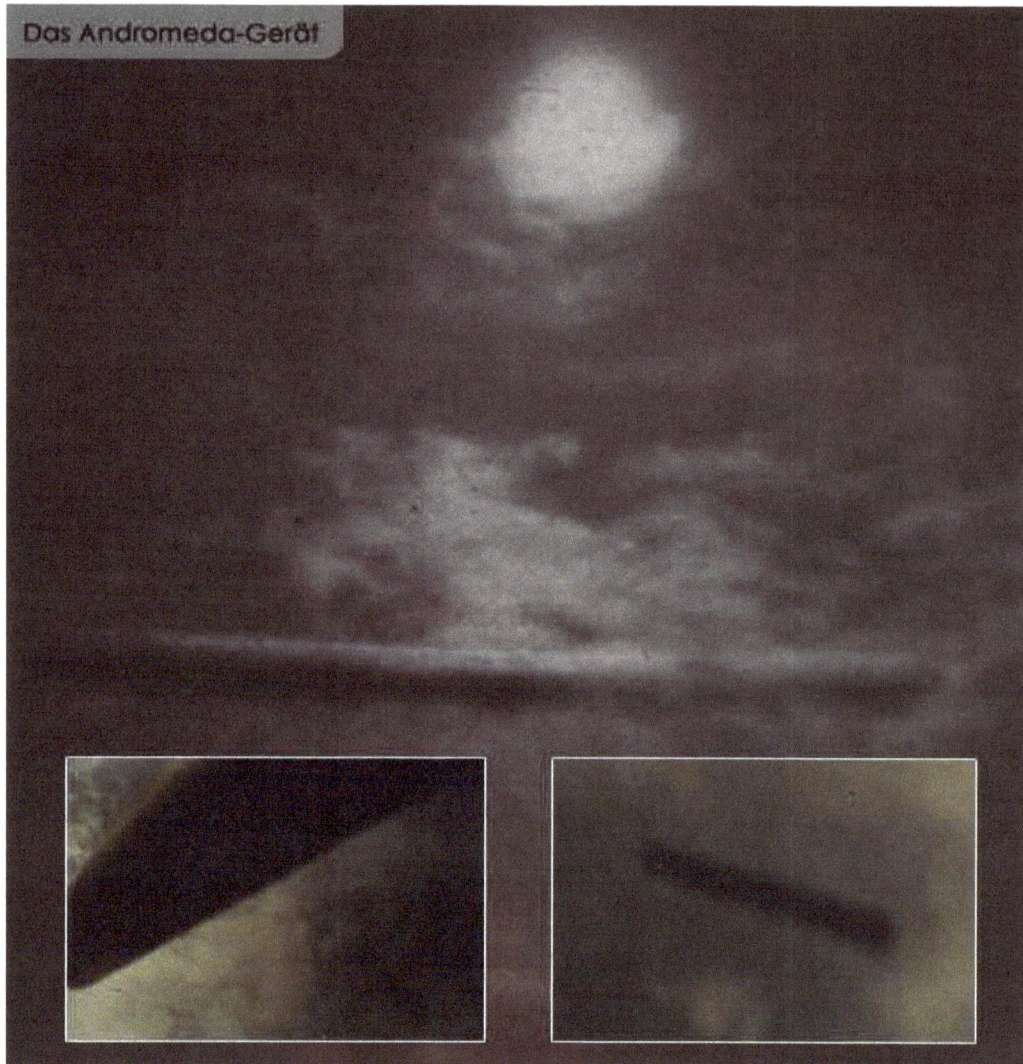

Das Andromeda-Gerät

Are these pictures of the Andromeda Device in flight? No date of when they were taken is available but, such objects have been reported over the post-war decades
http://www.paraportal.org/7696-reichsflugscheiben.html

Could such craft even have existed? According to the Van Helsing's book, *"Secret Societies and their Power in The 20th Century"*, a former CIA member and Green Beret by the name of **Virgil Armstrong**, claims that World War II German flying machines were capable of vertical take-off and landing and could perform right angle flight maneuvers. They were recorded flying at speeds of 3,000 km/h and had a laser-mounted weapon (probably the so-called KSK power ray gun) that could pierce four-inch armour.

It was also noted by **Ufologist Professor J. Hurtak**, author of: *"The Keys of Enoch"* that the German were in the process of building what the Allies referred to as "the wonder weapon system". Hurtak's documents described two events; the erection of a space city at Peenemunde and the enlistment and transport to the U.S. of the best German technicians and scientists from Germany after the war. **Secret Societies and their Power in the 20th Century by Jan Van Helsing (Pseudonym); copyrighted 1995; ISBN 3-89478-654 –X; Translated by**

164

EWERTVERLAG S.L.; P.O. Box 35290, Playa del Ingles, Gran Canaria, Spain
http://www.bibliotecapleyades.net/sociopolitica/secretsoc_20century/secretsoc_20century06
.htm#CHAPTER%2033

Some confusion arose in the evaluation by the CIA and the British secret service who knew already in 1942 about the construction and the use of these flying objects, but they thought the two different objects were actually only one object. Foo Fighters was the Allies' name for all glowing German flying machines. Probably there were two inventions that actually corresponded to the term **Foo Fighters**: the **"Flying Turtle"** and the **"Soap Bubble"** *which may have been Schauberger's remote controlled saucer devices!* [My italics added]

"The **"Flying Turtle"** was so called because its outer shape resembled the shell of a turtle. It was an unmanned probe that could cause disruption in the electric ignition systems of the enemy aircraft. They carried advanced **"Klystronohren" (klystron tubes)** that the SS called death rays. The initial ignition disrupter of the craft did not work perfectly but later improvements did and Ufologists would later confirm that the disruption of vehicle ignition systems, the cutting of electrical power to equipment was a typical sign that a UFO was nearby.

The late , U.S. Air Force pilot during the Second World War describes the Foo Fighters as sometimes gray-green and then, changing to red-orange. He reported that they would approach his aircraft to about five meters and then stayed there and no amount of evasive maneuvers away from them could shake them off, nor shooting at them could bring them down. Often the squadrons were forced to turn back and land.

"The **Soap Bubbles** that were also called Foo Fighters were something completely different, though. They were simple balloons in which there were metal spirals that disturbed enemy radar. The success probably was very limited, disregarding the psychological effect."

Seemingly, not knowing any limits to their imagination, the SS regime in 1943 began planning the construction of a large cigar-shaped mother ship in the Zeppelin works hangar which would allow an elite group of SS officers, soldiers, and personnel to travel to **Aldebaran**. The craft known as **Andromeda Device** had a length of 139m designed to transport several saucer-shaped craft in its body for flights of long duration (interstellar flights).

By Christmas 1943 an important meeting of the **VRIL-Gesselschaft (Vril Society)** took place in the seaside resort of Kolberg. Attending the meeting were the two Vril Society mediums, **Maria Ortic** and **Sigrun**, who were channellers of telepathic information from the **Aryan** race, human-like beings from the star system of Aldebaran in the constellation Taurus.

"The main item on the agenda was the **Aldebaran Project.** The mediums had received precise information about the habitable planets around the sun Aldebaran and one began to plan a trip there. At January 22, 1944, meeting between Hitler, Himmler, Kunkel (of the Vril Society) and **Dr. Schumann** this project was discussed. It was planned to send the **VRIL-7** large-capacity craft through a *"dimension channel independent of the speed of light"* **[Italics added for emphasis]** to Alderbaran.

"According to **Ratthofer**, a first test flight in the dimension channel took place in the winter of 1944. It barely missed disaster, for photographs show the Vril-7 after the flight looking "as if it had been flying for a hundred years". The outer skin was looking aged and was damaged in several places." **"Secret Societies and their Power in the 20th Century" by Jan Van Helsing** and
http://www.bibliotecapleyades.net/sociopolitica/secretsoc_20century/secretsoc_20century06.htm#CHAPTER%2033

Discovery of secret archives of the SS by the British and the Americans the during their occupation of Germany in early 1945 revealed photographs of the **Haunebu II** and the Vril 1 crafts as well as of the Andromeda device. **President Truman** in March 1946 upon hearing of Allies finds gave the immediate order to the war fleet command of the U.S. to start collecting as much material of the German high technology experiments as quickly and thoroughly as possible.

Under this order, **"Operation Paperclip"** commenced in earnest with German scientists who had worked on all manner of secret projects being sequestered to the U.S., most notably among them was **Viktor Schauberger** and **Wernher Von Braun.** It was a race against time and against the invading forces of the Russians army to remove as much technology as possible before the other side got to it first.

A short summary of the developments that were meant to be produced in series:

The first project was led by Prof. Dr. W.O. Schumann of the Technical University of Munich. Under his guidance seventeen disk-shaped flying machines with a diameter of 11.5m were built, the so-called **VRIL-1-Jager** (Vril-1-fighters), that made 84 test flights. At least one VRIL-7 large capacity craft apparently started from Brandenburg – after the whole test area had been blown up – towards Aldebaran with some of the Vril scientists and lodge members.

The second project was run by the SS-IV development group. Until the beginning of 1945 they had three different sizes of bell-shaped space gyros built:

The **Haunebu I**, 25m diameter, two machines built that made 52 test flights (speed 4,800 km/h).

The **Haunebu II**, 32m diameter, seven machines built that made 106 test flights (speed 6,000 km/h). The Haunebu II was already planned for series production. Tenders were asked from the Dornier and Junkers aircraft manufacturers, and at the end of March 1945, the decision was made in favor of Dornier. The official name for the heavy craft was to be **DO-STRA (Dornier stratospheric craft).**

The **Haunebu III**, 71m diameter, only one machine built that made at least 19 test flights (speed 7,000 km/h).

The ***Andromeda Device*** existed on the drawing board; it was 139m long and had hangars for one Haunebu II, two Vril I's and two Vril II's There are documents showing that the VRIL-7 large capacity craft had started for secret, still earth-bound, missions after it was finished and test flown by the end of 1944. There was a landing at the Mondsee in the Salzkammergut of Austria with dives to test the pressure resistance of the hull.

Probably in March and April 1945 the VRIL-7 was stationed in the "Alpenfestung" (Alpine Fortress) for security and strategic reasons, from whence it flew to Spain to get important personalities who had fled there, safely to South America and **Neuschwabenland** (explanation follows) to the secret German bases erected there during the war. Immediately after this, the VRIL-7 is said to have started on a secret flight to Japan about which however nothing further is known.

We know from the historical evidence presented so far that the Germans were developing and improving on saucer designs and had flown a number of prototypes as proof of concept and aeronautical performance thus, it can be concluded with a little logical reasoning, that a small number of the Haunebu II may have been built. Several photographs of UFOs that emerged after 1945 with the typical features of these German aeronautical marvels indicate that their reality was a certainty, in particular, further evidence of their existence appeared with the Adamski photos o the early 50s showing bell disc-shaped craft and also, of the cigar-shaped "mother ship, that the Germans dubbed the **"Andromeda Device."**

The cigar shaped craft was believed to have been constructed in a hidden camouflaged and modified Zeppelin hangar near Berlin and then test flown on December 8, 1942. The craft measured 300 meters across and was capable of carrying a crew of 200 people with enough food and water for trips lasting up to three years. The "carrier ship" interior compartments held 2 scout saucers, 2 attack ships, and one 200ft. diameter battle saucer. The propulsion system consisted of 3 rows of "Coler Converters" that interacted with 50 electro-magnetic inducers. A static electric charge was directed to the onboard Tesla coils, which generated a near 100% zero-point energy conversion. The craft was used to locate new territory for the Third Reich, in case things went wrong for the Nazi Regime.

It was alleged that this craft had completed a total of 11 trips to Mars between 1942 and 1945, including 21 trips to (galaxies?) over 3 light years from Earth. The Extraterrestrial biological entities brought back from these missions were eventually stored at Area 51 in the Nevada Desert.

For Vril, a Society formed by women based on metaphysics, the thought, and act of joining with other powerful male occult groups was almost unthinkable. Yet in 1919 Vril met with Thule and DHvSS members in an effort to obtain funding for their vision by 1921.

Thule had the most industrial influence and financial resources to fund such a project so an agreement was reached to join Societies in an effort to build the strange machine which became known as the **"Jenseitsflugmaschine" (JFM)** or Otherworld flight machine.

Under the code letters, J-F-M the machine was constructed in secret in Munich. It was constructed in 1922 in a barn and rolled out into a field for channelled flight testing. Despite two years of attempting to achieve channelled flight through high powered frequency field oscillations produced by strong electromagnetic forces within the craft, the occultists could not open what they termed a "white hole" in space/time and pass through; reaching the **Aldebaran system** and the **Aryan Sumerian aliens** that had contacted them.

When the casual reader or UFO researcher investigates these claims of advanced Nazi weapons technology, one has to ask is any of this information for real or is it post –war disinformation by a defeated Germany or a cover-up by the victorious Allies, particularly by the Americans to gain control over newly acquired technology via **Operation Paperclip**?

As incredible as this information of interstellar travel taking place, well before the current space age by Americans, Russians, and other nations, there may actually be some tangible proof to this Third Reich claim.

Professor W.O. Schumann led the JFM project but decided to scrap all research in 1924. The JFM was hurriedly dismantled and sent to Augsburg for storage at Messerschmitt's facility, where it was either destroyed or later moved up to Peenemünde and reassembled for further study. No one can confirm either. Professor Schumann did, however, managed to develop a levitator unit from it along with practical experience with generating increasingly intensified electromagnetic fields.

Thule and Vril Gesellschafts got a second chance in 1933 after Adolf Hitler became chancellor in Germany, to further develop their craft with a state approved **RFZ (RundFlugZeug, or Round Aircraft)** program also led by Professor Schumann.

The program started in 1937 and produced seven models. Himmler's SS had already become involved with the Thule Haunebu development program in 1935 before the RFZ series began. In 1939 the SS had used Hans Colers adapted gravitic battery and the applied ideas of Tesla to come up with a revolutionary new engine pioneered by Thule - the Thule Triebwerk (also known as a Tachyonator-7). Vril also perfected the SM-Levitator and invented their own **Triebwerk (Thrustwork)** in 1941 with the RFZ-7.

This brought about the abandonment of the RFZ series and introduced the Haunebu and Vril series of craft.

The Haunebu I, briefly listed as an RFZ design was dropped in 1939 and the RFZ-7 was re-designated Vri 1 Jäger (Hunter) in 1941. Both Societies then came under control of the **SS E-IV (Entwicklungsstelle 4)** Unit in 1941 under Hitler's order that all secret societies be banned. This unit was the technical branch of the SS which was tasked with developing alternative energies. Not only did this include alternative power through synthetic fuels but also with generating power fields that could be used in war. Thus, the occult discs were seen as a potentially rewarding war weapons.

Of course, Thule and Vril went along with the SS despite their intention of only producing a **Raumschiff (Spaceship).** Like von Braun, who also envisioned powered space flight with the A-series rockets, the Societies kept pioneering their own craft with full knowledge that any of the discs might be made into a war weapon at any time.

168

Fortunately, both types of discs were incredibly complex and difficult to control in flight. A special celestial navigation system built by both Siemens and AEG had to be developed to fly the machines which at first could only make turns of 22.5, 45, and 90 degrees. Control was only possible through Magnetic Field Shifting using Impulsers that transferred power to different parts of the electromagnetic rings rapidly, forcing very violent rotating electromagnetic fields inside the craft to move accordingly, changing instant direction of the craft. Maximum velocities in these craft proved to be a problem as well and required special SS metallurgists (some called them modern Alchemists) to create heat-resistant armour called Victalen that could withstand the heat friction of speeds well over 7,000 km/hr!

With both the Haunebu and Vril craft reaching their full potential by 1943/1944 it was decided that Vril would once again attempt to build a functioning **Raumschiff**; but this time, it would be built by the SS with all the resources of the SS state-within-a-state, including slave labour and safe facilities. Due to the immensity of this project, a special unit was formed to build just this type of craft --- the SS E-V (E-5) Unit.

The Raumschiff was to be called the **Andromeda-Gerät (Andromeda Device).** Work on two of these massive 139 meter long and 30 meter diameter "flying cigars" probably began with great effort in early 1943 in above ground heavily camouflaged shelters similar to that of the old Zeppelin hangars. These crafts were designed to hold both a **Haunebu II** or IV disc in one large bay with two other smaller Vril 1 or 2 discs in a secondary bay - both accessed from the side of each craft. Each craft was to be crewed by 130 and were to be armoured with a quadruple layered Victalen hull. The E-V unit nicknamed them as male and female "Freyr" and "Freya" after the Old Norse God and Goddess.

The propulsion systems located at both the fore and aft sections of the Andromeda craft would be beyond the last Haunebu-type Thule Tachyonator 7c drive. These crafts were to have four massive power units, two Tachyonator 11 at the front and two at the rear with four additional large SM-Levitator units located in pairs on the top and bottom of each craft which sat on a series of large underbody skids to support the massive weight. It is generally assumed that those engines were of the same **EMG (Electro-Magnetic-Gravitic)** type but other Allied Intelligence officers believe that they might have been photonic based on witness accounts of one large craft flying with a massive bright light source emitted from the rear. Thule might have simply used the Tachyonator term to describe an evolution of energy drives that might have varied in the latter models. Although designed as an armed vessel with provision for five turrets armed with powerful **KSK (KraftStrahlKanone, or Strong Ray Cannon)** it is doubtful that any armament actually made it onto either of the two prototypes under construction.

When the US Army came upon one of the uncompleted machines in 1945 the craft was first mistaken as some sort of radio tower due to the strange antennas protruding from the craft. The immense size of the uncompleted machine didn't reveal its purpose until someone actually climbed up on top of it and immediately spotted a cockpit with very thick glass that was shattered. The semi-completed craft painted gray with no markings was heavily damaged as the Germans retreated, taking with them all of the prototype's sensitive navigation equipment and the strange propulsion system. This caused damage to the craft except for the two open exposed bays - which were empty of any Haunebu or Vril discs.

169

Photos seem to indicate that the other prototype "Freya", or one similar to it, was completed and flew sometime in 1944/1945. The exotic craft was designed for a speed of 300,000 km/hr!

If one did escape in March 1945, that one most likely had the Vril Chef on board and achieved channelled flight.

Their final destination? The **Aldebaran system.**
http://greyfalcon.us/restored/Secret%20flying%20discs%20of%20the%20Third%20Reich.htm and https://www.youtube.com/watch?v=IClVotk8ff8 and https://www.youtube.com/watch?v=YoX_3kPCQAA

Is postwar flying cigar "Andromeda Device" of German Nazi design or a US copy?
http://greyfalcon.us/restored/Secret%20flying%20discs%20of%20the%20Third%20Reich.htm and https://tineye.com/

Neuschwabenland (New Swabia) and "Operation High Jump"

So, what became of these advanced flying machines after the war?

There were unverifiable reports that claimed that some of saucers had been sunk in the Austrian Mondsee, still, others maintain that some were flown to South America, or brought there in parts or even, made their way to a secret German base in Antarctica. The premise for this viewpoint is credible as many German immigrants have long resided there and many of the Spanish inhabitants of Argentina, Paraguay and Uruguay would have welcomed and provided a safe haven to any fleeing Nazi or scientist from the fatherland. Evidenced for this was proven time and again by the various Nazi hunters who, long after the war have captured many high ranking Nazi war criminals in the South American continent. There has even been an all out assault on Antarctica by many nations lead by the U.S.A.

From current UFO reports, there have long been many accounts of UFO sightings in these countries which may well indicate that saucer programs continue to this day. It is certain though that if these crafts didn't get to South America, the plans for them did allow for new ones to be built and flown there.

But, it appears that the chief escape route of the defeated Third Reich echelon was directly due south to the coldest continent on the planet, Antarctica. Prior to the outbreak of the Second World War in 1938, a German expedition to the Antarctic was made with the aircraft carrier Schwabenland (Swabia). Approximately 600,000 km2 of an ice-free area with lakes and mountains were declared German territory, the **Neuschwabenland (New Swabia).**

Hearing that Germany had laid claim to huge portions of the Antarctic continent as a part of German sovereignty and that an impregnable military base/settlement had been built for advanced weapons testing sent alarm bells ringing in the command centres of the Allied forces particularly in America, Britain, and Russia. If indeed Germany had surrendered unconditionally after the war in Europe, it did not appear to be so in the South Polar region of Antarctica, where German presence had been historically active since the 1870s with research and whaling.

Further expeditions took place in the early years of the twentieth century in 1911 under the command of **Wilhelm Filchner** with his ship the '*Deutchland*' and again, between the wars, the Germans made a further voyage in 1925 with a specially designed ship for the Polar Regions, the '*Meteor*' under the command of **Dr. Albert Merz**. But, it was Hitler who had authorized several expeditions to the poles shortly before WWII. Their stated objective was to either to rebuild and enlarge Germany's whaling fleet or test out weaponry in severely hostile conditions.
http://www.bibliotecapleyades.net/tierra_hueca/esp_tierra_hueca_6c.htm

The South Polar Continent of Antarctica
https://en.wikipedia.org/wiki/New_Swabia

Then, in the years directly preceding the Second World War, the Germans laid claim to parts of *Antarctica* in order to set up a permanent base there. Given that no country actually "*owned*" the continent and it couldn't exactly be conquered as no-one lived there during the winter months at least, it appeared to the Germans that the most effective way to "*conquer*" part of the continent was to physically travel there, claim it, let others know of their actions and await any disagreements.

Captain Alfred Ritscher was chosen to lead the proposed strike. He had already led expeditions to the North Pole and had proved himself in adverse and critical situations.

172

The Schwabenland and one of the flight boats, a 10 ton Dornier Super "Wals"
http://www.bibliotecapleyades.net/tierra_hueca/esp_tierra_hueca_6c.htm

For the mission Ritscher was given the '*Schwabenland*'; a German aircraft carrier that had been used for transatlantic mail deliveries by special flight boats, the famous 10 ton *Dornier Super 'Wals'* since 1934. The *Schwabenland* left the port of Hamburg on 17th December 1938 and followed a precisely planned and determined route towards the southern continent. In little over a month the ship arrived at the ice covered *Antarctica*, dropping anchor at 4° 30' W and 69° 14' S on January 20[th,] 1939. **The Antarctic Enigma from Violations.org Website; Book IV - Antarctica Enigma**
http://www.bibliotecapleyades.net/tierra_hueca/esp_tierra_hueca_6c.htm

The expedition then spent three weeks off Princess Astrid Coast and Princess Martha Coast off Queen Maud Land. During these weeks, the two *Schwabenland aircraft*, the '*Passat*' and '*Boreas*', flew 15 missions across some 600,000 square kilometers of Antarctica, taking more than 11,000 pictures of the area.

Nearly one-fifth of *Antarctica* was reconnoitered in this way and, for the first time, ice-free areas with lakes and signs of vegetation were discovered. This area was then declared to be under the control of the German expedition, renamed **'*Neu-Schwabenland*'** and hundreds of small stakes, carrying the swastika, were dumped on the snow-covered ground from the '*Wals*' to signal the new ownership.

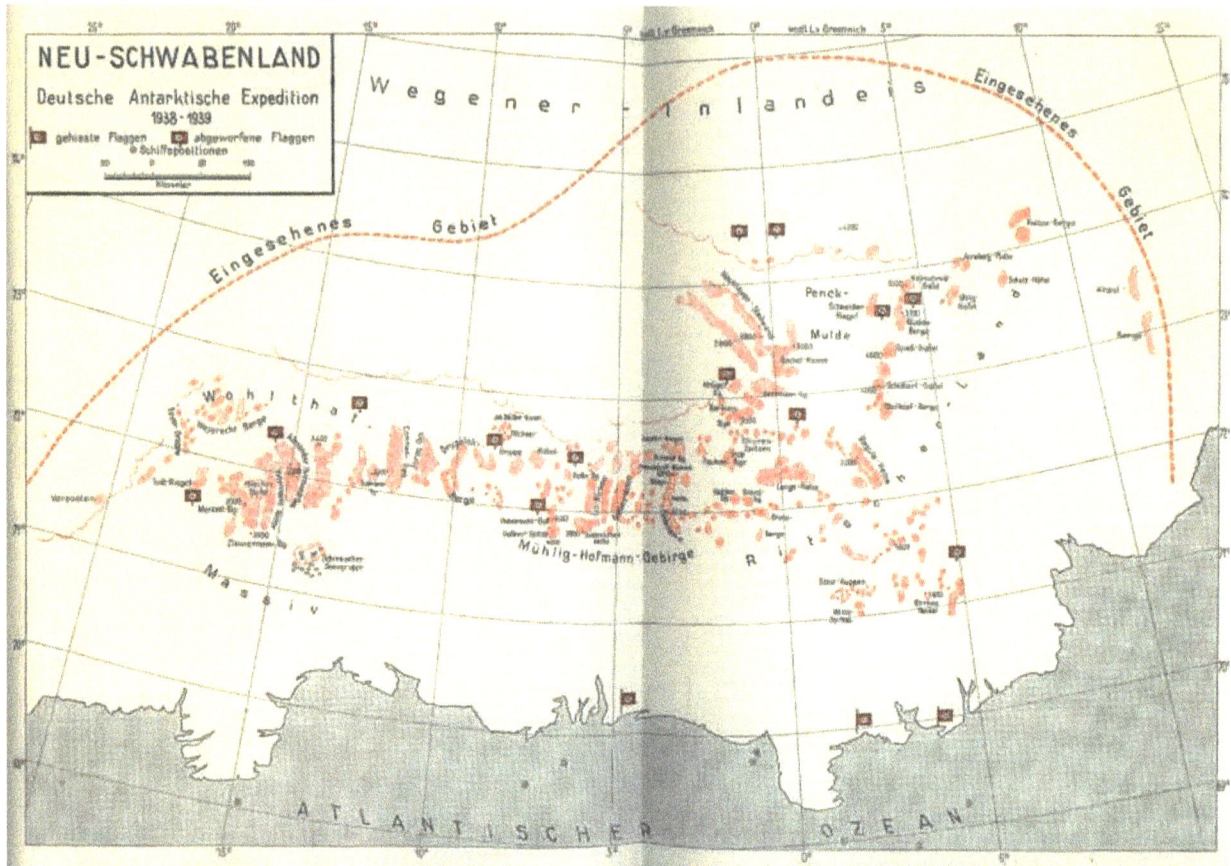

'Neu-Schwabenland' (New Germany) the Region of Antarctica claimed by Germany
http://www.thelivingmoon.com/47brotherthebig/03files/Neuschwabenland_the_Last_German_Colony.html

Ritscher and the *Schwabenland* left their newly claimed territory in the middle of February 1939 and returned to Hamburg two months later, complete with photographs and maps of the new German acquisition.

The true purpose of this expedition has never been satisfactorily explained; we are merely left with a series of puzzles, related reports and snippets of information that are no longer open to verification. Given that the seizing of **Neu-Schwabenland** occurred on the very eve of the war, it can only be concluded that that the polar expedition was of major importance and significance to the goals and development of the planned 1000-year Third Reich. **The Antarctic Enigma from Violations.org Website; Book IV - Antarctica Enigma**
http://www.bibliotecapleyades.net/tierra_hueca/esp_tierra_hueca_6c.htm

There was to be an invasion of the continent of *Antarctica* under the code name of **"Operation High Jump"** and comprised of some 4700 military personnel, six helicopters, six Martin PBM flying boats, two seaplane tenders, fifteen other aircraft, thirteen US Navy support ships and one aircraft carrier; the *USS Philippine Sea.*

Its purpose was highlighted with the official instructions issued by the then Chief of Naval Operations, **Chester W. Nimitz**, which were:

174

(a) train personnel and test material in the frigid zones

(b) consolidate and extend American sovereignty over the largest practical area of the Antarctic continent

(c) to determine the feasibility of establishing and maintaining bases in the Antarctic and to investigate possible base sites

(d) to develop techniques for establishing and maintaining air bases on the ice, (with particular attention to the later applicability of such techniques to Greenland)

(e) amplify existing knowledge of hydrographic, geographic, geological, meteorological and electromagnetic conditions in the area

Little other information was released to the media about the mission, although most journalists were suspicious of its true purpose given the huge amount of military hardware involved. The US Navy also strongly emphasized that *Operation High Jump* was going to be a navy show; Admiral Ramsey's preliminary orders of 26th August 1946 stated that "the Chief of Naval Operations only will deal with other governmental agencies" and that "no diplomatic negotiations are required. No foreign observers will be accepted." Not exactly an invitation to scrutiny, even from other arms of the government.

The operation was also launched with incredible speed, "a matter of weeks." I could be said that the Americans had some unfinished business connected with the war in the polar region. Indeed this was later confirmed by other events and the operation's leader, **Admiral Richard Byrd**, himself. However, the task force itself remained strictly under the military command of **Rear Admiral Richard Cruze.**

This is undeniable fact! "But... the part of the story that is seldom told, at least in "official" circles, is that Byrd and his forces encountered heavy resistance to their Antarctic venture from **"flying saucers"** and had to call off the invasion. This aspect of the story was pushed forward, again, a few years ago, when a retired Rear Admiral, allegedly living in Texas, who had been involved in the "invasion" said he was "shocked" when he read material from a documentary, entitled "Fire From The Sky". He allegedly claimed that he knew there had been "a lot of aircraft and rocket shoot-downs", but did not realize the situation was as serious as the documentary presented it." http://www.rense.com/general35/op.htm
"How High Can You Jump? Operation "Highjump & The UFO Connection by Erich J. Choron www.wintersteel.com kommissar@mtu-net.ru hmerik16@yahoo.com

"Operation High Jump" was basically an invasion of the Antarctic that consisted of three Naval battle groups, a British-Norwegian force, a Russian force, and some Australian and Canadian forces were also involved.

Following its arrival at Antarctica, the force began a reconnaissance of the continent from Little America, their base of operations in Antarctica. Six R4-D aircraft flew over as much of the continent as they could in the short three month "summer" period, with spy cameras and trailing a magnetometers mapping and recording magnetic data. The magnetometers, they carried would show anomalies in the Earth's magnetism, i.e. if there is a "hollow" place under the surface ice or ground, it will show up on the meter.

On one of the last mapping missions, Admiral Byrd's plane returned three hours late. "Officially", it was stated that he had "lost an engine" and had had to throw everything overboard except the films themselves and the results of magnetometer readings in order to maintain altitude long enough to return to Little America. If we are to believe the published and private accounts of what actually took place, this is almost certainly the time when he met with representatives of the "Aryan" Extraterrestrials, and a contingent of the German scientists working on the reverse engineering and construction of "flying discs"'.

Then the mission that had been expected to last for between 6-8 months, came to an early and faltering end. The Chilean press reported that the mission had "*run into trouble*" and that there had been "*many fatalities*". (The official record, though, states that one plane crashed killing three men; a fourth man had perished on the ice; two helicopters had gone down although their crews had been rescued and a task force commander was nearly lost.)

The Chilean claims to one side, it is known that the **Central Group** of **Operation High Jump** were evacuated by the Burton Island ice-breaker from the Bay of Whales on 22nd February 1947; the *Western Group* headed home on 1st March 1947 and the *Eastern Group* did likewise on 4th March, a mere eight weeks after arrival.

Contrary to what had gone on has never been made a matter of public record, however, it is known that Byrd was immediately summoned to Washington and interrogated by the Security Services on his return after being initially 'welcomed back' by **Secretary of War James Forrestal** on 14th April 1947. (Forrestal was later to commit suicide by jumping from a hospital window, although there are researchers who investigated his death, who strongly feel that he was murdered by being pushed out the window).

On 5th March 1947 the '*El Mercurio*' newspaper of Santiago, Chile ran the headline *'On Board the Mount Olympus on the High Seas'* which quoted Byrd in an interview with Lee van Atta:

"Adm. Byrd declared today that it was imperative for the United States to initiate immediate defense measures against hostile regions. The Admiral further stated that he didn't want to frighten anyone unduly but it was a bitter reality that in case of a new war the continental United States would be attacked by flying objects which could fly from pole to pole at incredible speeds.

Admiral Byrd repeated the above points of view, resulting from his personal knowledge gathered both at the north and south poles, before a news conference held for **International News Service.**"

Bearing in mind that all this occurred (the search for a craft that could fly from pole to pole at incredible speeds) a year after the war had ended with Germany defeated, makes it all the more intriguing.

So who was *the enemy* that owned or flew these flying objects that could "fly from pole to pole

176

with incredible speeds?" Rumors began to circulate that whilst Germany had been defeated, a selection of military personnel and scientists had fled the fatherland as Allied troops swept across mainland Europe and established themselves at a base on Antarctica from where they continued to develop advanced aircraft based on extraterrestrial technologies. (It is interesting to note that at the end of the war the Allies determined that there were 250,000 Germans unaccounted for, even taking into account casualties and deaths.)

Incredible as it may sound, there is considerable supporting evidence for these claims about a German base for, on the very eve of the Second World War, the Germans themselves had invaded part of *Antarctica* and claimed it for the Third Reich.
"Over the next four weeks, the planes spent 220 hours in the air, flying a total of 22,700 miles and taking some 70,000 aerial photographs. Then the mission that had been expected to last for between six to eight months came to an early and faltering end. The Chilean press reported that the mission had "run into trouble" and that there had been "many fatalities".

"However... the official record, states that one plane crashed killing three men; a fourth man had perished on the ice; two helicopters had gone down although their crews had been rescued and a task force commander was nearly lost".

"It is an indisputable fact that the *Central Group of Operation Highjump* were evacuated by the Burton Island ice-breaker from the Bay of Whales on 22 February 1947; the Western Group headed home on 1st of March 1947 and the Eastern Group did likewise on 4 March, a mere eight weeks after arrival".

In the end, the task force came steaming back to the United Sates with their data, which then, immediately became classified "top secret"'.
And this '*invasion*' was certainly not the end to German activity in the area; rather the prelude, providing support for the idea that Germany might have established a base on the apparently frozen wasteland.
That German activity continued around Antarctica through the war years is a matter of historical record.

It was reported that **Captain Bernhard Rogge** of the raider ship '*Atlantis*' made an extended voyage in the South Atlantic, Indian and South Pacific Oceans during the period of 1939 – 1941 and also visited the *Iles Kerguelen* (named the '*Most Useless Island In the World*' in 1995) where he had the sad duty of burying a seaman at Bassin de la Gazelle. This island of Kerguelen was also prominent in Nazi war plans as a meteorological weather station in 1942.

It was stated by Captain Rogge that the '*Atlantis'* was known to have been visited by an **RFZ-2 saucer craft** which had been in service as a reconnaissance aircraft since the 1940.

It is not unusual that both sides during the war would engage in normal patrols and in covert missions, and German submarines did engage in all sorts of covert activities. **Submarine U-859** for example, left on a mission in 4th April 1944, carrying 67 men and 33 tons of mercury sealed in glass bottles in watertight tin crates.

Karl Dönit, Commander of the German Navy

The submarine was later sunk on 23rd September by a British submarine (*HMS Trenchant*) in the Straits of Malacca and although 47 of the crew died, 20 survived. Some 30 years later one of these survivors spoke openly about the cargo and divers later confirmed the story on rediscovering the mercury. The significance being that *mercury* is usable as a *fuel source for certain types of aerospace propulsion.* Why would a German submarine be transporting such a cargo so far from home?

After Germany surrendered unconditionally to the Allies in 8th May 1945, there were events after that date which suggested something was happening that did not form a part of recognized world history. Something fuelled by a statement made by **Karl Dönitz**

Dönitz (16th September 1891 – 24th December 1980) had become *Oberbefehlshaber der Kriegsmarine* (Commander of the Navy) on 31st January 1943 and he led the German U-Boat fleet until the end of World War II. (Dönitz also has the distinction of briefly becoming head of the German state for 20 days after Hitler's death until his own capture by the Allies on 23rd May 1945.) His contribution to the mystery of post-war Antarctic activity came in a statement he

178

made in 1943 when he declared that the German submarine fleet had rebuilt "*in another part of the world a Shangri-La land – an impregnable fortress.*" Could he have been referring to the alleged base in Antarctica [*or was it propaganda nonsense or disinformation designed to throw off the Allied forces up to the bit end of defeat]?* (My italics added).
http://www.bibliotecapleyades.net/tierra_hueca/esp_tierra_hueca_6c.htm

Certainly, there are records of continued German naval activity in the area after the war had apparently ended.

Could it be possible that other German U-boats, in addition to U-530 and U-977 were continuing to operate in the area following the war? There are no formal records of such activity, however, it is known that 54 German U-boats '*disappeared*' during the war, of which only 11 are likely to have met their fate at the hands of mines).

The future may well reveal that fate of more of these submarines, however, given the French and South American reports, and the number of missing U-boats, it may not be unreasonable to conclude that at least some of them relocated to the South Polar area.

However, for *Operation High Jump* to have been an attempt to ferret out a remaining *Nazi base* on the *Antarctic* continent, there would have been two prerequisites. Firstly, *Operation High Jump* would have to provide evidence that the mission included a reconnaissance of *Neu-Schwabenland* and secondly, there would have to be an area of the frozen continent that could allow such a base to exist throughout the year. And indeed both criteria are met.

There is no conclusive evidence of a *Nazi base on Antarctica*, however, that something untoward was happening on, or around, the frozen continent appears, on balance of probabilities, to be likely. The evidence is there:

i) The Germans invaded and claimed part of *Antarctica* on the very eve of the war when all of their activity was geared towards the war machine and the establishment of a 1000-year Reich, *(and hundreds of small stakes, carrying the swastika, were dumped on the snow-covered ground from the 'Wals' to signal the new ownership, this is a documented fact).*

ii) There was ongoing ship and submarine activity in the *South Atlantic and polar regions* throughout and after the war had apparently ended.

iii) The US invaded the continent itself with considerable naval resources leaving mainland America exposed and vulnerable as the world edged into the Cold War. The task force limped home as if defeated only weeks later, and the local *South American press* wrote of such a defeat.

iv) Admiral Byrd spoke of *objects that could fly from pole to pole* at incredible speeds being based on Antarctica.

v) Hundreds of thousands of Germans and numerous U-boats were missing at the end of the war.

UFO Sightings in Antarctica

Yet such claims would have died out had they not been based on at least some real events. That something strange was happening around the foreboding continent took an interesting turn in the 1960s when the Argentine Navy was charged with the official investigation into strange sightings in the sky. A 1965 official report prepared by **Captain Sanchez Moreno** of the Naval Air Station, Comandante Espora in Bahia Blanca stated:

"Between 1950 and 1965, personnel of Argentina's Navy alone made 22 sightings of unidentified flying objects that were not airplanes, satellites, weather balloons or any type of known (aerial) vehicles". These 22 cases served as precedents for intensifying that investigation of the subject by the Navy.

Following a series of sightings at Argentine and Chilean meteorological stations on Deception Island, Antarctica, in June and July 1965, **Captain Engineer Omar Pagani** disclosed at a press conference that, *"the unidentified flying objects do exist. Their presence in Argentine airspace is proven. Their nature and origin are unknown and no judgment is made about them."*

More details of these UFO sightings were given in a report in the Brazilian newspaper '*O Estado de Sao Paulo*' in its 8th July 1965 edition.
http://www.bibliotecapleyades.net/tierra_hueca/esp_tierra_hueca_6c.htm

For the first time in history, an official communiqué has been published by a government about the flying saucers. It is a document from the Argentine Navy, based on the statements of a large number of Argentine, Chilean and British sailors stationed at the naval base in Antarctica.

The communiqué declared that the personnel of Deception Island naval base saw, at nineteen hours forty minutes on 3 July, a flying object of lenticular shape, with a solid appearance and a colouring in which red and green prevailed and, for a few moments, yellow. The machine was flying in a zig-zag fashion, and in a generally western direction, but it changed course several times and changed speed, having an inclination of about forty-five degrees above the horizon. The craft also remained stationary for about twenty minutes at a height of approximately 5,000 meters, producing no sound.

In March 1950 Commodore Augusto Vars Orrego of the Chilean Navy shot still pictures and 8mm movie footage of a very large cigar shaped flying object that hovered over and maneuvered about in the frigid skies above the Chilean Antarctic. Orrego stated, "during the bright Antarctic night, we saw flying saucers, one above the other, turning at tremendous speeds. We have photographs to prove what we saw".
http://www.bibliotecapleyades.net/tierra_hueca/esp_tierra_hueca_6c.htm

There were other Chilean sightings to add to the growing mystery in the south polar region of the planet. During January 1956 an event was witnessed by a group of Chilean scientists who had been flown by helicopter to Robertson Island in the Wendell Sea to study geology, fauna, and other features. This experience was the subject of a later article entitled '*A Cigar-Shaped UFO over Antarctica.*'

180

At the beginning of January 1956, during a period of stormy weather, the party suddenly became aware of something which, in other circumstances, could have been very grave for them. This was that their radio had mysteriously ceased to function. Prior arrangements alleviated any serious concerns as a helicopter would return to take them off again from the island.

A doctor, whose habit was getting up in the night to observe anything of meteorological interest, awakened another of the group, a professor from a deep sleep to come outside and witness an unusual aerial event. Looking upwards, almost overhead, they beheld two 'metallic' cigar-shaped objects in vertical positions, perfectly still and silent, and flashing vividly the reflected rays of the sun. They were joined by two other members, an assistant and a medical orderly around 7:00am. The group watched the two craft.

"At about 9:00am object No. 1 (the nearest to the zenith) suddenly assumed a horizontal posture and shot away like a flash towards the west. It had now lost its metallic brightness and had taken on the whole gamut of visible colours of the spectrum, from infrared to ultra-violet.
Without slowing down it performed an incredible acute-angle change of direction, shot off across another section of the sky and then did another sharp turn as before. These vertiginous maneuvers, the zigzagging, abrupt stopping, instantaneous accelerating, went on for some time right overhead, the object always following tangential trajectories with respect to the Earth and all in the most absolute silence.

"The demonstration lasted about five minutes. Then the object returned and took up position beside its companion in almost the same area of the sky as before, but now it was the turn of No. 2 to show its paces and do a weird zigzagging dance. Shooting off towards the east, it performed a series of ten disjointed bursts of flight, broken by brusque changes of direction, and marked by the same colour changes when accelerating or stopping, and so on. After about three minutes of this, object No. 2 returned and took up its station near its companion, and reassumed its original solid and metallic appearance.

Photos taken of the UFO seen at an outcrop of rock near Trindade Island
http://www.openminds.tv/trindade-ufo-case-205/5060

"The scientists had with them two *Geiger-Miller counters* of high sensitivity, one of the auditory and the other of the flash-type. When the two objects had finished their dance and reassumed their stations in the sky, someone discovered that the flash-type Geiger counter now showed that radioactivity around them had suddenly increased 40 times – enough to kill any organism subjected long enough to it. The discovery greatly increased the anxiety felt by the four men …
"Although they had no telescopic lens, they did, however, have cameras with them, and they took numerous photographs of the objects, both in colour and black and white. We are not told in the report what became of these photographs."
http://www.bibliotecapleyades.net/tierra_hueca/esp_tierra_hueca_6c.htm

Five years later, there was another documented account of a UFO sighting over Antarctica, this time by Rubens Junqueira Villela, a meteorologist, and the first Brazilian scientist to participate

182

in an expedition to the white continent, now a veteran of eleven expeditions to Antarctica. Whilst on board the *US Navy icebreaker Glacier* which had set sail from New Zealand at the end of January 1961, Villela claims that he witnessed a UFO event in the skies over Antarctica which he immediately recorded in his diary. During 16th March 1961 and after a fierce storm had forced the expedition to retreat to Admiralty Bay in the King George Isles, a strange light suddenly crossed the sky.

Crew members on board started shouting that they thought it was a missile or a meteor. There was growing wide-spread excitement in trying to describe the light which appeared over Almirantado Bay. The colors, the configuration, and contours of the object, as a body of light, had geometric form but, Villela sensed that it wasn't from this world or who could have possibly produced it.

Villela stated that "the object was multi-colored and had a luminous body – oval-shaped. It left a long tube-like orange/red trail. Suddenly, it split into two pieces, as if it had exploded. Each part shone even more intensively, with white, blue and red colors projecting 'V' shaped rays behind it. Quite quickly they moved away and could be seen 200 meters above the ground … Throughout the sighting, no noise was heard by any of the witnesses."
http://www.bibliotecapleyades.net/tierra_hueca/esp_tierra_hueca_6c.htm

The South Atlantic area was also host to another sighting on 16th January 1958 when the Brazilian naval vessel *Almirante Salddanha* was escorting a team of scientists to a weather station on Trindade Island. As the ship approached the island (or rather an outcrop of rock) a UFO reportedly swooped past the ship, circled the island then, flew off in front of dozens of witnesses.

One of these witnesses, the expedition photographer, took a number of photographs of the object, and later the film was handed over to the military by the Captain. After analysis, the Brazilian government released the film stating that they were unable to account for the images.

Whatever was going on in the Antarctic region, it certainly wasn't happening in isolation.
http://www.bibliotecapleyades.net/tierra_hueca/esp_tierra_hueca_6c.htm

The Brazilian naval vessel Almirante Salddanha

The 1936 Bavarian Crash

To understand this dedication, it is necessary to go back, before the outbreak of the Second World War, to an isolated section of the Bavarian Alps in the Schwarzwald (Black Forest) near Freiburg, It was there, in the summer of 1936, that an unidentified flying object, crewed by a distinctly human, and Aryan appearing race, made a forced landing, very similar to the one which was to occur, some ten years later, in the desert, near Roswell, New Mexico, in the United States.

While the occupants of the crashed saucer events were completely unrelated, the technology involved seems to have been strikingly similar. Also, the outcome of the recovery effort, undertaken by Germany, just as a similar recovery effort was undertaken by the United States, had strikingly different results.

The **Bavarian crash of 1936**, seems to have yielded a functioning, or almost functioning and repairable (with the technology of the time) power plant, and a near completely destroyed, or unrepairable airframe. The **Roswell crash of 1947** resulted in exactly the opposite... a nearly

184

intact airframe and a ruined power plant. Because of this, the German research, which was to follow, took a vastly different turn from that which was undertaken in the United States, some ten years later... Germany needed an airframe which was capable of supporting the "engine" (for lack of a better term), while the United States would eventually need an "engine" capable of giving maximum performance to the airframe.

This, of course, would explain the vast array of "experimental" aircraft... of extremely "unique" design... to literally pour out of the **design bureaus of Messerschmidt, Focke Wulf, Fokker** and a multitude of smaller firms in the period between 1939 and 1945. The most notable, of course, is the **Sänger "Flying Wing"** which was later copied by the United States, and is, of course, the ancestor of today's "stealth" bomber and fighter designs... notably, the B-2 Heavy Bomber. The United States of course would deny or debate that they too were also working on a flying wing design aircraft at the same time period.

It is also beyond doubt that both unidentified flying object recoveries are the initial impetus for the long-standing and ongoing research in "anti-gravity" propulsion seen in work of current aircraft manufacturers such as **Boeing** and **Lockheed** in the United States, and **PanAvia** in Europe.

In any case, it was the work on "reverse engineering" the downed Bavarian UFO that was the catalyst for the "exodus" to the South in the final days of the Second World War. Germany was in ruins, and the research was viewed, by those conducting it, as vital... vital enough to risk packing up all that they had and risking a perilous submerged crossing of the Atlantic... to an isolated experimental research base on a frozen continent... Antartica!

Granted, by modern standards... even by the standards of the day... U-Boats were small and cramped. They had very little cargo capacity. Still, a tiny fleet of them... ten to twelve boats... could easily transport the essential equipment, making several "runs", and serve to supply and, later re-supply the Antarctic bastion of the research.

Speculation exists, with much to support it, that at least one of the boats in the valiant little fleet contained the biggest prize of all... at least one living survivor of the 1938 crash... an Extraterrestrial... a literal Human Being... not a "Grey"... born on a distant planet. The best evidence indicates that there were several survivors of the crash and that they worked, and are most likely still working, with the original German scientists and engineers, or their descendants, in an effort to construct a viable "flying disc".

These are not the **"Grey Aliens" of Roswell**. These beings, biologically, completely human, are described as **"Aryan"** in appearance, and completely human, although at least two to three generations more advanced, technologically than Earth-born human beings. While their technology is similar to that of the Grays in general theory, it is somewhat different, apparently, in application.

This would tend to indicate that Earth technology and science is, at most, only one "major breakthrough" away from parity with the extraterrestrial cultures in question, and also explain the "urgency" of the project, as viewed by the German (and undoubtedly United States, as

well...) scientists and engineers involved in such research.
http://www.bibliotecapleyades.net/antarctica/antartica11.htm. **"Operation Highjump and The UFO Connection" by Erich J. Choron from Grey Falcon Website**

There can be little doubt left that history as we currently know it is not accurate and is always written from the viewpoint of the victor and when it comes to the history of the Second World War, we are left with a lot of contradictory information, suppressed records and documents (some only now coming to the light of public disclosure) as regards the technology of the super secret weapons of the Third Reich. We have tried to peel back some of the layers of secrecy surrounding the development of Germany's flying saucer and atomic bomb programs and of some it most brilliant scientists and engineers to reveal that these programs were very real and as Hitler often referred to as Germany's "**wunderwaffen** (*Wonder Weapons) that would snatch victory from the Allies in the last minutes of the war"*. Potentially had the Allied Forces not have won some of the more decisive battles of WWII , Hitler's statement would have come true and these wonder weapons, so radical in their concepts and designs would have made the difference in the outcome of the war, in favour of a Nazi victory.

The intelligence that was obtained by Allied Forces during the war of these advanced German technologies spurred a race by the British, American and Russian forces to push ever deeper into German-held territory, toward secret underground factories and facilities in order to be the first to lay claim to the secret documents, the weapons and assorted hardware that the Germans were working on as well as their scientists.

The Allied forces found and stripped these underground factories of everything they could lay their hands on but, the American got the lion's share of war booty and technology and most of the German scientists who worked on the weapons. Through **Operation Paperclip**, the American sequestered these scientists to the American southwest where they were forgiven their "sins of transgression"... expunging their Nazi war records), given comfortable accommodations, and fully stocked labs and equipment, in exchange they were to continue their work on the captured German super secret (**wunderwaffen**) weapons.

One of the curious aspects of the German development of flying saucer and the creation of a "*tall, blond, blue-eyed*" supposedly superior **Aryan** race through **eugenics** research and development has been the reports that have cropped up throughout the States and elsewhere on the planet particularly during the early 50s via contactee experiences. That being the similarity of alien saucer designs to the German saucer designs and the tall, long blond haired and blue-eyed aliens with the tall, blond, blue-eyed Aryan race of the German eugenics program. Some contactees , in particular, **George Adamski**, claiming that some of the aliens, now referred to as **Nordics,** spoke with a *"German accent"* while others communicated through telepathy.

The question that arises is, are these reports of human-like Aryan/Nordic alien beings real Extraterrestrial beings from another star system or just the product of good old fashion German or even, an American eugenics program?

A positive statement in support of either the Aryan eugenics creation or that a human looking Extraterrestrial being, who could be our distant cosmic cousin had come to pay us a visit has

186

profound repercussions on the whole question of whether we are indeed being visited by ETI or that earth-based science is a lot further advanced than we have currently been lead to believe.

The answer has become muddled and a point of contentious debate in support of either hypothesis but, there is a third alternative perspective and the answer to the question is yes! Both assumptions are correct!

We know that the German saucer programs were real as were their **eugenics program** to develop a "master race, an Aryan race. In Nazi Germany, a so-called mixed marriage of an "Aryan" with a "sub-human (defective human e.g. retarded person) was forbidden. To maintain the purity of the Germanic master race, eugenics was practiced. In order to eliminate "defective" citizens, the T-4 Euthanasia Program was administered by Karl Brandt to rid the country of the mentally retarded or those born with genetic deficiencies, as well as those deemed to be racially inferior. Additionally, a program of compulsory sterilization was undertaken which resulted in the forced operations of hundreds of thousands of individuals. Many of these policies are generally seen as being related to what eventually became known as the Holocaust.
http://en.wikipedia.org/wiki/Master_race#Nazi_beliefs_about_the_Aryans

Selection of the best qualities in human genetics and body types was fostered and several generations of such people resulted from the program. Some of the descendants from this national program are living in countries like Sweden, even today. Unfortunately, many suffer from prejudices by some European nations who are reminded of their wartime history by their very presence.

Could some of the Aryan people been recruited and trained in a covert program of disinformation or some other hidden agenda to discredit genuine UFO sightings and ET contacts by flying German built saucers in America and in some European countries? The evidence that this may indeed be the case is very strong and this aspect will be explored shortly in the next subsection of this book.

The focus for many decades of UFO research has been on the hardware of Extraterrestrial technology or the possibility of manmade saucer technology originating out of World War II Germany while almost ignoring that nazi scientists were involved in some of the most advanced biological experiments and some of the most cruel and grisly experiments upon prisoners of concentration camps.

Much of the research upon these poor prisoners involved injections of chemicals, gases, viruses and all manner of diseases to see the effects upon the body or being subjected to extremes of heat and cold, high atmospheric pressures and vacuums. There was also, live vivisection on people, removing of limbs and re-attachments, as well as genetic research, gene splicing, and engineering. Much of this information was to see how humans could survive in space or in other environments. Such unusual and cruel experiments were designed to push beyond the limits of human endurance and unfortunately, almost all POWs died from these experiments.

What is critical to our understanding of this research is that through **Operation Paperclip**, the American military scientists used this captured German medical research to help develop and advance their own space program and to understand and anticipate the likely conditions that humans in space might encounter.

Now, consider the reports of ET beings who appeared to be very human-like but, who spoke in some unknown strange language or had telepathic abilities to communicate to Earth beings. They were not always described as blue-eyed or blond but, sometimes had gold coloured eyes and were brunette or dark haired and who seem to be working along with other smaller or stranger looking ET beings.

UFO/ETI accounts like the one that **Travis Walton** experienced or the many **Nordic** type beings seen in past and more recent times by contactees like **George Adamski, Howard Menger, George Van Tassel or Billy Meier**, etc. force us to re-examine the possibly that life in the universe may actually follow some basic scientific laws of evolution. Laws of biology that are universal, immutable, and similarly reproducible to life on this planet but, with unique qualities of individuality peculiar to each planet or star system.

These two opposing concepts require further research not only by Ufologists but, by scientists in the fields of astronomy, sociology, and exo-biology to make sense of this dichotomy.
https://www.youtube.com/watch?v=DkH4TcVD_CM

CHAPTER 41
POSTWAR SCIENTISTS, SECRET PROJECTS AND PROGRAMS -THE BIRTH OF THE MILITARY INDUSTRIAL COMPLEX IN AMERICA

Operation Paperclip

Operation Paperclip was the **Office of Strategic Services (OSS)** program used to recruit the scientists of Nazi Germany for employment by the United States in the aftermath of World War II (1939–45). It was executed by the **Joint Intelligence Objectives Agency (JIOA)**, and in the context of the burgeoning Soviet–American Cold War (1945–91), one purpose of Operation Paperclip was to deny German scientific knowledge and expertise to the USSR and the UK.

Although the JIOA's recruitment of German scientists began after the European Allied victory (8 May 1945), US President Harry Truman did not formally order the execution of **Operation Paperclip** until August 1945. Truman's order expressly excluded anyone found "to have been a member of the Nazi Party, and more than a nominal participant in its activities, or an active supporter of Nazi militarism." Said restrictions would have rendered ineligible most of the scientists the **JIOA** had identified for recruitment, among them rocket scientists **Wernher von Braun** and **Arthur Rudolph,** and the physician **Hubertus Strughold**, each earlier classified as a "menace to the security of the Allied Forces".

A group of 104 rocket scientists (aerospace engineers) at Fort Bliss, Texas. Part of the 1500 German scientists that came to America during Operation Paperclip (Wernher von Braun is 7th from the right)
https://en.wikipedia.org/wiki/Operation_Paperclip

To circumvent President Truman's anti-Nazi order, and the Allied Potsdam and Yalta agreements, the JIOA worked independently to create false employment and political biographies for the scientists. The JIOA also expunged from the public record the scientists' Nazi

Party memberships and régime affiliations. Once "bleached" of their Nazism, the US Government granted the scientists security clearance to work in the United States. *Paperclip,* the project's operational name, derived from the paperclips used to attach the scientists' new political personae to their "US Government Scientist" JIOA personnel files.
http://en.wikipedia.org/wiki/Operation_paperclip

Unable to defeat the Russians in a number crushing winter battles, Nazi Germany found itself at a logistical disadvantage. German resources had been depleted from the failed conquest and its **Military-Industrial Complex** was unprepared to defend the Greater German Reich) against the Red Army's westward counterattack. By early 1943, many scientists, engineers, and technicians were recalled from the front lines of combat by the German government to work in research and development to bolster German defense for a protracted war with the USSR. Among them were 4,000 rocketeers who were sent back to Peenemünde, in north-east coastal Germany.

Reinstatement of these scientists and engineers by the Nazi government for scientific work required identifying and locating the scientists, engineers, and technicians, then ascertaining their political and ideological reliability. **Werner Osenberg**, the engineer-scientist heading the **Military Research Association**, recorded the names of the politically cleared men to the **Osenberg List**, thus reinstating them to scientific work.

However, whether through a deliberate intelligence operation or through bungling or stupidity, someone let this Osenberg List become retrievable which eventually fell into the hands of the Allied Forces!

In March 1945, at Bonn University, a Polish laboratory technician found pieces of the Osenberg List stuffed in a toilet; the list subsequently reached **MI6**, who transmitted it to US Intelligence. Then US Army Major Robert B. Staver, Chief of the **Jet Propulsion Section of the Research and Intelligence Branch** of the **U.S. Army Ordnance Corps**, used the Osenberg List to compile his list of German scientists to be captured and interrogated; **Wernher von Braun**, Nazi Germany's premier rocket scientist headed Major Staver's list.

Most of the engineers worked at the Baltic coast German Army Research Center Peenemünde, developing the V-2 rocket; after capturing them, the Allies initially housed them and their families in Landshut, Bavaria, in southern Germany.

Beginning on 19 July 1945, the **US Joint Chiefs of Staff (JCS)** managed the captured ARC rocketeers under a program called **Operation Overcast**. However, when the "Camp Overcast" name of the scientists' quarters became locally known, the program was renamed **Operation Paperclip** in March 1946. Despite these attempts at secrecy, later that year the press interviewed several of the scientists. http://en.wikipedia.org/wiki/Operation_paperclip

T-Force (Target Forces)

The **Supreme Headquarters Allied Expeditionary Force (SHAEF)** under General Eisenhower issued a directive to create **T-Forces** soon after the Normandy Landings. The success of 30 Assault Unit, a unit that had been created by **Ian Fleming** whilst working in Royal Navy intelligence was a key factor in the decision to create **'Target Force'**, normally referred to as T-Force. Fleming sat on the committee that selected targets for the unit, helping to create what were known as the **'Black Books'** which were issued to officers of the unit. The unit's most notable coup was the advance on the German port of Kiel where it captured the research centre where the engines for German rockets, missiles, jet fighters and high-speed U-Boats had been designed. Ian Fleming used elements of this story in his 1955 **James Bond** novel *"Moonraker"*.

T-Force (Target Forces) was an elite British Army force but, the Americans also had their T-Forces of intelligence officers which operated during the final stages of World War II.

In 1944 with the push into German-occupied France by the Allies and with the ever advancing American troops pushing back the German lines, small groups of intelligence officers called T-Forces followed to secure and exploit targets that could provide valuable intelligence of scientific and military value, and they were later tasked with seizing Nazi German scientists and businessmen in the aftermath of VE Day. It so became apparent, however, that in order to understand this new technology, the brains behind its development would also be needed as well. http://en.wikipedia.org/wiki/T-Force

President Roosevelt at the time not wanting any Nazi war criminals working in the US flatly refused to sign any papers allowing for such operations. Fearing that without these scientists the US would be fatally crippled with the upcoming conflict with Russia, the Joint Chiefs of Staff decide to overlook the President's refusal and implemented **Project Overcast (a.k.a. Operation Overcast),** a non-presidential sanctioned project.

This project might have eventually become a problem had it not been for President Roosevelt's death in June12, 1945. President Truman then took over the Whitehouse Office being totally unprepared or even briefed of an impending Soviet Russia problem. The Joint Chiefs of Staff approached Truman to expand on **Operation Overcast** and in august 1945 he consented signed into being Operation Paperclip.

The Joint Intelligence Committee which was the intelligence branch of the Joint Chiefs of Staff set up its own subcommittee the JIOA for the sole purpose of directing Operation Paperclip with the Exploitation Branch under the Army Intelligence of the War Department General Staff in charge of the physical execution of the project.

Through this is tangled web of sister agencies and subcommittees the government was able to effectively hide, twist, and erase many parts of the Nazi scientists pasts. Truman, like his predecessor, Roosevelt before him, though, not wholly in favour of Nazi scientists working in the US was unable to verify anything that these agencies did. Now, the hunt for Nazi scientists had official sanction. http://www.youtube.com/watch?v=bh37r3alzzs

Operation Alsos

Operation Alsos (or **Alsos Mission**) was an effort at the end of World War II by the Allies (principally Britain and the United States), branched off from the Manhattan Project, to investigate the German nuclear energy project, seize German nuclear resources, materials and personnel to further American research and to prevent their capture by the Soviets, and to discern how far the Germans had gone towards creating an atomic bomb. The personnel of the project followed close behind the front lines of the Allied invading armies, first into Italy, and then into France and Germany, searching for personnel, records, material, and sites involved.

Major General Leslie M. Groves was the military director of the **Manhattan Engineering District (the Manhattan Project)**, the Allied wartime effort to develop an atomic bomb (which itself was sparked out of fears of a German weapon). Groves was the major impetus behind the project, in part because of his desire to make sure that German technology and personnel did not fall into Soviet hands, so as to prolong the anticipated American monopoly on nuclear weapons as much as possible.

All of these special operation projects and forces had the sole objective of ensuring that Britain and the US had the lion's share of German scientists, engineers, research and development documents and actual weapons hardware before the soviets had a chance to get their hands on it. Operation Paperclip was to deny German scientific knowledge and expertise to the USSR, even from the UK.

As had been stated earlier, there had been a prior agreement among the Allied Forces, that captured intelligence and technology from Germany would be shared back and forth between the USA, Canada, and Britain after the war. The reality was that only a small percentage of information flowed equally to each country's government, military and intelligence agencies. United States, however, reneged on the deal or at least never fully lived up to its portion of the agreement. The U.S. kept much of what was captured and released only the stuff that was not considered super-secret or highly advanced retaining any saucer documents, blueprints, saucer prototypes, nuclear weaponry as well as Nazi scientists and engineers, even those tried and found convicted or war crimes came to America through Operation Paperclip! By accepting these people into the country, the US had chosen a policy of technology over justice! It appears that there is no honour among thieves or the victors of war!

Little wonder that President Eisenhower came out with his last famous public speech, to beware of the Military Industrial Complex, that it bears repeating part of it:

*"... In the councils of government, **we must guard against the acquisition of unwarranted influence, whether sought or unsought, by the military-industrial complex**. The potential for the disastrous rise of misplaced power exists and will persist. We must never let the weight of this combination endanger our liberties or democratic processes. We should take nothing for granted. Only an alert and knowledgeable citizenry can compel the proper meshing of the huge industrial and military machinery of defense with our peaceful methods and goals so that security and liberty may prosper together."*
http://www.hnet.org/~hst306/documents/indust.html

In historical perspective, I believe that Operation Paperclip will be viewed as an ill-conceived project that has cost the American public its identity, its democracy, its true sense of freedom and has sidetracked its destiny down a dark road which parallels the rise of the Third Reich in pre-war Nazi Germany. The US military along with its partners from the private industrial sector had

192

deliberately imported a "cancer" in the form of Nazi scientists, war criminals and political idealists into America. A cancer so pernicious, that it has metastasized through the body politic and the very fabric of American society. Yet, those in power will justify their position that it was necessary in order to stay ahead of the rising power of Communism in the world and their influence in the buildup of the Eastern Bloc in Europe after the Second World War.

We will prove that this has been a carefully orchestrated covert agenda that is still evolving and being played out on the world stage and the source of this information comes from no less a person than, Wernher von Braun, himself!

Further to this source of Nazi-inspired machinations, the ramifications nationally for the US has been a subdued, yet seething public distrust and unrest that threatens to erupt violently to the surface in the form of public protests, demonstrations of anti-governmentalism, open hostility toward the controlling corporate wealthy elite, and outright lawless anarchism bordering on second national revolution. The implications of which could easily spill over into other nations making the global social unrest movement of the 60s looking more like a ticket tape parade.

Internationally, the continued suspicion and distrust for American foreign policies, their hidden economic agendas and their deliberate interference with the affairs and operations of other nations' governments has created global agitation and social unrest, economic hardship and political strife, and numerous global "hot-spot" wars particularly in the Middle East, all in the name of stamping out global terrorism, defending democracy and maintaining American economic and foreign interests.

Instead of being the spiritual leaders of a new era of peace and justice leading to a global civilization, America has found itself, along with her partners retrogressing the world toward further chaos and conflict, impotent in its ability to create a stable, peaceful and sustainable world commonwealth. Its sole objective is the development of a world order based on materialism devoid of any spiritual values, with absolute power and control by the few ruling elite over the masses.

Operation Paperclip was a program to bring top German Nazi scientists into the country and continue to develop secret wartime Nazi technology under the umbrella of the National Security "clause" to further American's global brinkmanship interests!

Dr. Herman Oberth

Hermann Julius Oberth (25 June 1894 – 28 December 1989) was an Austro-Hungarian-born German physicist and engineer. He is considered one of the founding fathers of rocketry and astronautics.Oberth was born to a Transylvanian Saxon family in Nagyszeben (German: Hermannstadt, today Sibiu, Romania), Austria-Hungary. By his own account and that of many others, around the age of 11 years old, Oberth became fascinated with the field in which he was to make his mark through reading the writings of **Jules Verne,** especially *"From the Earth to the Moon"* and *"Around the Moon"*, re-reading them to the point of memorization. Influenced by Verne's books and ideas,

In the autumn of 1929, Oberth conducted a static firing of his first liquid-fueled rocket motor, although it lacked a cooling system, it did run briefly. He was helped in this experiment by an 18 year old student Wernher von Braun, who would later become a giant in both German and American rocket engineering from the 1940s onward, culminating with the gigantic Saturn V rockets that made it possible for men to land on the Moon in 1969 and in several following years. Indeed Von Braun said of him:

"Hermann Oberth was the first, who when thinking about the possibility of spaceships grabbed a slide-rule and presented mathematically analyzed concepts and designs.... I, myself, owe to him not only the guiding-star of my life, but also my first contact with the theoretical and practical aspects of rocketry and space travel. A place of honor should be reserved in the history of science and technology for his ground-breaking contributions in the field of astronautics."
https://en.wikipedia.org/wiki/Hermann_Oberth

Dr. Herman Oberth
https://en.wikipedia.org/wiki/Hermann_Oberth

In 1938, the Oberth family left Sibiu, Romania, for good, to first to settle in Austria, then in Nazi Germany, then in the United States, and finally back to a free Germany. Oberth moved to **Peenemünde, Germany,** in 1941 to work on Nazi German rocketry projects, including the **V-2 rocket** weapon. Oberth later worked on solid-propellant anti-aircraft rockets at the German WASAG military organization near Wittenberg.

Around the end of World War II in Europe in May 1945, the Oberth family moved to the town of Feucht, near Nuremberg, Germany, which became part of the American Zone of occupied Germany, and also the location of the high-level war-crimes trials of the surviving Nazi leaders. Oberth was allowed to leave Nurmberg to move to Switzerland in 1948, where he worked as an independent rocketry consultant and a writer.

During the 1950s and 1960s, Oberth offered his opinions regarding unidentified flying objects (UFOs). He was a supporter of the extraterrestrial hypothesis for the origin of the UFOs that were seen at the Earth. For example, in an article in *The American Weekly* magazine of October 24, 1954, Obert stated, "It is my thesis that flying saucers are real and that they are space ships from another solar system. I think that they possibly are manned by intelligent observers who are members of a race that may have been investigating our earth for centuries..."

"UFOs are conceived and directed by intelligent beings of a very high order, and they are propelled by distorting the gravitational field, converting gravity into useable energy. There is no doubt in my mind that these objects are interplanetary craft of some sort. I and my colleagues are confident that they do not originate in our solar system, but we feel that they may use Mars or some other body as sort of a way station. They probably do not originate in our solar system, perhaps not even in our galaxy."

"We cannot take the credit for our record advancement in certain scientific fields alone. We have been helped." In a statement made to a group of reporters after his retirement in 1960, one reporter asked, *"By who?"* He replied, *"The people of other worlds."*

Oberth returned to the United States to view the launch of STS-51J, the space Shuttle Discovery launched October 3, 1985.

Oberth died in Nuremberg, Germany, on 28 December 1989, just shortly after the fall of the **Iron Curtain** that had for so long divided Germany into two countries.
http://en.wikipedia.org/wiki/Hermann_Oberth

Wernher von Braun

Wernher Magnus Maximilian Freiherr von Braun (March 23, 1912 – June 16, 1977) was a German rocket scientist, engineer, space architect, and one of the leading figures in the development of rocket technology in Nazi Germany and the United States during and after World War II.

Wernher von Braun is perhaps regarded by American historians as the single most notable Nazi German rocket scientist to enter into the US through Operation Paperclip and though many

German scientists contributed to the development of the first cruise missile, the **V1 ("Doodle Bug or the "Buzz Bomb")** and the **V2 rocket**, the first ballistic missile, Wernher von Braun contributions were undeniable.

A former member of the Nazi party commissioned Sturmbannführer of the paramilitary SS and decorated Nazi war hero, von Braun would later be regarded as the preeminent rocket engineer of the 20th century in his role with the United States civilian space agency NASA. In his 20s and early 30s, von Braun was the central figure in Germany's rocket development program, responsible for the design and realization of the deadly V-2 combat rocket during World War II. After the war, he and some of his rocket team were taken to the U.S. as part of the then-secret Operation Paperclip. Von Braun worked on the US Army **intermediate range ballistic missile (IRBM)** program before his group was assimilated by NASA, under which he served as director of the newly-formed Marshall Space Flight Center and as the chief architect of the **Saturn V launch vehicle**, the super booster that propelled the Apollo spacecraft to the Moon. According to one NASA source, he is "without a doubt, the greatest rocket scientist in history. His crowning achievement was to lead the development of the Saturn V booster rocket that helped land the first men on the Moon in July 1969." In 1975 he received the National Medal of Science.

Wernher von Braun
https://pt.wikipedia.org/wiki/Wernher_von_Braun

Dr. von Braun developed the idea of a **Space Camp** that would train children in fields of science and space technologies as well as help their mental development much the same way sports camps aim at improving physical development.

He had since gone on to start the American space program in conjunction with NASA and is the chief architect of the Saturn V launch vehicle and his work on rockets lead to the space shuttle and the **International Space Station (ISS).**

On June 20, 1945, **U.S. Secretary of State Cordell Hull** approved the transfer of von Braun and his specialists to America; however, this was not announced to the public until October 1, 1945. Von Braun was among those scientists for whom the U.S. **Joint Intelligence Objectives Agency** created false employment histories and expunged Nazi Party memberships and regime affiliations from the public record. Once "bleached" of their Nazism, the US Government granted the scientists security clearance to work in the United States. **"Paperclip,"** the project's operational name, derived from the paperclips used to attach the scientists' new political personæ to their "US Government Scientist" personnel files.

On June 20, 1945, **U.S. Secretary of State Cordell Hull** approved the transfer of von Braun and his specialists to America; however, this was not announced to the public until October 1, 1945. Von Braun was among those scientists for whom the U.S. **Joint Intelligence Objectives Agency** created false employment histories and expunged Nazi Party memberships and regime affiliations from the public record. Once "bleached" of their Nazism, the US Government granted the scientists security clearance to work in the United States. **"Paperclip,"** the project's operational name, derived from the paperclips used to attach the scientists' new political personæ to their "US Government Scientist" personnel files.

Finally, von Braun and his remaining Peenemünde staff were transferred to their new home at Fort Bliss, Texas, a large Army installation just north of El Paso. While there, they trained military, industrial and university personnel in the intricacies of rockets and guided missiles. As part of the **Hermes project**, they helped to refurbish, assemble and launch a number of V-2s that had been shipped from Germany to the **White Sands Proving Ground** in New Mexico. They also continued to study the future potential of rockets for military and research applications. Since they were not permitted to leave Fort Bliss without military escort, von Braun, and his colleagues began to refer to themselves only half-jokingly as **PoPs (Prisoners of Peace) ."**

In 1950, at the start of the Korean War, von Braun, and his team were transferred to Huntsville, Alabama, his home for the next 20 years. Between 1950 and 1956, von Braun led the Army's rocket development team at Redstone Arsenal, resulting in the **Redstone rocket**, which was used for the first live nuclear ballistic missile tests conducted by the United States.

As director of the Development Operations Division of the **Army Ballistic Missile Agency (ABMA)**, von Braun, with his team, then developed the **Jupiter-C**, a modified Redstone rocket. The Jupiter-C successfully launched the West's first satellite, **Explorer 1**, on January 31, 1958. This event signaled the birth of America's space program.

Despite the work on the Redstone rocket, the twelve years from 1945 to 1957 were probably some of the most frustrating for von Braun and his colleagues. In the Soviet Union, **Sergei Korolev** and his team of scientists and engineers plowed ahead with several new rocket designs and the **Sputnik program**, while the American government was not very interested in von Braun's work or views and only embarked on a very modest rocket-building program. In the meantime, the press tended to dwell on von Braun's past as a member of the SS and the slave labor used to build his V-2 rockets. Von Braun's Nazi past would come back repeatedly to haunt him whenever the US military industrial complex needed to remind him who was in control and whenever; they required his full cooperation in matters of space technology and weapons development.

Von Braun envisioned a space station using recoverable and reusable rockets to construct it. It would be a toroid structure or giant orbiting wheel shape craft capable of generating its own gravity through rotational spin. The ultimate purpose of the space station would be to provide an assembly platform for manned lunar expeditions.

Von Braun developed and published his space station concept during the very "coldest" time of the Cold War when the U.S. government for which he worked put the containment of the Soviet Union above everything else. The fact that his space station – if armed with missiles that could be easily adapted from those already available at this time – would give the United States space superiority in both orbital and orbit-to-ground warfare did not escape him. Although von Braun took care to qualify such military applications as "particularly dreadful" in his popular writings, he elaborated on them in several of his books and articles

The U.S. Navy had been tasked with building a rocket to lift satellites into orbit, but the resulting Vanguard rocket launch system was unreliable. In 1957, with the launch of Sputnik 1, there was a growing belief within the United States that America lagged behind the Soviet Union in the emerging Space Race. American authorities then chose to utilize von Braun and his German team's experience with missiles to create an orbital launch vehicle, something von Braun had originally proposed in 1954 but had been denied. NASA was established by law on July 29, 1958. One day later, the 50th Redstone rocket was successfully launched from Johnston Atoll in the south Pacific as part of Operation Hardtack I. Two years later, NASA opened the **Marshall Space Flight Center** at Redstone Arsenal in Huntsville, and the ABMA development team led by von Braun was transferred to NASA. In a face-to-face meeting with Herb York at the Pentagon, von Braun made it clear he would go to NASA only if development of the Saturn was allowed to continue. Presiding from July 1960 to February 1970, von Braun became the center's first Director.

The Marshall Center's first major program was the development of Saturn rockets to carry heavy payloads into and beyond Earth orbit. From this, the Apollo program for manned moon flights was developed. During Apollo, he worked closely with former Peenemünde teammate, Kurt H. Debus, the first director of the **Kennedy Space Center**. His dream to help mankind set foot on the Moon became a reality on July 16, 1969, when a Marshall-developed Saturn V rocket launched the crew of **Apollo 11** on its historic eight-day mission. Over the course of the program, Saturn V rockets enabled six teams of astronauts to reach the surface of the Moon.

198

During the late 1960s, von Braun was instrumental in the development of the **U.S. Space & Rocket Center in Huntsville**. The desk from which he guided America's entry in the Space Race remains on display there.

After a series of conflicts associated with the truncation of the Apollo program, and facing severe budget constraints, von Braun retired from NASA on May 26, 1972. Not only had it become evident by this time that his and NASA's visions for future U.S. space flight projects were incompatible; it was perhaps even more frustrating for him to see popular support for a continued presence of man in space wane dramatically once the goal to reach the moon had been accomplished.

NASA it seems was heading in a different direction than was its original mission statement to the American public at least it was played up that way in the news media. **Big Media** gave the impression that the public had lost interest in moon landings, that they were becoming ordinary and somewhat boring. Had Big Media actually interviewed the public, they would have found the opposite was true. Space exploration has always been of extreme interest to most people with only a few detractors who viewed it as a waste of time, with no financial payback to the public.

After leaving NASA, **Wernher von Braun** became Vice President for Engineering and Development at the aerospace company, **Fairchild Industries** in Germantown, Maryland on July 1, 1972.

A young Dr. Carol Rosin beside an aging Wernher von Braun
http://www.visioninconsciousness.org/UFOs_ETs_32.htm

In 1973 a routine health check revealed he had cancer, which during the following years could not be controlled by surgery. Depending on the news source some newspapers reported he had kidney cancer (renal cell carcinoma) while others stated that he had pancreatic cancer, either one would eventually have been the cause of his eventual passing. Von Braun continued his work to the extent possible, which included accepting invitations to speak at colleges and universities as he was eager to cultivate interest in human spaceflight and rocketry, particularly with students and a new generation of engineers.

His deteriorating health forced him to retire from Fairchild on December 31, 1976. When the 1975 National Medal of Science was awarded to him in early 1977 he was hospitalized and unable to attend the White House ceremony.

On June 16, 1977, **Wernher von Braun** died of cancer in Alexandria, Virginia, at the age of 65. He was buried at the Ivy Hill Cemetery in Alexandria, Virginia.
http://en.wikipedia.org/wiki/Wernher_von_Braun

Many scientists have an interest in the subject of life in the universe and that UFOs may be alien spacecraft visiting the Earth, although such public comments were usually tempered with discrete carefully worded statements. Wernher von Braun was no exception to these points of view. In fact, he knew far more than was generally commented in the public news media! He has been linked with other top scientists of his time of taking part in the recovery and examination of the Roswell saucer crash of 1947and its alien crew.

Throughout his career, after the WWII while working for the US military's technical research and development division and later with NASA, Wernher von Braun became aware of a **"Big Lie"** that was being perpetrated on the American public and the rest of the world. He was not able to reveal outright what the **Big Lie** was all about to the American public due to threats and intimidations by the US government and military regarding his former Nazi associations and possible war crimes.

He was forced to keep these secrets to himself for much of his remaining life until he was told that he had inoperable cancer at which time he did not want to take the Big Lie with him to his grave so, he confided in one individual. In early 1974 while still at Fairchild industries, he met and befriended an award-winning educator, author, leading aerospace executive and space and missile defense consultant, **Dr. Carol Rosin**. It was at Fairchild Industries where they first met at which time Von Braun was dying of cancer. Von Braun revealed to her the nature of the Big Lie and what was really going on within the US military and the NASA space program!

Dr. Carol Rosin

Dr. Carol Rosin became the protégé of Wernher von Braun and she was his spokesperson until Von Braun's eventual passing. Dr. Rosin was also one of the many witnesses of **Dr. Steven Greer's Disclosure Project**.

200

Author's Rant: I had the pleasure to meet and talk with Dr. Carol Rosin in Vancouver during Canada's first Disclosure Project event held at Simon Fraser University on September 9. 2001.

Much of what has come out from Dr. Rosin came through interviews with Dr. Greer as he as he gathered videotaped testimony from first-hand eye-witness for disclosure on matters of UFOs, ETI and **Free Energy generation**. According to Rosin's accounts, Von Braun spent the last years of his life explaining to her his position that space-based weapons are dangerous, politically destabilizing, financially too draining on the US, completely unnecessary, and unworkable. He explained to Rosin that there were available alternatives. He asked Dr. Rosin to be his spokesperson and to appear on occasions when he was too ill to speak.

Her first speech for him was to the National Education Association where she introduced satellites as a tool for teachers to 18,000 educators. He also asked her to take on the challenge of educating about the need to ban ALL space-based weapons by educating decision-makers and the grassroots about how the military industrial complex can feasibly be transformed into a peaceful world cooperative space exploration complex...creating a global cooperative space program that will be large enough to replace the entire war game and mindset...and build a security system based on collaboration and information sharing, a stimulated economy that will provide more jobs and profits (and training programs) than during any hot or cold wartime, applied technologies and information that can provide solutions to urgent and potential man-made or natural disasters and solve problems of human needs, our common environment, and new energy.

Author's Rant: Carol Rosin told me personally and in her lectures that she was repeatedly told by Von Braun over and over again that a "Big Lie" was being perpetrated upon naive public and that not all that we are aware of is truly what it appears to be.

The Big Lie focused upon the buildup of military force and armaments. It began as a need for a greater military budget to defend against the buildup of nuclear weapons, missiles, rockets and other weapons of mass destruction that the ***"Russians (the USSR) and the Communist East Bloc"*** after the second world war. The Russians and her allies were considered the enemy. But that was a lie because, after the fall of Communist in Russia and Europe, it was found that the Russian military force was nowhere near as great as had been made out in the public media.

The next card to be played to get the public and the government to infuse greater financial resources into the military machine was to have another enemy and this time, it would be the "***drug cartels***" followed quickly by the "***Middle East Islamic regimes and insurgents***". This too would be a lie as most of the drug problems entering into the States come from direct involvement by intelligence agencies like the CIA who had financed rogue elements to act on their behalf. The money gained from the sale of drugs fuelled by the drug culture's addiction with its associated miseries and dependency would finance other CIA operations and agendas.

This would be followed next by the war on terrorism originating from ***"Nations of Concern"*** which would be spun off by the earlier Middle Eastern turmoil and political upheavals inflicted by US foreign policies and by the insatiable need for oil found in abundance in Arab nations like

Iraq, Saudi Arabia, Kuwait, Afghanistan, etc. If need be, a scenario similar to the Germans blowing up their own buildings like the Reichstag and blaming it on the Communists prior to the outbreak of WWII would also be used to convince the public for both moral and financial support of the Military Industrial complex. This too would be a lie. We all know about the grim day of 9/11 and the outcome from that tragedy leading to further wars in the Middle East and the buildup of more sophisticated weapons.

Finally, the last card to be played, as Dr. Rosin would explain and in which Wernher von Braun would hammer home over and over again to her was never to forget the real reason for all this military buildup and the dread drain on the county's financial resources would be to place weapons of mass destruction into space under the guise of **"*nuking inbound asteroids, large meteors, and comets"*** that approached too close to the Earth or on a direct trajectory with the Earth. This too would be a lie, the **Big Lie** as it will be discovered is that many of the weapon satellites were pointed outward and not down upon enemy nations wishing to strike at the US. When it is discovered, the answer given will be, to safeguard the Earth and humanity from asteroids, etc. but the reality is far more sinister. The US has had a program in place since the mid-60s or early 70s to monitor, track, target and *"shoot down extraterrestrial spacecraft"* using missiles or high-velocity particle beam scalar weapons. This will be the final card of deception in the Big Lie!

According to Rosin, Von Braun often spoke of the existence of **Off Planet Cultures (OPCs) - (Extraterrestrials)** and about how these **OPC's** are going to be identified as being *"enemies,"* when they are not enemies. http://en.wikipedia.org/wiki/Dr._Carol_Rosin, and **Carol Rosin. (2001-05-09),** *National Press Club Conference*, **[Video Recording], Washington DC: Disclosure Project.**

Wernher von Braun was well aware that the earth was being visited by Extraterrestrials since the time of the 1947 **Roswell Saucer Crash** when he was called upon to examine the wreckage and the alien bodies and he may have even been aware of the German recovery of a crashed ET craft back in **1938 in Bavarian** Alps of Germany. His knowledge and involvement in the existence of ET craft and their pilots were paramount in the understanding by the US Military Industrial Complex which has allowed it to leap ahead technologically of other nations.

He knew that a grand deception was being played out by the MIC and it was important to get that information out to the public through a trusted spokesperson that already had the ear of the public media. **Dr. Carol Rosin** was that spokesperson!

If there are a few top-level scientists who had insider knowledge and involvement with the retrieval and study of alien craft and bodies like von Braun, could there be other scientists, also well known to the American public who had insider knowledge of UFOs and ETI? As it turns out, there certainly were and it appears there's a growing list of scientists willing to throw their hats into the arena of UFO investigation.

Robert Oppenheimer

Julius Robert Oppenheimer (April 22, 1904 – February 18, 1967) was an American

202

theoretical physicist and professor of physics at the University of California, Berkeley. He is often called the "father of the atomic bomb" for his role in the **Manhattan Project,** the World War II project that developed the first nuclear weapons. The first atomic bomb was detonated on July 16, 1945, in the **Trinity test** in New Mexico; Oppenheimer remarked later that it brought to mind words from the **Bhagavad Gita:** *"Now, I am become Death, the destroyer of worlds."*

The public story is fairly well known --- leftist leanings in Berkely in the 1930s, marriage to a former member of the American Communist party, failure to report a contact with someone seeking to gain access for a third party to scientific information to be sent to Russian scientists, his opposition to the hydrogen bomb. These and these alone were the stated reasons for Oppenheimer to lose his **Q-clearance** for top secret work. Yet, these reasons were known to the military and the government before Oppenheimer was chosen to head the atomic bomb project. http://www.amazon.com/review/R1C5F8WI3FLI5M

Robert Oppenheimer, <u>c.</u> 1944
https://en.wikipedia.org/wiki/J._Robert_Oppenheimer

After the war, he became a chief adviser to the newly created United States Atomic Energy Commission and used that position to lobby for international control of nuclear power to avert nuclear proliferation and an arms race with the Soviet Union. After provoking the ire of many politicians with his outspoken opinions during the Second Red Scare, he had his security clearance revoked in a much-publicized hearing in 1954. Though stripped of his direct political influence he continued to lecture, write and work in physics. A decade later **President John F. Kennedy** awarded (and **Lyndon B. Johnson** presented) him with the Enrico Fermi Award as a gesture of political rehabilitation.

Oppenheimer's notable achievements in physics include the Born–Oppenheimer approximation for molecular wavefunctions, work on the theory of electrons and positrons, the Oppenheimer–Phillips process in nuclear fusion and the first prediction of quantum tunneling. With his students, he also made important contributions to the modern theory of neutron stars and black holes, as well as to quantum mechanics, quantum field theory, and the interactions of cosmic rays. As a teacher and promoter of science, he is remembered as a founding father of the American school of theoretical physics that gained world prominence in the 1930s. After World War II, he became director of the Institute for Advanced Study in Princeton.
http://en.wikipedia.org/wiki/J._Robert_Oppenheimer

But there is another story, one told in Dr. Burleson's book, ***Oppenheimer and UFO Crash Retrievals: A Possibility*** (as told by Robert B. Lelieuvre in a review of his book) offers reasons, some of which are on the record, while others are more speculative but hardly far-fetched. Accept as evident Oppenheimer's arrogance and directness, and its role in making enemies of **Edward Teller, Lewis Strauss, Edward Condon** as widely known elements; at least two of these men would come back to repay him later.

What is not widely known, and open to some debate, is Oppenheimer's involvement in at least two UFO crash retrievals --- Roswell, NM, in 1947 and Aztec, NM, in 1948. Dr. Burleson focuses on the following scenario. What would President Truman likely do when he received information on one or the other or both crashes? Send military and scientific teams to the sites. Who would make up the second team?

Here is where the story gets more complicated and rests on acceptance of several assumptions and suppositions --- that an off-hand, speculative comment about Oppenheimer being at Roswell is true; that **Majestic-12, in fact,** did exist; that four members of **Majic-12, Drs. Vannevar Bush, Lloyd Berkner, Detlev Bronk, and Jerome Hunsaker,** accompanied Oppenheimer to Aztec; that **Gordon Gray**, a member of **MJ-12** was selected to chair the AEC's Personnel Security Board for the reasons often cited and not other reasons (Gray's papers related to the Oppenheimer hearing, held in the Eisenhower Library, are permanently withheld from private research or public scrutiny); that Strauss had personal reasons for deepening **William Borden**'s antipathy toward Oppenheimer which led to Borden's letter to the FBI; that Teller's animosity toward Oppenheimer influenced **Sidney Sours** to recommend to President Eisenhower not to reappoint Oppenheimer to the AEC's General Advisory Committee.

Even with the above qualifying assumptions and suppositions, Dr. Burleson provides a cogent and plausible summary of the events between 1947 and 1954. He presents, in detail, information about the Smith-Sarbacher-Steinman correspondence and discussions that resulted in Steinman's book, "UFO Crash at Aztec," the first to offer the tantalizing possibility of Oppenheimer's involvement in two UFO crash retrievals. He also makes a strong case for the proceedings

204

against Oppenheimer being at best trumpery, at worst little more than a kangaroo court. With the board's two to one vote and the AEC's four to one vote, "J. Robert Oppenheimer was out of the loop. His government service was a thing of the past" (p. 75).

Oppenheimer said very little in public after the decision. Once he said that the hearing was a "train wreck" (p. 83), and it has been stated that he told a reporter that "there was a story behind the story" (p. 83). And, it is this story behind the story that Dr. Burleson argues is the real reason for the fall from grace. http://www.amazon.com/review/R1C5F8WI3FLI5M

Oppenheimer and UFO Crash Retrievals: A Possibility, November 3, 2008, By Robert B. Lelieuvre (Book Review)

Oppenheimer would later regret some of his decisions in the development of the atomic bomb and an advocate for de-escalation in the nuclear arms race with Russia, Teller on the other hand was the chief proponent of the nuclear arms race. Oppenheimer's conciliatory nature and demeanour were tempered by spiritual insight particularly from eastern philosophies and the Hindu religion (he was able to read the **Bhagavad Gita** in the original Sanskrit language and later he cited it as one of the books that most shaped his philosophy of life). Teller's hatred of **Communism** made him one of America's pre-eminent and vehement fighters of Communism next to **Joseph McCarthy**, the US Senator from Wisconsin who lead an all out witch hunt during the 1950s that became known as **McCarthyism** upon anyone in the American public who may have had any remote leanings toward socialism.

Edward Teller

Edward Teller, the "father of the American H-bomb," died at the age of 95. This person stands together with such glorious names as Kurchatov, Einstein, Sakharov, Ioffe and many others. Edward Teller was a pioneer of the nuclear epoch.

This American scientist also dealt with the UFO sightings. A meeting took place in Los-Alamos on February 16th, 1949, devoted to green balls of fire flying above top secret nuclear stations in New Mexico. Edward Teller was the most famous scientist at the meeting. The discussion started with meteorites, their descriptions, and peculiarities, and so on. The meeting lasted for several hours. Probably, the most interesting part of the discussion was connected with the following facts: the number of meteorite observations in December, January, and February was average in the USA on the whole, but there was no message to report green balls of fire beyond Los Alamos, Las Vegas, and Western Texas. In addition, the balls were noiseless, which confused Dr. Teller a lot. Edward Teller suggested the balls could be nonmaterial optical objects. Several members of the meeting believed, the balls had to be categorized as an inexplicable natural phenomenon, although Teller and others thought, the balls were a part of a new process that could be possibly connected with plasma or electricity in the atmosphere. In addition, Teller said the green balls phenomenon was close to ionized fluorescence.

Edward Teller could access the data that could help him find out if the balls of fire were the result of secret military experiments. The scientist assured everyone, they were not of the American technology. When the meeting was over, all questions remained unanswered. Despite the negative result, the scientist continued consulting the US government on the mysterious issue. http://www.cosmicparadigm.com/ufonews/edward-h-bomb-teller-consulted-re-ufos/

Dr. Teller led a one-man crusade within the White House pushing for SDI. Dr. Edward Teller was also the scientist, who by the late 1980's, was being named by many researchers as a key figure in the world of UFOs. His connections to UFO stories goes back a long way.

Edward Teller, Oppenheimer's former colleague
https://commons.wikimedia.org/wiki/File:Edward_Teller_(1958)-LLNL-cropped.png

Dr. Teller's first encounter came in the early days of the UFO mystery during the Truman Administration. On February 16, 1948, Dr. Edward Teller, along with **Dr. Lincoln La Paz**, a University of New Mexico astronomer, was part of a secret 1948 **Conference on Aerial Phenomena** that was held at Los Alamos to discuss the UFO phenomena. The particular interest of the conference was the so-called 'green fireballs' which were then being widely reported in the area. This green fireball investigation was also known as **"Project Twinkle".** Dr. Teller had commented during the conference that he felt the phenomenon was an electro-optic phenomenon rather than material phenomena due to the lack of noise.

In 1958 Teller expressed interest about possible life on Mars. In testimony before the Senate Preparedness Subcommittee on November 25, 1958, he stated that even though the Moon and Mars were inhospitable places, Teller felt there would be a search for "any kinds of traces of life."

The most dramatic tie-in to the world of UFOs for Teller came in the mid to late 80's when a story began to surface that the United States government was *"test flying"* and *"back engineering"* flying saucers at an area in Nevada known as **Area-51**. The main person to advance the theory that Area-51 housed flying saucers was **Robert Lazar**, who claimed to be a physicist from Las Vegas.

Lazar claimed to have worked at a spot within **Area 51** known as **S-4**. There he claimed he had worked on captured flying saucers and had seen one of the nine objects there during a test flight outside the underground hanger.

Many of the stories Lazar told could not be confirmed, and many items about Lazar's background seemed to be shaky at best. One item, however, seemed to check out. This item was June 28, 1982, meeting between Robert Lazar and Dr. Edward Teller, the same person who had recommended Keyworth as Reagan's top science man.

On June 28th Dr. Teller had been in Los Alamos, where Robert Lazar worked. Teller was there to give a speech. In an interview with **George Knapp** from a Las Vegas television station Lazar explained what happened:

"I had built a jet car, and they put it in the local newspaper on the front page. As I walked up to the lecture hall, I noticed Teller was outside sitting on a brick wall reading the front page.

I said, "Hi, I'm the one you're reading about there." He said, "That's interesting." I sat down and had a little talk with him."

Then in 1988, when searching for a job, Bob Lazar stated that he sent a copy of his resume to Dr. Edward Teller. Dr. Teller, just as he had recommended Dr. Keyworth for Reagan's science advisor, appeared to have recommended Lazar for a job inside **Area-51**. On November 29, 1988, Teller phoned and gave Lazar a name of someone at **EG&G**, a company believed to be involved in the flying saucer work. Lazar went to an interview, totally unaware of what the job would entail. Soon he was working at **S-4**.

When the *"Lazar story"* about working on flying saucers at **Area-51** broke, Dr. Teller was confronted by a TV reporter asking if he had gotten the job for Lazar and if he knew what was going on at Area-51. Dr. Teller responded to the reporter,

"Look, I don't know Bob Lazar. All this sounds fine. I probably met him. I might have said to somebody I met him and I liked him after I met him, and if I liked him. But I don't remember him . . . I mean you are trying to force questions on me that I simply won't answer." **Excerpts from The Presidents UFO Web Site - Star Wars, UFOs, and Dr. Edward Teller http://www.bibliotecapleyades.net/exopolitica/esp_exopolitics_F_h.htm**

Albert Einstein

It appears that **Albert Einstein** had very little to do with the investigation into the UFO phenomenon whether from a lack of personal interest or from any recruitment by the Military Industrial Complex to analyze retrieved crash saucer debris or alien bodies. His specialty if it was necessitated by the military would be in the area of theorizing the Extraterrestrial Intelligence's ability to traverse the vast interstellar distances with spacecraft that would have had to travel faster than the speed of light.

Einstein's **Special Theory of Relativity** which describes the motion of particles moving at close to the speed of light and his theory of a **Unified Field Theory (UFT)** which incorporated the theory of everything may have been invaluable to the US Military in understanding UFOs and ETI. Certainly, Einstein knew and worked with other scientists such as Von Braun, Oberth, Oppenheimer, Teller and many, many others, especially since their work on rocketry and the atomic bomb overlapped and contributed to their successful development.

Albert Einstein was the epitome of the very best in scientific brilliance yet, years later his appearance portrayed the image of the quintessential "Mad Scientist"

Since these scientists have all had involvement with UFO research of one kind or another, it would not be a long stretch of the imagination to consider for a moment that Einstein may also had his share of secret UFO involvement, even though there is no documented proof uncovered as yet, to support this conclusion.

CHAPTER 42

THE FORMATIVE AGE OF ETI CONTACT AND
THE MODERN AGE OF UFO SIGHTINGS

We have established early in this book that there is overwhelming evidence that indicates we have been visited by Extraterrestrial beings throughout ancient times up to the present age. It would appear that ETI have had an interest in mankind for millennia as they pop up throughout history as if monitoring the social evolutionary development of man on this planet. Perhaps, as the Old Testament claims, they are the Watchers!

Mankind's first extraterrestrial contact and their influence upon our civilizations many thousands of years ago may be regarded as the **formative age of ETI contact**. Their presence amongst men caused them to be venerated as gods, their intervention in the affairs of men awakened the development of humanity's first societies and cultures, seemingly guiding us on a journey toward an unknown destination which currently, we can only guess or speculate as to its nature. Then, as if some agenda or mission had been completed, they mysteriously vanished from the Earth in the same mysterious way in which they had appeared, leaving mankind to carry on by itself. Yet, it appears that they had not completely left in totality as some lingering outpost of their kind remained perhaps, as guardians to watch over us through the ages. Their surveillance of the Earth became routine perhaps, even boring and somewhat tedious, their appearance becoming more sporadic with merely the occasional interposition into the affairs of men, particularly during our times of war or exploration. We had forgotten who they were. A form of collective amnesia seem to settle over the body of mankind as the millenniums past, the memory of them ever fading into the distance becoming a part of a lost legend that was barely kept alive with shamanic traditions, religious scriptures, and practices. Yet, something of them did remain that reminded us of another intelligent presence beside ourselves, something that iconically was representative of them as a people from beyond the clouds, as star visitors... it was their ability of aerial flight in strange disc-shaped craft!

The **modern age of Ufology** should rightfully be considered as having begun in the mid-1800s and not the late 1940s. This timeline runs counter to the accepted historical ufological view which claims that the modern UFO era began with the account made by **Kenneth Arnold** back in July of 1947 with his sighting of nine saucer-like craft seen over **Mount Rainer** in Washington. This is an incorrect perception and understanding that was generated by the news media of the time and which has quickly gained popular acceptance in the national public consciousness and among ufologists. It also dismisses the strange sightings of unusual flying objects seen in other countries around the same time period.

Historically, the turning point by which the modern age of ufology began was during the height of the **industrial revolution** where it became the **technological age or modern age** beginning in the 1840s and '50s first in Britain, then in Europe and America and eventually throughout the rest of the world. Globally, the nineteenth century witnessed many military wars and conflicts that grew out of the ideology of imperialism fuelled by false notions of political, racial and military superiority and by the dogma of outdated religious beliefs which sadly to this day, still seem to be operating and inflicting a constant global unrest!

It is within the realm of possibility that ETI visiting our planet would have been monitoring humanity for much of our early development; it stands to reason that this monitoring would continue and even increase as we entered a technological renaissance during the mid-1840s! This technological development was taking place globally but, nowhere was it accelerating at such a rapid pace as in the USA and Europe.

This was an interesting time period, as many people in the government patent offices across America were about to close their doors for good feeling strongly that everything that been invented up to that time had already been invented and could only be slightly improved upon. The government saw no need to keep the offices open for the very few new inventions that were trickling into the patent offices. Then almost as if a dam had burst, the patent offices were literally inundated with new inventions from the rich fertile minds of their fellow countrymen! http://www.gilderlehrman.org/history-by-era/jackson-lincoln/essays/technology-1800s

The recorded technological and socio-economic growth of humanity has had a slow steady growth for nearly 10,000 years with only a slight rise in development occurring every thousand years or so. One such event happened around 622 AD with the appearance of **Mohammed** and the birth of **Islam** in fulfillment of Christian prophecy, sparking an acceleration in social and cultural development, politics, architecture, medicine, mathematics, language, and astronomy, etc. resulting in the **"creation of nationhood" (nation building)** and all the while Europe remained in the **Dark Ages** of ignorance, superstition, prejudice, religious suppression and intolerance controlled by feudal law and the Holy Church of Rome...the Vatican.

The next evolutionary rise in civilization came in 1844 with the appearance of **Baha'u'llah** and the birth of the **Baha'i Faith** in prophetic fulfillment to all religions including Judaism, Christianity, and Islam; a growth that witnessed a quantum leap forward socially, spiritually, geopolitically and technologically. So staggering, is the sociological and technological growth that globally, its impact is on an exponential scale, the likes of which has never been seen before and has not stopped since that time! Beginning in the 1850s, the advancement in technology has been doubling every hundred, then every ten years and now, it is doubling every two months or less, with no end in sight.

Such events in human history are usually so profound that the impetus of these sociological changes is often accompanied by a spiritual re-birth. In fact, down through recorded time, the promise and fulfillment of religious prophecy occur with the appearance of a manifestation of God. It is these manifestations of God who become the pivotal points in human history, who impart a renewed spirit into the world and it is they who advance civilization forward. The mid 19th Century is no different from any other age of social, spiritual and technological enlightenment with the unique exception that this is a **Universal Cycle** whose impact will be seen as being stronger and its influence more enduring, across the unborn reaches of time for thousands of centuries of human civilization upon this planet! It is destined to last for 500,000 years!!! **God Passes By; by Shoghi Effendi; 1944; published by the Baha'i Publishing Trust, Wilmette, Illinois, USA; ISBN0-87743-92—020-9**

Technological advancements in science have always had its counterpart in the spiritual world thus, maintaining the balance in the equation of life. It is important to understand the concept that science and religion are the **twin pillars of knowledge** that flow from one divine and infinite source. Science is knowledge made through discovery by the mind of man and religions is knowledge offered to man through inspiration or revelation by a manifestation of God. Most people, however, seems to be very ignorant of this fact, that science and religion are one and the same thing which is a reality that is pervasive throughout the universe and not just here, on Earth.

If there were an acceptance by scientists and the ecclesiastics of religion to this basic truth in a spirit of true harmony and unity, to perceive with new eyes, the world nay, the very universe itself would open up in a way never before seen to reveal all its hidden mysteries. This world would be transformed to a higher order of civilization almost within the blink of an eye! Such is the power and potential of this new age!

To any visiting **Extraterrestrial Intelligence** to Earth that kept an ever watchful eye on humanity's development, this time period would have been an incredible point in history as mankind would appear to be on the verge of a global civilization marked by technological achievements but also, increasing cross-border conflicts and wars. We needed closer monitoring and surveillance to see what direction we were headed in our social development.

The 19th Century witnessed many wars and conflicts appearing like multiple lightning strikes and ensuing wildfires that issued from the global storm of political and social unrest that enveloped many nations of the world. Even, the **American Civil War** caught the attention of many people around the world not only for its politically correct and moral position to free the black people from slavery and injustice but it was also, symbolic of freeing all mankind from the yoke of prejudice, oppression, and slavery on many moral and social levels.

It was not only in America that this social upheaval and change was occurring but, also, in the hotspots of Europe, the Middle East, and Far East. It was as if the world had gone mad, it had become spiritually bankrupt and moribund and in dire need of a divine physician to cure the world of its spiritual ills. To such an extent was this imperialistic attitude rampant in the world that many religious groups felt that this age was the **Day of Judgment**, the **End of Days**, and the time of **Armageddon** between the **forces of light and darkness**.

This time period was also, an age of religious fulfillment. People of the nineteenth century from many different cultures and religious beliefs had anticipated the return of their holy prophets. **Christians** anticipated the second coming of **Christ, Jews** expected the return of the **Messiah, Buddhist** awaited the second coming of **Buddha, Hindus** looked for the fifth return of **Krishna** and **Muslims** expected the coming of the **Twelfth Imam**. It was a time of great expectation and many people and religious groups actively searched for the **Promised One** not only in their own countries but, many searched the holy lands of the Middle East.

For the followers of the **Baha'i Faith**, the promise of all ages was fulfilled by the appearance of **Baha'u'llah,** the **Manifestation of God** for this day and age! To **Baha'is**, the locus of all knowledge for this day and age emanates from the personage of **Baha'u'llah** in fulfillment of prophecy from all the world's great religions.

This fact is important in understanding of why science has also made such a giant leap forward in the mid-1800s in comparison with other periods of history. This time period is a pivotal point in mankind's evolvement on the planet. The spiritual energy emanating from the teachings of **Baha'u'llah** will have a lasting influence upon mankind culminating in a global civilization that is destined to last for 500,000 years!

God Passes By, Shoghi Effendi, © 1944 by the National Spiritual Assembly of the United States of America, Sixth Printing 1970, SBN: 0-87743-020-9

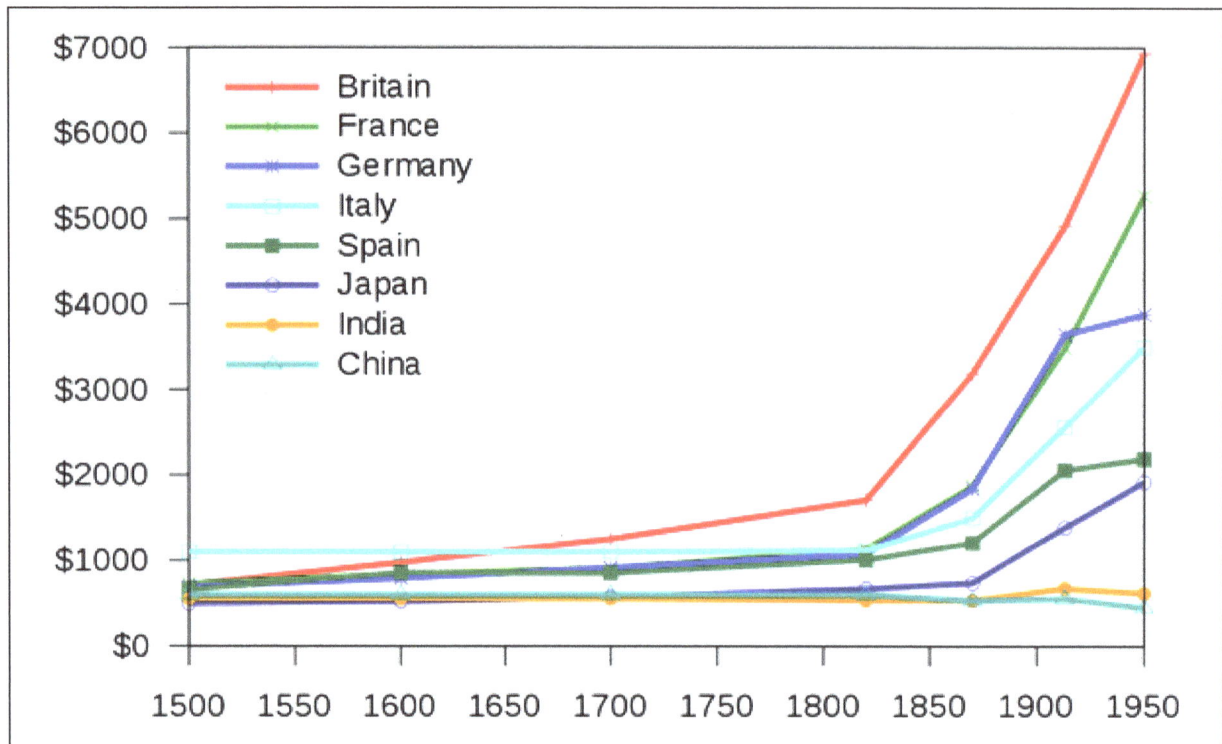

The effect of industrialization shown by rising income levels since the 1500s. The graph indicates the gross domestic product per capita between 1500 and 1950 reflected in 1990 international dollars for selected nations. Note the industrial growth for many countries turns sharply upward around 1825 to 1850!

http://en.wikipedia.org/wiki/Industrialization

Mystery Airships of 1896-'97

Small wonder then, in relationship to our subject matter that the spirit of this new age that was coursing throughout the planet and into the global consciousness of humanity had found expression in a technology that had been only dreamed of in past ages, that of manned flight.

Given the premise that the Earth has been under constant surveillance by ETI from times immemorial and with the unprecedented increase in wars, revolutions, economic meltdowns, the changes in the socio-political landscape and national boundaries, plus the technological developments and spiritual resurgence during the 19th Century, these factors may have been more than enough to triggered an extraterrestrial response to find out what the hell was going on

212

down on planet Earth. Perhaps, humanity was progressing faster than had been anticipated or in a direction not expected.

The response by ETs would be to fly over key industrialized cities, political hot spots and war zones to determine the current path humanity was moving along. The most advanced nations of the time that would have been monitored were America, Canada and Britain, most of Europe, Russia, China, Australia and New Zealand. These were exactly the places that were experiencing UFO or mysterious airship sightings during 1896-97, the US, in particular, became a lightning rod for a wave of mysterious airships that travelled across the country.

There were many famous sightings of dirigible or blimp-shaped UFOs that buzzed the southwestern states in small communities like and which travelled up the west coast to Washington and into Victoria, British Columbia. Eventually, reports were made of sightings seen over the mid-western states and on the eastern seaboard of America even, Europe had its share of unusual flying objects. Reports filled newspapers across the country and the offices of the local constabulary. Sightings by ranchers, town folk and the public, in general, reached a crescendo of activity by1897 ranging from curiosity and amazement to near public hysteria. The public was demanding a response from the government and military to get to the bottom of the mystery with some concrete answers.

Hundreds of newspapers and journals of the time reported that people around the globe genuinely believed that what they were seeing in the skies was not from this Earth but, possibly from Mars. Others speculated that there were a few inventors designing and constructing huge dirigible airships for their personal private use or for the military as a new weapon. **Thomas Edison**, the most well-known inventor of the time was even considered to be the man behind the construction of these mystery aircraft, although this has been proven not to be the case.

Could it have been a new military airborne weapon that was being tested with the purpose of a transcontinental flight to establish its airworthiness for distance and maneuverability as well as payload characteristics? This possibility certainly made sense at the time given the fact that manned flight technology may actually be a couple of thousand years old.

A slight detour is required to review a brief history of manned flight technology and the consequences of its use in order to illustrate the military mindset and interest in the power of flight. https://www.youtube.com/watch?v=RStc-wncVQ4

The Birth of Manned Airflight

The hot air balloon **Kongming lantern** was developed for military communications around the 2nd or 3th Century AD in China [**There have been a rash of UFO sightings in China and the USA during the writing of this book which have been attributed to this type of hot air balloon, (which is easily identifiable) whether these sightings are genuine UFOs or Chinese hot air paper lanterns remains to be determined.]** . It is thought that some ancient civilizations may have developed manned hot air balloon flight. For example, the **Nazca lines** (which are best seen from the air) allegedly presuppose some form of manned flight, such as a balloon

A modern Kongming Lantern

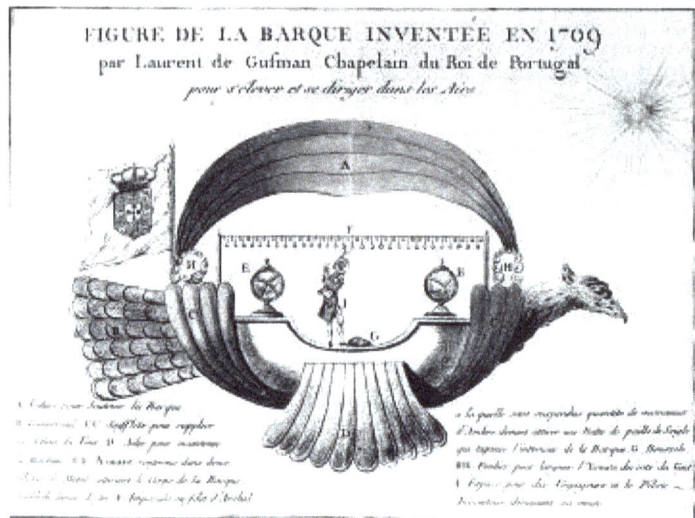

**Bartolomeu de Gusmão's airship *Passarola*
flew in 1709 in Lisbon.**

A 1500 year period of technological atrophy in manned flight befell the thinking of human endeavour until the 1700s when in1709 in Lisbon, **Bartolomeu de Gusmão** made a balloon filled with heated air rise inside a room. He also made a balloon named ***Passarola*** (English: ***Big bird***) and attempted to lift himself from Saint George Castle in Lisbon, but only managed to harmlessly fall about one kilometre away. This claim is not generally recognized by aviation historians outside the Portuguese speaking community, in particular, the FAI.
http://en.wikipedia.org/wiki/Balloon_(aircraft)

Following **Henry Cavendish**'s 1766 work on hydrogen, **Joseph Black** proposed that a balloon filled with hydrogen would be able to rise in the air.

The first recorded manned flight was made in a hot air balloon built by the **Montgolfier brothers** on November 21, 1783. The flight started in Paris and reached a height of 500 feet or so. The pilots, **Jean-François Pilâtre de Rozier** and **François Laurent d'Arlandes**, covered about 5½ miles in 25 minutes.

The history of the ***first recorded manned air flight*** began in earnest with the Montgolfier brothers, **Joseph-Michel Montgolfier** and **Jacques-Étienne Montgolfier** of France. They were the inventors of the ***montgolfière***-style hot air balloon, ***globe aérostatique***. They flew this unmanned craft on 4 June 1783, as their first public demonstration at Annonay in front of a group of dignitaries.

On October 15, 1783, **Étienne Montgolfier** became the first human to lift off the earth, making at least one tethered flight in the Faubourg Saint-Antoine. A little while later on that same day, Pilâtre de Rozier became the second to ascend into the air, to an altitude of 80 feet (24 m), which was the length of the tether. On 21 November 1783, the first free flight by humans was made by **Pilâtre**, together with an army officer, the **marquis d'Arlandes.**
http://en.wikipedia.org/wiki/Montgolfier_brothers

**First public demonstration in Annonay, June 4, 1783. A 1786 depiction
(left) of the Montgolfier brothers' historic balloon**

https://en.wikipedia.org/wiki/Montgolfier_brothers

It seems remarkable yet, in true to form human tradition with anything new that may benefit mankind, the Montgolfier hot air balloon was conceived and built initially to have military applications for siege assault warfare. The fact that it was even thought to have any potential public use such aerial journeys over vast distances was of secondary consideration.

Only a few days later, on December 1, 1783, **Professor Jacques Charles** and **Nicholas Louis Robert** made the **first gas balloon flight**, also from Paris. Their hydrogen-filled balloon flew to almost 2,000 feet (600 m), stayed aloft for over 2 hours and covered a distance of 27 miles (43 km), landing in the small town of Nesles-la-Vallée.

The **first aircraft disaster** occurred in May 1785 when the town of Tullamore, County Offaly, Ireland was seriously damaged when the crash of a balloon resulted in a fire that burned down about 100 houses, making the town home to the world's first aviation disaster. To this day, the town shield depicts a phoenix rising from the ashes.
http://en.wikipedia.org/wiki/Balloon_(aircraft)

Blanchard made the **first manned flight of a balloon in America** on January 9, 1793. His hydrogen filled balloon took off from a prison yard in Philadelphia, Pennsylvania. The flight reached 5,800 feet (1,770 m) and landed in Gloucester County, New Jersey. President George Washington was among the guests observing the takeoff.

215

By 1850 many nations were experimenting with hot air and hydrogen gas balloon flight, even Persia (now Iran) was testing their new "manmade wings". It seems that mankind could no longer keep his feet on the ground; he was destined to leave the Earth and ascend into the heavens and beyond!

Balloon landing in Mashgh Square, Iran (Persia) (left) in 1850 at the time of Nasser al-Din Shah Qajar. (Right) The Civil War balloon Intrepid used from 1860 – 1865
https://en.wikipedia.org/wiki/History_of_ballooning and
http://www.civilwar.org/education/history/civil-war-ballooning/ballooning-during-the-seven.html

The American Civil War was a war of many firsts in the use of advanced military technology in battle. Hot air and hydrogen gas-filled balloons such as the *Union, Intrepid, Constitution, United States, Washington, Eagle, and Excelsior* were used by both the Union and Confederate armies for reconnaissance. So successful were their use that the military ordered balloon corps developed employing aerial telegraphy for surveillance missions of the enemy's position. Flat coal barges became the world's first aircraft carriers complete with a coning observation tower for launching reconnaissance balloons.

The **first steerable balloon** (also known as a **dirigible)** was flown by **Henri Giffard** in 1852. Powered by a steam engine, it was too slow to be effective. As it did with heavier-than-air flight, the internal combustion engine made dirigibles – especially blimps – practical, starting in the late 19th century. http://en.wikipedia.org/wiki/Balloon_(aircraft)

216

A model of the Giffard Airship at the London Science Museum
https://en.wikipedia.org/wiki/Giffard_dirigible

The first ascent of LZ1 over Lake Constance (the *Bodensee*) in 1900
https://commons.wikimedia.org/wiki/File:First_Zeppelin_ascent.jpg

In Europe, and particularly in Prussia, the achievements of certain weapons technology had also caught the attention of the Prussian army who sent **Count Ferdinand von Zeppelin** to America to learn what he could from this kind of warfare. Count Von Zepplin returned from war-torn America inspired and would later develop the world famous line of Zepplin dirigibles that would carry passengers to America and throughout Europe.
http://www.centennialofflight.gov/essay/Lighter_than_air/Civil_War_balloons/LTA5.htm

Recent investigations of these late nineteenth century accounts have turned up a number of reports that were hoaxed events for publicity sake or as misidentifications of hot air balloons and dirigibles. Were the reports of the late 1890s nothing more than the misidentifications of new inventions of terrestrial aircraft that suddenly began appearing over the skies of the US and around the world? Certainly, the American military would have been developing and refining their hydrogen balloon technology during and after the Civil War working toward heavier-than-air flying machines.

Could these military built dirigibles secretly being tested over the countryside be what everyone was seeing in the skies over America resulting in the many eyewitness accounts of the **mysterious airships of 1896 and 1897**? It is highly plausible that this is the main explanation of all those late nineteenth century reports.

Perhaps, other individuals within the USA were also inspired by the use of balloons during the Civil War that they too set about developing and constructing larger balloons for carrying many people across the country. The cigar-shaped balloon or dirigible design appeared to be the best aerodynamic shape for an aerial craft and the accounts of these airships seem to reflect that basic aircraft design.

"Was there really a mysterious inventor who secretly built an airship and flew it around the country? Certainly, the public had been primed to accept such a story. Science fiction in this era often used the "mystery inventor" as a character. **Jules Verne's** *20,000 Leagues Under the Sea* featured a mystery inventor, Captain Nemo, who constructed a submarine. Verne's later book, *Robur the Conqueror*, featured a mystery inventor that built an airship and the similarities between the book and some of the airship stories are uncanny. Robur was published about ten years before the wave of airship sightings. http://www.unmuseum.org/airship.htm

Throughout the history of man's attempt at flight, the military mindset of any nation's army has viewed this particular invention as a potential weapon. It should not come as any surprise that those inventors of such balloon-type aircraft would find that the military had an interest in its weaponization. A nation's military strength could be ensured or advanced by either the seizure of the technology or by enlistment of the inventor to further the research and development of the aircraft.

Recall, that the 1850s witnessed a global phenomena unique in the annals of history, it was the beginning of a huge military arms buildup by many nations which has lasted into the 21st Century. Never has the world been witnessed to such epochal event in military growth and

218

weapons development which would become an impediment to the realization of a world civilization.

Is this then, "case closed" on this particular chapter in UFO history and is it score one more point for the side of the skeptics and debunkers? It would appear at first glance to be the case, however, keep in mind that UFO sightings and reports of alien beings have been recorded from prehistoric times as cave wall paintings to more recent history in the art and literature of many ancient cultures.

Was it possible that the aerial crafts sighted throughout America and Europe in the late 1890s, the invention of manmade design or the alien visitations of spacecraft from Mars or elsewhere in the universe? The question has its supporters on either side as it would appear that no matter what the time frame in mankind's history, Ufologists and skeptics would debate the pros and cons of the phenomenon. At this particular time, man's first serious attempts at flight would also be surrounded by controversy.

Ufologist have argued that mankind was incapable of flight, let alone of developing sophisticated and controllable powered aircraft that were mounted with powerful searchlights, a technology that was still nearly a decade away from becoming reality. These dirigible craft were constructed of metal according to most reported cases and the sightings were global in nature; not just restricted to America.

It was strongly thought that one of these aerial craft crashed in Aurora, Texas in 1897 and that its extraterrestrial occupant was buried in the community's cemetery. The debris of the spacecraft was disposed of down an empty well or possibly buried as well. Many Ufologists and the curious have tried to exhume the gravesite in order to determine if an alien body was buried in Aurora's cemetery. Even, a search of the old water well for the crash debris was attempted. In both cases, the search produced no evidence for their existence.

Skeptics, however, argued that there were aerial flights by a few ingenious men who initially flew in hot air balloons as far back as 1783 with the first manned flight by the Montgolfier brothers of France.

Some argued that the airship reports were genuine accounts. Steerable airships had been publicly flown in the US since the *Aereon* in 1863, and numerous inventors were working on airship and aircraft designs (the idea that a secretive inventor might have developed a viable craft with advanced capabilities was the focus of **Jules Verne's** 1886 novel *Robur the Conqueror*). In fact, two French army officers and engineers, **Arthur Krebs** and **Charles Renard**, had successfully flown in an ***electric-powered airship*** called the *La France* as early as 1885, making no fewer than seven successful flights in the craft over an eleven month period. Also, during the period of 1896-1897, Bosnian inventor **David Schwarz** built an aluminum-skinned airship in Germany that successfully flew over Templehof before being irreparably damaged during a hard landing. Both events clearly demonstrated that the technology to build a practical airship existed during the period in question, though if reports of the capabilities of the California and Midwest airship sighted in 1896-97 are true, it would have been considerably more advanced than any airship built up to that time.

Several individuals, including Lyman Gilmore and Charles Dellschau, were later identified as possible candidates for being involved in the design and construction of the airships, although little evidence was found in support of these ideas.
http://en.wikipedia.org/wiki/Mystery_airship

It would indeed have been strange if there had been no parallel activities in the U.S.A. at that time. Was there a possibility of a machine being the production of some far-sighted inventor with the ability, wealth, and resources to build and fly and also keep the whole project secret?
Such a person with a genius engineering ability and wealth was **Edward J. Pennington**, who seemed to be up to the challenge. A characteristic of Pennington was the secrecy he achieved to protect his projects and his habit of quietly dropping one idea in favor of another with little regard to the financial outcome.

One of his new companies produced **'Freight Elevators'** which may have been a euphemism for load-carrying Airships actually built a four cylinder radial engine..."for the propulsion of an aerial vessel". He also let it be known, that he was "readying a vessel to fly from Mount Carmel to New York".

In 1891 he exhibited a captive airship some thirty feet long and six feet in diameter. It flew in a circle propelled by an airscrew turned electrically. The current was conveyed by wires in the tethering cable.

By the mid to late eighteen hundreds, there were reports of airships that were gas filled with hydrogen and powered either electrically or by simple combustion piston engines.
The source for this file is the June 1970 issue of Flying Saucers magazine, published and edited by Ray Palmer. Reprinted from BUFORA Journal, British U.F.O. Research Association and http://www.unexplainable.net/artman/publish/article_782.shtml

New evidence suggests that there were actually five human built airships being flown around the Midwest in the period of 1896-1897. Amongst the evidence cited are patents, photographs, old newspaper accounts, and even census records used to find the inventors. The researcher who found this evidence, an author of a book called "Solving the 1897 Airship Mystery", also claims the Aurora, Texas crash was one of these airships. He points out that it could not have been mass hysteria that caused the sightings because of the quantity and quality of witnesses, and that they could not have been of extraterrestrial origin because they were built using 19th Century technologies, an interpretation he based on eyewitness testimony found in many of the newspaper articles of that time.

Witnesses of the airship were often men of excellent reputation for veracity and often crowds of onlookers were able to compare experiences. The descriptions tallied to a remarkable degree. It seems clear also that some of the sightings of night flying objects were of quite a different category and to present-day Ufologists may be recognized as being the result of 'normal' UFO activity.

From the reports still in existence, it is possible to build up a very good idea of the type of

220

dirigible involved and there is no doubt that in many respects it is similar to airships already built and flying in Europe particularly in France.

Von Zepplin had also, begun work on the famous Zepplin line of dirigibles in the 1860s in German. These were constructed of a heavy metal frame covered in a thick fabric and filled with Hydrogen, a highly flammable gas that was capable of lifting the huge craft into the air with easy.

Though the objects were technically UFOs in that they were unidentified and flying, many witnesses reported that the cigar-shaped object was not of extraterrestrial origin at all, but rather the result of multiple advanced manned craft designed to be held aloft by unknown means and kept secret to avoid the stealing of the plans for the vessels. The lights emitted from such an object would have not necessarily served to illuminate the ground for the crew members, but would have more likely served as beacons so that the "chasers" on the ground could track and follow the object. But throughout the entire series of incidents there was never a single explanation as to the real origins of these strange and mysterious craft - leading some to suggest that the objects were not of Earthly origin, but rather disguised alien ships cloaked as anachronistic dirigibles.
http://ufomania.proboards.com/index.cgi?board=history&action=display&thread=715

The official story, however, does not have much in the way of evidence pointing to an alien presence. Instead, hundreds came forward suggesting that the ships would regularly stop to ask about their location and even look for ways to resupply their ships. With the constant reports of beings lowering themselves down from ships held aloft in the sky, little information as to their origin came through. What little information did come about was almost as enigmatic as the ships themselves in its mundanity and the candid manner they interacted with people.

Witnesses reported that the men were friendly but reserved, and wore clothing that seemed typical of gentleman's dress at the time. The figures would not explain where they were from or what the purpose of their journey was, instead simply telling those who asked that they were travelers exploring the countryside in airships. Of course this ambiguity today may seem like precisely the sort of thing an alien race would suggest if it were attempting to gather intelligence on Earthlings. The fact that the men required supplies, however, also points to another often overlooked factor.

A traditional hot air balloon lacks the capacity for much weight to be carried into the air. Such travelers would indeed have to stop quite frequently for supplies and water as a single week's worth of supplies could, in theory, require as much weight as a full grown man if it were to accommodate several crew members. And if the airship were to have a mechanical system of propelling it forward against the wind, as many witnesses would suggest, the crew members would have even less weight to work with.
http://ufomania.proboards.com/index.cgi?board=history&action=display&thread=715

Speculation as to the origin of the 'Airship' reported over the central States of the U.S.A. in 1897 has resulted in many theories and at least one of these attributes the sightings to the activities of a peculiar antique UFO. The reason that the craft looked very much like the current airship design already flying in Europe is that the UFO denizens wished to present their ship to the natives in a manner that would be acceptable and understandable. However, the airship in question did not seem to be at all anxious to present itself, operating as it did almost exclusively by night and skulking during daylight hours in out of the way places.

Jacques Vallee proposes that down through history, the UFO and Extraterrestrial phenomenon has been with humanity seemingly interacting or intervening in the affairs of mankind disguising itself in the images of fairy folk or leprechauns or as angels and demons and in this current age as "gray aliens". They purposefully do this by staying 3 to 4 steps ahead of any human social and technological development, always elusive, sometimes within sight but, never within grasp to be interrogated. Their existence Vallee posits is on the order of interdimensionality rather than physicality and not extraterrestrial in nature. Their craft are capable of morphing into contemporary shapes and modes of transportation thus; many of the airships sighted in America, Europe and the rest of the world in 1896-97 reflect a global phenomenon that is beyond the capability of a coordinated timeline of human initiated activity for the purpose of dumbfounding with awe their fellow humans. Jacque Vallee's theory of ETs adapting to human perceptions of conventionality is certainly plausible in explaining many of the inconsistencies that are offered by the skeptic's explanation.

This explanation, however, was not well received by most Ufologists and skeptics alike, as they considered it as just too far out there and strained credulity beyond the limit of reason bordering on areas best left relegated to the paranormal, the mystical, the mythological and the superstitious. Vallee's thinking if nothing else is definitely radical, his investigation are usually logical based on hardcore data which makes him a true pioneer trending in areas where few brave scientists fear to go.

Passport To Magonia On UFOS, Folklore, and Parallel Worlds; 1969 by Jacques Vallee; published by Contemporary Books Inc.; ISBN: 0-8092-3796-2

Author's Rant: Personally, I believe that the answer to the airship mystery lies somewhere in the middle of rational explanation and deductive reasoning.

If the airship mystery of the late 1890s was ambiguous at best with pundits equally divided on the question of the existence of UFOs and ETI, by the turn of the 20th Century, there was a growing momentum in understanding toward the possibility, at least in the minds of the general public, that we were not alone in the universe. There was real possibility that not only did Extraterrestrials exist but, they were visiting the Earth.

When we consider manned flight, we usually think about hot air or hydrogen balloons and dirigibles as the only type of aircraft being designed and built in the 1850s. The reality, however, is quite different as gliders and fixed winged craft were developed, built and flown far earlier than this time but, serious flights withs heavier than air machines being engineered and flown occurred in the mid 19th Century.

What follows is a list of recorded manned flights from gliders, fixed or variable controlled winged craft and heavier than air engine powered aircraft fuelled by steam or combustion. Also, note that the **Wright Brothers** of America were not the first successful manned flight in a

222

controlled engine-powered aircraft contrary to the hype of the press and the historians in the US. That distinction goes to a Russian. The Wright brothers did nonetheless; contribute significantly toward the advancement of aeronautics with their successful flights in the **Wright Flyers** and the famous **Kittyhawk** aircraft.

There is much debate and controversy as to who and which nation accomplished this first manned powered flight. It matters not in the scheme of things who was first, as it was destined that man should fly into the air beside the birds of the earth and up into the heavens and beyond with angels!

Pre-19th Century Airflight

Archytas, Ancient Greece
According to Aulus Gellius, **Archytas, the Ancient Greek** philosopher, mathematician, astronomer, statesman, and strategist, was reputed to have designed and built the first artificial, self-propelled flying device, a bird-shaped model propelled by a jet of what was probably steam, said to have actually flown some 200 meters. This machine, which its inventor called *The Pigeon* (Greek: Περιστέρα "Peristera"), may have been suspended on a wire or pivot for its flight.

Bartolomeu de Gusmão, Brazil and Portugal, an experimenter with early airship designs
In 1709 Bartolomeu de Gusmão demonstrated a small airship model before the Portuguese court but never succeeded with a full-scale model.

Pilâtre de Rozier
Pilâtre de Rozier made the first trip by a human in a free-flying balloon (the Montgolfière): 9 km covered in 25 minutes, 21 November 1783, near Paris.

Professor Jacques Charles and Les Frères Robert, (Anne-Jean and Nicolas-Louis)
1. *Le Globe*, the first hydrogen gas balloon flew on 26 August 1783.
2. On 1 December 1783, *La Charlière* piloted by Jacques Charles and Nicolas-Louis Robert made the first manned hydrogen balloon flight.
3. On 19 September 1784, *La Caroline*, an elongated craft that followed Jean Baptiste Meusnier's proposals for a dirigible balloon, completed the first flight over 100 km from Paris to Beuvry.

19th Century Airflight

Hans Andreas Navrestad, Norway — 1825
Allegedly flew manned glider.

John Stringfellow, England — 1848
First heavier than air powered flight, accomplished by an unmanned steam-powered monoplane of 10-foot (3.0 m) wingspan. In 1848, he flew a powered monoplane model a few dozen feet at an exhibition at the Crystal Palace in London.

George Cayley, England — 1853
First well-documented Western human glide. Cayley also made the first scientific studies into the aerodynamic forces on a winged flying machine and produced designs incorporating a fuselage, wings, stabilizing tail and control surfaces. He discovered and identified the four aerodynamic forces of flight - weight, lift, drag, and thrust. Modern

airplane design is based on those discoveries including cambered wings. He is sometimes called the "Father of aviation".

Matias Perez, Havana, flight in 1856

Matias Perez was a Portuguese pilot, canopy maker and Cuban resident who, carried away with the ever increasing popularity of aerostatic aircraft, disappeared while attempting an aerostatic flight from Havana's "Plaza de Marte" (currently Parque de la Fraternidad) on June 1856.

Jean-Marie Le Bris, France, flight in 1856

Jean-Marie Le Bris was the first to fly higher than his point of departure, by having his glider pulled by a horse on a beach, against the wind.

Jan Wnek, Poland — controlled flights 1866 - 1869.

Jan Wnek controlled his glider by twisting the wing's trailing edge via strings attached to stirrups at his feet.[9] Church records only—Kraków Museum unwilling to allow verification.

Goodman Household, South Africa, 1871

Goodman built and flew his own glider over one hundred meters. The story is that he crashed breaking both glider and a leg. The event took place in the Kwazulu Natal Midlands near Curry's Post in 1871 and is recorded variously in legend and local literature.

Félix du Temple de la Croix, France, 1874

First take-off of a manned and powered aircraft, from a downsloped ramp, resulting in a brief hop a few feet above the ground.

Victor Tatin, France, 1874

First airplane to lift itself under its own power, the *Aeroplane* was an unmanned plane powered by a compressed-air engine.

John Joseph Montgomery, United States of America 1883

First controlled glider flight in the United States, from a hillside near Otay, California.

Alexander Feodorovich Mozhaiski, Russian Empire — 1884

First powered hop by a manned multi-engine (steam) fixed-wing aircraft, 60–100 feet (20-30 meters), from a downsloped ramp.

Clément Ader, France — October 9, 1890

He reportedly made the first manned, powered, heavier-than-air flight of a significant distance (50 m) but insignificant altitude from level ground in his bat-winged, fully self-propelled fixed wing aircraft with a single tractor propeller, the Ader Éole . Seven years later, the Avion III (a different machine) was said to be flown upon 300 metres (in fact just lifted off the ground, and lost control). The event was not publicized until many years later, as it had been a military secret. The events were poorly documented, the aeroplane not suited to have been controlled; there was no further development. Later in life Ader claimed to have flown the Avion II in 1891 for over 200 meters.

Otto Lilienthal, Germany — 1891

The German "Glider King" was a pioneer of human aviation—the first person to make controlled untethered glides repeatedly and the first to be photographed flying a heavier-than-air machine. He made about 2,000 glides until his death August 10, 1896, from injuries in a glider crash the day before.

Chūhachi Ninomiya, Japan - 1894

Japanese inventor who developed several small powered models including an early tailless aircraft.

Lawrence Hargrave, Australia—November 12, 1894

The Australian inventor of the box kite linked four of his kites together, added a sling seat and flew 16 feet. By demonstrating to a skeptical public that it was possible to build a safe and stable flying machine, Hargrave opened the door to other inventors and pioneers. Hargrave devoted most of his life to constructing a machine that would fly. He believed passionately in open communication within the scientific community and would not patent his inventions. Instead, he scrupulously published the results of his experiments in order that a mutual interchange of ideas may take place with other inventors working in the same field, so as to expedite joint progress.

Hiram Stevens Maxim, United Kingdom — 1894

The American inventor of the machine gun built a very large 3.5 ton flying machine that ran on a track and was propelled by powerful twin naphtha fueled steam engines. He made several tests in the huge biplane that were well recorded and reported. On July 31, 1894, he made a record-breaking speed run at 42 miles per hour (68 km/h). The machine lifted from the 1,800-foot (550 m) track and broke a restraining mechanism, crashing after a short uncontrolled flight just above the ground.

Samuel Pierpont Langley, United States — May 6, 1896

First sustained flight by a heavier-than-air powered, unmanned aircraft: the Number 5 model, driven by a miniature steam engine, flew half a mile in 90 seconds over the Potomac River near Washington, D.C. In November the Number 6 flew more than five thousand feet. Langley's full-size manned powered Aerodrome failed twice in October and December 1903.

Octave Chanute, United States — Summer 1896

Designer of first rectangular wing strut-braced biplane (originally tri-plane) hang glider, a configuration that strongly influenced the Wright brothers. Flown successfully at the Indiana shore of Lake Michigan, U.S. by his proteges, including Augustus Herring, for distances exceeding 100 feet (30 m).

Carl Rickard Nyberg, Sweden — 1897

Managed a few short jumps in his Flugan, a steam powered, manned aircraft

Gustave Whitehead, United States — 1899

Reportedly flew a steam-powered monoplane about half a mile and crashed into a three-story building in Pittsburgh in April or May 1899, according to a witness who gave a statement in 1934, saying he was the passenger

Percy Pilcher, England — 1899

Pioneer British glider/plane builder and pilot; protege of Lilienthal; killed in 1899 when his fourth glider crashed shortly before the intended public test of his powered triplane. Cranfield University built a replica of the triplane in 2003 from drawings in Philip Jarrett's book "Another Icarus". Test pilot Bill Brooks successfully flew it several times, staying airborne up to 1 minute and 25 seconds.

Augustus Moore Herring, United States — 1899

Claimed a flight of 70 feet (21 m) by attaching a compressed air motor to a biplane hang glider. However, he was unable to repeat said flight with anyone present.

20th Century Airflight

Dr. Wilhelm Kress, Austria — 1901
> Tested Drachenflieger, tandem monoplane seaplane similar to Samuel Langley, which made brief airborne hops but could not sustain itself.

Gustave Whitehead, United States — August 14, 1901
> First publicized account of a flight by an aeroplane heavier than air propelled by its own motor — Whitehead No. 21. Reports were published in the *New York Herald*, and the *Bridgeport (CT) Herald*. The event was reportedly witnessed by several people, one of them a reporter for the *Bridgeport Herald*. Children and youngsters who were present signed affidavits about 30 years later about what they saw. Reports said he started on the wheels from a flat surface, flew 800 meters at 15 meter height, and landed softly on the wheels. Other reports said he never flew.

Lyman Gilmore, United States — May 15, 1902
> Gilmore claimed to be the first person to fly a powered aircraft (a steam-powered glider). No witnesses. But he was an able inventor, rotary snow plow, 8-cylinder rotary motor, etc.

Gustave Whitehead, United States — January 17, 1902
> Whitehead claimed two spectacular flights on January 17, 1902, in his improved Number 22, with a 40 Horsepower (30 kilowatt) motor instead of the 20 hp (15 kW) in the Number 21 aircraft and aluminum instead of bamboo. In two published letters that he wrote to *American Inventor* magazine,[12] he said the flights took place over Long Island Sound and covered distances of about two miles (3 kilometers) and seven mi (11 km) at heights up to 200 ft (61 m), ending with safe landings in the water by the boat-like fuselage. Some later affidavits assert he flew, others that he never flew.

Orville & Wilbur Wright, United States — October 1902
> Completed development of the three-axis control system with the incorporation of a movable rudder connected to the wing warping control on their 1902 Glider. They subsequently made several fully controlled heavier than air gliding flights, including one of 622.5 ft (189.7 m) in 26 seconds. The 1902 glider was the basis for their patented control system still used on modern fixed-wing aircraft.

Richard Pearse, New Zealand — March 31, 1903
> Several people reportedly witnessed Pearse make powered flights including one on this date of over 100 feet (30 m) in a high-wing, tricycle undercarriage monoplane powered by a 15 hp (11 kW) air-cooled horizontally opposed engine. Flight ended with a crash into a hedgerow. Although the machine had pendulum stability and a three-axis control system, incorporating ailerons, Pearse's pitch and yaw controls were ineffectual. (In the mockumentary Forgotten Silver, director Peter Jackson recreated this flight, supposedly filmed by New Zealand filmmaker Colin McKenzie. The film was so convincing, Paul Harvey reported it as genuine on his syndicated *News and Comment* program).

Karl Jatho, Germany — August 18, 1903
> On August 18, 1903, he flew with his self-made motored gliding aircraft. He had four witnesses for his flight. The plane was equipped with a single-cylinder 10 horsepower (7.5 kW) Buchet engine driving a two-bladed pusher propeller and made hops of up to 200 ft (60 m), flying up to 10 ft (3 m) high.

Orville & Wilbur Wright, United States — December 17, 1903

226

First recorded controlled, powered, sustained heavier than air flight, in Wright Flyer. In the day's fourth flight, Wilbur Wright flew 279 meters (852 ft) in 59 seconds. First three flights were approximately 120, 175, and 200 ft (61 m), respectively. The Wrights laid particular stress on fully and accurately describing all the requirements for controlled, powered flight and put them into use in an aircraft which took off without the aid of a catapult from a level launching rail, with the aid of a headwind to achieve sufficient airspeed before reaching the end of the rail.

John Joseph Montgomery and Daniel Maloney, United States 1905
First high altitude flights with Maloney as pilot of a Montgomery tandem-wing glider design. The glider was launched by balloon to heights up to 4,000 feet (1,200 m) with Maloney controlling the aircraft through a series of prescribed maneuvers to a predetermined landing location in front of a large public gathering at Santa Clara, California.

Wilbur Wright, United States — October 5, 1905
Wilbur Wright pilots Wright Flyer III in a flight of 24 miles (39 km) in 39 minutes (a world record that stood until Orville Wright broke it in 1908) and returns to land the plane at the takeoff site.

Traian Vuia, Romania — March 18, 1906
Fully self-propelled, fixed-wing aircraft using a carbonic acid gas engine and a single tractor propeller. He flew for 12 meters in Paris without the aid of external takeoff mechanisms, such as a catapult, a point emphasized in newspaper reports in France, the U.S., and the UK. The possibility of such unaided heavier-than-air flight was heavily contested by the French Academy of Sciences, which had declined to assist Vuia with funding

Jacob Ellehammer, Denmark — September 12, 1906
Built monoplane, which he tested with a tether on the Danish Lindholm island.

Alberto Santos-Dumont, Brazil — October 23, 1906
The "14 Bis" at Bagatelle field, Paris. The Aero Club of France certified the distance of 60 meters (197 ft); height was about 2–3 meters (6–10 ft). Winner of the Archdeacon Prize for first official flight of more than 25 meters. Described by some scholars as the first "sportsman of the air". As reported in previous years and months by Ader, Whitehead, Pearse, Jatho and Vuia, the 14-Bis biplane flew and landed without a rail, catapult, or the presence of high winds, propelled by an internal combustion engine.

TABLE OF FLYING MACHINES

Designer/maker	Nationality	Title or specialty	Year	Status/Description
Roger Bacon	British	*Secrets of Art and Nature*	c. 1250	ornithopter design
Leonardo da Vinci	Italian	The Ornithopter	c. 1490	design, literature
Emanuel Swedenborg	Swedish	Flying Machine	1714	design, literature

Designer/Maker	Nationality	Machine name/description	Year	Description
Sir George Cayley	British	On Aerial Navigation	1809-1810	Technical literature. This work laid the ground rules for all later aircraft
Le Comte Ferdinand Charles Honore Phillipe d'Esterno		On The Flight Of Birds (*Du Vol des Oiseaux*)	1864	technical literature
Louis Pierre Mouillard	French	The Empire Of The Air (*L'Empire de L'Air*)	1865	literature
Otto Lilienthal	German	Birdflight as the Basis of Aviation (*Der Vogelflug als Grundlage der Fliegekunst*)	1889	literature
James Means	American	The Problem of Manflight, Aeronautical Annual	1894 - 1897	literature
Octave Chanute	American (born in France)	Progress in Flying Machines	1894	His technical articles collected in a book
Wilbur Wright	American	Some Aeronautical Experiments	1901	Published speech to Western Society of Engineers, Chicago
Martin Wiberg	Swedish	"Luftmaskin"	1903	Received a patent for a design powered by a liquid fuel rocket

MORE THAN DESIGN OR LITERATURE

Designer/Maker	Nationality	Machine name/description	Year	Claimed	Achieved
John Childs	American	"Feathered glider"	1757	Three successful flights in two days	Reports suggest that this was a fairground trick, involving sliding down a tethered rope. He had claimed to have performed the same stunt many times earlier in Europe
William Samuel Henson	British	Aerial Steam Carriage, "modern"-looking monoplane with "cabin", tail	1842		Models only, publicity illustrations

228

		and twin pusher propellers		
John Stringfellow	British	The Stringfellow Machines	1848, 1868	Indoor flights by fixed-wing steam-powered models
Sir George Cayley	British	"Governable Parachute"	1849-1853	Child- and man-carrying glides, both towed and free-flying
Rufus Porter	American	The New York to California Aerial Transport	1849	Uncompleted steam-powered dirigible
Jean Marie Le Bris	French	The Artificial Albatross	1857, 1867	Towed gliding flight
Felix and Louis du Temple de la Croix	French	Du Temple Monoplane, aluminum construction, steam-powered	1857-1877	Powered manned hop from ramp
James William Butler, Edmund Edwards		The Steam-Jet Dart	1865	Patented superposed wing design
Francis Herbert Wenham	British	"Aerial Locomotion" (academic paper)	1866	(biplane, mulitplane); invented wind tunnel
Jan Wnęk	Polish	glider	1866-1869	Controlled flights from local church tower
Frederick Marriott		Marriott flying machines	1869	
Alphonse Pénaud	French	Planophore, Pénaud Toy Helicopter	1871	Rubber-powered fixed-wing and helicopter models
Thomas Moy	British	Moy Aerial Steamer, tandem wings, 120 lb (55 kg), 15 ft (4.6 m) wingspan, 3 horsepower, twin fan-type propellers	1875	Lifted 6 inches (0.15 m) from ground at London Crystal Palace
Enrico	Italian	Demonstration in	1877	rose to 13 meters

229

Forlanini		Milan, Helicopter, unmanned, steam-powered.		(40 feet) for 20 s duration: first heavier than air self-powered machine to fly
Thomas Moy	as above	The Military Kite	1879	
Charles F. Ritchel	American	Ritchel Hand-powered Airship	1878	
Victor Tatin	French	Tatin flying machines	1879	
J. B. Biot	French	The Biot Kite	1880	
Alexandre Goupil	French	Goupi Monoplane, La Locomotion Aerienne	1883	
John Joseph Montgomery	American	Montgomery monoplane, Tandem-wing Gliders	1883-1911	A pre-1900 foot-launched manned glide; balloon-launched after 1900
Aleksandr Fyodorovich Mozhaiski	Russian	Mozhaiski Monoplane, multi-engine, steam	1884	Powered manned hop from ramp
Massia and Biot		Massia-Biot Glider	1887	
Pichancourt		Mechanical Birds	1889	
Lawrence Hargrave	British immigrant to Australia	Hargave flying machines and Box Kites	1889-1893	influential designs
Clément Ader	French	Eole, Avion, bat-wing, steam-driven	1890-1897	Manned, powered hops from level surface
Chuhachi Ninomiya	Japanese	The Tamamushi (model)	1891	
Otto Lilienthal	German	Bat-wing hang gliders, mono- and biplane	1891-1896	2,000 manned glides, dozens photographed
Horatio Frederick Phillips	British	Multiplanes	1893-1907	Multiple-wing test machines; successful flights in 1904 (50 feet) and 1907 (500 feet)
Hiram Stevens	British (born in America)	Maxim Biplane, a behemoth machine:	1894	Broke from restraining rail and

Name	Nationality	Craft	Date	Notes
Maxim		145 ft (44.2 m) long, 3.5 tons, 110 ft (33.5 m) wingspan, two 180 hp steam engines driving two propellers.		made uncontrolled manned flight. Total flying distance, 1,000 ft (305 m) while restrained, 924 ft (282 m) free flight. Total 1,924 ft (586 m)
Pablo Suarez		The Suarez Glider	1895	
Percy Sinclair Pilcher	British	Bat, Beetle, Hawk bat-wing hang gliders	1896-1899	Manned glides; fatal crash before planned public test of powered triplane; modern replica flown
Octave Chanute and Augustus Herring	American (Chanute born in France)	Hang gliders, "modern" biplane wing design	1896	Manned glides
William Paul Butusov, with Chanute group	Russian immigrant to U.S.	Albatross Soaring Machine	1896	unmanned unpowered uncontrolled hop from ramp
Samuel Pierpont Langley	American	Langley Aerodrome, Tandem wings, unmanned, steam-powered.	1896	5,000 ft. (1.7 km), photographed
William Frost	Welsh	Frost Airship Glider	1896	Manned, 500 meters, possibly with balloon assist
Carl Rickard Nyberg	Swedish	Flugan	1897 and on	Hops
Edson Fessenden Gallaudet	American	Gallaudet Wing Warping Kite	1898	
Lyman Wiswell Gilmore, Jr.	American	Gilmore Monoplane, steam-driven	1898	Too little info
Gustave Whitehead	German (Emigrated to U.S.)	Monoplane with pilot and passenger, steam powered	1899	Flew 500 m, crash
Wilhelm	Austrian	Kress Waterborne	1901	Long hops

Kress		Aeroplane			
Gustave Whitehead	as above	Whitehead Albatross, glider	1901		
Gustave Whitehead	as above	No. 21, bat-wing, 20 hp motor, twin tractor propellers	1901	800 m, 4 flights, body shifting control	Modern replica successfully flown
Gustave Whitehead	as above	No. 22, 40 hp motor, twin tractor propellers	1902	Flew 10 km circle; control by variable propeller speed and "rudder"	
Richard William Pearse	New Zealand	Pearse Monoplane	1903	150 m, believed controllable but unstable - numerous witnesses	
Karl Jatho	German	The Jatho Biplane	1903	.	70 m powered hop, unstable
Wright Brothers	American	Wright Flyer, level launch rail, headwind for sufficient airspeed	1903	.	Four flights, longest 852 feet (260 m), 59 s, controlled
Guido Dinelli		Dinelli Glider, Aereoplano	1903	70 m, no motor	
Wilbur Wright	American	Wright Flyer III, catapult launch	1905		24 miles (39 km), circling, max height about 50 feet (15.2 m)
Louis Blériot, Gabriel Voisin	French	Blériot-Voisin floatplane glider, Blériot-Voison biplane	1905		Towed up, 600 m
Alberto Santos-Dumont	Brazilian living in France	14-bis, Hargrave-style box-cell wings, sharp dihedral, pusher propeller, internal combustion. (Demoiselle in 1909, tractor monoplane with wing-warping)	1906		Controlled, rose off flat ground with no external assistance, 200 meters, 21 s, first official European flight
Jacob	Danish	Monoplane,	1906,		Tethered powered

Inventor		Aircraft	Year	Accomplishment
Ellehammer		helicopter	1912	fixed-wing flight
Traian Vuia	Romanian, flight experiments in France	Vuia I, Vuia II monoplanes, Carbonic acid engine on Vuia I, internal combustion engine on Vuia II	1906-1907	Powered manned hops
Glenn H. Curtiss and A.E.A.	American	June Bug, biplane with wingtip ailerons	1908	First official 1 km U.S. flight
Louis Blériot	French	Blériot XI monoplane, tractor propeller	1909	Crossed the English Channel, France to Britain, 23 miles (37 km)
Aerial Experiment Association (A.E.A)	American	Silver Dart	1909	First controlled powered flight in Canada
Edvard Rusjan	Slovenian	EDA 1	1909	
Ivan Sarić	Croatian	Sarić 1	1910	
Duigan Brothers, John and Reginald	Australian	Duigan Pusher Biplane	1910	

HISTORIC RECORDS

Inventor	Accomplishment or Claim	Year
Zhuge Liang	Kongming lantern, first hot air balloon	2nd or 3rd century
'Abbas Ibn Firnas	Single flight of manned ornithopter; ended in crash and injury.	875[13][14]
Eilmer of Malmesbury	Single flight of manned glider.	1010
Unknown Chinese	Manned kites are common. Reported by Marco Polo	1290
Lagari Hasan Çelebi	First manned rocket flight	1633
Bartolomeu de Gusmão	First lighter-than-air airship flight	1709
John Childs	Unnamed flying device flew 700m three times over two days. Documentation suggests that he glided down along a 700m rope and landed where the rope was fixed to the ground.	1757
Montgolfier	Modern hot air balloon	1783

brothers

Diego Marín Aguilera	Single flight of manned-glider-wings	1793
William Samuel Henson	Aerial Steam Carriage, flight of model	1842
John Stringfellow	Stringfellow Machines	1848, 1868
Henri Giffard	Non-rigid airship, hydrogen-filled envelope for lift, powered by steam engine	1852
Sir George Cayley	Cayley Glider, flight of manned glider. Investigating many theoretical aspects of flight. Many now acknowledge him as the first aeronautical engineer.	1853
Rufus Porter	New York to California Aerial Transport, an early attempt at an airline	1849
Jean Marie Le Bris	Artificial Albatross	1857, 1867
Félix du Temple de la Croix	Monoplane (1874) Maybe first powered manned fixed-wing flight, a short hop, from a downward ramp.	1857 - 1877
James William Butler and Edmund Edwards	Steam-Jet Dart Patented a prophetic design that of a delta-winged jet-propelled aircraft, derived from a folded paper plane.	1865
Francis Herbert Wenham	Wenham's Aerial Locomotion	1866
Jan Wnęk	Loty glider, many flights	1866
Frederick Marriott	Marriott flying machines, as well as an attempt at an early airline	1869
Alphonse Pénaud	Planophore, Pénaud Toy Helicopter	1871
Thomas Moy	Moy Aerial Steamer,	1875
Thomas Moy	The Military Kite	1879
Charles F. Ritchel	Ritchel Hand-powered Airship	1878
Victor Tatin	Tatin flying machines	1879
Massia and Biot	Massia-Biot Glider	1879? 1887?
Alexandre Goupil	Goupi Monoplane, La Locomotion Aerienne	1883
John J. Montgomery	Montgomery Monoplane and Tandem-Wing Gliders	1883 - 1911
Aleksandr Fyodorovich Mozhaiski	Mozhaiski Monoplane	1884
Charles Renard|Arthur	The first fully controllable free-flight was made with the La France	1884

234

Constantin Krebs		
Pichancourt	Mechanical Birds	1889
Lawrence Hargrave	Hargrave flying machines and Box kites	1889 - 1893
Clément Ader	Éole, Avion, short, manned and powered, flights	1890 - 1897
Chuhachi Ninomiya	Karasu model, Tamamushi model	1891 ,1895
Otto Lilienthal	Derwitzer Glider, Normal soaring apparatus and others, many flights	1891 - 1896
Horatio Phillips	Phillips Flying Machine	1893, 1906
Hiram Stevens Maxim	Maxim Biplane	1894
Pablo Suarez	Suarez Glider	1895
Octave Chanute and Augustus Herring	Chanute and Herring Gliding Machines	1896
William Paul Butusov	Albatross Soaring Machine	1896
William Frost	Frost Airship Glider	1896
Percy Sinclair Pilcher	Pilcher Hawk Based on the work of his mentor Otto Lilienthal, in 1897 Pilcher built a glider called The Hawk with which he broke the world distance record when he flew 250 m (820 ft)	1897
Samuel Pierpont Langley	Langley Aerodromes	1896 - 1903
Carl Rickard Nyberg	Flugan, very short manned flight	1897
Edson Fessenden Gallaudet	Gallaudet Wing Warping Kite	1898
Gustave Whitehead	A purported steam engine powered, 500-1000m flight, ending in collision with a three-story house, according to affidavit 37 years later by Louis Darvarich, self-described passenger.	1899
Count Ferdinand von Zeppelin	Zeppelin airship LZ 1. The first Zeppelin flight occurred on July 2, 1900, over the Bodensee lasted 18 minutes. The second and third flights were in October 1900 and October 24, 1900, respectively, beating the 6 m/s velocity record of the French airship La France by 3 m/s.	1900
Wilhelm Kress	Kress Waterborne Aeroplane hops	1901
Gustave Whitehead	A newspaper reported a manned, powered, controlled 800m flight. Whitehead claimed four flights on the same day in the aircraft, designated Number 21.	1901

Alberto Santos-Dumont	Santos-Dumont gained fame by designing, building, and flying dirigibles. On 19 October 1901, he won the Deutsch de la Meurthe prize of 100,000 francs by taking off from Saint-Cloud, flying his steerable balloon around the Eiffel Tower, and returning.	1901
Gustave Whitehead	He claimed a manned, powered, controlled 10 km flight, a circle over Long Island Sound, one of two flights the same day, landing in the water twice without damage to the plane, designated Number 22.	1902
Lyman Gilmore	Gilmore Monoplane Built a steam-powered airplane and claimed that he flew it on May 15, 1902.	1902
Richard William Pearse	Pearse Monoplane. First flight March 31, 1902, Waitohi, New Zealand. Evidence exists that on 31 March 1903 Pearse made a powered, though poorly controlled, flight of several hundred metres and crashed into a hedge at the end of the field. The aircraft had a tricycle type landing gear and primitive ailerons.	1903-1904
Wright brothers	Completed development of the three-axis control system with the incorporation of a movable rudder connected to the wing warping control on their 1902 Glider. They subsequently made several fully controlled heavier than air gliding flights, including one of 622.5 ft (189.7 m) in 26 seconds.	1902
Karl Jatho	Jatho Biplane 10 hp 70m hops	1903
Guido Dinelli	Dinelli Glider, Aereoplano	1903, 1904
Wright brothers	Wright Flyer I, Successful, manned, powered, controlled and sustained flight, 259m, in 59 seconds, according to the Federation Aeronautique International and Smithsonian Institution. Preceded by three other flights, each less than 200 feet.	1903
Ferdinand Ferber and Gabriel Voisin	Archdeacon glider	1904
Wright Brothers	Wright Flyer III Wilbur Wright pilots a flight of 24 miles (39 km) in nearly 39 minutes on Oct. 5, a world record that stood until Orville Wright surpassed it in 1908.	1905
Louis Blériot and Gabriel Voisin	Blériot-Voison floatplane glider, biplane	1905
Traian Vuia	Vuia I, Vuia II, Several short powered flights. August 1906, 24m flight. July 5, 1907, Flew 20m. and crashed.	1906 - 1907
Jacob Ellehammer	Ellehammer monoplane September 12, 1906, became the second European to fly an airplane (after Traian Vuia). He made over 200 flights in the next two years using many different machines. No distance data found.	1906 - 1907
Alberto Santos-	First official European flight on 23 October 1906 in aircraft	1906

236

Dumont	designated 14-bis or Oiseau de proie ("bird of prey"). On 12 November 1906, he flew the 14-bis 220 metres in 21.5 seconds. He won the Archdeacon Prize founded by the Frenchman Ernest Archdeacon in July 1906, to be awarded to the first aviator to demonstrate a flight of more than 25 m.	
Glenn H. Curtiss	AEA June Bug First official U.S. flight exceeding 1 kilometer (5,360 ft (1,630 m).	1908
Louis Blériot	Blériot V, Blériot XI On July 25, 1909, Louis Blériot successfully crossed the Channel from Calais to Dover in 36.5 minutes, 35 km	1909
Aerial Experiment Association (AEA)	Silver Dart on 10 March 1909, McCurdy flew the aircraft on a circular course over a distance of more than 35 km (20 mi).	1909
Aurel Vlaicu	Vlaicu 1909, Vlaicu I, Vlaicu II, Vlaicu III	1909-1910
Henri Fabre	Le Canard, First seaplane.	1910
Duigan Brothers	Duigan Pusher Biplane	1910
John William Dunne	With the Dunne D.5 tailless Biplane, the fifth in a series of tailless swept-wing designs, Dunne was among the first to achieved natural stability in flight in the same year.	1910.

http://en.wikipedia.org/wiki/Early_flying_machines

It would also be safe to say that the military mindset in most countries gave very little thought to the question of ET visitations, until sometime after the First World War. Their focus was on the development of other aspects of combat such as manned flight and aerial bombardment, larger long range cannons, machine guns and rifles, close hand to hand combat and other wartime tactics and strategies.

CHAPTER 43

BIG BROTHER SEES, LISTENS, MONITORS, TRACKS, TARGETS, SHOOTS DOWN, AND RETRIEVES ALL THINGS UFO

Ground and Airborne Radar Track Records

Radar is one of the touchstone technologies for the military to track, monitor and communicate with any of its military aircraft in the world as well as keeping track of other nation's military aircraft, even commercial airliners and small private aircraft. It is used to penetrate very deep underground mineral deposits and anomalies and is used as "over the horizon" surveillance and for deep space detection of astronomical and extraterrestrial objects!

Over the many generations of development since its invention, radar has improved to the point that it can, in the right hands of a professionally trained radar operator, be used to identify almost any aircraft from most meteorological phenomenon and nowadays, it is unusual to misinterpret common weather patterns from a flock of geese or from an enemy missile, rocket, unauthorized aircraft or even an orbiting satellite.

Radar has become a sophisticated tool for the military! It has been interfaced with HAARP and with "Scalar (Longitudinal) Weaponry" (it is what brought down the saucers in 1947 near Roswell, New Mexico...it wasn't just solely radar).

To understand radar better, a little history of its development is in order.

The first practical application of radar (radio detection and ranging) was developed prior to the Second World War by Britain and her allies as a means to detect the approach of German aircraft invasions of Britain from as far away as 160 km (100 mi) and to enable pilots to fly at night and detect enemy aircraft before engaging in aerial combat. Radar sites were initially setup on the south and east coastlines of England and later in the interior and northern regions of Britain. Not long after ground radar was established Allied aircraft were fitted with airborne radar units to help in night-time bombing sorties and detection of other aircraft, and radar was also employed in naval vessels. By 1940 Germany too had radar in use. Radar gave the advantage of early warning of inbound enemy planes because they could be picked on radar scopes or monitors while still hundreds of miles away and allowing for planes to get into the air and engage the enemy out over sea instead of over land where possible downed aircraft could cause potential damage to civilian or industrial areas or military bases. As ground radar improved, range increased to 6000 miles and it was possible to distinguish altitude, azimuth, speed of an object, size and the number of aircraft present. Radar was manned around the clock, thus the element of surprise was eliminated from the German aircraft invading England. The Germans needed to get pass the British radar and so, Germany developed and built the V1 and V2 rockets to fly high above the radar detection capability and re-enter in a steep trajectory aimed for London and selected targets. By the time radar was able to detect the inbound rockets, it was too late to scramble aircraft into the air to target and destroy them before they reached their impact target. Since guidance systems were not invented and built in to the rockets by the German scientists, destruction in London and surrounding areas was random relying merely on latitude and longitude trajectory settings from the point of launch. The fact that damage to buildings did occur in London by V1 and V2 rockets speaks highly of German engineering and accuracy.

"British physicist James Clerk Maxwell developed equations governing the behaviour of electromagnetic waves in 1864. Inherent in Maxwell's equations are the laws of radio-wave reflection, and these principles were first demonstrated in 1886 in experiments by the German physicist Heinrich Hertz. Some years later, a German engineer Christian Huelsmeyer proposed the use of radio echoes in a detecting device designed to avoid collisions in marine navigation. The first successful radio range-finding experiment occurred in 1924, when the British physicist Sir Edward Victor Appleton used radio echoes to determine the height of the ionosphere, an ionized layer of the upper atmosphere that reflects longer radio waves.

The first practical radar system was produced in 1935 by the British physicist Sir Robert Watson-Watt, and by 1939 England had established a chain of radar stations along its south and east coasts to detect aggressors in the air or on the sea.

Early military radar system used during World War II by the British (Home Chain Radar) and Germans (Würtzburg Radar)
http://spitfiresite.com/2010/04/deflating-british-radar-myths-of-world-war-ii.html/6 and
http://scholarsarchive.jwu.edu/cgi/viewcontent.cgi?article=1011&context=ac_symposium and
http://spitfiresite.com/2010/04/deflating-british-radar-myths-of-world-war-ii.html/6

In 1935 Watson-Watt wrote a paper entitled "The Detection of Aircraft by Radio Methods". This was presented to Henry Tizard, the chairman of the Committee for the Scientific Survey of Air Defence. Tizard was impressed with the idea and on 26th February 1935, Watson-Watt demonstrated his ideas at Daventry. His idea was based on the bouncing of a radio wave against an object and measuring its travel to provide targeting information. It was called radar (radio detection and ranging). As a result, he was appointed head of the Bawdsey Research Station in Felixstowe.

By the outbreak of the Second World War in September 1939, Watson-Watt had designed and installed a chain of radar stations along the East and South coast of England. In the same year, two British scientists were responsible for the most important advance made in the technology of radar during World War II. The physicist Henry Boot and biophysicist John T. Randall invented an electron tube called the resonant-cavity magnetron. This type of tube is capable of generating high-frequency radio pulses with large amounts of power, thus permitting the development of microwave radar, which operates in the very short wavelength band of less than 1cm, using lasers. Microwave radar, also called LIDAR (light detection and ranging) is used in the present day for communications, to measure atmospheric pollution and for very deep radar penetration of the ground for mineral deposits and unusual "anomalies". During the Battle of Britain, these stations were able to detect enemy aircraft at any time of day and in any weather conditions.

Radar was also used by ships and aircraft during the war. Germany was using radar by 1940 but Japan never used it effectively. The United States had a good radar system and it was able to predict the attack on Pearl Harbor an hour before it happened.

Britain tended to have the best radar system during the early stages of the war and in 1940, the invention of the Magnetron cavity resonator enabled more centimetric waves to be transmitted. It also enabled more compact high-frequency sets to be used by aircraft in the Royal Air Force."
www.ob-ultrasound.net/*radar***.html**

Radar had many problems early in its development such as false or "ghost" echoes from distant fixed targets; low pulse repetition frequency (PRF) problems, poor beam transmission of radar to the target, false radar returns from reflections off of birds, low clouds, terrain such as mountains, hills and lakes, "noise" either internally or externally produced, chaff from small objects used to create radar interference, etc. were aspects that needed to be worked out. Radar like any new technology was a gradual process of evolvement.

In peace time and in wars, over the years up to the present time, radar has become a very sophisticated and an indispensable tool with all its modern diverse applications from simple short range hand held radar guns to satellite equipped radar to large, long range, deep space radio telescopes capable of bouncing signals off the Moon and planets or receiving radio signals from billions of light years away. Like home video recording equipment, today's radar can record flight information about any aircraft in the air that is tracked by radar so that civilian and military operators, technicians and the **FAA (Federal Aviation Administration)** in the US and the **TC (Transport Canada)** can have access to any pertinent information that is required. It's the law of aviation. Whether used by civilian agencies or the military services, radar operators need to clearly and correctly understand and interpret the data received on their radar monitors. Split

240

second decision making is required on a daily, hourly, minute by minute and by the second because lives depend upon it in our modern world. Radar operators are under extreme stress from their jobs and it can have it toll emotionally and physically with a high percentage of employee turn-over. You are staring at a screen monitor for hours on end and fatigue can become a problem on body, mind and spirit, (although, this has changed somewhat, over the years where breaks and shift relief are taken between scheduled work periods). This profession has been rated as one of the most stressful jobs in the world to perform. There is little to no room for error from the radar equipment or the operator. Both must function as one. If an error does occur, it will usually be from the operator. The reason behind all this basic historic and somewhat technical information which is merely scratching the surface of a voluminous subject, is to remember that split second decision making and lives depend on its accurate data collecting and understanding of that data and that by law all radar track records must be kept and maintained.

So, if a person or persons have an unexplained aerial sighting of a strange aerial object unidentified (UFOs) or otherwise, by law there must be a track recording of the object witnessed indicating the time period of the event. The problem of course, is being able to get a copy of or the original radar track recording of the event or even an acknowledgement that something unusual was picked up on civilian or military radar. By FAA law, TC law and military law radar track records must be kept and maintained. They must exist! In fact all countries on the planet have them and each country has a government department or agency that manages them.

When it comes to the question of the existence of UFOs or ETVs, **Radar Track Records** provide crucial evidence and support to their existence! There are no "Ifs, Ands, Buts, Maybes or Wherefores" about it! This type of evidence will stand up in any court, in any country.

Why? Because it's the **LAW!** But sometimes the law becomes subverted or distorted!

Japan Air Lines flight 1628 UFO Incident

A remarkable case that illustrates this point came to light back in November 17, 1986 which was more than just another commercial airline pilot's UFO report, this sighting had been tracked by both ground and air radar. This UFO sighting was unusual in that it was the first real UFO account that had been tracked by both ground and air radar as well as witnessed by experienced airline pilots, and confirmed by a FAA Division Chief. The full story however, didn't come out until May 2001, twenty-five years later at the **Washington Press Club** in Washington, DC. It was one of twenty UFO **Disclosure Project Witness** testimonies brought to the public and the new media attention by **Dr. Steven Greer.** It is the famous case of the **Japan Air Lines flight 1628 UFO Incident.**

This sighting caught the attention of the world news media in late December and early January 1987 when the **Federal Aviation Administration (FAA)** announced that it was going to officially investigate this sighting because the Air **Route Traffic Control Center** in Anchorage, Alaska, had reported that the UFO had been detected on radar and the FAA needed to tell the news media something as they were hounding **RTCC** for details about this unusual radar tracked UFO sighting. http://www.ufoevidence.org/documents/doc1316.htm

There are two principle people in this UFO account, **Captain Kenju Terauchi** who flew the Boeing 747 JAL cargo plane and **FAA Division Chief John Callahan,** who reviewed the radar tapes of the UFO event. The testimony of these key witnesses confirms each other's story; however, Capt Terauchi testimony was not confirmed until 25 years later by FAA Division Chief Callahan as a **Disclosure Project Witness.**

Captain Terauchi testimony recounts how JAL Flight 1628 were settling in an otherwise routine flight pattern over northern Canada having left Iceland for the Anchorage leg of their flight originating from Paris to Tokyo with a cargo of French wine. It was when they crossed the Canada – Alaskan border that their flight no longer became routine as the flight crew saw and tracked three unidentified objects. On the night of November 17, 1986, the sighting of at least one of the UFOs was initially confirmed by FAA and U.S. military ground radar. http://www.ufocasebook.com/jal1628surfaces.html

Just after they crossed into Alaska, at 5:09 PM local time, Anchorage Air Traffic Control contacted them on the radio to report initial radar contact. The Anchorage flight controller asked them to turn 15 degrees to the left and head for a point known as Talkeetna on a heading of 215 degrees. They were at 35,000 feet and traveling at a ground speed of about 600 mph.

Japan Air Lines pilot Kenju Terauchi describes an alleged encounter with three UFOs over Alaska during a 1986 JAL flight. Terauchi's sketch (opposite page) shows how the largest craft dwarfed his plane. It also estimates the UFOs' positions relative to the plane, and on his radar screen, when they were first sighted.

**Japan Air Lines Captain Kenju Terauchi describes
the encounter with three UFOs over Alaska.**
http://www.ufoevidence.org/cases/case287.htm

According to **Captain Kenju Terauchi**, **First Officer Takanori Tamefuji** and **Flight Engineer Yoshio Tsukuda**, at 5:11 PM over eastern Alaska, Captain Terauchi noticed two craft to his far left, and some 2,000 ft (610 m) below his altitude, which he assumed to be military aircraft. These abruptly rose from below and closed in to escort their aircraft pacing his flight path and speed. Each had two rectangular arrays of what appeared to be glowing nozzles or thrusters, though their bodies remained obscured by darkness. When closest, the aircraft's cabin was lit up and the captain could feel their heat in his face. http://www.ufoevidence.org/cases/case287.htm

At 5:18 or 5:19 PM the two objects abruptly veered to a position about 500 ft (150 m) or 1,000 ft (300 m) in front of the aircraft, assuming a stacked configuration.

In doing so they activated "a kind of reverse thrust and [their] lights became dazzlingly bright". To match the speed of the aircraft from their sideways approach, the objects displayed what Terauchi described as a disregard for inertia: "The thing was flying as if there was no such thing as gravity. It sped up, then stopped, then flew at our speed, in our direction, so that to us it [appeared to be] standing still. The next instant it changed course. ... In other words, the flying object had overcome gravity." The "reverse thrust" caused a bright flare for 3 to 7 seconds, to the extent that captain Terauchi could feel the warmth of their glows. http://www.ufocasebook.com/jal1628surfaces.html

Air traffic control was notified at this point (i.e. 5:19:15 PM), who could not confirm any traffic in the indicated position. After 3 to 5 minutes the objects assumed a side-to-side configuration, which they maintained for another 10 minutes. They accompanied the aircraft with an undulating motion, and some back and forth rotation of the jet nozzles, which seemed to be under automatic control, causing them to flare with brighter or duller luminosity.

Each object had a square shape, consisting of two rectangular arrays of what appeared to be glowing nozzles or thrusters, separated by a dark central section. Captain Terauchi speculated in his drawings, that the objects would appear cylindrical if viewed from another angle, and that the observed movement of the nozzles could be ascribed to the cylinders' rotation. The objects left abruptly at about 5:23:13 PM, moving to a point below the horizon to the east. http://en.wikipedia.org/wiki/Japan_Air_Lines_flight_1628_incident and http://www.ufocasebook.com/jal1628surfaces.html

As these two craft departed before a third, much larger disk-shaped object started trailing them. **Captain Terauchi** now noticed a pale band of light that mirrored their altitude, speed and direction causing the pilots to request a change of course. Setting their onboard radar scope to a 25 nautical miles (46 km) range, he confirmed an object in the expected 10 o'clock direction at about 7.5 mi (13.9 km) distance, and informed **Anchorage ATC (Air Traffic Control)** of its presence. Anchorage found nothing on their radar, but Elmendorf ROCC, directly in his flight path, reported a *"surge primary return"* after some minutes.

As the city lights of Fairbanks began to illuminate the object, Captain Terauchi believed to perceive the outline of a gigantic spaceship on his port side that was *"twice the size of an aircraft carrier"*. It was however outside **First Officer Tamefuji's** field of view. Terauchi immediately requested a change of course to avoid it. The object however followed him "in formation", or in the same relative position throughout the 45 degree turn, a descent from 35,000 to 31,000 ft, and a 360 degree turn. The short-range radar at Fairbanks airport however failed to register the object.

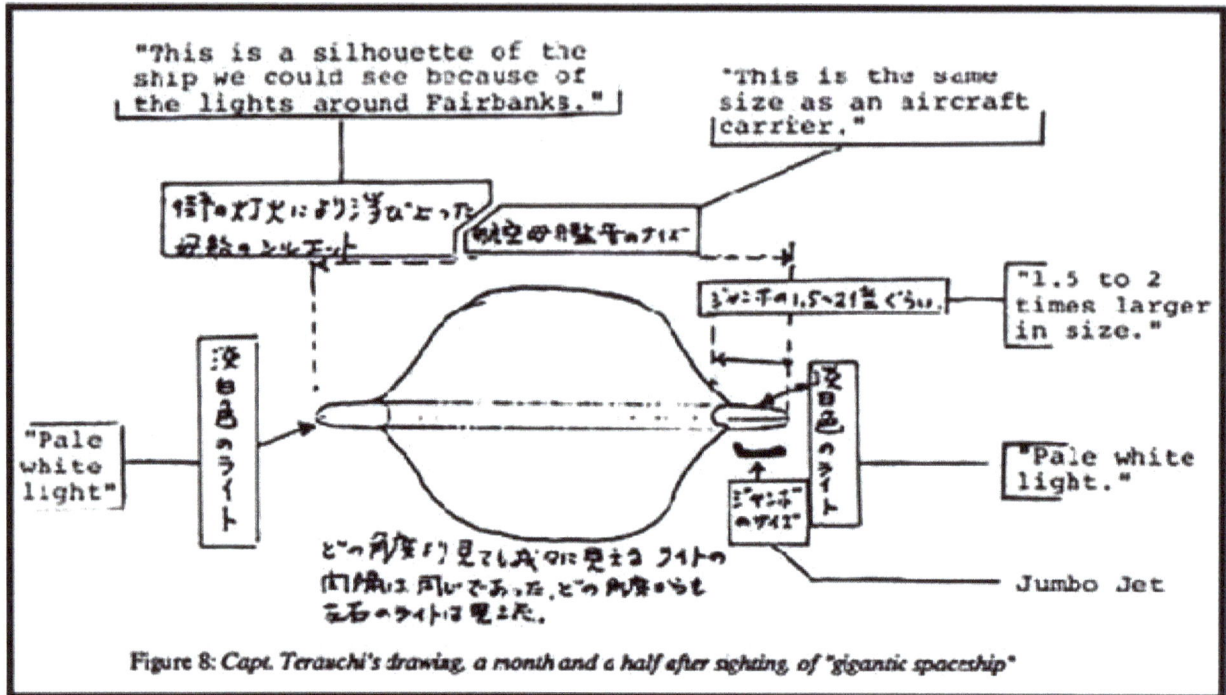

Figure 8: Capt. Terauchi's drawing, a month and a half after sighting, of "gigantic spaceship"

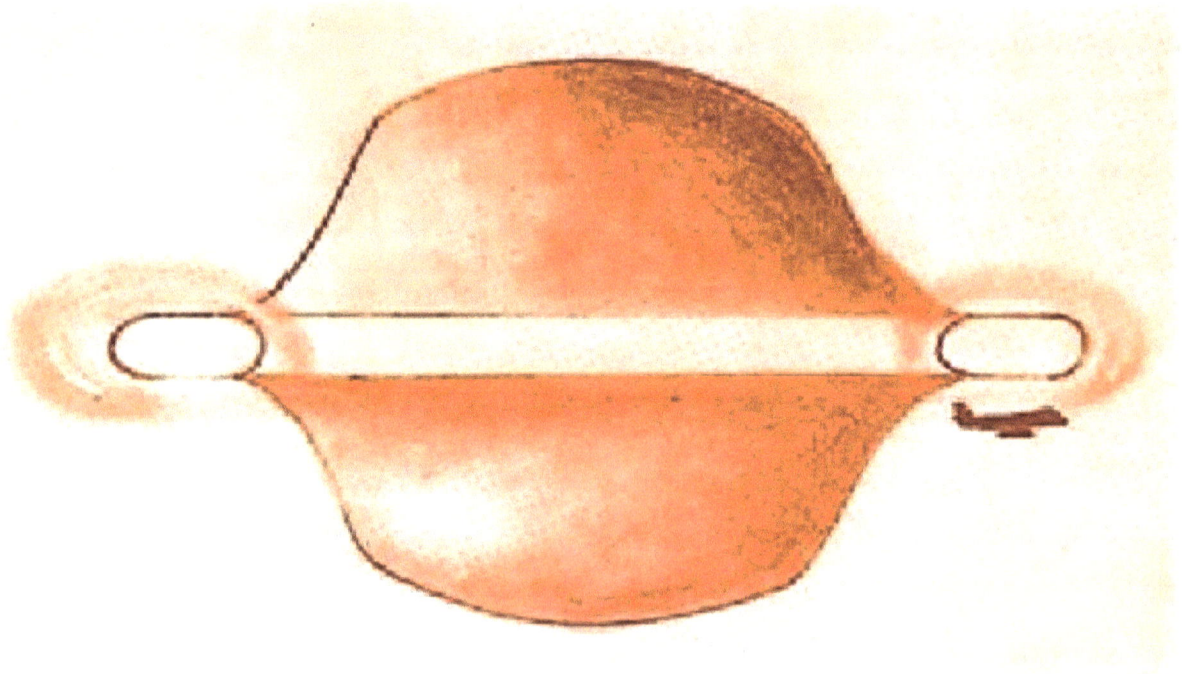

Capt. Terauchi drawings of the gigantic spaceship that paced his JAL 1628 cargo plane

244

Anchorage ATC offered military intervention, which was declined by the pilot, due to his knowledge of the **Mantell Incident.** Anchorage Air Traffic Control obliged and requested an oncoming United Airlines flight to confirm the unidentified traffic, but when it and a military craft sighted JAL 1628 at about 5:51 PM, no other craft could be distinguished by which time JAL 1628 had also lost sight of it. JAL 1628 arrived safely in Anchorage at 18:20 after the sighting of 50 minutes ended in the vicinity of Mt. McKinley.
http://en.wikipedia.org/wiki/Japan_Air_Lines_flight_1628_incident and
Shincho, Shukan. "Terauchi's London interview of December 1986". *JAL Pilot's UFO Story Surfaces after 20 Years.* **Japan Today, UFO casebook. Retrieved 2011-05-08. Japan Air Lines Flight 1628, Loy Lawhon Terauchi, Kenju.**
http://www.ufoevidence.org/cases/case287.htm **"Statement of Captain Terauchi, Pilot of JAL Flight 1628".** *Concerning JAL Flight 1628's Sightings of Unidentified Air Traffic on November 17, 1986.* **FAA official document. Retrieved 2011-05-08. Maccabee, Bruce (Mar– Apr 1987), "The Fantastic Flight Of JAL 1628",** *International UFO Reporter,* **12 (2) Maccabee, Bruce. "The Fantastic Flight Of JAL 1628". UFO Evidence. Retrieved 7 May 2011.**

The FAA at first confirmed the claims that several of its radar traffic controllers tracked the 747 and the large object, and that U.S. Air Force radar did as well. Later official statements hedged on this, and tried to ascribe the radar targets to weather effects. At the end, however, an FAA spokesman stated, *"We are accepting the descriptions of the crew, but are unable to support what they saw."*
http://www.bibliotecapleyades.net/ciencia/ufo_briefingdocument/1986b.htm

Captain Terauchi cited in the official Federal Aviation Administration report that the object was a UFO. In December 1986, Terauchi gave an interview to two Kyodo News journalists.
He was also featured on numerous radio and TV programs and in People Magazine. Within a few months of these events JAL grounded him for talking to the press, apparently for his indiscretion of reporting a UFO, even though he was a senior captain with an excellent flying record. He was moved him to a desk job and several years later he was reinstated to pilot status and eventually retired in north Kanto, Japan.

Because of the report of unusual traffic the crew was interviewed immediately by FAA official Jack Wright and by agents James Derry and Ronald Mickle. Wright recorded the details as outlined above but also all the specific times of tower control and pilot conversations back and forth, course corrections, turns, speed and altitude changes, including individual pilot testimonies of the events, etc.

Although the Anchorage FAA alerted the FAA Security Office in Washington, D.C., no further action regarding the reported traffic was taken. This is unfortunate since careful debriefing at the time of the event possibly could have uncovered details which had been forgotten by the time the crew was interviewed again in January 1987.

It is quite likely that the JAL sighting would never have been known to the general public, nor would it have been investigated, if it hadn't been for the interest by the American news media in an odd story out of Japan. On December 29, 1986, the Kyoda Press in Japan published a news story about a Japan Airlines crew that had seen a UFO while flying over Alaska and that the UFO had been picked up by military radar. The Kyoda press got the story as a result of the air crew talking to their friends, etc., and someone alerted the press. On the 29th after the Christmas holidays that story must have been printed somewhere in Japan because **United Press International (UPI)** reporter **Jeff Berliner** broke the story in the United States on December 29.

Numerous newspapers reprinted the story and the FAA headquarters in Washington, D.C. learned about it from the press reports, especially the January 1 report in the staid (***"never report a UFO story, and, if you must, be sure it is unfavorable!"***) **Washington Post**. When the FAA headquarters called Anchorage for the full story, it learned that the ***radar data tape had been saved*** (which was unusual since the tapes were generally reused after 30 days). http://www.ufoevidence.org/Cases/CaseSubarticle.asp?ID=288

This information about the sighting must have come as a complete surprise to the FAA headquarters. On January 1 the FAA reopened its inquiry into the sighting. Capt. Terauchi was interviewed on January 2 [1], at which time he supplied his written testimony [2]. On January 4 the national press reported that the FAA had reopened the inquiry and numerous news stories followed.

The FAA released portions of the information through **Mr. Steucke**, the Public Information Officer at Anchorage, as the information became available. Unfortunately the FAA did not have a complete and accurate story to report and consequently the early news stories on the UFO incident contained errors. The FAA conducted further investigations of the incident, and did not issue its final report until March 5.

In the meantime while waiting on the FAA to give its verdict of what took place on the JAL 1628 cargo flight, **CSICOP's (Committee for the Scientific Investigation of Claims of the Paranormal) Phil Klass** in his usual effort to debunk all things related to any UFO story issued a premature statement on January 22 claiming that ***the UFOs were the planets Jupiter and Mars*** - an impossible solution because the UFO was seen in a part of the sky opposite the position of these planets and because the UFOs moved from positions one above the other to side by side. http://www.ufoevidence.org/Cases/CaseSubarticle.asp?ID=288

Not knowing when to quit with its ridiculous explanations, CSICOP later issued a second explanation that the UFO was light reflecting off of clouds of ice crystals - also unlikely because the sky was clear at the reported altitude of the UFO. The FAA attributed the radar images received by ground radar to a "split radar return from the JAL Boeing 747."

You really have to wonder, whether Phil Klass and CSICOP really expected the public to buy into their drivel or if they knew something about the news media that the general public didn't. Could it be that all major news media are forced into a phony dutiful response from orders that originate from the news offices of spin-meisters, whose policy is to down-play or ridicule all UFO stories, in order to covey a public sense of open fair-mindedness from the "fifth estate" by portraying both sides of the story in an alleged judicial manner?

246

Fortunately, we don't always have to rely upon false statements and deliberate disinformation to get to the truth and the heart of the real story of the JAL 1628 UFO incident.

Although, the FAA could not substantiate the pilot's eye-witness accounts, the radar tracking tapes don't lie and they tell another part of this intriguing story. As it turns out, 25 years later, a new fascinating revelation to this tale with explosive implications which buttresses the reality of the UFO phenomenon comes from no less than the testimony of a former executive of the National FAA!

John Callahan, a former Division Chief of the FAA executive gave testimonial remarks at the **National Press Club** presentation in May 2001 under the directive auspices of **Dr. Steven Greer**'s **Disclosure Project Witnesses Program**:

"What I am about to tell you now, is about an event that never happened!"

"I am John Callahan, and I was Division Chief of the Accidents, Evaluations and Investigations Division of the FAA in Washington, for 6 years during the 1980s.

In early January of 1987, I received a call from the Air Traffic Quality Control Branch in FAA's Alaskan regional office, requesting guidance on what to tell the media personnel who were overflowing their office. The media was requesting information about the UFO that chased a Japanese 747 across the Alaskan sky for some 31 minutes at flight levels between 31000 and 35000 ft. on November 17, 1986. Somehow, the word had got out."

"What UFO; when did this take place; why wasn't Washington Headquarters informed?" I asked.

"Hey," he replied, "who believes in UFOs? I just need to know what to tell the media to get them out of here."

The answer to that question was easy: "Tell them it's under investigation. Then, collect all the data - the voice tapes and data disc for both the facility and the military. Send then overnight to the FAA Tech Center in Atlantic City, New Jersey."
http://www.theufotimes.com/contents/News_358%20Alaska%20Japan%20airlines.html

"Japan Air Lines flight 1628, a cargo jet with a pilot, co-pilot and one crew member, was just north of Anchorage, and it was about 11 p.m. when the pilot first reported the UFO. He described it as a huge ball with lights running around it. He said it was about 4 times bigger than a 747, and he saw it with his own eyes. So did the other two-crew members.

Over the course of 31 minutes, the UFO jumped miles in a few seconds, changing places to different spots around the 747, in one sweep of the radarscope.

*After receiving the data from Alaska almost two months later, I briefed my boss **Harvey Safer** and the **FAA Administrator Admiral Engen**, and Safer and I went to the FAA Tech Center to observe the playback.*

The FAA had developed a program capable of recreating the traffic on the controller's scope, or plan view display. I instructed the FAA specialist to synchronize the voice tapes with the radar data so we could hear everything the controller and pilot said. I videotaped the radar display.

Later that day, I requested that the FAA automation specialists plot the radar targets along the route of flight and explain what each target was doing. I videotaped the resulting chart.

The print out and radar play back displayed primary targets in the vicinity of the 747. These target returns were displayed about the same time and place as the pilot advised viewing the UFO.

Both the radar and manual controller observed the primary target. The military controllers also viewed the primary target on their radar and identified it as a double primary. The pilot and crew viewed the target on their radar and were able to see the huge UFO at the same time as it approached their aircraft.
http://www.theufotimes.com/contents/News_358%20Alaska%20Japan%20airlines.html
If this craft had been a Learjet or military aircraft at the wrong attitude, that would have been clear. The FAA has procedures that cover tracking unidentified aircraft violating another's airspace - but it has no procedures for UFOs.

*Back at FAA headquarters, I gave **Administrator Engen** a quick briefing and showed him the video. He set up a briefing with **President Reagan's** scientific staff, and told me my function was to give them a dog-and-pony show and hand this event off to them, since the FAA does not deal with UFOs.*

*I brought along **a copy of the video and all the data print outs** available at the time.*
One of the scientists asked a number of questions, such as, what is the range of the radar, what is the frequency of the radar, what is the band width, what is the formula for the height equipment, etc.

In a videotaped interview with **Dr. Steven Greer (SG), John Callahan (JC)** explains in detail:

 *"Then the next day I got a call from someone with the Scientific Study Group [for President Reagan], or the CIA, I'm not sure who it was, the first call. And they had some questions about the incident. And I had said, I don't know what you are talking about, you probably want to call the **Admiral [FAA Administrator Engen]**.*

Well a few minutes later the Admiral calls down and says, I have set up a briefing tomorrow morning at 9:00 am in the round room. Bring all the stuff you have. Bring everybody up there and give them whatever they want. We want to get out of it. Just let them do whatever they want. So I brought all the people from the Tech Center. We had all kinds of boxes of data that we had them print out; it filled up the room. They brought in three people from the FBI, three people from the CIA, and three people from Reagan's Scientific Study team — I don't know who the rest of the people were but they were all excited...

When they got done, they actually swore all these other guys in there that this never took place. We never had this meeting. And this was never recorded.

SG: Who said that? Who was saying that?

JC: This was one of the guys from the CIA. Okay? That they were never there and this never happened. At the time I said, well I don't know why you are saying this. I mean, there was something there and if it's not the stealth bomber, then you know, it's a UFO. And if it's a UFO, why wouldn't you want the people to know? Oh, they got all excited over that. You don't even want to say those words. **He said this is the first time they ever had 30 minutes of radar data on a UFO. And they are all itching to get their hands onto the data and to find out what it is and what really goes on. He says if they come out and told the American public that they ran into a UFO out there, it would cause panic across the country. So therefore, you can't talk about it. And they are going to take all this data…**

Well when they read the reports that came through the FAA decided it had to protect themselves — you can't say you saw a target, even though this is what he said. So they made him change his report to say "position symbols," which makes it sound like it wasn't really a target. Well if it's not a target then a lot of the other position symbols that we are separating [on radar] aren't targets either. And when I read that, I thought oh, there is something fishy here, that somebody is worried about something or other and they are trying to cover up.

When the CIA told us that this never happened and we never had this meeting, I believe it was because they didn't want the public to know that this was going on. Normally we would put out some type of a news release that such and such happened...

Well, I've been involved in a lot of cover-ups with the FAA. When we gave the presentation to the Reagan staff I was behind the group that was there. And when they were speaking to the people in the room, they had all those people swear that this never happened. But they never had me swear it never happened. And it always bothered me that we have these things going on and when you see something or you hear something on the radio or TV, the news that it's put down as it's not there. I have a hard time saying nothing.

It still bothers me that I've seen all this, I know all this, and I'm walking around with the answer, and nobody wants to ask the question to get the answer. And it kind of irritates me a little bit. And I don't believe our Government should be set up that way. I think when we have something like this, that you can probably find out more about what's going on in the world [by not covering it up]. If they [the UFOs] can travel that far, that distance with that type of machinery, who knows what they could do here for the health of the nation, the people, the food they could give them, the cancers we could cure. They have to know more than us to be able to travel at that speed.

For those people that say that if these UFOs existed, they would someday be on radar and that there'd be professionals who would see it, then I can tell them that back in 1986 there were enough professional people that saw it. It was brought down to headquarters, FAA headquarters, Washington D.C. The Administrator saw the tape of it. The people that we were debriefing, they've all seen. Reagan's Scientific Study team, three of those professors, doctors, they've seen it. As far as I was concerned they were the ones that verified my own thoughts about it. They were very, very excited about the data. They had said that this was the only time a UFO was ever recorded on radar for any length of time where it is 30 some minutes. And they have all this data to look at...

What I can tell you is what I've seen with my own eyes. I've got a videotape. I've got the voice tape. I've got the reports that were filed that will confirm what I've been telling you. And I'm one of those, what you would call the high Government officials in the FAA. I was a Division Chief. I was only three or four down from the Admiral...
http://www.ufoevidence.org/Cases/CaseSubarticle.asp?ID=290

Again, from his Disclosure Witness testimony, Callahan goes on to further state*:*

I was impressed with the response from the FAA experts. At the end, one of the three people from the CIA said, **"This event never happened; we were never here; we're confiscating all this data and you are all sworn to secrecy."**

"What do you think it was?" I asked the CIA person.

"A UFO and now they have over 30 minutes of radar to go over," *he responded.*
"Well let's get a TWIX out and advise the American public that we were visited by a UFO," I suggested.
"No way, if we were to tell the American public there are UFOs they would panic," *he informed me.*

Some time after the briefing, the detailed FAA report which included extensive interviews with the pilot and crew, the chart prepared at the Tech Center; and facility voice tapes arrived at my office and were placed on a small table waiting for the CIA to request more data. (Meaning these documents, charts and radar tracking tapes were in the possession of **FAA Division Chief, John Callahan** up and until 2001, when they became evidence in the **Disclosure Project Witnesses** testimonies!)

The material stayed there until I retired 2 years later, and I've had it ever since.

Most people including FAA controllers really aren't familiar with how the FAA radar system works and why all aircraft traveling through our airspace are not caught on radar or displayed on the controllers PVD. But I am.

The system and organization of the FAA is not configured to identify and track these types of performances."
http://www.theufotimes.com/contents/News_358%20Alaska%20Japan%20airlines.html

Let's jump back in history once again toward the end of World War II to see how American military advancements along with some unusual extraterrestrial circumstances better known as flying saucer crash retrievals began to shape the US military and indirectly America's political

250

position to become the most powerful nation the world has ever seen.

At the close of the Second World War, radar and many other technological advancements had played a significant role in helping the Allies win the war over the Axis Powers and now, the world was tallying its victories and losses in human carnage, 66 million people had died. From the beginning of the 20th century to the end of WWII about 130 million people were killed in both world wars and various smaller wars. By the end of the 20th century, conservatively over 180 million people had been killed in wars, conflicts, and atrocities either in attacking or defending their particular ideologies. http://users.erols.com/mwhite28/war-1900.htm

In comparison to the 18th Century where an approximate 18 million people died in wars or unnatural deaths or in the 19th Century where many bloody wars and conflicts resulted in approximately 45 million dead, by end of the 20th Century killing seemed to have been honed to a fine art form. http://necrometrics.com/20c5m.htm and http://users.erols.com/mwhite28/warstat8.htm

By the end of WWII, America found itself well position upon the seat of power and security in the world fairing much better than most nations who suffered the devastations of war in their lands. Most nations struggled to get back on their feet, both economically and politically with the rebuilding of their country's infrastructures. In America, rather than downsizing their military forces or merely maintaining them, they decided to jump in bed with big industrial business to increase the size of their military against the wishes of post-war president Eisenhower.

At the turn of the 21st Century, humanity having not learned a damn thing about peace and global stability, the world continues to venture onwards in its violent ways with America seemly at the helm of the warship. This is not to say that Russia or China are absent from the helm of their own warships but, it does seem that America and her allies were trying to steer their warship into even rougher waters of conflict, in pursuit of some clandestined global mission that is destined to bring about a potential third world war or some other form of chaos and horror upon the world if it is not halted immediately.

White Sands Missile Range - White Sands, New Mexico

The US was moving forward after the successful completion of Operation Paperclip. Nazi German scientists were now firmly embedded into the American way of life under control by the US military and were cooperating fully in the reverse engineering of captured Nazi super secret weapons from the war. Many of the German scientists were sent to Fort Bliss near White Sands Proving Grounds in New Mexico to work on the V2 rocket and other advanced concepts of rocketry.

White Sands Proving Grounds is a rocket range of almost 3,200 square miles (8,300 km^2) in parts of five counties in southern New Mexico. (The site was officially renamed to **White Sands Missile Range (WSMR) o**n May 1, 1958.) The largest military installation in the United States, WSMR includes the Oscura Range and the WSMR Otera Mesa bombing range. WSMR and the 600,000-acre (2,400 km^2) McGregor Range Complex at Fort Bliss to the south, form a contiguous swath of territory for military testing.

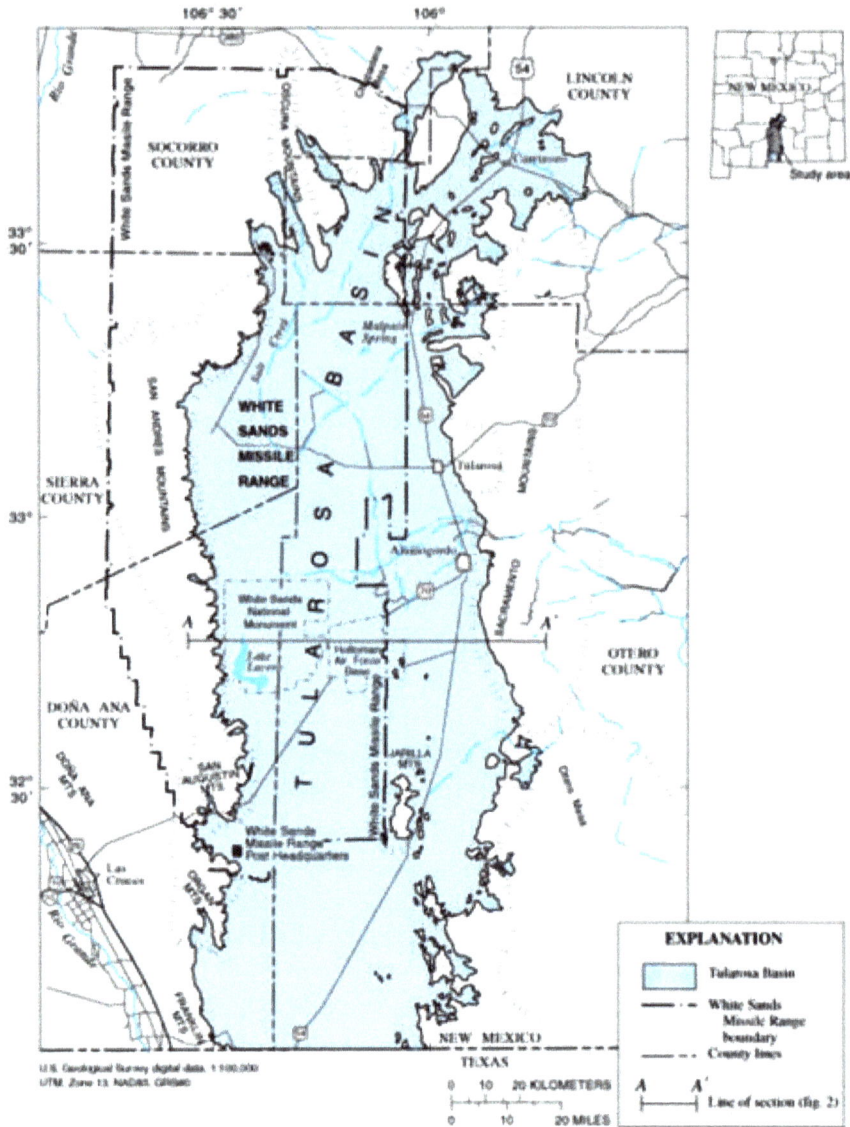

Outline of WSMR at the Tularosa Basin
https://en.wikipedia.org/wiki/White_Sands_Missile_Range

The creation of atomic weapons arose out of political and scientific developments of the late 1930s. The rise of fascist governments in Europe, new discoveries about the nature of atoms and the fear that Nazi Germany was working on developing atomic bombs converged in the plans of the United States, the United Kingdom, and Canada to develop powerful weapons using nuclear fission as their primary source of energy. The **Manhattan Project**, as the American nuclear physics effort was called, culminated in the test of a nuclear weapon at what is now called the **Trinity Site** on July 16, 1945, and the atomic bombings of Hiroshima and Nagasaki just a few weeks later. http://en.wikipedia.org/wiki/Trinity_(nuclear_test)#First_deployment

On July 16, 1945 deep in the deserts of Socorro, New Mexico, Pandora's Box was opened and the Atomic Age was born with the first nuclear explosion.
The only surviving color photograph of the Trinity explosion.
http://www.atlasobscura.com/places/trinity-atomic-bomb-site Photo by Jack Aeby

Oppenheimer was one of the chief scientists in the development of the atomic bomb and would later state that, while watching the test, he was reminded of a line from the Hindu scripture the **Bhagavad Gita:** *"Now I am become Death, the destroyer of worlds"*

The **Atomic Age** is usually considered to have begun on July 16, 1945, with the first nuclear weapons test of an atomic bomb, code named **Trinity**. Trinity was a test of an implosion-design plutonium device. This test was conducted by the United States Army in the Jornada del Muerto desert about 35 miles (56 km) southeast of Socorro, New Mexico, at the new White Sands Proving Ground, which incorporated the Alamogordo Bombing and Gunnery Range. The Trinity detonation produced an explosive power equivalent to the explosion of about 20 kilotons of TNT. http://en.wikipedia.org/wiki/Trinity_(nuclear_test)#First_deployment

By executive order of President Harry S. Truman, the U.S. dropped the nuclear weapon **"Thin man" (aka. Little Boy)** on the city of **Hiroshima** on Monday, August 6, 1945, followed by the detonation of **"Fat Man"** over **Nagasaki** on August 9, 1945.
http://en.wikipedia.org/wiki/Trinity_(nuclear_test)#First_deployment

For six months before the atomic bombings, the United States intensely fire-bombed 67 Japanese cities. Together with the United Kingdom and the Republic of China, the United States called for a surrender of Japan in the **Potsdam Declaration** on July 26, 1945. The Japanese government ignored this ultimatum. By executive order of **President Harry S. Truman**, the U.S. dropped the nuclear weapon **"Little Boy" (aka. Thin man)** on the city of **Hiroshima** on Monday, August 6, 1945, followed by the detonation of **"Fat Man"** over **Nagasaki** on August 9th.

Six days after the detonation over Nagasaki, on August 15, Japan announced its surrender to the Allied Powers, signing the **Instrument of Surrender** on September 2, officially ending the Pacific War and therefore World War II. Germany had signed its Instrument of Surrender on May 7, ending the war in Europe. The bombings led, in part, to post-war Japan's adopting **Three Non-Nuclear Principles**, forbidding the nation from nuclear armament. The role of the bombings in Japan's surrender and the U.S.'s ethical justification for them, as well as their strategic importance, is still debated.

It is a point of debate and conjecture that the bombing of Hiroshima and Nagasaki by the US with nuclear weapons was even necessary in order to get the Japanese government to accept surrender. With all the previous fire bombings of many Japanese cities by American planes, Japan was on the verge of surrender as they had lost the war in the Pacific to America and the Allied Forces also, the Soviet Union had just declared war on Japan as well. It appears that Japan's hold out on surrender was merely their coming to terms with defeat in the war. The dropping of the atomic bombs was unnecessary in getting the Japanese to surrender; this was a blatant use by the US of a new more deadly force to send a clear message across the Pacific and into the heart of Asia as well as into Europe that the USA was not a nation to be trifled with.

The US knew of the destructive power of the atomic bomb but, they wanted to see its devastating effects first hand on buildings and people first hand. It is the reason why, they sent teams of doctors and scientists along with tens of thousands of soldiers into these cities, the first day after dropping the bomb, to see the results of their handy work. The "experiment" was not quite complete and the need for further data had to be collected and analyzed.

The atomic bombings of Hiroshima and Nagasaki killed at least 148,000 people immediately and many more over time. By 1950, the death toll was over 340,000. They were followed days later by the surrender of Japan. Debate over the justification of the use of nuclear weapons against Japan persists to this day, both in scholarly and popular circles.

From the first testing of atomic weapons in New Mexico to their use upon Japan at the end of the Second World War, an unusual signal had been unwittingly sent across the universe to other more advanced civilizations. Someone had manufactured extremely destructive weaponry and had used it in a global war. It was already understood by military forces on both sides of the war that non-terrestrial devices were following the air raid sorties of aircraft, they were monitoring

254

ground combat and observing naval battles. Before the war had ended, it had been determined by military officials and political leaders that the Earth was under surveillance by Extraterrestrial Intelligences, their existence was real and not visionary.

Exploding atomic bombs was a good way to say to everyone in the galactic neighbourhood, "Hey! Here we are! Come and check us out!" It was an invitation, perhaps unintentional, to anybody listening or observing to respond with a much closer inspection of the species called homo sapien.

Imagine a race of interstellar explorers travelling past our planet a few thousand years ago perhaps, to some other distant reaches of the universe, may have decided to have a casual layover visit to examine the various types of flora and fauna on such a pristine blue planet. By happenchance they observed a rather young primitive humanoid species whose favourite pastime seems to be using sticks and stones and other rudimentary weapons to settle localized tribal conflicts. These battles and skirmishes may have been viewed as quaint and short-lived except that such bloody barbarism resulting from these tribal conflicts never diminished but grew in size with each encounter. Nevertheless, with a non-interference protocol into the affairs of other worlds, these ETs departed the planet and continued on with their primary mission. They did, however, mark the location of this pretty blue world on their star charts with the promise to themselves to stop by again, on their way back home to see if this new species was at peace with itself and their planet.

Then, about 2 thousand years later, one of their scientists quite unexpectedly picks up a seismic tremor rippling across all the EM bands from the Infrared to the Gamma wavelength ends of the spectrum and into the multi and hyperdimensional wavelengths of physical and quantum space. From his initial discovery he is not successful in pinpointing its location having been caught off guard by its unique suddenness, however, his investigation and persistence is rewarded when two more strong energy bursts are detected. To his horror, he discovers that the energy bursts emanate from that pretty blue planet where they found the primitive humanoid life forms.

He reports his findings to his commanding officer and immediately small armadas of spacecraft are dispatched to the Earth. Upon their arrival, they are greeted by other spacecraft piloted ET races of the galactic federation, who have been monitoring this major global conflict. It seems that in this most recent global war, the kids on this planet have discovered the matches to the big explosives and now, they threaten to burn down the planet and themselves with it. This young primitive species that are barely out of the trees have taken their predilection for tribal violence and have turned it into global warfare, not just once, but twice! They have even built rockets to blast into space! This deadly combination of technology is a potential threat to the galactic neighbourhood! What would these kids be contemplating next?

An immediate council of the planetary federation was convened and an agreement was discussed, voted upon, and unanimously passed. **General Order 19** was to be implemented. **Quarantine the Earth** and her people! Remove the Zone of Quarantine Order only when humans have demonstrated maturity and their ability to live together in peace and unity in a global society and not until then.

Sounds like a great plot line for a Spielberg sci-fi movie, if it wasn't for the sad truth that it is true and not science fiction!

1941 Cape Girardeau, Missouri Saucer Crash

One of the earliest accounts of a crashed UFO with alien bodies predated the well known Roswell incident by some six years. **Leo Stringfield** first brought this case to fellow UFO investigators in his book "UFO Crash / Retrievals: The Inner Sanctum" and he recounted Mann's story in the July 1991 issue of his "Status Report," a monthly publication on UFO activities and investigations.

He recounts the story of a UFO crash retrieval under military control, the details of the case which were sent to him in a letter from one **Charlette Mann**. She relates in her minister grandfather's deathbed confession how he had been summoned to pray over alien crash victims outside of **Cape Girardeau, Missouri** in the spring of 1941.

As an evangelist for many years, **Reverend William Huffman** had taken the resident minister post at the Red Star Baptist Church in early 1941. His employment there is substantiated by church records during that time period.

Huffman Reverend had received a call to perform last rites duty presumedly from police or military personnel and was immediately driven on a 10-15 mile journey to some woods outside of town. Upon arriving at the scene of the crash, he saw policemen, fire department personnel, FBI agents, and photographers already mulling through the wreckage. His first thoughts were probably that this was a plane crash where the aircrew had died in the disaster.

He was soon asked to pray over three dead bodies. As he began to take in the activity around the area, his curiosity was first struck by the sight of the craft itself.

Expecting a small plane of some type, he was shocked to see that the craft was disc-shaped, and upon looking inside he saw hieroglyphic-like symbols, indecipherable to him.

He then was shown the three victims, not human as expected, but small alien bodies with large eyes, hardly a mouth or ears, and hairless.

Upon completion of the last rites over the alien crash victims, he and other witnesses at the scene were soon sworn to secrecy by military officials who had taken charge of the crash area with dire warnings given to them not to reveal what they had seen.

Arriving back at his home, Huffman Reverend's mind was reeling, playing over and over the events that had just transpired. He needed to unburden his soul to his wife and his sons and tell them, the story he had participated in a few hours ago. It would be this late night family discussion that Charlette Mann would hear from her grandmother in 1984, as she lay dying of cancer at Charlette's home while undergoing radiation therapy.

Although Charlette had heard bits and pieces of this story before, she knew her grandmother was

256

very near the ends of her days, so she needed to get the full details of this intriguing story before the eventual death of her grandmother.

She also learned that one of the members of her grandfather's congregation thought to be Garland D. Fronabarger, had given him a photograph taken on the night of the crash. This picture was of one of the dead aliens being helped up by two men.
http://ufocasebook.com/missouricrash.html

In an interview for a television documentary, Charlette Mann retells the account of what she knew. It should be kept in mind that this is a third-hand witness account of the crashed saucer event that her grandfather experienced. Some of that account is given here:

"I saw the picture originally from my dad who had gotten it from my grandfather who was a Baptist minister in Cape Girardeau Missouri in the Spring of '41. I saw that [picture] and asked my grandmother at a later time she was at my home fatally ill with cancer so we had a frank discussion.

"She said that grandfather was called out in the spring of 1941 in the evening around 9:00-9:30, that someone had been called out to a plane crash outside of town and would he be willing to go to minister to people there which he did."

"Upon arrival, it was a very different situation. It was not a conventional aircraft, as we know it. He described it as a saucer that was metallic in color, no seams, did not look like anything he had seen. It had been broken open in one portion, and so he could walk up and see that.

"In looking in he saw a small metal chair, gauges and dials and things he had never seen. However, what impressed him most was around the inside there were inscriptions and writings, which he said he did not recognize but were similar to Egyptian hieroglyphics."

"There were 3 entities, or non-human people, lying on the ground. Two were just outside the saucer, and a third one was further out. His understanding was that perhaps that third one was not dead on impact. There had been mention of a ball of fire, yet there was fire around the crash site, but none of the entities had been burned and so father did pray over them, giving them last rites.

"There were many people there, fire people, photographers, and so they lifted up one, and two men on either side stood him up and they stretched his arms out, they had him up under the armpits and out here. http://ufocasebook.com/missouricrash.html

As I recall from the picture I saw, he was about 4 feet tall, appeared to have no bone structure, soft looking. He had a suit on, or we assume it was a suit, it could have been his skin, and what looked like crinkled, soft aluminum foil. I recall it had very long hands, very long fingers, and I think there were three but I cannot swear to that."

"My grandfather upon arrival, said there were already several people there on the scene, two that he assumed were local photographers, fire people, and so not long after they arrived, military just showed up, surrounded the area, took them off in groups separately, and spoke to each of them.

"Grandfather didn't know what was said to the others, but he was told 'this didn't happen, you didn't see this, this is national security, it is never to be talked about again.'

"My grandfather was an honorable man, being a preacher, that's all that needed to be said to him. And so he came home and told the story to my dad, who was there, and my grandmother and my uncle. Now, my mother was expecting at the time, so she was off in the bedroom."

"My sister was born May 3, 1941, so we are assuming this was the middle to the last of April. And he never spoke of it again. But about two weeks later, one of the men who had a personal camera that he had put in his shirt pocket, approached grandfather and said I think someone needs a copy of this.

"I have one and I would like you to keep one. So that's how it came about that grandfather had the picture, to begin with. But he never spoke of it again. The other people seem to be very intimidated and very frightened and paranoid."

Other living supporting witnesses include Charlette Mann's sister who confirmed her story in a notarized sworn affidavit, and the living brother of the Cape Girardeau County sheriff in 1941, Clarance R. Schade. He does remember hearing the account of the crash, yet does not have many details. He does recall hearing of a "spaceship with little people."
http://ufocasebook.com/missouricrash.html

There are also Fire Department records of the date of the crash. This information does confirm the military swearing department members to secrecy, and also the removal of all evidence from the scene by military personnel.

Guy Huffman, Charlette's father also told the story of the crash and had in his possession the photograph of the dead alien. He showed the picture to a photographer friend of his, Walter Wayne Fisk.

He has been contacted by Stanton Friedman, but would not release any pertinent information.

Charlette had no luck in getting Fish to return calls or answer letters. It has been rumored that Fisk was an advisor to the President, and if this was the case, would account for his silence on the facts of the Missouri crash.

Another UFO investigator, James Westwood of Centreville, Virgina, a retired Navy man and engineer interviewed Charlette Mann about the 1947 incident in which the government allegedly recovered and then covered up a UFO crash in New Mexico.

His assessment is that this case ends like many others, but appears by all indications to be authentic. All who have come in contact with Charlette Mann found her to be a trustworthy person who is not given to sensationalism and has sought no gain from her account.

Huffman died in 1959. His wife, who died in 1984, told Mann the story.

Woodward also speculates the military officers on the scene may have been called in from an Army Air Corps base in Sikeston at the time. If the crash happened, the military and police wouldn't have known what they were looking at, Westwood said, because Roswell and the other early UFO sightings hadn't happened.

And the incident may have been covered up for military security reasons since the U.S. was gearing up for World War II, he said. "It wouldn't be implausible" for the incident to have been reported as an airplane crash," Westwood said.

There is still research being done on the Missouri crash, and hopefully, more information will be forthcoming to validate this remarkable case. http://ufocasebook.com/missouricrash.html

What is very interesting about this case is the number of alleged witnesses to the event who have yet to come forward to tell of their account of the incident given the fact that they had ample opportunity to take in a great amount of detail and to have examined closely the saucer debris wreckage and the non-human bodies, all before the military officials showed up to sequester and cordoned off the area to further public scrutiny.

This would indicate the inexperience of the US military to coordinate and handle this type of foreign materiel retrieval scenario and their inability to get to the scene of the crash before the general public or other government officials and intelligence agencies.

When we examine the military handling of the **Battle of L.A. Incident of 1943** and the **Roswell Saucer Crash of '47**, there was a definite progression of protocols, rapid mobilization of equipment and men as well as highly tightened security around the event impact areas. Certainly by the time the **Kingman and Aztec saucer crash events** took place, the US Army and Air Force were operating like a well-greased piece of machinery, national security measures had been setup, sworn oaths of national allegiance trumpeted and enforced backed up by intimidations and threats, all had become standard practices in crash retrievals and in securing public cooperation.

1942 Battle of Los Angeles

World War II was the Earth's bloodiest global conflict to date and it was a well-known fact that the military forces on both sides of the war knew that they were under surveillance by strange aerial craft not of this world, before the war ended. **Eisenhower and Churchill** had received reports from their commanding officers and pilots in active war zones and had discussed the matter and concluded that these crafts were Extraterrestrial in nature but, felt the need to classify the subject above top secret, to prevent any further alarm by the general public who were already on edge from the aftermath of war.

US Military programs like the **Interplanetary Phenomenon Unit** (or **IPU**) were instituted to get to the bottom of the matter with the objective to determine if this was a new manmade weapon developed by the Russians from captured Nazi technology or a possible Extraterrestrial threat to national security and if necessary to shoot them down. The Interplanetary Phenomenon Unit was established by at least 1947, possibly as early as 1942 due to the **"West Coast Air Raid"** in which an unidentified object over L.A. resulted in a massive anti-aircraft barrage, by the late 1950s IPU was dissolved.

If as Ufologists have stated that the Earth has been monitored for hundreds or even thousands of years then when the world implodes sociologically and politically from within resulting in global war, not once but twice, the probability that ETI are going to investigate the on-goings of humankind is no longer in question.

Once again, imagine a very late fall night on 24th February and the early hours of 25th February 1942 when the city of Los Angeles received an aerial visit from a very large orange coloured object. An object that was picked up by army coastal radar and a total blackout was ordered. Air raid sirens sounded out across the County at 2.25 a.m. as a warning of an impending raid.

With the outbreak of the war with Japan and the rising fear of a Japanese air attack, or even invasion of the West Coast, 12,000 thousand residents volunteered to become Air Raid Wardens for wartime duties for the sprawling city of Los Angeles and surrounding communities.

This image of a UFO was on the front page of the Los Angeles Times in February 1942.
The searchlights are targeting the object while artillery shells
explode showing up as smaller spots of light.
http://www.huffingtonpost.com/jason-apuzzo/the-time-a-ufo-invaded-lo_b_6749734.html

After the air raid warning sounded, pilots from the 4th Interceptor Command prepared their aircraft to intercept the UFOs, but no order to intercept was given. The aircraft remained on the ground. http://www.where-is-area-51.com/ufo-sightings-west-coast-air-raid.html
The U.S. Army anti-aircraft searchlights by this time had the object completely covered. **Jeff Rense**, who runs an internet radio program and website (http://www.rense.com) reports in an article on the Battle of L.A. that Katie (last name unknown), one of the Air Raid Wardens and a close observer of the object as it appeared nearly over her house is insistent about the use of planes in the attack on the object. *"They sent fighter planes up (the Army denied any of its fighters were in action) and I watched them in groups approach it and then turn away. There were shooting at it but it didn't seem to matter."*

At 3.16 a.m. the **37th Coast Artillery Brigade** started firing 12.8 pound anti-aircraft shells at the UFO, which was illuminated by bright searchlights. http://www.where-is-area-51.com/ufo-sightings-west-coast-air-raid.html

The planes were apparently called off after several minutes and then the ground cannon opened up. *"It was like the Fourth of July but much louder. They were firing like crazy but they couldn't touch it."* The attack on the object lasted over half an hour before the visitor eventually disappeared from sight. Many eyewitnesses talked of numerous "direct hits" on the big craft but no damage was seen done to it. http://www.rense.com/general27/battle.htm

In the photograph above note, the convergence of the searchlights and you will clearly see the shape of the visitor within the illuminated target area. Note also, that the beams from the searchlights do not emerge on the other side of the BIG object. They appear to be absorbed or redirected to some other point in space. The object seems completely oblivious to the hundreds of AA shells bursting on or adjacent to it which caused it no evident damage.

Artillery fire continued occasionally for nearly an hour and stopped at 4.14 a.m. There were at least 6 people who died as a direct result of the Army's attack on the UFO, three of which died of heart attacks attributed to the stress of the hour-long bombardment. The UFO slowly made its way down to and then over Long Beach before finally moving off and disappearing.

Several buildings had been damaged by friendly fire before the all-clear was sounded and the blackout order lifted at 7.21 a.m.

The UFO sightings made front-page news along the U.S. Pacific coast and earned some mass media coverage throughout the United States. One Herald Express writer who observed some of the incidents insisted that several antiaircraft shells had struck one of the objects, and he was stunned that the object had not been brought down.

It should be noted that on the TV series, **Ancient Aliens** in one of its episodes which covered in part, this "Battle over L.A.", other witnesses who lived on Long Beach recalled that the object appeared almost overhead when the artillery started its bombardment. They stated that the large object which appeared elliptical in shape slowly moved off down toward Redondo Beach which was then quickly followed by three to five interceptor aircraft in pursuit.

So, here we have two sets of witnesses, who stated that they remembered Army Air Force aircraft flying overhead in investigative pursuit, contrary to the denial that the US Army said that none of the their aircraft were in the skies at the time of the event.

'In spite of the heavy anti-aircraft fire, Army Defense Command flatly denied any reports of any hits or any bombs dropped in this area. However, persistent eye-witness reports were that a plane was downed at 185th Street and Vermont Ave. The area was roped off and spectators were not permitted to approach what appeared to be a wrecked plane.' *Long Beach Independent EXTRA February 25, 1942*

Secretary of the Navy, Knox announced that the entire incident was a false alarm due to anxiety and "war nerves". The press was outraged, some suspected a cover-up: the Long Beach Independent wrote, "There is a mysterious reticence about the whole affair and it appears that some form of censorship is trying to halt discussion on the matter." Others speculated that the incident was a ruse designed to give coastal defense industries an excuse to move further inland. And if there truly was nothing to the incident, the possibility that Navy personnel had fired heavy artillery shells for nearly an hour at nothing at all, killing three civilians in the process seemed to suggest that the men of the U.S. Navy were dangerously incompetent.

262

ARMY SAYS ALARM REAL

U.S. Flyers Reap Indies Victories

Sink Two Transports and Destroy Three Planes; MacArthur Breaks Lull by Successful Attacks; American-British Airmen Blast Foe in Burma

Storm Grows Over Delay in Alien Ouster

Telegraphic Plea Sent to Olson Urging Action; Navy Speeds Evacuation

INFORMATION, PLEASE

(Editorial)

Five Deaths Laid to Raid Blackout

Traffic Accidents and Heart Attacks Take Lives of Quintette

Roaring Guns Mark Blackout

Identity of Aircraft Veiled in Mystery; No Bombs Dropped and No Enemy Craft Hit; Combat Report Seeing Planes and Balloon

Rangoon Aces Bag 30 Planes

Americans Sink Jap Ships

IN THE 'TIMES' TODAY

Knox Indicates Raid Just Jittery Nerves

Rubber in Corsets Ordered Banned

Attention Subscribers!

Front page story of the LA Times on February 26, 1942 as witnessed by hundreds of thousands of LA residents

https://forum.termometropolitico.it/71900-la-battaglia-di-los-angeles-24-02-1942-a.html

The Long Beach Independent reports that eye-witnesses had seen a downed aircraft in the L.A. area after the air raid

On March 5, 1942, **George C. Marshall** writes a top-secret memo to the President, which states: *"regarding the air raid over Los Angeles it was learned by Army G2 that Rear Admiral Anderson... recovered an unidentified airplane off the coast of California... with no bearing on conventional explanation... This Headquarters has come to the determination that the mystery airplanes are in fact not earthly and according to secret intelligence sources they are in all probability of interplanetary origin."* Marshall goes on to state: *"As a consequence, I have issued orders to Army G2 that a special intelligence unit be created to further investigate the phenomenon and report any significant connection between recent incidents and those collected by the director the office of Coordinator of Information."*

The memo bears correct **Office of Chief of Staff (OCS)** file numbers and has **"Interplanetary Phenomenon Unit" (IPU)** typed on it at a later time by a different typewriter. It is logical to believe that this is the order that sets up the IPU.
http://www.majesticdocuments.com/documents/pre1948.php

"It is noteworthy that for thirty years until the release of the Marshall memorandum, the Department of Defense claimed to have no record of the event. Five years before Roswell, the military was already learning to clamp down on UFOs." UFOs and the National Security State' By Richard M. Dolan Keyhole Publishing, 2000

A **Freedom of Information Act** request was made in 1974 to gain access to a memorandum regarding the incident, access was granted and documentation released. Written by **General George C. Marshall** for **President Franklin Roosevelt**, and dated February 26, 1942, the memo contradicts Knox's assertion that the incident was due only to "war nerves," and proves that officials took the event seriously.

Photo enhancements and analysis made by **Dr. Bruce Macabee** on July 24, 2010, show definitively that a huge craft is caught in the converging spotlights. Some investigators have argued that the object seen in the spotlights and which was fired upon by artillery shells was a large weather balloon. However, no balloon debris was ever found and it certainly wouldn't require 1430 rounds of artillery fire to bring down a balloon. One or two direct shots would have done the job. As already stated many eyewitnesses saw the large object take numerous "direct hits" but no damage was seen done to it. Nor was this air raid a scheduled late night army battery practice, this event was taking place in real time.

Everything about this event strongly indicates that unknown object that flew over Los Angeles was not identified as an enemy aircraft of Japan or any other country as it appeared non-hostile. Its behaviour seems inquisitive and almost casual or ambivalent in response to the open hostility it received from the many gun batteries on the ground firing round after round of artillery shells. The army was unable to bring the craft down during its fierce barrage leaving gun crews and the populace in absolute bewilderment as to what had taken place.

After meticulous but, unsuccessful efforts by UFO researchers and other investigators to duplicate and video record a re-enactment of the event with accurate positioning of artillery guns, searchlight placements and actual open shelling upon balloon targets it was concluded that this event was anomalous and probably extraterrestrial in nature. No manmade aerial devices could operate and behave in the same manner that this unknown object was able to without being destroyed by artillery fire.

Some documents, perhaps of dubious authenticity, have suggested that the Los Angeles Air Raid-inspired the formation of the **Interplanetary Phenomenon Unit (IPU).** While the IPU undoubtedly existed (official U.S. Military sources have confirmed its reality), little is known of the unit, and any connection to the Los Angeles Air Raid must be regarded as unproved.
http://www.where-is-area-51.com/ufo-sightings-west-coast-air-raid.html

It would be logical that events of anomalous flying objects seen and engaged with during and after the war would by reason of a national security issue necessitate an investigation like IPU to be setup to get a handle on this problem.

Unidentified flying objects continued to plague the US Military after the war and it was discovered that radar which was developed during WWII seem to have an adverse effect upon these alien craft as well as being able to detect them from other friendly aircraft. It appeared that radar affected the navigation and stability of flight of alien flying saucers. This was a very important fact that would be utilized by the military if the opportunity arose to engage these alien craft. It wasn't long when such an opportunity did arise.

July 2, 1947, Roswell Saucer Crash – The "Holy Grail" of All Saucer Crashes

The annals of UFO literature are strewn with the accounts UFO crash cases found in almost every country on the planet but, the review of this particular aspect of UFO literature illustrates an inherent problem with almost all of them. The problem is that the physical proof of debris wreckage of a crashed disk or the remains of alien bodies, if they were real during the initial eyewitness account of the event, that evidence quickly became the illegitimate possession of the military usually taken by well armed force or was hauled away by some covert black ops group under the control of a governmental agency. In the Ufology community, this was a clear problem of 'corpus delicti' and 'habeas corpus'.

Many of these crash retrieval cases read like a great Sci-Fi script produced by some Hollywood movie mogul, the only difference is that these are real accounts and not from some science fiction script.

In 1945, America sent a signal into space with the explosion of a few atomic bombs then, by 1947 that signal had been received by Extraterrestrial Intelligences, who had grave concern for the Earth and its people. Their response came with increased surveillance particularly in the one region of the planet that had nuclear capability: Roswell and White Sands, New Mexico., USA.

White Sands Proving Grounds had been the area where the first nuclear weapon had been detonated and not too far away to the east was Roswell, New Mexico and the **Roswell Army Air Field**, home of the **509th Bomb Group of the Eighth Air Force, RAAF** which was the armament staging point for the deployment of atomic bombs. During 1945, it was the only air base of its kind in the world to have this nuclear capability.

266

For an Extraterrestrial civilization in search of the unique signature that comes from the radioactive material of an atomic weapon or from the radiation waste debris found at a nuclear detonation site, Roswell, New Mexico stood out like the proverbial sore thumb. It wasn't difficult to locate and with a few flyovers of the air base and the surrounding area, the ETI visitors would soon have enough data collected to tell the story of what was going on. This may sound highly speculative and no one really knows the mindset or the agenda of another sentient intelligence, particularly one visiting from another star system but, it is intriguingly synchronistic, that the one area of the planet possessing weapons of mass destruction should have alien spacecraft flying in the skies over its air base.

The telling of the Roswell saucer crash of July 1947 has become the **"Holy Grail"** of all flying saucer crash stories and the "touchstone" by which all other UFO accounts are measured. It has become the darling golden little "Oscar" to which Hollywood Sci-Fi writers and producers aspire to emulate in their movies or the **"Ebie" trophy** that Ufologist and documentarians try to achieve as a hallmark of UFO investigative excellence.

Small wonder then, that there is chaos, misinformation, and disinformation within the field of Ufology whose corrupt database resembles the debris field of the saucer wreckage found by **Mac Brazel** on the Foster homestead ranch back in 1947. No other single post wartime event has done more to create and foster the concept of what a conspiracy theory can develop into in the minds of the general public than has the Roswell incident of 1947.

The story is a confusing, twisting, chaotic jungle of people, places, timelines and events that have miraculously sprung up from the dry harsh desert of the southwest to become an ever-growing mythology of monumental proportions that has taken on a life of its own, outside of the actual event. The Roswell saucer crash is a unique historical event that is truly out of this world but, has now become an annual celebration that is commemorated with weeklong festivities of parades, strangely costumed alien party-goers and curiosity seekers. It is littered with UFO lecturers espousing their latest current theories and conspiracies while promoting their research in timely UFO books, articles, CDs and DVDs and all officially sanctioned by the Roswell city fathers in perpetuity of tourism, the cottage industry and good old fashion US of A free enterprise.

How then, do you make any rational sense from a interplanetary event that if the US government and the military had only been more open, forthcoming and responsive to the national interests of the people, the world's destiny from such a momentous revelation may have been more hopeful and positive, instead of one that is cloaked in clandestine agendas.

Author's Rant: It is not my intent to give a detailed and in depth recount of the Roswell incident as this would take up many books on the subject. Suffice it to say that many great authors like Stanton Friedman, Don Berliner, Kevin Randle, Donald Schmitt and Jessie Marcel M.D. etc. have written far better books dealing with every aspect and nuance of this historic case than I could ever do justice to. To these gentlemen and their books, I refer the reader to investigate further and do their homework. I shall merely give a brief overview of the more important aspects and highlights to wet the thirst of the reader leaving him to discover the details for himself from other sources.

It is now strongly thought that development of long range high powered radar utilizing **Scalar Physics technology** from several ground stations in and around the **Roswell, Corona and Socorro** areas and possibly the Alamogordo area were used on the night of July4, 1947. Reports had been coming to the military and the local police stations as well as the various newspapers of the southwest with accounts of eyewitness testimony of UFOs seen or photographed. The military may well have been on high alert for any unidentified aircraft moving about the countryside that did not follow prescribed flight protocols.

It has been theorized that powerful long range radar dishes from the ground base stations were rotated into a triangulation configuration focusing and emitting the radar energy beams which caused havoc with the navigational systems of the alien craft thus, causing them to become unstable and to crash.

A Brief History of Scalar Energy

This theory has been debunked as implausible as conventional aircraft and electronic of that time are not affected by radar as they routinely fly through radar all the time unaffected. However, some researchers think that these radar sites were re-configured with **Scalar Technology** to emit microwave-like beams based upon the early theory and work of **Nikola Tesla**. Decades later, it was demonstrated that it was possible to remotely fly aircraft with microwave energy beams and it has also been proven that focused microwave beams could also be used to fry electronic circuitry from a distance. Much of this came out of radical scientific developments during the Second World War from both the Allied and Axis Forces.

The history of scalar energy and scalar physics is not well known in the general public consciousness unless, you happen to be, either a physicist or student at university studying physics. If you've ever attended a whole life expo, you may have heard something about scalar energy from someone who gave a presentation on UFO physics or someone practicing alternative medicine and health. It is a science that is gaining popularity as more people become informed of its theory and application in everyday life, particularly among the alternative health community and the science of alternative energy generation, even though, the basic knowledge of it has been around for quite some time.

James Clerk Maxwell, a theoretical physicist and mathematician born in Scotland in 1831 originally discovered in the mid-1850's a new form of energy that became known as **Scalar Energy.**

Not long after Maxwell's discovery of this new energy, Nikola Tesla, a brilliant Yugoslavian mathematical and mechanical engineer in 1856, began work on a provable and demonstrable application of the energy's existence. Tesla referred to Scalar energy as **standing energy (universal waves)** but more commonly, he referred to it as **"-"** and he was able to collect it without the use of any cables and wires. http://www.slideshare.net/scalarenergyproducts

Albert Einstein also noted and made reference to Scalar Energy during his work in the 1920's, but the question asks, how is it different? Our everyday electromagnetic waves have wave action and frequencies, which are easily measurable in the unit of Hertz. Scalar energy, on the other

268

hand, is a static, stationary energy that has no frequencies and cannot be measured in Hertz like electromagnetic waves. It is a very special and unique type of energy and only now are we beginning to understand its importance and possibilities.

Although only discovered within the last 150 years, scalar energy has existed since the beginning of time. Scalar energy is found within in the vacuum of empty space, ranging anywhere from the empty atmosphere in the earth to the small spaces between your bodily cells. In nature, it can be found in desolate hills, rain forests and is generated by crashing ocean waves.

Perhaps, the most promising benefit of scalar energy is its healing properties. It has been known to reduce and can possibly eliminate many forms of disease, but since its discovery more 150 years ago, knowledge on scalar energy has been held in secrecy by a select few *(most notably by the military industrial complex)* who were able to understand its complexity *(and its incredible far reaching usage)*. [Italics added by author for emphasis].
http://www.slideshare.net/scalarenergyproducts

What is scalar energy? Simply put, it created when two opposing forces of common electromagnetic waves collide. As soon as the two opposing forces collide, they immediately cancel one another out, creating a stationary or static form of energy we know as scalar energy.

Two of the most important properties of scalar energy is that one, it is not **Hertzian**, meaning it cannot be measured as regular electromagnetic waves can, and two, it is non-linear like regular electromagnetic waves. This energy is in a class of its own with many distinguishing properties. One of the more fascinating properties of scalar energy is that unlike most forces that occur in the world such as waves, rolling objects, or any directional force, scalar energy does not decay or diminish over time. Also, the motion of scalar energy differs from that of regular electromagnetic waves. Regular waves tend to be sent out in beams, or running through cables. Scalar energy radiates outwards from the point of collision with another wave, which resembles circles.
http://www.slideshare.net/scalarenergyproducts

Another extremely rare and unique quality of scalar energy is that the space the energy equals is not a vacuum, meaning, the energy is unbounded to anything and can freely move through any solid object. The space that scalar energy fills tends to be networks of harmoniously balanced energies. Yet another rare quality of scalar energy is that it is freely created throughout the universe, or can be manually created via the collision of electromagnetic waves.

As previously mentioned, a distinguishable property of scalar energy is that, unlike regular electromagnetic waves, scalar energy cannot be measured in **Hertz**. The only way to be able to view the effect that scalar energy has or to measure its concentration is through techniques such as **Gas Discharge Visualization (GDV), or Kirlian photography**.
http://www.slideshare.net/scalarenergyproducts

Another theory considers a more natural phenomenon, the weather that night was stormy with many lightning strikes in the area and that the UFOs which crashed became the unfortunate victims of Mother Nature's own natural fury.

First-hand eyewitness accounts support the evidence that two ET spacecraft crashed that evening, one in the Corona area northwest of Roswell and the other on the Plains of San Agustin southwest of Socorro. **The White Sands Missile Test Range** and the Trinity site where the first atomic bomb was detonated are located just south of Socorro and west of Roswell.*(These areas should be kept in mind as they come up frequently throughout UFO literature as places of re-occurring UFO sightings.)* [My italics added].

An office memorandum from the US Government during an FOI search indicates that "three saucers" crashed in New Mexico (not two) with each having three humanoid bodies on board. The evaluation from investigators concluded that high powered radar interfered with the navigation system of the flying saucers causing them to crash.

The first crash area or "site #1" is the field debris discovered by "Mac" Brazel on the Foster Ranch NW of Socorro, where an alien craft came down hard due to loss of its navigational controls from interference or "jamming" from high powered military radar. It hit the ground hard scattering structural debris across a half mile of the ranch's landscape whereupon it bounced back up in a long but shallow trajectory either under semi-control or in a non-responsive flight where it came down in a final impact site located approximately 40 miles away north of Roswell. This is "site #2" where the alien bodies were found.

The second saucer crash was on the Plains of San Agustin where more alien bodies were found amongst the wreckage. This is "site #3".

Now, it is possible after revision and correction of testimony by one of the witnesses that site #3 is actually site #2 and that site #2 is actually site #3 but, only the US Army would know this for sure as their modus operandi was crash retrieval and only the forensics from both crashes would determine the certainty of that part of the event.

270

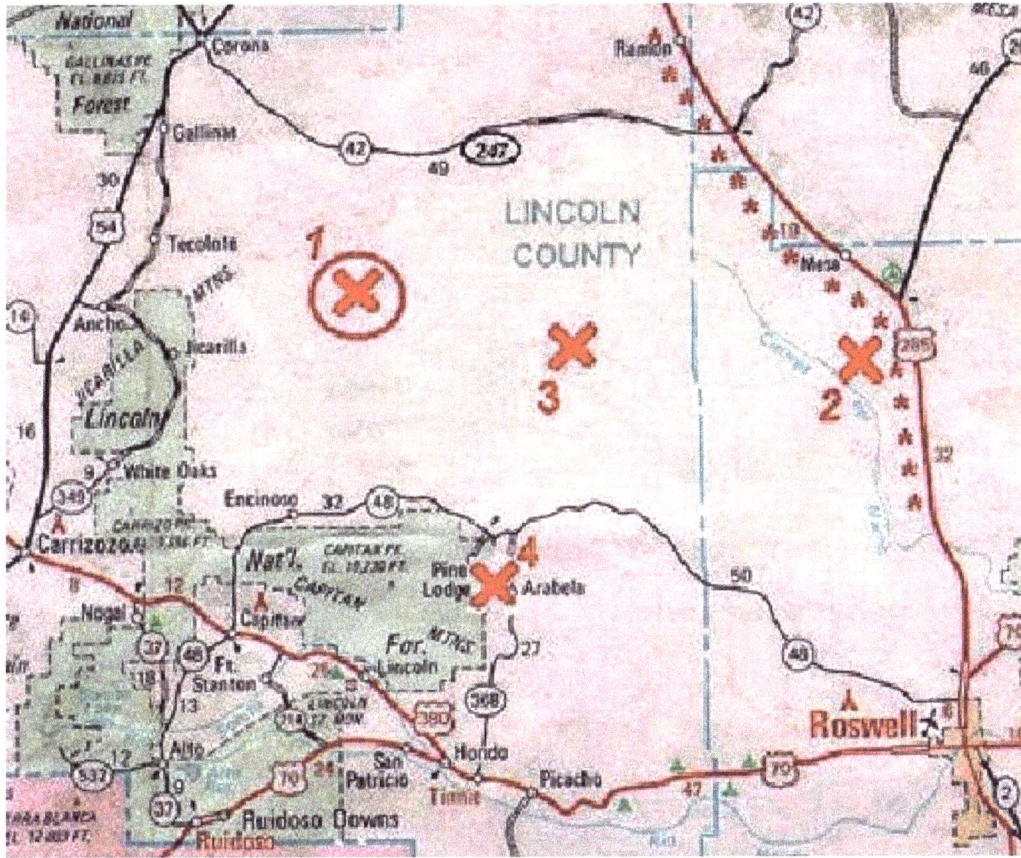

Saucer crash sites based upon witness testimony as compiled by David Rudiak from his website.

http://www.roswellproof.com/cordon.html

Depending upon who was interviewed and the testimony of the eyewitness, there is as many as six possible different crash sites for the same Roswell incident. Some sites may be considered as misidentification of the event locations, some sites may be mere hoaxing or

prevarication, while others are considered as the legitimate crash site locations.

Here, then is the beginning of some of the confusion in this landmark event that many Ufologists believe is the best UFO/ET case ever documented. Only time and government disclosure will tell if indeed it is the best case ever. If nothing else, it certainly is the most widely publicized and best-known case in the UFO literature.

Somewhere in the data and testimony is the correct information of the crash sites. It would appear that there were two crash sites based upon the independent research work of Freidman, Randle, and Schmitt.

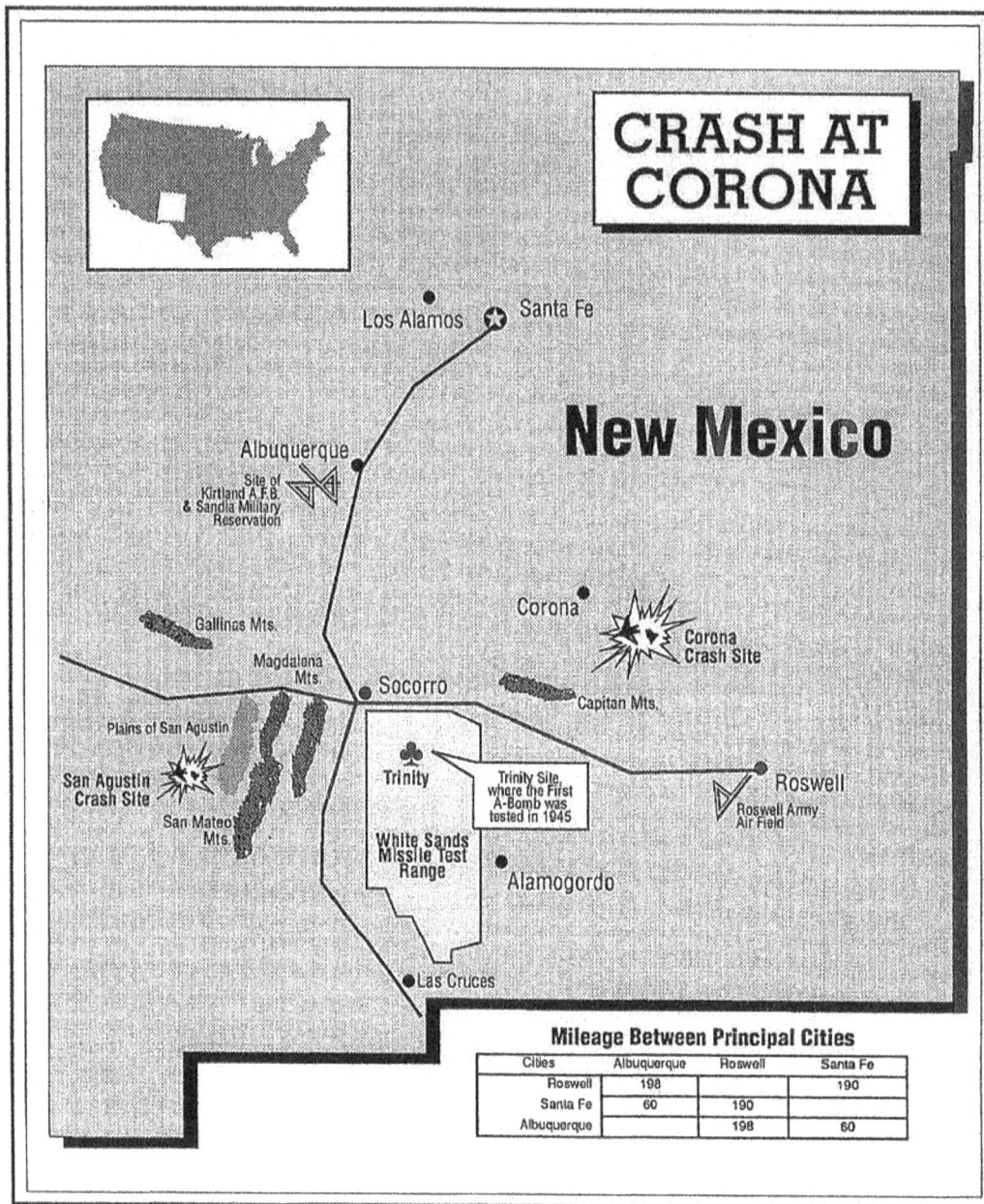

The crash sites as determine by Stanton T. Friedman and Don Berliner's investigation from his book "Crash At Corona"

The Current Accepted Version of the 1947 Roswell Saucer Crash

Here then, in a nutshell, is the story heard around the world which drew press coverage and public inquiry from every corner of the planet to the small southwestern town of Roswell, New Mexico in June 1947. It is a synopsis taken from Randle and Schmitt's book *"The Truth About the UFO Crash at Roswell"*. It provides key witnesses to various aspects of the event, public interaction with military counter-interaction and cover-up to prevent further public inquiry and scrutiny.

The event unfolds in **July 1, 1947** with radars in the White Sands, Alamogordo and Roswell area tracking an object which maneuvers and speeds around in an unconventional manner suggesting that it is not manmade, but of possible extraterrestrial origin. Checks of the radar equipment prove to be operating properly with no malfunctions.

The next day, **July 2nd.,** two witnesses, **Mr. and Mrs. Dan Wilmot** report seeing a typical saucer shape craft, "like two inverted saucers face mouth to mouth" fly over their house in Roswell moving at a high rate of speed toward the northwest.

On **July 3rd.** **Steve MacKenzie** sees and unidentified flying object flash overhead in the skies at the White Sands Proving Ground while investigating a V2 rocket accident.

July 4th., **William Woody** and his father watch a flaming object, white with a red tail fall toward the ground north of Roswell.

Re-creation of crash site in the Corona area of New Mexico with one surviving alien being. Note that the craft is triangular in shape with wings and stabilizer fins
http://www.crystalinks.com/roswell.html

During and intense thunderstorm near Corona, New Mexico, William W. "Mac" Brazel hears a tremendous thunderclap that more sounds like an explosion than thunder with reports from other witnesses stating the same phenomenon.

Radar sites in the area continue to monitor the unknown object which pulsates on the radar screen a few times, then explodes into a starburst image. Radar technicians conclude that the object has now crashed.

About the same time, **Jim Ragsdale and Trudy Truelove** spot a bright flash of light and heard a roaring sound as it passes overhead. Ragsdale senses that something has struck the ground near to their campground.

Events begin to rapidly unfold as witness after witness to various aspects of the event become involve in the otherworldly incident and begin to recognize that something fantastic has just occurred with the military and various intelligent departments trying desperately to cover up

and secure the crash site while shutting down any public information release. On **July 5th** a sheepherder and friend of Mac Brazel in central New Mexico comes across the remains of a crashed saucer. Not wishing to tell anyone what he had found, it isn't until many years later that he reveals what he had discovered. **The Truth About the UFO Crash at Roswell by Kevin D. Randle and Donald R. Schmitt; 1994; Published by M. Evans and Company, New York, New York; ISBN 0-87131-761-3**

W. Curry Holden, an archaeologist working the sites about 35 miles north of Roswell along with other archaeologists stumble across the impact site where the object has crashed and discover small humanoid bodies **[*shades of Indiana Jones!!!*]** (My italics added for effect).

One of the men heads out to find the nearest phone to call **Sheriff George Wilcox** to report that his team has found a crashed aircraft of some kind. Wilcox, in turn, calls the fire department and a fire truck with **Dan Dwyer** on board escorted by a Roswell police vehicle eventually meet up with the other civilians at the impact site.

Early in the morning, the military move into the crash site with a selected team highly trained in the recovery of the craft to find that civilians are already investigating the saucer and alien bodies. The civilians are quickly escorted off the area with dire warnings to keep what they have seen a secret. Five alien bodies are recovered and placed into lead-lined body bags. Only the highest clearance personnel are allowed at ground zero. Guards are ordered to face outward around the impact site to keep the curious away. The site is cleared and secured by mid afternoon.

At the same time that the military are securing area near the archaeology work site, Brazel is up early riding out with the young son of **William D. Proctor** to inspect the pastures around the ranch house after the rain storm the night before. Brazel during his inspection comes upon a large debris field about three-quarters of a mile long and thick enough to cause the sheep to refuse crossing over the field thereby forcing them to travel over a mile around it to reach water.

274

The debris consists of metal, plastic like beams, pieces of lightweight material, foil and string like substance.

John McBoyle, a reporter for a Roswell radio station KSWS tries to get to the crash site and see a crushed "dishpan" and while phoning Lydia Sleppy at a parent radio station in Albuquerque, she overhears a heated argument between McBoyle and someone else. Mcboyle is forced to tell Sleppy to forget the story but she decides to put out a message on teletype which is immediately intercepted by the FBI in Dallas and is ordered not to complete the transmission.

Melvin E. Brown, a guard at the crash site is ordered into the back of a truck and told not to look under the tarp. But, as it is human nature to sometimes do the opposite to what one is told, he looks under the tarp when everyone's back is turned and sees the bodies of the alien flight crew. He describes them as small beings with large heads with yellow to orange coloured skin. The Truth About the UFO Crash at Roswell by Kevin D. Randle and Donald R. Schmitt; 1994; Published by M. Evans and Company, New York, New York; ISBN 0-87131-761-3

Glenn Dennis, the mortician at the Ballard Funeral Home in Roswell receives a call from the military base mortuary officer asking about small caskets. A second call from the same base mortuary officer is made to Dennis, this time asking questions about various chemical would do to blood and tissue and the procedures for a body that has lain out in the elements.

Dennis also operates the ambulance from the funeral home and gets another phone call from downtown Roswell that an airman has injured himself and needs to be taken to the airbase right away. He has no problems being waved through the base security gate to the hospital. Parking in the rear at the loading dock, Dennis sees small canoe-like devices and strange debris being unloaded next to three other ambulances.

Two officers confront Dennis, one is a red-haired captain that threatens Dennis that he has seen nothing and heard nothing and should he open his mouth to anyone, they will be picking his bones out of the desert sand.

The alien bodies arrive in the base hospital for examination where **Dr. Jesse Johnson** pronounces them dead. Two other doctors arrive by special flight at the base who then begin a preliminary autopsy. Melvin Brown along with other soldiers stand guard outside the hangar and deny entry to anyone even including Brown's commanding officer who wants to get a look inside.

In the meantime, Brazel takes a few scraps of the strange material over to the home of his nearest neighbours, **Floyd and Loretta Proctor** to show them is recent find. They try to burn and cut some of the material but, are unable to do so and the Proctors suggest that Brazel take it into town to the sheriff.

The military is on high alert along the West Coast with fighters on airborne alert should flying disks show up looking for their comrades. Some of the Oregon and Washington planes have gun cameras on board.

Brazel returns to the debris field that evening and retrieves a large, circular piece of material that he hauls away with his truck to a nearby livestock shed three miles away where he stores it.

The childlike bodies of the strange beings are sealed into large crates which are then taken to a hangar [*is this, the infamous Hangar 18, so often reported with the Roswell Incident?]* (My italic added) and MPs stand guard overnight, kept only company by a spotlight focused upon the crates.

Early Sunday morning, **July 6th.** Brazel drives into Roswell after completion of his chores and heads to Sheriff Wilcox's office. Wilcox is excited about the find and suggests that the next best course of action is to notify the Roswell Army Air Field.

William Woody and his father try to drive out to the area where they saw the object coming down but all the roads to the area from Vaughn to the west are blocked and guarded by military police allowing no one to pass. **The Truth About the UFO Crash at Roswell by Kevin D. Randle and Donald R. Schmitt; 1994; Published by M. Evans and Company, New York, New York; ISBN 0-87131-761-3**

The air intelligence office, **Jesse A. Marcel** under orders from **Colonel William Blanchard**, commanding officer of the 509th Bomb Group investigates the report from Sheriff Wilcox. He drives to the sheriff's office interviews Brazel and examines the pieces of material. He then decides to visit the debris field for a closer examination and takes some of the debris with him back to the base and reports to Blanchard.

Blanchard is convinced that he is now in possession of something unusual possibly Soviet in origin and alerts higher command. At no time is the material considered any type of balloon.

Marcel returns to the sheriff's office with a senior counterintelligence agent assigned to the base, **Captain Sheridan Cavitt**. They escort Brazel back to his ranch and examine the debris field. The distance to the ranch is long and over rough roads and they do not arrive until after dark. They stay at the "Hines" house near the debris field and wait for first daylight. During that time Marcel checks the large piece of wreckage that Brazel had stored with a Geiger counter but it is found to be non-radioactive.

Blanchard acting on orders from **Major General Clements McMullen**, deputy commander of **Strategic Air Command (SAC)** obtains more debris from sheriff's office. It is sealed in courier pouch and loaded on an airplane destined to Fort Worth Army Air Field and handed over to **Colonel Thomas J. DuBose** where it is transported to Washington, D.C. and General McMullen.

Sheriff Wilcox Deputies having been sent out to check the debris field return to say that no debris field had been found but, observed a burned area in one of the pastures stating that the sand in the area had turned to glass and appeared blackened as if some circular object had touched down

By Monday, **July 7th.,** a 2:00 am flight leaves for Andrews AAF in Washington, D.C. with some of the debris and the bodies on board.

276

Brazel has taken the two military officers to the crash site which was an area three-quarters of a mile long by two to three hundred feet wide. There was a gouge at the northern end of it extending four to five hundred feet toward the other end. It appeared as if something had touched down and then skipped along. The largest piece of debris was recovered at the southern end of the gouge.

For most of the day, the debris is collected and loaded into the Marcel's car and Cavitt's jeep.

The wreckage was thin like newsprint but incredibly strong. There was foil found in the debris that when crumpled would unfold itself without a wrinkle, I-beams that flex slightly had strange symbols on them and there was material that resembled Bakelite. By dusk that evening the men head back into Roswell.

News has gotten out into the general public somewhat sporadically and news reporters smell a story in the air. **General Carl Spaatz** commander of the army air forces on "vacation" in the Northwest tells reporters that he knows nothing about flying disks or any local units out in search of them.

Lieutenant General Nathan F. Twining, commander of the Air Materiel Command, the parent organization at Wright Field, Ohio changes his plans and flies into Alamogordo.

At 2:00 am on **July 8th** Marcel stops at his home on his way to the base. He awakens his wife Viaud and his son, Jesse Jr; to show them the unusual material he has collected from the Proctor ranch. For the next hour the debris is examined on the kitchen floor and Marcel Sr. states that it is from a flying saucer. Jesse Jr. is intrigued with the strange symbols on the "I" beams and remembers that night many, many years later. The material is gathered up and reloaded into the car and Marcel heads back towards the base.

Blanchard receives the report from Marcel and Cavitt of what they observed and now takes control of the situation by ordering the provost marshal to post guards on the roads around the debris field with no access to anyone not on official business. Brazel is to direct MPS to the crash site.

Blanchard calls the eight Air Force headquarters and advises them of the new find. The message goes up the chain of command to SAC headquarters. No one at this time considers the material is from a Soviet device.

Blanchard convenes an early staff morning meeting to discuss the new find and its disposition. In attendance are Marcel, Cavitt, **Lt. Col. James I' Hopkins**, the operations officer; **Maj. Patrick Saunders**, the base adjunct; **Maj. Isidore Brown**, personnel officer; **Lt. Col. Charles W. Horton, Lt. Col. Fernand L. Andry, Lt Walter Haut,** and possibly **MasterSergeant, Lewis Rickett** was also there in attendance.

Walt Whitmore Sr. of KGFL having been intrigued by the story of fellow KGFL reporter and announcer Frank Joyce's interview with Mac Brazel wished to interview him further. On July8th, Whitmore had taken Brazel to the military base and then returns to the radio station to broadcast the wired recorded interview. Before he can even do so, he receives a phone call from Washington telling not to air the interview with Mac Brazel. Breaching this order would cause the station to lose its Broadcast license.

Blanchard confers by phone **Brigadier General Roger Ramey** who orders Marcel to Fort Worth.

Mac Brazel is interrogated by military officers and is kept on base for several days staying at a guest house.

Walter Haut finishes a press release as he'd been ordered to write and heads into town taking it to both radio stations in Roswell and both daily newspapers.
Wondering what is happening at the crash site, sheriff Wilcox has two more deputies go back out to the site but they run into the cordon thrown up by the military and are turned back as unauthorized personnel.

By 2:00pm the story is out on the AP wire announcing: *"The army air forces here today announced a flying disc has been found."*

Phones start ringing at the base and an irritated Blanchard unable to get a phone line out, orders Haut to do something about the incoming calls but there is nothing he can do about incoming calls. The Truth About the UFO Crash at Roswell by Kevin D. Randle and Donald R. Schmitt; 1994; Published by M. Evans and Company, New York, New York; ISBN 0-87131-761-3

FBI director J. Edgar Hoover receives a teletype message from a previous inquiry telling him that a balloon was responsible for the reports and it is now on its way to Dayton for examination by Army air force experts.

278

Publicity photo of Major Jesse Marcel as he poses with rawin weather balloon debris which had replaced the real saucer disc debris which he had brought into General Ramey's office earlier

United States Air Force/AFP/Getty Images http://www.crystalinks.com/roswell.html

By late evening, AP breaks into its last message with a bulletin telling the world that the Roswell flying disk was nothing more than a balloon. Ramey re-enforces this bulletin by announcing that the officers at Roswell had been fooled by misidentifying a weather balloon and appears on Fort Worth-Dallas station WBAP. On **July 9th** newspapers and wire services are now trumpeting the "flying saucer" found in Roswell is nothing more than a weather balloon as "Gen. Ramey Empties Roswell Saucer" story. **The Truth About the UFO Crash at Roswell by Kevin D. Randle and Donald R. Schmitt; 1994; Published by M. Evans and Company, New York, New York; ISBN 0-87131-761-3**

However, the wreckage is still being cleaned up and sent to Roswell Army Air Forces base where it is being loaded onto C-54's for Kirkland air base and then onto Los Alamos.

Mac Brazel is seen by **Floyd Proctor** and **Lyman Strickland** escorted by three military officers into Roswell and uncharacteristically ignores them.

279

A nurse that Dennis has been trying to see for some time, who works on the air base, finally arranges to meet with him for lunch. She reveals to him what has happened and gives him drawings of the alien bodies. She then states she feels sick and leaves for the barracks.

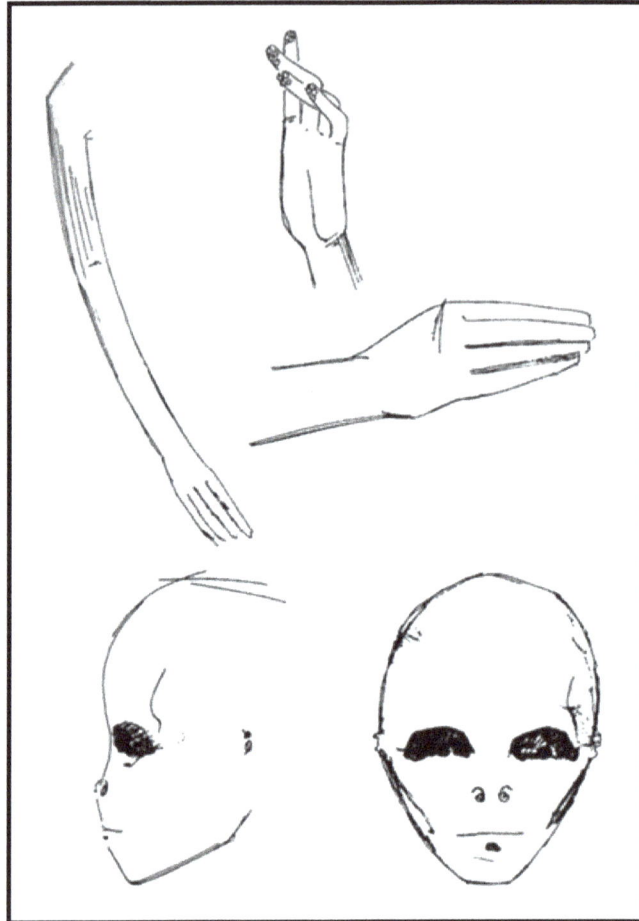

Drawing given to Glenn Dennis by a nurse friend
at the Roswell Army Air Force base
http://theudericus.free.fr/Ufologie/Ufologie.htm

Mac Brazel is taken to the Roswell Daily Record where he states that the debris was found on June 14 but, that he has found balloon wreckage before on two occasions and this debris was like no weather balloon he had seen before.

Copies of Lt. Haut's press release are retrieved from the radio station and news presses by an officer from the base. **The Truth About the UFO Crash at Roswell by Kevin D. Randle and Donald R. Schmitt; 1994; Published by M. Evans and Company, New York, New York; ISBN 0-87131-761-3**

Diversionary tactics are used on some officers and MPs including Marcel of a departure and arrival an aircraft of wreckage and alien bodies but, this deliberate ploy to confuse military personnel who may leak information. The bodies had already been sent to Andrews.

280

Mac Brazel calls on **Frank Joyce** to inform him that the story he gave on Sunday was different from the one he now gives stating that it "would go hard on him" if he didn't tell the new story. It becomes apparent that Brazel has been intimidated and/or threaten by military officers to tell a different story from his original testimony. Brazel is still under house arrest on the military base as a "guest" to ensure his full cooperation and to keep him away from other reporters.

Sheriff Wilcox is asked to surrender the stored material he has at his office, he does so without protest. Glen Dennis is also warned to keep silent on what he knows and Sheriff Wilcox is visited by a military sergeant to ensure he gets Glenn's silence.

From **July 11th** until the end of 1947 the military set about debriefing its personnel and everyone else associated with the recovery to forget what had happened and to never speak of it again.

Glenn Dennis is not able to contact his nurse friend who seems to have been transferred from the base to parts unknown. He hears from another nurse later that she had been killed in a plane crash while stationed in London, England.

Everyone in Roswell who knows something about the incident is asked never to talk about what happened, some even been threatened with death should they speak to anyone.

Mac Brazel is forced to take an oath not to reveal what he knows but is very bitter about the whole situation but honour his oath.

Bits and pieces of foil–like material are found by Bill Brazel, Mac Brazel's son and he shows it to Sally Strickland (later Tadolini). **The Truth About the UFO Crash at Roswell by Kevin D. Randle and Donald R. Schmitt; 1994; Published by M. Evans and Company, New York, New York; ISBN 0-87131-761-3**

In **September of 1947**, Lewis Rickett is assigned to assist Dr. Lincoln La Paz from the University of New Mexico to determine the possible speed and trajectory of the craft when it hit. They discover a touchdown point five miles from the debris field where the sand has been crystallized from some heat source and they find more of the foil –like material. La Paz concludes that the object was an unoccupied probe from another planet, unaware that alien bodies have already been recovered. He remains steadfast in this belief, years later. He goes on to work at various secret government projects, such as the "Ghost Rockets of Sweden."
Bill Brazel continues to discover more scraps of the alien craft which are confirmed by his father as the same material he had found two years earlier. Captain Armstrong and three other military personnel get wind of the recent find and show up demanding they hand over the material that Bill found reminding him of his father's patriotic oath of secrecy and cooperation. Bill surrenders the material to Armstrong.

From 1961 to 1991 various witnesses, both military and public (about three hundred in all) start coming forward to reveal their first-hand accounts of the crashed saucer event of 1947 to investigative reporter and UFO researchers.

By 1978 researchers like **Leonard Stringfield** and **Stanton Friedman** interview **Jesse Marcel** and his involvement in the event. In 1980, **Charles Berlitz** and **William Moore** publish *"The Roswell Incident"* and in 1989, **Dr. J. Allen Hynek** from the **Center for UFO Studies** makes the first scientific expedition to the crash site.

Randle and Schmitt have their book published by Avon Books in 1991, *"UFO Crash at Roswell"*. The Truth About the UFO Crash at Roswell by Kevin D. Randle and Donald R. Schmitt; 1994; Published by M. Evans and Company, New York, New York; ISBN 0-87131-761-3

What can we take away from the Roswell saucer crash incident of 1947?

From the research work of UFO investigators into the Roswell incident, many people were witnesses to a unique extraterrestrial event leading up to and including the actual crash of two alien spacecraft along with their alien pilots, both dead and alive, as well as to the aftermath process of wreckage retrieval and its transportation to various military bases all cloaked in clandestined secrecy.

At every possible turn in the events that unfolded where some public or military witness saw something, heard something or tried to reveal something of the unusual event to someone else or through the news media, the military were there from the top brass, down through the recovery teams and the intelligence officers "closing the doors" and "shutting down" all information leading to or coming out of the extraterrestrial event. No leaks as to what had happened were permitted, only those with a need to known were included in the "loop", all others were denied access or told in outright bold lies that that nothing happened or given some other plausible explanation that only a crashed **Rawin balloon** had been recovered.

So effective was this campaign of denial and secrecy surrounding the 1947 Roswell saucer crash that it took over 30 years before anyone could discover what the truth was behind the ET incident.

When witnesses came forward foregoing their oath to remain quiet as true patriotic citizens in the name of national security, the military strong arm of cover-up came into play again by debunking these first-hand witnesses. When that failed to silence their testimony, the military came out and re-stated that only a weather balloon had been found and that the military made an error in identifying it initially as a crashed disc. This did not have the effect they had hope and then, they decided that they would reveal the "real truth" of what they found on that day in July 1947. They had recovered one of their own Mogul high altitude balloons used to monitor possible radiation from atomic explosions that the Russian may have been testing.

The US Air Force felt that the Condon Committee Report on Unidentified Flying Objects had given them a way out of the UFO investigation business through **Project Blue Book** but, this was a short-lived victory. It did not satisfy Ufologists or the general public and mistrust for the military grew with each passing year.

The Roswell story would not die of natural causes as more damning evidence came out about the recovery of alien bodies and other alien craft. **The Majestic 12 Documents (aka. MJ 12 or**

282

Majic 12 documents, etc) investigated by Stanton Friedman sent everything over the top new evidence of top scientists and military involvement into the UFO phenomenon and the **"Canadian Connection"** via letters by **Wilbur Smith** to Sarbacher was the proverbial "nail in the coffin". **Crash At Corona by Stanton T. Freidman; 1992; Paragon House Publishers, New York, NY; ISBN 1-55778-449-3**

Add to this the digital blow up of the letter in the handoff General Ramey when the photos of him and Jesse Marcel holding the balloon debris were taken, they revealed and confirmed that a crashed disc had been recovered by the military in July 1947.

In the on-going public battle between the military and the public, notably with Ufologists, the military were determine to convince everyone that they had a firm grip on the truth. Two Air Force reports on the **Roswell UFO incident**, published in 1994/5 and 1997, form the basis for much of the skeptical explanation for the 1947 incident, the purported recovery of aliens and their craft from the vicinity of Roswell, New Mexico

The first report, *"The Roswell Report: Fact versus Fiction in the New Mexico Desert,"* identified a secret military research program called **Project Mogul** as the source of the debris reported in 1947. The second report, *"The Roswell Report: Case Closed"* concluded that reports of alien recoveries were likely misidentified military programs or accidents. http://en.wikipedia.org/wiki/Air_Force_reports_on_the_Roswell_UFO

A news article appeared in the Las Cruces local paper in 1997 by Phillip Klass stating that the Air Force was about to release a new report on the Roswell Crash of 1947 two weeks before the town would commemorate the 50th anniversary of the event. The report was aimed at addressing the complaints that a 1994 Air Force report, which explained alleged flying saucer debris as the remains of top-secret high-altitude balloons, did not explain away reported sightings of alien bodies.

The new report, drafted by **Air Force historian Capt. James McAndrew** suggested that purported witnesses have confused the 1947 incident with late 1950s tests. It will say the so-called saucer crash victims were dummies dropped from the sky. http://www.subversiveelement.com/roswell_crash_dummies.html

The military also suggested that people's memories and their recollections were at fault due to a unique phenomenon known as **"time compression of events"** which could affect people by confusing their recall. The so-called hairless alien bodies seen by witnesses were nothing more that "crash test dummies" used to test high-altitude free-fall accidents resulting from parachuting and that the alien craft that was found was an experimental landing craft that was somewhat saucer-shaped which was being tested and developed for future Mars landings.

It appeared that the military had a very reasonable explanation but, upon closer inspection, it would seem that it was the military who were suffering from the time compression syndrome. Their testing of anthropomorphic dummies occurred in the early 1950s but, the Roswell events happen in 1947! Many of the Roswell witnesses have stated years after the event that their memories of that day were intact, that there was nothing faulty with their recall of the facts. In the minds of the general public, it was the Military who were the dummies for coming up with such a lame-brained excuse. It would appear that the Air Force publicist had not done his homework or got his facts right. Perhaps, he had been tossed out on his head from a great height, one too many times without a parachute!!!

No doubt many Ufologists and other researchers must have had a good laugh around the water cooler when the new military explanation for Roswell hit the media airwaves. For some strange notion, the US Air Force still believes that the masses of the public are unintelligent and are incapable of thinking rationally. Their final explanation on the Roswell matter came in a public report released as "The Roswell Report: Case Closed". It did not have the effect of settling the issue in the minds of most Americans, but, rather had the opposite effect, it further distant public trust for the military and re-enforced the concept of cover-up and secrecy.

In fact, the public demanded more investigation be done into the Roswell incident. Polls, like a 1997 CNN/*Time* poll, suggested that a strong majority of Americans believed that the government was hiding evidence of the existence of aliens, and specifically, that the Roswell incident involved the recovery of aliens. **Poll: U.S. hiding knowledge of aliens - UFO Evidence**

In that context, many were demanding answers from their government on what really happened at Roswell in 1947, so in January 1994, **Congressman Steven Schiff** requested that the United States Congress' investigative branch, the **General Accounting Office (GAO),** look into the matter. The next month, the Air Force was informed of the GAO's planned formal audit. The Air Force was not the sole agency to be investigated, but it was the focus of the investigation as it had been consistently identified as most involved with the alleged cover-up. (The US Army Air Forces became the US Air Force in September 1947 and inherited all personnel, equipment, records etc.) The Secretary of the Air Force subsequently ordered an investigation to locate any information it had on the incident.

The result, published in 1994 and 1995 was a near-1000 page report entitled ***"The Roswell Report: Fact versus Fiction in the New Mexico Desert."*** The report was significant for identifying for the first time a likely source of the debris found on the Foster ranch: the remnants of a balloon train from a secret military program called Project Mogul. Though several others had suggested a Mogul program balloon as a possible candidate previously, the report had specific information that had never been revealed before about the program that led many to conclude that the "incident" had been explained.
http://en.wikipedia.org/wiki/Air_Force_reports_on_the_Roswell_UFO_incident

The problem with this report was that it was reminisce of the **University of Colorado's Condon Report on Unidentified Flying Objects** which was also sponsored by the USAF and this set alarm bells ringing and red flags waving in the UFO community. The Condon Report was also a thousand page tome of pre-ordained foregone conclusions, that there was nothing to the UFO phenomenon and so it would be with the **Roswell Incident of 1947**. One thousand page whitewash in 1969 was followed by yet another thousand page whitewash in 1994. Did the Air Force really believe that the public was still gullible enough into buying another prefabricated UFO report this time centered on a specific incident?

284

US Congressman Steve Schiff began to become interested in the Roswell Incident early in the 1990's when his constituents requested he find out what really happened in Roswell, New Mexico (July 1947). In January of 1994, Congressman Schiff announced to various members of the press that he had been getting the 'run-around' when trying to press the US Defense Department for information on the Roswell Incident. Soon afterward, he requested that the GAO investigate the government's involvement in the Roswell Incident.

The GAO requested information on the Roswell Incident from the White House, the FBI, the Department of Energy and the National Security Council. The GAO discovered that all administrative records and outgoing messages from the **Roswell Army Air Field (RAAF)** between March 1945 and December 1949 had been destroyed. The GAO estimates that the documents were destroyed 40 years ago and were unable to determine who destroyed them.
http://www.abductee.ca/seven/roswell_article4.html

UFO researchers found it alarming that the documents pertaining to the Roswell Incident were destroyed. Researchers also believed that the GAO did not press hard enough to get all the facts and documentation from US government organizations associated with the Roswell Incident.

The release of the GAO audit on the Roswell Incident was unable to put an end to the rumours of a massive government cover-up of a crashed extraterrestrial ship. In fact, this audit merely added to the mystery surrounding the Roswell Incident.
http://www.abductee.ca/seven/roswell_article4.html

When Congressman Schiff ordered the GAO to investigate and get to the bottom of the matter, the GAO found that:

(1) in 1947, Army regulations required that air accident reports be maintained permanently and although none of the military services filed a report on the Roswell incident, there was no requirement in 1947 to prepare a report on the weather balloon crash;

(2) although some of the records concerning Roswell activities had been destroyed, there was no information available regarding when or under what authority the records were destroyed;

(3) only two government records originating in 1947 have been recovered regarding the Roswell incident;

(4) a 1947 Federal Bureau of Investigations record revealed that the military had reported that an object resembling a high-altitude weather balloon with a radar reflector had been recovered near Roswell; and

(5) a 1947 Air Force report noted the recovery of a flying disc that was later determined by military officials to be a radar-tracking balloon. http://www.fas.org/sgp/othergov/roswell.html.

[The following story in part is based partly on an Associated Press release dated March 25, 1998.]

"**Congressman Steve Schiff**, a five-term Republican from Albuquerque, New Mexico's 1st Congressional District, died on March 25, 1998, of skin cancer. He was 51.

Schiff had been fighting aggressive squamous-cell skin cancer for more than a year. He had been absent from Washington, D.C. since last April but had continued to conduct congressional business from home while undergoing treatment that included surgery, radiation, and chemotherapy.

A spokesman for Schiff had denied as late as last month that the disease was terminal and said the congressman was responding well to treatment. Caught early, squamous-cell carcinoma reportedly has a 95% cure rate. But doctors treating Schiff recently expressed concern that his cancer was unusually aggressive.

Though Schiff had a long and distinguished career in public service, he will be remembered by readers of CNI News primarily for his courageous and unprecedented efforts to uncover secret government information pertaining to the Roswell incident of 1947. Responding to concerns of his constituents, Schiff made inquiries at the Pentagon and several government archives in 1993 before concluding that he was getting "the runaround."

Angered by what appeared to be stonewalling on the part of Pentagon officials, Schiff pushed the **General Accounting Office (GAO),** the investigative arm of Congress, to look into possible improprieties related to Roswell. An 18-month investigation by the GAO concluded that many records of government activity in Roswell during the summer of 1947 were mysteriously missing. The apparent destruction of those records was unusual and unauthorized, the GAO said, but without them, the investigation came to a standstill.

No government official ever pushed harder or more publicly for disclosure of Roswell-related government records than Congressman Steve Schiff. And, although there is absolutely no evidence of a connection between his GAO investigation and his cancer, UFO researchers are compelled to wonder why this particular Congressman came down with an unusually vicious disease so soon after his Roswell investigation ended." http://www.anomalies.net/archive/cni-news/CNI.0999.html

From the Associated Press, Congressman Schiff gives additional statements on his GAO investigation:

Subject: Roswell Documents Already Gone
Copyright: 1995 by The Associated Press
Date: Sat, 29 Jul 95 14:50:05 PDT

ALBUQUERQUE, N.M. (AP) -- Key military documents on the so-called Roswell Incident, cited by UFO buffs as an alien crash, were apparently destroyed without authorization decades ago, a congressman said Saturday.

Rep. Steve Schiff of New Mexico said a General Accounting Office report shed no new light on the 1947 crash and showed that important documents are missing.

286

"Documents that should have provided more information were destroyed," Schiff said. ``The military cannot explain who destroyed them or why."

Schiff said the GAO estimates the information was destroyed more than 40 years ago.

The Air Force has said that the wreckage was probably a balloon launched as part of a classified government project to detect Soviet nuclear weapons.

The GAO report, released Friday, said that two government documents are the only official records remaining of the crash near what was then the Roswell Army Air Force Base.

For nearly half a century, the mysterious crash has fueled speculation about aliens in the New Mexico desert, Cold War secrecy and a government cover-up.

"The debate on what crashed at Roswell continues," the GAO report said.

It said the Roswell base's administrative records from March 1945 through December 1949 and its outgoing messages from October 1946 through December 1949 were destroyed.

Those messages, internal military communications, would have shown how military officials in Roswell explained what happened to their superiors, the Republican congressman said.

``My understanding is that these were permanent records which should not have been destroyed," Schiff said.

Scientists and Pentagon officials have said an experimental aerial surveillance balloon crashed northwest of Roswell in 1947, but UFO buffs have contended that was a cover up for the crash of an alien spaceship.

The GAO report includes an FBI teletype and a reference to a ``radar tracking device," or weather balloon.

The weather balloon story has since been discredited by the Air Force itself, which last year said the wreckage was probably a balloon launched as part Project Mogul. The project was a highly classified effort to detect Soviet nuclear weapons using balloons that carried radar reflectors and acoustic sensors, the GAO report said. http://www.v-j-enterprises.com/gao.html

It appeared to Congressman Schiff that the US Air Force was stonewalling him and giving the GAO the runaround on the Roswell investigations. Some UFO researchers believe that his tenacity in the investigation leads to his eventual demise through covert action by some department or arm of the Air Force.

This is a difficult aspect to prove but, at the same time Steven Schiff came down with an aggressive form of facial skin cancer, Dr. Steven Greer, director of **CSETI** and his assistant **Shari Adimak** came down with aggressive forms of cancer as well. Dr. Greer had melanoma from which he recovered through treatments but, Shari had breast cancer which spread to other

287

parts of her body eventually taking her life. In the latter two cancer cases, Dr. Greer concluded that he and Shari had been targeted with **microwave/scalar weapons** that induced their cancers. Dr. Greer feels strongly that Congressman Schiff had also been a target by covert operatives because of his fierce determination in getting to the truth of the Roswell Incident. The technology to induce disease into people has existed for decades and obviously, bacteria warfare development is big business within the military circles so, targeting and inducing cancer or any other disease in people is definitely achievable.

Before leaving this historical UFO account, two other aspects have arisen to greatly complicate this story and add still further controversy to an already controversial event. One has since been proven to be a cleverly hoaxed piece of film of a supposed alien autopsy, known as the **Ray Santilli Alien Autopsy** and the other the **Majestic 12 Documents** which name names of people and scientists involved in the **Roswell Incident** cover-up and although, very controversial in its own right seems to have more than enough evidential weight behind to be authentic.

The Roswell saucer crash of 1947 was an opportunity for the US Military to shakedown its crash retrieval teams to get the bugs out of their operations. The US government and its military have known since the earlier part of World War II that the Earth was being monitored by Extraterrestrial Intelligence operating saucer shape craft. The acquisition of first-hand evidence in the form of alien craft and bodies was paramount to understanding their technology and their occupants. The alien spacecraft that crashed onto terra firm in Roswell was the opportunity the US Military had been looking for even if that recovery of an alien craft was less than stellar in covering up the evidence of their retrieval program. Part of the weakness in maintaining a suppression of information was that many of the military people on the retrieval team were young men with no experience for this type of recovery scenario. Contrary to how much and how often they were told by their superior officers to keep the information of what they saw to themselves, many either before their deaths or upon their deathbeds revealed to their families or to some UFO investigator what had happened back on that historic day, this was something that they could not take with them to their graves.

We shall see that in other saucer crash retrievals that occurred in the USA and in other nations friendly to America, a tighter level of security and information control or suppression around the event sites was maintained. The problem, however, in any crash retrieval, it appears that there is always one or more witnesses to the event, whether this be from the general populace of a neighbouring community or from an inside source who participated in the actual retrieval incident. Sparsely populated or remote regions of the planet or the passage of time are not guarantees that news of the alien crash will not go unnoticed or undiscovered or find its way eventually to the attention of the public.

May 24, 1948, Aztec Saucer Crash

The **Aztec Saucer Crash** on a desert plateau in Aztec, New Mexico on May 24, 1948, had been a well-kept secret for 40 years. There is substantial evidence that a 100-foot diameter spacecraft from another world piloted by intelligent human-like beings with technologies far superior to our own, lost control and made a forced landing in the remote area of Aztec, N.M. It was

288

immediately recovered by a special military unit who were trained and prepared to conduct such operations in great secrecy.

The original story found its way to **Dorothy Kilgallen**, whose brief report from her investigations appeared in the Hearst Syndicate newspapers, this was picked up by **Walter Winchell** in New York and **Frank Edwards** in Washington. But it was **Frank Scully**, a columnist out west, who pursued another lead on the story in a more systematic investigation by contacting first-hand witnesses to the event. His account of the case came out in a book titled "Behind the Flying Saucers". Much later in 1981 **William S. Steinman** and the late **Colonel Wendelle C. Stevens** collaborated and re-investigated this suppressed UFO account to obtain a more detailed understanding of what took place back in 1948. Their investigative report appears in their book "UFO Crash at Aztec: A Well Kept Secret".

For years many Ufologists considered Scully's account of the Aztec Crash to be a hoax because of the duplicitous behavior two oilfield con men, **Silas Newton and Leo GeBauer**, who were trying to peddle a "**doodlebug device**" which supposedly could find oil, precious metals, and UFOs. These two men knew something of the crashed-saucer incident which came to them through leaked second or third-hand informants. Although Frank Scully's investigations were not thorough by today's standards of UFO investigation, his information of the incident is essentially correct. Interestingly, the character of Dana Scully from the popular sci-fi series, the "**X-files**" is named after him.

Late breaking news: the "X-Files" series is to resume after a twenty-year hiatus with filming back in Vancouver and interior of B.C. beginning in spring 2015!

Frank Scully (1892-July 23, 1964) was an author in the 1940s and 1950s, who wrote for the show business publication *Variety*.
https://en.wikipedia.org/wiki/Frank_Scully

Scully claimed in 1950 when his book came out that up to that time there had been four such recoveries, one of which was alleged to have taken place around Aztec, New Mexico when sixteen humanoid bodies were recovered together with their undamaged craft. According to Scully's informants, the disk that landed near Aztec was 99.99 feet in diameter, its exterior made of a light metal resembling aluminum but so durable that no amount of heat (up to 10,000 degrees was applied) or diamond-tipped drill had the slightest effect. The disk apparently incorporated large rings of metal which revolved around a central, stabilized cabin, using an unfamiliar gear ratio. There were no rivets, bolts, screws or signs of welding. Investigators were eventually able to gain entry. Scully was told, because of a fracture in one of the portholes, which they enlarged, revealing a knob inside the cabin which when pushed (with a pole) caused a hidden door to open. Sixteen small humanoids, (there were actually 14 bodies) ranging in height from 36 to 42 inches, were supposedly found dead inside the cabin, their bodies charred to a dark brown color. Scully was told that the craft landed undamaged, having landed under its own guidance.

290

The craft was eventually dismantled, the investigators having discovered that it was manufactured in segments which fitted in grooves and were pinned together around the base. The complete cabin section, measuring 18 feet in diameter was lifted out of the base of the saucer, around which was a gear that fitted a gear on the cabin. These segments, together with the bodies, were then transported to **Wright Field (Wright-Patterson AFB).** Some of the bodies were later dissected and examined by the Air Force and were found to be similar in all respects to human beings, with the exception of their teeth, which were perfect. UFO Crash at Aztec: A Well Kept Secret" By William S. Steinman, Contribution by Wendelle C. Stevens; Copyright 1986 by UFO Photo Archives; Privately Published by Wendelle C. Stevens; Distributed by America West Distributors; ISBN 0-934269-05-X

Steinman discovered that the Aztec disk came to earth having been detected by three separate radar units in the southwest, one of which was said to have disrupted the craft's control mechanism. The spacecraft like the Roswell saucer that crashed in 1947 had lost its navigation control due to high powered ground radar interference. The argument against this being the reason for loss of navigational control is that our own aircraft are not affected by radar so, why would a more superior technology be adversely affected? This line of reasoning is anthropocentric and shows the flaw in human thinking as we have no idea of the type of technology used by Extraterrestrials in their spacecraft, which may or may not be affected by radar; it is after all by definition alien in origin!

The area of impact was calculated by triangulation and this information was immediately relayed to Air Defense Command and Gen. George C. Marshall, then Secretary of State, who allegedly contacted the MJ-12 group as well as the Interplanetary Phenomenon Unit (IPU) of the Army Counterintelligence Directorate. The IPU operated out of Camp Hale, Colorado, at this time, Steinman claims, and its main function was to collect and deliver disabled or crashed disks to certain specified secret locations. The craft was recovered within hours by the IPU scout team about 12 miles northeast of Aztec. General Marshall ordered Air Defense Command to go off alert status, and the radar units were advised that there had been a false alarm. Marshall then gave orders to the commander of the IPU to organize a recovery team and contacted Dr. Vannevar Bush - the head of MJ-12 to gather together a team of scientists to accompany the IPU to the crash site. Steinman has named these scientists as follows:

Dr. Lloyd Berkner,   **Dr. Detlev Bronk,**

Dr. Carl A. Heiland,   **Dr. Jerome Hunsaker,**

Dr. John von Neumann,   **Dr. Robert J. Oppenheimer,**

Dr. Merle A. Tuve,   **Dr. Horace B. van Vandenberg**

Four of these scientists, it will be noted, were members of the original MJ-12 panel set up in September 1947. **Dr. Carl A. Heiland** was a geophysicist and magnetic sciences expert who was the head of the Colorado School of Mines and according to Steinman leaked details of the recovery to one of Scully's sources, **Leo GeBauer**. **Dr. Horace B. van Vandenberg** was an inorganic chemist associated with the University of Colorado. **Dr. Merle A. Tuve** worked for the Office of Scientific Research and Development during World War II and is chiefly remembered as a geophysicist for his techniques of radio wave propagation of the upper atmosphere. **Dr. Robert J. Oppenheimer** distinguished himself primarily as the leader of the Los Alamos atomic bomb project, commanding the allegiance of the world's top physicists. He was the Director of the Institute of Advanced Studies at Princeton from 1947 and became Chairman of the General Advisory Committee of the Atomic Energy Commission. **Dr. John von Neumann**, the famous Hungarian-born mathematician, became a consultant on the atomic bomb **(Manhatten Project)** in 1943. His main area of expertise lay in the design and development of computers. The scientists, according to Steinman, were told by Dr. Bush to assemble at Durango Airfield, Colorado, 35 miles to the north of Aztec, with the minimum delay. All those involved in the recovery were sworn to an above top secret oath. UFO Crash at Aztec: A Well Kept Secret" By William S. Steinman , Contribution by Wendelle C. Stevens ; Copyright 1986 by UFO Photo Archives; Privately Published by Wendelle C. Stevens; Distributed by America West Distributors; ISBN 0-934269-05-X

The IPU convoy used a route to the site that avoided main roads, and on arrival, road blocks were set up at strategic points within two miles of the recovery area. The owner of a ranch and his family were allegedly held incommunicado and told never to discuss the matter (cf. the Roswell incident). Equipment hauling trucks were camouflaged to look like oil drilling rigs during the operation.

A team of scientists arrived at the site a little later than the IPU team and began dissecting the disk. According to Steinman, they entered the craft one by one, entry having been gained via a fractured porthole as described in Scully's account. The portholes themselves looked metallic and only appeared translucent on close inspection. Inside the craft, they found two humanoids, about two feet in height, slumped over an instrument panel charred deep brown. Another 12 bodies lay sprawled on the floor in a chamber within the cabin, making a total of 14 bodies (not 16 as Scully had been told).

An instrument panel supposedly had several push buttons and levers with hieroglyphic-type symbols, as well as symbols illuminated on small display screens. Bush and von Neumann discovered that the control panel had drawers which rolled out, but no wiring could be detected. A book composed of parchment-like leaves with the texture of plastic also contained the strange hieroglyphs - similar to Sanskrit, Oppenheimer thought. This was given to General Marshall, who then passed it on to two leading cryptological experts for analysis, **William F. Friedman and Lambros C. Callihamos** (who both later led distinguished careers in the National Security Agency).

Dr. Bronk, a physiologist, and biophysicist, examined the bodies and asked Bush to get hold of cryogenic equipment with which to preserve them. **Cryogenics specialist Dr. Paul A. Scherer**, a colleague of Bush's, was contacted and advised Bush to obtain some dry ice. Meanwhile, another small group of scientists and military personnel examined the craft and were eventually able to dismantle it when several interlocking key devices were found which opened up seams at specific points. UFO Crash at Aztec: A Well Kept Secret" By William S. Steinman, Contribution by Wendelle C. Stevens; Copyright 1986 by UFO Photo Archives; Privately

Published by Wendelle C. Stevens; Distributed by America West Distributors; ISBN 0-934269-05-X

Three days later the segments were loaded onto three trucks, together with the bodies, and transported with a tarpaulin marked "Explosives". The convoy headed at night by the least conspicuous and often most laborious route to the restricted Naval Auxiliary Airfield Complex at Los Alamos, arriving one week later. Here they remained for over a year, Steinman claims, before being transported to another base.

Dr. Paul A. Scherer eventually obtained special preservation containers for the least damaged bodies, Steinman relates. One of the companies which supplied equipment was the Air Research Corporation, of which Scherer was Director of Research and Development; it supplied the liquid nitrogen pump, circulation system, and refrigeration units. Other specimens were given a complete autopsy, by a team headed by Dr. Bronk, of biophysicists, histochemists and pathologists. The results were put in a report, part of which, Steinman claims, appeared in the **"*Air Force Project Sign (Grudge) Report No. 13*"** which has never been released.

According to the report, the bodies were described as averaging 42 inches in length. The facial features strongly resembled "mongoloid Orientals" in appearance, with disproportionately large heads, large "slant" eyes, small noses and mouths. The average weight was about 40 pounds. The torsos were very small and thin, with very thin necks. The arms were long and slender, reaching the knees, with hands containing long and slender fingers with webbing between them. There was no digestive or gastrointestinal tract, no alimentary or intestinal canal, and no rectal point. No reproductive organs were apparent. Instead of blood, there was a colorless liquid with no red cells which smelled similar to ozone. *[At no time is it reported that these beings were the common "Grey" type aliens that have become so popular in the UFO literature]*. (My italics added for emphasis) UFO Crash at Aztec: A Well Kept Secret" By William S. Steinman, Contribution by Wendelle C. Stevens; Copyright 1986 by UFO Photo Archives; Privately Published by Wendelle C. Stevens; Distributed by America West Distributors; ISBN 0-934269-05-X

Veteran researcher Leonard Stringfield, a former Air Force intelligence officer, who is the world's leading specialist on what he calls **"Retrievals of the Third Kind,"** has uncovered further evidence himself. **Captain V. A. Postlethwait**, who was on detached service with **Army G-2 (Intelligence)** in 1948, told Stringfield that he was cleared to see a top secret cable describing the crash of a saucer-shaped craft 100 feet in diameter and 30 feet high, with one porthole broken, causing suffocation to the five occupants - who had turned blue as a result. The bodies were about four feet tall with relatively large heads, Postlethwait recollects. The metallic skin of the saucer was too tough to penetrate, although as thin as newspaper. The incident was said to have occurred near White Sands, New Mexico. Aside from a few discrepancies, there are some significant parallels with the Aztec case. Postlethwait revealed to Stringfield, for example, that private property was purchased to facilitate transporting the craft.

Leonard Stringfield has also spoken with **Dr. Robert Spencer Carr**, a retired University of South Florida professor who claims to have testimonial evidence from five sources, including a nurse and a high-ranking Air Force officer who participated in the recovery of a crashed UFO and occupants in 1948 - presumed to be the one at Aztec (although there was another alleged recovery that year, just across the Mexican border near Laredo, Texas.

According to Bill Steinman, two of Carr's sources were aeronautical engineers who provided important information regarding the saucer's construction and propulsion. A source now named is **Arthur Bray** (not to be confused with the Canadian researcher), a security guard involved with the recovery project. Carr also interviewed a woman whose father was present during the recovery.

At the still fenced-off crash site on a plateau twelve miles northeast of Aztec, Bill Steinman has uncovered charred and scraped-off rocks of various sizes as well as some metal bracing struts that might possibly have been used for supporting the craft. On one of his visits to the area, he was shadowed by two unmarked helicopters.

Steinman has traced at least four people who knew where the crash site was located, one of whom, "V.A.," recalls that sometime between 1948 and 1950 a huge disk-shaped flying object with a dome on top skimmed about 100 feet above the ground not far from him. The witness pointed out to Steinman a cliff jutting above the Animas River.

"That thing, or flying saucer, tried hard to clear that cliff, but it hit the very corner up there, shooting sparks and rocks in every direction," he claims. "Finally, it made a right-angle turn in midair and headed straight north in the direction of the alleged crash site at Hart Canyon. That's the last I saw of it. I ran into the house and called the military in Albuquerque. I never heard from them about it." http://www.aztecufo.com/crash.htm

May 21, 1953, Kingman Saucer Crash

The earliest reference of a crash near Kingman has been made to **MUFON** researcher **Richard Hall** in April 1964. A UFO is alleged to have crashed 8.1 miles NW of Kingman AFB, now Kingman airport (see map below). He was told the story by a future Vietnam commander. The case of the **Kingman UFO retrieval** was then brought to the public attention by **Raymond Fowler**, a respected UFO researcher, in June 1973.

*"**Fritz Werner" (Arthur Stanzel),** this member of the official USAF investigation team to the recovery operation signed an affidavit which contained the following statement:*

"I, Fritz Werner, do solemnly swear that during a special assignment with the U.S. Air Force, on May 21, 1953, I assisted in the investigation of a crashed unknown object in the vicinity of Kingman, Arizona. The object was constructed of an unfamiliar metal which resembled brushed aluminum. It had impacted twenty inches into the sand without any sign of structural damage. It was oval and about 30 feet in diameter. An entranceway hatch had been vertically lowered and opened. It was about 3-1/2 feet high and 1-1/2 feet wide. I was able to talk briefly with someone on the team who did look inside only briefly. He saw two swivel seats, an oval cabin, and a lot of instruments and displays.

294

A tent pitched near the object sheltered the dead remains of the only occupant of the craft. It was about 4 feet tall, dark brown complexion and had 2 eyes, 2 nostrils, 2 ears, and a small round mouth. It was clothed in a silvery, metallic suit and wore a skull cap of the same type material. It wore no face covering or helmet.

I certify that the above statement is true by affixing my signature to this document this day of June 7, 1973.
Signature: Fritz Werner   *Witnessed By: Raymond E. Fowler*
Date Signed: June 7, 1973   *Date Signed: June 7, 1973*
http://arizonaufosightings.com/kingman-arizona-1953-ufo-crash.html

It involved an engineer who took preliminary measurements to assess the momentum of a crashing craft, measurements useful to any reverse engineering efforts. The engineer who brought this story to light was **Arthur G. Stancil** (Stanzel) (previously known by the pseudonym "Fritz Werner"). Stancil graduated from Ohio University in 1949 and was first employed by **Air Material Command** at Wright-Patterson Air Force Base in Dayton, Ohio as a mechanical engineer on testing Air Force aircraft engines.

Dr. Eric Wang, who was suspected of leading a reverse engineering team on alien craft, headed the Installations Division within the Office of Special Studies where Arthur worked. Stancil signed a legal affidavit vouching to the honesty of his testimony, which was released by Ray Fowler in *UFO Magazine*, April 1976.

Stancil told that he was loaned out to the Atomic Energy Commission and was designated as a project engineer on some atomic bomb tests referred to as **"Operation Upshot Knothole"**.

The location of these tests was at Frenchman's Flats at the southern end of the Nevada Test Site. The test director was a **Dr. Ed Doll**.

Kingman as it appears today

On May 21, 1953 Stancil was called away by his boss and told to report for a special assignment at the **Indian Springs Air Force Base** where he was joined by 15 other specialists. They were flown by military plane to Phoenix where they boarded a bus with blacked-out windows and rode for an estimated four hours. When they arrived at their destination somewhere southeast of Kingman in one of the washes of the Hulapai Mountains, they were met and briefed by an Air Force Colonel who told them they were to investigate the crash of a super-secret test vehicle. He and the others on the bus were told not to speak to each other under any circumstances. Stancil's job was to determine the forward and vertical velocities of the vehicle when it impacted in the sand. Stancil was escorted to the site by military police. Two military arc-lights illuminated the saucer, which appeared to be two convex oval plates inverted over each other approximately 30 feet in diameter. The saucer was embedded in the sand about 20 inches. From this Stancil had determined that the saucer crashed at a velocity of 100 knots yet it had no dents, marks, or scratches on its burnished aluminum surface. It was constructed of dull silver metal like brushed aluminum.

Another specialist had gotten a look inside the craft as a 1.5 x 3.5 foot hatch was open revealing an oval interior cabin with two swivel seats and many instruments. Stancil saw one body recovered from the crash. It was humanoid, about 4 feet tall, with brown skin and wearing a silver metallic flight suit.

Whilst they were back on the bus and being taken back they were made to sign the **"Official Secrets"** act and were told never to tell anyone about this incident.

296

Fowler made several checks as to the integrity of Stancil and everyone who knew him said that he was a man of considerable integrity and scientific ability.
http://arizonaufosightings.com/kingman-arizona-1953-ufo-crash.html

Another story supporting the 1953 crash near Kingman came to UFO researcher Len Stringfield in 1977. A soldier in the National Guard at Wright-Patterson claimed that he was a witness to a delivery from a "crash site in Arizona" in 1953. He said that 3 bodies had been recovered and were packed in dry ice. They were 4ft tall, with large heads and brownish skin.
http://thechurchofufology.blogspot.com/2010/03/ufo-crash-near-kingman-az

According to UFO researcher **Kevin Randle,** Stancil's involvement in a top-secret crash retrieval of a wrecked saucer somewhere in the desert near Kingman, Arizona is the best account available to date even, though Stancil's credentials appear excellent, his embellishments on multiple retellings of the story particularly after a few drinks, however, tarnished his account of the event.

After some years of investigation, Randle in his book *Crash: When UFOs Fall From the Sky* tells of one **Judie Woolcott,** whose husband, a professional military officer was on duty at an airbase control tower somewhere in the Kingman vicinity. They had been tracking something on radar when it lost altitude, seemed to disappear and then, in the distance, there was a bright flash of white light.

According to Woolcott, something crashed and the Military police began talking about something intentionally being brought down and they drove out in the general direction of the flash where they discovered a domed disk had struck the ground with some force embedding itself in the sand. There was no apparent damage to the exterior of the craft or any strewn wreckage on the ground.

Before they had a chance to advance, a military convoy appeared. Woolcott and those with him were stopped before they could get close to the disk. They were ordered away from it and then escorted from the site. They were taken back to the base, where they were told that the event had never happened and that they had never seen anything. Just as others have been, they were sworn to secrecy. Woolcott never saw any bodies from the craft but, did hear talk about non-human casualties from some military police.
The details of the story came from a letter that Woolcott's husband had sent her during the Vietnam War while on a tour of combat duty. He indicated that he knew more, but he didn't want to write it down. Perhaps, it was part of his last will and testimony that he wanted to get off his conscience should he not return home after the war. A week after receiving the letter, Jodie Woolcott heard that her husband had been killed.

The letter would be an important piece of documented evidence of the event as described by her husband, however, the letter by itself would not be absolute proof that the crash had actually happened. Unfortunately, the letter was lost and only Woolcott's memory of it remains. This means that this witness's testimony is unreliable at best as it is a second-hand evidence. **Crash:**

This leaves only the Stancil story as the only strong case for this event with a couple of second-hand reports to flesh out this account. As far as crash retrieval accounts go, this one is not a very strong case but, has the potential to become an important one if, additional witnesses can be tracked down who can then, corroborate the story.

1955 Del Rio, Texas Saucer Crash

This next UFO crash (**Del Rio, Texas Saucer Crash**) was reported by a retired U.S. Air Force colonel and former fighter jet pilot who says that he chased a UFO across West Texas and watched it crash along the Rio Grande River near Del Rio, Texas. Decorated World War II and Korean War veteran **Colonel Robert B. Willingham**, 82, discloses his strange 1955 encounter in a new book titled "The Other Roswell: UFO Crash on the Texas-Mexico border," written by UFO researchers **Noe Torres and Ruben Uriarte**.
http://www.ufodigest.com/news/0508/theotherroswell.html

The story of Willingham's 1955 encounter with the crashed UFO began earlier in the day while flying an F-86 fighter jet during a Cold War simulated bombing run out of Carswell Air Force Base in Fort Worth, Texas.

A radio message warned Willingham and the others about a fast-moving UFO that was approaching Texas from the northwestern U.S. "Suddenly it came into their view, looking like an intensely bright light – like a bright star seen through a telescope," Torres said. "It blazed across the sky past them, and everyone in all the planes saw it. But, because of the location of Willingham's jet, he was in the best position to see what happened after the object flew by. The large, bright object moved generally west to east at about 2,000 miles per hour before suddenly executing a 90-degree turn and heading south toward Del Rio, Texas, located on the border with Mexico.

Willingham requested and received permission to pursue the object "to find out what the hell it was and give a full report to the commander back at base."

Turning his F-86, he followed the object's contrail down toward the Rio Grande River and observed that the UFO had begun wobbling uncontrollably and descending rapidly. He watched as the streak of light nearly took the roof off a house on the Texas side of the border before impacting on the south bank of the Rio Grande, where it skidded for 300 yards prior to coming to rest against a small hill. http://www.ufodigest.com/news/0508/theotherroswell.html

Determined to return to the crash site in a smaller plane that he could land along the riverbank, Willingham returned to his squadron and asked permission to go back to base. After returning to Carswell AFB, Willingham turned in his jet and picked up a small, two-seater training plane, which he and a man named **Jack Perkins**, now deceased, piloted back down to where the UFO had crashed.

298

Artist's Conception of the UFO crashing on the Texas-Mexico border
http://roswellbooks.com/delrio/?page_id=17 © Torres-Uriart

Retired Colonel Robert B. Willingham, 84, recounts that upon arriving at the crash site, he and Perkins found a damaged silver orb, roughly disc-shaped, "sticking in the side of a hill." They also noticed a 300-yard gouge in the earth, where the UFO had skidded before stopping. The object, which was still intensely hot, was being guarded by a large detachment of Mexican Army personnel, including several officers - all of them armed. The Mexicans would not allow Willingham to get very close to the smoldering wreckage, and a lieutenant told him that they were waiting for the U.S. Air Force to arrive in order to turn the object over to them.

It was while talking to the lieutenant, whose name he remembers only as "Martinez," that Willingham caught a glance of the three dead creatures inside the wrecked UFO through a hole in the ruptured hull of the ship.

The strange beings, two of which were badly mangled, appeared to have died in the crash. "It didn't look like humans to me," Willingham said during a recent interview on the Jeff Rense radio program. The retired aviator said the beings he saw fit the common description of UFO occupants with large heads, slit-like mouths, and arms "like broomsticks."

Asked how the dead entities were clothed, Willingham replied, "They weren't dressed at all." He said that despite the mangled condition of the bodies, he does not recall seeing traces of blood around the creatures. "But man, I just got to glance in there, because he [a Mexican Army lieutenant] wouldn't let me go look in it." Even though he got only a momentary glance inside the wrecked UFO, what he saw has haunted him for the past 55 years.

Told by the Mexicans that they needed to leave, Willingham and Perkins reluctantly got back in their small plane, but not before Willingham picked up a chunk of metallic debris about the size of a man's hand and stuck it in a pocket of his flight suit. Upon later examination back in Fort Worth, Willingham found that the metallic debris had properties unlike any metal known to earth science. Attempts to burn it, cut it with a torch, and deform it in any manner were totally unsuccessful. The curved piece of silvery-gray metal was highlighted by a honeycombed pattern of holes on each end and metallic ridges along the edges. "It was a piece of something not of this world," Uriarte said. http://www.ufodigest.com/news/0508/theotherroswell.html

Col. Robert V. Willingham, Circa 1960
http://roswellbooks.com/delrio/?page_id=17

Willingham possessed the unearthly metal for several days before he was ordered to surrender it to a Marine metallurgic laboratory in Maryland, from where it mysteriously "disappeared" days later in an obvious military intelligence cover-up. Subsequently, Willingham received threatening phone calls from intelligence operatives warning him to never disclose any of what he had seen "down on the border."

Willingham is a highly credible witness who served in World War II and was at one time a personal assistant to **General George S. Patton**. He was a pilot in the Korean War and sustained a battlefield injury in December 1950. Willingham later joined the Reserve and became fascinated by the widespread UFO sightings occurring throughout North America in the 1950s

and 1960s. www.ufodigest.com/news/0310/images/del-rio.jpg and
https://www.youtube.com/watch?v=9u7yd9j1qR0

Laredo, Texas Saucer Crash July 7, 1948

The **Laredo, Texas UFO Crash** is a case in which at least two U.S. military aircraft allegedly chased a 90-foot-diameter (27 m) silver disc-shaped **unidentified flying object (UFO)** across Texas before watching the object crash approximately 30 miles (48 km) south-southwest of Laredo, Texas on July 7, 1948. U.S. servicemen were reportedly dispatched from a nearby military base to cordon off the UFO crash site until a special U.S. retrieval team arrived to examine the wreckage and carry it away to a military base in San Antonio, Texas. Supposedly, the badly burned body of a non-human entity was recovered from the crash site.

Texas Monthly magazine recently included the Laredo UFO Crash on a list of the eight most significant UFO cases in Texas history. Interestingly, this case is said to have occurred almost exactly one year after the more famous Roswell UFO Incident. Rumors about this case first began circulating in the 1950s, although details were not widely known until 1977. This case shares similarities with the **Del Rio, Texas UFO Crash of 1955** and the Coyame UFO Incident of 1974, both of which reportedly also occurred along the Texas-Mexico border.

According to *Texas Monthly*, talk of a UFO crash near Laredo first surfaced in the 1950s, with additional details being released in 1978 by the late Leonard Stringfield, one of the first UFO researchers to advocate serious investigation of reported UFO crashes. Stringfield wrote, "In the Fall of 1977, new word of a 1948 crash came to me from a well-informed military source. His information, however, was scanty. He had heard from other "insider" military sources that a metallic disc had crashed somewhere in a desert region. His only details indicated that the craft had suffered severe damage on impact and was retrieved by military units."

Also in 1977, Stringfield received more information about the case from another UFO researcher, the late **Todd Zechel**. Stringfield wrote, "Formerly with the **National Security Agency**, Zechel stated that a United States Air Force technician told him that his uncle, then a Provost Marshall at Carswell Air Force Base near Fort Worth, Texas, had taken part in the 1948 recovery of a crashed UFO, which was described as a metallic disc, 90 feet in diameter."

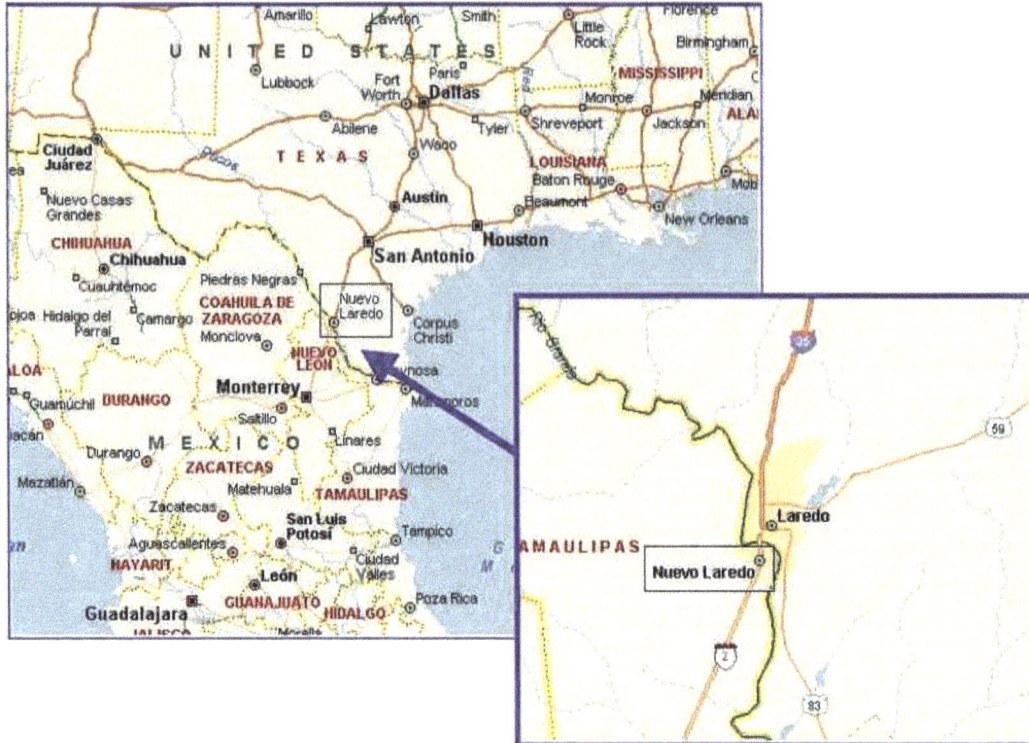

Map Showing Location of Alleged 1948 UFO Crash
http://holidaymapq.com/map-of-laredo-texas.html

In a report presented at the **Mutual UFO Network (MUFON) Symposium** on July 29, 1978, Stringfield stated that "one dead alien was found aboard the craft, which was described as about 4 feet, 6 inches tall, completely hairless, with hands that had no thumbs."[3]

In December 1978, two photographs fitting Stringfield's description of the dead alien suddenly appeared in Maryland. The photos, along with a brief note about them, were received in the mail by **Willard F. McIntyre**, founder of a civilian UFO group called the **Mutual Anomaly Research Center and Evaluation Network (MARCEN)**. The photos showed the badly burned body of a small biped with a large head and claw like hands. The photos were purportedly sent by a retired U.S. Navy photographer from Tennessee who claimed to have taken them at a UFO crash site along the Texas-Mexico border in 1948.

McIntyre corresponded by mail with the unnamed former Navy photographer from 1978 through 1981 and learned more details about the Laredo crash, which McIntyre later disclosed to numerous civilian UFO organizations. McIntyre claimed that MARCEN had thoroughly checked out the photographer's military service record and had verified that he was who he claimed to be. McIntyre further claimed that the Eastman Kodak company and the UFO group **Ground Saucer Watch (GSW)** had both independently verified that the negatives of the photos given to McIntyre in 1978 were approximately 30 years old.

The photos were first released to the media in April 1980 by Charles Wilhelm, director of the now-defunct **Ohio UFO Investigators League (OUFOIL),** were picked up by the Associated

302

Press, and were published in a number of U.S. newspapers, including the Cincinnati Enquirer (on April 29, 1980).

Based on a number of accounts published in the late 1970s and early 1980s, a large UFO was spotted in the airspace above Albuquerque, New Mexico on the afternoon of July 7, 1948, moving at approximately 2,000 miles per hour. Stringfield said that the object "was tracked on radar screens" and other sources stated that the object at one point made a 90-degree turn before heading toward southwest Texas.

It is important to note that in the 1980s some UFO researchers confused this story with that of the Del Rio, Texas UFO Crash of 1955, which also occurred along the Texas-Mexico border. Some of the early accounts of the Laredo crash, therefore, contain inaccuracies based on this confusion.

A number of UFO investigators have stated that, prior to crashing near Laredo, the UFO was chased across the skies of Texas by at least two military aircraft. Neither the type of aircraft nor the base from which they were dispatched is known. **Ron Schaffner** referred to the aircraft as Lockheed F-94 Starfire jets; however, the F-94 was not in use until 1949. It is possible that the actual aircraft involved were Lockheed P-80 Shooting Star jets, of which the F-94 was a variant.

It is also not known if the pursuing aircraft might have contributed to the downing of the UFO by firing upon it or otherwise causing it to fall. However, there are numerous other documented cases of U.S. military aircraft firing upon UFOs during this time period.

Stringfield wrote that the UFO crashed "about 30 miles inside the Mexican border across from

Laredo, Texas, and was recovered by U.S. troops after it was tracked on radar screens." Ohio UFO investigator Ron Schaffner wrote, "At 1410 hours, other pilots in pursuit said the object was slowing down and was wobbling in flight. By 1429 hours, the object disappeared from all radar screens. Using triangulation from all the radar installations, it was determined that the object must have gone [sic] down in Mexico, approximately 30 miles south of Laredo, Texas.

The location of the crash was given more specifically in *UFO Crash at Aztec: A Well-Kept Secret*, a 1986 book by William S. Steinman and Wendelle C. Stevens. They wrote, "This site was about 30 miles SSW of Laredo, not far from the highway to Mexico City, and near where the Rio Sabinas joins the Rio Salado before they empty into the Rio Grande, in the Sierra Madre Oriental."

The crashed UFO was described by Steinman and Stevens as follows: "As best the source could ascertain, the craft was nearly perfectly circular and was about 90 feet in diameter and about 28 feet [8.5 m] in thickness at the center and tapering off to about 5 feet [1.5 m] thick at the perimeter. There appeared to be five or six levels in the center of the craft and they were told some sort of instrumentation and machinery were removed before they had arrived. No propulsion system or mechanism was apparent to the source."

According to Stringfield's account of the Laredo case, a provost marshal stationed at Carswell Air Force Base near Fort Worth admitted that he had taken part in a mission to cordon off and secure the site where a UFO crashed to the Earth near Laredo, Texas in 1948. The marshal told his nephew, who later told UFO researcher Todd Zechel. Stringfield later wrote, "Zechel learned from his sources that the troops involved in the retrieval were warned that if they said a word about the incident, they would be the sorriest people around."

Steinman and Stevens later identified the eyewitness as **John W. Bowen**, who they said: "was sent over to take immediate charge of cordoning off and controlling the area." According to the authors, after Bowen's group secured the area, a team was flown in from the missile range at White Sands, New Mexico to photograph the crash site, and later, a convoy of large Army transport trucks removed the wreckage, taking it the San Antonio Air Depot for further study.

Early in 1978, Stringfield described the humanoid found at the crash site as "about 4 feet, 6 inches tall, completely hairless, with hands that had no thumbs." That description seemed to fit the body shown in two photographs that were mailed to MARCEN founder Willard F. McIntyre in December 1978. Shaeffner later described the body shown in the photos: "There was one body found within the craft. The photographers managed to get a series of pictures even though there was intense heat. When the object cooled down, the body was removed to a hillside and another series of pictures were taken. The body was said to be 4 feet 6 inches [1.4 m] long with a head extremely large compared to the torso."

In 1986, Steinman and Stevens added, "They [the military photographers] only saw and photographed one body but rumors were floating around the site that two or more creatures had been blown out of the vehicle and were captured and taken away injured severely but still alive. Our source said he had no confirmation of this aspect of the case. The body they photographed was 4' 6" long. Its head was extremely large for the body size by human proportions. The eyes were gone from the fire but the eye sockets were much larger than in humans and were almost wraparound as if to give 180-degree vision. There were no visible ears or nose, but there were openings where ears and nostrils would have been in humans. There were no lips and the mouth was just a sort of slit with no teeth or tongue. There were two legs of normal proportions with short feet having no discernible toes. The two arms were longer than in humans and the hands had four claw-like fingers each with no apparent thumbs. The arms and legs appeared to have joints in approximately the same places as in humans."

Sketch of "Tomato Man." Body is face down amidst debris. Left arm and claw-like hand are visible. Head is very large. Skin is mostly burned off.

Original black and white photos showing the charred Remains of the "Alien Being"

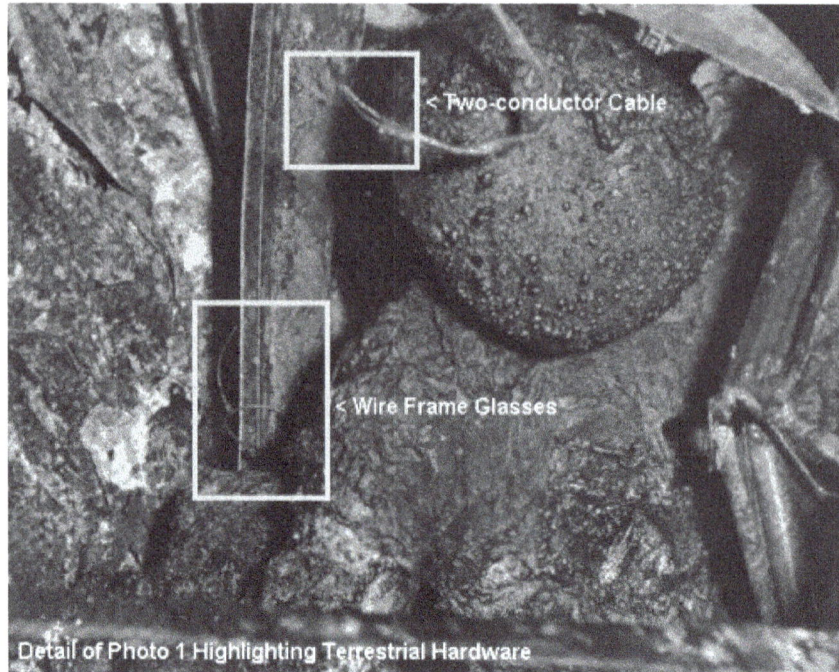

A closer examination of the photograph reveals artifacts of conventionality indicating a more terrestrial explanation rather than an extraterrestrial one.

http://www.ufosightingsdaily.com/p/ufo-crashes.html

Shaeffner wrote, "Army doctors arrived on July 8 and performed an examination of the body. They could not find any reproductive organs. They compared the gray skin to the texture of a human female breast. The bone structure was more complicated than a human and no muscle fiber was discovered within the torso."

The body depicted in the photos sent to McIntyre has over the years come to be known as the "Tomato Man" due to its large, roundish head. Many UFO researchers, including Shaeffner and Kevin Randle, believe the body is that of a human pilot who was badly disfigured by intense heat following a plane crash. They argue that one of the photos shows a pair of eyeglasses, such as a human pilot would wear, near the body. Randle has classified the entire "Tomato Man" story as a "hoax."

Other researchers believe the body might be that of a monkey used as a test subject in a missile experiment. Still, other UFO researchers argue that even if the "Tomato Man" photos are fakes, that does not necessarily invalidate the UFO crash incident itself, knowledge of which preceded the appearance of the photos. Aftermath

Steinman and Stevens looked into a rumor that notable U.S. scientist **Luis W. Alvarez**, (now deceased) may have been involved in an investigation of the site where the Laredo UFO Crash occurred. Supposedly in July 1948, Alvarez and other top U.S. scientists were taken under circumstances of complete secrecy to a location in the **Sierra Madre Oriental Mountains** of Nuevo Leon, Mexico, which is the general vicinity of the alleged Laredo UFO Crash. Their mission was to "examine the residue on site of a crashed 100-foot-diameter [30 m] circular flying vehicle of unknown origin." As a scientist, Alvarez was noted for applying scientific principles

306

to paranormal subjects. Steinman and Stevens contacted Dr. Alvarez in the late 1980s and asked whether he was involved in any investigations of crashed UFOs, but he refused to make any comment to them.

Physicist Luis W. Alvarez.
http://www.reformation.org/secrets-of-the-great-pyramid-revealed.html

UFO researcher Wendelle C. Stevens, whose 1986 book *UFO Crash at Aztec: A Well-Kept Secret* included a section about the Laredo crash, now believes that the UFO was a top-secret U.S. experimental aircraft and that the burned body was that of a large rhesus monkey. In a 2009 interview, Stevens said that, although he believes many UFO incidents do involve extraterrestrial spacecraft, he thinks the 1948 Laredo crash was really a secret experiment that originated at the White Sands, New Mexico missile range. **www.ufodigest.com/news/0310/images/del-rio.jpg**

Kecksburg, Pennsylvania Saucer Crash Dec 9, 1965

What exactly soared through the late afternoon skies of Canada, Michigan, Ohio, and Pennsylvania on December 5, 1965? Eye witnesses described the unknown object as a "fireball," but it seemed to be under some type of intelligent control, as it veered somewhat in Ohio toward the Quaker State.

One of the first official reports of that day came from **Frances Kalp**, who phoned in her experience to radio station WHJB in Greensburg at 6:30 P.M. She related seeing a fiery object crash into a wooded area near her home in Westmoreland County. Kalp and her children had approached the site within a half-mile, and there they saw an odd object resembling a "four-pointed star."

Radio station employee **John Murphy** immediately phoned in the report to the Pennsylvania State Police Department. The Police phoned Kalp and arranged to meet her in Kecksburg.

Murphy also raced to the site of the alleged crash. He interviewed Kalp and her children for his report, while the State Police searched the woods for the crashed object. Murphy eagerly awaited the return of the searchers. When they finally completed their search, Murphy was unable to get any clear information from either **Carl Metz or Paul Shipco**, who headed the search. They only stated that they were calling in the Military to handle the case.

Undaunted, Murphy made phone contact with Captain Dussia at State Police Headquarters in Greensburg. Murphy was instructed to visit the office to receive an official statement on the search party results. Upon arriving at headquarters, Murphy noticed that the Military had already arrived in force. Murphy was startled when he received the "official" statement;

"The Pennsylvania State Police have made a thorough search of the woods. We are convinced that there is nothing whatsoever in the woods."

By this time, Murphy was convinced that there was a cover-up of some kind. If there was nothing in the woods, why would the Military be in force at Pennsylvania's Police headquarters? After turning in his report to the radio station, Murphy overheard one of the policemen involved in the search describing a "pulsating blue light" in the forest.

Murphy was told that Officer Metz and the Military were going back to the woods, and to his surprise, he was given permission to join the second search. Murphy's excitement soon turned to disappointment when after accompanying the party to the outskirts of the woods, he was kept from going any farther.

Murphy was an eye witness to the Military sealing off the area, and banning all civilians from the scene and its immediate surroundings. The story of the crash soon made newspapers and television, and the area was quickly overflowing with people wanting to get first-hand information on what could possibly be a historic event.

What had actually crashed into the woods? Was it just a fragment of a comet? Or space debris reentering the Earth's atmosphere? If the explanation was so simple, however, why had the Military cordoned off the area? Could it have been a secret Military craft? Or something far more mysterious?

It soon became common knowledge that some eager, interested civilians had made a trip into the woods before the Military gained control of the area. These few individuals were interviewed by Stan Gordon and told an amazing story.

308

They stated that they saw a copper-bronze colored, saucer-shaped object crashed in the woods. This craft was anywhere from 9-12 feet in length and bore a gold band around its bottom.

Some of the witnesses described writing on the craft which resembled Egyptian hieroglyphics. These few witnesses were quickly whisked away when discovered by Military personnel. Later that night, witnesses claimed that they observed a flatbed truck toting a large object, covered by a tarpaulin.

Shortly after the departure of the flatbed, many of the Military personnel vacated the search area. Could this have been a craft from another world being taken down our highways?

An extremely intense day of activity was followed by a quiet morning, and it seemed that what had occurred was just a dream. The Air Force concluded their investigation with the "official" statement that a meteorite was responsible for the report of a glowing craft and subsequent crash in the woods.

The media as a whole accepted this explanation, and the matter seemed for all intent and purposes, closed. Had it not been for a 1990 television program, the **Kecksburg saucer crash** would have been just another fancied report by a few overly excited witnesses.

The area of Kecksburg would again become a beehive of controversy after a dramatization of the events on **"Unsolved Mysteries"** in 1990. The citizens of the area seemed to be equally divided on the value of airing the segment, with some accepting the "official" explanation, and others claiming "cover-up."

Even before the segment aired, some protesters promoted a petition to stop the network from airing the Kecksburg story.

The actual witnesses of the event prevailed, stating that the petitioner's list did not include any eye witnesses to the original event. Opposite sides were also taken by those officially investigating the facts of Kecksburg.

After the television show ran, two new witnesses came forward. One was a USAF officer at Lockbourne AFB (near Columbus, Ohio). In the early hours of December 10, a truck arrived by the little used back gate of the base and he was ordered to patrol it. It was a flat-bed with a large tarpaulin on the surface covering a conical object.

He was told to shoot anyone who tried to get too close. He was advised the truck was bound for Wright-Patterson AFB, which is the reputed home of other crashed saucers.

The other witness was a building contractor who was asked two days later to take a load of 6,500 special bricks to a hangar inside Wright-Patterson. When he sneaked a look inside the hangar he saw a bell-shaped device, some 12 feet high sitting there. Several men wearing white anti-radiation style suits were inspecting the object.

After he had been escorted out he was told that he had just seen an object that would become common knowledge in 20 years time.

Investigator Robert Young, for one, became an advocate of the Military's official explanation. Stan Gordon, on the other hand, believed the eyewitness accounts, and the cover-up theory. Gordon took his findings to the next level, producing a 92 minute documentary video of his findings, titled "Kecksburg: The Untold Story."

As to be expected, alternative theories were put forward by skeptics and debunkers, such as the reentering of the Russian VENUS probe, but this scenario was rejected by the American and Russian governments.

One interesting footnote to this haunting tale came from the widow of John Murphy, who shortly after his death, publicly stated that her husband had been one of the first on the scene, and had actually photographed the strange craft.

Supposedly, these valuable snapshots were confiscated by Military officers, and Murphy was instructed not to discuss what he knew unless he wanted to suffer severe consequences.

In January 1980 UFO investigator **Clark McClelland** interviewed the assistant fire chief of Kecksburg, **James Mayes**, and **Melvin Reese**, another fireman. They reported that their team had come within sixty meters of the object. They had seen an object on the ground that had smashed its way through the trees.

Mayes explained how the military had cordoned off the woods and had established a temporary base, complete with telecom link. Fire chief **Robert Bitner** would later confirm this story. He also said he had seen an object that was about 6 feet high, 6 feet wide, and some 15 feet long, clearly not an aircraft. It was resting at an angle on the ground as if it had impacted nearly horizontally. Another fire officer, 'Pete', stated he had seen a ring of bumpers around it into which were described some pictorial symbols.

In 1990, researcher Stan Gordon from the **Pennsylvania Association for the Study of the Unexplained**, traced an apparent first-hand witness, **James Romansky**. He recalled seeing the object on the ground some 25 years previously when he was an 18 year old firefighter. He had been called to duty that night following concerns that an airplane had crashed.

He described the object as bronze coloured and shaped like an acorn some 12 feet long and 25 feet in diameter; it had slightly raised "blunt" end and strange markings.

"It had writing on it, not like your average writing, but more like ancient Egyptian hieroglyphics. It had sort of a bumper on it, like a ribbon about six to 10 inches wide, and it stood out. It was elliptical the whole way around and the writing was on this bumper.
It's nothing like I've ever seen, and I'm an avid reader. I read a lot of books on Egypt, the Incas, Peruvians, Russians and I've never to this day come across anything that looked like that."

310

The Kecksburg crash, in many ways similar to the Roswell case, remains an extremely interesting one and is still considered "unsolved. http://ufocasebook.com/Kecksburg.html

On 9th December 1965 hundreds of witnesses in Michigan, Ohio and Pennsylvania observed a UFO crash. It first appeared to be nothing more than a spectacular meteorite but after 30 years, it is still a source of much controversy amongst UFO researchers.

In March 1966, UFO researcher **Ivan Sanderson** compiled a detailed account from various eyewitnesses and soon realized that there was more to this story than that of a simple meteorite.

His findings indicated that the object showed a clear trajectory, moving from northwest to southeast. Its total visible journey lasted no more than 6 minutes, which indicated a speed far too slow for a meteor. Sanderson calculated a speed of only around 1,000 mph.

Not all the eyewitnesses were located on the ground at the time of the sighting, there were also several pilots who spoke of being buffeted by shock waves as the large bright object sped by. This was strange as most meteorites are observed several thousands of feet above most commercial aircraft. There were also claims of shock waves and sonic booms reported from witnesses on the ground.

The vapour trail left by the object was so intense that they lasted for more than 20 minutes in which time they were filmed by several people.

Several bits of silvery debris were found on the ground at Lapeer, Michigan and these were assumed to be from the object. Later analysis of this material indicated that they were indeed chaff which are pieces of aluminum foil released by aircraft to fool radar operators.

However then most convincing aspect that Sanderson discovered to rule out the possibility of a meteorite was that the object appeared to change direction and head in an easterly direction.

The object finally came to rest in a wooded area in the town of Kecksburg. The object was initially witnessed by 2 children who reported that a 'start that had caught fire had crashed in the woods'. Their mother who's first thought was that her children had witnessed a plane crash called the state police and the fire service.

As soon as she had finished phoning she went out into the woods and to her surprise found that a military unit had beaten them to the crash site. This unit took command and told all civilians, police and fire department officials to leave the site immediately.

The military unit reported to the police that they had found nothing, and they left. This remained the case for around 15 years when some UFO researchers re-opened the case.

After some investigation, they found out that the fire service had come within 200ft of the object before being turned away by the military. They reported seeing blue flashing lights and noticing that the tops of several of the trees nearby were broken as if an object had come crashing through.

The investigators also reported that they had found witnesses who had seen a large flat-bed truck leaving the area of the woods with a large oval object covered with sheets. Another witness indicated that he observed the military loading the object onto the truck. He described the object as looking like a large acorn, with 'bumpers' on the base. He also noticed that there was strange hieroglyphics on the craft surface.

The case was reported in the **Project Blue Book** files and indicated that "a three man team has been dispatched to investigate and pick up an object that started a fire". This three man team is now known to of been part of the then highly classified **'Project Moon Dust'**.

The official report was that the UFO was simply a meteorite.

In 1990 a new witness came forward who claimed to be part of the military team that was sent in to retrieve the object. He claims that he was given orders to "shoot anyone who got too close". He also revealed that the object was being transported to the **Wright Patterson base.**

The last major discovery related to a worker at Wright-Patterson who claims that a strange object was shipped in on the 16th December the same year, just days after the events at Keckburg. He described the object almost identically as the other witnesses had described. Whilst he was observing the object a guard escorted him out of the hanger and told him "that you have just seen an object that will be common knowledge in 20 years time".

List of Alien Crashes

CSETI's interest in this matter stems from the fact that Earth is being visited by ET's with benign intentions who are currently being tracked, targeted and shot down by covert forces operating outside democratic controls.

This assessment of the situation and the ET's intentions is endorsed by CSETI military and intelligence witnesses, some of whom have been on the retrieval teams.

Events listed include those where ET's may have been captured (dead or alive), but craft were not retrieved. Events involving only the recovery of artifacts are listed as well. In some instances, the reports may be duplicates covering the same incident.

Some events probably also involve mishaps with **ARV's (Alien Reproduction Vehicles)**, or weapons platforms, of terrestrial manufacture, and some may have been **SDI (Star Wars)** or scalar electromagnetic (Tesla) weaponry episodes. These are banned by treaty but are being conducted at an escalating rate.

At least three nations on Earth now have their own electro-gravitic "flying saucers". Some may also have been events involving captured ET craft. Hence the numerous Press Releases about "meteors" by various National Laboratories.

Some incidents may have even been psychological warfare experiments involving holographic projections - and some, too, may be outright disinformation.

312

A thought of remembrance, though, for those beings of all species who perished in the crashes, and for those who have been killed or brainwashed to keep these events secret from humanity.

Extraterrestrial Space Vehicle (ETVs) crashes as listed by date:

1. **10,000 BC - Sino-Tibetan Border**
2. **2,000 BC - Grand Canyon, AZ**
3. **840 - Lyon, France**
4. **18th Century**
5. **1864 September - Cadotte Pass, Missouri**
6. **1884 Early June - Holdredge, Nebraska**
7. **1884 Dec. 13 - Sorisole, nr. Bergamo. Italy**
8. **1897 April 17 - Aurora. Texas**
9. **1907 - Burlington, Vermont, USA**
10. **1908 June 30 - Tunguska River, USSR**
11. **1909 Dec 22 - Chicago**
12. **1910/1915 - Puglia Italy**
13. **1923 - Quetta, Pakistan**
14. **1925 - Chevy Chase, Maryland, USA**
15. **1925 Sept/Oct - Polson, Montana**
16. **1930 - Mandurah, West Australia**
17. **1933 - Italy**
18. **1936 - Black Forest, Germany**
19. **1938 summer - Czernica, Poland**
20. **1941 - west of San Diego, Ca**
21. **1941 Spring - Cape Girardeau, Missouri**
22. **1941 July 4 - Tinian Island, Oceania**
23. **1945 approx. - UK**
24. **1945 - Mataquescuintla, Guatemala**
25. **1946 - Date and location unknown**
26. **1946 - Magdalena, NM**
27. **1946 July 9 - Lake Barken, Sweden**
28. **1946 July 10 - Bjorkon, Sweden**
29. **1946 July 18 - Lake Mjosa Sweden**
30. **1946 July 19, noon - Lake Kolmjarv, Sweden**
31. **1946 August 12 - SW Sweden**
32. **1946 August 16 - Malmo Sweden**
33. **1946 mid-October - Southern Sweden**
34. **1947, January - Papagos Indian Reserve. AZ**
35. **1947 May - Spitzbergen, Norway**
36. **1947 May 31- Socorro, NM**
37. **1947 July - nr. St. Joseph, MO**
38. **1947 July 4 Roswell, NM**
39. **1947 July 5 - Plains of San Augustin, NM**
40. **1947 July 31 - Maury Island, Tacoma**
41. **1947 Aug. 13 - Hopi Reservation, AZ**

42. 1947 October - Paradise Valley, AZ
43. 1947 October 20 - San Diego
44. 1948 - Kingman, AZ
45. 1948 February 13 - Aztec, NM
46. 1948 April - 12 mi. outside Aztec, NM
47. 1948 March 25 - White Sands, NM
48. 1948 7/8 July - Laredo/30mi inside Mexico
49. 1948 August - Laredo, Texas
50. 1949 - Roswell, NM
51. 1949 August 19 - Death Valley, California
52. 1950 (before) - nr. Mexico City, Mexico
53. 1950 January - Mojave Desert, Ca
54. 1950s (mid) - Birmingham, Alabama
55. 1950 May 10 - Bahia Blanca Province, Argentina
56. 1950 Dec 6 - El Indio/Guerrero area, Tx/Mex border
57. 1952 June - Spitzbergen, Norway
58. 1952 July - Washington, DC
59. 1952 August - Ohio
60. 1953 - Brady, Montana
61. 1953 May 20 - western Utah
62. 1953 May 20/21 - Kingman, Az
63. 1953 Summer - Fort Polk, LA
64. 1954 (Spring) - Mattydale, NY
65. 1955 July - Vestra Norrland, Sweden
66. 1957 Sept 14 - Ubatuba, Brazil
67. 1957 Nov 21 - Reasty Hill, Scarborough, Yorks
68. 1958/1959 - Woomera, Australia
69. 1958 - Utah desert
70. 1959 Jan 21 - Gdynia, Poland
71. 1959 Sept 17 - Wormer nr Amsterdam
72. 1959 Undated - Italy, North of Rome
73. 1960s - offshore Spain
74. 1960s - Great Sand Dunes, Co
75. 1960 March - New Paltz, NY
76. 1961 - Timmensdorfer, Germany
77. 1961 April 28 2am. - Lake Onega, USSR
78. 1961 June 9 - Nr. Woodbridge AFB, UK NEW
79. 1962 - Otero County, NM
80. 1962 April 18 - Las Vegas
81. 1963/1972 - Australia, 12 recoveries
82. 1963 - Atlantic Ocean
83. 1963 July 16, 6 am. - Charlton, Wiltshire
84. 1963 Oct. 10/15 - Xerekena, Congo
85. 1963 Dec 10 - RAF Cosford nr Wolverhampton
86. 1964 Feb/March - Penkridge, UK
87. 1964 April 13 - London

88. 1964 Dec 10 - Fort Riley, Kansas
89. 1965 Jan - San Miguel, Argentina
90. 1965 Dec 9 - Kecksburg, Penn.
91. 1965 Dec 17 - Llandegla, Wales
92. 1966 June/July - Topolevka, West Siberia
93. 1967 - nr. Avon Park AFB, Florida
94. 1967 January - Southwest Missouri
95. 1967 Feb. 7 - General Teran, N.L., Mexico
96. 1967 August 17 - Sudan
97. 1967 Oct. 6 - Shag Harbor Nova Scotia, Canada
98. 1968 Feb 12 - Orocue, Columbia
99. 1968 March 25 - Nepal
100. 1969 March - Sverdlovsky, USSR
101. 1969 July 13 evening - Tunisia
102. 1970s - vicinity of Wichita Falls, Tx
103. 1970s (mid to late, or early 80s) - Kanakee, IL
104. 1970 Oct 7 - nr. Lai, Chad
105. 1971 Summer - Elk Mountain, Wyoming
106. 1971 Summer - nr. Edwards AFB
107. 1972 - Rossendale Valley, UK
108. 1972 April 7 - New Zealand
109. 1973 June - Pacific Ocean
110. 1973 Sept. - Great Lakes Naval Base
111. 1974 - nr. Detroit, MI
112. 1974 Jan - Cannock Chase, Staffs, UK
113. 1974 Jan 23 - Llandrillo. N. Wales
114. 1974 May - Ramstein AFB, Germany
115. 1974 May 17 - Chili, NM
116. 1974 July 2 - Santa Catarina, Brazil
117. 1974 July 15 - Spain
118. 1974 Aug 25 - Chihuahua, Mexico
119. 1974 Nov. 9 - Carbondale, PA
120. 1974 Undated - USAF base nr. Savonna, Italy
121. 1975 - upstate Pennsylvania
122. 1975 Spring - Nr. Ohio-Michigan border
123. 1975 Summer - Millard, Nebraska
124. 1976/77/78 - White Sands, NM
125. 1976 - upstate New York
126. 1976 February 14 - Cubatao, Brazil
127. 1976 March (?) - Simi Valley, Ca
128. 1976 May 12 - Australian Desert
129. 1977 - West Australia
130. 1977 Jan 10 - Wakefield, NH
131. 1977 April 5 - nr Xenia, Ohio
132. 1977 May 26 - Corinth, Heard County, Georgia
133. 1977 June 22 - NW Arizona

134.　1977 July - Minley Manor Woods, UK
135.　1977 July - Puebla, Mexico
136.　1977 August 17 - Tabasco, Mexico
137.　1977 September - Ocotillo, Ca
138.　1977 November 16 - Ellsworth AFB
139.　1978 - Isle of Wight, UK
140.　1978 - upstate Pennsylvania
141.　1978 Undated - Soviet Union
142.　1978 January - McGuire AFB, NJ
143.　1978 May - Argentina
144.　1978 May 6 - Tarija, Bolivia
145.　1979 Feb 12 - Pocono Mts. Pennsylvania
146.　1979 Feb 24 - 2am. Stacksteads, Lancs, UK
147.　1979 June 27 - Oregon-Idaho border
148.　1979 Aug 20/21 - La Paz Bolivia
149.　1979 Nov 25 - Grays Harbor, Washington
150.　1980s early - Massachusets
151.　1980 Dec 26 - Rendlesham Forest, UK
152.　1981 Aug 22 - Argentina
153.　1982, Jan. 12 - offshore Atlantic City, NJ
154.　1982 Aug 24 - over New York
155.　1983 Jan 12 - Gallup, New Mexico
156.　1984 Feb 19 - Coca Falls, Puerto Rico
157.　1988 - Mt. Orab, OH (Brown County)
158.　1988 Nov. 21 - Gulf Islands Ntl. Seashore, FLA
159.　1988 Dec 26 - nr Dayton, Ohio
160.　1989 May 7 - Kalahari Desert
161.　1989 Sept 16 - Zaostrovka, USSR
162.　1989 Sept 28 - Smith's Point Beach, L.I.
163.　1989 Nov 4 - Carp,
164.　1990 Jan 9 - Colombia
165.　1990 Aug 22 - Ebenezer, P.E.I., Canada
166.　1990 Sept 2 - Megas Platanos, Central Greece
167.　1991 Jan - 1st Gulf War Crash
168.　1991 Jan - 2nd Gulf War Crash - Saudi Arabia
169.　1991 August 18 - Second Carp incident, nr. w. Carleton, Canada
170.　1991 August/Sept. - Tien-Shan, Russia
171.　1992 Nov 24 - Long Island
172.　1992 Dec 14 - Arizona
173.　1993 July - Italy
174.　1993 Nov - Fylingdales, Yorks, UK
175.　1993/1996 - Exact date unknown. Canada
176.　1994 Jan 12 - Greeny Mt. Colo.
177.　1994 March 6 - Mount Mutria, Guardairegia, Italy
178.　1994 May 15 - Kharkov, Ukraine
179.　1994 September - off Monterey, California

316

180. 1994 Sept 24 - Eureka, Ca
181. 1994 Oct 31 - Hepton Hill, UK
182. 1995 Undated - Chile
183. 1995 August 17 - Salta, Argentina
184. 1995 September 15 - Lesotho
185. 1995 September 28 - Central Israel
186. 1995 November - Itum-Kale, Chechnya
187. 1995 December 5/7 - Somalia
188. 1996 Spring - Lansing, MI
189. 1996 Jan 14 - North of New Orleans
190. 1996 Jan 20 - Varginha, Brazil
191. 1996 Jan 28 - Mt. Popocatepetl, Mexico
192. 1996 October 3 - El Paso, Tx video
193. 1996 October 3 - Simi Valley, Ca
194. 1996 October 3 - So. California
195. 1996 Oct. 26 - Isle of Lewis, Scotland
196. 1996 Feb 15 - Isle of Jura, Scotland
197. 1996 March 30 - Hoosier National Park, Indiana
198. 1996 May - nr. Boyle, Eire
199. 1996 June 23 - Pinedale, AZ
200. 1996 November 21 - Annency, France
201. 1996 November 29 - Franklin, Ohio
202. 1996 December 19 - China
203. 1997 - Date Unknown - NM?
204. 1997 March 15 - Wegorzewo, Poland
205. 1997 April - Peru
206. 1997 May 5 - Puerto Rico
207. 1997 June 18 - Cumpas, Sonora, Mexico
208. 1997 July 1 - Teresopolis, Brazil
209. 1997 Sept 22 - offshore Scotland
210. 1997 Oct 9 - El Paso, Tx
211. 1997 November - Grand Prairie, Tx
212. 1997 Dec. 9 - off E. Greenland
213. 1997 Dec. 17 - Rogersville, Mo
214. 1998 January 8 - Kennewick, Wa
215. 1998 January 12 - Rocky Mtns., Co
216. 1998 January 18 - Vechne, Ukraine
217. 1998 January 27 - Breckenridge, Co
218. 1998 January 27 - Hanna, Wyoming
219. 1998 February 22 - Puerto Rico
220. 1998 March 1 - Bonnybridge, Scotland
221. 1998 March 18 - nr. Khartoum, Sudan
222. 1998 July 11 - Azle, TX
223. 1998 August 5 - Tracy, CA
224. 1999 March 13 - N of Haig, Australia
225. 1999 May - Kentucky

226. 1999 Oct. 9 - Parah State, Brazil
227. 1999 7 Dec. - Guyra, NSW, Australia
228. 2000 25 January - N. Argentina
229. 2000 May 5 - Munchengladbach, Germany
230. 2000 August 27 - Pakistan
231. 2000 October 13 - nr. Elk City, OK
232. 2000 October 18 - Midwest USA
233. 2000 April 27 - Worchester, West Cape, South Africa
234. 2000 August 27 - Baluchistan, Pakistan
235. 2004 May 15 – Puerto Ordaz, Venezuela
236. 2006 May 20 – Port Shepstone, South Africa
237. 2006 November - Bahia, Brazil
238. 2006 December 1 – Krasnoyarsk, Siberia
239. 2007 January 1 – Lephalale, South Africa
240. 2008 May 14 – Needles, California
241. 2008 May 27 - Phu Quoc, Vietnam (possibly an aircraft.)
242. 2009 June 20 – Gin-Gin Queensland, Australia
243. 2009 July 27 – Ottawa, Ontario, Canada
244. Undated: Veneto, Italy
245. Undated: British Columbia, Canada
246. Undated: Australia - various
247. Undated: Alabama Military Base
248. Undated: Antarctica
249. Undated: White Sands, NM
250. Undated: Wright-Patterson AFB
251. Undated: Montauk, NY
252. Undated: Area S4, NV - Bob Lazar "Sport Model"
253. Undated: Area S4, NV - Bob Lazar "Jello Mold"
254. Undated: Area S4, NV - Bob Lazar "Top Hat"
255. Undated: Area S4, NV - Bob Lazar "Damaged disc"
256. Undated: Area S4, NV - Bob Lazar craft #5
257. Undated: Area S4, NV - Bob Lazar craft #6
258. Undated: Area S4, NV - Bob Lazar craft #7
259. Undated: Area S4, NV - Bob Lazar craft #8
260. Undated: Area S4, NV - Bob Lazar craft #9
261. Undated: Area 51, NV - David Adair craft
262. Undated - nr. Malad City, Idaho
263. Undated - Central Russia
264. Undated - Estonia
265. Undated: Brazil #1
266. Undated: Brazil #2
267. Undated: Madagascar

http://www.cseti.org/crashes/crash.htm

Crashed UFO on the Moon? Refer to the appropriate section in this book for full details.

Crashed UFO on Mars? Refer to the appropriate section in this book for full details.

For reasons of consistency, we are not referring to any of the above as "alleged", or "possibly hoaxed" etc. At such time as the events are definitively proved to be so, they will be removed from the list.

We are indebted to:

The Late *Leonard Stringfield*
Stanton Friedman
Kevin Randle - books recommended: "A History of UFO Crashes", Avon Books, 1995 and "Crash – When UFOs Fall From the Sky", New Page Books, 2010
Jenny Randles - book recommended: "UFO Retrievals", Blandford Books, 1995
"Sightings" - TV show
Chris O'Brien - book recommended: "The Mysterious Valley"
Dr. Steven Greer
"Beyond Reality: The UFO & Paranormal Files" – Web Site and all other researchers and anonymous sources who have assisted. http://www.cseti.org/crashes/crash.htm and https://www.youtube.com/watch?v=HYv6TFuB0A4

Alien Bodies Recovered From UFO Crash Retrievals

The following is a preliminary list of UFO crashes throughout the world. Currently, little is known about most of these crashes

Most physical evidence of these crashes has been confiscated by the government; however, not all evidence is in their possession. In the past, the government's central archive of UFO wreckage was Wright-Patterson Air Force Base, Dayton, Ohio. It has been reported that most of the alien bodies have been sent to Great Britain for storage and examination.

1939-46	Spitzenbergen, Norway	?
4 July 1947	Roswell, New Mexico	4 Bodies
13 Feb 1948	Aztec, New Mexico	12 Bodies
7 July 1948	Mexico South of Laredo, TX	1 Body
1949	Roswell, New Mexico	1 ET Living
1952	Spitzenbergen, Norway	2 Bodies
14 Aug 1952	Ely, Nevada	16 Bodies
10 Sep 1950	Albuquerque, New Mexico	3 Bodies
18 Apr 1953	S.W. Arizona	No Bodies

20 May 1953	Kingman, Arizona	1 Body
19 June 1953	Laredo, Texas	4 Bodies
10 July 1953	Johofnisburg S.Africa	5 Bodies
13 Oct 1953	Dutton, Montana	4 Bodies
5 May 1955	Brighton, England	4 Bodies
18 July 1957	Carlsbad, New Mexico	4 Bodies
1961	Timmensdorfer, Germany	12 Bodies
12 June 1962	Holloman AFB, New Mexico	2 Bodies
10 Nov 1964	Ft.Riley, Kansas	9 Bodies
27 Oct 1966	N.W. Arizona	1 Body
1966-1968	5 crashes IN/KY/OH Area	3 Bodies Disk Intact
18 July 1972	Morroco Sahara Desert	3 Bodies
10 July 1973	NW Arizona	5 Bodies
25 Aug 1974	Chihuahua, Mexico	? Bodies Disk Intact
12 May 1976	Australian Desert	4 Bodies
22 June 1977	NW Arizona	5 Bodies
5 Apr 1977	SW Ohio	11 Bodies
17 Aug 1977	Tobasco, Mexico	2 Bodies
May 1978	Bolivia	No Bodies
Nov 1988	Afghanistan	7 Bodies
May 1989	South Africa	2 ET Living
June 1989	South Africa	2 ET Disk Intact
July 1989	Siberia	9 ET Living
2 Sept 1990	Megas Platanos, Greece	?
Nov 1992	Long Island NY, New York	?

http://ufocasebook.conforums.com/index.cgi?board=general&action=display&num=1
http://www.angelfire.com/journal/alienseek/ufocrash.html

Most of the crashes listed were identified in the Summer 1980 issue of The New Atlantean Journal on page 54.

This is a list of the UFO crashes up to 1992 with recovery information: researched by the **Phoenix Foundation**. https://www.youtube.com/watch?v=WgMY8wTNsdU

CHAPTER 44

THE US MILITARY'S PROACTIVE APPROACH TO UFOs

When one reflects on the number of alien spacecraft that have crashed upon terra firma over the past couple of centuries, one has got to ask the question, why have there been so many crashes of Extraterrestrial spacecraft upon the Earth?

It seems ironic to be an interstellar explorer who has ventured away from one's home planet in another star system or from another galaxy and has travelled hundreds or thousands or perhaps even, millions of light years so successfully, only to end your long space journey by crashing and dying on an unknown alien world or to possibly languish in misery for the remainder of one's days on an inhospitable and hostile planet as a military prisoner. However, these are the possibilities and realities of space flight that needs to be considered by every space-faring civilization that sets out to explore the universe.

Could such an unforeseen demise be the outcome of so many alien spaceship disasters chalking it up to dumb luck and poor spacecraft design or inadequate interstellar navigational skills?

Alien mortality obviously indicates that these are physical beings subject to all the same laws of a physical universe that we humans confront on a daily basis. This fact alone would seem to answer the question as to the nature and reality of Extraterrestrial lifeforms. It would eliminate the **Interdimensional Hypothesis** and the other hypotheses that these particular Extraterrestrial beings that have crashed upon the Earth are beings from the same physical universe that we inhabit. However, a caveat to this assessment: it does not eliminate completely the Interdimensional Hypothesis that there may be beings existing in other dimensions or realms of existence. The universe or multiverse is, after all, a complex reality beyond human understanding at this time in our existence.

But is this really the reason for so many alien flying saucer crashes? No doubt, such may be the reason for a few untimely and fatal alien disasters; however, there is a more reasonable down to earth explanation for so many crashes. These interstellar explorers are being deliberately shot down by aggressive military forces here on Earth!

It was rumoured that a secret and an elite group of top notch Nazi scientists were experimenting with unusual and highly advanced aircraft designs and though German military officials after the War stated that they were unaware of such advanced aircraft programs by their own country, the Allies took nothing for granted and left nothing to chance. Answers were few and questions were many and although, still unsure what the armed forces were dealing with during the World War II, the Americans, British, Russians and their allied partners began to shoot down any foreign or alien craft flying over their country. Their success rate, however, was fortunately for the alien visitors, a pitiful failure for the military.

The aerodynamics and performance of these unusual craft were beyond any human capability or science and thus, a startling realization became evident, that these strange crafts were Extraterrestrial in nature and therefore, vis-a-vis, we were no longer alone in the universe!

Allied intelligence had acknowledged that flying saucers or UFOs were not being built by either side during wartime. Recall, from a previous section in this book that recently released British documents on UFOs revealed that Eisenhower and Churchill conferred on the highly unusual matters of strange aerial craft seen over the European and Pacific theatres of War. From the many reports that started to come in from the debriefings of pilots after each mission into enemy territory, they realized that they were be monitored by an alien intelligence.

With the end of WWII, the Cold War began between the USA and the newly formed Union of Soviet Socialist Republic. Early in the 1950s, there were widespread fears among some US military leaders that some nations within the newly formed Soviet bloc, would try to create chaos in the United States by causing psychological, social and political upheaval with a **"War of the Worlds"** scenario that would have the US Air Force wasting it time chasing shadows. Never knowing, if what they were observing visually or tracking on radar was anything real.

In this infamous 1938 radio broadcast, **Orson Wells** staged a fictitious Martian attack that reached millions of homes across the country. This fictitious attack caused widespread hysteria to such an extent that some people even committed suicide believing the alien attack was real. Wells was forced to go on the air and say that the whole thing was made up, a kind of Hallowe'en prank, in order to quell the panic that had ensued as a result.

At the height of the Cold War, there were widespread fears that the American public would once again go into a mass panic over a fictitious radio program, or even from the rapidly growing television market. **McCarthyism** had also become a part America's cold war paranoia to search and root out anyone with Communism or socialist affiliations of any kind. As a result, the **Robertson Panel (a.k.a. Robertson Report or Durant Report)** was convened to study the "UFO Question" and its possible impact on the public as a psychological weapon with its heavy burden on military involvement.

The Robertson Report was released on January 14, 1953, and it said the military and CIA should take measures to reduce interest in UFOs and should debunk all UFO reports with any explanation that is considered feasible. In order to debunk UFO reports, the military should use ridicule, character assassination, media manipulation and even threats in order to prevent evidence UFOs from being studied or reported. The report also stipulated that all UFO study groups be closely monitored "because of their potentially great influence on mass thinking..."

One member of the Robertson Panel even went so far as to say *"people who report UFOs are as dangerous to society as drug addicts."*

AFR 200-2 and JANAP 146

The Air Force quickly developed two regulations that re-enforced the restrictions for all military and commercial pilots on reporting UFOs, the first was **JANAP 146 (Joint Army Navy AirForce Publications)** with all its revisions. JANAP 146 regulations made the reprinting or publication of UFO reports by pilots a crime under the Espionage Act, punishable by a fine up to $10,000 and 1 to 10 years in jail.

On Aug. 12, 1954, shortly after JANAP 146 was issued, **AFR 200-2 (Air Force Regulations AFR 200-2)** made all UFO reports classified and prohibited release of UFO information unless it had been "positively identified" by the Air Force. A 1958 revision of AFR 200-2 stipulated that the Air Force would give the names of people who report UFOs to the FBI so these people can be closely monitored. It was deemed that these people were "illegally or deceptively bringing the subject of UFOs to public attention."

With the **FOIA (Freedom of Information Act),** potentially any information on almost any subject matter was made accessible to anyone in the general public but subject like UFOS was definitely off limits. To such extent that when there was a Congressional investigation into the UFO phenomenon about to occur, the Air Force along with the CIA desperately tried to prevent it. This collaboration resulted with the Air Force releasing a sanitized, one-page summary of the **Robertson Report** that concealed the CIA's involvement. This one-page summary said nothing about the measures the Air Force and CIA would take in order to silence UFO reporters.

Furthermore, whenever anyone requests a copy of the Robertson Report under FOI, they may receive a variety of versions of the report from one or two pages, with a variety of font sizes and line spacing or with information or paragraphs omitted from other versions and clearly with the absence of signatures of the Robertson Panel members from what is considered an authentic official report. The public really has no idea how many different versions of the report that are being circulated and it leads one to suspect that other "official" FOI documents regarding UFOs, may also be mere versions of the original reports and are not authentic.

AFL 200-5 and **JANAP 146** and all their revisions were regulations to ensure that all sightings of unidentified flying objects made by US and Canadian pilots were to be reported to an appointed debriefing officer, who would collect these accounts for further analyze.

Reproduced below is the classic order from the Inspector General of the Air Force to all air base commanders in the continental United States on December 24, 1959.

"Unidentified flying objects - sometimes treated lightly by the press and referred to as "flying saucers" - must be rapidly and accurately identified as serious USAF business in the ZI. As AFR 200-2 points out, the Air Force concern with these sightings is threefold: **First of all, is the object a threat to the defense of the US? Secondly, does it contribute to technical or scientific knowledge? And then there's the inherent USAF responsibility to explain to the American people through public information media what is going on in their skies.** [Bold font added for emphasis by author]

The phenomenon or actual objects encompassing UFO's will tend to increase, with the public more aware of goings on in space but still inclined to some apprehension. Technical and defense considerations will continue to exist in this area.

Published about three months ago, AFR 200-2 outlines necessary orderly, qualified reporting as well as public information procedures. This is where the base should stand today, with practices judged at least satisfactory by commander and inspector:

Responsibility for handling UFO's should rest with either intelligence, operations, the Provost Marshal or the Information Offices - in order of preference, dictated by limits of the base organization;

-A specific officer should be designated as responsible;

-He should have experience in investigative techniques and also, if possible, scientific or technical background;

-He should have authority to obtain the assistance of specialists on the base;

-He should be equipped with binoculars, camera, Geiger counter, magnifying glass and have a source for containers in which to store samples.

What is required is that every UFO sighting be investigated and reported to the **ATIC (Air Technical Intelligence Center)** at Wright-Patterson AFB and that explanation to the public be realistic and knowledgeable. Normally that explanation will be made by the OSAF information Officer."

Flying Saucers: Top Secret; copyright © 1960 by Donald E. Keyhoe, U.S. Marine Corps, Ret.; Published by G. P. Putnam's Sons; New York

These regulations, however, appear to the general public as if the Air force knows little to almost nothing about this phenomenon and is trying to get a handle on the situation by getting as much information from pilots while controlling the information release to the public in a perceived satisfactory manner. But, it has already been stated that the Air Force knew exactly what this phenomenon was before the end of the Second World War and that these regulations were nothing more than censorship on its pilots and the public. The military does not like competition into anything new that has incredible potential for military application in weapons and aircraft R&D. Their enemies certainly didn't need to know for national security reasons, least of all the public had absolutely no need to know, whatsoever!

But, a closer look at the situation seems to reveal a much deeper covert understanding of the phenomenon that not everyone in the Air Force or for that matter, in any other military branch was aware of, namely what was really going on or what exactly was the true nature of the UFO phenomenon.

On January 22, 1958, on a televised program of the 'Armstrong Circle Theatre' which was to interview **Major Donald Keyhoe** on "UFOs – Enigma of the Skies ", Keyhoe was suddenly cut off the air. Censorship was imposed by the producer before Keyhoe had the opportunity to give his testimony for fear of possible negative Air Force reactions. In his place the Air Force Assistant Secretary, **Richard E. Horner** appeared and read an Air Force approved-script stating:

"During recent years, there had been a mistaken belief that the Air Force had been hiding from the public information concerning unidentified flying objects. Nothing could be further from the truth. And I do not qualify this in any way…There is no evidence at hand that objects popularly known as 'flying saucers' actually exist."

Less than twenty–four hours later, the Air Force in an unguarded moment and unaware of Horner's broadcast openly admitted through their PIO (Public Information Officer) out of Langley Air Force Base to one of the **NICAP** members, Larry W. Bryant:

"The public dissemination of data on unidentified flying objects is contrary to Air Force

policy and regulations…specifically, Air Force Regulations 200-2".

In Section B-9 of AFR 200-2, only UFO sightings reported within the vicinity of an Air Force base 'may be released to the press or general public by the commander of the Air Force base concerned, if it had been *positively identified as a familiar or known object'.* [Air Force italics added]

The official position of the Air Force about UFOs is found in Section A-3 of AFR 200-2:

'Since the possibility cannot be ignored that UFOS reported may be hostile or new foreign air vehicles of unconventional design, it is imperative that sightings be reported rapidly, factually, and as completely as possible.' **Flying Saucers: Top Secret; copyright © 1960 by Donald E. Keyhoe, U.S. Marine Corps, Ret.; Published by G. P. Putnam's Sons; New York**

Here is a glaring contradiction, a denial to Horner's broadcasted "official" Air Force statement earlier which is nothing more than an outright **LIE**, one which would be repeated over and over again to the press and the general public!

As **Timothy Good** points out in his book "Need to Know" with regard to AFR 200-2, "The US Air Force may have an 'inherent responsibility to explain to the American people…what is going on in their skies', yet it is manifestly obvious that they were doing nothing of the sort. The public, evidently, did not have a need know." **Need To Know - UFOS, the Military, and Intelligence" by Timothy Good; 2007; Published by Pegasus Books LLC; New York; ISBN 978-1-933648-38-5**

Concurrently, with the release of these Air Force regulations, the Air Force was engaged in an aggressive shoot down program upon any flying object that was determined as unidentified. Pilots were already leaking their accounts to the news media that they were routinely chasing after UFOs and even, firing upon them when so ordered. The Air Force needed protocols and regulations to control the information leaks that were spilling out to the news media or to investigating UFO researchers. JANAP 146 and AFR 200-2 was the Air Force's way of controlling the information.

The concerns of the Air Force with UFO sightings in AFR 200-2 and JANAP 146 appears to pose the question as to the reality of UFOs and its possible impact on national security. The conclusions and the recommendations of the supposed 'legitimate' UFO research projects like the Robertson Panel Report, Project Sign, Project Grudge and Project Bluebook, as well as the findings of the biggest UFO research con job ever, the USAF sponsored Condon Committee Report on the "Scientific Study of Unidentified Flying Objects" out of the University of Colorado seems to answer the question about UFOs once and for all. Not only does the Condon Report provide an answer to the UFO question but re-enforces the conclusions and recommendations of all previous US Air Force UFO projects before it.

For comparison here are the conclusions reached by Project Grudge, Project Bluebook and particularly the Condon Committee Report which bears repeating once more in this book because it has become the foundation on which the Air Force has gullibly lead the news media and the general public to believe that there is nothing to the whole UFO phenomenon.

The assessments are repetitive and re-enforce the previous project's conclusions and are

carefully worded which gives the impression that each project investigation was up-front, honest and scientific in its approach but, actually cloaks the truth in deception permitting the assessments to have another meaning, other than the obvious one stated.

Project Sign (In Brief)

Project Sign is the exception to the other UFO research projects which followed it by unofficially concluding UFOs are real alien spacecraft but, in its official report, came to no conclusion about UFOs with their final report stating that the existence of "flying saucers" could neither be confirmed nor denied. However, prior to this, Sign officially argued that UFOs were likely of extraterrestrial origin, and most of the project's personnel came to favor the extraterrestrial hypothesis before this opinion was rejected and Sign was dissolved. Project Sign was not the conclusion that the Air Force wanted to release to the public and so its successor took the negative position in keeping with its project name: **Project Grudge**.

Project Grudge (In Brief)

Project Grudge report's conclusions stated:

> *A. There is no evidence that objects reported upon are the result of an advanced scientific foreign development; and, therefore they constitute no direct threat to the national security. In view of this, it is recommended that the investigation and study of reports of unidentified flying objects be reduced in scope. Headquarters AMC [Air Materials Command] will continue to investigate reports in which realistic technical applications are clearly indicated.*
> *NOTE: It is apparent that further study along present lines would only confirm the findings presented herein. It is further recommended that pertinent collection directives be revised to reflect the contemplated change in policy.*
>
> *B. All evidence and analyses indicate that reports of unidentified flying objects are the result of:*

1. Misinterpretation of various conventional objects.

2. A mild form of mass hysteria and war nerves.

3. Individuals who fabricate such reports to perpetrate a hoax or to seek publicity.

4. Psychopathological persons.

Project Blue Book (In Brief)

Project Blue Book had two goals:

1. to determine if UFOs were a threat to national security, and
2. to scientifically analyze UFO-related data.

Thousands of UFO reports were collected, analyzed and filed. As the result of the **Condon Report,** which concluded there was nothing anomalous about UFOs, Project Blue Book was ordered shut down in December, 1969 and the Air Force continues to provide the following summary of its investigations:

1. No UFO reported, investigated and evaluated by the Air Force was ever an indication of threat to our national security;
2. There was no evidence submitted to or discovered by the Air Force that sightings categorized as "unidentified" represented technological developments or principles beyond the range of modern scientific knowledge; and
3. There was no evidence indicating that sightings categorized as "unidentified" were extraterrestrial vehicles.

Condon Committee Report (In Brief)

Condon Committee Report assessment states briefly:

"Our general conclusion is that nothing has come from the study of UFOs in the past 21 years that has added to scientific knowledge. Careful consideration of the record as it is available to us leads us to conclude that further extensive study of UFOs probably cannot be justified in the expectation that science will be advanced thereby."

Let's look at what is being said in Blue Book's carefully worded findings which are upheld by the Condon Committee Report:

1. "No UFO reported, investigated and evaluated by the Air Force was ever an indication of threat to our national security".

This is because the US Air Force had already made the assessment that no hostility has ever come from these flying objects, whether in peacetime or wartime unless, fired upon by US pilots, in which case the alien craft either evaded the aggressive aircraft or had no choice but to return fire, because they felt that eminent danger to themselves was unavoidable.

No aggressive alien invasion has ever taken place in the USA or anywhere else upon the Earth and although almost every country has had their borders penetrated by alien spacecraft which may be viewed as a violation of a country's airspace and a possible national security breach, no hostility has ever taken place against planet Earth. In fact, aggression has always been the first response by humans to these alien visitors, who are perceived as uninvited guests to our planet.

2. *"There was no evidence submitted to or discovered by the Air Force that sightings categorized as "unidentified" represented technological developments or principles beyond the range of modern scientific knowledge"*, this is nothing but a bold face lie!

Most Allied military forces during the Second World War realized very quickly that these strange aerial craft were no longer "unidentified" having demonstrated a highly advanced technology in design and performance to anything that was being built on Earth. These alien objects were identified as Extraterrestrial in origin and their advanced spacecraft represented a highly desirable technology that was beyond the range of current scientific knowledge. Having such alien technology in one's possession would mean a huge advantage over countries on Earth, particularly one's enemies. The evidence is undeniable and the potential for scientific advancement is pregnant with possibilities. The only reason to make such an absurd statement was to sidetrack other nations and the public interest away from any further investigation, while the US seize the advantage of making some major breakthroughs through captured spacecraft and be light years ahead of anyone else before they discovered some of the same answers to the UFO mystery.

3. *"There was no evidence indicating that sightings categorized as "unidentified" were extraterrestrial vehicles"*. This statement too is another bold face lie!

The evidence from visual observations, both air and ground radar trackings, photos and videos, physical landing traces and actual contact with these craft and alien beings from all over the world is insurmountable and undeniable proof that these unidentified flying objects are Extraterrestrial vehicles. Many countries governments through public release of UFO documents and an ever growing list of scientists are now publicly stating that we are being visited by Extraterrestrial life forms not originating from our Solar System. People are becoming more aware of the things of the Earth, the elements of nature, even when observed under unusual circumstances and they are able to recognize what is manmade and what objects under intelligent control are truly out of this world.

So, once again, another **BIG LIE** in the guise of what is touted as truth to an unwitting news media, the government and the science community, who have accepted blindly the final conclusions and recommendations that have come from each successive Air Force sponsored UFO investigative program.

Those intelligent enough to question the findings from these projects will see that there is an official cover-up being perpetrated upon the public as UFO sightings and case evidence accumulates each year to mountainous proportions. The official military position according to **Condon Report** is that there is no secrecy concerning UFO reports. What may be perceived as secrecy is the "intelligence policy of delay in releasing data so that the public does not become confused by premature publications or incomplete of studies of reports." Yet, another **BIG LIE!!** Edward W. Condon, Director, and Daniel S Gillmor, Editor; Final Report of the Scientific Study of Unidentified Flying Objects; Bantam Books, 1968

The Condon Report went so far as to state that schools should discourage courses and student study of this phenomenon, that it was a wasted pursuit of knowledge which would be better channelled in the study of natural sciences like astronomy, meteorology, etc.

"A related problem to which we wish to direct public attention is the miseducation in our schools which arises from the fact that many children are being allowed, if not actively encouraged, to devote their science study time to the reading of UFO books and magazine articles of the type referred to in the preceding paragraph. We feel that children are educationally harmed by absorbing unsound and erroneous material as if it were scientifically well founded. Such study is harmful not merely because of the erroneous nature of the material itself, but also because such study retards the development of a critical faculty with regard to scientific evidence, which to some degree ought to be part of the education of every American.

Therefore, we strongly recommend that teachers refrain from giving students credit for school work based on their reading of the presently available books and magazine articles. Teachers who find their students strongly motivated in this direction should attempt to channel their interests in the direction of serious study of astronomy and meteorology, and in the direction of critical analysis of arguments for fantastic propositions that are being supported by appeals to fallacious reasoning and false data." Edward W. Condon, Director, and Daniel S Gillmor, Editor; Final Report of the Scientific Study of Unidentified Flying Objects; Bantam Books, 1968

330

CHAPTER 45

U.S. MILITARY'S UFO INTERCEPTION AND SHOOT DOWN PROGRAM WITH AIRCRAFT CASUALTIES AND LOSSES

The acquisition of alien technology suddenly became the primary goal of the US, British, and the Allied Forces having determined that these strange aerial craft were not being built by any military power on Earth at the time. Over the years, the Russians, Chinese, South Africans, and other emerging superpowers from South America have also become actively involved in the hunt for alien spacecraft in a catch-up program to keep pace with the Americans and her partners.

The initial justification for such aggressive action toward these alien visitors was that they could be mistaken as German or Japanese invading forces against Allied countries. UFOs behave unexpectedly and differently when tracked on radar and they did not respond to any communication to identify themselves when tracked or visually observed. In the minds of the military such non-compliance to air flight protocols within a nation's borders without communication or identification is usually considered an act of hostility of airspace intrusion by a possible enemy. In the minds of the military, these alien spacecraft represented a real potential threat to the national security of America.

These mysterious craft also demonstrated a technology far beyond any earthly science in aircraft design, propulsion, and performance. To have one of these craft in one's possession for closer inspection and possible reverse engineering would enable America to establish a quantum leap ahead of any other country in military technology and power. The exploitation of the almost limitless potential that could be derived from Extraterrestrial technology became the real reason for why the US Air Force and the other US military branches wanted.

How does one acquire something that is out of this world? The most obvious way would be to shoot one of these alien spacecraft down when they enter your airspace, a task easier said than done. There had been no successes in downing one of these spacecraft before the War ended mainly in part due to the fact that no official regulation was in place by any of the Allied Forces. The problem was the inability of wartime aircraft technology to compete with these saucers that were able to out fly and outmaneuver both in speed and altitude any aircraft built.

On June 27, 1947, the Air Force still not knowing what UFOs were ordered their pilots to down a UFO for examination either, by shooting at it or if necessary, by ramming at the craft and then, bailing out. There has been a number of incidents in which military aircraft were ordered to pursue, close within range and either film the strange craft with gun-cameras or in many cases to open fire upon the alien craft and bring it down when it failed to respond to radio communications to land.

"The 1950's became the military's 'shoot down' era and according to **Stanton Friedman** and **Frank Feschino Jr**. during a **Coast to Coast** radio interview with **George Noory** stated that the military had standing orders to shoot down any unidentified craft that didn't land when instructed --- and it appears that on occasion UFOs shot back! There is no question, however, that our planes were the aggressors, Friedman commented. http://www.coasttocoastam.com/

There were many pilot deaths and mysterious military plane incidents during the early to mid-1950s. In fact, the New York Times described jets as "disintegrated and disappeared" in their coverage, Friedman reported. Feschino detailed how an F-86 jet fighter plane crashed in South Glastonbury, CT on August 5, 1952, under mysterious circumstances and connected it to a UFO flap that was occurring that summer. **Project Blue Book** contained 1500 reports from 1952, with over 300 of them classified as unidentified, he continued.

The Flatwoods Monster case also took place in the summer of '52-- there were sightings over 11 states the night the curious craft/robot set down in Braxton County, Feschino noted. On that same night, thirty objects were seen coming in over the Eastern Seaboard, and appeared to be following a craft that was damaged, he added. Friedman suggested that the US military eventually gave up on their shootdown policy, and instead began simply observing UFOs with their instruments." http://www.coasttocoastam.com/show/2007/12/06

Throughout the years when the Flying Saucer/UFO phenomenon caught and gripped the public's imagination, the military branches, and some scientists began to realize that UFO sightings appeared to come in **"flaps"** and **"waves"**. Was it possible that alien ships were making reconnaissance runs over our planet during frequent time periods? Some scientist thought that it was during the **perihelion** (nearest approach) of the planet Mars in its orbit to Earth. This serendipitous fact drew speculation that Mars had intelligent life which was visiting the Earth in flying saucers on some hidden mission. There have been many flaps and waves of UFO activity since records of these strange objects were being documented globally. 1942, 1947, 1948, 1952, 1954, 1957, 1962, 1965, 1967, 1972 were just some of years when waves of UFO activity occurred and of particular interest in those years are the military chases or **"dog fights"**. These dogfights and chases would be repeated time and time again over the next six decades, some ending in tragedy for the pilots.

1948 Mantell UFO Incident

One of the most famous cases of US military pursuit of a UFO that ended tragically for the pilot was the Mantell UFO Incident.

The **Mantell UFO Incident** is one of the most significant UFO sightings of all time, mainly because it resulted in the death of a fighter pilot and subsequent change in public attitude towards UFOs.

On January 7th, 1948, Kentucky State Highway Patrol began receiving calls regarding an unidentified object hovering above Maysville, Kentucky. The object was said to be 250 to 300 feet in diameter and heading west. The reports were brought to the attention of military police and eventually **Colonel Guy F. Hix** of the Godman Army Air Field at Fort Knox.

For nearly an hour and a half, personnel at the airfield observed the object but were unable to identify it. Witnesses also watched from surrounding areas. Most witnesses described the object as motionless or moving very slowly.

Captain Thomas F. Mantell
http://www.ufocasebook.com/Mantell.html

At approximately 2:45 p.m., a flight of four F-51 Mustang fighters flew into the area. Flight leader, **Captain Thomas F. Mantell**, was asked to investigate the unidentified object. The object was at a much higher altitude so the fighters had to climb. By 22,500 feet, three of the four fighters had abandoned the pursuit due to low fuel or oxygen. Captain Mantell was also instructed to level off but continued to climb regardless.

There is some dispute over exactly which words were spoken over the radio network during the final few moments. Some airfield staff claimed Mantell had described the object as "metallic and of tremendous size". Others have quoted him saying "[it's] above me and appears to be moving about half my speed".

Mantell's plane was last seen climbing in altitude. Lacking oxygen equipment, he apparently blacked out and his plane spun out of control, crashing on a farm near Franklin, Kentucky. Mantell was found dead at the scene a couple of hours later. By this time the UFO was no longer visible.

Rumors soon spread about the crash, including claims that the plane had been shot down by a UFO, the plane was full of near-microscopic holes, and mysterious wounds had been found on Captain Mantell. Although there was never any evidence for the mysterious wounds claim, two factors may have helped fuel speculation about the plane's destruction:

1. The aircraft broke apart before hitting the ground. This is a normal result of an aircraft operating well outside its design parameters, but it is easy to see how witnesses may have misinterpreted the breakup.
2. The aircraft landed horizontally, rather than nose-first as some people believe should have been the case.

Despite a severe lack of evidence, conspiracy and cover-up theories found a ready audience. The initial explanation was that the object was Venus, which is known to create a similar type of illusion. However, this was later ruled out by astronomers who said that Venus would not have created this effect at the time of the incident.

The final official explanation was that the object was a **US Navy Skyhook weather balloon**. This is by far the most plausible explanation because:

- The weather balloon project was secret—no one involved in the sighting knew about it.
- A number of Skyhook balloons had been launched the same day in Clinton County, Ohio, approximately 150 miles from Fort Knox.
- The UFO's appearance and behavior matched the weather balloons (made of reflective aluminum, about 100' in diameter).

Following the Mantell incident, public opinions about UFOs inevitably changed. Whereas UFOs had previously been considered light entertainment, the Mantell incident was announced by such alarming headlines as "Flier Dies Chasing A Flying Saucer" (The New York Times). From this point on, UFOs were no longer assumed to be harmless. The UFO phenomena had changed forever.
http://www.paranormal-encyclopedia.com/u/ufo/sightings/1948/mantell.html

The death of World War II ace Captain Thomas Mantell while chasing a flying saucer was admitted by the Air Force and forced them to set up a secret investigative unit called Project Sign, later Project Grudge and then Project Blue Book

In 1949 **General Hoyt Vandenburg**, Air Force Chief of Staff refused to accept the spaceship conclusion, warning project officers of the effect on the public. The policy was changed to one of explaining away sightings as mistakes, hallucinations or hoaxes. But according to Major Donald Keyhoe of NICAP, an intelligence summary was apparently released by error to the public which suggested possible observations of our planet by Extraterrestrial beings:

"Such a civilization might observe that on Earth we now have atomic bombs and are fast developing rockets. In view of the past history of mankind, they should be alarmed. We should, therefore, expect at this time above all to behold such visitations."

334

Photograph of Mantell wreckage, retouched to remove scratches

Part of Captain Thomas Mantell F-51 Mustang fighter
http://www.ufocasebook.com/Mantell.html

The summary was quickly withdrawn from the press but not until a few newsmen at the Pentagon had examined it. But official Air force policy in 1950 was to view every report as originating from crackpots, religious cranks, publicity hounds or malicious pranksters. However, this policy did nothing to stem the tide of hundreds of global UFO reports that were coming in from pilots and trained observers.

Many saucer sightings by 1952 though not considered hostile were still unexplained and so, the Air Force once again ordered its pilots to fire on UFOs. It was a policy of "shoot first and ask questions second". When this information went public, protests caused the order to be revoked. The public's approach to the problem was to demonstrate a peaceful inquiry with force as a last measure to open hostility. The US Air Force may have backed off when UFOs retaliated to their aggressive behavior but, it didn't stop the Air Force from chasing them around the skies and taking the occasional potshot at them if the pilots deem them hostile!

By November 1953 with the tightening of censorship, the Air force had more worries than the constant leaks of UFO sighting to the press or the public when an F-89 jet with its two man crew went missing while pursuing an unidentified flying object. This was the **Kinross Incident** in which Traux Air Force Base had tracked the jet by radar until it merged with an object over Lake Superior. The object was never identified, nor was any trace of the jet and its crew ever found.

Flying Saucers: Top Secret by Major Donald E. Keyhoe; © 1960; Published by G.P. Putnam's Sons; New York

1948 George Gorman UFO Chase

In the skies above Fargo, North Dakota, on October 1, 1948, **Lieutenant George F. Gorman** of the North Dakota Air National Guard had an experience he would never forget. After flying a cross-country flight with this Guard squadron, he elected to do some night flying. He flew over the football stadium and then circled the city before deciding to land at Hector Airport.

It was 9:00 PM when he received clearance from the control tower to make his landing. The tower warned him of a Piper Cub in his vicinity. Gorman could see the Cub about 500 feet below him. All of a sudden, he saw what he thought were the tail lights of a different plane. Radioing the tower, he was informed that the tower only had Gorman's F-51 and the Piper Cub on the radar. Gorman decided to investigate the uncharted plane.

He got to within 1,000 feet of the flying object and got a good look at it. Gorman would later make this statement:

"It was blinking on and off. As I approached, however, the light suddenly became steady and pulled into a sharp left bank. I thought it was making a pass at the tower. I dived after it and brought my manifold pressure up to sixty inches but I couldn't catch up with the thing. It started gaining altitude and again made a left bank; I put my F-51 into a sharp turn and tried to cut the light off in its turn. By then we were at about 7,000 feet."

The unknown object suddenly made a sharp turn. Suddenly, the two planes were heading straight toward each other. Gorman went into a dive, missing the unknown object as it passed at about 500 feet above him. The UFO then went to 1,000 feet, Gorman still in pursuit. The UFO had turned and was again heading toward the F-51. Was it playing games with him? What was this object? Just before the two fliers hit head-on, the UFO made an unbelievable move, streaking straight upward.

Gorman made one final attempt to approach the UFO, but his F-51 stalled at 14,000 feet. The object was nowhere in sight. Time elapsed-exactly 27 minutes. Gorman, an experienced flier, was shaken to the point that he struggled to control his plane. He recalled that he saw no exhaust at all from the disc-shaped UFO, and heard not a sound. He had flown his F-51 at 400 mph, pushing it to its limits, yet was no match for the unknown object.

Traffic controllers Lloyd D. Jensen and H. E. Johnson, both in the tower during the dogfight, witnessed Gorman's ordeal. The two had seen the Piper Cub and the UFO independently, eliminating any possibility that Gorman had somehow chased the Piper.

Testimony by the Piper Cub pilot **Dr. A. E. Cannon, and his passenger, Einar Nielson** also validated the flight of the UFO. They both stated that they had witnessed the unknown object

336

streaking through the sky, at the same time they were communicating with the tower. The tower had previously asked them what the "third plane" (UFO) was.

Gorman's final word on the dogfight was:

"I am convinced that the object was governed by the laws of inertia because its acceleration was rapid but not immediate, and although it was able to turn fairly tight at considerable speed, it still followed a natural curve." http://ufos.about.com/od/bestufocasefiles/a/gorman.htm

1951 Newfoundland, Canada Giant UFO Encounter

In Newfoundland, Canada, on February 10, 1951, Atlantic-Continental Air Transport Squadron 1, a US Navy flight, encountered a giant UFO.

The flight took off from Keflavik, Iceland, about 90 miles west of Gander, Newfoundland. The pilot of the craft was US Naval Reservist **Lieutenant Graham Bethune**, who was the first crew member to spot a large, unknown flying craft.

Bethune could quickly see that the 300 ft. diameter UFO was on a collision course with the plane. Members of the crew in the cockpit were all scurrying to see the object, which resulted in several of them sustaining minor injuries.

In an official statement, Bethune stated:

"I observed a glow of light below the horizon about 1,000 to 1,500 feet above the water. We both [the pilot as well] observed its course and motion for about 4 or 5 minutes before calling it to the attention of the other crew members... "

Suddenly, the UFO seemed much larger... it was accelerating, estimated at up to 1,000 mph, and the crew was extremely concerned with a possible head-on collision. Then the object's color began to change.

To the relief of the crew, the UFO's angle of flight suddenly changed. At this time, the crew was able to clearly see that the object's shape was circular, and the UFO was of a red-orange color.

The object had reversed its course, tripled its speed, and soon disappeared over the horizon. Later, it would be verified by radar that the UFO had come to within 5 miles of the plane. The UFO was tracked by **DEW Line Radar** at Goose Bay, Labrador.

The plane arrived safely, and soon the entire crew was interrogated by Air Force Intelligence. It was also reported the captain of the plane was shown a photograph of a UFO by what was believed to be the CIA. The UFO was similar to what the crew had encountered.

In a subsequent video interview, Bethune would add that the UFO was under intelligent control and that he had disengaged his automatic pilot, to attempt to fly under the object, and therefore avoid what seemed to be an imminent collision.

1953 Kinross UFO Incident

Is this a case of aliens abducting aircraft right out of the air? The account has a familiarity to the 1953 Sc-Fi movie "This Island Earth" where a light aircraft while in flight is tractor-beamed aboard a large flying saucer.

Maj. Donald Keyhoe of NICAP recounts this event in several of his books as he had a strong interest in military jet pursuits of UFOs:

"One of the strangest cases on record occurred in 1953. Though it has received considerable publicity, some of the follow-up developments are not generally known.

First Lieutenant Felix Eugene Moncla, Jr. disappears
while pursuing a UFO over Lake Superior

On the night of November 23, 1953, an F-89 all-weather interceptor was scrambled at Kinross AFB, to check on a UFO flying over the Soo Locks. The jet had a crew of two – **Lt. Felix Moncla**, the pilot, and **Lt. R. R. Wilson**, the radar observer. Guided by an AF **GCI (Ground Control Intercept) radar station**, Moncla followed the unknown machine out over Lake Superior, flying at 500 miles an hour.

Minutes later, a GCI controller was startled to see the blips of the jet and the UFO, suddenly merge on the radar-scope. Whatever had happened, one thing was certain: The F-89 and the UFO were locked together.

As the combined blip went off the scope, the controller hurriedly radioed Search and Rescue. Moncla and Wilson might have bailed out before the collision. Both had life jackets and self-inflating life rafts – even in the cold water, they could survive for a while.

All night, U.S. and Canadian search planes with flares circled low over the area. At daylight, a score of boats joined the hunt, as the pilots crisscrossed the lake for a hundred miles. But no trace was found of the airmen, the jet or the UFO."

The search was still on when Truax AFB gave the Associated Press this official release:

"The plane was followed by radar until it merged with an object 70 miles off Keweenaw Point in upper Michigan."

In view of AF secrecy, this was a surprising admission. The statement appeared in an early edition of the Chicago Tribune, headed JET, TWO ABOARD, VANISHES OVER LAKE SUPERIOR. (Photocopy in author's possession.) Then AF Headquarters killed the story.

Denying the jet had merged with anything, the AF said that radar operators had misread the scope. The reported UFO, it stated, had been an off course Canadian airliner which the F-89 had intercepted and identified. After this, the AF speculated, the pilot evidently had been stricken with vertigo and the jet had crashed in the lake.

The Canadian airlines quickly denied any flights in the area. Expert pilots also hit at the AF explanation: Moncla could have switched on the automatic pilot until the vertigo passed; also Wilson could have taken over temporarily.

As customary, the AF sent two officers to the families of the lost airmen to give them official messages of sympathy. According to letters which a relative of Moncla sent me, here is what followed. Explaining the accident, the AF representative told Moncla's widow that the pilot had flown too low while identifying the supposed Canadian airliner and had crashed in the lake.

By some headquarters mix up, a second AF officer was sent to offer condolences to the Moncla family. When Moncla's widow asked if her husband's body might be recovered the officer said there was no chance – the jet had exploded at a high altitude, destroying the plane and its occupants." **Aliens From Space – The Real Story of Unidentified Flying Objects By Major Donald E. Keyhoe; © 1973; Published by Doubleday & Company, Inc; New York**

From the above account, it is clear that the Air Force seems to routinely engage their mouths before they engage their brains when it comes to sharing what it knows or doesn't know about UFOs in general. This results in endless news media and public speculation of possible Air Force cover-ups and ulterior motives. When the truth of the situation is unintentionally announced and the wrong thing is stated, damage control kicks in to correct the news story because someone on the USAF 'team' has not been briefed on Air Force policy regarding UFOs.

 Lies, deception, and plausible deniability become the game plan of damage control by which the military falls back on whenever there is a report of a truly unidentified flying object that is encountered by one of their pilots.

F-98C 51-5855 of 433d Fighter Interceptor Squadron in B&W and colour photo.
The aircraft involved in the Kinross incident was nearly identical.
http://www.nicap.org/reports/kinross3.htm

There are also other agencies and military branches with their own intelligence officers who often overlap with other agents in the field and quite often their style of investigation conflict with each other when investigating the same UFO case. This is quite telling as it indicates that there are other agencies or covert groups interested in the UFO phenomenon that are competing with each other and that their agendas are not necessarily the same! This aspect of competing covert groups will be examined a little further in this book. In the above Kinross Incident, the mix-up in sending out a second AF officer to Moncla's widow and family to offer condolences and expressing a different consequence of the air tragedy was nothing but Air Force incompetence in handling the matter. Remarkably and very nobly, Moncla's widow accepted the contradiction in the air tragedy story stating that she realized that the Air Force may have its security reasons for keeping the whole matter a secret.

"Unlike the Mantell incident, the Kinross case attracted minimal newspaper coverage; also unlike Mantell, Kinross has never been satisfactorily explained. Later, after aviation writer **Donald E. Keyhoe** broke the story in his best-selling *The Flying Saucer Conspiracy* (1955), the Air Force insisted that the "UFO" had proved on investigation to be a **Royal Canadian Air Force** C-47. The F-89C had not actually collided with the Canadian transport plane, but something unspecified had happened, and the interceptor crashed. Aside from implying woeful incompetence on the radar operators' part, this "explanation" -- still the official one -- flies in the face of the Canadian government's repeated denials that any such incident involving one of its aircraft ever took place."

"In 1958 Keyhoe got hold of a leaked Air Force document that made it clear that officialdom considered the Kinross incident a UFO encounter of the strangest kind. The document quoted these words from a radar observer who had been there: 'It seems incredible, but the blip apparently just swallowed our F-89.'"**http://science.howstuffworks.com/space/aliens-ufos/ufo-government4.htm**

Off the record, those close to the case believed that the loss of the F-89 Scorpion jet was a direct result of an encounter with the unknown object, clearly seen on radar. In recent months, the Michigan Diving Company has made assertions that they may have found the jet and the UFO, but this finding is still being investigated. Current belief amongst Ufologists is that the Michigan Diving Company's find is a hoax as the company can no longer be found operating anywhere.

British Pilots Chase UFOs

It would seem that the US Air Force weren't the only ones chasing UFOs, as the British had its share of military planes chasing after strange disc shaped objects and large balls of light over the British countryside. According to Nick Pope, who used to run the Ministry of Defence's UFO project, pilots have apparently fired upon the unidentified objects without success since the 1980s.

In the Sun newspaper, he told: "There was a faction in the MoD who said 'We want to shoot down a UFO and that will resolve the issue one way or another', we know of cases where the order has been given to shoot down - with little effect to the UFO."

Mr. Pope claimed that the RAF only attempted to engage when the mysterious objects were perceived to be a threat.

He said: *"In the case of UFOs, whether the object is causing a threat is very much a pilot's judgement call. The public won't know unless it comes down in a heavily populated area."*
By Daily Telegraph Reporter; 8:14AM GMT 26 Jan 2009

1957 Milton Torres UFO Encounter

The Milton Torres 1957 UFO Encounter is a case reported by a former jet pilot in the United States Air Force, **Milton John Torres**, of Miami, Florida. On May 20, 1957, the 25-year-old Torres was ordered on a mission to intercept and shoot down a large UFO that had been picked up by radar in the skies over East Anglia, United Kingdom. Torres, flying a North American Aviation F-86D Sabre jet approached the object and prepared to fire upon it. As he locked on his weapons, the UFO suddenly zoomed away at a high rate of speed. The details of this case were first made public in a release of UFO-related case files by the British government in October 2008.

This case shares some similarities with the Del Rio, Texas UFO Crash of 1955, which also involved an aerial encounter between an F-86 Sabre and a UFO.

Milton Torres was a lieutenant with the 514th Fighter Interceptor Squadron in the 406th Air Expeditionary Wing based at RAF Manston in Kent. He was the pilot of one of two North American Aviation F-86D jets that were on five-minute alert at the end of the runway at RAF Manston airfield, awaiting the signal to scramble. The two American pilots were on standby at the "alert shack" at RAF Manston and their jets were fully powered for immediate take-off when they received instructions to take to the air.

**American fighter pilot Milton Torres was ordered to shoot down
a suspected UFO in 1957 over East Anglia** Source: Daily Mail

342

"The initial briefing indicated that the ground was observing for a considerable time a blip that was orbiting the East Anglia area." Torres was then told that the object of his pursuit "was an unidentified flying object with very unusual flight patterns" and that "the bogey [UFO] actually was motionless for long intervals."

Having difficulty believing that he had been ordered to open fire on the unknown object, Torres requested authentication. "I was only a lieutenant and very much aware of the gravity of the situation." When the authentication code response came back, Torres struggled to match the numbers with those in his daily code sheet. "It was totally black, and the lights were down for night flying. I used my flashlight, still trying to fly and watch my radar." The authentication code matched.

Torres, who was in the lead position, received orders to push his jet to maximum power by using full afterburner and to proceed to an altitude of 32,000 feet on a heading that would intercept the UFO. The F-86D moved to Mach .92, (more than 700 miles per hour). The maximum speed of the F-86D is 761 miles per hour.
http://www.nationalmuseum.af.mil/factsheets/factsheet.asp?id=362

Torres and his wingman were vectored by Met Sector to a position over the North Sea, just east of East Anglia. Asked to report any visual observations about the UFO, Torres told ground control, "I'm in the soup, and it is impossible to see anything." High altostratus clouds and night flying conditions prevented him from making any visual contact.

F-86D Sabre fighter jet similar to that flown by Milton Torres in 1957.
http://www.nationalmuseum.af.mil/Visit/MuseumExhibits/FactSheets/Display/tabid/509/Article/198076/north-american-f-86d-sabre.aspx

Aware that he was reaching the upper limit of his jet's capabilities, Torres requested to come out of afterburner, but ground control denied his request. At that time, he prepared to execute the order to "fire a full salvo of rockets at the UFO." The F-86D carried 24 rockets. The National Museum of the U.S. Air Force describes the aircraft as having a "retractable tray of 24 rockets" and adds that "the effect of these weapons would have been devastating to an enemy bomber" because each rocket contained the power of a 75mm artillery shell.
http://www.nationalmuseum.af.mil/factsheets/factsheet.asp?id=362

As Torres prepared all 24 of his rockets for firing, he could hear the other F-86D pilot also responding to orders from ground control. "I wasn't paying too much attention ... but I clearly remember him giving a 'Roger' to all the transmissions. I can only suppose he was as busy as I was."

Torres was given a final turning maneuver to execute and was told to look 30 degrees to port. His radar screen displayed the UFO at 30 degrees and about 15 miles distant. "The blip was burning a hole in the radar with its incredible intensity. It was similar to a blip I had received from B-52s and seemed to be a magnet of light."

Map of England highlighting location of 1957 UFO encounter.
http://www.thelivingmoon.com/49ufo_files/03files2/1957_May_20_Milton_Torres.html

344

According to Torres, the radar return "had the proportions of a flying aircraft carrier." He added, "By that I mean the return on the radar was so strong that it could not be overlooked by the fire control system on the F-86D... the larger the airplane the easier the lock on. This blip almost locked itself... it was the best target I could ever remember locking on to. I had locked on in just a few seconds, and I locked on exactly 15 miles, which was the maximum range for lock on."

Suddenly, he noticed that the object on the radar screen was moving. "Do you have a Tally Ho?" Ground control radioed, wanting to know if he had made visual contact with the UFO. Torres replied that he was still "in the soup" and could see nothing.

"By this time, the UFO had broke lock, and I saw him leaving my 30-mile range. Again I reported that he was gone, only to be told that he was now off their scope as well. With the loss of the blip off their scope, the mission was over. We were vectored back to home plate (Manston) and secured our switches. My last instructions were that they would contact me on the ground by land line."

Torres was left with the impression that the UFO was moving at no less than Mach 10 (over 7,000 miles per hour) when it disappeared. "This thing had to be going double-digit mach making turns that I didn't think were possible, breaking all the rules of physics," Torres said. According to Torres, the UFO "didn't follow classic Newtonian mechanics. It made a right turn almost on a dime. The (Royal Air Force radar) scope had a range of 250 miles. And after two sweeps, which took two seconds, it was gone. And I was flying almost at Mach 1, at .92."

After the mission, Torres was advised that the mission he had just completed was classified and that he would receive further information about it from an "investigator." Torres later wrote, "I had not the foggiest idea what had actually occurred, nor would anyone explain anything to me."

On the day after the incident, Torres was ushered into a room of the squadron operations area, where a well-dressed American civilian in a dark blue trenchcoat was waiting for him. The civilian, who looked to Torres like an "IBM salesman" and was in his late 30s or early 40s, had come from London to interview Torres. He waved a **National Security Agency (NSA)** ID at Torres and immediately began asking questions about the UFO sighting. "He advised me that this would be considered highly classified and that I should not discuss it with anybody, not even my commander. He threatened me with a national security breach if I breathed a word about it to anyone. He disappeared without so much as a goodbye, and that was that, as far as I was concerned." The interview took only about 30 minutes, but Torres was "significantly impressed" by it that he did not speak to anyone about the UFO incident until recent years.

Torres told the *Air Force Times* that the NSA agent threatened to revoke his flying privileges and end his Air Force career if he talked about the mission. "He told me I would lose my pilot's license and it would be the end of my flying days...."
http://www.airforcetimes.com/news/2008/10/airforce_ufo_shootdown_102008/

Since Torres never saw the UFO with his own eyes, some have suggested that what he and the ground radar observers saw was only a radar anomaly. According to inquiries by the **British Ministry of Defence (MoD)** in 1988, a possible explanation for the blip on Torres' radar screen

345

is that "an experiment in 'electronic warfare' was taking place by means of a bogus radar pulse being transmitted to create an illusion on the Pilot's radar of a solid moving target and that at the last minute before the Pilot was due to release his salvo of rockets the target was very swiftly removed off his radar screen by some technical means creating the further illusion that the target had outrun the Pilot."

According to the online edition of the British newspaper *The Times*, the experiment may have been part of a secret **Central Intelligence Agency (CIA)** program codenamed **Palladium**. The program used advanced electronic equipment to create simulated radar blips close to Soviet airspace. However, recently released CIA documents show that the Palladium program did not actually begin until the early 1960s.

The MoD began releasing declassified UFO-related files in 2005 as requests were made under **Britain's Freedom of Information Act (FOIA).** According to the **British National Archives** website, "In 2008 MoD announced their intention to transfer all their remaining records on UFOs to The National Archives before 2010. This transfer has now begun with 27 UFO files dating from 1979 to 1991 opened to the public via The National Archives UFO website during 2008."

Documents released in October 2008 included a 1988 letter from Torres to the MoD relating his UFO experience and expressing his desire to know more about what really happened on that day. http://en.wikipedia.org/wiki/Milton_Torres_1957_UFO_Encounter

Britain's military said it had no record of the incident, according to the files. Neither did the U.S. military. The second pilot's account, also included in the files, paints a somewhat different picture of events, saying there were not one but several "unknowns" and that he did not remember being contacted by anyone about staying quiet. He did not mention the targets' size.

"I know this is not a very exciting narrative but it is all I can recall," the second pilot said. His name, like his colleague's, was redacted from the files.

David Clarke, a UFO expert who has worked with the National Archives on the document release, said it was one of the most intriguing stories he had culled from the batch of files released Monday.

He said that the CIA once had a program intended to create phantom signals on radar and that this may have been an exercise in electronic warfare. Whatever the case, Clarke argued that "there's no doubt something very unusual happened."

Clarke said the batch of files released Monday — which include witness accounts, investigations, and sketches — was part of a three- to four-year program intended to make a total of 160 UFO-related files available to the public. By Raphael G. Satter of AP Associated Press; Updated 10/20/2008 3:50:07 PM ET

346

1976 Tehran UFO Incident

For a UFO account to reach the United States from Iran, a country that is currently a hostile nation to the U.S. it has to have had a tremendously high caliber of authenticity. The 1976 Tehran UFO Incident was a radar and visual sighting of an unidentified flying object over Tehran, the capital of Iran, during the early morning hours of 19 September 1976. The incident is particularly notable for the **electromagnetic interference effects** observed upon aircraft coming within close proximity to the object: two **F-4 Phantom II jet fighters** independently lost instrumentation and communications as they approached, only to have them restored upon withdrawal; one of the aircraft suffered temporary weapons systems failure while preparing to open fire.

The incident, extensively recorded in a four-page U.S. **Defense Intelligence Agency (DIA)** report distributed to at least the White House, Secretary of State, Joint Chiefs of Staff, National Security Agency (NSA) and Central Intelligence Agency (CIA), remains one of the most well-documented military encounters with anomalous phenomena in history, and various senior Iranian military officers directly involved with the events have gone on public record stating their belief that the object was not of terrestrial origin.
http://en.wikipedia.org/wiki/1976_Tehran_UFO_incident

In September of 1976 word reached the west that an Iranian military pilot had engaged in a dogfight with a UFO over Tehran. On 19 September 1976, the Imperial Iranian Air Force command post at Tehran received four reports by telephone, from civilians in the Shemiran city district, of unusual activity in the night sky. Some callers reported seeing a bird-like object; others reported a helicopter with a bright light.

When the command post found no helicopters airborne to account for the reports, they called **Assistant Deputy Commander of Operations, General Yousefi**. General Yousefi at first said the object was only a star, but after conferring with the control tower at Mehrabad International Airport and then looking for himself to see a very bright object larger than a star, he decided to scramble one F-4 Phantom II jet fighter from Shahrokhi Air Force Base in Hamadan, west of Tehran.

At 0130 hours, the **F-4**, piloted by **Captain Mohammad Reza Azizkhani** was proceeding to a point 40 nautical miles north of Tehran. It was noted that the object was of such brilliance that it could be seen from 70 miles away. When the aircraft approached to approximately 25 nautical miles from the object, the jet suddenly lost all instrumentation and communications capabilities, prompting Azizkhani to break off the intended intercept and turn back toward Shahrohki. As soon as the plane dropped its original course, all of the plane anomalies disappeared and both systems resumed functioning. This implies that the UFO was affecting the plane's functions as it neared the object. http://en.wikipedia.org/wiki/1976_Tehran_UFO_incident

At 0140 hours, a second F-4 was scrambled, piloted by **Lieutenant Parviz Jafari**. Jafari would eventually retire as a general and participate on 12 November 2007, at **a National Press Club** conference demanding a worldwide investigation into UFO phenomena. Jafari's jet had acquired a radar lock on the object at 27 nautical miles range. The radar signature of the UFO resembled that of a Boeing 707 aircraft. Closing on the object at 150 nautical miles per hour and at a range of 25 nautical miles, the object began to move, keeping a steady distance of 25 nautical miles from the F-4. So intense was the UFO's glow that the actual shape and size of the object could not be discerned visually, even as the jet bridged the distance between it and the unknown object. http://www.theufochronicles.com/2007/11/transcipt-of-witness-declarations-from.html

On 19 September 1976, Lieutenant Parviz Jafari of the Imperial Iranian Air Force pilots one of two F-4 Phantom II jet fighters scrambled to investigate a UFO near Tehran, Iran.
Google Images

General Parviz Jafari, Iranian Air Force (Ret.)
http://www.educatinghumanity.com/2011/03/ufo-government-documents-taken-very.html

The lights of the object were alternating blue, green, red, and orange, and were arranged in a square pattern. The lights flashed in sequence, but the flashing was so rapid that they all could be seen at once. Suddenly, the UFO began to move away, and the F-4, even flying above Mach 1, could not keep up with the UFO. The crew members stated that the UFO's sudden burst of speed was unbelievable.

While the object and the F-4 continued on a southerly path, a smaller second object detached itself from the first and advanced on the F-4 at high speed. The members of the F-4 crew were now fearful for their own lives. The F-4 had only one course of action left, Lieutenant Jafari, thinking he was under attack, tried to launch an AIM-9 sidewinder missile, but he suddenly lost all instrumentation, including weapons control, and all communication. He later stated he attempted to eject, but to no avail, as this system also malfunctioned. Jafari then instituted a turn and a negative G dive as evasive action. The object fell in behind him at about 3 to 4 nautical miles distance for a short time, then all of the F-4 functions returned to normal as the UFO turned and sped away to rejoin the primary object.

Once again, as soon as the F-4 had turned away, instrumentation and communications were regained. The F-4 crew then saw another brightly lit object detach itself from the other side of the primary object and drop straight down at high speed. The F-4 crew expected it to impact the ground and explode, but it came to rest gently. The F-4 crew then overflew the site at a decreased altitude and marked the position of the light's touchdown. Jafari would later comment that the object was so bright that it lit up the ground and he could see rocks around it. The object had touched down nearby **Rey Oil Refinery** on the outskirts of Iran. Then they landed at Mehrabad, noting that each time they passed through a magnetic bearing of 150 degrees from Mehrabad, they experienced interference and communications failure.

A civilian airliner that was approaching Mehrabad also experienced a loss of communications at the same position relative to Mehrabad. As the F-4 was on final approach, they sighted yet another object, cylinder-shaped, with bright, steady lights on each end and a flashing light in the middle. The object overflew the F-4 as they were on approach. Mehrabad tower reported no other aircraft in the area, but tower personnel were able to see the object when given directions by Jafari. Years later, the main controller and an investigating general revealed that the object also overflew the control tower and knocked out all of its electronic equipment as well

The next day, the F-4 crew flew out in a helicopter to the site where they had seen the smaller object land. In the daylight, it was determined to be a dry lake bed, but no traces could be seen. They then circled the area to the west and picked up a noticeable "beeper" signal. The signal was loudest near a small house, so they landed and questioned the occupants of the house about any unusual events of the previous night. They reported a loud noise and a bright light like lightning.

Further investigation of the landing site, including radiation testing of the area was apparently done, but the results were never made public. Since this event occurred before the fall of the Shah, any records in Tehran itself may be lost.
http://en.wikipedia.org/wiki/1976_Tehran_UFO_incident

There was not to be any conventional explanation for the F-4 and UFO encounter that occurred over Tehran. There is, however, ample evidence and documentation to state categorically that an unknown flying object was seen visually by the F-4, made radar return on the plane's radar, and was confirmed by ground radar and ground visual observation.
http://ufos.about.com/od/bestufocasefiles/p/iran1976.htm

Alongside the four-page report, there was a form from the DIA which assessed the quality of the report. The form indicated in checked boxes that the content was of high value, that the report was confirmed by other sources, and that the utility of the information was potentially useful to them. The form from the DIA also stated the following:

> "An outstanding report. This case is a classic which meets all the criteria necessary for a valid study of the UFO phenomenon:
> a) The object was seen by multiple witnesses from different locations (i.e., Shamiran, Mehrabad, and the dry lake bed) and viewpoints (both airborne and from the ground).
> b) The credibility of many of the witnesses was high (an Air Force general, qualified aircrews, and experienced tower operators).
> c) Visual sightings were confirmed by radar.
> d) Similar electromagnetic effects (EME) were reported by three separate aircraft.
> e) There were physiological effects on some crew members (i.e., loss of night vision due to the brightness of the object).
> f) An inordinate amount of maneuverability was displayed by the UFOs."

http://en.wikipedia.org/wiki/1976_Tehran_UFO_incident
Jerome Clark, *The UFO Book: Encyclopedia of the Extraterrestrial*, 1998, Visible Ink Press, ISBN 1-57859-029-9
Lawrence Fawcett and Barry J. Greenwood, *The UFO Cover-up* (formerly titled *Clear Intent*), Fireside Book/Simon & Schuster, ISBN 0-671-76555-8
Timothy Good, *Above Top Secret*, 1988, William Morrow and Co., ISBN 978-0-688-09202-3 (contains copy of full DIA report of incident)
Timothy Good, *Beyond Top Secret*, 1996, Pan Books, ISBN 0-330-34928-7 (contains copy of MIJI Quarterly report of incident, also below)

NOW YOU SEE IT, NOW YOU DON'T! (U)

Captain Henry S. Shields, HQ USAFE/INOMP

Sometime in his career, each pilot can expect to encounter strange, unusual happenings which will never be adequately or entirely explained by logic or subsequent investigation. The following article recounts just such an episode as reported by two F-4 Phantom crews of the Imperial Iranian Air Force during late 1976. No additional information or explanation of the strange events has been forthcoming; the story will be filed away and probably forgotten, but it makes interesting and possibly disturbing, reading.

* * * * *

Until 0030 on a clear autumn morning, it had been an entirely routine night watch for the Imperial Iranian Air Force's command post in the Tehran area. In quick succession, four calls arrived from one of the city's suburbs reporting a series of strange airborne objects. These Unidentified Flying Objects (UFOs) were described as 'bird-like', or as brightly-lit helicopters (although none were airborne at the time). Unable to convince the callers that they were only seeing stars, a senior officer went outside to see for himself. Observing an object to the north like a star, only larger and brighter, he immediately scrambled an IIAF F-4 to investigate.

Approaching the city, the F-4 pilot reported that the brilliant object was easily visible 70 miles away. When approximately 25 NM distant, the interceptor lost all instrumentation and UHF/Intercom communications. Upon breaking off the intercept and turning towards his home base, all systems returned to normal, as if the strange object no longer regarded the aircraft as a threat.

DECLASSIFY ... 4 Dec 81
by: HLS/A HQUSA

32

US Air Force files on the incident (page 1)
https://en.wikipedia.org/wiki/1976_Tehran_UFO_incident

A second F-4 was scrambled ten minutes after the first. The backseater reported radar-lock on the UFO at 27 NM/12 o'clock high position, and a rate of closure of 150 knots. Upon reaching the 25 NM point, the object began rapidly moving away to maintain a constant separation distance while still visible on the radar scope. While the size of the radar return was comparable to that of a KC-135, its intense brilliance made estimation of actual size impossible. Visually, it resembled flashing strobe lights arranged in a rectangular pattern and alternating blue, green, red, and orange. Their sequence was so fast that all colors could be seen at once.

As the F-4 continued pursuit south of Tehran, a second brightly-lit object (about one-half to one-third the size of the moon) detached from the original UFO and headed straight for the F-4 at a high rate of speed. The pilot attempted to fire an AIM-9 missile at the new object but was prevented by a sudden power loss in his weapons control panel. UHF and internal communications were simultaneously lost. The pilot promptly initiated a turn and negative-G dive to escape, but the object fell in behind the F-4 at 3-4 NM distance. Continuing the turn, the pilot observed the second object turn inside of him and then away, subsequently returning to the primary UFO for a perfect rendezvous.

The two UFOs had hardly rejoined when a second object detached and headed straight down toward the ground at high speed. Having regained weapons and communications systems, the aircrew watched the third object, anticipating a large explosion when it struck the ground. However, it landed gently and cast a bright light over a two-three kilometer area. The pilot flew as low over the area as possible, fixing the object's exact location.

Upon return to home base, both crewmen had difficulty in

DECLASSIFY ON: 4 Dec 81
by: ACS/I, HQ USAF
33

US Air Force files on the incident (page 2)
https://en.wikipedia.org/wiki/1976_Tehran_UFO_incident

adjusting their night vision devices for landing. The landing was further complicated by excessive interference on UHF and a further complete loss of all communications when passing through a 150 degree magnetic bearing from the home base. The inertial navigation system simultaneously fluctuated from 30 to 50 degrees. A civil airliner approaching the area also experienced a similar communications failure, but reported no unusual sightings.

While on a long final approach, the F-4 crew noted a further UFO. This was described as a cylinder-shaped object (about the size of a T-33 trainer) with bright steady lights on each end and a flasher in the middle. It quickly approached and passed directly over the F-4. In answer to the pilot's query, the control tower reported no other air traffic in the area, although they subsequently obtained a visual sighting of the object when specifically directed where to look.

The following day, the F-4 crew was flown by helicopter to the location where they believed the object had landed. This turned out to be a dry lake bed, but nothing unusual was noticed. As the helicopter circled off to the west, however, a very noticeable beeper signal was received, and eventually traced to a nearby house. They immediately landed and asked the inhabitants if anything strange or unusual had occurred the previous night. Yes, they replied, there had been loud noises and a very bright light, like lightning. The helicopter returned to base and arrangements were made to conduct various tests, such as radiation checks, in the vicinity of the house. Unfortunately, the results of such tests have not been reported.

DECLASSIFY ON: 4 Dec 81,
BCS/1, HQ USAF

34

US Air Force files on the incident (page 3)
https://en.wikipedia.org/wiki/1976_Tehran_UFO_incident

353

2005 - Fighter Jet Chases Triangle UFO over Oklahoma

A witness who described himself as a competent, educated scientist of sound mind and character, made an extraordinary claim of an event he witnessed on May 30, 2005. In the city of Noble, Oklahoma, the witness was taking a nightly walk when he observed a triangular UFO moving across the sky from north to south.

The sky was clear with several miles of visibility as the craft moved at about 500 mph, making not a sound. The UFO was not alone, as the witness saw a plane several miles behind the object, evidently in full pursuit. The plane was traveling at an estimated speed of 600-700 mph.

Both objects were traveling toward the witness, and as they came closer, he could hear the jet's engines. He could clearly see that the UFO was toying with the plane - as the plane got closer to the UFO, the object would speed up and distance itself from the jet.

This scenario played out several times, and as the two objects flew almost directly over the witness' head, the UFO suddenly made a sharp right turn - a maneuver the plane could not duplicate, as it made a much wider looping curve to change direction in pursuit of the UFO.

Soon, the UFO sped away from the jet completely. The jet turned to the north-northeast and flew out of sight. The witness made the observation that the three lights on the triangle were not an equal distance from each other. The lights of the object were a milky white, while the jet's lights were a bright red, green, and white.

The **National UFO Reporting Center** investigated the sighting, and the only object they found that could have accounted for the UFO would be an F-22 "Raptor" from Tinker Air Force Base, but it bears the traditional navigational lights, and can, therefore, be ruled out.
http://ufos.about.com/od/ufohistory/a/decade_2.htm

For over six decades, the USAF has denied any involvement in a "shoot down order" on UFOs, after all, their position has always been that there is nothing to the UFO phenomenon. They instituted military investigative programs like Project Sign, Grudge, and Bluebook to determine any validity to the many sighting reports from citizens and when the public demanded a truly unbiased scientific investigation into the subject, the Air Force sponsored a group of scientists out of the University of Colorado known as the Condon Committee to investigate the phenomenon.

At the same time that these programs were being carried out, the US Air Force were engaged in their shoot down protocols. The success ratio of shoot downs, however, between the military and UFOs, favoured the UFOs over the Air Force. It wasn't that Extraterrestrial Intelligences were hostile as there has never been any aggression against humans down through history but, as an adolescent humanity emerged into a technological era at an unprecedented rate, their moral compass did not keep pace with these developments, therefore, there may have been circumstances to not only defend themselves against human hostility but, in those rare situations return fire with the purpose to disarm rather kill. It would seem that military jets were incapable

354

of matching the speed or the maneuverability of alien spacecraft or even bringing them down with just aircraft gunfire and missiles.

With numerous aircraft encounters with UFOs continually occurring throughout the world from many nations, the more advanced militaries of the world particularly the United States, became aware that ET spacecraft had some sort of defensive electromagnetic shielding surrounding their craft. It became eminently practical and necessary for any interstellar intelligence visiting the Earth to have a type of defensive electromagnetic shielding which they could generate as a defensive measure against possible micrometeorite collision during space travel or as a defensive measure in combat situations. It also became apparent to the US that ETI, even have the ability if they encounter aggressive hostility from military aircraft to interfere with that aircraft avionics by either jamming or completely disarming flight control and power. The unfortunate result for the aircraft, unless it disengaged and pulled away from the encounter, it essentially became an uncontrollable "flying brick" and often **plummeting from the skies.**

This was not merely science fiction as scientists have known about magnetism for hundreds of years. By the late eighteenth century Michael Faraday, an English chemist, and physicist made a major contribution to the fields of electromagnetism and electrochemistry. It was on account of his research regarding the magnetic field around a conductor carrying a DC electric current that Faraday established the basis for the concept of the electromagnetic field in physics, which was subsequently enlarged upon by **James Clerk Maxwell**. Faraday established that magnetism could affect rays of light and that there was an underlying relationship between the two phenomena. He similarly discovered the principle of electromagnetic induction, diamagnetism, and the laws of electrolysis. His inventions of electromagnetic rotary devices formed the foundation of electric motor technology, and it was largely due to his efforts that electricity became viable for use in technology. http://en.wikipedia.org/wiki/Michael_Faraday

This inability of the US Air Force to compete with alien spacecraft in a dogfight with simple weapons did not stop the Air Force from trying. Air Force regulations and protocols like **JANAP 146** and **AFR 200-2** kept pilots reigned in from going to the press with reports of their strange aerial encounters. Yet there were many in the military forces that had no idea of what was going on, what was the nature and origin of the UFO phenomenon and many service men never spoke about the subject among themselves at least not openly. Pilots were routinely ordered to scramble their jets to the skies to intercept, target and engage any hostiles whether earth-bound or extraterrestrial in nature. In the case of alien spacecraft, pilots were ordered to shoot them down regardless, if they knew that they were extraterrestrial in origin.

Some pilots were lucky in their shoot down of a UFO and according to Boyd Bushman, a retired Lougheed engineer and scientist during a YouTube interview with UFOTV stated that the Roswell Crash of 1947 was the first successful military shoot down of a UFO. This statement made by Bushman certainly needs to be confirmed by other UFO researchers but, it would appear on the surface to fit with the general practices of aerial engagement. The story of the Roswell Incident may need to be revisited. The pilot was either lucky or the ETs underestimated human ability. Not all shoot downs, however, were successful and some ended in tragedy for the pilots or the planes similarly disappeared completely out of the skies and off the radar scopes, as if absorbed or beamed aboard the ET spacecraft.

The unfortunate downside to all this terrestrial – extraterrestrial confrontation was the possibility of fulfilling the late **General Douglas McArthur's** prediction, that the next world war would be fought in outer space against extraterrestrials. It would seem that the United States and her allies along with other coerced nations seem eager to throw the world into a war of unimaginable proportions, the likes of which has never been witnessed before in history as per the dire warning to **Dr. Carol Rosin** from the late **Wernher von Braun**.

General Benjamin Chidlaw
http://content.time.com/time/covers/0,16641,19541220,00.html

Timothy Good writes, "The destruction or disappearance of military aircraft during interceptions of UFOs continued apace." As General Benjamin Chidlaw, former commanding general of the Air Defense Command told Robert C. Gardener (ex USAF) in 1953: *"We have stacks of reports of flying saucers. We take them seriously when you consider we have lost many men and planes trying to intercept them."* Leonard Stringfield, the former Air Force intelligence officer was told by a reliable source in the 1950s that the "Air Force was losing about a plane a day to the UFOs. According to US Defense Department figures, from 1952 until the end of October 1956, there were 18,662 major accidents of military aircraft, broken down as follows:

Year - Air Force - Navy - Total Accidents			
1952	2,274	2,086	= 4360
1953	2,075	2,325	= 4400
1954	1,873	1,911	= 3784
1955	1,664	1,566	= 3230
1956	1,530	1,358	= 2888

Of this astonishing four-year total of 14,302 US aircraft losses, most involved fast new jets (such as those scrambled in UFO interceptions), of which 56.2 percent were found to be caused by pilot error; 8.1 percent by ground-crew or other personnel failure; 23.4 percent by failure of parts and equipment in the aircraft; 2.8 percent by various unsafe conditions, and ¨9.5 percent (1,773) were due to unknown factors. **Thanks to Timothy Good new book, "*Need to Know*" P.172**

In the excitement of the chase, it is easy to exceed aircraft capabilities and often some part of the aircraft may fail. Many stories are told of firing missiles and making direct hits on the UFOs that were unharmed only to have them return the fire and destroy the interceptor.

One well-known example of aircraft loses to a UFO was the F-89 with tail number 51-5853A, piloted by 1st Lt. Felix Moncla, Jr, with Second Lt. Robert L. Wilson as the Radar Intercept Officer. I spoke with a radar operator who worked in the Air Defense Command radar site on November 23, 1953. During the evening, radar picked up an unidentified target over Lake Superior and alerted the 433rd Fighter Interceptor Squadron at Truax Field, in Wisconsin. Lt. Moncla in his F-89 C all-weather interceptor was scrambled and instructed to intercept the target. The F-89 diving steeply down intercepted the target at 8,000 feet, and radar operators watched as the "blips" of the UFO and the F-89 merged on their scopes, in an apparent collision, and disappeared. No trace of the plane was ever found. You can visit Lt. Moncla's memorial headstone at Sacred Heart Catholic Cemetery in Moreauville, Louisiana, which reads:

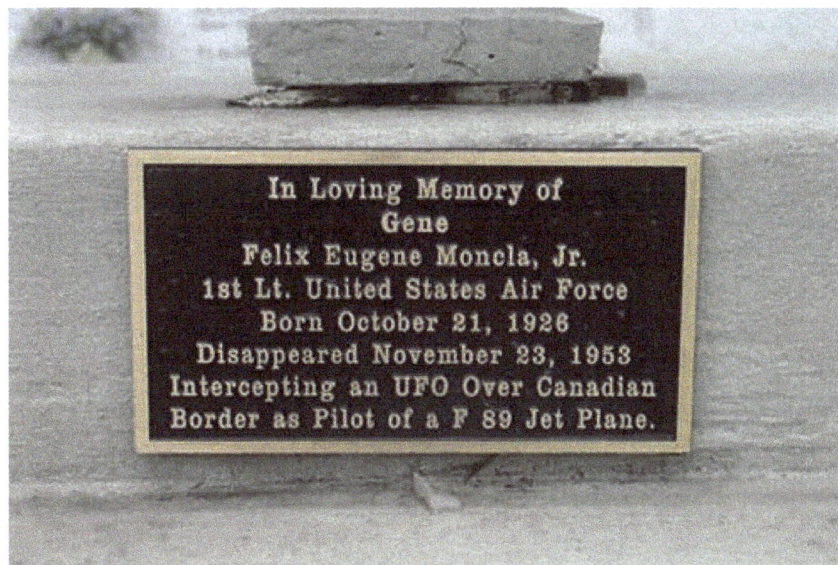

Memorial plaque to 1st Lt. Gene Felix Moncla, Jr. USAF
https://en.wikipedia.org/wiki/Felix_Moncla

During the summer of 1952, Americans reported a record amount of UFO sightings to Project Blue Book, the Air Force group responsible for investigating unidentified flying objects. Author/Illustrator Frank C. Feschino, Jr. has written an exciting book entitled, "Shoot Them Down" that encompasses the summer of 1952 and America's overwhelming UFO problem during that time. The author documents the numerous aerial encounters and confrontations that occurred between United States fighter planes and UFOs over America.

Feschino writes about the overwhelming amount of mysterious fighter plane accidents and vanished jets that occurred during the 1950s. "Shoot Them Down" is an important historical record of the deaths of brave men who were ordered to engage UFOs. Not only to hunt them down, but as Lt. Col. Moncel Monts is quoted in the book; ***"The jet pilots are, and have been under orders to investigate unidentified objects and to shoot them down if they can't talk them down"***.

Frank proceeds to lay his documented cases of pilot disappearances' as a result of actual aerial dogfights with documented sightings of UFOs. **President Barack Obama** laid a wreath at the Tomb of the Unknowns at the Arlington National Cemetery this Memorial Day, perhaps those who sacrificed their lives chasing Unknowns has a new meaning. Those who died intercepting UFOs has been ignored. Most of the men who died in our wars have been honored for their sacrifice, now it is time to honor those who died for their country fighting UFOs.
http://www.amazon.com/Shoot-Them-Down-Flying-Saucer/dp/0615155537

Marine Major Donald Keyhoe, states: *"In 1949 the Air Force told me they had been ordered to 'get' a flying saucer by any possible means. This was admitted by an intelligence officer at the Pentagon ... Major Jere Boggs. In front of General Sory Smith, Boggs told me that one Air Force pilot had fired at a saucer over New Jersey."*

"In the late 1950s, as a number of futile US chases mounted, some pilots were convinced that the UFOs were immune to gunfire and rockets. Several Intelligence analysts believed the aliens might be using some negative force linked with gravity control to repel or deflect bullets and missiles. But the top control group disagreed. In a special evaluation of US and foreign reports, they found evidence that UFOs were not invulnerable. Some had been temporarily crippled, apparently from power or control failures, and a few others had been completely destroyed by strange explosions. In one or two cases, it appeared that missiles or rocket fire could have been the cause." http://www.flatwoodsmonster.com/

Many cases involved the disappearance of aircraft and aircrews that are still missing.

The Fullerton News-Tribune of California wrote in their newspaper of July 26, 1956...

"The United States Navy will not publicly admit that it believes in flying saucers, but it has officially ordered combat-ready pilots to 'shoot to kill' if saucers are encountered, OCNS [Orange County News Service] has learned. The information was first learned when Navy pilots navigating trans-Pacific routes from the United States to Hawaii were ordered in a briefing session to engage and identify 'any unidentified flying objects.' If the UFOs (saucers) appeared hostile the briefing officer told the pilots of Los Alamitos Naval Air Station reserve squadron VP 771, they are to be engaged in combat... It was found that the orders are not unusual."

May 5, 1953, off the coast of Atlantic City, New Jersey, the USN reported, "3 Navy fliers missing" that were aboard a "Douglas Sky Raider, a night attack bomber." The Navy reported, "A search by twenty-seven aircraft from daybreak to sundown, failed to find any trace of the men or plane."

Stanton T. Friedman states, *"I have personally heard of seven instances in which military planes chasing flying saucers were never heard from again. The chasing isn't surprising, in view of the official USAF 1952 orders to shoot them down if they don't land when instructed to do so."*

"If I have heard of seven such events, then surely there have been more. (More details are given in SHOOT THEM DOWN by Frank Feschino, Jr.). One can perhaps understand the reluctance of the government to admit that such losses have occurred (even though Major General Roger Ramey admitted in 1952 that more than 300 interceptors have been scrambled)." "Impotence in the face of intruders is not something that governments want to admit." Excerpt from, Flying Saucers and Science - A Scientist www.amazon.co.uk/Flying-Saucers-Science-Scientist

"The reported operating characteristics such as extreme rates of climb, maneuverability (particularly in roll), and action which must be considered evasive when sighted ... lend belief to the possibility that some of the objects are controlled ..." **--General Nathan Twining, Head of Air Material Command (AMC), 1947.**

Major General Robert B. Landry writes, *"I was called one afternoon [in 1948] to come to the Oval Office ¨C the President wanted to see me.... I was directed to report quarterly to the President after consulting with Central Intelligence people, as to whether or not any UFO incidents received by them could be considered as having any strategic threatening implications ."*

**Major General Robert B. Landry, Air Force Aide to
President Harry S. Truman getting another star**

https://www.trumanlibrary.org/oralhist/landryr.htm

It is sad to know that many governments kept a lid on these engagements. The deaths of many airmen were attributed to pilot error or engine failure when in reality they had died in air combat with craft, not of this world. It is time to honor these men who died in aerial battles with UFOs after being scrambled to hunt them down.

"For we are not fighting against people made of flesh and blood, but against principalities, against powers, against the rulers of the darkness of this world, against spiritual wickedness in high places." - **Ephesians 6**
http://www.nationalufocenter.com/artman/publish/article_280.php

Author's Rant: A personal perspective is necessary at this point. This last quote is most telling as a possible insight into the mental mindset of the military and the private industrial sector based upon archaic Christian scripture. Such a belief system justifies the aggressive behavior and agenda to shoot upon a foe without knowing anything about the alleged enemy other than it is not of this Earth. Such philosophy is premised upon poorly misunderstood religious writings which have not only been divisive for most of humanity but, has now become the rallying cry to falsely unite mankind against an extraterrestrial intelligence stemming from primitive anthropocentric thinking that is not only immature but, outright dangerous for the whole planet. Rational heads must prevail at this critical juncture to re-assess the potential dangers from aggressive encounters with

360

Extraterrestrials as well as the potential benefits to be gained by a more peaceful communicative approach to these visitors from space.

Project Moon Dust and Operation Blue Fly: Crash Retrieval Teams

Once a "successful shoot down" is confirmed and its location is established, special **Rapid Mobilization Response Teams (RMRT)** who are always on stand-by alert are deployed and mobilized to retrieve any unusual foreign or unknown space material. The impact and surrounding area is cordoned off, sometimes for hundreds of square miles, witnesses are interviewed, sequestered and threatened with oaths of secrecy in the name of national security and if needed, if witnesses are uncooperative, they may be verbally threatened, subject to physical injury or worse suffer an "accidental" death.

These rapid mobilization response teams were formerly known by the code names **Project Moon Dust** and **Operation Blue Fly** although currently, the names of such teams have been changed, even though the Air force has stated that such recovery teams no exist longer.
http://www.nicap.org/moondust.htm

"Although ostensibly two projects involved in the recovery and exploitation for the US Government of foreign space debris such as crashed satellites, rocket boosters and so on, there is intriguing data at our disposal showing that both projects have been involved in the recovery of far more exotic items – including possibly crashed UFOs and UFO debris. A 1961 US Air Force document states that:

In addition to their staff duty assignments, intelligence team personnel have peacetime duty functions in support of such Air Force projects as Moon Dust, Blue Fly, and UFO, and other AFCIN directed quick reaction projects which require intelligence team operational capabilities.

Unidentified Flying Objects (UFO): Headquarters USAF has established a program for investigations of reliably reported unidentified flying objects within the United States.

Blue Fly: Operation Blue Fly has been established to facilitate expeditious delivery to Foreign Technology Division of Moon Dust or other items of great technological intelligence interest.

Moon Dust: As a specialized aspect of its overall material exploitation program Headquarters USAF has established Project Moon Dust to locate, recover, and deliver descended foreign space vehicles.

An example of one 1965 report from the nearly 1000 pages of official documentation on Moon Dust and Blue Fly that have now been released into the public domain by the Department of State, Air Force, Defense Intelligence Agency and CIA, is titled: "FRAGMENT METAL, RECOVERED IN THE REPUBLIC OF THE CONGO, ORIGIN BELIEVED TO BE AN UNIDENTIFIED FLYING OBJECT." This report is just one of many confirmations that have come out from a top secret covert ops group that UFOs are real and not imaginary.

It should be noted too that Project Moon Dust is referenced in the 1-page CIA paper pertaining to crashed UFOs, alien bodies, the late actress Marilyn Monroe and the Kennedy brothers John and Robert."
http://www.majesticdocuments.com/official.investigations.projectmoondust.bluefly.php

On September 23, 1947, Lt. **General Nathan Twining**, Commander of Air Material Command at Wright-Patterson Air Force Base, wrote an opinion concerning "Flying Discs" to **Brig. General George Schulgen**, Chief of the Air Intelligence Requirements Division at the Pentagon.

Twining wrote:

"The phenomena is something real and not visionary or fictitious…The reported operating characteristics such as extreme rates of climb, maneuverability (particularly in roll), and action which must be considered evasive when sighted or contacted by friendly aircraft and radar, lend belief to the possibility that some of the objects are controlled either manually, automatically or remotely."

He described the objects as metallic or light-reflecting, circular or elliptical with a flat bottom and domed top, and usually silent. Twining stated that "due consideration must be given" to "the lack of physical evidence in the shape of crash recovered exhibits which would undeniably prove the existence of these objects.

"He recommended that Army Air Forces assign "a priority, security classification and Code Name for a detailed study of this matter…"

In 1953, the Air Defense Command created the 4602d Air Intelligence Service Squadron (AISS) and assigned it to the official investigations of UFOs. The squadron was headquartered at Ent Air Force Base, CO and soon moved to Fort Belvoir, VA with field units throughout the country. All UFO reports were to go through the 4602d AISS prior to any transmission to **Project Blue Book**, a public relations project with no access to reports above the Secret level. The 4602d AISS dealt with more sensitive cases of national security concern requiring a higher classification. Thus, many UFO reports bypassed Blue Book altogether.
http://www.bibliotecapleyades.net/sociopolitica/esp_sociopol_mj12_3k.htm

In 1954, Air Force Regulation 200-2 ("Unidentified Flying Objects Reporting") stated that the Air Defense Command has, "a direct interest in the facts pertaining to UFOBs" and will conduct all field investigations, "to determine the identity of any UFOB."

General Nathan Twining, Head of the U.S. Air Force Material Command

It stated that the ADC will investigate the reports through the 4602d AISS, a highly mobile unit composed of "specialists trained for field collection and investigation of matters of air intelligence interest." The document outlined collection responsibilities for this unit.

According to an Air Force Intelligence Letter (**"Betz Memo"**) of 13 Nov 1961, the 4602d had three peacetime functions:

UNIDENTIFIED FLYING OBJECTS (UFO- A program for investigation of reliably reported unidentified flying objects within the United States.

PROJECT MOONDUST - A specialized aspect of the U.S. Air Force's overall material of the exploitation program to locate, recover, and deliver descended foreign space vehicles.

363

OPERATION BLUE FLY – [A unit] to facilitate expeditious delivery to the **Foreign Technological Division (FTD)** of Moon Dust and other items of great technical intelligence interest.

The memo stated that all three functions involve, "employment of qualified field intelligence personnel on a quick reaction basis to recover or perform field exploitation of unidentified flying objects, or known Soviet/Bloc aerospace vehicles, weapons systems, and/or residual components of such equipment."

A classified 1969 Air Force document terminating **Project Blue Book ("Bolender Memo")** made it clear that existing operations would continue to investigate UFOs even though the Air Force was closing Blue Book. The memo established that UFO reports affecting national security,

"are made in accordance with JANAP 146 or Air Force Manual 55-11, and are not part of the Blue Book system" and that "the defense function could be performed within the framework established for intelligence and surveillance operations." It stated that "reports of UFOs which could affect national security would continue to be handled through the standard Air Force procedures designed for this purpose."

As far as the public was concerned, the termination of Project Blue Book meant the end of the Air Force investigation into UFOs. The Air Force stated publicly two months after the issuance of the classified **Bolender Memo** that the continuation of Project Blue Book could not be justified on the grounds of national security since no UFO has ever presented a threat to national security. The Air Force misinformed the public by not acknowledging its continuing, secret investigation of UFOs independent of Blue Book, and it's very real national security concerns.
http://www.bibliotecapleyades.net/sociopolitica/esp_sociopol_mj12_3k.htm

Ongoing efforts to retrieve fallen objects are described in a 1970 State Department telegram to its embassies and consulates around the world requesting that they post any "reports or sightings of entry into atmosphere or landing of 'space debris.'" (Quotes around "space debris" are throughout.) The offices are instructed to follow leads "as expeditiously as possible" without informing the local government or making public comment.

"Recovery of any material from such space debris would be of great scientific interest to USG," the telegram states.

A 1973 Secret State Department Airgram confirms that "the designator '**MOONDUST**' is used in cases involving the examination of non-US space objects and objects of unknown origin." Beginning in 1989, **Sergeant Clifford E. Stone**, US Army ret., of New Mexico wrote to fourteen intelligence agencies for records on Project Moon Dust and Operation Blue Fly under the Freedom of Information Act. Many of the documents cited here were obtained through his efforts.

The responses from numerous agencies were inconsistent and evasive. In 1990, the U.S. Air Force told **Stone,**

"We do not have any records responsive to your request." The U.S. Air Force Intelligence Service stated, "We have made a thorough search of our records and found none responsive to your request." Four months later, the Air Force reversed their initial position stating, "we have two records responsive to your request. However, they are exempt from disclosure because the information is properly classified."
http://www.bibliotecapleyades.net/sociopolitica/esp_sociopol_mj12_3k.htm

The Defense Intelligence Agency stated that information pertaining to **Project Moondust** is classified and that the agency located no records on "Project Bluefly." Fifteen months later, the **DIA** acknowledged that the State Department had eight DIA documents, but that two were not releasable. (There are numerous references in Moon Dust documents to **DIA** participation in retrievals.)

In a 1991 letter, the Air Force told **Stone**,

"We can neither confirm nor deny the existence or nonexistence of records responsive to your request regarding Projects or Operations known as Blue Fly, Moon Dust…"

On **Stone**'s behalf, New Mexico **Senators Jeff Bingaman and Pete Domenici** agreed to make inquiries to the Air Force about Project Moon Dust and Operation Blue Fly. In response to a letter from Senator Bingaman in 1992, the Air Force told the Senator that, "there is no Project Moon Dust or Operation Blue Fly. These missions have never existed."
When the Senator responded with documents challenging this, the Air Force "amended" it's previous statement, acknowledging the existence and function of Moon Dust and Blue Fly with regards to UFOs.

In 1994, **Senator Domenici** requested eleven Air Force documents pertaining to Moon Dust and Blue Fly that were in State Department files but were denied **Stone** in 1991. (At that time, the Air Force had not been willing to "confirm nor deny the existence or non existence" of these documents.) The Air Force responded in December of 1994 that "the projects, as such no longer exist, nor do their files. Classified reports that existed, if any, presumably were destroyed."

Yet the Air Force informed a New Jersey citizen in 1998, in response to an independent request on Project Moon Dust and Operation Blue Fly, that "the information relating Project Moon Dust remain classified" and is being withheld. This contradicts the earlier statement by the same office that the files did not exist and were likely destroyed.

Why did the US Air Force state that the documents had been destroyed when they were requested by a US Senator?

Why did it tell a second Senator that Moon Dust and Blue Fly never existed?

Why the great concern about releasing information concerning fallen space debris collected decades ago?

In a letter dated February 28, 1994, New Mexico **Congressman Joe Skeen** told Stone that the "House Government Operations Committee has taken an interest in this matter… Congressional hearings may be held on this matter later this session." Skeen said he would also share Stone's report with the House Intelligence Committee.

In April 1997, the Air Force acknowledged to Stone that Operation Blue Fly's mission included, "space objects and unidentified flying objects (UFOs) if any were reported available for recovery."

It goes on to state that no Soviet Bloc planes were ever downed in the US, and "no UFOs were ever reported downed or recovered in the United States or anywhere else."

These statements are patently false. Air Force Intelligence files show that Operation Blue Fly was assigned to the intelligence exploitation of a Soviet-built Cuban helicopter in Florida. More importantly, with respect to UFOs, official documents released through **FOIA** directly contradict the 1997 Air Force statement.
http://www.bibliotecapleyades.net/sociopolitica/esp_sociopol_mj12_3k.htm

In 1965, a three-man team was sent to recover an object of unknown origin reported downed in **Kecksburg, PA**. (Witnesses state an object was recovered; the Air Force says nothing was found.)

In August 1967, an object described as a satellite crashed and was recovered in the **Sudan** under Moon Dust. (The description on the **DIA** document released by the State Department does not fit that of a satellite.)

In 1968, Project Moon Dust recovered four unknown objects in **Nepal**.

Also in 1968, a "dome-shaped object" with no identification marks was retrieved underwater off **Cape Town, South Africa**. The metal object had been subjected to extreme heat and showed no signs of corrosion. **NASA** determined it was made of "almost pure aluminum" and stated that the NASA analysis of the sample and photographs "does not otherwise provide a clue as to its origin or function although it is possible it is a space object of US origin."

In 1970, Moon Dust investigated a metal sphere that fell "with three loud explosions and then burned for five days" in South America. It had "ports" which had been melted closed.

A May 1970 State Department document describes a fallen, unidentified object in **Bolivia**, depicted in the newspapers as metal and egg-shaped. The Department expresses a desire to assist the Bolivian Air Force in the investigation. "The general region had more than its share of reports of UFOs this past week," the document notes.

It says that **Panama** and **Paraguay** checked with appropriate government agencies and "no direct correlation with known space objects that may have reentered the earth's atmosphere near May 6 can be made."

All the documents on the above events represent raw, unprocessed field intelligence data. The public, however, is not privy to the final determinations of these investigations. Where are the finalized intelligence products? Where are the recovered fragments?
Our government will not disclose what these objects were. In fact, the Air Force denies these

366

events ever happened, even though official documents show otherwise. What is the purpose of keeping this information classified? The search for this information must pick up where **Sgt. Stone**, **Senators Domenici and Bingaman,** and **Congressman Skeen** left off.

The Kecksburg incident is an ideal focal point for further inquiry into Moon Dust and Blue Fly since it is already well documented. The object fell on American soil. There were witnesses to the object on the ground and its removal by an Army vehicle. Countless others saw the Army cordon off the area, blocking access.

Project Blue Book files state that no object was found in Pennsylvania. They also acknowledge that no space debris entered our atmosphere that day and that "aluminum type" fragments were retrieved in Michigan. (Where are they now?). It is likely that **Blue book** was not informed about the retrieval of this object since it would have been classified higher than Secret. In short, the documentation shows that the United States Air Force has continued to conduct a highly classified UFO investigation program in conjunction with other government agencies.

Under this program, Project Moon Dust and Operation Blue Fly have recovered objects of unknown origin. We, as citizens, have been denied knowledge of what they were. Physical evidence in the possession of the U.S. Government could shed light on the UFO question as would nothing else. The Kecksburg case also has the potential to generate more documentation on Project Moon Dust and Operation Blue Fly, which hold the key to other cases involving downed objects of unknown origin.
http://www.bibliotecapleyades.net/sociopolitica/esp_sociopol_mj12_3k.htm and
https://www.youtube.com/watch?v=pBSTT7qZGig

CHAPTER 46

UFO INTRUSIONS AND OVER-FLIGHTS OF MISSILE BASES AND WEAPONS STORAGE SITES

Missile and Advanced Weapon Shoot Downs of UFOs

Extraterrestrial Intelligences adapted quickly to military dogfights as evidenced from the above aircraft – UFO engagement accounts. When alien spacecraft evaded or fail to plummet to terra firma from gun-firing military jets, the outcome frequently suffered by the pursuing plane was either electrical or navigational malfunction or sometimes a deadly crash from such encounters. Most militaries, however, would see this outcome as still further reason to escalate any future encounters with a more aggressive posture.

Many nations' militaries escalated these types of encounters with the use of missiles, laser cannons, particle beam and microwave weapons and even advance scalar and psychotronic technology, all these have been utilized at one time or another in the hope of gaining some advantage over the high-flying alien technology.

We know that many militaries have had some success in their shoot down programs as indicated by the list of over 260 crash retrievals from around the world. Not all these UFO crashes were the result of US military track and targeting of inbound objects but, they certainly have lead the way with the lion's share of successes. A few accounts will illustrate the efforts of some militaries to fire upon ET craft and the retaliatory response by ETI to this aggression by humans.

1961 – Voronezh (Rybinsk) Incident: Multiple Missile Salvos Launched Against UFO

Richard Haines in his book **CE-5 Close Encounters of the Fifth Kind** describes an account which he states is impossible to verify, where the Russian military fired several missile salvos at a UFO which had hovered over the Voronezh missile base. Even, though this story is hard to verify it obviously has been around awhile since the '80s as it is recounted in other UFO books and magazines and it is certainly deserving of further investigation.

According to Haines there are two versions of this event which have some differences and similarities both large and small but, the common elements to both accounts reveal that the personnel of an anti-aircraft missile battery saw a huge UFO which "swooped down over the city," a number of smaller objects (referred to as "scout craft") came out of it. The commander ordered (apparently without authorization) missiles to be launched.

"A salvo of anti-aircraft missiles roared up – only to explode harmlessly when still two kilometers from [the object]." It was as if the missiles had hit an invisible wall. He ordered a second salvo which suffered the same fate as the first. The commander was about to order a third salvo of missiles be launched when the commander's superiors told him to forget about the third attempt as the entire base's electrical system malfunctioned for some unexplained reason. It was obvious that the missiles were having no effect upon the large UFO hovering over the base. The smaller objects seemed to return to the large object in a leisurely fashion and then all flew away.

In one report *(Saga)* the event took place near Voronezh while in the other *(Flying Saucer Review),* near Rybinsk, 150 kilometers north of Moscow. This article also claimed that the smaller objects descended toward the rocket batteries and somehow "stalled the electrical apparatus of the whole missile base. Creighton G.W.; 1962; *Amazing News from Russia.* Flying Saucer Review 8, no. 6 (Nov. –Dec.): 27-28
Richard F. Haines Ph.D.; CE-5 - Close Encounters of the Fifth Kind, *Case No. 112;* 1999; Sourcebooks, Inc.; ISBN 1-570571-427-4

1989 - Kapustin Yar Incident: Multiple Witness Case at Russian Missile Base

Very little was known about UFO investigations in Russia and the republics of the former USSR during the communist era. UFOs were officially labeled "capitalist propaganda" in the 1950s and1960s. A few scientists such as Professor Felix Zigel and Yuri Fomin documented UFO incidents, but their results were rarely published and circulated mostly in **samizdat** form. Ufology began to prosper in the early 1980s when **"Commissions on Anomalous Phenomena"** were established under the patronage of a few academicians.

Stories of secret military UFO investigations began to spread with **glasnost,** increasing with the break-up of the USSR. Retired military and intelligence officers were now speaking up and offering documents. One collection, covering a 10-year period of military UFO investigations between 1978 and 1988, was sold by its former director, Colonel Boris Sokolov, to American journalist George Knapp and to ABC News.124 In 1991, the **Committee of State Security (KGB)** declassified 124 pages of documents of "Cases of Observations of Anomalous Occurrences in the Territory of the USSR, 1982-1990," covering a total of 17 regions.

One of the most interesting cases in the KGB file is a multiple-witness **CE-I (Close Encounter of the First Kind)** at an army missile base in the district of Kapustin Yar, Astrakhan Region, on the night of July 28-29, 1989.

Kapustin Yar is a Russian rocket launch and development site in Astrakhan Oblast, between Volgograd and Astrakhan. Known today as Znamensk , it was established 13 May 1946 and in the beginning used technology, material, and scientific support from defeated Germany. Numerous launches of test rockets for the Russian military were carried out at the site, as well as satellite and sounding rocket launches.
The file is surely incomplete but still offers an interesting glimpse into the maneuverability of UFOs. The dossier consists of the depositions of seven military witnesses (two junior officers, a corporal, and four privates) plus illustrations of the object by the observers, and a brief case summary by an unnamed KGB officer. (Neither the author nor the department are identified, but the document is at the beginning of the KGB file on the **Kapustin Yar Incident**). It states in part:

"Military personnel of the signal center observed UFOs in the period from 22:12 hrs. to 23:55 hrs. on 28 July 1989. According to the witnesses' reports, they observed three objects simultaneously, at a distance of 3-5 km. [2-3 mi.]."

A nearby base reported the UFO from 23.30 hrs. on July 28 until 1.30 hrs. on July 29. The report continues:

"After questioning the witnesses, it was determined that the reported characteristics of the observed UFOs are: disc 4-5 m. [13-17 ft.] diameter, with a half-sphere on top, which is lit brightly. It moved sometimes abruptly, but noiselessly, at times coming down and hovering over ground at an altitude of 20-60 m. [65-200 ft.]. The command of [censored] called for a fighter... but it was not able to see it in detail, because the UFO did not let the aircraft come near it, evading it. Atmospheric conditions were suitable for visual observations."

The KGB file on the case is obviously incomplete, since there is no data on the jet scramble mission or whether ground or airborne radar detection was also reported. Nevertheless, the hand written descriptions by the seven witnesses from the signal center do provide interesting reading about the flight behavior exhibited by the UFOs. The most detailed communication was submitted by the Officer-on-Duty, **Ensign Valery N. Voloshin**. A Captain from the telegraph center informed him at 23:20 hrs. that "an unidentified flying object, which he called a flying saucer, was hovering over the military unit for over an hour." After confirming the sighting with the operation signal officer on duty, Ensign Voloshin, and **Private Tishchayev** climbed the first part of an antenna tower.

According to his deposition:

"One could clearly see a powerful blinking signal which resembled a camera flash in the night sky. The object flew over the unit's logistics yard and moved in the direction of the rocket weapons depot, 300 meters [1,000 ft.] away. It hovered over the depot at a height of 20 meters [65 ft.]. The UFO's hull shone with a dim green light which looked like phosphorous. It was a disc, 4 or 5 m. [13-17 ft.] in diameter, with a semispherical top.

"While the object was hovering over the depot, a bright beam appeared from the bottom of the disc, where the flash had been before, and made two or three circles, lighting the corner of one of the buildings... The movement of the beam lasted for several seconds, then the beam disappeared and the object, still flashing, moved in the direction of the railway station. After that, I observed the object hovering over the logistics yard, railway station, and cement factory. Then it returned to the rocket weapons depot and hovered over it at an altitude of 60-70 m. [200-240 ft.]. The object was observed from that time on, by the first guard shift and its commander. At 1:30 hrs., the object flew in the direction of the city of Akhtubinsk and disappeared from sight. The flashes on the object were not periodical, I observed all this for exactly two hours: from 23:30 to 1:30."127

A drawing of the UFO was attached:

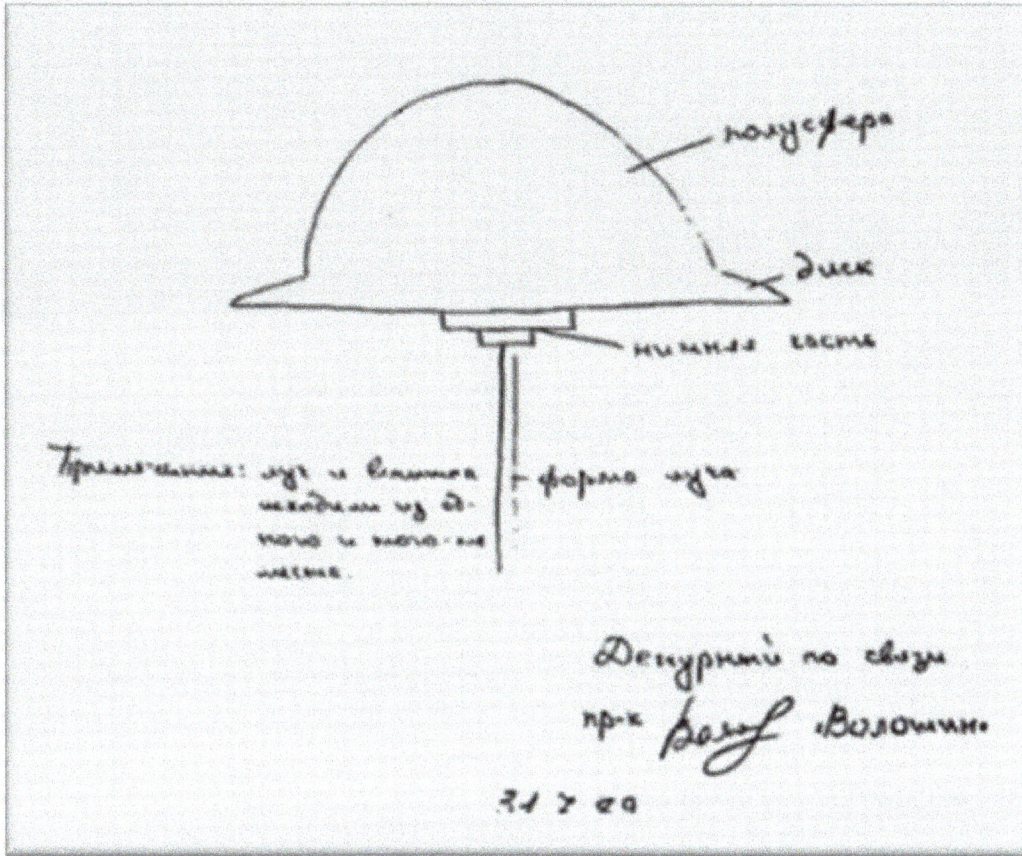

Drawing by Ensign Voloshin of UFO with beam seen over a Russian missile base in Kapustin Yar in July 1989.

Private Tishcahayev essentially confirmed Ensign Voloshin's testimony. The guard shift of Corporal Levin and Privates Bashev, Kulik and Litvinov basically tell the same story. They were all alerted by **1st Lt. Klimenko** and they all saw up to three UFOs performing fantastic acrobatics in the sky, such as:

"Suddenly, it flew in our direction. It approached fast and increased in size. It then like divided itself in three shining points and took the shape of a triangle. Then it changed course and went on flying in the same sector."

"After veering, it began to approach us and its speed could be felt physically. (It swelled in front of our eyes). Its flight was strange: no aircraft could fly in this manner. It could instantly stop in the air (and there was an impression that it wobbled slightly up and down); it could float (exactly that: float, because the word 'fly' would not be adequate, it was as if the air was holding it, preventing it from falling). At all times that I observed it, it was blinking, blinking without any order and constantly changing colors (red, blue, green, yellow). The point itself was not blinking but something above it."

371

Reconstruction of the event published in the Moscow magazine Aura-Z.
Both illustrations courtesy of Antonio Huneeus/Aura-Z.
http://www.ufoevidence.org/cases/case663.htm

"Here is what I observed: there was a flying object, resembling an egg, but flatter. It shone brightly alternating green and red lights. This object gathered a great speed. It accelerated abruptly and also stopped abruptly, all the while doing large jumps up or down. Then, appeared a second and then a third object. One object rose to low altitude and stopped. It stayed there in one place and was gone. Later a second object disappeared, and only one stayed. It moved constantly along the horizon. At times, it seemed it landed on the ground, then it rose again and moved."

All the testimonies coincided with the appearance of a jet fighter attempting to intercept the UFOs. The fighter made a first pass above the object apparently without seeing it. Then, according to Lt. Klimenko's deposition, "the airplane, which could be identified by its noise, approached the object, but the object disengaged so fast, that it seemed the plane stayed in one place."

It is difficult to make a final evaluation of the Kapustin Yar CE-I since no information about the scramble mission and possible radar tracking has been released by the KGB. But the detailed testimony of seven military witnesses, who were familiar with rocket launches and various aircraft because of their post (Kapustin Yar is somewhat equivalent to the White Sands Proving Grounds in New Mexico), appears to confirm the unusual flight characteristics and extraordinary maneuverability displayed by UFOs in many instances. Moreover, as in the SAC flap of 1975 and the Bentwaters affair in England in 1980, the UFOs seemed capable of "demonstrating a clear intent in the weapons storage area," as described in a 1975 declassified teletype concerning Loring AFB in Maine.

372

One of the official milestones of Soviet/Russian ufology occurred less than a year later, as a result of a radar-visual and jet scramble incident on the Pereslavl-Zalesskiy region, east of Moscow, on the night of March 21, 1990. A statement issued by Colonel-General of Aviation Igor Maltsev, Chief of the Main Staff of the Air Defense Forces, was published in the newspaper *Rabochaya Tribuna*. Unit commanders compiled "more than 100 visual observations" and passed them on to Gen. Maltsev, who stated:

"I am not a specialist on UFOs and therefore I can only correlate the data and express my own supposition. According to the evidence of these eyewitnesses, the UFO is a disc with a diameter from 100 to 200 meters [320 to 650 feet]. Two pulsating lights were positioned on its sides... Moreover, the object rotated around its axis and performed an 'S-turn' flight both in the horizontal and vertical planes. Next, the UFO hovered over the ground and then flew with a speed exceeding that of the modern jet fighter by 2 or 3 times... The objects flew at altitudes ranging from 100 to 7,000 m. [300 to 24,000 ft.]. The movement of the UFOs was not accompanied by sound of any kind and was distinguished by its startling maneuverability. It seemed the UFOs were completely devoid of inertia. In other words, they had somehow 'come to terms' with gravity. At the present time, terrestrial machines could hardly have any such capabilities." http://www.bibliotecapleyades.net/ciencia/ufo_briefingdocument/1989.htm

It is clear that when threatened with hostile action, UFOs will not only defend themselves but will respond with a retaliatory action. A response which given the hostile environment is non-escalating and non-aggressive in action. Military aircraft and their weapons are usually neutralized and made non-functional whenever possible while trying at all costs to avoid harming humans in the process. This is a dangerous cat and mouse game that is repeated all too often, over and over again around the world at military bases, on the high seas, under the oceans, during military exercises by either NATO or by Russia and her allies or during any wartime conflict that may be occurring anywhere in the world at any given time.

Throughout recorded history, down through the ages of mankind's existence on Earth, whenever there has been sightings and interaction with unidentified flying objects, on no occasion has there been an alien invasion by Extraterrestrials bent on taking over the Earth, or enslaving mankind or the exploitation and illegal acquisition of Earth's natural resources. In reality, Alien visitors to this planet seem to have been very tolerant and incredibly patient with such an immature race as ours. In fact, in almost every encounter whether public or military in nature with a few exceptions of course, when there may have been open peaceful contact between humans and ETI, humans have displayed repeatedly an aggressive posture towards these visitors. We are barely out of the trees in terms of evolutionary time-scale, trying to develop societies and a global civilization, yet our behaviour is still primitive and ape-ish! If there is a race of **"Klingons"** as portrayed in the TV series **Star Trek**, somewhere out in the universe then, we need not look beyond this planet for them, for we are by temperament and behaviour those very same Klingons!

Huge Triangle Shaped UFO in Cabo Rojo Makes Two U.S. Jet Fighters Disappear December 28, 1988

From Puerto Rico comes an astounding account of two Navy F-14 'Tomcat' fighters that engaged a gigantic triangular shaped UFO in the Cabo Rojo area.

This encounter took place December 28, 1988, in the area of the towns of Lajes and Cabo Rojo in the western end of Puerto Rico. *(see January 1990 issue of the MUFON UFO Journal for details).*

According to one set of reports most of the witnesses (there are more than 115 witnesses that have surfaced up to this moment), there seemed to be three jet fighters involved with the incident, two of them disappeared in mid air as they intercepted and closed in on the huge triangular UFO, and the third one fled the area flying to the north, being chased by several big red balls of light that came out from the UFO.

Eyewitnesses alleged the 'Tomcats' intercepted the object and then were somehow 'trapped' or taken aboard this huge craft. The UFO then split into two separate triangular objects in a silent flash of light, after which one of the objects or sections flew off very fast to the north and the other one flew to the east, disappearing.
www.iraap.org... and http://www.iraap.org/Martin/PR.htm

Quoting Timothy Good from his book Alien Liaison: "Although the FAA denied this incident had occurred, investigator, Jorge Martin subsequently obtained confirmation from a U.S. Navy source who said there were radar-tapes showing what had happened which were immediately classified and sent to Washington D. C."

Numerous residents of the area reported seeing jet fighters flying through the area, starting at 6 p.m. A week earlier, some people had seen jet fighters chase a small UFO over the Sierra Bermeja mountain ridge and Laguna Cartagena (a lagoon), sites where a number of UFO sightings had been reported since 1987.

At 7:45 this evening, residents saw a large triangular UFO flying over the Sierra Bermeja. It seemed to have some kind of extended appendage on its frontal section with many brilliant colored lights constantly blinking on and off.

It was slightly curved at its rear end, had a gray metallic structure and had a large central yellow light that was being emitted from a big bulging luminous circular concave appendage.
At the triangle's right "wing tip" were brilliant yellow lights and on the left were red lights.
As the people watched, two jet fighters tried to intercept the object.

They passed in front of it, at which time the UFO veered to the left and made a turn back, reducing its speed. The jets tried to intercept it three times, and that's when the UFO decreased its speed, almost stopping in mid-air.

One jet stationed itself near the UFO's right side and the other at its rear. Suddenly, the jet in the back just disappeared on top of or inside the UFO. One witness who was watching with binoculars said he never saw the jet emerge from beyond the UFO.

The second jet remained very close to the right side of the object, looking very small in comparison. As the UFO flew a little to the west, the second jet also disappeared, apparently

374

inside the UFO, and its engine noise stopped.

The object then began descending and came down very close to the ground over a small pond known as Saman Lake. It stood still in mid-air for a moment, then straightened its corners and gave off a brilliant flash of yellow light from the central ball of light, like an explosion, but without making any noise.

It then divided into two different and distinct triangular sections. The triangle to the right was illuminated in yellow and the other in red.
Then both shot away at great speed, one to the southeast, and the other to the northwest. Red sparks could be seen falling when the object divided itself.

A retired Army veteran living in the area said that at 8:20 p.m., a bunch of black helicopters arrived and flew over the Sierra Bermeja and the Laguna Cartagena areas without lights until midnight and appeared to be searching for something.

UFO researcher **Jorgé Martin** checked with all Puerto Rican and U.S. Government agencies that might have knowledge of the incident but all denied knowing anything about what happened. Martin said that a week later a U.S. Naval officer whom he could not identify confirmed all that had happened and said radar tapes had been sent to Washington, D.C. and that a lid of secrecy had been placed on the whole affair. http://www.alien-ufos.com/ufo-alien-discussions/20220-unidentified-aerial-phenomena-2.html

These remarkable photos were taken in Peurto Rico by Amaury Rivera in 1988. The pictures show a US Navy F-14 "Tomcat" intercepting a large disk-shaped object
http://www.ufoevidence.org/photographs/section/topphotos/Photo145.htm

Amaury Rivera - Abductee

http://www.ufoevidence.org/photographs/section/topphotos/Photo145.htm

Subsequent scientific analysis of the photographs concluded that both objects were between 3 and 5 miles away from the camera and that the jet was moving quickly, whilst the disk was moving relatively slowly or not at all.

https://www.youtube.com/watch?v=mIM2s6YqNmQ

**The US Navy F-14 "Tomcat" has circled around the large
flying saucer approaching ever closer to it**

**The intentions of the F-14 appear potentially hostile to the UFO
It is believed the jet disappeared inside the ET craft**

377

British Jets Chase UFO over Brum – 2012

Recently this year, an unknown cameraman took footage of two military aircraft chasing a shiny orb or more precisely a flying saucer over the M5 over Brum in the Midlands.

The 30-second clip is believed to have been taken from a West Midlands service station car park. The jet fighters were filmed trailing the saucer as they streaked past the cameraman's location in pursuit of the UFO.

Expert Nick Pope, who probed UFO sightings for the MoD, said: "This is one of the best videos I've seen. It could be a new drone — that might explain the military jets.

"But you don't normally test-fly secret projects in daylight. Alternatively, this could be the real thing — a UFO in our airspace and military aircraft scrambled to intercept, probably due to it being tracked on radar."

The MoD refused to comment on the alleged sighting, but confirmed it would scramble jets to meet an air threat.

West Midlands Police said: "We are not aware of any reports of unidentified aircraft near the M5." v.soodin@the-sun.co.uk

Close-up ... 'UFO'

Rapid first and second passes by jets as UFO remains stationary

List of Aircraft-UFO Incidents and Encounters

The following list should be considered only as a partial list of aircraft and UFO encounters as obviously there are certainly many unreported encounters of this kind that are not reported to the public, particularly if they involve military aircraft in pursuit of UFOs. Other UFO encounters may involve private planes that see UFOs sharing the same area of the sky but, for reasons of possible public ridicule these accounts never get reported into the public domain. The actual number of aircraft - UFO encounters probably number conservatively in the hundreds to thousands.

1942

- On 25 March, over the Zuiderzee an RAF Vickers Wellington from No. 301 Polish Bomber Squadron at RAF Hemswell commanded by 2nd Lt Roman Sobinski was followed by an orange-coloured luminous disc on a journey back from a mission over Essen in the Ruhr Valley. The disc followed the bomber for five minutes at a distance of 150m and height of 14,000 ft and was fired on by the tail gunner, then disappeared at a very high speed. Such sightings were later christened *foo fighters* and many were seen in November 1944 on bombing missions over Germany.

1947

- Kenneth Arnold UFO sighting – on 24 June, Kenneth Arnold saw up to 20 objects over Mount Rainier in Washington State.

1948

- Mantell UFO incident – on 7 January an object was seen by four P-51 Mustangs over Fort Knox in Kentucky
- Chiles-Whitted UFO Encounter – on 24 July an object collided with an Eastern Air Lines Douglas DC-3 near Montgomery, Alabama.
- Gorman Dogfight – on 1 October an object was seen by North Dakota National Guard P-51 Mustang over Fargo, North Dakota.

1950

- Sperry UFO case – on 29 May an object was seen by an American Airlines Douglas DC-6 near Mount Vernon, Maryland.

1952

- Nash-Fortenberry UFO sighting – on 14 July, eight objects were seen by two commercial pilots over Chesapeake Bay in Virginia.
- Carson Sink UFO incident – on 24 July an object was seen by two pilots in a B-25 Mitchell over Carson Sink in Nevada.
- Operation Mainbrace – from 14 to 25 September many UFOs were observed. On 19 September at 11am a silver disc-shaped object followed a Gloster Meteor returning to RAF Topcliffe and seen by observers on the ground. It rotated whilst hovering. It then travelled towards the west at high speed. On 21 September, six RAF planes followed a spherical object over the North Sea. It followed one of the planes back to the base.

1953

- Ellsworth UFO sighting – on 5 August an unknown object over Bismarck, North Dakota was seen by people on the ground and by two F-84 Thunderjet pilots.

1954

- On 14 October Flt Lt James Salandin of the Royal Auxiliary Air Force, flying in a No. 604 Squadron RAF Gloster Meteor F8 from RAF North Weald, narrowly missed two UFOs over Southend-on-Sea at around 4.30pm at 16,000 ft. The objects were circular with one being coloured silver and the other gold. He narrowly avoided having a head-on collision with the silver object.

1956

- Lakenheath-Bentwaters incident – on 13 August, 12 to 15 objects were picked up by USAF radar over East Anglia. One object was tracked at more than 4,000 mph by USAF GCA radar at RAF Bentwaters. The objects sometimes travelled in formation, then converged to form a larger object and performed sharp turns. One object was tracked for 26 miles which then hovered for five minutes then flew off. One object at 10pm was tracked at 12,000 mph. RAF de Havilland Venoms from RAF Waterbeach had sightings of the objects.

1957

- West Freugh incident – on 4 April, a large object was seen on radar at RAF West Freugh near Stranraer at 50,000 ft which was stationary for 10 minutes over the Irish Sea. It moved vertically to 70,000 ft and was also tracked by radar at Ardwell. The object did an 'impossible' sharp turn and was described as being as large as a ship, bigger than a normal aircraft.
- Milton Torres 1957 UFO Encounter – on 20 May, a USAF F-86D Sabre based at RAF Manston intercepted an object over East Anglia.
- UFOs seen by Portugal Air Force – on 4 September an object was seen by four Portuguese Air Force pilots

1959

- On 24 February an American Airlines Douglas DC-6 was followed for 45 minutes by three saucer shaped objects on its flight from Newark to Detroit. It was seen by the crew and the 35 passengers and by Captain Peter Killian.

1965

- On 18 March a Convair CV-240 from Toa Airways heading from Osaka to Hiroshima was followed over Iejima by an oblong 15 metre long green luminescent object, forcing the pilot, Yoshiharu Inaba, to make a 60 degree turn to avoid a collision. The object stopped and flew alongside the plane for about three minutes, and affected the Automatic Direction Finder.

1970

- On April 24, a Soviet bomber disappeared on its flight from Moscow to Vladivostok without a trace. On the same day, several UFOs were observed in the area of the Soviet-Chinese border, which could not be shot down by the Russian military.
- William Schaffner on 8 September intercepted an unknown object over the North Sea. His BAC Lightning was later retrieved from the sea.
- On September 21st at 20:01pm, the Emergency controller at West Drayton was informed of two reports of a great ball of fire seen by pilots flying over the North Sea.

1974

- Coyame UFO Incident – on 25 August an object collided with an aircraft near Coyame, Chihuahua in Mexico near the USA border.

1976

- On 30 July at 8pm, 42 year old Captain Dennis Wood and his crew in a British Airways Hawker Siddeley Trident from Edinburgh to Faro saw objects near Lisbon, seen by all passengers on board. They looked "due west and saw an incredibly brilliant white object", which "took off vertically at very high speed and disappeared". The pilots were alerted by the radar operator at Lisbon who picked up two unknown objects. It was observed for around eight minutes. An astrophysicist, Dr. David Ramsden, from the University of Southampton believed it was a large research balloon that was released on 29 July and drifted from Italy to America.

- 1976 Tehran UFO incident – on 19 September over an area 40 miles north of Tehran two F-4 Phantom IIs from Hamedan Air Base (Shahrokhi) attempted to intercept an unknown object seen on ground radar. When closing in on the target, all the plane's communications systems would not work, which regained their function when the plane turned away from the unknown object. Another object emerged from the main unknown object and headed towards the intercepting planes. The pilots tried to discharge an AIM-9 Sidewinder missile but when engaging their weapons system, it would momentarily not lock-on to the target or even function until the planes changed to a less hostile course.

1977

- On 7 March a Dassault Mirage IV piloted by Major René Giraud and navigated by Captain Jean Paul Abraham was returning to Luxeuil Air Base when at 8pm they encountered an unknown object over Chaumont, Haute-Marne. It was reported to the air traffic control at Contrexéville and it briefly followed the aircraft twice, although nothing was noticed on radar.

1978

- Valentich disappearance – on 21 October, 20 year old Frederick Valentich was piloting a Cessna 182 over Bass Strait in Australia when he was "buzzed" by an unknown object with four bright lights about 1000 ft above him and reported this to air traffic controllers at Melbourne with his last message being *it's not an aircraft*.

1979

- Manises UFO incident – on 11 November a TAE Sud Aviation Caravelle from Salzburg to Las Palmas was forced to make an emergency landing at Manises in Spain.

1986

- Japan Air Lines flight 1628 incident – on 16 November on a Japan Airlines flight in a Boeing 747 from Paris to Tokyo whilst near Alaska, an object was seen that was twice the size of an aircraft carrier. In subsequent months, two other aircraft reported unidentified objects over Alaska.

1990

- Belgian UFO wave – on 31 March two Belgian Air Component F-16 Fighting Falcons flew from Beauvechain Air Base to intercept a UFO seen near Wavre. They were unable to maintain a radar lock for their missiles because the object changed position too quickly. Such rapid changes of position would involve g-forces fatal to human pilots.
- On 4 August at 9pm a 75 ft wide diamond object was seen by an RAF Harrier 20 miles north of Pitlochry in Perth and Kinross. It appeared to hover for 10 minutes above the A9 near Calvine.

1991

- On 21 April, pilot Achille Zaghetti from Grosseto, Tuscany, in an Alitalia McDonnell Douglas MD-80 on a flight from Milan to Heathrow saw a three-metre-long khaki-coloured object over Lydd in Kent at 22,000 ft about 300m away. It was seen on radar.

1994

- On 28 January on Air France Flight AF3532 in an Airbus A320 flying from Nice to London, Jean-Charles Duboc saw an unusual object near Paris at around 1pm which looked like a huge red-brown disc at 35,000 ft of approximately 800 ft in diameter. The object was reported to air traffic control at Reims in Champagne-Ardenne and was determined to be close to Taverny Air Base, the headquarters of the French Air Force. It was stationary in the sky for around one minute then disappeared in around 10–20 seconds.

1995

- On 6 January pilots aboard a Boeing 737 on British Airways Flight BA5061 from Milan saw an object on their descent to Manchester at 4,000 ft when over the southern Pennines.

1996

- Westendorff UFO sighting – on 5 October a large object was seen over Lagoa dos Patos in Brazil.

2007

- 2007 Alderney UFO sighting – two huge craft observed near Alderney and Guernsey respectively.
- 2007 Romania UFO incident - a Romanian military plane taking off from Gherla almost hit 2 solid entities which could not be identified.

On many occasions aircraft - UFO encounters took on the form of an aerial "cat and mouse game" (from the perspective of the ET pilot) or a mock "dogfight" (from the perspective of the military pilot) which sometimes ended in a deadly outcome usually for the terrestrial pilot.

A cat and mouse chase was essentially just that, where the jet aircraft would pursue a UFO trying vainly to catch up with it. **[I have personally witnessed this type of encounter on at least two occasions]**. The alien flying saucer would hover for awhile and just when the aircraft approached to within gun or missile firing range, the saucer would speed off a short distance or zigzag into a new direction and then hover again, waiting for the aircraft to close the distance. Then, when the pilot was within range again, the UFO would once more repeat the same aerial maneuvers to the frustration of the jet pilot. This cat and mouse game could go on for an hour or more or until either the pilot just decided to give up or was nearly out of fuel or the UFO decided to end the game by merely zipping off into the distance or flying straight up into space at incredible speed and out of sight.

Sometimes, the UFO would fly wing side along with the military aircraft at the same speed or maneuver above the cockpit as in the movie "Top Gun" starring Tom Cruise or fly on the tail end of the aircraft and sometimes, depending on the size of the military aircraft and the size of the flying saucer, the ET craft would slip itself in between the forward and tail wings close to the fuselage in a show of superior aeronautical flight control! Such an event was witnessed by the late Colonel Wendell Stevens and his crew up in Alaska and many other veteran pilots of WWII had shared the same experiences with **Foo Fighters** on numerous sorties into Europe.

Needless to say, such aerial acrobatic maneuvers by ET craft left many pilots freaked out and overwhelmed by the experience thinking that a mid-air collision was inevitable between themselves and the UFO. Sometimes, the UFO would interfere with the on –board electrical systems of the plane or cause stalling of the engines. The alien craft, however, would simply slide back out from between the wings and fly off into the distance; the aircraft electrical and power systems would return to normal leaving it none the worse for wear.

CHAPTER 47

IS THE NAVY REWRITING THE UFO NARRATIVE

When we think of the recent news story that broke in the *New York Times* in December 2017 that the Pentagon acknowledged a secret UFO program, we find, like most of the public confused about the news story, wanting more.

Stephen Pope wrote an interesting article on the website *"Flying"* in which he stated that there were legitimate reasons to be skeptical of the *New York Times* UFO story "which reads like science fiction" leaving a person with more questions than answers. Almost immediately a sensationalistic amount of worldwide media reporting began, yet most failed to recognize that the *Times* original reporting had glaring major problems with the story. Here are five conspicuous reasons which Steve Pope brings up:

1. The Pentagon didn't release those UFO videos, an official connected to a Las Vegas company who resigned in October did.

The *New York Times* story states that the release of the naval videos came from high-level government officials while glossing over the actual disclosures came from an official who led the Pentagon's small UFO investigation program and who has since resigned to join a Las Vegas company called **To the Stars Academy of the Arts and Sciences** while initiating private funding for more UFO research. This fact appears to be ignored by most news media since the original *Times* news article publicly broke.

2. The Pentagon UFO program's prime contractor claims to possess mysterious metal alloys that exhibit mystical powers — but won't show them to anyone.

The Nevada-based **Bigelow Aerospace** company founded by billionaire **Robert Bigelow** collects unusual alloys from UFOs that physically affected people who interacted with them and receives most of the Pentagon funding for the UFO research project. This has created obvious conflicts of interest between Bigelow Aerospace company and his friend, former Nevada Sen. Harry Reid. Interestingly, it was Reid who pushed for the $22 million in funding for the government's UFO program in the first place. So, why is there all the continued secrecy surrounding these alloys and other UFO findings? https://www.flyingmag.com/five-reasons-to-be-skeptical-about-that-new-york-times-ufo-story/

3. One of the authors of the *Times* article wrote a book about UFOs and doesn't even work for the paper.

When the UFO story appeared *Times's* Sunday edition, they divulged that a non-employee of the paper, UFO hunter **Leslie Kean** and author of a book "UFOs" with a forward by **John Podesta**, UFO conspiracy theorist who worked in the White House during the Clinton years, had written the byline to the story. In fact, not only did Kean help write the article, she is the one who pitched the story to the *Times* in the first place. At best, shouldn't she merely have been interviewed for the story? Why was she permitted to be a part of the writing and vetting process?

4. In one of the videos, a Navy airman says, "That's a [bleeping] drone." Why are we doubting him?

In the only UFO video released with pilot audio accompanying it, a Navy airman describes the object he is following as a "drone." The object does not appear to do anything particularly unusual in this video. Nobody claims that the object did anything unusual. The Pentagon won't even say where and when the video was shot. Was it a drone?

5. There are plausible explanations for these videos and eyewitness accounts. Where is the rest of the evidence?

Far from proving that our planet has been visited by aliens in aerodynamically advanced space craft, the videos and eyewitness accounts that emerged in the *Times* story and elsewhere merely provide a starting point for further investigation and inevitably lead to more questions. That's what makes all the secrecy and intrigue still surrounding this story so frustrating for the public. The *New York Times* has dangled this carrot in front of the world, and now leaves us all wondering if the people who claim to know more will ever divulge that information, or merely continue to direct us to the "Donate Here" buttons on their websites.
.https://www.flyingmag.com/five-reasons-to-be-skeptical-about-that-new-york-times-ufo-story/

The Times reveals in its news story, organizations and names of people that require further investigation to fully understand what is going on. Is the Military, especially the US Navy providing open and honest disclosure on the UFO subject or is there a new narrative being re-written to convince the public that UFOs now known as UAPs (Unidentified Aerial Phenomenon) could be potentially hostile?

At this point, what we need is an examination of what these questions imply as it pertains to the Pentagon, the Navy, and the pilot's sighting and the three Navy videos, along with what is **To The Stars Academy of Arts and Sciences** (aka. **To The Stars Academy**) and who are some of its major members, **The Advanced Aerospace Threat Identification Program,** and its agenda. the **Pentagon's** position and the three Navy videos released into the public domain.

To The Stars Academy of Arts and Sciences (TTAASA aka. TTSAd)

The company was founded in 2017 as a public benefit corporation by **Jim Semivan**, a former senior Intelligence Officer with the CIA; **Harold E. Puthoff**; and **Tom DeLonge**. The Entertainment Division was created by acquiring DeLonge's previous media company, "**To the Stars, Inc.**" which publishes albums, books, TV shows and films.

The science and aerospace divisions are devoted to the "outer edges of science" such as investigating unidentified flying objects. Harold E. Puthoff described their goals as *"imagine having 25th-century science this century."* One of their potential projects is an "**electromagnetic vehicle**."

386

https://en.wikipedia.org/wiki/To_the_Stars_(company)

Some sources have opined that To the Stars is responsible for reinvigorating public imagination of UFOs, including **Jan Harzan** of **Mutual UFO Network** and **Dan Zak** of *The Washington Post.* For his work at To the Stars, Tom DeLonge was named UFO Researcher of the Year in 2017 by the UFO hunting organization **Open Minds**.

Despite the company's work being primarily associated with ufology, **Luis Elizondo** has stated: *"None of us at TTSA consider ourselves 'Ufologists' or part of the 'Ufology culture,' in fact, most of us come from a U.S. government background (both Defense and Intelligence)."*

Virtual Analytics UAP Learning Tool (VAULT)

The "**Virtual Analytics UAP Learning Tool" (VAULT)** is a public-facing database of UFO sightings. The VAULT team collects, analyses and provides their authentication of UFO sightings, most famously reported in the media as having been obtained through declassified government materials.

Three videos from the VAULT taken during the **USS Nimitz UFO incident** and the **USS Theodore Roosevelt UFO incidents** were publicly confirmed by the US Navy in September 2019 as authentic videos taken by Navy pilots. The videos were part of a campaign by former intelligence officer Luis Elizondo, who now works for To the Stars, who said that he wanted to shed light on the **Advanced Aerospace Threat Identification Program**, a secretive **Department of Defense** operation to analyze reported UFO sightings. In April 2020, the same footage released by the company was subsequently declassified and officially released by the Navy.[1]

The company, with assistance from **Chris Mellon**, who worked formerly for the **Senate Intelligence Committee** and the Department of Defense, engaged congress and arranged classified congressional hearings with the pilots involved in the incidents aimed at understanding potential threats to aviators. **https://en.wikipedia.org/wiki/To_the_Stars_(company)**

Entertainment

The entertainment leg of To the Stars, often referred to as **To the Stars Media**, publishes albums, books, TV shows and films. *Sekret Machines* non-fiction companion series *Gods, Man & War* were co-authored by DeLonge and Peter Levenda.

A major TV project was the **History Channel** series *Unidentified: Inside America's UFO Investigation* in 2019. To the Stars has also started production on a documentary TV series and a feature film for the *Sekret Machines* franchise.

Science and Aerospace

The science and aerospace divisions are devoted to the "outer edges of science" such as investigating unidentified flying objects. **Harold E. Puthoff** described their goals as "imagine having 25th-century science *this* century. One of their potential projects is an "electromagnetic vehicle."

Vice reported that the company would participate "in the investigation of UFOs and other fringe science projects" and that "many of the technologies or phenomena being researched by the company are based on highly speculative theories that toe the line of pseudoscience". **https://en.wikipedia.org/wiki/To_the_Stars_(company)**

ADAM Research Project

The company's **ADAM Research Project** is promoting what they believe to be an "extraterrestrial" metal for commercial and military applications.

In July 2019, the company stated it had acquired and was studying "potentially exotic materials" as part of its **Acquisition & Data Analysis of Materials (ADAM)** research project. **Steve Justice**, To The Stars's COO and former head of **Advanced Systems at Lockheed Martin's Skunk Works** said in a statement that "*the structure and composition of these materials are not from any known existing military or commercial application"* and that the materials would be studied in an attempt to reverse engineer them.

Regarding the origin of the materials, he stated: *"they've been collected from sources with varying levels of chain-of-custody documentation, so we are focusing on verifiable facts and working to develop independent scientific proof of the materials' properties and attributes."* In its SEC filing, the company is recorded as having paid $35,000 for several items including "six pieces of Bismuth/Magnesium-Zinc metal" and a piece of aluminum. According to the company, the metals are from an unidentified flying object, and were previously "retained and studied" by ufologist **Linda Moulton Howe**. Moulton Howe claimed in 2004 that the metals become a

"lifting body" when subjected to electromagnetic radiation. Today, however, she claims she has had the samples tested by **Carnegie Science's Department of Technical Magnetism** in 1996 and again by **Harold E. Puthoff** and others on several occasions. According to a letter from Puthoff in 2012 the tests were unable to prove the alien origin of the samples or any "interesting/anomalous outcome" but suggested that one additional test was remaining that required special equipment which was not readily available.
https://en.wikipedia.org/wiki/To_the_Stars_(company)

On October 17, 2019 the company announced it entered into a cooperative research and development agreement with the **United States Army Combat Capabilities Development Command**. The five-year contract will focus on "inertial mass reduction, mechanical/structural meta materials, electromagnetic meta material wave guides, quantum physics, quantum communications, and beamed energy propulsion."

According to the U.S. Army, no public funding will go the group but at least $750,000 will be provided in support and resources for developing and testing To the Stars technologies. The contract states that **To the Stars** will provide samples in its possession of **"metamaterials"**, any data or "obtained vehicles" that use "beamed energy propulsion," and any information or technology related to "active camouflage" for testing and analysis of potential application on Army ground vehicles.

Doug Halleaux, a spokesperson for the **U.S. Army Combat Capabilities Development Command Ground Vehicle Systems Center**, has stated that the US government has approached To the Stars since *"If materials represented in the TTSA ADAM project are scientifically evaluated and presented with supporting data as having military utility by the TTSA, it makes sense to look deeper here".* According to Halleaux, the Army is also interested in the results of a collaboration between To the Stars and *TruClear Global*, a company that creates custom video screen billboards, aimed at providing "advanced technology solutions to United States Government clientele." https://en.wikipedia.org/wiki/To_the_Stars_(company)

The Advanced Aerospace Threat Identification Program (AATIP)

The **Advanced Aerospace Threat Identification Program (AATIP)** was an unclassified and unpublicized investigatory effort funded by the United States Government to study **unidentified flying objects (UFOs) or unexplained aerial phenomena (UAP)**. The program was first made public on December 16, 2017. The program began in 2007, with funding of $22 million over the five years until the available appropriations were ended in 2012. The program began in the U.S. **Defense Intelligence Agency.**

According to the **Department of Defense**, the AATIP ended in 2012 after five years, however reporting suggested that U.S. government programs to investigate UFOs continued. This was confirmed in June 2020 with the acknowledgement of a similar military program, the unclassified but previously-unreported **Unidentified Aerial Phenomenon Task Force**

A group including **Luis Elizondo**, who was an AATIP program director, founded a public-benefit corporation named **To The Stars Academy of Arts & Science** in 2017.

Initiated by then **Senate Majority Leader Harry Reid** (D-Nevada) as the **Advanced Aerospace Weapon Systems Applications Program (AAWSAP)** to study **unexplained aerial phenomena (UAP)** at the urging of Reid's friend, Nevada billionaire and governmental contractor Robert Bigelow, and with support from the late senators **Ted Stevens** (R-Alaska) and **Daniel Inouye** (D-Hawaii), the program began in the DIA in 2007 and was budgeted $22 million over its five years of operation.

AATIP was headed by **Luis Elizondo**. Elizondo resigned from the **Pentagon** in October 2017 to protest government secrecy and opposition to the investigation, stating in a resignation letter to **US Defense Secretary James Mattis** that the program was not being taken seriously. Elizondo, said on December 19, 2017, that he believed there was **"very compelling evidence we may not be alone".** While the United States Department of Defense stated that the program was terminated in 2012, after acknowledgement of AATIP in 2017 the exact status of AATIP and its termination remained unclear Elizondo commented that, while the effort's government funding ended in 2012, the program continued with support from Navy and **CIA** officials even after his resignation Reports in 2020 confirmed Elizondo's statement with the reporting of the U.S. government successor to the AATIP, the **Unidentified Aerial Phenomenon Task Force'**

The materials studied by AATIP have been the subject of classified congressional hearings aimed at understanding and identifying the potential threat to the safety and security of aviators. The Navy has confirmed that, in response to inquiries by members of Congress, they have provided a series of briefings by senior naval intelligence officials as well as testimony from *"aviators who reported hazards to aviation safety"*. The contents of those briefings are classified, but **Senator Mark Warner**, the vice chairman of the **Senate Intelligence Committee,** who participated in one of those briefings, released a statement requesting further research into "unexplained interference in the air" that could pose safety concerns for naval pilots.

According to *Popular Mechanics*, **Senate Intelligence Committee Brigadier General Richard Stapp**, **Director of the DoD Special Access Program Central Office**, testified the mysterious objects being encountered by the military were not related to secret U.S. technology.

On July 23, 2020, *The New York Times* reported that while former **Senator Harry Reid** *"believed that crashes of objects of unknown origin may have occurred and that retrieved materials should be studied; he did not say that crashes had occurred and that retrieved materials had been studied secretly for decades."* News reports also repeated a claim made by **Eric W. Davis**, a former employee of **Harold E. Puthoff** (co-founder of UFO-promoting company *To the Stars*) that an "off-world vehicle" might be in the possession of the US government.
https://en.wikipedia.org/wiki/Advanced_Aerospace_Threat_Identification_Program

Author's Rant: It is this author's opinion that AATIP real agenda is not about full public disclosure that UFOs and by that acknowledgement, that ETI also exist, but to spin or re-write the narrative on this UFO/ETI subject. By re-writing the narrative to make it acceptable to the younger generation who are highly impressionable, especially when one of the ATTAAS members is a former rock musician who is recognized for his music, more so,

than the older generation, who are more demanding of honest full disclosure to the UFO question which has never been adequately addressed, other than with denials, lies, coverups, suppression, misinformation, disinformation. and obfuscation!

The general secrecy behind UFOs is the coverup of decades of US military involvement from all the branch services that come crash retrievals of UFO shootdowns since the early 1940s, the research and development of alien technology, especially their spacecraft, and their medical research of alien DNA, Cloning and Hybridization of alien biome! This latter R&D has led to reverse engineered alien craft referred to as ARVs (Alien Reproduction Vehicles) and PLFs (Programmable Life Forms) which has evolved into a secret space force and mutated into a potential future false flag alien invasion scenario!

The re-write of the narrative is to portray Extraterrestrial intelligence as hostile, interfering with commercial air traffic and especially with military aircraft, and engaging with military jets in dog-fights where military aircraft have either been disabled or tragically shot down by ET craft. In such tragic outcomes, it has been the military who hostility and aggressiveness by initiating cat and mouse chases because of national air intrusions over the US or over military and nuclear weapon bases or in dog fight engagements, all in the quest of acquiring alien technology. These militaristic aggressive actions are not solely in the US but are global in many nations!

Tom DeLonge

Thomas Matthew DeLonge is an American musician, singer, songwriter, author, record producer, actor, and filmmaker. He is the lead vocalist and guitarist of the rock band "Angels & Airwaves", which he formed in 2005, and was the co-lead vocalist, guitarist, and co-founder of the rock band **"Blink-182"** from its formation in 1992 until his dismissal from the group in 2015.

DeLonge has been a believer in aliens, UFOs, and conspiracy theories since his youth, well before founding **Blink-182**.

In 2015, DeLonge founded an entertainment company called To The Stars, Inc. which, in 2017 he merged into a larger To the Stars Academy of Arts & Sciences. Aside from the entertainment division, the new company has aerospace and science divisions dedicated to ufology and the fringe science proposals of To the Star's co-founder, **Harold Puthoff**.

In 2019, the company produced the **History Channel** television show *Unidentified: Inside America's UFO Investigation*, about the **USS Nimitz UFO incident**, which also features DeLonge.

In April 2020, the **Pentagon** declassified three videos which had been captured of UFOs. DeLonge had previously released these videos through his company, back in 2017.
https://en.wikipedia.org/wiki/Tom_DeLonge

Tom DeLonge

"Unidentified" reveals the findings of the Pentagon's top secret $22m UFO Task Force, which investigated the threat of UAPS around the globe. With DeLonge's help, military footage of these incidents has now been released into the public domain. ***While the visual evidence is restricted to grainy radar footage***, the scary part is that the pilots, military officials and other eyewitnesses interviewed in Unidentified all give the same story: that of multiple sightings of giant white Tic-Tac-shaped craft moving at speeds and trajectories that seem impossible to man. (Bold Italics added by author emphasis)

When interviewed by The Guardian, DeLonge comes across as 99% enthusiast/expert and 1% eccentric. He can't answer some of my questions *("What's the most convincing piece of evidence you've seen as to the existence of aliens?")* due to *"national security issues"*. But he clearly knows something, or thinks he knows something. But what level of government cover-up are we talking about? Are aliens (2-1 to him by now) really living among us like in Men in Black?
https://www.theguardian.com/tv-and-radio/2020/sep/15/star-of-bethlehem-spaceship-tom-delonges-new-career-ufo-expert-blink-182

In November 1996, there was a **CSETI expedition** to **Joshua Tree National Park**, California that is a favourite spot for **Dr. Steven Greer** to do **CSETI CE-5 contact work** by engaging with **UFOs (ETVs)** and **ETI** as this area has had a history of UFO sightings and ETI encounters.
On this expedition, **Jamie (James) Cromwell** the actor from the **Star Trek** movie "First Contact" and "Babe" and **Tom DeLonge**, the founder of the rock band **"Blink 182"** were in attendance for a few evenings, as both men had been following the work of CSETI for years.
"Contact: Countdown to Transformation - The CSETI Experience 1992 - 2009" by Steven M. Greer; September 2009; ISBN 9780967323831; Printed in USA; Produced by 123Printfinder, Inc.

Tom has had an interest in the UFO subject as a youngster, however, it wasn't until he met Dr. Greer and witnessed sightings of UFOs at Joshua Tree, that the whole phenomenon gelled together into actual reality.

Tom prior to this expedition had flown **Dr. Steven Greer** and his daughters out to attend his concert and to hear more about UFOs where they kindled their friendly relationship. Then after a phone call, he flew to Greer's ranch with a music producer from MTV Diary where they heard more information on UFOs and ETI. At that time, Dr. Greer entrusted copies of his **Disclosure Witness Project** video tapes to Tom for safe keeping, before he initiated the Disclosure Witness event at the **Washington, DC Press Club** in May 2001. DeLonge was one of very few people who had in his possession copies of these **Witness Disclosure tapes** which he hid away in his home. It wasn't long after that time afterward his phone started to be bugged!

Tom was even inspired from his time with Dr. Greer and wrote the song "Aliens Exist" and the influence of Tom DeLonge's rock music on the younger generation was apparent with crowds numbering 100,000 at some concerts. Tom wanted to have another meeting with Dr. Greer along with Dr. Carol Rosin and some show biz people out west in Los Angeles to spread the message that UFOs and ET were visiting the Earth.

It became apparent after phone tapping of Tom's home and probable surveillance that the US military people from the **Pentagon** took an immediate interest in **Tom Delonge**.

He was soon approached by Pentagon military who let him know *"the real truth"* and the potential threat behind UFOs entering US airspace. From the narrative of the official Naval military position, which he believed was insider information on the UFO phenomenon, he went along with them and their desire to let the public know what they had been investigating for 12 years.

Author's Rant: I had in that same year, June of 1996 been on a week-long CSETI seminar called "Ambassadors to the Universe" in Crestone, Colorado with Dr. Greer where we had UFO sightings and ETI encounters nightly!
We saw military jets chasing after the ETVs that we were communicating with, even to the point that one jet fired a green flare toward the ground where there were several ET craft to light up the area from which they were engaging in photonic communication with me and our CSETI group! Initially, my first thought was that the military jet had fired a missile into the group of UFOs, but when there was no explosion, I discovered the next day from an ex-military man that this was probably flare which was a standard practice to light up the ground to get visibility of what was there. The lights of the ETVs dimmed out completely and only came back on after the jet left the area to continue our photonic communication. To say that the military such as the USAF or the Navy knows nothing about what UFOs are is ludicrous.
It is worth noting that when you are in the public spotlight as a singer, or actor, or politician, etc. the media seeks you out to hear of your story. As an example, more recently the singer Demi Lovato's UFO public announcement of having a UFO sighting with Dr. Greer at Joshua Tree in October 2020. Since then, more singers and Hollywood movie stars

have come out stating their experiences and beliefs in extraterrestrial life visiting the Earth. The entertainment media are always eager to interview such people for their shows.

Luis Elizondo

Elizondo, who is also on the Zoom call, headed up a top secret $22m project - officially called the **Advanced Aerospace Threat Identification Program (AATIP)** - tasked with investigating the possible threat of UFOs, from 2007 to 2012.

He says the topic is *"fraught with taboo and stigma… because most people jump immediately to the conclusion of tin-foil hats and, quote-unquote, 'Elvis being on the mothership.'"*

He continues: *"All of a sudden, lo and behold, the world finds out we actually had a real programme with $22 million of taxpayer money to actually look at UFOs, but nobody wanted to talk about it."*

Luis Elizondo being interviewed on CCN.
https://edition.cnn.com/2017/12/18/politics/luis-elizondo-ufo_pentagon/index.html?sr=fbCNN121817luis-elizondo-ufo-pentagon0959PMVODtopLink

Elizondo says DeLonge is *"not afraid to take a topic that is controversial and hit it head on"* and that his involvement is helping to get people talking about it. https://news.sky.com/story/tom-delonge-on-ufo-research-i-wouldnt-have-left-blink-182-for-something-pie-in-the-sky-12061013 and https://www.youtube.com/watch?v=Jmhf_TQ_w2M&feature=youtu.be

In the $600 billion annual Defense Department budgets, the $22 million spent on the **Advanced Aerospace Threat Identification Program** was almost impossible to find. Which was how the Pentagon wanted it.

394

For years, the program investigated reports of unidentified flying objects, according to Defense Department officials, interviews with program participants and records obtained by The New York Times. It was run by a military intelligence official, Luis Elizondo, on the fifth floor of the Pentagon's C Ring, deep within the building's maze.

The Defense Department has never before acknowledged the existence of the program, which it says it shut down in 2012. But its backers say that, while the Pentagon ended funding for the effort at that time, the program remains in existence. For the past five years, they say, officials with the program have continued to investigate episodes brought to them by service members, while also carrying out their other Defense Department duties.

Senator Harry Reid

The shadowy program — parts of it remain classified — began in 2007, and initially it was largely funded at the request of **Harry Reid**, the (Nevada Democrat) who was the Senate majority leader at the time and who has long had an interest in space phenomena. Most of the money went to an aerospace research company run by a billionaire entrepreneur and long-time friend of Mr. Reid's, **Robert Bigelow,** who is currently working with NASA to produce expandable craft for humans to use in space.

Working with Mr. Bigelow's Las Vegas-based company, the program produced documents that describe sightings of aircraft that seemed to move at remarkably high velocities with no visible signs of propulsion, or that hovered with no apparent means of lift.

Officials with the program have also studied videos of encounters between unknown objects and American military aircraft — including one released in August of a whitish oval object, about the size of a commercial plane, chased by two Navy F/A-18F fighter jets from the aircraft carrier Nimitz off the coast of San Diego in 2004.

Harry Reid, the former Senate majority leader, has had a long-time interest in space phenomena. Credit...Al Drago/The New York Times

Mr. Reid, who retired from Congress this year, said he was proud of the program. *"I'm not embarrassed or ashamed or sorry I got this thing going,"* Mr. Reid said in a recent interview in Nevada. *"I think it's one of the good things I did in my congressional service. I've done something that no one has done before."*

Two other former senators and top members of a defense spending subcommittee — **Ted Stevens,** an **Alaska Republican,** and **Daniel K. Inouye,** a **Hawaii Democrat** — also supported the program. Mr. Stevens died in 2010, and Mr. Inouye in 2012.

While not addressing the merits of the program, **Sara Seager**, an astrophysicist at M.I.T., cautioned that not knowing the origin of an object does not mean that it is from another planet or galaxy. *"When people claim to observe truly unusual phenomena, sometimes it's worth investigating seriously,"* she said. But, she added, *"what people sometimes don't get about science is that we often have phenomena that remain unexplained."*

The following videos were featured in "*Unidentified: Inside America's UFO Investigation*", a 2019 History Channel series executive produced by **Tom DeLonge**.
https://en.wikipedia.org/wiki/Pentagon_UFO_videos

Commander David Fravor's Testimony

A former U.S. Navy fighter pilot, now retired, **Commander. David Fravor** recalls the strange encounter he had 16 years ago in 2004 as the commanding officer of the VFA-41 Black Aces, a U.S. Navy strike fighter squadron of F/A-18 Hornet fighter planes the squadron was doing an exercise some 60 to 100 miles off the coast between San Diego and Ensenada, Mexico, in advance of a deployment to the Persian Gulf for the Iraq War.

Fravor said he was told by the command that there were some unidentified flying objects descending from 80,000 feet to 20,000 feet and disappearing; he said officials told him they had been tracking a couple dozen of these objects for a few weeks.

Fravor says, he was dispatched to investigate radar anomalies. When he and his squadron arrived closer to the point where the object had descended, they saw the object, flying around a patch of white water in the ocean below. As he got closer and pulling back up on the nose of his plane, the object accelerated and it was gone, faster than anything he had seen in his life. He describes an encounter with a white "**Tic Tac**" (named after the breath mint because its similar appearance) shaped UFO about the same size as a Hornet, 40 feet long with no wings, and believes the flying object committed an "***act of war***."

Cmdr. Fravor later described the Tic Tac-shaped object as able to turn on a dime and make itself invisible to radar. The object moved rapidly and unlike any other thing he had ever seen in the air. It was *"Just hanging close to the water"* before it accelerated away.

He was followed by other pilots who managed to catch it on video. Clips were leaked in 2017 by a UFO research group **To the Stars Academy of Arts and Sciences (TTSAAS)**, founded by punk singer **Tom DeLonge** of **Blink 182,** and formally declassified in 2020 by **the Pentagon,** according to the station.

https://www.washingtonpost.com/news/checkpoint/wp/2017/12/18/former-navy-pilot-describes-encounter-with-ufo-studied-by-secret-pentagon-program/

A video shows a 2004 encounter near San Diego between two Navy F/A-18F fighter jets and an unknown object. It was released by the Defense Department's **Advanced Aerospace Threat Identification Program (AAATIP).** Credit: U.S Department of Defense.

Fravor retired from the Navy in 2006 and nothing really came of his sighting until 2009 when a government official he declined to name contacted him while doing "an unofficial investigation." Fravor declined to give more details about the official, but said he was later contacted by **Luis Elizondo**, an intelligence officer who ran the secretive program at the Department of Defense that was just disclosed, who was also a consultant to **TTSAAS.**

Elizondo, who has since left the government to work for a private company that is hoping to promote UFO research for both scientific and entertainment purposes, is a large part of why the story about UFOs and the government's program are in the news; he quietly arranged to secure the release of three videos of UFOs from the Pentagon, including *"the one shot"* the same day and place as Fravor's.

James E. Oberg, a former NASA space shuttle engineer was also doubtful. *"There are plenty of prosaic events and human perceptual traits that can account for these stories,"* Mr. Oberg said. *"Lots of people are active in the air and don't want others to know about it. They are happy to lurk unrecognized in the noise, or even to stir it up as camouflage."* Still, Mr. Oberg said he welcomed research. *"There could well be a pearl there,"* he said.

https://www.washingtonpost.com/news/checkpoint/wp/2017/12/18/former-navy-pilot-describes-encounter-with-ufo-studied-by-secret-pentagon-program/

"FLIR" Video
FLIR1 Official UAP Footage from the USG for Public Release.webm

"GIMBAL" Video
Gimbal The First Official UAP Footage from the USG for Public Release.webm

"GOFAST" Video
Go Fast Official USG Footage of UAP for Public Release.webm

Author's Rant: Having had private written debates with Oberg in the past, his reputation is also, well known as a professional debunker and he is usually unwilling to budge from his highly skeptical position or acknowledge that UFOs are possibly not from this world yet, this military pilot seems to have made him alter his perception. This is probably, only because he may have known in advance that the Navy is coming out on their admission that some Tic Tacs or UAPs (Unidentified Aerial Phenomenon) are extraterrestrial in origin.

What Oberg has not acknowledged is that some UFOs, UAPs or Tic Tacs are in fact, (ARVs – Alien Reproduction Vehicles) manmade from reverse-engineering alien spacecraft. This may also be an identification problem with Cmdr. Fravor's sighting and his inability to distinguish between what is advanced manmade craft from what are ETVs (Extraterrestrial Vehicles).

The Pentagon

In response to questions from *The Times*, **Pentagon** officials this month acknowledged the existence of the program, which began as part of the **Defense Intelligence Agency**. The United States Pentagon has finally decided to initiate disclosure by providing *a slow, scripted release of information* through **Tom Delonge** and the "**To The Stars Academy**" **(TTSA)**. Officials insisted that the effort had ended after five years, in 2012.

But according to Mr. Elizondo, all that ended was the funding. From that time on, he worked with officials from the Navy and C.I.A. and the program remains in existence today. From then on, Mr. Elizondo said in an interview, he worked with officials from the Navy and the C.I.A. He continued to work out of his Pentagon office until this past October, when he resigned to protest what he characterized as excessive secrecy and internal opposition.

Most of the funding went to billionaire entrepreneur **Robert Bigelow** who has stated that he is convinced that extraterrestrials exist and are visiting our planet. In 2010, The FAA and **MUFON** started funneling all sighting reports through **Bigelow Aerospace Advanced Space Studies (BASS)**. Mr. Bigelow formed a research team of scientists to interview experiencers involved in significant UFO contact cases. A special area was also created to house various materials.

"Why aren't we spending more time and effort on this issue?" Mr. Elizondo wrote in a resignation letter to **Defense Secretary Jim Mattis**. Mr. Elizondo said that the effort continued and that he had a successor, whom he declined to name.

Pentagon officials say the program ended in 2012, five years after it was created, but the official who led it said that only the government funding had ended then.
Credit: Charles Dharapak/Associated Press

UFOs have been repeatedly investigated over the decades in the United States, including by the American military. In 1947, the Air Force began a series of studies that investigated more than 12,000 claimed UFO sightings before it was officially ended in 1969. The project, which included a study code-named **Project Blue Book**, started in 1952, concluded that most sightings involved stars, clouds, conventional aircraft or spy planes, although *701 remained unexplained.* (Bold italics added by author)

Robert C. Seamans Jr., the secretary of the Air Force at the time, said in a memorandum announcing the end of Project Blue Book that it *"no longer can be justified either on the ground of national security or in the interest of science."* (Bold italics added by author)

Mr. Reid said his interest in U.F.O.s came from **Mr. Bigelow**. In 2007, Mr. Reid said in the interview, Mr. Bigelow told him that an official with the **Defense Intelligence Agency** had approached him wanting to visit Mr. Bigelow's ranch in Utah, where he conducted research.

Mr. Reid said he met with agency officials shortly after his meeting with Mr. Bigelow and learned that they wanted to start a research program on U.F.O.s. Mr. Reid then summoned Mr. Stevens and **Mr. Inouye** to a secure room in the Capitol.

"I had talked to **John Glenn** a number of years before," Mr. Reid said, referring to the astronaut and former senator from Ohio, who died in 2016. Mr. Glenn, Mr. Reid said, had told him he thought that the federal government should be looking seriously into U.F.O.s, and should be talking to military service members, particularly pilots, who had reported seeing aircraft they could not identify or explain.

The sightings were not often reported up the military's chain of command, **Mr. Reid** said, because service members were afraid they would be laughed at or stigmatized.

The meeting with **Mr. Stevens** and **Mr. Inouye,** Mr. Reid said, *"was one of the easiest meetings I ever had."*

He added, "Ted Stevens said, *'I've been waiting to do this since I was in the Air Force.'"* (The Alaska senator had been a pilot in the Army's air force, flying transport missions over China during World War II.)

During the meeting, Mr. Reid said, Mr. Stevens recounted being tailed by a strange aircraft with no known origin, which he said had followed his plane for miles.

None of the three senators wanted a public debate on the Senate floor about the funding for the program, Mr. Reid said. *"This was so-called black money,"* he said. *"Stevens knows about it, Inouye knows about it. But that was it, and that's how we wanted it."* Mr. Reid was referring to the Pentagon budget for classified programs.

"Internationally, we are the most backward country in the world on this issue," **Mr. Bigelow** said in an interview. "Our scientists are scared of being ostracized, and our media is scared of the stigma. China and Russia are much more open and work on this with huge organizations within their countries. Smaller countries like Belgium, France, England and South American countries like Chile are more open, too. They are proactive and willing to discuss this topic, rather than being held back by a juvenile taboo."

By 2009, **Mr. Reid** decided that the program had made such extraordinary discoveries that he argued for heightened security to protect it. "Much progress has been made with the identification of several highly sensitive, unconventional aerospace-related findings," Mr. Reid said in a letter to **William Lynn III**, a deputy defense secretary at the time, requesting that it be designated a "restricted special access program" limited to a few listed officials.

A 2009 **Pentagon** briefing summary of the program prepared by its director at the time asserted that "what was considered science fiction is now science fact," and that the United States was incapable of defending itself against some of the technologies discovered. Mr. Reid's request for the special designation was denied.

Mr. Elizondo, in his resignation letter of Oct. 4, said there was a need for more serious attention to "the many accounts from the Navy and other services of unusual aerial systems interfering with military weapon platforms and displaying beyond-next-generation capabilities." He expressed his frustration with the limitations placed on the program, telling Mr. Mattis that "there remains a vital need to ascertain capability and intent of these phenomena for the benefit of the armed forces and the nation."

Mr. Elizondo has now joined **Mr. Puthoff** and another former Defense Department official, **Christopher K. Mellon,** who was a deputy assistant secretary of defense for intelligence, in a new commercial venture called **To the Stars Academy of Arts and Science.** They are speaking publicly about their efforts as their venture aims to raise money for research into U.F.O.s.

In the interview, Mr. Elizondo said he and his government colleagues had determined that the phenomena they had studied did not seem to originate from any country. *"That fact is not something any government or institution should classify in order to keep secret from the people,"* he said.

For his part, Mr. Reid said he did not know where the objects had come from. *"If anyone says they have the answers now, they're fooling themselves,"* he said. *"We do not know."* https://www.nytimes.com/2017/12/16/us/politics/pentagon-program-ufo-harry-reid.html

The History Channel's "Unidentified: Inside America's FO Investigation" TV Series

If anyone has seen the new **History Channel** TV series *"Unidentified: Inside America's UFO Investigation",* they will know that it is about "The Pentagon's Mysterious UFO Program", a clandestine U.S. government program that had been investigating UFOs. For eight years, the secret program was run by **Luis Elizondo**, the main star of the docu-entertainment series with occasional appearances by other members of To The Stars Academy.

The **History Channel** show is being promoted as *"ground-breaking nonfiction,"* goes on to follow Elizondo as he re-investigates strange UFO incidents he says he learned of when he was at the Pentagon running the **Advanced Aerospace Threat Identification Program (AATIP).**

Whatever the truth about otherworldly UFOs (cue a collective eye-roll from scientists), there is one crucial detail missing from "Unidentified," as well as from all the many stories that have quoted Elizondo since he outed himself nearly two years ago to a wide-eyed news media: ***"There is no discernible evidence that he ever worked for a government UFO program, much less led one."***

Yes, AATIP existed, and it *"did pursue research and investigation into unidentified aerial phenomena,"* Pentagon spokesperson Christopher Sherwood stated. He added: *"Mr. Elizondo had no responsibilities with regard to the AATIP program while he worked in* **OUSDI (**the **Office of Under Secretary of Defense for Intelligence)***, up until the time he resigned effective 10/4/2017."*

The email was sent over a year ago by **Kari DeLonge**, a public relations representative for To the Stars, to **John Greenewald**, a UFO researcher who runs an online archive of **Freedom of Information Act** obtained government documents on a website called the **Black Vault**. At the time, Greenewald had become frustrated at the lack of tangible information about AATIP and Elizondo's role; additionally, Elizondo had spurned Greenewald's interview requests.

"Hi John – Thanks for reaching out," DeLonge wrote. *"The program was initially run out of [the Defense Intelligence Agency] but when Lue took it over in 2010 as Director, he ran it out of the Office for the Secretary of Defense (OSD) under the Under Secretary of Defense for Intelligence (USDI). Hope that clarifies."*

Greenewald tried contacting Elizondo multiple times via email and his cellphone. He has not responded. It's not as if he is on retreat somewhere; I noticed that in the run-up to his star turn on the new History Channel show, he has been speaking to everyone from the New York Times to UFO media personalities and military bloggers.

Indeed, judging by all the UFO stories lighting up the internet this week, the self-described "career spy" is having another big moment in the media spotlight. The timing is either an auspicious coincidence or the "flying saucers are here" brigade's well-oiled PR machine is working overtime.

Mellon, like Elizondo, works for To the Stars neglected to mention this connection, along with the fact that *the History Channel show was made by the company Elizondo and Mellon work for.*.

who Luis Elizondo, **Chris Mellon** are like the A-Team of former, high-level government officials who still have security clearances, still have networks in Washington, still are in the business, if you will.

"We know that UFOs exist," **Chris Mellon**, a deputy assistant secretary of defense for intelligence in the Clinton and George W. Bush administrations, pronounced on the show. *"This is no longer an issue. The issue is why are they here? Where are they coming from? And what is the technology behind these devices that we are observing?"*

In the UFO community, in which conspiracy theories have flourished like a mutating virus, some suspect that DeLonge is being played like a useful idiot or patsy — and that his To the Stars Academy is a front for some kind of black ops project.

DeLonge's connection with Mellon and Elizondo is an odd and certainly makes no sense and is bizarre when you consider the interview he did with **Joe Rogan** on his podcast show where DeLonge claimed to be the military's chosen vessel for UFO disclosure.

Author's Rant: This is fascinating when considering that Joe Rogan has had the Father of Disclosure, Dr. Steven Greer on his program several times. Yet, Dr. Greer will tell you that no military or government senator or politician has come knocking at his door to offer their services or to bankroll his ongoing efforts to get the word out to America or the world. Much of Dr. Greer's success has come from the average citizens support, who believe that he is in the know on the UFO and ETI subject and that he has the ears of high-ranking military, government officials, royalty, scientists, and various religious orders, etc.

Long time UFO researchers have been troubled finding out what the program exactly did, as well as the scope of Elizondo's role. FOIA requests were also turning up dry.

Elizondo alludes to internal opposition at the Pentagon to investigate UFOs that he wrote had menaced Navy Pilots and posed an "existential threat to our national security." He was leaving, he strongly implied in his letter, because the Pentagon wasn't taking that threat seriously.

Even the much-ballyhooed Navy's UFO/UAP grainy video footage of tiny, darting objects, combined with Elizondo's claims of *"compelling evidence"* for the *"phenomena"* that he couldn't identify for some made great television. The problem is that the there are much better video in the public sector and even the Navy and the Air Force has much clear videos and photographs of UFOs but these will never see the light of day without an act of God forcing their hand!

On the Times's podcast, "The Daily," **Helene Cooper,** the newspaper's Pentagon correspondent, described Elizondo as a *"spooky, secretive guy"* but added that he was *"completely credible."* He showed her documents, pictures, and military videos of potential UFOs, which appeared fantastic to her, but also persuasive. *"I did believe him,"* Cooper said on the podcast. *"It seemed completely credible to me in the moment."*

Later on,, after she left the hotel room, Cooper acknowledged that doubts crept in. In the end, though, she decided that what mattered most was whether the Pentagon's UFO program was real. https://theintercept.com/2019/06/01/ufo-unidentified-history-channel-luis-elizondo-pentagon/

He offered the following statement regarding the footage release:

"In these uncertain times, it is more important than ever that governments speak the truth to their citizens, as it is the most important pillar of a Democracy. We commend the leadership at the Department of Defense for sharing the truth and TTSA is optimistic that they will continue to share more information transparently as it becomes publicly available."

"At To The Stars Academy, we do not fear skepticism and will continually work to decrease doubt around the topic of the existence of UAPs. We are fueled by the Pentagon's significant actions and hope this encourages a new wave of credible information to come forward. We acknowledge and stand by those who have the courage to always speak the truth, no matter how difficult or complex the issue is."

Elizondo is a principal cast member of HISTORY's hit nonfiction series "**Unidentified: Inside America's UFO Investigation,**" which reached 19.1 million people during season one. Elizondo appeared alongside **Commander David Fravor**, U.S. Navy (Ret), Commanding Officer of Strike Fighter Squadron 41, who encountered the UAP in the Navy's declassified footage. Throughout the series, Elizondo and his colleagues are profiled for their work at To the Stars Academy of Arts & Science collaborating with public and private entities around the world to shine a spotlight on the truth and produce tangible evidence to build the most indisputable case for the existence and threat of UAPs ever assembled. Season two of the series is set to return this summer and will continue to follow Elizondo and the TTSA team as they further their work with retired and active military to expose **Unidentified Aerial Phenomena** sightings worldwide.

To The Stars Academy Team Members

Here are the senior team members in the To The Stars Academy which include **Tom DeLonge, Jim Semivan, Dr. Hal Puthoff, Steve Justice, Luis Elizondo, Chris Mellon,**

Dr. Garry Nolan, Dr. Paul Rapp, Dr. Norm Kahn, Dr. Colm Kelleher, and Dr. Adele Gilpin. There are other board members in a lesser role and these include **J. Christopher Mizer** and **C. Chris Herdon.**

Pay attention to the short bios of each senior member and note their involvement with intelligence organizations and/or covert intelligence operations which are highlighted.

Tom DeLonge
President & CEO

Tom DeLongeis the co-founder., President and interim CEO of To The Stars Academy of the Arts and Sciences. As President of the company's subsidiary, To The Stars, since 2015, he was primarily engaged within the entertainmwnt sector. His career spans over two decades, selling over 25 million records worldwide with the bands he founded, Blink182 and Angels & Airwaves. Prior to forming the TTS Academy, Mr. DeLonge do-founded Really Likeable People. (RLP),the parent company of international cosumer lifestyle brands including Atticus Clothing and Macbeth Footwear, and technology monetization platform, Modlife. Mr. DeLonge has taken his award-winning creative content that spans music, books, and film and built To The Stars Inc. as a vertically integrated entertainment business that develops, produces and distributes multi-media and merchandise world-wide.

Jim Semivan
Vice President Operations

Jim Semivan is the co-founder and Vice President Operations of TTS Academy. In 2007, Mr. Semivan founded the conssulting firm, Jim Sem LLC, after his retirement as a senior intelligence service member of the Central Intelligence Agency. Since retirement, Mr. Semivan primarily worked as a consultant for the Intelligence Community (IC) on classified topics including IC leadership training, CIA tradecraft training and IC programs for countering weapons of mass destruction. Mr. Semivan retired from the Central Intelligence Agency's Directorate of Operations after 25 years as an operations officer, both overseas and domestically. Mr. Semivan holds a bachelor's degree from The Ohio State University and a M.A. from San Francisco State University.

https://www.auforg.ca/recent-posts/disclosure/

404

Dr. Hal Puthoff
Vice President Science & Technology

Dr. Hal E. Puthoff is the co-founder and Vice President of Science and Technology of TTS Academy. Since 1985, Dr. Puthoff has served as President and CEO of Earth Tech International, Inc. (ETI), and Director of the Institue for Advanced Studies at Austin (IASA). He has published numerous papers on electron beam devices, lasers, communications, and energy fields. Dr. Puthoff's professional background spans more than five decades of research at General Electiric, Sperry, the National Security Agency, Stanford University and SRI International. Dr. Puthoff regularly advises NASA, the Department of Defense and intelligence communities, corporations and foundations on leading edges technologies and future technology trends. He earned his Ph.D. from stanford University in 1967 and won Who's Who Lifetime Achievement in 2017 that recognizes individuals that have achieved greatness in their industry and have excelled in their field for at least 30 years.

Steve Justice
Areospace Division Director

Steve Justice is the Director of the TTS Academy Division, tasked with leading the effort to examine the possibilities of emerging sciences and technologies. This team will work to define advanced systems exploiting radical technologies, prototypes promising concepts, and develop operational systems that shatter conventional thinking. He entered the defense aerospace industry in 1978 after graduating from Georgia Institute of Technology. After 31 years, Stephen is the recently retired Program Director for avanced Systems from Lockheed Martin Advanced Development Programs - better known as the "Skunk Works". Stephen's industry experience brings to TTS Academy a deep understanding of strategy definition, breakthrough technology development, advanced concept design, prototyping, system fielding, and program planningand execution using a leadership style that inspires innovation.

https://www.auforg.ca/recent-posts/disclosure/

Luis Elizondo
Director of Global Security & Special Programs

Luis Elizondo is a career inteligence officer whoise experience includes working with the U.S. Army, the Department of Defense, the National Counterintelligence Executive, and the Director of National Security. As a former Special Aggent In-Charge, Luis conducted and surpervised highly sensitive espionage and terrorism investigation around the world. As an Intelligence Case Officer, he ran clandestine source operations throughout Latin America and the Middle East. Most recently, Luis managed the security for certain sensitive portfolios for the U.S. Government as the Director for the National Programs Special Management Staff. For nearly the last decade, Luis also ran a sensitive aerospace threat identification Program focusing on unidentified aerial technologies. Luis' academic background includes Microbiology, Immunology and Parasistology, with research experiencenin tropical diseases, Luis is also an inventor who hold several patterns.

Chris Mellon
National Security Affairs Advisor

Christopher Mellon is aprivate equity investor, political commentator and the Chair of there Science Committee at the Carnegie Museum of Natural History. He served 20 years in the federal Government, including as the Deputy Assistant Secretary of Defense for Intelligence in the Clinto and Bush Administrations. In addition, he worked many tears on Capitol hill including as the Minority Staff director of the Senate Select committee on Intelligence. As an aide to Senator William s. Cohen, he drafted the legislation that established the US Special Operations Command. He is the author of numerous articles on politics and national security, and the recipient of multiple awards from the Department of Defense and agencies of the US Intelligence Community. He holds a B.A. in economics from Colby College and an M.A. in international affairs from Yale University.

https://www.auforg.ca/recent-posts/disclosure/

406

Dr. Garry Nolan
Genetics Technology Consultant

Dr. Garry Nolan is a Rachford and Carlota A. Harris Professor in the Department of Microbiology and Immunology at Stanford University School of Medicine. He trained with Leonard Herzenberg (for his Ph.D. and Nobel recipient Dr. David Baltimore (postdoctoral work). He has published more than 200 research articles and is holder of 20 US patents, has been honored as one of the top 25 inventors at Stanford University and is the first recipient of the Teal Innovator Award (2012) from the Department of Defense. Dr. Nolan was the founder and has served on the Board of Directors for several successful biotechnology companies. He holds a B.S. in genetics from Cornell University, a Ph.D. in gentics from Stanford University.

Dr. Paul Rapp
Brain Function & Consciousness Consultant

Dr. Paul Rapp is a Professor of Military and Emergency Medicine ath the Uniformed Services University and Director of the Traumatic Injury Research Program. He also holds a secondary appointment as a Professor of Medical and Clinical Psychology. He is a past editor of Physica and served on the editorial boards of the International Journal of Bifurcation and Chaos, Chaos and Complexity Letters, and cognitive Neurodynamics. His past honors include a Certificate of Commendation from the Central Intelligence Agency for significant contributions to the missions of the Office of Research and Development. Dr. Rapp attended the University of Illinois and earned degrees in physiology (minor in Chemistry, Summa cum Laude) and engineering physics (Summa cum Laude). He received a Ph.D. from Cambridge University, working under the supervision of Professor Sir James Lighthill in the Department of Applied Mathematics and Theoretical Physics.

https://www.auforg.ca/recent-posts/disclosure/

Dr. Norm Kahn
National Security &Program Management Consultant

Dr. Norm Kahn is currently a consultant on national security for the US Government,with a focus on preventing the use of biological weapons of mass destruction/disruption. Dr. Kahn has over 30-year career with the Central Intelligence Agency, culminating in his developement direction of the Intelligence Community's Counter-Bilogical Weapons Program. Dr. Kahn is the recipient of the Agency's Distinguished Career Intelligence Medal and the Director of National Intelligence's National Intelligence Distinguished Service Medal.. Dr. Kahn holds a B.S. degree in bilogy from the City College of New York and a Ph.D. in oceanography from the University of Rhode Island.

Dr. Colm Kelleher
Biotech Consultant

Dr. Colm Kelleher is a biochemist with a twenty-eight year research career in cell molecular biology currently working in senior management in the aerospace sector. He served as Laboratory Director at biotech company, Prosetta Corporation, leading serveral small molecule drug discovery programs focused on viruses of interest to the US Department of Defense. He worked for eight years as Deputy director of the National Institute for Discovery Sciences (NIDS), a research organiztion using Forensic science methodology to unrevel scientific anomalies. From 2008 -2011, he served as Deputy Administrator of a US government funded threat assessment focused on advanced aerospace technology. Dr. Kelleher has authored more than forty peer reviewed scientific articles in the cell and molecular biology, immunology and virology as well as two best selling books, " Hunt for the Skinwalker" and "Brain Trust". He holds a Ph.D. in biochemistry from the University of Dublin, Trinity College.

https://www.auforg.ca/recent-posts/disclosure/

408

https://www.auforg.ca/recent-posts/disclosure/

It should be obvious to the reader the connections that most of these people have with either one or with both the **US Department of Defense** and the **Central Intelligence Agency** holding either a directorship, a certificate, a commendation or a medal for distinguished service in those service branches.

Each person is highly intelligent with University degrees from one or more universities or from medical colleges specializing in various medical disciplines or in the aerospace or oceanographic sciences or having direct involvement in intelligence agency operations either within the US or overseas in Latin America or the Middle East.

All, except for one person, **Tom Delonge**, who is in the entertainment business as a rock musician writing music or in writing books, producing movies and promoting clothing. His main interest has been in UFOs since he was a child and this interest grew more intense after spending some time with Dr. Greer and being trusted by Dr. Greer to look after a copy of tapes on disclosure Witnesses. This may have made him an easy target to be coerced with inducement to feed his ego, both in the entertainment business but also, his interest in Ufology and his close association with Dr. Greer and his work.

It would strongly appear to any seasoned UFO researcher that DeLonge was being unwittingly played as a patsy and coached into the negative narrative of potential hostile ETI piloting UFOs from other star systems with a hostile intent against humanity or at least against the military.

DeLonge has come out on the **Joe Rogan** podcast show as stating that he was in the position of being the military's emissary to disclose the reality of UFOs to humanity. This boast by Delonge ignores the fact that there is a growing global network of CSETI and CE-5 groups who are already ambassadors for humanity. DeLonge also ignores the information that Dr. Greer has postulated that ETI are essentially benevolent or neutral and that very few are actually hostile and these particular ETI are not permitted to interfere with human affairs or its evolution.

So, why is **Tom DeLonge** going down this negative narrative path? Why is he in support of the military position that these **UFOs/UAPs** area potential threat as proclaimed by the Navy's **Advanced Aerospace Threat Identification Program (AAATIP)?** Does DeLonge know something about UFOs that the rest of the US populace and the world doesn't know or is he been spoon-fed only what the military wish for him to know?

Close scrutiny of the military's negative narrative on this show a consistency of lies, cover-ups, suppression of information, all in a propaganda campaign of public deception that dates back to World War Two when **Allied Forces and Axis Powers** were reporting **Feu (Foo) Fighters** following and interacting with their aircraft and naval ships. To such an extent were these reports that **General Douglas MacArthur** reported his finding to then **President Roosevelt** and even **Winston Churchill** and **General D. Dwight Eisenhower** conferred many times on this strange phenomenon fearing that the release of such news would put more undue stress onto an already fearful public stemming from a world war.

What does all this stuff about flying saucers amount to?   What can it mean?   What is the truth   Let me have a report at your convenience.

W.S.C.

A letter from Winston Churchill to the Secretary for Air, dated July 28, 1952, requesting an explanation on flying saucers. *The National Archives UK and* https://www.history.com/news/uk-ufos-mainbrace-nato

They felt the best course of action was not to tell the public within their nations but instead to track reports coming in from pilots and naval servicemen and to investigate this strange Feu Fighter phenomenon. Under the establishment o**f General Douglas MacArthur**, the secret **Interplanetary Phenomenon Unit** was organized to compile reports and assessing the situation. This was the beginning of a true UFO investigation under military control.

To place things into perspective, for why the Interplanetary Phenomenon Unit was instituted, here is a couple of examples of what the Navy really knows about UFOs:

The USS New York UFO Incident of 1945

 After a successful invasion of Iwo Jima n February 1945, the **USS New York** returned home for a refit and once again, it was deployed along with two destroyers to the Pacific for the massive invasion of Okinawa in April of 1945. However, between March and that time, near New Guinea, off the Admiralty Islands, the radar room picked up unidentified flying object headed their way. Moments later the object appeared over the USS New York as extremely shiny and silver in colour which was witnessed by 2000 sailors and marines, including the ship's commanding officer **Captain K.C. Christian.** This was a daylight sighting.

As the ship moved at 12 knots, so did the strange object keeping pace with the ships' movements, ruling out the possibility of it being a plane. Everyone was in amazement thinking maybe, it was possibly a secret Japanese weapon. Then, Captain Christian ordered the ship's gun to open fire on the object fearing a security risk. But the non-hostile object appeared to be completely unaffected by the furious gunfire. The Captain ordered a ceasefire seeing that they may be wasting artillery on the strange object that mysteriously remained unscathed by the naval gunfire.

As soon as the guns ceased their firing, the object quickly disappeared leaving everyone below was astonished by what had just taken placed! **"UFOs in Wartime What They didn't Want You to Know"; 2011; Published by Berkley Publishing Group a division of the Penguin Group, New York, New York; ISBN: 978-0-425-24011-3**

Exercise (Operation) Mainbrace of 1952

The second naval incident with UFOs which will only be recounted here in brief as it is discussed elsewhere under Naval encounter with UFOs, was the first large scale massive naval exercise known as the **Exercise Mainbrace (aka. Operation Mainbrace).**

Exercise Mainbrace was undertaken by the newly established **Allied Command Atlantic (ACLANT)**, one of the two principal military commands of the **North Atlantic Treaty Organization (NATO)**. It was part of a series of NATO exercises jointly commanded by **Supreme Allied Commander Atlantic Admiral Lynde D. McCormick, USN,** and **Supreme Allied Commander Europe General Matthew B. Ridgeway, U.S. Army,** during the Fall of 1952.

With over 200 naval ships (Aircraft Carriers, Battleships, Cruisers, Escorts, MCM, Submarines, Torpedo Boat Squadrons, Motorized Naval Trawlers) with 700 aircraft, 1000 aircraft and 80,000 sailors and soldiers from nine NATO nations along with involvement by other countries.
https://www.history.com/news/uk-ufos-mainbrace-nato

The UFO sightings incidents interrupted **Exercise Mainbrace** and forced the UK to take 'Flying Saucers' more seriously.

One of three photos taken by Metropolitan Group news photographer Wallace Litwin on September 20, 1952 while aboard the USS Roosevelt during Operation Mainbrace.
https://www.saturdaynightuforia.com/html/articles/articlehtml/itn52part43.html

The first Mainbrace encounter came on September 13 when the captain and crew of a Danish destroyer spotted a triangular-shaped object moving through the night sky at alarming speeds.

412

The unidentified craft emitted a blue glow and was estimated by **Lieutenant Commander Schmidt Jensen** to be traveling upward of 900 miles per hour.

On September 20, an American newspaper reporter named **Wallace Litwin** was aboard the **USS Franklin D. Roosevelt**, an aircraft carrier participating in the Mainbrace exercises, when he saw a commotion on deck: several pilots and flight-crew members pointing at a silver sphere in the sky that appeared to be following the fleet. Litwin quickly shot four color photos of the round object, which he assumed was a weather balloon.

Some years later in a letter to a UFO investigator, the ship's executive officer, informed Litwin that no weather balloons had been released that day. The officer then radioed the Midway, the only other ship in the vicinity, which also confirmed that no weather balloons were in the air or unaccounted for.

Later this story would become a part of **Captain Edward J. Ruppelt** and the **Project Blue Book** team followed up with the Navy and interviewed members of the flight-deck crew.
https://www.history.com/news/uk-ufos-mainbrace-nato

It would also come under investigation in the Colorado University, USAF sponsored **Scientific Study on Unidentified Flying Objects** headed up by **Edward U. Condon.**

There were so many UFOs sightings witnessed by the 80,000 naval sailors occurring over the vast international naval armada that UFOs were seen emerging out of the ocean, flying over the ships and then zipping off toward space. Sailors were snapping photographs such a degree that the commanders of various ships ordered the men to stay below deck unless they had duties on deck that required their presence.

When the US naval ship eventually returned home most film taken by sailors were confiscated and everyone was told not to leak out information to anyone, or to reporters or the families and friends. They were even told not to speak about the Naval exercise or the UFO sighting amongst themselves!

The evidence that the US Navy and the US Air Force including Britain and most of the international participants in Operation Mainbrace were very much aware of the existing UFO phenomenon and had determined that it was of an extraterrestrial nature. There has been numerous military study programs in place during and since that time period, and the fact is that when the Navy stated it only had a short span of five years when the secret investigation ended in 2012, this is an out-right, bold-face LIE!

But hey, did any serious UFO researcher think that this would be otherwise? Were we really believing that the military was becoming completely transparent on what it knows about the UFO phenomenon? Is this the beginning of a carefully controlled public release of information on the reality that UFOs (aka, ETVs) are in fact extraterrestrial in origin? Or is there a far greater sinister agenda afoot that we may soon see unleashed on an unsuspecting humanity?

Time will tell!

CHAPTER 48

AMERICAN, SOVIET AND CHILEAN NAVAL ENGAGEMENTS WITH UFOS

Operation Mainbrace (NATO Naval Exercise) – September 1952

Exercise Mainbrace was the first large-scale naval exercise undertaken by the newly established **Allied Command Atlantic (ACLANT),** one of the two principal military commands of the **North Atlantic Treaty Organization (NATO).** It was part of a series of NATO exercises jointly commanded by Supreme Allied Commander AtlanticSt Admiral Lynde D. McCormick, USN, and Supreme Allied Commander Europe General Matthew B. Ridgeway, U.S. Army, during the Fall of 1952.

The strategic importance of control of Norway and the adjacent Norwegian and Barents seas was recognized by Anglo-American naval planners as early as the First World War. The invasion and occupation of Norway by Nazi Germany Norway during World War II confirmed the importance of the region, as Germany was able to establish bases for submarine and air operations against Allied convoys bound for the Soviet seaport of Murmansk.

Following the Second World War, several former allied navies executed a number of individual and multilateral naval exercises, including:

- **Operation Frostbite,** a 1946 naval exercise involving U.S. Navy Task Group 21.11 led by the aircraft carrier USS *Midway* (CVB-41) that operated in the Davis Straits between Labrador and Greenland;
- **Exercise Verity**, a 1949 combined naval exercise involving the British, French, and Dutch navies which carried out naval bombardment, convoy escort, minesweeping, and Motor Torpedo Boat attack evolutions;
- **Exercise Activity**, a 1950 Dutch-led naval exercise to refine combined communications and tactical procedures; and
- **Exercise Progress**, a 1951 French-led combined naval operation with Belgian, French, Danish, Dutch, Norwegian, and British naval units that participating in antisubmarine warfare operations, air defense maneuvers, minesweeping operations, and convoy exercises.

Initially with his "long telegram" and subsequently in his article "The Sources of Soviet Conduct" that appeared in the July 1947 issue of *Foreign Affairs*, American diplomat **George F. Kennan** argued that **Joseph Stalin** would not (and moreover could not) moderate the determination of the Soviet Union to undermine and overthrow Western governments following World War Two, noting:

"... the main element of any United States policy toward the Soviet Union must be a long-term, patient but firm and vigilant containment of Russian expansive tendencies... Soviet pressure against the free institutions of the Western world is something that can be contained by the adroit and vigilant application of counterforce at a series of constantly shifting geographical and

political points, corresponding to the shifts and 415 expansion of Soviet policy, but which cannot be charmed or talked out of existence."

Kennan's advocacy for a policy of containment against the Soviet Union formed the basis of American foreign policy during the **Cold War**, giving rise to the **Truman Doctrine**, the **Marshall Plan**, and the formation of **NATO**.

The Soviet Union characterized **Operations Mainbrace and Holdfast,** and other NATO military exercises as "war-like acts" by NATO, with particular reference to the participation of Norway and Denmark, while the USSR was preparing for its own military maneuvers in the Soviet Zone. http://en.wikipedia.org/wiki/Exercise_Mainbrace

Operation Mainbrace part of the Cold War (1947–1953) NATO Exercises
http://www.thinkaboutit-ufos.com/operation-mainbrace/

Author's Rant: Here, we see the beginning of the Cold War between Democracy and Communism and the gradual expansion of American foreign policies toward other nations, acting as if it had been bestowed with the mantle of being the world's police force to protect and fight against communism and global terrorism, all the while strongly voicing the values of democracy and the free-enterprise system while maintaining American global interests abroad.

In a world emerging from the devastation of two world wars in less than twenty-five years, people wanted to get back to rebuilding their countries. Democratic nations were reluctant to allow the United States, who had come out more powerful after WWII, to act as policemen to fight the growing communist threat which had now invaded the East Bloc of Europe. Such unofficially sanctioned world police force operated by one nation without the checks and balances by other nations would come with a high price in cold war tensions while permitting American self-interests to take precedence over the interests of other nations. Thus, this would inevitably create suspicion and distrust toward a post-war American government, its leaders, and its foreign policies which were ultimately controlled and manipulated by the wealthy corporate elite.

Jim Marrs may have said it best in his book, "The Rise of the Fourth Reich" that *"Germany may have been defeated but, Nazism won the war"* (and is alive and well in America!) Where Hitler was openly arrogant and falsely confident that his political war machine was invincible, yet still lost the war, in America that same transplanted Nazi agenda found renewed life, steadily advancing in a more methodical covert development.

Despite what the rest of the world thought during the Cold War years, America was on a roll and unbeknownst to its people, she had an agenda!

Operation Mainbrace was conducted over twelve days between September 14–25, 1952, and involved nine navies: United States Navy, the British Royal Navy, French Navy, Royal Canadian Navy, Royal Danish Navy, Royal Norwegian Navy, Portuguese Navy, Royal Netherlands Navy, and Belgian Naval Force operating in the Norwegian Sea, the Barents Sea, the North Sea near the Jutland Peninsula, and the Baltic Sea. Its objective was to convince Denmark and Norway that those nations could be defended against attack from the Soviet Union. The exercise featured simulated carrier air strikes against "enemy" formation attacking NATO's northern flank near Bodø, Norway, naval air attacks against aggressors near the Kiel Canal, anti-submarine and anti-ship operations, and U.S. marines landing in Denmark.
http://en.wikipedia.org/wiki/Exercise_Mainbrace

With a demonstration of such military presence and strength in the Northern Atlantic, an area that had witnessed more wartime carnage and suffering during the Second World War than any other area on the planet, such an exercise once again drew the attention of the Extraterrestrial presence. To the ETI who shadowed this naval operation, this may have looked like the beginnings of another global conflict!

What follows is a collection of UFO reports were reported in the vicinity of the **"Operation Mainbrace"** NATO maneuvers that were held in September 1952. The maneuvers commenced September 13th and lasted twelve days. According to the U. S. Navy, "units of eight NATO governments and New Zealand participated, including 80,000 men, 1,000 planes, and 200 ships; in the vicinity of Denmark and Norway." The operation, directed by **British Admiral Sir Patrick Brind**, "was the largest NATO maneuver held up until that time."

- September 13 – The Danish destroyer Willemoes, participating in the maneuvers, was north of Bornholm Island. During the night, **Lieutenant Commander Schmidt Jensen** and several members of the crew saw an unidentified object, triangular in shape, which moved at high speed toward the southeast. The object emitted a bluish glow. Commander Jensen estimated the speed at over 900 mph.

  Within the next week, there were four important sightings by well-qualified observers. (Various sources differ by a day or two on the exact dates, but agree on details. There is no question about the authenticity of the sightings; the British cases were officially reported by the Air Ministry, the others are confirmed by reliable witnesses. All occurred on or about September 20).

- September 19 – A British Meteor jet aircraft was returning to the airfield at Topcliffe, Yorkshire, England, just before 11 A. M. As it approached for landing, a silvery object was observed following it, swaying back and forth like a pendulum. **Lieutenant John W. Kilburn** and other observers on the ground said that when the Meteor began circling, the UFO stopped. It was disk-shaped and rotated on its axis while hovering. The disk suddenly took off westward at high speed, changed course, and disappeared to the southeast.

- About September 20 – Personnel of the **U.S.S. Franklin D. Roosevelt,** an aircraft carrier participating in the Mainbrace maneuvers, observed a silvery, spherical object which was also photographed. (The pictures have never been made public). The UFO was seen moving across the sky behind the fleet. Reporter Wallace Litwin took a series of color photographs, which were examined by Navy Intelligence officers. The Air Force project chief, **Captain Ruppelt** stated: "[The pictures] turned out to be excellent...judging by the size of the object in each successive photo, one could see that is was moving rapidly." The possibility that a balloon had been launched from one of the ships was immediately checked out. No unit had launched a balloon. A poor print of one of the photographs appears in the Project Blue Book files, but with no analysis report.

- September 20 – At Karup Field, Denmark, three Danish Air Force officers sighted a UFO about 7:30 P.M. The object, a shiny disk with metallic appearance, passed overhead from the direction of the fleet and disappeared in clouds to the east.

- September 21 – Six British pilots flying a formation of RAF jets above the North Sea observed a shiny sphere approaching from the direction of the fleet. The UFO eluded their pursuit and disappeared. When returning to base, one of the pilots looked back and saw the UFO following him. He turned to chase it, but the UFO also turned and sped away.

- September 27/28 – Throughout Western Germany, Denmark, and southern Sweden, there were widespread UFO reports. A brightly luminous object with a comet-like tail was visible for a long period of time moving irregularly near Hamburg and Kiel. On one occasion, three satellite objects were reported moving around a larger object. A cigar-shaped object moving silently eastward also was reported.
  http://www.nicap.dabsol.co.uk/mainbrace.htm source: Richard Hall

During Operation Mainbrace, the aircraft carrier, USS *Franklin D. Roosevelt* (CVB-42) was buzzed several times by flying saucers at close range as were other ships in the naval exercise. Many sailors and airmen on board these ships in this international naval operation reported these objects at close range during the two weeks at sea.

In one event, there had been a "sonar contact" by three or four destroyers which was an unknown to the fleet which was moving with them and it was determined to be a quarter mile wide (that is over 1250 feet in diameter)! It then came out of the water and "dinged" one of the ships and flew off into the sky and out of sight! In an official cover-up of the event, it was later reported as a collision between two destroyers which is not what happened according to one witness on board the USS *FDR* who reported the event to one of the History Channel's "Alien Hunters" investigators. http://www.youtube.com/watch?v=tBXOBAiglIc

USS Franklin D. Roosevelt (CVB-42) with Carrier Air Group 4 (CVG-4)
as it may have looked during Operation Mainbrace in 1952
http://www.navsource.org/archives/02/42.htm

It was even reported that General Dwight Eisenhower was on board the USS *FDR* as a guest and saw one of these objects first hand for himself. Some reports have been released into the public

domain regarding Operation Mainbrace, but many more remain classified.
http://www.metacafe.com/watch/576930/operation_mainbrace_ufos/

A recent update has come out on January 28, 2012, that states that it may be no coincidence that flying saucers were monitoring the Mainbrace Exercise and in particular, the Franklin D Roosevelt aircraft carrier which may have been the first and only aircraft carrier or ship in the 1950s, equipped with nuclear weapons! http://www.realufos.net/2012/01/operation-mainbrace-ufo-encounters-1952.html

Newspaper clipping of NATO maneuvers, code name "Operation Mainbrace"

Many of the first nation's militaries continue to keep a tight lid on the subject regarding these objects over the last 60 years because they treat them as threats. This is particularly apparent when these objects can penetrate our airspaces with almost virtual impunity when they can go into secure areas and out-fly our fastest jets then, it no wonder these unidentified flying objects could be viewed as being potentially hostile by the military.

From the Sunday Dispatch:

"FLYING SAUCER" JOINS IN "MAINBRACE"

"A Flying saucer entered exercise Mainbrace today following a report from the RAF station at Topcliffe (York's), of a white object having been seen, which accelerated at a speed in excess of a shooting star.
RAF officers at the Mainbrace headquarters at Pitreavie (Fifeshire) would not state what view was being taken of the report.
An RAF spokesman refused to comment himself whether the report was regarded as a serious one. He added, "It is being investigated."
The report led to a NAAFI saucer being placed on the plot of the exercise. The saucer, fastened with ribbons, was placed on the plot alongside the name Topcliffe, and the RAF duty controller at Pitreavie open up a new signal file, headed "SAUCER SIGHTINGS AND MOVEMENTS," where he could file further sightings of the object.

Five Miles Astern
The object was seen by about 10 RAF officers and men, the crews of Shackleton aircraft operating from Topcliffe. The signal reporting what they had seen was passed by the intelligence officer at Topcliffe to the maritime headquarters at Pitreavie and to coastal command headquarters at Northwood, Middlesex. The signal said that the object was seen at 10:53 am yesterday when a Meteor aircraft was flying at 5000 ft and was descending. A white object was seen five miles astern of the aircraft at approximately 15,900 ft.

Silver in Colour
It was moving at a comparatively slow speed on a course similar to that of the aircraft. The object was silver in colour and circular. It maintained a slow forward speed before beginning to descend, swinging like a pendulum.

Not a Parachute
It was thought by those who saw it to be a parachute or cowling from a meteor, but an RAF spokesman said today that no cowling or parachute had fallen in the vicinity of the station. The signal reported that the aircraft turned towards Dishforth and the object, while still descending appeared to follow.

Incredible Speed
It then began a rotary motion about its own axis, but it suddenly accelerated at incredible speed in a westerly direction, then turned to a south-easterly course.
Those who saw it stated that its movements were not identifiable with anything they had seen in the air and that the acceleration was in excess of that of a shooting star.

420

The duration of the incident was between 15 and 20 seconds. The object was seen by Flt.-Lt. Kilburn and Flt.-Lt. Cybulski, both captains of aircraft, by FO Paris, Master Signaller Thompson, and by about 6 other aircrew members.”
http://www.aliensthetruth.com/UFO_sightings_famous.php?view=1&ID=33

Sunday Dispatch

151st Year. No. 7,872.   2½d.   SEPTEMBER 21, 1952.

What Intruded Into 'Exercise Mainbr

'SAUCER' CHASED R JET PLANE Say 6 Airmen

NOT Smoke-Ring Or Weather Balloon, Says Pilot

By Sunday Dispatch Reporter

SERIOUS investigation was being made last night by the R.A.F. into the mystery of a silvery-white object that chased a Meteor jet-plane over Yorkshire during " Exercise Mainbrace."

It was seen by two R.A.F. officers and three aircrew as they stood near Coastal Command Shackleton Squadron H.Q. at Topcliffe.

They had just landed after a flight and were watching a Meteor coming in to land at the neighbouring Dishforth R.A.F. station.

One of them, Flight Lieut. John W. Kilburn, 31, of Egremont, Cumberland, then spotted "something different from anything I have ever seen in 3,700 hours flying in a variety of conditions."

He told me last night:

"It was 10.53 a.m. on Friday. The Meteor was coming down from about 5,000ft. The sky was clear. There was sunshine and unlimited visibility.

"The Meteor was crossing from East to West when I noticed the white object in the sky.

"This object was silver and circular in shape, about 10,000ft. up some five miles astern of the aircraft. It appeared to be travelling at a lower speed than the Meteor but was on the same course.

Here are five of the R.A.F. men who saw the Flying Saucer. Left to right standing : L.A.C. Grime, Sgt. T. B. Dewys, Master Sigs. A. E. Thompson, Flight-Lieut. M. Cybulski, Flight-Lieut. J. W. Kilburn.

— GOLF MATCH OF THE CENTURY !

More 1952 British Headlines about Exercise Mainbrace
https://drdavidclarke.co.uk/secret-files/operation-mainbrace-ufos/

Headline from Yorkshire Evening Press, September 1952

Navy Flight Encounters Gigantic UFO - 1956

The next account comes from **Admiral Delmar S. Fahrney**, former Navy missile chief in a brief message to Major Donald E. Keyhoe of a naval flight in 1956. Cruising at 19,000 feet, a Navy R7V-2 transport Super-Constellation was flying west across the Atlantic Ocean heading for a stop-over at Gander, Newfoundland before its final destination at Naval Air Station in Patuxent, Maryland.

The night was clear, visibility unlimited. In the senior pilot's seat, **Commander George Benton** was checking the dim-lit instruments. At thirty-four, Benton had a decade of Navy flying behind him.

He had made the Atlantic crossing more than two hundred times. Back in the cabin were two extra Navy aircrews, en route home from foreign duty.

Most of these men were asleep. Including Benton's regular and relief crews, there were nearly 30 airmen-pilots, navigators and flight engineers aboard the Constellation.

422

As Commander Benton finished his cockpit check, he glanced out at the stars. Then he leaned forward, puzzled. A few minutes before, the sea below had been dark. Now there was a cluster of lights, like a village, about twenty-five miles ahead.

Benton looked over at his co-pilot, **Lieutenant Peter W. Mooney**. "What do you make of those lights?" Mooney peered down, startled.

"Looks like a small town!"

"That's what I thought." Benton quickly called the navigator, Lieutenant Alfred C. Erdman. "We must be way off course. There's land down there." "It can't be land." Erdman hurried forward from his map table. "That last star sight shows..." He broke off, staring down at the clustered lights. "Well?" said Benton. "They must be ships," said Erdman. "Maybe a rendezvous for some special operation."

"They don't look like ships," said Benton. He called Radioman John Wiggins. No word of any unusual ship movements, Wiggins reported. And no signals from the location of the lights. If they were ships, they were keeping radio silence. "Wake up those other crews," Benton told Erdman.

"Maybe somebody can dope it out." In a few moments, two or three airmen crowded into the cockpit. Benton cut off the automatic pilot, banked to give them and the men in the cabin a better view.

As the transport began to circle, the strange lights abruptly dimmed. Then several colored rings appeared, began to spread out. One, Benton noticed, seemed to be growing in size.

Behind him, someone gave an exclamation. Benton took another look. That luminous ring wasn't on the surface - it was something rushing up toward the transport.

"What the devil is it?" said Mooney. "Don't know," muttered Benton. He rolled the Constellation out of its turn to start a full-power climb. Then he saw it was useless. The luminous ring could catch them in seconds.

The glow, he now saw, came from the rim of some large, round object. It reached their altitude, swiftly took shape as a giant disc-shaped machine.

Dwarfing the Constellation, it raced in toward them. "It's going to hit us!" said Erdman. Benton had known normal fear, but this was a nightmare. Numbed, he waited for the crash.

Suddenly the giant disc tilted. Its speed sharply reduced, it angled on past the port wing. The commander let out his breath. He looked at Mooney's white face, saw the others' stunned expressions.

Watching out the port window, he cautiously started to bank. He stopped as he saw the disc.

It had swung around, was drawing abreast, pacing them at about one hundred yards. For a moment he had a clear glimpse of the monster.

Its sheer bulk was amazing; its diameter was three to four times the Constellation's wing span. At least thirty feet thick at the center, it was like a gigantic dish inverted on top of another.

Seen at this distance, the glow along the rim was blurred and uneven. Whether it was an electrical effect, a series of jet exhausts or lights from opening in the rim, Benton could not tell. But the glow was bright enough to show the disc's curving surface, giving a hint of dully reflecting metal.

Though Benton saw no signs of life, he had a feeling they were being observed. Fighting an impulse to dive away, he held to a straight course. Gradually, the strange machine pulled ahead.

Tilting its massive shape upward, it quickly accelerated and was lost against the stars.

Commander Benton reached for his microphone, called Gander Airport and identified himself. "You show any other traffic out here?" he asked the tower. "We had something on the scope near you," Gander told him. "But we couldn't get an answer."

"We saw it," Benton said grimly. "It was no aircraft." He gave the tower a concise report, and back at Gander teletype messages were rushed to the U.S. Air Defense Command, the Commanding Officer, Eastern Sea Frontier, the Director of Air Force Intelligence and the Air Technical Intelligence Center.

When the Constellation landed at Gander, Air Force intelligence officers met the transport. From the start, it was plain they accepted the giant disc sighting as fact.

For two hours, Benton and the rest were carefully interrogated [debriefed], separately and together: How close did the object come? What was its size... estimated rate of climb... any electrical interference noted... what happened to the other luminous rings?

From the answers to scores of questions, the majority opinion emerged. The flying disc was between 350 and 400 feet in diameter, and apparently metallic. No interference with ignition noted; instruments not observed and radio not operating during this brief period.

Time for the giant disc to climb to the transport's altitude, between five and eight seconds, indicated speed between 1,400 and 2,200 knots; the disc had accelerated above this speed on departure.

Not all the men in the cabin had seen the luminous rings. Of those who had, most were watching the huge disc approach and did not see the "rings" disappear. If they, too, were flying discs, in a rendezvous as some suggested, they apparently had raced off while the other one was checking on the Constellation.

424

At one point, an Intelligence captain asked Benton if he had seen any indication of life abroad the disc.

"No, but it was intelligently controlled, that's certain. Benton looked at him closely. "That size, it would hardly be remote-controlled, would it?" "I couldn't say," replied the Air Force man.

Nor would he tell what the Gander Airport radar had shown about the disc's speed and maneuvers. "What's behind all this?" demanded Mooney. "Up to now, I believed the Air Force. You, people, say there aren't any flying saucer..."

"Sorry, I can't answer any questions," said the captain. "Why not? After a scare like that, we've got a right to know what's going on." The Intelligence officer shook his head. "I can't answer any questions," he repeated.

As quickly as possible, intelligence reports with full details were flashed to the four Defense commanders already notified, with an extra message for the Director of Naval Intelligence.

After the Constellation reached Patuxent, the air crews were interviewed [debriefed] again, by Navy order. Each man made a written report, with his opinion of what he had seen.

Five days later, Commander Benton had a phone call from a scientist in a high government agency. "I'm informed you had a close-up UFO sighting. I'd like to see you."

Benton checked, found the man was cleared by the Navy. Next day, the scientist appeared, showed his credential, listened intently to Benton's report. Then he unlocked a dispatch case and took out some photographs.

"Was it like any of these?" At the third picture, Benton stopped him. "That's it!" He looked sharply at the scientist. "Somebody must know the answers if you've got photographs of the things."

The other man took the pictures. "I'm sorry, Commander." He closed his dispatch case and left.

At the time when I (Donald Keyhoe) learned of this case, I had served for two years as Director of the National Investigations Committee on Aerial Phenomena.
Flying Saucer Review, Volume 49/2, Summer 2004, pp. 21-23 From the NICAP records, by Major Donald E. Keyhoe and http://www.ufocasebook.com/navy1956.html

Warships Shoot Down UFO During Persian Gulf War January 24, 1991

U.S. warships patrolling the Persian Gulf shot down a UFO during a battle with Saddam Hussein's troops during operation desert storm. Now a top secret mission to recover the UFO debris is underway.

The U.S. navy downed a UFO on January 24, 1991, according to a classified report detailing allied cooperation in the war to liberate Kuwait. It was detected by radar for nearly 30 seconds and was flying in an erratic manner. The order to down the aircraft was given when it buzzed several American ships, including the USS Wisconsin, USS England and USS O'Brien and two British frigates, the HMS Battleaxe and HMS Jupiter.

All vessels in the area fired on the alien spacecraft, which was described by officers as a chromium plated air- craft which emitted a high-pitched piercing sound, unlike conventional jets. The UFO was most likely downed by a missile, although every ship in the U.S. Task force deployed in the Persian Gulf emptied its conventional armaments to destroy it.

News of the alien involvement in the Persian Gulf War was first revealed when Pentagon intelligence reports of the U.S. armed forces performance were partially disclosed in Washington. For security reasons, much of the sensitive information is still classified top secret.

But rumors of the UFO incident were verified by London-based reporter Anthony Edens, who gained access to the joint American-British accounts of the UFO incident revealed in London.

The UFO skirmish is by far the most pertinent military knowledge gained from the Persian Gulf War, says Edens. The sophisticated military hardware of the United States was not only superior to Saddam Hussein's feared Republican Guard troops but also to the UFO.

The hunt for the UFO debris was reportedly delayed until the Memorial Day weekend when president bush issued an executive order to recover the downed spacecraft. Intelligence analysts are relying on space satellite transmissions of the Persian Gulf to race the whereabouts of the UFO debris.

Respected ufologist **Craig Shopley** says if the UFO incident resulted in the death of any aliens, the results could be devastating for all mankind.

Alien spacecraft are probably monitoring the Earth to learn more about us, he says.
Conspiracy Shack website by Unknown - Reformatted by Kidd 11/2000
http://j_kidd.tripod.com/b/140.html

Soviet Submarines Pursued by UFOs

North Americans by no means have a monopoly on UFO encounters, nor do land dwellers. In 2009 the Russian Navy declassified a vast amount of documents on encounters they have experienced with UFO activity. These encounters have been so prevalent that reports as far back as the Soviet era have been made on a weekly basis to Naval High Command. The department in charge of these reports was headed up by Deputy Commander Admiral Nikolai Smirnov.

According to Captain Vladimir Azhazha, former Deputy Section Chief of Exploration for the Oceanographic Commission of the Academy of Sciences of the USSR, *"50 percent of UFO related with the ocean, 15 percent - from lakes. So the UFO apparently gravitate to the water element. Therefore, the naval collection of data on UFOs has special value."*

426

While the first reported encounter with a UFO took place in the Atlantic Ocean, 44% of the Russian sightings have been in the Pacific, 16% in the Atlantic, 10% in the Mediterranean Sea, and the remaining 30% in various other bodies of water, including 15% in lakes, especially Lake Baikal in Siberia.

One report that stands out in the Russian Navy's UFO encounters involves a submarine of the Pacific Fleet. It had made sonar contact with six unknown objects that shadowed the vessel. Being unable to lose the pursuit the submarine's Commander ordered his ship to surface, a gross violation of standard combat rules of conduct. Upon reaching the surface, he reported the six objects not only surfaced with his vessel but also then launched themselves into the sky and disappeared. http://www.alieneight.com/alien-encounters-the-russian-navy-and-aquatic-ufos.htm

Russian Navy nuclear-powered cruise missile submarine OMSK (K-186)
http://www.nationalufocenter.com/artman/publish/article_291.php **Credit: US Dept of Defence**

Azhazha said that the captain of this sub assumed the objects were American subs. When he couldn't shake them by maneuvering he decided to surface under the assumption that they would also and there would be communication during which time the captain could tell the Americans to "get lost." Instead, when the captain and a crew member reached the sail (top) of the surfaced submarine he was astonished to see the objects rise out of the water and fly away. http://www.nationalufocenter.com/artman/publish/article_291.php

Rear Admiral Yuri Beketov, a former nuclear submarine commander, also reported repeated encounters while stationed in the Bermuda Triangle. He said that often they recorded material (solid) objects traveling underwater at "unimaginable" speeds, sometimes reaching speeds of 230 knots (ca 400 KM per hour), which given the pressure resistance of water, he said was contrary to any known laws of physics. With the large number of objects detected, along with the depth (8,742 meters) of the Atlantic-Puerto Rican Rift, he stated that the possibility of an unknown, advanced civilization residing deep under the ocean was a distinct possibility which should be further examined. Veteran Intelligence Navy Captain Igor Barclay stated that such UFO activity often congregated around areas where the NATO fleets would be in operation.

There are watery accounts of UFOs out of the former Iron Curtain motherland, one of these involves Russia's Lake Baikal which is the oldest and deepest lake on the planet at 25-30 million years old, and the second largest following the Caspian Sea. It has had its own share of water based UFO encounters. In 1982 Russian military divers were on a training mission in the Lake when they were confronted with a group of other beings. The alien divers were described as being humanoid but about nine feet (3 meters) tall, encased in tight silvery body suits with no apparent diving equipment other than spherical helmets. This was at the depth of 50 meters (over 160 feet). The Russian divers attempted to pursue the strange beings but ultimately gave up the attempt after three of their numbers were killed and the other four injured.

A few of their reports concerning unidentifiable craft were taken from clandestine surveillance reports of US Navy Fleet activities. During a training exercise off the coast of Puerto Rico, involving an aircraft carrier and five escort ships and submarines, one of the subs broke ranks to pursue an underwater sonar contact. It was traveling at 150 knots, nearly three times the greatest possible speed for a terrestrial submarine, the fastest of which have a top speed closer to 45 knots.

People have been spotting UFO's diving into and out of the oceans for centuries. Even the modern navies of this world can attest that one does not have to just keep their eyes to the sky to see strange and inexplicable craft moving about planet Earth, behaving as nothing we are capable of building can do. http://www.alieneight.com/alien-encounters-the-russian-navy-and-aquatic-ufos.htm

Chilean Navy UFO Incident - October 24, 1969

Just after midnight, on or about, a Chilean Navy destroyer, a week out of dry dock at Talcahuano Port (where the ship's axle had been removed and replaced), was navigating at 20 knots and heading north 20 degrees portside from NNW). The incredible events that followed took place over the next eight minutes.

At 12:43 A.M. the radar officer reported a long-range flying contact. A minute later the "contact" was at 400 miles. Because of the "object's" speed, the operator suspected a malfunction in his equipment. In the next minute, the contact was approximately 150 miles away closing from 331 degrees of true north. But the operator and officer in charge during the late night duty (an officer of second-class rank) speculated that the contact was a "plane flying southeast" --but at 213 miles in a minute: 12,780 mph!

The officer in charge advised the ship's commander, who asked that an eye be kept on the object. At 12:47 the contact was only 12 miles away. Suddenly the single contact became six "targets." The "thing spread ... Little dots appeared in the fire of the light on the radar...."

The officer in charge informed the commander just as the objects making the returns came into view. Both eyewitnesses and radar were reporting six objects approaching the ship. The

428

commander came up to the ship's bridge to see what was going on.

It was a fantastic sight. One massive object and five small objects were approaching at high speed. The "big thing looked like a big box [with] semicircles in the side" looking as if they had been scooped out. It was bathed in brilliant light. "The thing must have been metallic," my informant told me. "It was bigger than the destroyer, which was about 110 metres long." He thought it might be at least twice the length of the ship but it was difficult to tell because "the light was bright."

The five smaller objects were egg-shaped and appeared to be no bigger than eight feet long and five to six feet wide. They were bluish in color.

At about 2000 yards out from the destroyer the smaller objects left the proximity of the larger object. Three went port side and two went starboard. The smaller objects at times seemed to be flying in "elliptical circles backwards and forwards between the big one and our vessel ...

"The main thing did not change direction -- if it had been in the sea there would have been a collision."

A humming noise was audible when the large object got to within 300 yards. It was at this point that the power went out. For a couple of seconds, all instruments went dead as the huge craft passed overhead.

At this point the commander came on to the bridge, asking, "What the hell is that?" My informant remarked that the commander "was very calm because probably he knew what he saw. We didn't.

"You could see the whole thing, the light was so strong. You could see the water, the funnel head, the head of the ship, the towers, everything... Everybody on the bridge was sort of listening to the noise. I don't know how long this thing took to go across."

The bright red lights underneath the main UFO seemed to be moving back and forth inside the "box," shining through a half-circle or crescent shape on the bottom of the object. On the side were what looked like "corn cobs," with green or turquoise pulsing lights. These lights, my informant said, "went right through your head."

The large object passed over the powerless ship. When it was 200 yards away, everything returned to normal. The smaller objects, never coming closer than 500 to 1000 yards, flew around the ship and joined up with the larger object on the other side.

The main object continued on its way until it was about two miles beyond the ship. At that point, along with (we assume) the smaller objects, it vanished. "It was like somebody opened a big door; then it closed and it was darkness. The only thing that was left behind was like floating little bits of metallic paper."(!)

At least eight minutes had passed. Three persons had tracked the UFOs on radar. Five persons saw the objects themselves. The ship's commander ordered everyone to keep silent about what they had seen.

The "cover-up" started immediately, According to my informant, the officer in charge (the second-class officer) had made several entries concerning the UFO incident in the ship's log.

The entries at midnight and on either side of the UFO entries were routine. Duty proceeded normally for the rest of the morning.

Awakening after he had retired from his shift, the officer rechecked the log, only to find normal entries in what appeared to be his own handwriting. The UFO incident was not mentioned. The officer had been left-handed but an accident had forced him to use his right hand. Because writing was difficult for him, he had to press down hard on the paper. He had to press down so hard, in fact;-that the outline of what he wrote was always visible on the page. But when he checked, it was not. In other words, someone had forged his handwriting but neglected this telltale physical detail.

At 6:45 that morning the destroyer arrived in Valparaiso Port. After eating breakfast, the men on duty during the UFO event were asked to report to the commander. He told them that some people were coming to talk with them and that they were to go with them. The visitors turned out to be two Chilean Navy officers and four Americans in civilian clothes. The Americans, who spoke Spanish, were identified as naval attaches with the U.S. embassy.

The six witnesses were taken to a mine- or torpedo storage area at the port. Each was taken to a separate room.

My informant was directed to explain what happened last night. When he told them, they said, "No, you didn't see that !"

"I said that's what we saw," my informant related to me. "They started getting very cranky. They said, 'No, you didn't see it. You didn't see anything. You know nothing.'

"I spoke to the highest-ranking officer there, a tactic commander in the Chilean Navy: 'Are we under arrest or what?' He said, 'No you are not.' "

Asked why they were there, the Chilean officer replied, 'You are under orders. These people just want to talk to you, to put you on the right track."

My informant claimed, "While the men never asked us to go through the story in detail, they knew what had happened by the questions they asked. They also had the radar-plot blueprint." They were more interested in securing his silence than in getting information about the event from him.

For two days the harassment continued. "Every two hours," the informant said, "this guy would come up. 'What do you think now?' he'd say... We had arguments. They didn't have any right to do what they were doing."

My informant threatened to complain when he got out. He was told that if he did so, that would be the end of him. He did not see the other witnesses at any point during the two-day interrogation. He was not allowed to sleep until he signed a document. Finally, after he had done so, he was put in a jeep and driven to another location. One week later he was transferred unexpectedly from destroyer duty. He never saw the others again.

This account was originally reported by **Bill Chalker,** one of Australia's leading UFO investigators from New South Wales. http://www.ufoevidence.org/cases/case1016.htm

The USS John F. Kennedy Atlantic Ocean UFO Incident - 1971

This incident occurred as the **USS John F. Kennedy** was returning to port in Norfolk, Virginia, following an "**Operational Readiness Exercise**" in the Caribbean. The carrier was sailing north in the Atlantic Ocean just a few hundred nautical miles off of the coast of South Carolina.

It was near sundown and the ship had just completed Flight Operations for the day, when at approximately 20:30 hours, the communications, navigational, and defense electronics started to function erratically.

As the officers and crew moved quickly and efficiently through checklists and well-rehearsed procedures to identify the source of the electronic disruptions, a voice from the signal bridge boomed over the ships intercom, "There's something hovering over the ship."

Within the next few seconds, the "General Quarters" alarm started going off. As crewmen scrambled to assume their assigned battle station, they encountered equipment that wasn't fully operational.

The USS John F. Kennedy returning to port in Norfolk, Virginia, following an "Operational Readiness Exercise" in the Caribbean
https://en.wikipedia.org/wiki/USS_John_F._Kennedy_(CV-67)

431

Radar screens glowed but did not function normally, the ships compasses moved erratically and weapons controls were inoperative. Attempts to launch two CAP F-4 Phantom jets to check out the phenomenon failed for reasons unknown.

According to officers and crew on the flight deck and "The Island" (the ships control tower), the object was a bright golden yellow pulsating light.

It looked to be a solid metallic orb but was also said to be completely silent. Estimates on its size and distance above the ship vary, but the most conservative estimate was that it was at least 300 feet in diameter.

It hovered over the USS John F. Kennedy for a little over twenty minutes, before rapidly accelerating straight up and out of sight. Almost immediately, the JFK's systems returned to normal operations.

Upon return to Norfolk, select members of the crew were de-briefed privately. The general population of the 'non-involved' crew were advised that "certain events" that occur on Naval Warships are classified and not to be discussed with anyone lacking proper "clearance."
http://www.zuko.com/Inexplicable/US_Navy_USO_Encounters.asp

The USS Lexington – Gulf of Mexico - 1965

The USS Lexington was on routine patrol in the Gulf of Mexico in the fall of 1965. During night time flight operations, an unknown contact was picked up on radar and attempts were made to quickly identify the contact. Support crew on the flight deck that night were also able to observe a bright orange glow in the sky.

Unable to identify the object with a radar profile, and unable to get any sort of radio response from the object, the Lexington launched two F-4 fighter jets to intercept and identify. The F-4 jets immediately went to afterburners (a method for jets to temporarily accelerate to a high rate of speed) but even at 1,500 miles per hour, the F-4s could not get close to the object. It simply accelerated away, vanishing from sight.

This game of cat and mouse played out the same way three times during that night. The F-4s would land, the light would appear. The jets would launch and the light would accelerate away. The fourth time the light appeared the Navy elected not to launch the F-4s and the light appeared to follow them harmlessly for nearly a half hour before vanishing for the final time.

Officers and crew were debriefed and reminded that the incident should not be talked about with any unauthorized individuals.
http://www.zuko.com/Inexplicable/US_Navy_USO_Encounters.asp

The USS Constellation Aircraft Carrier - 1994

This account has many similarities to the encounter of 1989 concerning the USS Midway. The USS Constellation was on patrol in an undisclosed location in the Indian Ocean in the early summer months of 1994.

432

The **Combat Direction Center (CDC)** picked up a low flying contact on their radar that was closing in on the ship at a very high rate of speed. Initially thought to be an incoming missile the ship braced for impact when the radar contact vanished suddenly.

As the ship went to a General Quarters alert, it was then reported that there was an object now hovering over the aircraft carrier's "Island." Witnessed only by those on the ship's superstructure, it was described as a large bright light. The object was visible for only a few seconds and with an eerie silence, shot straight up into the sky until it was completely out of sight.

Officers and crew were debriefed and reminded that the incident should be forgotten.
http://www.zuko.com/Inexplicable/US_Navy_USO_Encounters.asp

The USS Midway – The Gulf of Oman - 1989

In 1989 the aircraft carrier USS Midway was on a tour of duty in the Indian Ocean. On a mission in the Gulf of Oman, the Combat Direction Center (CDC) watch officer and his crew were electronically monitoring the positions and movement of all surface contacts to ensure a safety margin of three miles or more from any potentially hostile vessels.

Suddenly the officer spotted three strong blips (contacts) on his radar screen grouped in a triangular formation. On the second sweep of the radar screen, the officer was stunned to see that the contacts had moved a considerable distance in a remarkably short period of time. While this CDC group's mission is to monitor and track ocean surface contacts, low flying aircraft are also periodically picked up on their low angled radar. These contacts, however, were moving faster than any aircraft the officer or his crew had ever observed. The next unnerving thought to occur to the CDC officer was that these were missiles. The officer checked with the airborne detection group within the CDC and they responded that they weren't picking up or tracking anything at all. There were no contact blips flashing on their screens.

The aircraft carrier: USS Midway
https://en.wikipedia.org/wiki/USS_Midway_(CV-41)#/media/File:USS_Midway;024105.jpg

Tensions mounted quickly, however, as the CDC Officer and his crew continued to track the three fast moving blips as they angled toward the ship. As the 'contacts' rapidly got closer, two of the blips made a startling ninety degree turn away from the ship, but the third turned directly toward the ship. The CDC officer excitedly barked over the ship's intercom to the observers on the ship's superstructure that something was coming right at them at a high rate of speed. Accelerating, it was coming in so fast there was no time to react. When the object got within one mile of the ship it suddenly disappeared from radar. The CDC Officer again called out to the lookouts high above in the superstructure, asking if they saw anything through their binoculars. They reported that they looked at the exact location but didn't see a thing.

The CDC Officer then reported the incident to his commander but was a bit taken aback by the commander's lack of reaction. He didn't react at all.

Within a few minutes, the CDC radar screens picked up the three blips again, very nearly in the same location where they first picked them up. Alerting the commander that they were back,

434

the sequence of events played out nearly as before, with neither, the airborne radar or the lookouts seeing anything.

After this second incident, the CDC Officer was more than a bit concerned that something had actually approached to within one mile of them without anything or anyone other than surface radar picking them up and tracking them. (The air defense protection radius for an aircraft carrier guarding against air attack is fifty miles.)

The CDC Officer calmly approached the commander and began discussing the close contact incident, commenting particularly on the incredible speeds and the virtually impossible 90 degree turns executed by the unknown contacts. The officer then commented that the only way he could have picked them up on his low angled radar was if they were flying 30 feet or lower over the ocean's surface. He then speculated that the only way they could have disappeared so suddenly, within one mile of the ship, would be to dive beneath the ocean's surface.

The commander then offered (in a rather hushed voice) that the CDC had, some months before detected an unidentified object in this general vicinity and had launched an 'Alert Fighter' to intercept and identify the object. The commander stated that the fighter pilot had just managed to get within visual range and report the object as a strange metallic glint when the object quickly accelerated away from the Navy jet and dove into the water. As the pilot flew over the point of the object's entry into the sea, he reported no wreckage or any sign of oil or fluids floating on the water.

The Navy immediately sent one of the aircraft carrier's escort ships over to the site to investigate, but no trace of a crash or of any type of debris was observed.

While detailed reports were filed by the Officers and personnel involved, the Navy has reserved comment or official position on this incident.
http://www.zuko.com/Inexplicable/US_Navy_USO_Encounters.asp and
https://www.youtube.com/watch?v=u3FTpJJkcbE and
https://www.youtube.com/watch?v=k92OKqporaE

FOR EARLY WARNING IN DEFENSE OF THE NORTH AMERICAN CONTINENT
MERINT
RADIOTELEGRAPH PROCEDURE

1. WHAT TO REPORT

Report immediately all airborne and waterborne objects which appear to be HOSTILE, SUSPICIOUS or are UNIDENTIFIED.

Surface warships positively identified as not U.S. or Canadian

Guided Missiles

Aircraft or contrails which appear to be directed against the United States, Canada, their territories or possessions

Submarines

Unidentified Flying Objects

2. SEND TO ANY

United States Naval Radio Station
Canadian Naval Radio Station
United States Coast Guard Radio Station
United States Commercial Radiotelegraph Station
Canadian Department of Transport Coastal Station

Receiving station will relay to military destination

3. HOW TO SEND

* MERINT MERINT MERINT (Coastal Station) DE (Own Signal Letters) K (Own Signal Letters) DE (Coastal Station) K
EMERGENCY (For U.S. or Canadian Naval or Coast Guard Radio Stations)   or
RAPID US GOVT COLLECT (For U.S. Commercial Coastal Stations)   or
RUSH COLLECT (For Canadian Dept of Transport Coastal Stations)

4. SEND TO ONE DESTINATION

ComWestSeaFron Navy SFran
NavyCharge Halifax
NavyCharge Esquimalt

Select destination nearest to your receiving station

5. SEND THIS KIND OF MESSAGE

Content—	Example—
a. Begin your message with the word "MERINT"	MERINT
b. Give the reporting ship's name and signal letters	SS TOLOA WHDR
c. Describe briefly the objects sighted	TWO UNIDENTIFIED SURFACED SUBMARINES
d. Give ship's position when objects are sighted, also TIME and DATE	5034N 4012W 071430 GMT
e. If objects are airborne, estimate altitude as "low", "medium", "high"	(not applicable)
f. Give direction of travel of sighted objects	HEADING 270 DEGREES
g. Estimate and give speed of sighted objects	15 KNOTS
h. Describe condition of sea and weather	SEA CALM
i. Give other significant information	ELONGATED CONNING TOWERS

6. SEND IMMEDIATELY

a. DO NOT DELAY YOUR REPORT DUE TO LACK OF INFORMATION
b. EVERY EFFORT SHOULD BE MADE TO OBTAIN ACKNOWLEDGMENT FROM RECEIVING STATION THAT MESSAGE HAS BEEN RECEIVED.

* The International urgency signal (XXX XXX XXX) may be used as an alternate to clear circuit.

Authorized by Secretary of the Navy

OPNAV 94-P-3B

A placard from the "Blue Book UFO Reports at Sea by Ships" of what do when hostile, suspicious or unidentified airborne or waterborne objects are sighted. Note the top right image shows a flying saucer and a spaceship as UFOs

CHAPTER 49

ETI SEND A CLEAR MESSAGE TO THE MILITARIES OF THE WORLD – "WE WILL NOT LET YOU DESTROY THIS PLANET!"

UFO "Alters" Tracking of Navy Polaris Test - January 10, 1961

This next case was investigated by **NICAP (National Investigation Committee on Aerial Phenomena)** by one of it committee members who was at the site when the event occurred. On Jan. 10, 1961 at a US Air Force Test Range in Cape Canaveral, Florida a Polaris missile was launched by the US Navy from a ground pad. As the missile steadily gained altitude, a flying disc suddenly appeared which then paced the missile's ascent then, it appeared to alter its tracking but did not block the missile in any way. The tracking system continued to follow the unidentified flying object and later returned to again to track the Polaris downrange.

The diameter of the flying saucer was close to the length of the Polaris, it was approximately 20 to 25 feet and it was about 6-8 feet thick at its center. In other words, it was a disc and not tubular like the body of the Polaris. It was visually lost as it continued downrange to ground observers and the primary witness, Clark McClelland, Florida NICAP member, who had watch the event unfold with his 10x50 binoculars at the Cape. The whole event was apparently filmed but to date, that film has yet to surface for closer scrutiny.
http://www.nicap.org/610110capecanav_dir.htm

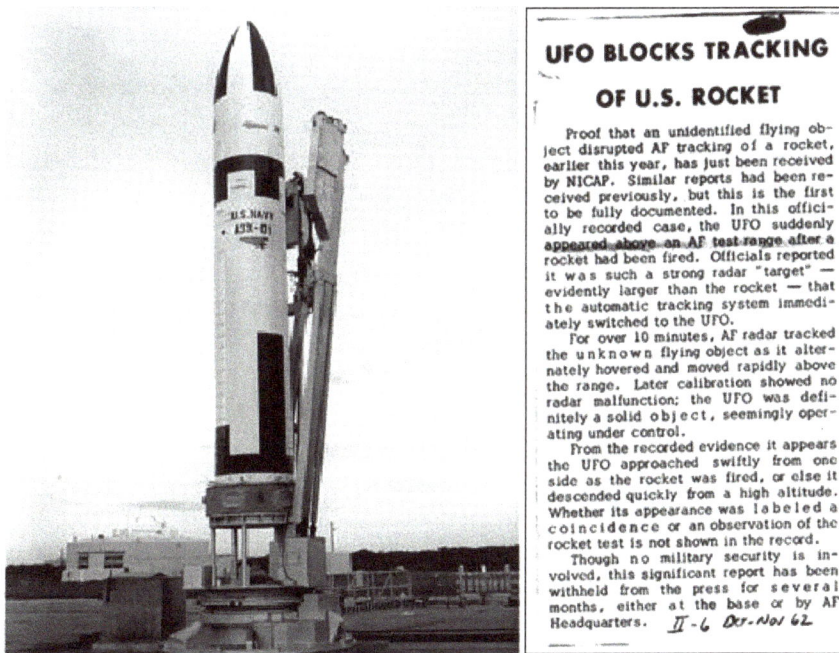

UFO BLOCKS TRACKING OF U.S. ROCKET

Proof that an unidentified flying object disrupted AF tracking of a rocket, earlier this year, has just been received by NICAP. Similar reports had been received previously, but this is the first to be fully documented. In this officially recorded case, the UFO suddenly appeared above an AF test range after a rocket had been fired. Officials reported it was such a strong radar "target" — evidently larger than the rocket — that the automatic tracking system immediately switched to the UFO.

For over 10 minutes, AF radar tracked the unknown flying object as it alternately hovered and moved rapidly above the range. Later calibration showed no radar malfunction; the UFO was definitely a solid object, seemingly operating under control.

From the recorded evidence it appears the UFO approached swiftly from one side as the rocket was fired, or else it descended quickly from a high altitude. Whether its appearance was labeled a coincidence or an observation of the rocket test is not shown in the record.

Though no military security is involved, this significant report has been withheld from the press for several months, either at the base or by AF Headquarters. II-6 Oct-Nov 62

**Polaris A-3 on launch pad at Cape Canaveral and
Newspaper clipping of the reported UFO event**
https://en.wikipedia.org/wiki/UGM-27_Polaris and http://www.nicap.org/reports/blocmiss1.htm

The Air Force evaluation for this particular sighting as it was reported to Project Blue Book was that it was the "...star Gamma Piscium." Hynek's response to the Air Force's explanation: "Gamma Piscium is a relatively faint star, and quite stationary. It is absurd to think that a person professionally qualified to track missile launches would be puzzled by one particular star out of a great many." The Hynek UFO Report p. 44; 1977; Dell Publishing Co., Inc.; ISBN 0-440-19201-3

UFO Fires Upon Missile Launched From Vandenberg Air Force Base - 1964

Professor Robert Jacobs, a Lieutenant in the US Air Force and **Major Mansmann**, his commanding officer, both witnesses of Dr. Steven Greer's **Disclosure Project** tell of another similar event involving a missile launch from Vandenberg Air Force base in California in 1964.

Prof. Jacobs was the office in charge of optical instrumentation at Vandenberg AF Base in the 1369th photo squadron. His duty was to supervise the instrumentation photography of **ICBM (inter-county ballistic missiles)** as many of them blew up on launch. Forensic analysis of sequential photography was the way for engineers to determine why they blew up. Most of these missile launches were testing dummy warheads that were designed to carry nuclear weapons, a precursor to **inter-continental ballistic missiles (ICBMs)**, which would come down on a target island in the Pacific Ocean.

On the day of the Atlas missile launch, everything appeared to go well as high powered telescopic cameras captured the successful launch and the separation of the booster rockets at each stage of flight.

About two days later, Major Mansmann calls Prof. Jacobs into his office where two civilians in gray suits are also present which upon reflection by Jacobs was unusual. He was asked to sit down on a couch and watch the film of the missile launch that occurred a couple of days earlier. As he watched, he saw that the film had recorded everything they saw with the naked eye on the day of the launch with the exception of something else the naked eye didn't see.

Into the frame of the that showed the dummy warhead could be seen an object that can only be described as a flying saucer, that it appeared like two saucers cupped together at the rim with a ping-pong shape cupola on the top from which it fires a beam of light at the warhead.

Keep in mind that this warhead is flying straight up 60 miles into subspace at 11,000 to 14,000 miles per hour when this UFO flies toward the missile at the same rate of climb fires a beam of light at the warhead, hits it and then, the UFO moves to the other side and fires another beam of light, then goes down and fires another beam of light and then, finally flies out the way it came in. At which point the warhead is seen to tumble out of space.

The film stops and the lights in the room come back on and immediately Major Mansmann now visibly upset asked Jacobs "were you guys screwing around up there"? Jacob replies that it appears that they had got a UFO and after some discussion with Major Mansmann that Jacobs is told to never to speak of this incident again, that this event never happened and that he should not

438

have to be reminded that there were dire consequences to himself of a national security breach, even if years later, he was asked under force by someone.

Should he be asked, he was to say that it was laser strikes, laser tracking strikes (possibly a type of laser distance ranging). Once again, Jacobs thought this statement was strange giving that lasers were still in their infancy of development used only in laboratories. Mansmann would conclude years later that the UFO was extraterrestrial in nature which may have fired a plasma beam weapon at the dummy warhead.

For as long as Jacobs had worked at Vandenberg Air Force base not even his Commanding Officer Major Lewis S. Clement, Jr. knew nothing about the event, nor did any of his superior officers, whether Captains or Lieutenants or Chief Ward Officers.

Now, this is very interesting as this demonstrates the concept and position of **USAPs (Unacknowledged Special Access Projects)** held by various officers and personnel in the military. Even, if a junior rank officer knew aspects of the UFO phenomenon intimately and was asked by a superior officer to inform or reveal detail of what they knew, the junior officer would reply: "Sir, there is nothing to it" or that he knew nothing about the subject. End of discussion! Basically, the junior ranked officer would lie even, if the other officer was a four-star General.

A Lieutenant or a Major could be "in the loop" but a Captain, Colonel, or General could be "out of the loop" with regard to UFO knowledge. They simply don't have a need to know regardless of how high up their rank. It could be the President of the United States, the supreme executive office of the country and still he may not have a need to know. Many Presidents since Truman and Eisenhower have tried to enter this most sacred of inner sanctums and have been shut out, *even under threat of assassination*! Promising to be a **President of Transparency** has meant nothing when in office and no amount of executive legal action or bill passing will open the doors on UFO secrecy and cover-up. [We will discuss this whole matter further in a later section.]

Years later, Jacobs did eventually reveal what he knew about the event but, someone else had already publically disclosed the event to the news media. This caused him many personal threats and harassment to himself and his family even, to the point of having his mailbox blown up. There were humiliating letters and phone calls belittling Jacobs from professional skeptics like **James O'Berg** at NASA and **Phillip J. Klass**, who is known within the UFO community as a paid informant for the US government.

Urologists and news reporters had tracked down a now retired **Major Florence J. Mansmann** who had become a rancher; he confirmed everything that Jacobs had stated about the missile launch and the UFO encounter.

Major Mansmann also reported that the two unidentified men in the gray suits in the room when the film was being watched were not CIA but were from another intelligence agency. They had taken scissors to the film and cut out the sections of the film that showed the UFO firing at the dummy warhead and left the remaining film with Major Mansmann with the same warnings and threats of not breaching national security. Disclosure: Military and Government Witnesses

Reveal the Greatest Secrets in Modern History; 2001; Publisher's Cataloging-in-Publication by Quality Books, Inc.; ISBN 0-9673238-1-9

Extraterrestrial Intelligences have been kept busy monitoring those nations who have researched, developed, tested, deployed and stockpiled nuclear weapons since the United States first exploded an atomic bomb back in July of 1945.
United States, Russia, Britain, France, and China are the original members of the "Nuclear Club" and the only members in the United Nations that have a veto power in decision making.

Though the US and Russia have gone through a nuclear arms race, a period of cold war escalation as bitter enemies between the 50' to the early 90s; they have emerged unscathed within a peaceful state of co-existence and growing partnership. Yet, their nuclear arsenal has only diminished minimally in size and now, other nations have entered into the **"Nuclear Club"** with hopes of being recoginzed on the world stage as a major global power. Nuclear-armed states currently include the United States, Russia, United Kingdom, France, China, Israel, India, Pakistan, North Korea and South Africa (former). The saber rattling of nations has not ceased!
http://www.guardian.co.uk/news/datablog/2009/sep/06/nuclear-weapons-world-us-north-korea-russia-iran

These nations are under constant monitoring by ETI and as the list of nuclear nations continues to grow, reports of UFO sightings within these countries also increase.

Although, Canada in partnership with Britain and the US developed the first atomic bomb, it is not a member of the nuclear-armed states and has been the only nation, never to have developed or detonated nuclear weapons for itself, but has instead, turned its attention towards the peaceful uses of nuclear energy like the Candu reactors for electrical power generation, rather than for the purposes of war.

Extraterrestrial Intelligences are very concerned about our predilection with weapons of mass destruction such as nuclear and scalar weaponry. The impact of weapons affects not only the people on our planet but it sends destructional waves or pulses of energy far beyond our planet not only into the physical realm of existence but, into inter-dimensional realms or higher planes of existence of which we are just starting to learn about. Life exists on many multiple and infinite levels of existence. Thermonuclear weapons when detonated resonate into these multiverses like so many ripples on a lake but, the affecting shockwaves are seismic in nature to lifeforms in these realms of existence. It is only natural that they become alarmed when these destructive seismic-like waves invade their home world causing them to investigate their source of origin.

Lakenheath-Bentwaters UFO Incident -1956

The **Lakenheath-Bentwaters Incident** was a series of radar and visual contacts with **Unidentified Flying Objects** that took place over airbases in eastern England on the night of 13 - 14 August 1956, involving both RAF and USAF personnel. The incident has since gained some prominence in the literature of ufology and the popular media.

The final Report of the Condon Committee, which otherwise concluded that UFOs were simple misidentifications of natural phenomena or aircraft, took an unusual position on the case: "In conclusion, although conventional or natural explanations certainly cannot be ruled out, the probability of such seems low in this case and the probability that at least one genuine UFO was involved appears to be fairly high". It has, however, also been argued that the incidents can be explained by false radar returns and misidentification of astronomical phenomena. **Condon Report, Case 2; p.387;**
http://en.wikipedia.org/wiki/Lakenheath-Bentwaters_incident

The commonly cited sequence of events is that recorded in the original Project Blue Book file by the US Air Force, subsequently analyzed by the Condon Committee's report and by atmospheric physicist Dr. James E. McDonald.

The incident began at the USAF-tenanted **RAF Bentwaters,** Suffolk, on the evening of 13 August 1956. This was a dry, largely clear night with, observers noted, an unusually large number of shooting stars, associated with the Perseid meteor shower. Radar operators at the base tracked a target, appearing similar to a normal aircraft return, approaching the base from the sea at an apparent speed of several thousand miles per hour. They also tracked a group of targets moving slowly to the north-east which merged into a single very large return (several times the strength of that from a B-36) before moving off the scope to the north, as well as a further rapid target proceeding east-west.

A T-33 trainer from the 512th Fighter Interceptor Squadron, crewed by **1st Lieutenants Charles Metz and Andrew Rowe,** was directed to investigate the radar contacts but saw nothing. No visual sightings of the objects were made from Bentwaters in this period with the exception of a single amber star-like object which was subsequently identified as probably being Mars, then low in the southeast.

At 22:55, a target was detected approaching Bentwaters from the east at a speed estimated around 2-4000 mph. It faded from the scope as it passed over the base (possibly suggesting anomalous propagation as a source for the target), reappearing to the west. However, as it passed overhead a rapidly-moving white light was observed from the ground, while the pilot of a C-47 at 4000 feet over Bentwaters reported that a similar light had passed beneath his aircraft. At this point, Bentwaters alerted the US-tenanted **RAF Lakenheath base**, 40 miles to the northwest, to look out for the targets. Ground personnel at Lakenheath made visual sightings of several luminous objects, including two which arrived, made a sharp change in course, and appeared to merge before moving off. The angular size of these objects was compared to that of a golf ball at arm's length, and they started to dwindle to pinpoint size as they moved away, an observation which seemed to rule out a bolide or bright meteor.

The final phase of the incident was described in some detail by **T/Sgt Forrest Perkins**, who was the Watch Supervisor in the Lakenheath Radar Air Traffic Control centre, and who wrote directly to the Condon Committee in 1968. Perkins claimed that two RAF De Havilland Venom interceptors were scrambled and directed towards a radar target near Lakenheath. The pilot of the first Venom achieved contact, but then found that the target maneuvered behind him and chased the aircraft for a period of around 10 minutes despite the latter's taking violent evasive action;

441

Perkins characterized the pilot as "getting worried, excited and also pretty scared". The second Venom was forced to return to its home station due to engine problems; Perkins stated that the target remained on their screens for a short period before leaving on a northerly heading.

Perkins' well-written letter convinced the Condon Committee to include the case in its analysis and a previously classified teleprinter message from 3910th Air Base Group to Air Defence Command at Ent AFB, transmitted three days after the incident confirmed the events of Perkin's letter, including the 'chase' episode.

RAF De Havilland Venoms. The NF.3 night fighter was involved in the Lakenheath incidents
http://www.platz-hobby.com/products/3367.html

Based on the information available, the Committee's researcher (Thayer) felt that while anomalous propagation was possible, the lack of other targets on radar scopes at the time made it unlikely. Focusing on the later phase of the incident at Lakenheath, he came to the remarkable conclusion that "this is the most puzzling and unusual case in the radar-visual files. The apparently rational, intelligent behavior of the UFO suggests a mechanical device of unknown origin as the most probable explanation of this sighting".
Thayer, Condon Report, Optical and Radar Analysis, p.246 and
http://en.wikipedia.org/wiki/Lakenheath-Bentwaters_incident

At this point, another aspect of this event which seems to be deliberately overlooked, if we consider this case as a genuine UFO Extraterrestrial incursion into military airspace then, why then did a UFO fly over the US-tenanted **RAF Lakenheath base** in the first place?

442

The United Kingdom was the first country to set up a nuclear program and, in 1952, became the third country to test an independently developed nuclear weapon. It is one of the five **Nuclear Weapons States** following its 1968 ratification of the **Nuclear Non-Proliferation Treaty.**

The UK is thought to retain a stockpile of around 225 thermonuclear warheads, of which 160 are operational, but has refused to declare the exact size of its arsenal.

Since the 1958 **US-UK Mutual Defence Agreement**, the United States and the United Kingdom have cooperated extensively on nuclear security matters. The special relationship between the two countries has involved the exchange of classified scientific data and materials such as plutonium.

The UK has not run a program to develop an independent delivery system since the cancellation of the Blue Streak missile in 1960. Instead, it has purchased US delivery systems for UK use, fitting them with warheads designed and manufactured by the **UK's Atomic Weapons Establishment** and its predecessor.
http://en.wikipedia.org/wiki/Nuclear_weapons_and_the_United_Kingdom

A little backtracking is necessary in order to get the full picture of the importance of RAF Lakenheath and to understand the implication of UFO sightings in the area.

RAF Lakenheath, (IATA: LKZ, ICAO: EGUL) is a Royal Air Force military airbase near Lakenheath in Suffolk, England. Although an RAF station, it hosts United States Air Force units and personnel. The host wing is the 48th Fighter Wing (48 FW), also known as the Liberty Wing, assigned to **United States Air Forces in Europe (USAFE).**
http://en.wikipedia.org/wiki/RAF_Lakenheath

The first use of Lakenheath Warren as a Royal Flying Corps airfield was in World War I, when the area was made into a bombing and ground attack range for aircraft flying from elsewhere in the area. It appears to have been little used and was abandoned when peace came in 1918.

In 1940, the Air Ministry selected Lakenheath as an alternative for RAF Mildenhall and was used as a decoy airfield. False lights, runways, and aircraft diverted Luftwaffe attacks from Mildenhall. Lakenheath Airfield was used by RAF flying units on detachment late in 1941. The station soon functioned as a Mildenhall satellite with Stirling bombers of No. 149 Squadron RAF dispersed from the parent airfield as conditions allowed.

In June 1943, No. 199 Squadron RAF was established as a second Stirling squadron. No. 149 Squadron ended its association with RAF Lakenheath the same month, taking its Stirlings to RAF Methwold. Between them, the two squadrons lost 116 Stirling bombers in combat while flying from Lakenheath.

The squadron exchanged its Vickers Wellingtons for Stirlings late in 1941. After becoming fully operational with its new aircraft, the squadron moved into Lakenheath in April 1942 and remained until 1944.

The reason for the departure of the two bomber squadrons was Lakenheath's selection for upgrading to a **Very Heavy Bomber** airfield. Lakenheath was one of three RAF airfields being prepared to receive United States Army Air Force **Boeing B-29 Superfortresses**, which were tentatively planned to replace some of Eighth Air Force's Third Air Division B-24 Liberator groups in the spring of 1945.

Cold War tensions with the Soviet Union in Europe began as early as 1946. In November, President Harry S. Truman ordered Strategic Air Command B-29 bombers to RAF Burtonwood, and from there to various bases in West Germany as a "training deployment". In May 1947, additional B-29s were sent to the UK and Germany to keep up the presence of a training program. (***These deployments were only a cover-up, as the true aim of these B-29s was to have a strategic air force permanently stationed in Europe.***) (Italics added for emphasis. This may also indicate part of the real hidden American agenda which is to establish US military bases across most countries of the world).

In April 1947, **RAF Bomber Command** returned to Lakenheath and had the runways repaired, resurfaced, and readied for operations by May 1948.

In response to the threat by the Soviet Union, by the 1948 Berlin blockade, President Truman decided to realign **USAFE** into a permanent combat-capable force. In July, B-29 Superfortresses of the **SAC 2nd Bombardment Group** were deployed to Lakenheath for a 90-day temporary deployment.

On 27 November 1948, operational control of RAF Lakenheath was transferred from the Royal Air Force to USAFE. The first USAFE host unit at RAF Lakenheath was the **7504th Base Completion Squadron**, being activated that date. The squadron was elevated to an **Air Base Group (ABG)** on 28 January 1950 and to a Wing (ABW) on 26 September 1950.

Control of RAF Lakenheath was allocated to Third Air Force at South Ruislip Air Station, which had command of SAC B-29 operations in England. Third Air Force was subsequently placed directly under USAF orders, with Strategic Air Command establishing the 7th Air Division Headquarters at RAF Mildenhall. The collocation of the two headquarters within the United Kingdom allowed HQ USAFE to discharge its responsibilities in England, while at the same time allowing **Strategic Air Command** to continue in its deterrent role while retaining operational control over flying activities at Lakenheath.

By 1950, Lakenheath was one of three main operating bases for the U.S. Strategic Air Command in the UK; the others were **RAF Marham and RAF Sculthorpe**. A succession of bombardment squadrons and wings, 33 in all, rotated through Lakenheath, the B-29s giving way to the improved B-50 Superfortresses and then, in June 1954, B-47 Stratojets.

On 1 May 1951, Lakenheath was transferred from USAFE to SAC, and placed under the **3909th Air Base Group**. By 1952, high-security perimeter fencing was erected for security. The 3909th moved to RAF Greenham Common in 1954 and was replaced by the 3910th Air Base Group.
http://en.wikipedia.org/wiki/RAF_Lakenheath

444

A note by U.S. President Dwight D. Eisenhower quoted him as saying:

> *"The United States Government welcomes the agreement to coordinate the strike plans of the United States and United Kingdom bomber forces, **and to store United States nuclear weapons on RAF airfields under United States custody** for release subject to decision by the President in an emergency. We understand that for the present at least these weapons will be in the kiloton range. The United Kingdom forces could obviously play a much more effective part in joint strikes if the United States weapons made available to them in an emergency were in the megaton range, and it is suggested that this possibility might be examined at the appropriate time."*

http://en.wikipedia.org/wiki/Project_E

Many SAC Squadrons had aircraft at RAF Lakenheath on a transitory basis without any recorded deployment to the base. (Italics added by author for emphasis).

For example, in January 1951, a detachment of Convair RB-36D Peacemaker intercontinental bombers from the 5th Strategic Reconnaissance Wing at Travis AFB, California arrived for a few days, and various tanker and transport aircraft also made periodic appearances at the base. Several of the temporary detachments included in-flight refueling tanker aircraft.

Meanwhile, on 30 April 1956, two Lockheed U-2s were airlifted to RAF Lakenheath to form CIA Detachment A. The first flight of the U-2 was on 21 May. The Central Intelligence Agency unit did not remain long, moving to Wiesbaden Air Base, West Germany on 15 June. (Italics added by author for emphasis).

A near nuclear accident occurred on 27 July 1956 - when a B-47 bomber crashed into a storage igloo at Lakenheath containing three MK-6 nuclear weapons while on a routine training mission. (Italics added by author for emphasis).

Although the bombs involved in the accident did not have their fissile cores installed, each of them carried about 8,000 pounds of high explosives as part of their trigger mechanism. The crash and ensuing fire did not ignite the high explosives and no denotation occurred. The damaged weapons and components were later returned to the Atomic Energy Commission. The B-47 involved in the accident, which killed four crewmen, was part of the 307th Bombardment Wing. http://en.wikipedia.org/wiki/RAF_Lakenheath

An upgrade to Lakenheath runways for Very Heavy Bombers such as the Boeing B-29 Superfortresses, the B-50 Superfortresses and which were then replaced in June 1954 by the B-47 Stratojets, followed by a beehive of American deployment of aircraft and arsenal to Lakenheath and other nearby airbases ***meant only one thing, that the United States Air Force was moving nuclear weapons into Britain!*** (Italics added by author for emphasis).

Until about 2006 the US continued to store nuclear weapons in the UK when approximately 110 tactical B61 nuclear bombs stored at RAF Lakenheath for deployment by USAF F-15E Strike Eagle aircraft were removed. However, well before that time, Britain had already developed its own thermonuclear weapons.

Suddenly, it becomes clear why ETI had an interest in the Lakenheath Air Base and the neighbouring military air bases and why aerial incursions in this area have taken place. ETI were monitoring the mobilization of nuclear weaponry from one continent to another and the movement of nuclear material was like a big red flashing beacon saying, here we are! In fact anytime you want to get the attention of Extraterrestrial Intelligences to a particular area, you merely have to "hang your radioactive laundry (hardware) out in the breeze" and they will show up to investigate!

Rendlesham Forest UFO Incident - 1980

RAF Lakenheath was not the only RAF air base to have over flights UFOs showed up particularly in this part of Britain. In December 1980, RAF Bentwaters and RAF Woodbridge Air Bases received worldwide attention when USAF military personnel tenanted there experienced a major UFO encounter event over a 2 – 3 day period that has become known as the **Rendlesham Forest Incident** or the **Bentwaters – Woodbridge UFO Incident.**

RAF Bentwaters, now known as **Bentwaters Parks**, is a former Royal Air Force station about 80 miles (130 km) northeast of London, 10 miles (16 km) east-northeast of Ipswich, near Woodbridge, Suffolk in England. The name was taken from two cottages ('Bentwaters Cottages') that had stood on the site of the main runway during its construction in 1943.

It was used by the RAF during World War II and by the United States Air Force during the Cold War, being the primary home for the 81st Fighter Wing under various designations from 1951 to 1993. For many years the 81st Fighter Wing also operated RAF Woodbridge, with Bentwaters and Woodbridge airfields being known by the Americans as the "Twin Bases".

Some Ufologists believe it is perhaps the most famous UFO event to have happened in Britain, along with the **Berwyn Mountain UFO incident**, ranking amongst the best-known UFO events worldwide. It has been compared to the Roswell UFO incident in the United States, and is sometimes referred to as "Britain's Roswell". *"Minister warned over 'UK Roswell'". BBC News. 2009-08-17. Retrieved 2012-05-21.*

The incident occurred in the vicinity of two former military bases - RAF Bentwaters, which is just to the north of the forest, and RAF Woodbridge which extends into the forest from the west and is bounded by the forest on its northern and eastern edges. At the time, both were being used by the United States Air Force and were under the command of wing commander **Colonel Gordon E. Williams**. The base commander was **Colonel Ted Conrad**, and his deputy was **Lieutenant Colonel Charles I. Halt**. Halt's memo to the Ministry of Defence on the incident and his personal involvement in the second night of the sightings has given the case credibility. http://en.wikipedia.org/wiki/Rendlesham_Forest_incident

The Rendlesham Incident occurred over a three-day span in 1980. It is uncertain whether the incident first occurred on December 26 or December 27, although witness statements suggest the 26th. The **"Halt Memo"** states the 27th, but Halt admits that it should be the 26th. On the 27th, the **"Halt Tape"** was recorded.

446

Map showing the locations of RAF Lakenheath and RAF Bentwaters within Suffolk where UFOs flew over these air bases in 1956 and 1980, respectively.

https://en.wikipedia.org/wiki/RAF_Woodbridge

Retired **Sgt. John Burroughs (LE)** states that the events took place over three successive nights (pm into am); 24–25, 25–26 and 26–27 December 1980. One of the key pieces of primary evidence suggests that the first sightings were on the 26th, rather than 25th. The memo was written almost two weeks after the event and its author later agreed that he had probably made a mistake in his recollection of the dates. This discrepancy in dates has not only confused subsequent researchers but also led to confusion at the time, for example in the MoD's investigation and analysis of contemporaneous radar records. Nevertheless, Halt's memo and his account of the events during that time period remain credible.
http://www.theblackvault.com/wiki/index.php/Rendlesham_Forest_Incident

The **Ministry of Defence (MoD)** denied that the event posed any threat to national security, and stated that it was therefore never investigated as a security matter. Later evidence indicated that there was a substantial MoD file on the subject, which led to claims of a cover-up; some interpreted this as part of a larger pattern of information suppression concerning the true nature of unidentified flying objects, by both the United States and British governments. One such person to take this view was eyewitness and **Deputy Base Commander Colonel Charles Halt**. When the file was released in 2001 it turned out to consist mostly of internal correspondence and responses to inquiries from the public. The lack of any in-depth investigation in the publicly released documents is consistent with the MoD's earlier statement that they never took the case seriously. http://en.wikipedia.org/wiki/Rendlesham_Forest_incident

447

Some U.S. airmen witnessed a conical metallic object suspended in a yellow mist hovering over a clearing in the trees, with a pulsating blue and red circle of light above. One serviceman, Sgt. Jim Penniston stated he encountered a "craft of unknown origin", made detailed notes of its features, touched its "warm" surface, and copied symbols on its body.
https://www.youtube.com/watch?v=a7Btl5Y-DJs

In typical debunking fashion, skeptics have ignored the hardcore facts and statements of the witnesses choosing to thrown out any of the detailed evidence they don't agree with like the

proverbial adage of "throwing the baby out with the bathwater" and simply regard the sightings as a misinterpretation of a series of nocturnal lights – a fireball, the Orford Ness lighthouse and bright stars. Other explanations for the incident have also included a downed Soviet spy device and a possible a nuclear incident similar to the one that occurred in RAF Lakenheath in July 1956.

Author's Rant: Personally, I think it's time for all professional debunkers and skeptics to stand aside and stop hindering the serious investigation into UFO accounts. Their weighing in on such matters only muddy's the research process and creates confusion in the minds of scientists and the general public. But then, that has always been their agenda when it comes to any serious investigation of any well documented UFO case that has all the right elements of evidence in place. They will either discredit the witness if they can't discredit the evidence or they will simply choose to ignore the critical core evidence! UFO cases like the Rendlesham Forest Incident demonstrate that regardless of a person's professional training as a ground or sky observer, when it comes to discrediting a witness's testimony, their professional training means nothing!]

448

The main events of the incident, including the supposed landing or landings, took place in the forest, which starts at the east end of the base runway or about 0.3 miles (0.5 km) to the east of the **East Gate of RAF Woodbridge**, where guards first noticed mysterious lights appearing to descend into the forest. The forest extends east about 1.0 mile (1.6 km) beyond East Gate, ending at a farmer's field, where additional events allegedly took place.

Orford Ness lighthouse, which skeptics identify as the flashing light seen off to the coast by the airmen, is along the same line of sight but 5 miles (8.0 km) further east of the forest edge. However, bear in mind that most if not all personnel on base would be familiar with this lighthouse at any time, day or night and its location so, it would not be an unusual sight to them if they were observing the horizon or night sky. These are trained ground observers and as such governments and society put their trust and confidence in their ability to know what they are looking at. Misidentifications are simply not going to cut it in the military when a country's security is on the line! **End of discussion, end of subject!!**

Around 3 a.m. on 26 December 1980 strange lights were reported by a security patrol near the East Gate of RAF Woodbridge apparently descending into nearby Rendlesham Forest. Servicemen initially thought it was a downed aircraft but, upon entering the forest to investigate, they saw strange lights moving through the trees, as well as a bright light from an unidentified object. Some U.S. airmen claim to have seen a conical metallic object suspended in a yellow mist hovering over a clearing in the trees, with a pulsating blue and red circle of light above.

One of the servicemen, **Sgt. Jim Penniston** later claimed to have encountered a "craft of unknown origin" and to have made detailed notes of its features, touched its "warm" surface, and copied the numerous symbols on its body. The object allegedly flew away after their brief encounter. Penniston also claimed to have seen triangular landing gear on the object, leaving three impressions in the ground that were visible the next day. The servicemen claimed that the object seemed clearly aware of their presence and moved away from them, forcing them to give chase. Later the servicemen were found in a dazed state, and local reports said that local farmyard and domestic animals had been behaving in a possessed state of fear and panic; some cows went on a stampede.

While undergoing regression hypnosis in 1994 Penniston subsequently claimed that the "craft" he encountered had come from our future, and was occupied by time travellers, not extraterrestrials. Sgt. Penniston's report made shortly after the incident contains no mention of physically encountering an unknown craft, nor of interacting with it. This report and associated sketches are neither signed nor dated, nor are they representative of AF Form 1169, Statement of Witness.

Shortly after 4 a.m. local police were called to the scene but reported that the only lights they could see were those from the Orford Ness lighthouse, some miles away on the coast. Some reports claim that local farmyard animals had been behaving in a state of fear and panic.

After daybreak on the morning of 26 December, servicemen returned to a small clearing near the eastern edge of the forest and found three small impressions in a triangular pattern, as well as burn marks and broken branches on nearby trees. Plaster casts of the imprints were taken and have been shown in television documentaries. At 10.30 a.m. the local police were called out again, this time, to see the impressions on the ground, which they thought could have been made by an animal.

The servicemen returned to the site again in the early hours of 28 December 1980 with radiation detectors, although the significance of the readings they obtained is disputed. The deputy base commander Lt. Col. Charles I. Halt investigated this sighting personally and recorded the events on a micro-cassette recorder. The site investigated by Halt was near the eastern edge of the forest, at approximately 52° 05' 20" N, 1° 26' 57" E.

It was during this investigation that a flashing light was seen across the field to the east, almost in line with a farmhouse. The Orford Ness lighthouse is visible further to the east in the same line of sight.

Later, starlike lights were seen in the sky to the north and south, the brightest of which seemed to beam down a stream of light from time to time. Halt described this light like a laser beam which was regarded as a warning to them not to proceed further or as a type of probe beam as it moved over the base's **weapon storage area (WSA).**

There are claims that the incident was videoed by the USAF; but, if so, the resulting tape has as yet, not been made public.

The first public report of the incident was published in the tabloid newspaper *News of the World*, on 2 October 1983, beneath the sensational headline *UFO lands in Suffolk – and that's official*. The story was based on an account by a former US airman, using the pseudonym Art Wallace (supposedly to protect himself against retribution from the USAF), although his real name was **Larry Warren**

The first piece of primary evidence to be made available to the public was a memorandum written by the deputy base commander, Lt. Col. Charles I. Halt, to the Ministry of Defence (MoD). Known as the **"Halt memo",** this was made available publicly in the United States under the **US Freedom of Information Act** in 1983. The memorandum (left), was dated "13 Jan 81" and headed "Unexplained Lights". The two-week delay between the incident and the report might account for errors in dates and times given. The memo was not classified in any way.

450

REPLY TO
ATTN OF: CD   13 Jan 81

SUBJECT: Unexplained Lights

TO: RAF/CC

1. Early in the morning of 27 Dec 80 (approximately 0300L), two USAF security police patrolmen saw unusual lights outside the back gate at RAF Woodbridge. Thinking an aircraft might have crashed or been forced down, they called for permission to go outside the gate to investigate. The on-duty flight chief responded and allowed three patrolmen to proceed on foot. The individuals reported seeing a strange glowing object in the forest. The object was described as being metalic in appearance and triangular in shape, approximately two to three meters across the base and approximately two meters high. It illuminated the entire forest with a white light. The object itself had a pulsing red light on top and a bank(s) of blue lights underneath. The object was hovering or on legs. As the patrolmen approached the object, it maneuvered through the trees and disappeared. At this time the animals on a nearby farm went into a frenzy. The object was briefly sighted approximately an hour later near the back gate.

2. The next day, three depressions 1 1/2" deep and 7" in diameter were found where the object had been sighted on the ground. The following night (29 Dec 80) the area was checked for radiation. Beta/gamma readings of 0.1 milliroentgens were recorded with peak readings in the three depressions and near the center of the triangle formed by the depressions. A nearby tree had moderate (.05-.07) readings on the side of the tree toward the depressions.

3. Later in the night a red sun-like light was seen through the trees. It moved about and pulsed. At one point it appeared to throw off glowing particles and then broke into five separate white objects and then disappeared. Immediately thereafter, three star-like objects were noticed in the sky, two objects to the north and one to the south, all of which were about 10° off the horizon. The objects moved rapidly in sharp angular movements and displayed red, green and blue lights. The objects to the north appeared to be elliptical through an 8-12 power lens. They then turned to full circles. The objects to the north remained in the sky for an hour or more. The object to the south was visible for two or three hours and beamed down a stream of light from time to time. Numerous individuals, including the undersigned, witnessed the activities in paragraphs 2 and 3.

CHARLES I. HALT, Lt Col, USAF
Deputy Base Commander

Letter from Lt. Col. Charles Halt

Former Air Force Col. Charles Halt was deputy base commander at a joint British/American airbase, RAF Bentwaters, in 1980. Halt saw a disc-shaped UFO directing beams of light down to the ground near him.

http://www.therendleshamforestincident.com/Halt_Memo_and_Tape.html

It has been rumored that small beings with domed heads left the craft, and that twin base **Commander Gordon Williams** claimed to have made sign language communication with the aliens. It must be stated that these rumours have no evidence to back them up and come from unconfirmed reports.

A short while after the incident, the Deputy Commander of Bentwaters air base, Lt. Col. Charles Halt, issued the following report to the **Ministry of Defence (MOD):** "The object was described as being metallic in appearance and triangular in shape, approximately three metres across the base and two metres high. It illuminated the entire forest with a white light ... As the patrolmen approached the object, it maneuvered through the trees and disappeared. At this time the animals on a nearby farm went into a frenzy. The next day three depressions were found where the object had been sighted on the ground."

452

The Scottish researcher James Easton succeeded in obtaining the original witness statements made for Col. Halt **by Fred A. Buran, 81st Security Police Squadron, Airman First Class John Burroughs, 81st LE, Airman Edward N. Cabansag, 81st Security Police Squadron, Master-Sergeant J. D. Chandler, 81st Security Police Squadron and Staff-Sergeant Jim Penniston, 81st Security Police Squadron.**

These documents describe the sightings of strange lights. Penniston, for instance, states that "directly to the east [of East Gate] about $1\frac{1}{2}$ miles [2.4km] in a large wooded area...a large yellow glowing light was emitting above the trees. In the center of the lighted area directly in the center ground level, there was a red light blinking on and off 5 to 10 sec intervals. And a blue light that was being for the most part steady." Burroughs, Penniston, and Cabansag drove into the forest in search of the source of the lights.

They heard strange noises, too. Burroughs reported a noise "like a woman was screaming" and also that "you could hear the farm animals making a lot of noises". Halt heard the same noises two nights later. In a CNN interview in January 2008, he said: "The livestock around the barn seemed to be going crazy". Such noise could also have been made by Muntjac deer in the forest, which are known for their loud, shrill bark when alarmed. Cabansag said: "We figured the lights were coming from past the forest since nothing was visible when we passed through the woody forest. We would see a glowing near the beacon light, but as we got closer we found it to be a lit-up farmhouse. We got to a vantage point where we could determine that what we were chasing was only a beacon light off in the distance." Burroughs' statement also states that "We could see a beacon going around so we went towards it. We followed it for about two miles [3 km] before we could [see] it was coming from a lighthouse."

Penniston's statement is the only one that positively identifies a mechanical object as the source of the lights. He states that he was within 160 feet (50 m) of the object and "it was definitely mechanical in nature". Penniston has subsequently claimed that contrary to his statement at the time, he actually encountered a landed craft in the forest which he circled, touched and made notes of for 45 minutes, although there is no corroborating evidence of this from other witnesses. Penniston has shown on television a notebook in which he claims to have made real-time notes and sketches of the object. The notebook is headed with the date 27 December and the time 12:20 (00:20 GMT), which does not accord with the date and time given by the other witnesses for the incident.

Penniston claims that he saw the object at a different landing site from the one investigated by Halt, much closer to RAF Woodbridge. This is inconsistent with his initial assessment that the light lay a mile and a half from East Gate.

The witnesses were unnerved by their experience and believed that they had witnessed something, as Buran expresses it, "out of the realm of explanation".

Also, in 1984, a copy of what became known as the **"Halt Tape"** fell into the hands of researchers. Unfortunately, because of static and the fact that the tape had been dubbed on an old machine, much of its background conversations could not be discerned. The Sci Fi Channel had later acquired the original recording, which documents Lt. Colonel Halt (USAF) and his patrol investigating a UFO sighting in Rendlesham Forest in December 1980. This tape not only reveals much more of the background conversations but features names that could not be heard on the poor-quality 1984 dub.

The British Government has released files to researchers, but the United States continues to remain silent despite the SciFi Channel-sponsored investigation entitled "UFO Invasion at Rendlesham", the History Channels "UFO Files - Britain's Roswell" and Coalition for Freedom of Information inquiries.

There has been much controversy over this account within consistencies in dates and details even among witnesses that British Ufologist Jenny Randles who originally brought the case to prominence, came to the conclusion that there were some puzzles about the case that remained, but felt that no unearthly craft were seen in **Rendlesham Forest** and that the events were a series of misperceptions of everyday things encountered in unusual circumstances.

On the other hand, the late **Lord Hill-Norton, (Admiral of the Fleet and former Chief of the Defence Staff of the UK)** believed that a UFO landed at Rendlesham and repeatedly questioned the UK Government on the issue.

Larry Warren who was the source of the original *News of the World* article has written extensively on the subject and is a firm believer in an extraterrestrial explanation. Warren was certainly a USAF airman at the **Woodbridge base,** but his own claims that he was a witness to the incident are disputed by others, notably by Col. Halt.

Newspaper clipping of the Bentwaters UFO Incident at Rendlesham Forest

http://www.telegraph.co.uk/culture/film/film-news/11415831/Rendlesham-Forest-UFO-incident-truth-conspiracies.html

One of the most prominent believers in the extraterrestrial origin of the Rendlesham UFOs is **Nick Pope** who worked for the MoD, researching and investigating UFO phenomena between 1991 and 1994. He discussed the **Rendlesham Forest Incident** in his various books and his articles and has gone on record as saying that *"the Rendlesham Forest Incident is bigger than Roswell"* (quoted on Sci Fi Channel).

Lieutenant Colonel (later Colonel) **Charles I Halt**, the former Deputy Base Commander of USAF Bentwaters and Woodbridge, who was a major witness to these events, believes he witnessed an extraterrestrial event that was then covered up. Halt was also a speaker at the **National Press Club** in Washington D.C. on 27 September 2010, one of half a dozen former Air Force officers testifying on the subject of *"U.S. Nuclear Weapons Have Been Compromised by Unidentified Aerial Objects".*

RAF Bentwaters is another air base tenanted by the USAF from the British government and is also known to have stored nuclear weapons on the base. UFOs, that is Extraterrestrial spacecraft with their overflights of military air bases or naval bases or missile sites compromise the military's location of nuclear weapons. In doing so, they point out to the military forces of any nation that, "We know where your weapons of mass destruction are located" and as we shall see, they were also able to neutralize the destructive capability of these weapons!

It is quite possible that many people can see the same thing and yet come away from the observation of the event with completely different perspectives. This may be the case with the witnesses involved in the Bentwaters UFO event and explains the differences and inconsistencies that have arisen between eyewitness accounts of the same event over a three day period. Keep in mind that these men are trained professional in their particular area of expertise. This case is a prime example of "**high strangeness** often referred to by the late **J. Allen Hynek** as associated with **Close Encounters of First, Second and Third Kind** that seem to transcend the physical reality of our perceptions. This is due in part to the reality of one's life experiences, their perceptions and interactions of that reality within their environment and the ability to process that information into something meaningful, a sort of Jungian archetype.

Recall earlier in this book, in which world-renown Swiss psychiatrist **Carl Jung** spoke about the proto-archetype of reality. An archetype at its essence is simply an expression of original energy, the ideal behind the idea. Jungian archetypes are similarly prototypical energies. They are the idea behind the experience.

An archetypal energy, or concept, precedes and forms the structure of physical manifestations. Sometimes referred to as "racial memory," the collective unconscious consists of the totality of human experience and wisdom. Archetypes, and what they symbolize, spring from the depths of collective experience and are usually universally recognized and understood.

Author's Rant: A personal example of this occurred when the members of Vancouver CSETI and I experienced high strangeness while demonstrating the CSETI protocols in front of a national television crew, who were filming a week of UFO related topics. After the interview and demonstration, our team left the work site in two cars; as we moved down a major highway, all of us witnessed a large triangular craft from our moving vehicles. What was interesting about this event was that everyone saw the object but, of the six people who saw the ET craft, only one lady reported that what she saw was a boomerang shape object while the rest of us saw a triangular craft. It was the same object yet, described differently based upon the individual life experiences and understanding of reality. It is possible of course that the ETI deliberately created altered manifestations of their craft's reality to each individual in our group based on our individual perceptions of reality. This would imply that ETI possesses transdimensional technology that interacts with our consciousness and physical reality which is outside of our three dimensional reality.

There are many reports from around the world where a group of people to a UFO event see an object but, some of the people in the group do not see the same object even when everyone in the group is all looking in the same direction where the object is reported to be located. The ability to see something physically may actually be the ability to see with the inner eye first which registers in the mind of the individual, who is open and receptive, before the outer eye physically acknowledges its existence.

456

This is true for example, when the first Europeans came to the shores of North and South America in their tall sailing ships, they were greeted by the indigenous people of America. Most Native Americans who witnessed the Europeans explorers coming to shore in small row boats could not "see" the tall sailing ship just anchoured off shore which was clearly visible, even when the explorers explained to these Native people that they had come across the ocean in these tall ships and not the small row boats. They simply could not see these ships as they had no point of reference to understand the concept of tall ships with large sails. This is a historical fact. It took awhile before a few Native people, perhaps the shaman was the first individual who could physically"see" these ships and then, eventually everyone in the Americas was able to "see" the sailing ships. It would seem that only a few who were more consciously evolved or receptive than the others were the first to understand this new paradigm of reality. They in turn culturally discriminated the new concepts to the rest of society then collectively, they were able to understand this new reality.

Regardless of the true nature of reality, one thing was certain that ET spacecraft were definitely interacting in a physical way with the nuclear missiles of most of the world's military air and naval bases. No nation's missile base silos were beyond the intrusive penetration of alien craft which were now monitoring the military movements and weapon storage bases of the planet.

Minot AFB B-52 UFO Incident - 1968

For over three hours on October 24th, 1968, numerous ground witnesses observed an unconventional aerial object maneuvering over the [Minuteman] ICBM missile fields surrounding **Minot Air Force Base**, North Dakota. During this time, an unidentified object paced a B-52 and was recorded on radar. Later, the crew of the B-52 overflew and observed a large glowing object on or near the ground.

As airmen on the ground were witnessing a brightly-lit object hovering above the ground, a B52 flying in the area was ordered to divert it flight penetration exercises to go and investigate what the airmen on the ground were seeing. The crew clearly saw a structured craft and they appear on camera to describe their experiences. The co-pilot, an Air Force captain, is certain that what he saw was an alien spacecraft. The navigator picked it up on his radar scope and we are shown photographs of the actual blip as it paced the aircraft. When it vanished from the scope, they turned the aircraft in an attempt to locate the UFO visually. They saw it hovering close to the ground. It was described as at least 200-feet in diameter, hundreds of feet long, glowing yellow, with a metallic cylinder that was attached.

The crew of the B52 and sixteen ground witnesses attested that they saw a UFO that night. **Blue Book** came to the astonishing conclusion that what they actually saw were nothing more than stars!

There were radar sightings from ground and weather's radar. There were visual sightings from the crew of the B-52, and an airborne radar sighting where the target traveled at 3,000 miles per hour. Scope photographs were taken. There were sightings made by S.Sgt. Bond the FSC at Nov. Flight, S.Sgt. Smith at Oscar-1, Julelt, and Mike Flight Team and a number of men in widely scattered locations. The object landed at location AA-43 and the entire observation lasted for 45 minutes. Fourteen other people in separate locations also reported the UFO.

At about 0300 hours (3:00 A.M.) local, a B-52 that was about 30 miles northwest of Minot AFB and making practice penetrations sighted an unidentified blip on their radars.

Initially, the target travelled approximately 2 1/2 miles in 3 sec. or at about 3,000 mph. After passing from the right to the left of the plane it assumed a position off the left wing of the 52. The blip stayed off the left wing for approximately 20 miles at which point it broke off. Scope photographs were taken.

When the target was close to the B-52 neither of the two transmitters in the B-52 would operate properly but when it broke off both returned to normal operation.

At about this time a missile maintenance man called in and reported sighting a bright orangish-red object. The object was hovering at about 1000 ft, or so, and had a sound similar to a jet engine. The observer had stopped his car, but he then started it up again. As he started to move, the object followed him, then accelerated and appeared to stop at about 6-8 miles away. The observer shortly afterward lost sight of it.

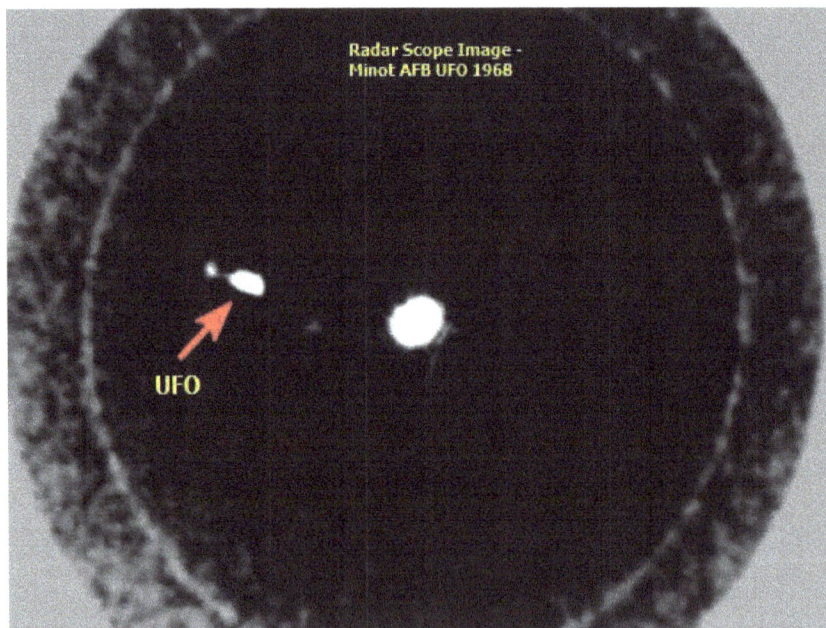

Here's the image of the object taken from the radarscope
http://ufo-blogg.blogspot.ca/2012/04/ufo-over-minot-afb-north-dakota-usa-1968.html

In response to the maintenance man's call, the B-52, which had continued its penetration run, was vectored toward the visual which was about 10 miles northwest of the base. The B-52 confirmed having sighted a bright light of some type that appeared to be hovering just over or on the ground.

458

Also, at this approximate time, security alarm for one of the sites was activated. This was an alarm for both the outer and inner ring. When guards arrived at the scene they found that the outer door was open and the combination lock on the inner door had been removed."

DOD, USAF, and CIA document reveal that during October, November, and December of 1975, reliable military personnel repeatedly sighted unconventional aerial objects in the vicinity of nuclear weapons storage areas, aircraft alert areas, and nuclear-missile control facilities at Loring Air Force Base, Maine; Wurtsmith AFB Michigan; Malstrom AFB, Montana; Minot AFB, North Dakota; and Canadian Air Forces Station, Falconbridge, Ontario. Many of the sightings were confirmed by radar. At Loring AFB, the interloper "demonstrated a clear intent on the weapons storage areas."

The incidents drew the attention of the **CIA**, the **Joint Chiefs of Staff**, and the **Secretary of Defense**. Though the Air Force informed the public and the press that individual sightings were isolated incidents, an Air Force document says that "Security Option III was implemented and that security measures were coordinated with fifteen (15) Air Force bases from Guam to Newfoundland. Another AF document reveals that the Air Force conducted an investigation into the incidents but found no explanation for their occurrence.

It appears Air Force "security measures" provided no protection against the "invasion." One month later, on January 21, 1976, UFOs "25 yards in diameter, gold or silver in color with a blue light on top, hole in middle, and red light on bottom" were observed, "near the flight line of Cannon AFB, N.M." Ten days later, on January 31, a UFO was observed "over the ammo storage area" at Fort Richie, Maryland.

From 1948 through 1950, an FBI document reveals, UFOs were sighted by "persons whose reliability is not questioned, "near highly sensitive military and government installations, including nuclear weapons design, construction, testing and stockpiling sites. Security officials were greatly alarmed by these incidents."

A CIA document reveals that in 1952 "sighting of unexplained objects at great altitudes and traveling at high speeds" were reported in the vicinity of major U.S. defense installations and posed a threat to national security.

Apparently, this wasn't the first time that Minot AFB had an Extraterrestrial visitation to see what was going on below. http://www.abovetopsecret.com/forum/thread546470/pg1

March 5, 1967; Minot AFB, North Dakota

Witnesses: Air Force security police and other base personnel.
Features: disc, radar-visual sighting, flashing body lights, Hover acceleration.

Air Defense Command radar tracked an unidentified target descending over the Minuteman missile silos of the 91st Strategic Missile Wing. Base security teams saw a metallic, disc-shaped object ringed with bright flashing lights moving slowly, maneuvering, then stopping and hovering about 500 feet off the ground. After a while, the object circled directly over the launch

control facility. F-106 fighter-interceptors were scrambled to investigate. At that moment the UFO "climbed straight up and streaked away at incredible speed." **(Fowler, 1981, p. 187)**
http://www.abovetopsecret.com/forum/thread546470/pg1

UFOs Intrude Into SAC Base Weapons Areas: October - November 1975

In Lawrence Fawcett and Barry J. Greenwood's book "The UFO Cover-up", the authors were seeking **Freedom of Information Act (FOIA)** requests on military UFO reports which came opportunistically by way of an article seen in the **National Enquirer** newspaper. The National Enquirer publication was not normally noted for its accuracy in UFO journalism based upon the past poor reporting of such subject matters instead most reports often appeal to sensationalism for the sake of its readership. However, the December 13, 1977, issue had an article entitled "UFOs Spotted at Nuclear Bases and Missile Sites" was different from other issues that reported UFO accounts, in that it gave times, places and some verifiable detail. Lawrence Fawcett and Barry J. Greenwood; The UFO Cover-up (Originally published as *Clear Intent*); 1984; Published by Simon and Schuster Inc.; ISBN 0-671-76555-8

FOIA case files came back to Fawcett and Greenwood from Loring AFB, Maine; Wurtsmith AFB, Michigan; and Malstrom AFB, Montana.

Over a period of about three weeks in October and November of 1975, several Strategic Air Command (SAC) bases in the northern tier states were placed on a high priority (Security Option 3) alert because of repeated intrusions of unidentified aircraft flying at low altitude over atomic weapons storage areas.

The Commander-in-Chief of North American Air Defense Command (NORAD) sent a four-part message to NORAD units on November 11, 1975, summarizing the events. Some excerpts follow:

"Since 28 Oct 75 numerous reports of suspicious objects have been received at the NORAD CU; reliable military personnel at **Loring AFB, Maine, Wurtsmith AFB, Michigan, Malmstrom AFB, Mt, Minot AFB, ND, and Canadian Forces Station, Falconbridge, Ontario, Canada** have visually sighted suspicious objects."

On October 27-28, 1975, **Staff Sgt. Danny K. Lewis**, 42nd Security Police Squadron, while on duty at the munitions storage area of Loring AFB, Maine, at 7:45 p.m. saw an apparent aircraft at low altitude along the northern perimeter of the base. Other witnesses were **Sgt. Clifton W. Blakeslee and Staff Sgt. William J. Long**. The craft had a red light and a pulsating white light.

A teletype message to the **National Military Command Center** in Washington, D.C., said: "The A/C [aircraft] definitely penetrated the LAFB [Loring Air Force Base] northern perimeter and on one occasion was within 300 yards of the munitions storage area perimeter."

Staff Sgt. James P. Sampley, 2192nd Communications Squadron, was on duty in the control tower when he picked up the craft on radar nearing the base. He tried to make radio contact to warn the presumed aircraft that it was entering a restricted area. The craft began to circle in the

460

vicinity of the nuclear weapons storage area at about 150 feet altitude. When it penetrated the nuclear weapons storage area at an estimated 300 feet altitude, Lewis reported it to the command post.

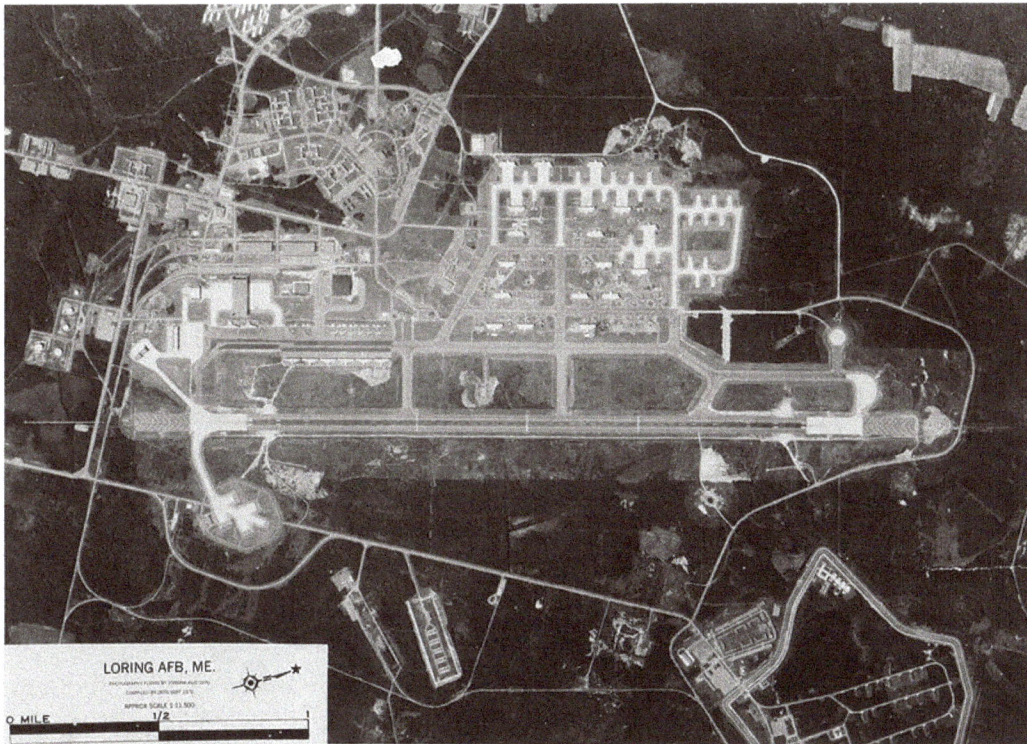

Loring Air Force Base - Limestone Air Force Base located at Limestone, Maine
https://en.wikipedia.org/wiki/Loring_International_Airport

At 8:45 p.m. **Sgt. Grover K. Eggleston** was on duty in the control tower when a call came from the command post requesting a radar track on the mysterious craft. For 40 minutes he observed the object on radar circling around, then it abruptly vanished from the screen as if it had either landed or dropped below the minimum level of radar coverage.

Witnesses observed the craft flying away toward New Brunswick. Radar tracked it again as it receded from the base until contact was lost in the vicinity of Grand Falls, 12 miles from Loring AFB.

Priority messages on the incident were sent to the National Military Command Center in Washington, D.C., the Air Force Chief of Staff, SAC headquarters, and other major commands. Loring remained on a high state of alert into the following morning. Efforts to identify the "aircraft" through the Maine State Police, local police departments, and the Federal Aviation Administration office in Houlton, Maine, were not successful.

The next night at 7:45 p.m. a similar craft with body lights again approached the base, tracked on radar, and maneuvered around in the vicinity for more than 30 minutes. Its speed and motions were similar to those of a helicopter. The craft would appear and disappear from view. Its lights went off and the craft reappeared over the weapons storage area at 150 feet (45 meters).

461

At about this time **Sgt. Steven Eichner**, a B-52 crew chief, **Sgt. R. Jones** and other crew members spotted an unidentified red and orange object over the flight line. It looked like a "stretched out football" (cigar-shaped) and was hovering in mid-air.

As the B-52 crew watched, the lights on the object went out and it disappeared, but soon reappeared over the north end of the runway, moving in a jerky, erratic fashion. When it stopped and hovered, Eichner and the others jumped into a truck and drove toward the object. As they turned onto the road that led to the weapons storage area, they encountered the object about 300 feet ahead hovering about five feet off the ground. Its length appeared to be equivalent to about four car lengths.

An eyewitness to the UFO presence on the second night, Sgt. Steven M. Eichner, described the unknown aerial intruder as an elongated football, as long as four cars, and reddish-orange in color, hovering just above the ground.

In a sworn affidavit, Eichner stated, "The object looked like all the colors were blending together, as [if] you were looking at a desert scene. You see waves of heat rising off the desert floor. This is what I saw. There were these waves in front of the object and all the colors were blending together. The object was solid and we could not hear any noise coming from it." Greenwood, Barry and Fawcett, Lawrence. Clear Intent: The Government Cover-up of the UFO Experience, Prentice-Hall Inc., 1984. pp. 41-45

The object again was tracked on radar as it departed towards New Brunswick. And once again priority messages about the intrusion were sent to higher commands. No explanation was ever found.

On October 30, 1975, at Wurtsmith AFB, Michigan, about 10:10 p.m., base personnel saw the running lights of a low-flying craft thought to be a helicopter, except that it hovered and moved up and down erratically near the base perimeter. One white light was pointed downward and two red lights were visible near the trailing edge.

Airman Michael J. Myers, an air policeman on duty near the main gate, saw several unidentified lights near the western edge of the base. The object turned north and appeared to lose altitude.

Between 10:15 and 10:25 p.m., base security police at the back gate of Wurtsmith reported to the command post that an unidentified "helicopter" with no lights had come over the back gate and hovered at low altitude over the weapons storage area. Radar was also tracking low-flying objects intruding into the base, and an incoming KC-135 confirmed seeing two unidentified craft that sped away each time they attempted to close in for identification.

A teletype message November 2nd from Loring Air Force Base, Maine, Office of Special Investigations (OSI) detachment to the National Military Command Center and OSI headquarters in Washington, D.C., reported another "unidentified helicopter sighted at low level over Loring AFB" over the past two nights (October 31-November 1). It also referred to the intruder as an "unknown entity."

462

Capt. Richard R. Fuhs, Operations Officer, 42nd Security Police Squadron (SPS), "...advised that there had been three verified sightings of an unidentified A/C [aircraft] flying at low level over and in the vicinity of LAFB [Loring Air Force Base]" during this period. The initial sighting was made by Staff Sgt. Michael D. Scott, 42nd SPS, on duty at 11:14 p.m. Tech. Sgt. David E. Mott, Flight Chief, 42nd SPS, spotted the object from a position near the East Gate just past midnight, flying from east to west.

At Malmstrom AFB, Montana, November 7, 1975, electronic sensors at the Minuteman missile sites triggered an alarm indicating a breach of the K-7 site security at about 3:00 p.m. A Sabotage Alert Team headed toward the site, and from a distance of about a mile saw a glowing orange object over the area. As they came closer they could see that it was an enormous disc, the size of a football field, whose light was illuminating the missile site. They reported this to the launch control facility and were ordered to proceed into the site. But they refused to do so because they were fearful of the consequences.

The object then began to rise and was picked up on NORAD radar at about 1,000 feet. Two F-106 interceptors were scrambled from Great Falls and sped toward the area, but the object continued to rise and disappeared off the radar screen at about 200,000 feet. As noted in the NORAD Commander-in-Chief's report cited above, the pilots were unable to obtain a visual sighting. Later investigation established that computer codes in the missile warhead had been altered. Several other UFO sightings at the missile sites later that evening and next day were documented in military records.

During the same time period as the SAC base intrusions, civilians, police officers, military officers, and NORAD radar saw and tracked UFOs that alternately hovered and darted around at high speed at Falconbridge Air Force Station, a radar site near North Bay, Ontario, Canada. The sightings occurred between 3:00 a.m. and 11:00 a.m. local time, November 11, 1975.

NORAD regional director logs obtained by **Larry Fawcett and Barry Greenwood** gave some indication of the events, with times expressed in Greenwich Mean Time (GMT) or "Zebra" (Z) time.

1205 GMT. "Unusual sighting report" made.

1840 GMT. Jet interceptors were scrambled, airborne at 1750Z "due to unusual object sighting...UFO report from Falconbridge."

At 0202Z on November 15, 1975, "Report sent to NCOC Surveillance, referred to **Assistant Command Director Space Defense Center**, and intelligence. These 3 individuals considered the report a UFO report and not an unknown track report."

A detailed NORAD report on the incidents and a subsequent press release based on it, both have been made public. **Raymond E. Fowler** quotes the NORAD report:

"Falconbridge reported [at 4:05 a.m.] search and height finder paints [radar targets] on an object 25 to 30 nautical miles south of the site ranging in altitude from 26,000 feet to 72,000 feet

463

[appearing visually as like a bright star]. With binoculars, the object appeared as a 100-ft. diameter sphere and appeared to have craters [sic] around the outside... To date, efforts by Air Guard helicopters, SAC helicopters and NORAD F-106s have failed to provide positive identification."

On November 13, NORAD issued a press release in Sudbury, Ontario, containing essentially the same information. The press release added that "Two F-106 aircraft of the U.S. Air Force Air National Guard's 171st Fighter Interceptor Squadron at Selfridge ANGB [Air National Guard Base], Michigan, were scrambled; but the pilots reported no contact with the object."

Seven Ontario police officers were among the witnesses to the UFOs that were also tracked on NORAD radar alternately hovering and darting around at high speed. Capt. Gordon Hilchie, director of public affairs for the 22nd NORAD Region Control Center at North Bay, Ontario, acknowledged: "Yes, we saw this so-called UFO at the same time people outside were seeing it too."

Lt. Col. Brian Wooding, Control Center Director, said: "We get quite a few UFO reports, but to my knowledge, this is about the only one we've actually seen on radar, and the only time we've gone to the point of scrambling interceptors. The jets were scrambled because the indications were there was something very evident to a large number of people, and because we did manage to get some sort of radar sighting."

Del Kindschi, spokesman for NORAD headquarters in Colorado Springs, said the UFO was tracked on radar intermittently for six hours, first spotted 25-30 miles south of the radar site. The object zoomed from 26,000 feet to 45,000 feet, "... stopped a while, and then moved up very quickly to 72,000 feet." The first visual sightings were at 3:00 a.m. from Sudbury, Ontario, as brilliant lights that hovered low in the sky, then suddenly shot straight up at tremendous speed.

Lawrence Fawcett and Barry J. Greenwood; The UFO Cover-up (Originally published as *Clear Intent*); 1984, pp. 27-31; Published by Simon and Schuster Inc.; ISBN 0-671-76555-8 24th NORAD Region Senior Director's log excerpt reproduced on page 29, and National Military Command Center memorandum reproduced on pp. 30-31.
Raymond Fowler, Casebook of a UFO Investigator, Englewood Cliffs, N.J., Prentice-Hall, 1981, pp. 190-191.
Fawcett and Greenwood; The UFO Cover-up (Originally published as *Clear Intent*); 1984, pp. 16-19, 46-47.
Copies of Government documents in Fund for UFO Research files. also:
http://www.ufocasebook.com/sacbaseweapons1975.html

It seems that, contrary to the Condon Report (The Scientific Study of Unidentified Flying Objects of the University of Colorado) conclusions on UFOs that *'The history of the past 21 years has repeatedly led Air Force officers to the conclusion that none of the things seen, or thought to have been seen, which pass by the name of UFO reports, constituted any hazard or threat to national security.'* and the often cited U.S. Government claim about *'UFOs being of no defense significance'*, there *was,* nonetheless, **considerable panic** about them over a period of three weeks in October and November, 1975. (Bold italics added by author for emphasis).

464

Several Strategic Air Command bases in the Northern U.S. were placed on a high priority alert due to repeated reports of UFOs flying at low altitude over atomic weapons storage areas and the Commander-in-Chief of North American Air Defense Command sent a four part message to NORAD about the 'intrusions' on Nov 11th. (see documents below).

http://www.abovetopsecret.com/forum/thread556334/pg1

```
                                                        0J:9/5

                                              AL6


                        EXTRACTS

              NORAD COMMAND DIRECTOR'S LOG (1975)

29 Oct 75/0630Z:   Command Director called by Air Force Operations
                   Center concerning an unknown helicopter landing
                   in the munitions storage area at Loring AFB, Maine.
                   Apparently this was second night in a row for this
                   occurrence.  There was also an indication, but not
                   confirmed, that Canadian bases had been overflown
                   by a helicopter.

31 Oct 75/0445Z:   Report from Wurtsmith AFB through Air Force Ops
                   Center - incident at 0355Z.  Helicopter hovered over
                   SAC weapons storage area then departed area.  Tanker
                   flying at 2700 feet made both visual sighting and
                   radar skin paint.  Tracked object 25NM SE over
                   Lake Huron where contact was lost.

 1 Nov 75/0920Z:   Received, as info, message from Loring AFB, Maine,
                   citing probable helicopter overflight of base.

 8 Nov 75/0753Z:   24th NORAD Region unknown track J330, heading SSW,
                   12000 feet.  1 to 7 objects, 46.46°N x 109.23W.  Two
                   F-106 scrambled out of Great Falls at 0754Z.  SAC
                   reported visual sighting from Sabotage Alert Teams (SAT
                   K1, K3, L1 and L6 (lights and set sounds).  Weather
                   section states no anomolous propagation or northern
                   lights.  0835Z SAC SAT Teams K3 and L4 report visual,
                   K3 reports target at 300 feet altitude and L4 reports
                   target at 5 miles.  Contact lost at 0820Z.  F-106's
                   returned to base at 0850Z with negative results.
                   0905Z Great Falls radar search and height had inter-
                   mittent contact.  0910Z SAC teams agains had visual
                   (Site C-1, 10 miles SE Stanford, Montana).  0920Z SAC
                   CP reported that when F-106's were in area, targets
                   would turn out lights, and when F-106's left, targets
                   would turn lights on.  F-106's never gained visual
                   or radar contact at anytime due to terrain clearance.
                   This same type of activity has been reported in the
                   Malmstrom area for several days although previous to
                   tonight no unknowns were declared.  The track will
                   be carried as a remaining unknown.

10 Nov 75          Apparently Minot AFB was reportedly "buzzed" by a
                   bright object.  The object's size seemed to be that
                   of an automobile.  It was flying at an altitude of
                   1000 to 2000 feet and was noiseless.  No further
                   information or description has been received by this
                   organization.


                           - 453 -
```

The Commander-in-Chief of North American Air Defense Command sent a four part message to NORAD about the 'intrusions' on Nov 11th.

http://www.abovetopsecret.com/forum/thread556334/pg1

465

b. 24th NORAD Region Senior Director's Log (Malmstrom
AFB, Montana).

7 Nov 75 (1035Z) — Received a call from the 341st Strategic Air
Command Post (SAC CP), saying that the following missile locations
reported seeing a large red to orange to yellow object: M-1,
L-3, LIMA and L-6. The general object location would be 10
miles south of Moore, Montana, and 20 miles east of Buffalo,
Montana. Commander and Deputy for Operations (DO) informed.

7 Nov 75 (1203Z) — SAC advised that the LCF at Harlowton, Montana,
observed an object which emitted a light which illuminated the
site driveway.

7 Nov 75 (1319Z) — SAC advised K-1 says very bright object to
their east is now southeast of them and they are looking at it
with 10x50 binoculars. Object seems to have lights (several)
on it, but no distinct pattern. The orange/gold object overhead
also has small lights on it. SAC also advises female civilian
reports having seen an object bearing south from her position
six miles west of Lewistown.

7 Nov 75 (1327Z) — L-1 reports that the object to their northeast
seems to be issuing a black object from it, tubular in shape.
In all this time, surveillance has not been able to detect any
sort of track except for known traffic.

7 Nov 75 (1355Z) — K-1 and L-1 report that as the sun rises, so do
the objects they have visual.

7 Nov 75 (1429Z) — From SAC CP: As the sun rose, the UFOs disappeared. Commander and DO notified.

2

466

8 Nov 75 (0635Z) - A security camper team at K-4 reported UFO with white lights, one red light 50 yards behind white light. Personnel at K-1 seeing same object.

8 Nov 75 (0645Z) - Height personnel picked up objects 10-13,000 feet, Track J330, EKLB 0648, 18 knots, 9,500 feet. Objects as many as seven, as few as two A/C.

8 Nov 75 (0753Z) - J330 unknown 0753. Stationary/seven knots/ 12,000. One (varies seven objects). None, no possibility, EKLB 3746, two F-106, GTF, SCR 0754. NCOC notified.

8 Nov 75 (0820Z) - Lost radar contact, fighters broken off at 0825, looking in area of J331 (another height finder contact).

8 Nov 75 (0905Z) - From SAC CP: L-sites had fighters and objects; fighters did not get down to objects.

8 Nov 75 (0915Z) - From SAC CP: From four different points: Observed objects and fighters; when fighters arrived in the area, the lights went out; when fighters departed, the lights came back on; to NCOC.

8 Nov 75 (0953Z) - From SAC CP: L-5 reported object increased in speed - high velocity, raised in altitude and now cannot tell the object from stars. To NCOC.

8 Nov 75 (1105Z) - From SAC CP: E-1 reported a bright white light (site is approximately 60 nautical miles north of Lewistown) NCOC notified.

9 Nov 75 (0305Z) - SAC CP called and advised SAC crews at Sites L-1, L-6 and M-1 observing UFO. Object yellowish bright round light 20 miles north of Harlowton, 2 to 4,000 feet.

9 Nov 75 (0320Z) - SAC CP reports UFO 20 miles southeast of Lewistown, orange white disc object. 24th NORAD Region surveil- lance checking area. Surveillance unable to get height check.

9 Nov 75 (0320Z) - FAA Watch Supervisor reported he had five air carriers vicinity of UFO, United Flight 157 reported seeing meteor, "arc welder's blue" in color. SAC CP advised, sites still report seeing object stationary.

9 Nov 75 (0348Z) - SAC CP confirms L-1, sees object, a mobile security team has been directed to get closer and report.

9 Nov 75 (0629Z) - SAC CP advises UFO sighting reported around 0305Z. Cancelled the flight security team from Site L-1, checked area and all secure, no more sightings.

TERRENCE C. JAMES, Colonel, USAF
Director of Administration - 451 -

Cy to: HQ USAF/DAD
HQ USAF/JACL

467

10 Nov 75 (0125Z) - Received a call from SAC CP. Report UFO sighting from site ..-1 around Harlowton area. Surveillance checking area with height finder.

10 Nov 75 (0151Z) - Surveillance report unable to locate track that would correlate with UFO sighted by K-1.

10 Nov 75 (1125Z) - UFO sighting reported by Minot Air Force Station, a bright star-like object in the west, moving east, about the size of a car. First seen approximately 1015Z. Approximately 1120Z, the object passed over the radar station, 1,000 feet to 2,000 feet high, no noise heard. Three people from the site or local area saw the object. NCOC notified.

12 Nov 75 (0230Z) - UFO reported from K01. They say the object is over Big Snowy mtn with a red light on it at high altitude. Attempting to get radar on it from Opheim. Opheim searching from 120° to 140°.

12 Nov 75 (0248Z) - Second UFO in same area reported. Appeared to be sending a beam of light to the ground intermittently. At 0250Z object disappeared.

12 Nov 75 (0251Z) - Reported that both objects have disappeared. Never had any joy (contact) on radar.

13 Nov 75 (0951Z) - SAC CP with UFO report. P-SAT team enroute from R-3 to R-4 saw a white lite, moving from east to west. In sight approx 1 minute. No determination of height, moving towards Brady. No contact on radar.

19 Nov 75 (1327Z) - SAC command post report UFO observed by FSC & a cook, observed object travelling NE between M-8 and M-1 at a fast rate of speed. Object bright white light seen 45 to 50 sec following terrain 200 ft off ground. The light was two to three times brighter than landing lights on a jet.
-----------LAST ENTRY PERTAINING TO THESE INCIDENTS------

3   - 452 -

468

National Enquirer

Dec. 13, 1977

UFOs Spotted at Nuclear Bases And Missile Sites

Mysterious UFOs flew near or hovered over nuclear bomb storage areas of two Strategic Air Command (SAC) bases and near at least 10 missile sites during a three-week period in 1975, The ENQUIRER has learned.

Several times the Air Force scrambled jet fighters and helicopters to chase down the UFOs but failed to intercept them.

By BOB PRATT

One pursuing helicopter was guided to within 1,000 feet of a UFO hovering over a SAC base, but the men aboard said they couldn't see the intruder object — even though men on the ground could see both craft.

In several instances the mysterious UFOs were first reported to be helicopters, but an Air Force spokesman admitted: "The overflights were not identified as helicopters. Unsuccessful attempts were made to trace the craft."

The ENQUIRER uncovered these UFO sightings by digging through official Air Force documents available under the Freedom of Information Act.

On October 27, 28 and 31, 1975, a UFO invaded the security of Loring Air Force Base in Limestone, Maine, hovering over the weapons storage area.

The first night the UFO was seen visually for 35 minutes and tracked on radar for 40 minutes more. The next night the object hovered only 150 feet over the weapons storage area. A helicopter was sent up and flew to within 1,000 feet of the UFO — but the crew could not see it.

On October 30, a UFO was also seen over the weapons storage area of another SAC base, Wurtsmith Air Force Base at Oscoda, Mich.

It was tracked by an Air Force plane on radar. Later the plane's crew reported "visual contact" with two UFOs.

But as the plane approached the UFOs they sped off.

On November 7, officers at a Minuteman missile launch control facility near Lewistown, Mont., were startled to see a large UFO with red and white lights hovering 10 to 15 feet off the ground.

The next night, after a UFO was spotted at eight missile sites in Montana, two F-106 jet fighters were scrambled. But as the jets got close to the UFOs, their lights went out — then went on again when the jets left.

On November 10, a bright object the size of a car passed silently 1,000 to 2,000 feet above a radar station at an Air Defense Command installation, Minot Air Force Station, Max, N. Dak.

Three people saw it, the Air Force reported.

Two days later, November 12, Air Force headquarters in Washington, replying to a question from the North American Air Defense (NORAD) headquarters in Colorado Springs, Colo., regarding "unknown air activity," instructed NORAD and SAC on how to handle queries about the sightings. The message said, in part:

"Unless there is evidence which links sightings or unless media queries link sightings, queries can best be handled individually at the source and as questions arise. Responses should . . . emphasize that the action was taken in response to an isolated or specific incident."

Thus, newsmen inquiring about UFO incidents at one base were not told about similar incidents at other bases.

Dear Mr. Hansen,

Finally found this, per

Mufon Journal Editor Bob Pratt's article for the National Enquirer found on the Alien –UFOs.Com website

Falconbridge Canadian Forces Radar Station, Sudbury, Ontario November 1975

One of the longest and most controversial UFO sightings ever recorded by radar personnel happened in the Sudbury area in northern Ontario. The appearance of unidentified craft on November 11, 1975, prompted NORAD officials to send military jet interceptors to investigate. Despite denials, this move clearly exposed the government's interest in exploring the phenomenon. The Sudbury sightings coincided with an unprecedented week-long flurry of UFO activity over key military installations in both Canada and the United States.

First to spot the objects were two **Sudbury Regional Police Constables, Bob Whiteside, and Alex Keable**. At about 5:00 AM, while patrolling the streets of western Sudbury they spotted four bright objects high up in the sky. Because of the brilliance of the craft, no shape could be discerned, but the officers agreed that the bizarre vessels, which produced no noticeable sound, were definitely not conventional aircraft. One object, brighter than the others, appeared in the southwest and seemed to be bobbing up and down like a ping-pong ball; a second one in the northeast remained stationary, while two others drifted aimlessly.

In the western part of the city, meanwhile, **Constable Gary Chrapynski and Policewoman J.B. Deighton** watched what were presumably the same four objects. They saw light rays being emitted which seemed to illuminate the clouds overhead. Viewed through binoculars, one of the objects looked long and cylindrical, similar in shape to a dirigible. Other police officers stationed at various locations in a thirty-mile radius around Sudbury also reported spotting various types of pulsating, circular craft, noiselessly maneuvering in the early morning sky.

At 6:15 AM, four officers at the **Canadian Forces Radar Station** at Falconbridge, ten miles north of Sudbury, similarly reported three unidentified targets on their Height Finder Radar and Search screens. One appeared to be a very bright stationary light at thirty thousand feet over the station, visible for thirty seconds. Another, spherical in shape, appeared to be rotating while ascending and descending thirty miles south of the station. This object apparently remained visible for over two hours, while maintaining elevations ranging from forty to seventy thousand feet. The third object appeared to be circular, brilliantly lit, with two black spots in the centre, moving upwards at high speeds from 42,000 to 72,000 feet. No circular movement viewed for fourteen minutes.

It was reported that a Major O. *(name of military officer not given to protect identity of witness, although it is presumed to be Major Oliver)* took pictures, but it is not sure whether they will turn out. [Italic added by author].

That same Tuesday, a report in the *Sudbury Star* confirmed that photographs of the mysterious objects had been taken.

In Ottawa, National Defence Headquarters confirmed that four people at the radar station, alerted by the police, saw three bright circles with two black dots about 6:15 AM. The objects were photographed by the base staff.

Later that afternoon, *Star* reporters were advised by the public information office at Defence Headquarters in Ottawa that the photos would soon be released to the press. This was corroborated by Falconbridge radar station personnel, who indicated that the developed prints would be available the following (Wednesday) morning.

When contacted the next morning, the station's commanding officer, **Major Oliver**, made the following surprise announcement: There have been no photographs taken, nor any messages sent to Ottawa that mentioned photographs! He said he had investigated and had found "no one had grabbed a camera."

**Canadian Forces Radar Station at Falconbridge, ten miles north of Sudbury
was the center for a lengthy UFO activity over the radar station
which was also picked up by NORAD**
http://www.noufors.com/cfs_falconbridge_incident_revisited.html

This sudden reversal was in direct conflict with statements issued earlier by Defence Headquarters. What's more, the Ottawa statements confirming the existence of the photos were based primarily on the Telex report sent *from* Falconbridge to Defence Headquarters, which specifically stated: *"Major 0. took pictures, but it is not sure whether they will turn out."* Even more bizarre is the mystery of why, for a period of over twenty-four hours, Defence

Headquarters and the **National Research Council (NRC),** as well as Sudbury Star reporters, were led to believe that the (non-existent) photos would be released to the public! Was this an intra-departmental communications breakdown, or a last-minute cover-up?

We may never know the reasons for the apparent secrecy, but there seems to be no doubt that UFOs were indeed present over Sudbury that morning. In fact, the objects were still in the neighbourhood six hours later, when NORAD officials decided to send up jet interceptors. The *Sudbury Star* reported that: "the fighters were scrambled from the U.S. Air Force base at Selfridge, Michigan, at 12:50 PM local time."

This was eventually confirmed by **Captain Rudy Miller,** public relations officer at the 22nd Division of NORAD in North Bay, who stated that the two F-106 interceptors of the United States Air National Guard Squadron "reported to have a lock on the object. The only thing the pilots reported encountering were sun reflections on ice crystals in the clouds."

It cannot be disputed that the pilots may indeed have observed sun reflections off cirrus clouds. What remains questionable is whether these reflections could account for the many reported sightings. The NORAD explanation clearly overlooked the fact that seemingly geometrical maneuvers were observed both visually and on radar by a variety of qualified witnesses.

Perhaps the most original explanation came from **Dr. Ian Halliday**, research officer at the Herzberg Institute of Astrophysics at the National Research Council, who commented that in all probability what the officers had seen was Venus or Jupiter.

Venus rises around 3 AM high in the southeast and is still bright and high in the sky after sunrise. Jupiter is also bright and sets about 4:30 AM.

As to the sightings registered on radar, Dr. Halliday ventured *"as near as we can tell, it is a coincidence. This sort of thing is not uncommon on radar. They just happened to see one at the same time."*

Meanwhile, area residents continued reporting sightings for the next few days.

More than three years later, the entire matter surfaced again with the release of previously "top secret" documents by the U.S. Air Force and the U.S. Defence Department. The documents, released under provisions of the U.S. Freedom of Information Act, following the successful court action by **Peter Gersten** of **Citizens Against UFO Secrecy (CAUS)** of New York, indicate that the UFO presence over military installations was far more widespread than initially reported. These disclosures were confirmed by National Research Council **(NRC)** officials in Ottawa on January 19, 1979.

According to Research Officer **Dr. Bruce McIntosh** of NRC's Planetary Sciences Section of the Herzberg Institute of Astrophysics, Canadian jets were scrambled to intercept UFOs on at least two occasions during the week-long wave of sightings. It seems that on the night of November 6, six days prior to the Sudbury occurrence, unidentified targets were also spotted on the radar screens at the North Bay NORAD Command base, seventy miles east of Sudbury. As in the

Falconbridge case, the prolonged presence of the targets on the radar screens prompted officials to send up Canadian interceptors later that morning. Nothing was found, according to Dr. McIntosh. During the same period, Canadian interceptors were again scrambled to intercept a UFO that was approaching the Canadian border after it had hovered over the missile launch area at Loring Air Force Base in Maine. The documents gave no indication whether or not the Canadian plane spotted the UFO. The U.S. records also reveal extensive UFO activity over other nuclear missile launch sites and bomber bases along the Canadian border in Maine, Montana and Michigan.

Once again, NRC downplayed the North Bay sighting. One possible explanation proposed by Dr. McIntosh was that the layers of high-density ice crystals could reflect radar beams onto aircraft over the horizon, creating a false radar signal. He also suggested that Venus, "sticking out like a sore thumb," could have accounted for the sighting.
http://noufors.com/falconbridge_canadian_forces_radar_station.html

Military Bases Suspected of UFO Activity

There were other military bases in a steady growing list of bases around the world that also

experienced UFO overflight intrusions, most of them directly over nuclear weapons storage facilities. These UFOs were not planetary or atmospheric phenomena that were interacting with radar ground control units as police officers, private citizens as well as military officials were seeing the same objects either with the naked eye or through binoculars. Private citizens may be forgiven for possible incorrect identification of what they were witnessing although, this is highly unlikely, however, police and military personnel including pilots are highly training in identifying both ground and air observations and not prone to misidentification.

Countries like Australia, Canada, England, Iran, Russia, United States were major target countries that experience sightings of UFO activity over their military bases. To this growing list may be added China, India, North Korea, Pakistan, Israel, Ukraine, France, Germany, Brazil, Argentina, and almost any country that has an R&D program for development and building a nuclear arsenal, complete with **Weapons Storage Areas (WSA).**

UNITED STATES

Andrews Air Force Base:

Location: Southeast of Suitland, Maryland. Very close to Washington, D.C.
Description: UFOs have been spotted over this base by military personnel and public.

Barksdale Air Force Base:

Location: West of Shreveport, Louisiana.
Description: This base was put on alert during the 1975 **Loring AFB intrusions.**

Bolling Air Force Base:

Location: Washington D.C.
Description: UFOs have been spotted at this major Air Force Base.
Brunswick Naval Air Station:

Location: West of Brunswick, in southern Maine.
Description: A UFO was sighted at this base just before the **Loring AFB intrusions** in 1975. A strange abduction occurred less than 40 miles from this base.

Cannon Air Force Base:

Location: East of Clovis, on the eastern edge New Mexico.
Description: UFOs have been sighted over this secretive air base.

Carswell Air Force Base:

Location: Fort Worth, Texas.
Description: It is believed that some of the UFO wreckage from the 1947 Roswell, New Mexico, crash was sent here.

Edwards Air Force Base:

Location: A large section of land east of Rosamond, California, that includes Rogers Lake (Dry) and Rosamond Lake (Dry).
Description: A well-known experimental aircraft testing range, this area has tested many saucer-shaped aircraft. A UFO landed here over a 2-day period in April of 1954. The occupants of the UFO referred to as the "**Etherians**", gave amazing demonstrations to a select group of individuals (including President Eisenhower). The Etherians allowed the military to inspect the UFO and even helped them.

Eglin Air Force Base:

Location: Located on the western edge of Choctawhatchee Bay, south of Valparaiso, Florida.
Description: In 1976, UFOs were spotted over the base's Armament Development and Test Center. This base is just east of Gulf Breeze, Florida, and is a major UFO hotspot.

Ellington Air Force Base (a.k.a. NASA Ellington Field):

Location: Near NASA's Johnson Space Center in Houston, Texas.
Description: Bob Oechsler (a former NASA Engineer and now a famous Ufologist) was flown (by helicopter), in 1990, to a NASA facility 20 miles southwest of Ellington AFB. The facility had a microgravity chamber that, Oechsler is convinced, was made using alien technology.

474

Ellsworth Air Force Base:

Location: South Dakota.
Description: UFOs have been spotted over this air base.

Fairchild Air Force Base:

Location: Washington.
Description: This base was put on alert during the 1975 **Loring AFB intrusions**.

Fort Ritchie:

Location: Maryland.
Description: During 1976, UFOs were spotted over Weapons Storage Facilities on the base.

Grand Forks Air Force Base:

Location: This base is located 16 miles west of Grand Forks, North Dakota.
Description: This base was put on alert during the 1975 **Loring AFB intrusions.** It experienced its own intrusions also during 1975.

Groom Lake Test Facility:

Location: About 120 Miles northwest of Las Vegas, Nevada, in the Nellis Air Force Range. The base is constructed on the edge of the Groom Lake salt bed in the dry Emigrant Valley between two jagged mountain ranges.
Description: Officially, the base that created the **U-2** and **SR-71 Blackbird Spyplanes** doesn't exist. The base cannot be seen on any public map. Old Government maps will list it as **Area 51**. It is a common hotspot for UFOs and it is believed that a new Mach-8 spyplane, nicknamed **Aurora**, has been developed there. It is also believed that an aircraft made from UFO technology, the **Human-Piloted Alien Craft (H-PAC),** was developed there. **Bob Lazar**, a nuclear physicist who claims he was hired by the government to study an alien propulsion system, said that another facility, called **S-4 (Section-4),** is located just south of Groom Lake at Papoose Lake (also dry). There have been many UFO sightings and cattle mutilations in and around the area. Base personnel have disappeared during night time UFO activity. Area 51 has a formidable security system (the area is protected by a group of Green berets). Recently, the USAF gained control over a key ridge **(Freedom Ridge)** near Groom Lake from which you could see the installation. A 1988 Soviet Satellite photo shows a high-security compound at one end of the base, purpose unknown.

Holloman Air Force Base:

Location: Approximately 15 km. southwest of Alamogordo, New Mexico.
Description: This air base is a well-known testing site for experimental aircraft. It is also near where the first A-bomb was tested. On April 25, 1964 (12 hours after the famous Socorro, New Mexico, UFO landing), a UFO (escorted by 2 others that remained in the air) landed at this base.

Three aliens got out of the craft and talked with base officials. This story was confirmed by Richard C. Doty (possible pseudonym), a retired counterintelligence officer with the **Air Force's Office of Special Investigations (OSI).**

Homestead Air Force Base:

Location: Florida.
Description: This base is suspected of keeping alien bodies in a top-secret, underground repository.

Hunter Army Air Field:

Location: Georgia.
Description: In 1976, 2 base security policemen spotted a UFO near this base.

Killeen air Force Base:

Location: Texas.
Description: At Killeen Base in Texas - one of the first nuclear bomb depots – (March 6, 1949). One U.S. Army intelligence report reveals that on the evening of April 28th of that year, a total of 12 guards and other personnel were involved in nine separate sightings of small lighted objects, maneuvering southeast of the weapons depot. One sighting involved a group of four lights; another formation was composed of eight to ten lights. **Clark, Jerome. The UFO Book, Visible Ink Press, 1998, p. 259**

Kinchloe Air Force Base:

Location: Michigan.
Description: This base was put on alert during the 1975 **Loring AFB intrusions**.

Kinross Air Force Base:

Location: Michigan.
Description: UFOs have been sighted over this base and jets have also been scrambled from this base to chase UFOs.
Kirtland Air Force Base:

Location: Albuquerque, New Mexico.

Description: This Airbase is a well-known landing site for space shuttles and is suspected to be harboring alien bodies. During 1980, UFOs were spotted over Kirtland AFB by Base Security Personnel. The UFOs would fly solo and would hover over an area known as Coyote Canyon (part of a restricted testing range on the base). This testing range is used by the Air Force Weapons Laboratories, Sandia Laboratories, Defense Nuclear Agency, and the Department of Energy. Another incident, as summarized in a USAF Office of Special Investigations (AFOSI or simply OSI) memo, involved a disc-shaped UFO that landed not far from the Manzano WSA, located east of Kirtland AFB on August 8, 1980.

Langley Air Force Base:

Location: Just Northeast of Hampton, Virginia.
Description: Also known as Langley Field, this highly-secretive base is suspected to be harboring alien bodies. UFOs have been spotted near this base. Jets have been scrambled from this base to intercept UFOs.

Loring Air Force Base:

Location: Just Northeast of Limestone, Maine.
Description: During 1975, a rash of UFO sightings occurred over this base. The base, along with others, was put on a "Security Option 3" and was told to prepare to "Defend against a helicopter assault". However, the UFOs were never identified as helicopters, and many witnesses said they looked like bright "stretched-football" crafts about the size of a car. Jets were scrambled to intercept the UFOs, but their attempts proved unsuccessful. This base has nuclear weapons installations.

Los Alamos Research Facility:

Location: Los Alamos (north of Santa Fe), New Mexico.
Description: This facility is owned and operated by the **Nuclear Regulatory Commission (NRC)**. It was the site of the famous "**Manhattan Project**" that developed the Atom Bomb. UFOs have been sighted over this facility. This facility is suspected to have analyzed UFO wreckage and EBEs from the **1947 Roswell Crash**. Between 1948 and 1952, UFOs were spotted over the facility by personnel.

Luke Air Force Base:

Location: Arizona.
Description: UFOs have been sighted over this air base.

Malstrom Air Force Base:

Location: Montana.
Description: This is a base that was put on alert in 1975 after the Loring AFB intrusions. In 1975, UFOs were spotted (both visually and by radar) hovering over the base's nuclear missile installations. Closer examination of the missiles revealed that the UFOs had somehow tampered

with the missiles. Jets were scrambled to intercept these UFOs, but the UFOs were far superior to any military aircraft and therefore could not be shot down.

March Air Force Base:

Location: California.
Description: This installation is a Strategic Air Command (SAC) base. UFOs have been spotted over this base.

Maxwell Air Force Base:

Location: Near Prattville, Alabama.
Description: Two base security policemen spotted a UFO over this base.

Minot Air Force Base:

Location: The base is located 13 miles north of Minot, North Dakota.
Description: This was another base put on alert after the **Loring AFB intrusions** of 1975. This base is a Strategic Air Command (SAC) base and has a large amount of land reserved for missile sites. In 1975 it experienced UFO intrusions however, it has had a history of UFO intrusions back in 1967 and 1968. There is something of real interest at this base to have attracted a lot of UFO attention over the years!

NORAD Headquarters:

Location: Inside Cheyenne Mountain, near Colorado Springs, Colorado.
Description: This highly-secretive base is headquarters to the **North American Air Defense (NORAD)** System. It is built inside a mountain, and can withstand a nuclear blast. Its **Deep Space Surveillance Center** is believed to have radar tracking of UFOs.

Norton Air Force Base:

Location: California.
Description: Fighter jets have been dispatched from this base to intercept UFOs.

Oakdale Armory:

Location: Near Greater Pittsburgh International Airport, Pennsylvania.
Description: This armory is home to the 662nd Radar Squadron. This squadron was the military group involved in the cover-up of the Kecksburg, Pennsylvania UFO crash of 1965.

Otis Air National Guard (formerly **Otis Air Force Base**):

Location: Located on the western portion of Cape Cod, in western Barnstable County, Massachusetts, United States.

478

Description: During the 1950s, Otis had quite a few instances of UFOs being associated with it. One of these instances allegedly involved an F-94C Starfire which was said to have vanished along with its radar operator over the base. The pilot reportedly escaped. The incident was mentioned on the **History Channel.** This alleged incident is controversial, and may never have actually occurred.

Pease Air Force Base:

Location: Near Portsmouth, New Hampshire.
Description: This base was put on alert during the 1975 **Loring AFB intrusions**. UFOs have been spotted over this installation.

The Pentagon:

Location: Arlington, Virginia.
Description: The Pentagon is The United States Military Headquarters. It is home to the **National Military Command Center (NMCC).** The NMCC is believed to have radar tracking of UFOs.

Peterson Air Force Base:

Location: Colorado Springs, CO
Description: This Air Base is the home of the **USAF Space Command Center (SCC).** The SCC is in charge of the **Defense Support Program (DSP) Satellites**. These satellites are highly classified. They were designed to warn the U.S. Military of **Inter-continental Ballistic Missile (ICBM)** launches. They are believed to have recordings of UFOs.

Plattsburgh Air Force Base:

Location: New York.
Description: This installation was another base that was put on alert during the 1975 **Loring AFB intrusions**. UFOs have been sighted at this base (especially over the base's Weapons Storage Area).

Sawyer Air Force Base:

Location: Michigan.
Description: This base was put on alert during the 1975 Loring AFB intrusions.

Air Force Base:

Location: Michigan.
Description: UFOs have been spotted over this base. Fighter jets have been scrambled from this base to chase UFOs.

Wright-Patterson Air Force Base:

Location: About 5 km. west of Fairborn, Ohio.
Description: An extremely secretive Air Base, security officials once refused entrance to the defense secretary. Wright-Patterson is home to the infamous **Hangar 18** which is suspected of harboring UFO wreckage and dead **Extraterrestrial Biological Entities (EBEs)**. The alien bodies and some of the UFO wreckage retrieved from the 1947 Roswell crash was sent to this base. UFO wreckage from around the world (Australia, etc.) has been sent here. W PAFB is headquarters to **Project Moondust**. Moondust is a "foreign space debris" analysis program of the **Air Force Systems Command's Foreign Technology Division (FTD).** Hangar 18 is known to have huge **Cryogenic freezing chambers**, purpose unknown.

Wurtsmith Air Force Base:

Location: Michigan.
Description: This was another base put on UFO alert during the 1975 **Loring AFB intrusions.** Many UFOs were spotted in and around this base during the 1975 UFO-flap (especially over the base's Weapons Storage Area). Fighter jets were scrambled to intercept the UFOs, but their attempts were useless.

F.E. Warren Air Force Base:

Location: Wyoming
Description: In the 1980s, UFOs were repeatedly observed hovering over the Weapons Storage Area at F.E. Warren AFB, according to former USAF Air Policeman Jay DeSisto.

AUSTRALIA

Pine Gap Research Facility:

Location: Located 12 miles from Alice Springs in Australia's Northern Territory.
Description: This base is described as a "Defense Space Research Facility" sponsored by both the American and the Australian defense departments. It is run by the American National Security Agency (NSA). This secret communications base serves primarily as a downlink for geosynchronous SIGINT (Signals Intelligence) Satellites. It is suspected to be monitoring UFO communications. UFOs have been sighted in and around the base. It is believed that there is an underground alien base on the facility's grounds.

CANADA

Falconbridge Air Force Station:

Location: North of Sudbury and Near North Bay, Ontario, Canada.
Description: In 1975, this Radar Station north of Sudbury experienced a rash of UFO sightings. American jets were scrambled to intercept the UFOs after the Canadian Forces at the 22nd Division of NORAD in North Bay requested assistance.

480

ENGLAND

Lakenheath - Bentwaters RAF Bases

Location: Lakenheath and Bentwaters, England
Description: The Lakenheath RAF Base and Bentwaters RAF Base were areas where a series of radar and visual contacts with Unidentified Flying Objects took place over these airbases. On the night of 13 - 14 August 1956, UFOs flew over and hovered above the nuclear storage facilities of newly acquired American weapons in eastern England, involving both RAF and USAF personnel. Reports were filed with Project Blue Book and the incident was also investigated by the Condon Committee which considered the account has a true unknown and has since gained some prominence in the literature of ufology, the popular media.

Bentwaters Air Force Base (a.k.a. RAF Bentwaters):

Location: Near Bentwaters, England.
Description: In 1980, UFOs were sighted (one hovering above the ground), at very close range, by base personnel in Rendlesham forest (a few miles from the base). The personnel was threatened into silence and the public was misinformed about the incident.

IRAN

Shahrokhi Air Force Base:

Location: In Hamadan, Iran.
Description: In 1976, jets were dispatched from this base to pursue UFOs. The UFOs, which were spotted by the public, and the pilots (and subsequently confirmed by radar), outmaneuvered these jets. Every time the jets approached the crafts, their systems failed. The smaller of the two UFOs engaged one jet, causing it to break off the chase. The smaller UFO landed behind a hill while the larger "Mother Ship" disappeared into the sky. Military Helicopters dispatched later to find the UFO that apparently landed reported no evidence of a landing.

RUSSIA

Plesetsk Military Cosmodrome:

Location: Plesetsk, in northwest Russia.
Description: This super-secret Russian military space center has been responsible for many UFO sightings in Russia. The base's satellite and missile launches along with experimental plane testing have often been mistaken for UFOs throughout Russia.

Voronezh (Rybinsk) Missile Base:

Location: near Voronezh near Rybinsk, 150 kilometers north of Moscow.

Description: an astounding UFO intrusion occurred in 1961 when personnel of a Russian anti-aircraft missile battery saw a huge UFO swoop down over the city releasing a number of smaller objects out of it. The commanding officer on duty ordered without authorization missiles to be launched at the UFO which had now hovered over the Voronezh missile base. Several missile salvos were fired, only to explode harmlessly two kilometers from the object as if the missiles had hit an invisible wall. The smaller objects descended toward the rocket batteries and disabled the electrical apparatus of the whole missile base. A third salvo of missiles to be launched was pre-empted by the commander's superiors as the entire base's electrical system had unexplainably malfunctioned. The smaller objects were seen to return to the larger object and then all the crafts flew away.

Kapustin Yar Army Missile Base:

Location: Volgograd and Astrakhan. Known today as Znamensk.
Description: This missile, satellite and rocket launching and testing site is equivalent to the White Sands Proving Grounds in New Mexico. Numerous Military personnel of the signal center observed three UFOs on July 28-29, 1989, one of which flew over the unit's logistics yard and moved in the direction of the rocket weapons depot. The object hovered over the depot where a bright beam suddenly appeared from the bottom of the disc lighting the corner of one of the buildings then, the beam disappeared and the object, still flashing, moved in the direction of the railway station. A jet fighter attempted to intercept the UFOs but, the object disengaged so fast, that it seemed the plane stayed in one place. The UFO then move toward the city of Akhtubinsk and disappeared from sight.

UFOs Filmed Hovering Over U.S. Air Force Nuclear Weapons Storage Area

Declassified U.S. government documents and military veterans have confirmed repeated UFO incursions at American nuclear weapons sites, decade after decade. Declassified Soviet documents and former Soviet Army personnel verify that such incidents occurred in the former U.S.S.R.

During September 27, 2010, at the "UFOs and Nukes" press conference in Washington D.C., UFO researcher **Robert Hastings**, one of the main speakers at the **National Press Club** conference stated that UFOs had been repeatedly seen by US military officials and servicemen hovering over weapons storage areas. These unidentified objects would send laser-like light beams down into the missile silos and storage areas shutting down and neutralizing the electronic launch systems of our missiles. Every country with nuclear capability has had UFO intrusion flights at one time or another into their sensitive weapons storage sites. The present of Extraterrestrial spacecraft continue to this day to monitor our military bases, fly circles around our aircraft and hover near our warships despite any challenges by us to confront them.

What happens when your investigations into the UFO phenomenon occurs almost next door to where you live and you capture the intrusion of UFOs over a military base and its sensitive WSAs on film? If your reaction is to inform anyone who will listen whether it be friends, fellow Ufologists, the news media or even, the Air Force, you will undoubtedly find that the military will take you seriously enough to ***seek you out!***

482

Paul Bennewitz Incident and the UFO Over Kirkland AFB

Hastings spoke about one incident, as summarized in a **USAF Office of Special Investigations (AFOSI or simply OSI)** memo, involved a disc-shaped UFO that landed not far from the Manzano WSA, located east of Kirtland AFB, New Mexico, on August 8, 1980. This incident also involved the late **Dr. Paul Bennewitz**, the poor hapless soul who became the deliberate target of a disinformation and smear campaign to discredit him in the eyes of his fellow Ufologists and the general public leaving him in a state of confusion and paranoia. **Clark, Jerome. The UFO Book, Visible Ink Press, 1998, p. 259**
Greenwood, Barry and Fawcett, Lawrence. Clear Intent: The Government Cover-up of the UFO Experience, Prentice-Hall Inc., 1984. pp. 41-45
http://www.wired.co.uk/news/archive/2010-10/06/mirage-men
http://www.sott.net/articles/show/245414-UFOs-Filmed-Hovering-Over-U-S-Air-Force-Nuclear-Weapons-Storage-Area

that this planet is being visited
by beings from another world

Robert Hastings speaks about UFOs hovering American Air Force Bases and the Dr. Paul Bennewitz affair.
https://www.sott.net/article/245414-UFOs-Filmed-Hovering-Over-U-S-Air-Force-Nuclear-Weapons-Storage-Area

Citing researcher **Christian P. Lambright's** book titled *X Descending,* Hastings tells of previously unpublished information and evidence - including photographic stills taken from motion picture films of the UFOs - relating to the real reason the Air Force launched a covert campaign of disinformation and harassment against the person who took those films, civilian systems engineer Dr. Paul Bennewitz.

After Bennewitz approached USAF commanders at Kirtland with his films, and subsequently began giving interviews about his sightings at Manzano to ufologists and the press, OSI agent **Richard Doty** was secretly ordered to neutralize his testimony and evidence through the use of dirty tricks, including providing Bennewitz with forged documents and other bogus information about U.S. government-sanctioned alien operations near Kirtland, and other bizarre subjects - which he was encouraged to publicize - in an effort to undercut his credibility with the media and the public. Bennewitz was also told that aliens were monitoring his own activities and, as a result, he began wearing a sidearm at home. The ensuing psychological assaults on the hapless Bennewitz resulted in his being temporarily hospitalized for paranoid psychosis and, ultimately, his complete, permanent mental collapse.

In short, Bennewitz became a casualty of the Air Force's determined efforts to direct attention away from the real story - repeated UFO incursions at the Manzano nukes storage site - evidence of which he had captured on film. While the tragic outcome, involving lasting damage to his mental health, was probably not deliberately planned, at least to that extreme degree, it nevertheless occurred.

Almost nothing relating to the genesis of the Bennewitz affair remains available for public scrutiny. While the Internet yields a lot of information about the unfolding of this sad, repugnant episode in the ongoing UFO cover-up by the U.S. government, almost all of the commentary about its underlying cause is inaccurate and wildly speculative. As author Lambright correctly notes, "Sadly, most of the information available today, in this case, focuses on the fallout, often just bizarre claims and interpretations promoted by others, some with little, if any, basis in fact."

While some of the information that Doty provided is factual - having already been exposed by various researchers, including me, years ago - the key facts about the reasons for the covert operation against Bennewitz were withheld, replaced by claims that Bennewitz had stumbled upon a National Security Agency (NSA) project to monitor Soviet satellite-related communications. As Mark Pilkington ironically told one interviewer:

"[As a result of disinformation provided to him by Doty,] Bennewitz believed that he was eavesdropping on the communications of ETs flying in and out of the base. These were apparently [electronic] emissions from an NSA project which he was [misguidedly] 'decoding' to find alien messages in. After being encouraged in his delusions for some years by Doty and others working out of Kirtland, Bennewitz developed serious paranoid psychosis and had to be institutionalized for a month. During this entire time Bennewitz was quite influential in the UFO field, and many of the ideas he promoted and was encouraged to believe by AFOSI - that there's an ET base next to Dulce, New Mexico, that the ETs have traded their technology with the U.S. government - are still central to the UFO lore. My sense is that Bennewitz was deliberately targeted to distribute nonsense into the UFO community."[3]

On this last point, Pilkington is entirely correct. Note, however, the absence of any mention of Bennewitz having taken films of UFOs over the Manzano WSA. Unwittingly, Pilkington and other authors have assisted in keeping the "classified NSA project" cover story alive, simply because they took Richard Doty at his word on the matter. Consequently, prior to the publication of *X Descending*, the true story of the origins of the Bennewitz affair remained hidden. Thankfully, Lambright's book nicely and convincingly fills that void with its credible, documented research. Clark, Jerome. The UFO Book, Visible Ink Press, 1998, p. 259

484

Greenwood, Barry and Fawcett, Lawrence. Clear Intent: The Government Cover-up of the UFO Experience, Prentice-Hall Inc., 1984. pp. 41-45
http://www.wired.co.uk/news/archive/2010-10/06/mirage-men
http://www.sott.net/articles/show/245414-UFOs-Filmed-Hovering-Over-U-S-Air-Force-Nuclear-Weapons-Storage-Area

Hasting's own investigation of Richard Doty in the mid-1980s revealed that among his officially-sanctioned capers, Doty successfully duped not only Bennewitz but also filmmaker **Linda Moulton Howe** regarding all manner of alleged covert developments, including secret alien-U.S. treaties, the supposedly-authorized building of underground alien bases on American soil, and so on. Of course, all of these tales were complete fabrications, meant to confuse the UFO cover-up issue, and the mythology that resulted from Howe's regrettable widespread dissemination of this disinformation is in evidence even today, strewn all over the Internet.

Hastings was suspicious about the real reason Bennewitz was targeted.
Years later, after researcher **Robert J. Durant's** article "Doty and the Body Snatchers" was published in October 2005, in the journal *International UFO Reporter*, I wrote to him, saying: "Despite Doty's recent public 'explanation' regarding the reasons for the campaign against Bennewitz, I am of the opinion that Bennewitz may have actually photographed and filmed bona fide UFOs over the Manzano Weapons Storage Area (WSA), which is located just east of Kirtland AFB. It was this nuclear weapons depot, now decommissioned, which directly bordered Bennewitz' subdivision, Four Hills. If you are familiar with some of the nuclear weapons-related UFO sightings - including those at ICBM sites and weapons research labs - then you may also be aware that a few of those sightings have occurred at WSAs."

"In view of these facts [about other UFO sightings at various WSAs], I have suggested the following scenario to other researchers: Bennewitz - a reputable businessman whose company held contracts to supply engineering components to various government agencies - photographed bona fide UFOs above the Manzano WSA, and then talked about it to anyone who would listen, including the Air Force, ufologists, and the media. Because nuclear weapons-related UFO incidents were/are extremely sensitive, a decision was made by the Air Force to undermine Bennewitz' credibility. Consequently, OSI at Kirtland formulated a disinformation scheme whereby the talkative Bennewitz would be provided with outrageous stories of alien visitations at Kirtland, underground alien bases in the Southwest, secret U.S./alien treaties, and all the rest of it. Once this 'inside information' had been passed along to others by the increasingly paranoid Bennewitz, the legitimate media - as well as the more rational members of ufology - would quickly lose interest in his claims, leaving only the most gullible to ooh and ahh at these amazing 'revelations'. The net result? The initial, bona fide UFO sightings at a highly sensitive nuclear weapons facility got lost in all of the hoopla and were only rarely, if ever, mentioned in the articles and news stories about Bennewitz' claims."

When viewing the hazy, indistinct photographic images below - individual frames excerpted from Bennewitz' 8mm motion picture films - it is important to remember that countless observers worldwide have mentioned the plasma-like quality of UFOs at night, perhaps due to the ionization of the surrounding atmosphere and/or some other phenomenon, which lends a shimmering, constantly shifting appearance to their outer hulls. For example, as Sgt. Eichner at

Loring AFB said, "The object looked like all the colors were blending together, as [if] you were looking at a desert scene. You see waves of heat rising off the desert floor. This is what I saw. There were these waves in front of the object and all the colors were blending together." **Clark, Jerome. The UFO Book, Visible Ink Press, 1998, p. 259**
Greenwood, Barry and Fawcett, Lawrence. Clear Intent: The Government Cover-up of the UFO Experience, Prentice-Hall Inc., 1984. pp. 41-45
http://www.wired.co.uk/news/archive/2010-10/06/mirage-men
http://www.sott.net/articles/show/245414-UFOs-Filmed-Hovering-Over-U-S-Air-Force-Nuclear-Weapons-Storage-Area

Additionally, Bennewitz' films were pre-digital and shot on traditional color film, which uses light-sensitive silver halide crystals in the emulsion. When developed in a darkroom, these clumps of interlocking crystals yield a highly-visible "grain" pattern when viewed at high magnification, such as the images above, which are zoomed-in close-ups of the UFOs that only filled a tiny part of the original picture frame. Therefore, the resulting ragged-edged appearance of the unidentified aerial objects is due in part to this effect. Indeed, in any pre-digital still or motion picture image, regardless of the subject being filmed, the same distortion would be apparent if one zoomed-in on any small object in the larger film frame.

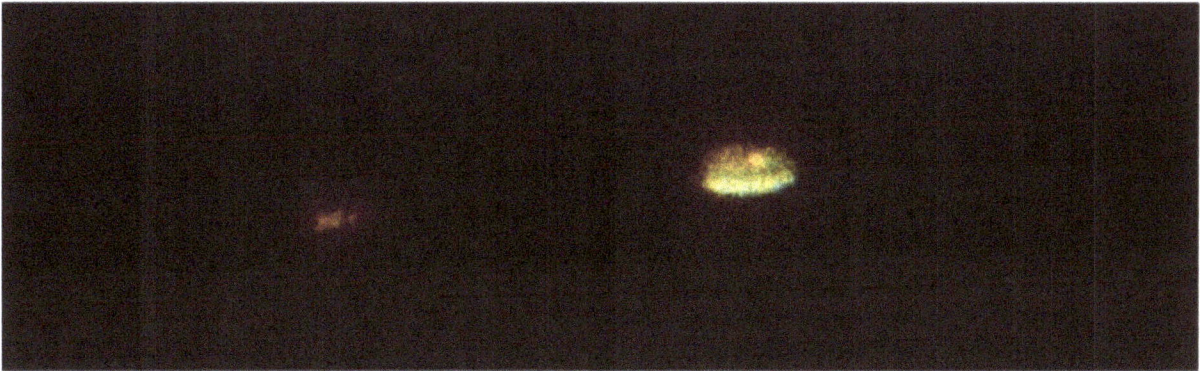

A unique view of the largest vehicle with a smaller one faintly visible in the distance.

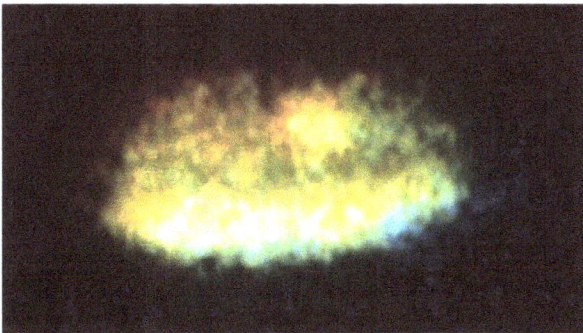

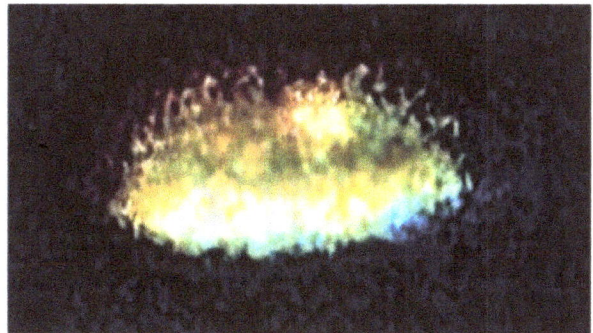

Enhancement shows unusual patterns in the upper energy field, possibly converging at the glowing node.

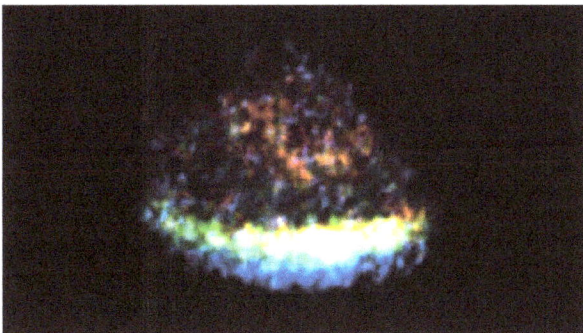

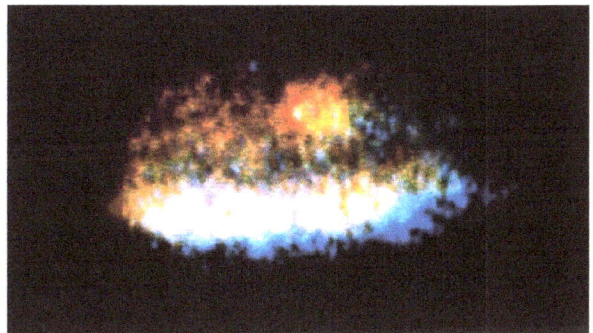

Above left is the vehicle in an apparent low-power state. Above right is a color-enhanced view of the vehicle in flight, the upper surface now engulfed in yellow-orange and the vehicle appears an intense white. The same iridescent blue-green glow stands out along the lower edge.

The above set of photos from an 8mm film are believed to have been taken by Dr. Paul Bennewitz of UFOs seen over the Manzano nuclear weapons storage site

https://www.sott.net/article/245414-UFOs-Filmed-Hovering-Over-U-S-Air-Force-Nuclear-Weapons-Storage-Area

The other intriguing UFO film discussed in **Lambright's** book was taken by **Ray Stanford**, in Corpus Christi, Texas in October 1985 and shows a procession of disc-shaped objects, each one enveloped in a pinwheel swirl of, well, something, with a bright spike of light extending from a tower in its radial center. Successive frames in the motion picture reveal the spikes to be pulsating, as well as growing and shrinking in length. Perhaps counter-intuitively, each vehicle is traveling with one face toward the direction of travel, rather than its more aerodynamic edge. Without going into a lot of technical detail here, **Lambright** devotes a significant part of *X Descending* discussing his exhaustive search for evidence of a technology in existence in 1985

487

which would account for the bizarre images captured on the Stanford film - only to come up empty-handed.

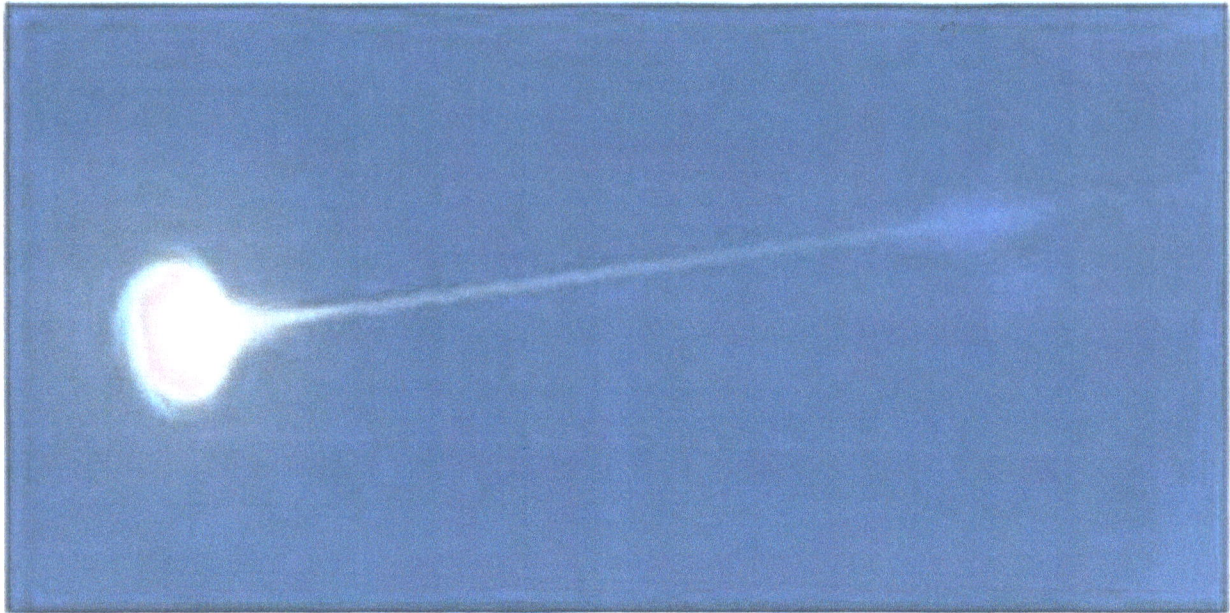

A Reproduction of one vehicle as seen in the Stanford film (c) 2012 Christian P. Lambright
https://www.sott.net/article/245414-UFOs-Filmed-Hovering-Over-U-S-Air-Force-Nuclear-Weapons-Storage-Area

He did, however, stumble upon the mid-1990s theoretical work of aerospace engineer Dr. Leik Myrabo, a professor at Rensselaer Polytechnic Institute, involving an advanced, ultra-high-velocity aircraft's use of directed-energy technology to improve performance. Intriguingly, as it turns out, Myrabo's **'Directed Energy Air-Spike'** concept - for the purpose of mitigating shock waves, reducing drag, and aiding in overall propulsion - apparently derived from his having visited Stanford's house in March 1987, where he viewed several still images from the film for himself.

Because Lambright's comprehensive research confirms, beyond a reasonable doubt, that no such disc-shaped aircraft utilizing directed-energy technology existed anywhere on Earth in 1985, the question becomes: If the objects photographed in Corpus Christi did indeed employ some already-functional variation of the Air-Spike system, just who was operating the craft captured on the Stanford film? Clark, Jerome. The UFO Book, Visible Ink Press, 1998, p. 259 and Greenwood, Barry and Fawcett, Lawrence. Clear Intent: The Government Cover-up of the UFO Experience, Prentice-Hall Inc., 1984. pp. 41-45
http://www.wired.co.uk/news/archive/2010-10/06/mirage-men
http://www.sott.net/articles/show/245414-UFOs-Filmed-Hovering-Over-U-S-Air-Force-Nuclear-Weapons-Storage-Area
Some military officials, scientists as well as some Ufologist speculate that ETs are gathering information about our defensive and first strike capabilities by monitoring our nuclear storage sites. Others see this action by ETs as an aggressive pre-emptive strike to disarm most nuclear-armed countries before beginning an all-out alien invasion of our planet, while still others,

488

downplay the whole subject as simply defects in outdated electronic firing systems in need of replacement or upgrading.

Nothing can be further from the truth! This is simply militaristic thinking incapable of a paradigm shift away from the possibility that not everything under the sun is hostile or has hostile intentions.

CHAPTER 50

THE AGENCY OF UFO SECRECY - MAJIC 12
(AKA. MAJESTIC 12, MJ 12)

As indicated earlier in this section of the book, the intrusion by UFOs into the weapons storage sites of many of the world's nuclear military bases has left no doubt in the minds of the military brass that we are not only being visited by Extraterrestrials but also, being vigilantly monitored in an on-going demonstration of deterrence from engaging in an all-out nuclear war on this planet by these visiting Extraterrestrials.

ETI over-flights and intrusions into our most sensitive military bases around the planet is not an act of invasion or first strike against us but rather a display of their grave concern for us and our continued proclivity for sabre rattling and nuclear brinkmanship which has destabilized the peace and security of the inhabitants of this planet. If ETs wanted to take over our planet, why wait until we are capable of defending ourselves with nuclear weaponry and other weapons of mass destruction? If, as some Ufologists, scientists and military officials have said, that they have been observing us for millennia then, they could have easy done it ages ago! Why now, instead of back then?

ETI over-flights and intrusions into military bases is definitely a clear demonstration of their superior ability and technology to intervene and prevent a nuclear holocaust or any other type of global conflict from happening. It should appear to any rational individual with an ounce of common sense, that ETI are not going to allow us to blow ourselves up or destroy this beautiful planet, we call home. Global or interstellar war is clearly not the objective of ETI visiting out planet.

The military know this to be the case but the military mindset is to defend the country, the interests and agendas of private industrialists and somewhere down the list, defend the general public from enemies from within and from without, by following a protocol of shooting first and asking questions second.

It has been shown that ETI have an interest in our military forces, our aircraft, our naval exercises, our warships, our rocketry and launch sites, our nuclear weapons storage bases, our missile test ranges, proving grounds and test facilities.

We are an adolescent species, destined to become a global civilization. We are coming of age and are about to enter into adulthood but as yet, we are still heedless of the age in which we live in.

Compounding this age of adolescence is our misplaced trust and reliance given over to those in authority, to those in political power, to those who captain our industries, to those who maintain our financial, medical and educational institutions, as well as those who supposedly uphold our most cherished religious belief systems. We believe that they will look after our well-being and our way of life. These are our leaders that we have elected, voted, appointed and placed into positions of power and authority. We expected them to be servants of the highest caliber in their

490

duty to advancing the goals of our society. We do not expect them to have self-vested interests for themselves or for some other party, political organization, association, or some quasi-elitist group or cell.

Unfortunately, the more we go down the rabbit hole into the covert black world, the more apparent it becomes that such groups within or outside of society's oversight and control exist with their own agendas, their own missions and their own path of destiny and ultimately bring about a future of their own design.

Understanding the nature of the UFO and ETI phenomenon is but one small aspect of this covert group. Their goal is to research, develop and utilize alien technology to fulfill their agendas and destiny and it does not include sharing the benefits with the rest of society or the world.

With an increasing interest in our world by Extraterrestrials and particularly in the military forces of the world, the US government and the military felt the need for a small carefully selected, well positioned group of scientists, military leaders, and government officials to form a secret committee whose sole function would be to investigate UFO activity in the aftermath of the alien spaceship crash near Roswell, New Mexico, in July 1947. A secret committee was formed to deal with such matters in 1947 by an executive order from U.S. **President Harry S Truman**. Its codename is known as **Majestic 12** (also known as **Majic 12**, **Majestic Trust**, **M12**, **MJ 12**, **MJ XII** or **Majority 12**, it is also known as **MAJI - Majority Agency for Joint Intelligence.**

The alleged purpose of the committee was to investigate UFO activity in the aftermath of the Roswell incident—the purported crash of an alien spaceship near Roswell, New Mexico, in July 1947. 39Majestic 12, as viewed by UFO conspiracy theorists, debunkers, skeptics and those outside of the pale of UFOs and ETI matters, see this as an important part of the of an ongoing government cover up of UFO information. The **Federal Bureau of Investigation (FBI)** has concluded that documents associated with the Majestic 12 committee are completely bogus". http://en.wikipedia.org/wiki/Majestic_12

The primary evidence for the existence of a group named Majestic twelve is a collection of documents that first emerged in 1984 and which have been the subject of much debate. The original MJ-12 documents state that: The Majestic 12 group was established by secret executive order of President Truman on 24 September 1947, upon recommendation by Dr. Vannevar Bush and Secretary of Defense James Forrestal.

The existence of MJ-12 has sometimes been denied by some agencies of the United States government, which insist that documents suggesting its existence are hoaxes. The FBI investigated the documents and concluded they were forgeries, based primarily on an opinion rendered by the **U.S. Air Force Office of Special Investigations, AFOSI**. Opinions among UFO researchers are divided: Some argue the documents may be genuine while others contend they are phony, primarily due to errors in formatting and chronology.

In 1985, another document mentioning MJ-12 and dating to 1954 was found in a search at the National Archives. Its authenticity is also highly controversial. The documents in question are rather widely available on the Internet, for example on the FBI website, where they are dismissed as bogus.

Since the first MJ-12 documents, thousands of pages of other supposed leaked government documents mentioning MJ-12 and a government coverup of UFO reality have also appeared, sometimes collectively referred to as the **"Majestic Documents."** All of them are controversial, with many disputing their authenticity. A few have been proven to be unquestionably fraudulent, usually retyped rewrites of unrelated government documents. The primary new MJ-12 document is a lengthy, linotype-set manual allegedly dating from 1954, called the MJ-12 **"Special Operations Manual (SOM)".** It deals primarily with the handling of crash debris and alien bodies. Objections to its authenticity usually center on questions of style and some historical anachronisms.

The MJ-12 documents are alleged to date from 1942 to 1997 and have been hotly debated in the UFO community. The documents include such matters as the conduct to be used when meeting an alien, diagrams and records of tests on UFOs, memos on assorted cover-ups, and descriptions of the President's statements about UFO-related issues. The documents contain supposed signatures of important people such as Albert Einstein and Ronald Reagan, creating a major debate in the conspiracy and UFO communities. No more documents have been leaked or released since 1997. Their authenticity remains uncertain, and some claim them to be entirely fake.

However, before the appearance of the various dubious MJ-12 Documents, Canadian documents dating from 1950 and 1951 were uncovered in 1978. These documents mention the existence of a similar, highly classified UFO study group operating within the Pentagon's **Research & Development Board (RDB)** and headed by **Dr. Vannevar Bush**. Although the name of the group is not given, proponents argue that these documents remain the most compelling evidence that such a group did exist. ***The bone of contention here, are these the MJ-12 documents or some other top secret documents that are being referred to?*** There is also some testimony from a few government scientists involved with this project confirming its existence. (Bold italics added by author for emphasis). http://en.wikipedia.org/wiki/Majestic_12#Arguments_for

There have been many UFO researchers, investigators, and authors who have delved into the murky waters of the secret group known as the Majestic 12. There are, however, some who have dug deeper than others and have come up with incredible documented evidence which if not part of a cleverly hoaxed disinformation campaign by some government agency, stand as incontestable and incontrovertible proof that the US government and military have known about the reality of UFOs since the early 1940s and possibly longer.

Robert M. Wood, Ph.D., his son Ryan S. Wood, Nick Redfern, Stanton T. Friedman, Timothy S. Cooper, Jim Marrs and **Mr. Jim Clarkson** comprise the **investigators of the Majestic documents**. Each of them has published books individually and having their own websites, together, they have a collective website at http://majesticdocuments.com from where the MJ 12 Documents can be viewed and downloaded.

The Majestic 12 Documents reveals a remarkable work of investigative journalism led by Robert and Ryan Wood in establishing the authenticity of top secret UFO government documents that tell the story of military and government cover-up for over 60 years. The documents describe presidential briefings and military action, authorization, and cover-up regarding crashed discs, and the recovered bodies of alien occupants. The MJ-12 investigators paid special attention to the forensic authentication issues of content, provenance, type, style and chronology leaving little doubt that the cover-up is real, shocking, and at times unethical.

Over 500 pages of newly surfaced documents are contained in the Wood's website many of which date years before the Roswell crash. The website tells of the U.S. government's work on retrieval and analysis of extraterrestrial hardware and alien life forms from 1941 to present.

The Majestic documents tell a mind-boggling story of deception, intelligence, and counterintelligence, revolutionary alien technology, missing nuclear weapons and compartmentalized secrecy spanning in time from the first crashed disc retrieval in 1941 until three days before President Kennedy's assassination in 1963.
http://majesticdocuments.com/documents/intro.php

From left to right and top to bottom:
Robert M. Wood, Ph.D. , his son Ryan S. Wood, Nick Redfern, Stanton T. Friedman, Timothy S. Cooper, Jim Marrs and Mr. Jim Clarkson.
http://majesticdocuments.com/team.php

Operation Majestic 12 was established by special classified presidential order on September 24, 1947, at the recommendation of Secretary of Defense James Forrestal and Dr. Vannevar Bush, Chairman of the Joint Research and Development Board. The goal of the group was to exploit everything they could from recovered alien technology.

Buried in a super secret "MAJIC EYES ONLY" classification that was above TOP SECRET long before the modern top secret codeword special access programs of today, Major General Leslie R. Groves (who commanded the Manhattan Project to deliver the atomic bomb) kept just one copy of the details of crashed alien technology in his safe in Washington, D.C.

Ambitious, elite scientists such as **Vannevar Bush, Albert Einstein,** and **Robert Oppenheimer,** and career military people such as **Hoyt Vandenberg, Roscoe Hillenkoetter, Leslie Groves,** and **George Marshall**, along with a select cast of other experts, feverishly and secretively labored to understand the alien agenda, technology, and their implications.
http://majesticdocuments.com/documents/intro.php

Einstein and Oppenheimer were called in to give their opinion, drafting a six-page paper titled "Relationships With Inhabitants Of Celestial Bodies". They provided prophetic insight into our modern nuclear strategies and satellites and expressed agitated urgency that an agreement be reached with the President so that scientists could proceed to study the alien technology.

The extraordinary recovery of fallen airborne objects in the state of New Mexico, from July 4 - July 6, 1947, caused the Chief of Staff of the Army Air Force's **Interplanetary Phenomena Unit**, Scientific and Technical Branch, Counterintelligence Directorate to initiate a thorough investigation. The special unit was formed in 1942 in response to two crashes in the Los Angeles area in late February 1942. The draft summary report begins At 2332 MST, 3 July 47; radar stations in east Texas and **White Sands Proving Ground (WSPG)**, N.M. tracked two unidentified aircraft until they both dropped off radar. Two crash sites have been located close to the WSPG. Site LZ-1 was located at a ranch near Corona, Approx. 75 miles northwest of the town of Roswell. Site LZ-2 was located approx. 20 miles southeast of the town of Socorro, at latitude 33-40-31 and longitude 106-28-29.

The first ever known UFO crash retrieval case occurred in 1941 in Cape Girardeau, Missouri. This crash kicked off early reverse engineering work, but it did not create a unified intelligence effort to exploit possible technological gains apart from the Manhattan Project uses.

The debris from the primary field of the 1947 crash 20 miles southeast of Socorro, New Mexico was called **ULAT-1 (Unidentified Lenticular Aerodyne Technology),** and it excited metallurgists with its unheard of tensile and shear strengths. The fusion nuclear (called neutronic at that time) engine used heavy water and deuterium with an oddly arranged series of coils, magnets, and electrodes, descriptions that resemble the cold fusion studies of today.

Harry Truman kept the technical briefing documents of September 24, 1947, for further study, pondering the challenges of creating and funding a secret organization before the CIA existed (although the **Central Intelligence Group (CIG)** did exist) and before there was a legal procedure of funding non-war operations.

In April 1954, a group of senior officers of the U.S. intelligence community and the Armed Forces gathered for one of the most secret and sensational briefings in history. The subject was Unidentified Flying Objects, not just a discussion of sightings, but how to recover crashed UFOs, where to ship the parts, and how to deal with the occupants. For example, in the **Special Operations Manual (SOM1-01)** Extraterrestrial Entities Technology Recovery and Disposal, MAJESTIC 12, red teams mapped out UFO crash retrieval scenarios with special attention given to press blackouts, body packaging, and live alien transport, isolation, and custody. The SOM-01 document has since been authenticated by **Bob and Ryan Wood,** long-time Majestic-12 documents, and UFO investigators, in December 2014 through research and specific evidence on their new website: www.SpecialOperationsManual.com
https://www.earthfiles.com/news.php?ID=2276&category=Environment

UFO related secret programs have consumed a significant part of America's black budget since the **Manhattan Project**. The 1997 government disclosed intelligence budget portion alone is $26 billion and according to Tim Weiner's 1990 book *Blank Check*, the total black budget was about $35 billion in 1990. Even the most sensational conspiracy of modern times, the Kennedy assassination is likely linked to the UFO cover-up and the military cabal, as several of the documents demonstrate.

RESTRICTED SOM1-01

TO 12D1—3—11—1
MAJESTIC—12 GROUP SPECIAL OPERATIONS MANUAL

EXTRATERRESTRIAL ENTITIES AND TECHNOLOGY, RECOVERY AND DISPOSAL

TOP SECRET/MAJIC EYES ONLY

WARNING! This is a TOP SECRET—MAJIC EYES ONLY document containing compartmentalized information essential to the national security of the United States. EYES ONLY ACCESS to the material herein is strictly limited to personnel possessing MAJIC—12 CLEARANCE LEVEL. Examination or use by unauthorized personnel is strictly forbidden and is punishable by federal law.

MAJESTIC—12 GROUP • *APRIL 1954*

MJ—12 4838B–Mar   270485⁰—54——1

The controversial SOM-01 document that specifically deals with retrieval and disposal of Extraterrestrial Intelligences and their technology
http://thewebmatrix.net/disclosure/1954.html

Overall, the United States UFO program grew out of necessity. First, to determine the alien threat, second to exploit their advanced technology in any way we could to gain a military, economic or even a psychological advantage and win World War II, and third to maintain power, authority, and control of both technology, governments, and world stability. Initially, to make the project public would have sent unpredictable turmoil into science, religion, politics, and global economics. http://majesticdocuments.com/documents/intro.php

There are other theories of the **MJ-12** group that are sometimes associated in recent UFO conspiracy literature with the more historically verifiable but also, deeply secretive **NSC 5412/2 Special Group**, created by **President Dwight D. Eisenhower** in 1954. Although the Special Group was not specifically concerned with UFOs and post-dates the alleged creation of MJ-12 in 1947, the commonality of the number '12' in the names of the two groups is cited as intriguing, as is the first chairman, **Gordon Gray**, being one of the alleged MJ-12 members. As the highest body of central intelligence experts in the early Cold War era (the Group was alleged to include the President but exclude the Vice President), the Special Group certainly would have had both clearance and interest in all matters of national security, including UFO sightings if they were considered a real threat.

Others have speculated that MJ-12 may have been another name for the **Interplanetary Phenomenon Unit**, an officially recognized UFO-related military group active from the 1940s through the late 1950s.

Another government group recently associated with MJ-12 was the CIA's **Office of National Estimates (ONE)**, a forerunner of the current **National Intelligence Council (NIC).** ONE was created in 1950 by **CIA Director Gen. Walter Bedell Smith**, alleged to have replaced **Secretary of Defense James Forrestal** on MJ-12 after his death.

A history of the **NIC** states that **ONE** was a type of super branch of the CIA "whose sole task was to produce coordinated **'National Intelligence Estimates**.'" Besides Smith, it apparently consisted of 11 other members. A recent article on the history of the CIA's involvement in UFO investigations states that ONE received a UFO intelligence briefing on January 30, 1953, immediately after the end of the CIA's UFO debunking study known as the **Robertson Panel.** Members of ONE at that time included **FBI director J. Edgar Hoover, William Bundy**, President Eisenhower's **chief of staff Admiral B. Bieri, and William Langer**, a Harvard historian, who was chairman. Referring to ONE as "super think tank" within the CIA, the article states, "ONE is as close as we get to a documented version of the rumoured Majestic-12 group." http://majesticdocuments.com/documents/intro.php

At the **Mutual UFO Network (MUFON)** 2007 Symposium in Denver Colorado, UFO researcher **Brad Sparks** presented a paper showing a connection to the **Pratt Documents** that describes the MJ12 documents as an elaborate disinformation campaign perpetrated by William Moore, Richard Doty, and other **Air Force Office of Special Investigations (AFOSI)** personnel. The sources for this information are files dating from 1981 (3 years before the first alleged MJ12 documents surfaced) that UFO researcher **Bob Pratt** gave MUFON before his death in 2005. The information lay hidden in MUFON's archives until they were digitized as part of **MUFON's Pandora Project** and made available to UFO researchers. The paper can be downloaded from the MUFON website.

Of interest will be the paragraph that has a handwritten date of 1/02/82 and states: "3. UFO project is Aquarius, classified Top Secret with access restricted to MJ 12. (MJ may be "magic").

497

This project began about 1966, but apparently inherited files of the earlier project."
http://en.wikipedia.org/wiki/Majestic_12

All the alleged original members of MJ-12 were notable for their military, government, and/or scientific achievements, and all were deceased when the documents first surfaced (the last to die was Jerome Hunsaker, only a few months before the MJ-12 papers first appeared).

The original composition was six civilians (mostly scientists), and six high-ranking military officers, two from each major military service. Three (Souers, Vandenberg, and Hillenkoetter) had been the first three heads of central intelligence. The Moore/Shandera documents did not make clear who was the director of MJ-12, or if there was any organizational hierarchy.

The named members of MJ-12 were:

- Rear Adm. Roscoe H. Hillenkoetter: first CIA director
- Dr. Vannevar Bush: chaired wartime Office of Scientific Research and Development and predecessor National Defense Research Committee; set up and chaired postwar Joint Research and Development Board (JRDB) and then the Research and Development Board (RDB); chaired NACA; President of Carnegie Institute, Washington D.C.
- James Forrestal: Secretary of the Navy; first Secretary of Defense (replaced after his death on MJ-12 by Gen. Walter Bedell Smith, 2nd CIA director)
- Gen. Nathan Twining: headed Air Materiel Command at Wright-Patterson AFB; Air Force Chief of Staff (1953-1957); Chairman of Joint Chiefs of Staff (1957-1961)
- Gen. Hoyt Vandenberg: Directed Central Intelligence Group (1946-1947); Air Force Chief of Staff (1948-1953)
- Gen. Robert M. Montague: Guided missile expert; 1947 commander of Fort Bliss; headed nuclear Armed Forces Special Weapons Center, Sandia Base
- Dr. Jerome Hunsaker: Aeronautical engineer, MIT; chaired NACA after Bush
- Rear Adm. Sidney Souers: first director of Central Intelligence Group, first executive secretary of National Security Council (NSC)
- Gordon Gray: Secretary of the Army; intelligence and national security expert; CIA psychological strategy board (1951-1953); Chairman of NSC 5412 committee (1954-1958); National Security Advisor (1958-1961)
- Dr. Donald Menzel: Astronomer, Harvard; cryptologist during war; security consultant to CIA and NSA
- Dr. Detlev Bronk: Medical physicist; aviation physiologist; chair, National Academy of Sciences, National Research Council; president Johns Hopkins & Rockefeller University
- Dr. Lloyd Berkner: Physicist; radio expert; executive secretary of Bush's JRDB

According to other sources and MJ-12 papers to emerge later, famous scientists like Robert Oppenheimer, Albert Einstein, Karl Compton, Edward Teller, John von Neumann, and Wernher von Braun were also involved with MJ-12. http://en.wikipedia.org/wiki/Majestic_12

In retrospect, these would be exactly the kind of people who, you would expect to be call upon to investigate a truly out of this world phenomenon, people who are at the top of their professions with impeccable credentials and fully experienced in their fields of expertise.

Members of Majestic 12 from the U.S. Government, Military and Science Communities:

1. Admiral Roscoe H. Hillenkoetter
2. Dr. Vannevar Bush
3. Secretary James V. Forrestal
4. General Nathan Twining
5. General Hoyt S. Vanderberg
6. Dr. Detlev Bronk

7. Dr. Jerome Hunsaker
8. Admiral Sidney W. Souers
9. Mr. Gordon Gray
10. Dr. Donald Menzel
11. General Robert M. Montague
12. Dr. Lloyd V. Berkner

Many of these men had reliably documented activities related to UFOs:

- **Vandenberg**, as Director of Central Intelligence in 1946, had overseen investigations into the so-called Ghost rockets in Europe and wrote intelligence memos about the phenomena. Later as Air Force Chief of Staff, both Vandenberg and Twining oversaw early U.S. Air Force UFO investigations, like Project Sign and Project Blue Book and made some public statements on UFOs.
- **Twining** had previously written a famous Secret memo on September 23, 1947 (the day before Truman allegedly set up MJ-12) stating that flying saucers were real and urged formal investigation by multiple government organizations such as the AEC, NACA, NEPA, Bush's JRDB, and the Air Force Scientific Advisory Board. This led directly to the creation of Project Sign in late 1947.
- **Vandenberg** met with Bush's JRDB in a suddenly called meeting on the morning of the Roswell UFO incident (July 8, 1947) and was reported in the press as handling the later public relations crises. Radm. Konstantine Konopisos most recently implicated to involvement with later years of the MJ12 activities as Vice JAG of the US NAVY.
- **Bush** was directly implicated in 1950-51 Canadian documents heading a highly secret UFO investigation within the **Research and Development Board (RDB);** this assertion was confirmed by Canadian scientist Wilbert B. Smith Immediately after the Roswell UFO incident, Bush made public statements denying any knowledge of UFOs or any relation they might have to secret government projects.
- **Berkner** was on the 1953 CIA-organized Robertson Panel debunking UFOs and helped establish **Project Ozma**, the first radio telescope search for extraterrestrial intelligence.
- **Menzel** filed a UFO report in 1949, later wrote several UFO debunking books, and was the most prominent public UFO debunker of his time. Many conspiracy theorists see this as a "cover-up" for alleged MJ12 - alien connections.
- **Hillenkoetter** was on the board of directors of the powerful civilian UFO organization NICAP and made public statements to Congress about UFO reality in 1960.
- The **National Advisory Committee for Aeronautics (NACA),** which Hunsaker chaired from 1941-1958, is also known from documents to have occasionally delved into UFO cases. Other **members of NACA** were **Bush** (1938-1948), **Compton** (1948-1949), **Vandenberg** (1948-1953), **Twining** (1953-1957), and **Bronk** (1948-1958).
- **Teller** was a member of a scientific panel in 1949 at **Los Alamos National Laboratory** looking into the UFO phenomenon known as the green fireballs.
  http://en.wikipedia.org/wiki/Majestic_12

Research has also shown that there were many social and professional connections between many of the alleged members of MJ-12. For example, **Bush, Hunsaker, Bronk, and Berkner** all sat on the oversight committee of the **Research and Development Board (RDB)**, which Bush had established and initially chaired. Other notables on the RDB oversight committee were **Karl Compton, Robert Oppenheimer, and Dr. H. P. Robertson**, who headed up the debunking Robertson Panel, of which Berkner was a member. As mentioned, 1950 Canadian documents indicated that Bush headed up a small, highly secret UFO study group within the RDB.

500

Various alleged MJ-12 members or participants would also naturally be part of the Presidential office's **National Security Council (NSC)**, created in 1947. This would include (depending on NSC composition, which evolved) various NSC permanent members:

Executive Secretary (Souers, Cutler), the Secretary of Defense (Forrestal), the Secretary of the Army (Gray), National Security Advisor (Gray), and the Air Force Chief of Staff (Vandenberg, Twining).

Other nonpermanent members who would attend NSC meetings as advisors and implement policy would be the **CIA director (Hillenkoetter, Smith),** the head of the **Research and Development Board (Bush, Compton)**, the **President's Special Assistant for National Security Affairs (Cutler, Gray)**, and the **Chairman of the Joint Chiefs (Twining).** Therefore the supposed members and participating personnel of MJ-12 would have served on many other high government agencies and could conceivably have influenced government policy at many levels. (See also list of supposed current MJ-12 members below, again indicating various individuals who have served in high government positions.)

The purported members were trusted and high-ranking officials and scientists, with a history of inclusion in important government projects and councils; they possessed a diverse range of skills and knowledge while also having a high-security clearance. At the same time that they possessed these qualities, however, they were not-so-public figures as to be instantly missed should they be called upon in an emergency one wished to keep secret. If such a group existed, these members would be the likely types of individuals to be part of its membership.
http://en.wikipedia.org/wiki/Majestic_12

Arthur Bray's Discovery (1978) – The Canadian Connection

In 1978, Canadian researcher **Arthur Bray** uncovered previously classified Canadian UFO documents naming **Dr. Vannevar Bush** as heading *a highly secret UFO investigation group* within the **U.S. Research and Development Board**. *No name for the group was given.* Bray published excerpts of the documents in his 1979 book, *The UFO Connection.*

Aquarius Document and the Paul Bennewitz Connection (1980)

The earliest citation of the term "MJ Twelve" originally surfaced in a purported U.S. Air Force Teletype message dated November 17, 1980. This so-called **"Project Aquarius"** Teletype message had been given to Albuquerque physicist and businessman Paul Bennewitz in November 1980, by U.S. Air Force Office of Special Investigations counterintelligence officer **Richard C. Doty** as part of a disinformation campaign to discredit Bennewitz. Bennewitz had photographed and recorded electronic data of what he believed to be UFO activity over and nearby Kirtland AFB, a sensitive nuclear facility. Bennewitz reported his findings to officials at Kirtland, including Doty. Later it was discovered the Aquarius document was phony and had been prepared by Doty.

One sentence in the lengthy Teletype message read:

The official US Government policy and results of Project Aquarius is [sic] still classified TOP SECRET with no dissemination outside channels and with access restricted to "MJ TWELVE.

As Greg Bishop writes, "Here, near the bottom of this wordy message in late 1980, was the very first time anyone had seen a reference to the idea of a suspected government group called **'MJ Twelve'** that controlled UFO information. Of course, no one suspected at the time, the colossal role that this idea would play in the 1980s and '90s Ufology, and it eventually spread beyond its confines to become a cultural mainstay."

As **Bennewitz** was the subject of a disinformation campaign, many investigators are automatically suspicious of any documents or claims made in association with the Bennewitz affair. Because the entire MJ-12 affair made its appearance only a year after Bray had made public the incriminating Canadian documents about the secret UFO committee, one theory is that the Project Aquarius teletype message was part of a counterintelligence hoax to discredit the information in the just-revealed Canadian documents. Thus the various MJ-12 documents could be fakes, but the secret committee described in the verified Canadian documents could still have been real. http://en.wikipedia.org/wiki/Majestic_12

This Canadian connection is the proverbial "fly in the ointment" because the Canadian documents that originated from Wilber Smith's contact with the Canadian Embassy in Washington, DC. It's a source of sensitive information that is outside of US information control, independent from all other sources of Majestic Twelve documentation. Someone had obviously leaked the information to the Canadian Embassy (as it is doubtful that there were any Canadian spies or secret agents running around the US looking for very unusual out of this world top secrets, so soon after the War), perhaps from a government "insider" !

These documents were released in 1978 Arthur Bray prior to any known Majestic 12 Documents came into public awareness. It may add validity to the Majestic 12 group because of the association of Dr. Vannevar Bush with MJ-12, however, the documents that Smith obtained from the Canadian Embassy did not name specifically Bush as being with MJ-12 and in fact, the name of the organization never arises in the documents. Simply put, the creditability of the MJ-12 documents is by association only, in that one person among twelve other people who were named in a document called Majestic Twelve. The intrigue of the MJ -12 documents whether real or hoaxed demonstrates that there is a cover-up on the UFO subject matter and shows the outright lying, the disinformation, and misinformation that runs from top levels of government agencies, from political and military officials down through to the individual agents of disinformation, in order to conceal the evidence of the existence of UFOs and ETI visiting our planet.

The Moore/Shandera Documents (1984)

What came to be known as the **"MJ-12 papers"** – detailing a secret UFO committee allegedly involving Vannevar Bush – first appeared on a roll of film in late 1984 in the mailbox of television documentary producer (and amateur ufologist) Jaime Shandera. Shandera had been collaborating with Roswell researcher William Moore since 1982.

Just prior to Bennewitz and the Aquarius document, Moore had been contacted in September 1980 by Doty, who described himself as representing a shadowy group of 10 military intelligence insiders who claimed to be opposed to UFO secrecy. Moore called them **"The Aviary"**.

In January 1981, Doty provided Moore with a copy of the phony Aquarius document with mention of MJ Twelve. Moore would later claim in 1989 that he began collaborating with AFOSI in spying on fellow researchers such as Bennewitz, and dispensing disinformation, ostensibly to gain the trust of the military officers, but in reality to learn whatever UFO truth they might have, and also to learn how the military manipulated UFO researchers. In return, Doty and others were to leak information to him.

Later it would turn out that some of the UFO documents given Moore were forged by Doty and compatriots, or were retyped and altered from the originals. Furthermore, the film mailed to Shandera with the MJ-12 documents was postmarked "Albuquerque," raising the obvious suspicion that the MJ-12 documents were more bogus documents arising from Doty and AFOSI in Albuquerque.

In 1983, Doty also targeted UFO researcher and journalist Linda Moulton Howe, revealing alleged high-level UFO documents, including those describing crashed alien flying saucers and recovery of aliens. Doty again mentioned MJ-12, explaining that *"MJ"* stood for *"Majority"* (not **"Majestic"**)

Moore soon showed a copy of the Aquarius/MJ 12 Teletype message given him by Doty to researchers Brad Sparks and Kal Korff. In 1983, Moore also sought Sparks' reaction to a plan to create counterfeit government UFO documents, hoping to induce former military officers to speak out. Sparks strongly urged Moore not do this. The previous year Moore had similarly approached nuclear physicist and UFO researcher Stanton T. Friedman about creating bogus Roswell documents, again with the idea of encouraging witnesses to come forward. Also, in early 1982, Moore had approached former *National Enquirer* reporter Bob Pratt (who had first published a story on Roswell in the *Enquirer* in 1980). Moore asked Pratt to collaborate on a novel called *MAJIK-12*. As a result, Pratt always believed that the Majestic-12 papers were a hoax, either perpetrated personally by Moore or perhaps by AFOSI, with Doty using Moore as a willing target. Moore, however, flatly denied creating the documents, but eventually thought that maybe he had been set up. Noted UFO skeptic Philip J. Klass would also argue that Moore was the most likely hoaxer of the initial batch of MJ-12 documents.

Unlike Pratt, who was convinced they were a hoax, Friedman would investigate the historical and technical details in the MJ-12 documents and become their staunchest defender.
http://en.wikipedia.org/wiki/Majestic_12

The 1984/1985 MJ-12 Papers - The Eisenhower Briefing Document

The film allegedly received by Shandera in 1984 consisted of two MJ-12 documents. The main document, dated November 18, 1952, was supposedly prepared by Rear Admiral Roscoe Hillenkoetter, the first CIA director, to brief incoming president Dwight Eisenhower on the committee's progress. The document lists all the MJ-12 members and discusses United States Air

Force investigations and concealment of a crashed alien spacecraft near Roswell, New Mexico, plus another crash in northern Mexico in December 1950.

Eisenhower did indeed receive extensive briefings November 18, 1952, including a briefing at the Pentagon by the Joint Chiefs of Staff, which would have included alleged MJ-12 members Twining and Vandenberg. However, Eisenhower's Pentagon briefing is still classified and thus the subject matter discussed remains speculative.

A recent document found by **James Carrion** in the **Truman Library** shows that **General Eisenhower** was briefed by **James S. Lay** and alleged MJ12 member **Walter Bedell Smith** on November 15th in Augusta, Georgia. This real briefing may have been the basis for fabricating the EBD story. http://en.wikipedia.org/wiki/Majestic_12

The Cutler/Twining Memo (1985)

In 1985, Shandera and Moore began receiving postcards postmarked "New Zealand" with a return address of "Box 189, Addis Ababa, Ethiopia." The cards contained a series of cryptic messages referring to "Reeses [sic] Pieces" and "Suitland" (among other terms) that Shandera and Moore assumed were a code; however, they were unable to "decode" the seeming message.

A few months later, a happenstance request from Friedman unlocked the mystery: busy due to previous obligations, Friedman asked Moore and Shandera to examine newly declassified Air Force documents at the **National Archives (NARA)** repository in Suitland, Maryland; the head archivist there was named Ed Reese.

After a few days in Suitland, Shandera and Moore discovered yet another MJ-12 document, the so-called **Cutler/Twining memo**, dated July 14, 1954. Interestingly enough, the memo turned up in "Box 189" of the record group. In this memo, NSC Executive Secretary and Eisenhower's **National Security Advisor Robert Cutler** informed **Air Force Chief of Staff** (and alleged MJ-12 member) **Nathan Twining** of a change of plans in a scheduled MJ-12 briefing.

The Cutler-Twining memo lacked a distinctive catalog number, leading many to suspect that whether hoaxed or genuine, the memo was almost certainly planted in the archives.

Moore and Shandera have been accused of hoaxing the memo and then planting it in the archives. However, Friedman notes that the memo, unlike the other early MJ-12 papers which were available only as photos, is on original onionskin paper widely used by the government at that time (1953 - early 70's) and unavailable in stationery stores. The document also has some subtle historical and other details that a civilian hoaxer would be unlikely to know, such as a red pencil declassification marking also found with the other declassified files. Furthermore, NARA security procedures would make it difficult for a visitor to the Archives to plant such a document; even the skeptical Klass argued that NARA security procedures made it highly unlikely that Shandera and Moore could have planted the Cutler-Twining memo in the archives. Instead, Friedman has argued that one of the many Air Force personnel involved in declassifying NARA documents could easily have planted the Cutler/Twining memo in with other unrelated documents.

504

However, most researchers have argued that various subtle details point to a forgery. For example, the date of the alleged MJ-12 meeting does not correspond to any known meeting of import (see Arguments against for more examples). However, this doesn't negate Friedman's point that the memo could have been planted by someone in the Air Force.
http://en.wikipedia.org/wiki/Majestic_12

The FBI investigation (1988)

The MJ-12 documents were first made public in 1987 by Shandera, Moore, and Friedman. Another copy of the same documents Shandera received in 1984 was mailed to British researcher **Timothy Good** in 1987, again from an anonymous source. Good first reproduced them in his book *Above Top Secret* (1988), but later judged the documents as likely fraudulent.

After the documents became widely known with the publication of Good's book, the **Federal Bureau of Investigation** then began its own investigation, urged on by debunker **Philip J. Klass**. The MJ-12 documents were supposedly classified as "Top Secret", and the FBI's initial concern was that someone within the U.S. government had illegally leaked highly classified information.

The FBI quickly formed doubts as to the documents' authenticity. FBI personnel contacted the U.S. Air Force Office of Special Investigations (counterintelligence), asking if MJ-12 had ever existed. AFOSI claimed that no such committee had ever been authorized or formed and that the documents were "bogus." The FBI adopted the AFOSI opinion and declared the MJ-12 documents to be "completely bogus."

However, when Stanton Friedman contacted the AFOSI officer, Col. Richard Weaver, who had rendered this opinion, Friedman said Weaver refused to document his assertion. Friedman also noted that Weaver had taught disinformation and propaganda courses for AFOSI and was principal author of the Air Force's debunking Roswell report in 1994. (Friedman, 110-115)

Timothy Good in *Beyond Top Secret* also noted that Weaver in 1994 was the Director of Security and Special Programs Oversight of AFOSI's Pentagon office, a very high-level organization within the Office of the Secretary of the Air Force. Good commented that AFOSI is "an agency whose work involves counterintelligence and deception, and which has a long record of deep involvement in the UFO problem." Within Weaver's office were "special planners." According to Good, "In Air Force parlance, the term 'special plans' is a euphemism for deception as well as for 'perception management' plans and operations." Conducting an interview with one Roswell witness, Weaver himself admitted, *"We're the people who keep the secrets." It is difficult to tell from interviews such as these, as the cold war tactics of deceptions within deceptions are intentionally vague as to where the disinformation and cover-up of espionage ends and the government's actual investigation into UFOs begins.* **[Italics added by author for emphasis].**

William Moore would later reveal that the whole New Mexico UFO disinformation scheme was run out of the Pentagon by a Colonel Barry Hennessey of AFOSI. When the Defense Department phone directory was checked, Hennessey was listed under the "Dept. of Special Techniques." Working under him at the time was the same Col. Weaver.

Friedman, therefore, raised the question as to whether Weaver rendered an objective intelligence opinion about the authenticity of the MJ-12 papers or was deliberately misleading the FBI as a counterintelligence and disinformation agent, much like Doty had done with Moore and Howe earlier.

Journalist **Howard Blum** in his book *Out There* (1990) further described the FBI's difficulty in getting at the truth of the matter. One frustrated FBI agent told Blum, ***"All we're finding out is that the government doesn't know what it knows. There are too many secret levels. You can't get a straight story. It wouldn't surprise me if we never know if the papers are genuine or not.*** http://en.wikipedia.org/wiki/Majestic_12

Whether it is the CIA, the FBI, the US Air Force or some other government agency calling the Majestic 12 Document as bogus or hoaxed, it is nothing more than a deliberate attempt to discredit the whole UFO/ETI subject as a quasi-science, while the real science of the "dark world" investigates this phenomenon behind secured locked doors away from the prying eyes of the public.

The arguments continue on to the present day over the authenticity of the MJ 12 Documents with many UFO researchers divided in their support of the Majestic Documents and some UFO researchers and professional debunkers opposing the documents as bogus and fabricated having bought into the Doty disinformation campaign with the unwitting support of Moore's participation and the deliberate support by the AFOSI. From the outside perspective of the private UFO researcher, it would appear that the US Air Force Office of Scientific Intelligence counterintelligence have accomplished their goal of throwing doubt into the authenticity of the Majestic Documents and causing another split in the UFO Community.

Stanton Friedman and William Moore, the researchers primarily pushing these materials out into the public eye, remain fairly controversial, even within UFO circles. The documents themselves have been the subject of endless analysis, with very little in the way of fruitful conclusions.

Moore was one of the first people to view the MJ-12 documents. While he has maintained their veracity, he has been involved in a couple of very public embarrassments which finally led to his "confession" that some information he had been peddling was wrong and might have been "disinformation" sent out by nefarious shadow government types seeking to discredit UFO researchers.

Friedman is viewed either as a dogged and persevering hero or as an intellectual fraud, depending mostly on the preconceptions of the person describing him. http://www.rotten.com/library/conspiracy/majestic-12

Into this controversial fray have entered the father and son team of Robert and Ryan Wood, who add to the Majestic mystique with the claim of authentication of FOIA documents removing any remaining doubt from the minds of even the staunchest debunker as to the legitimacy of the Majestic Documents.

Arguments for Authenticity of the MJ 12 Documents

Some of the arguments for the authenticity of the MJ 12 Documents are listed below:

- The National Archives contain one document relating to MJ-12, found in 1985, which has been interpreted as corroborative evidence for the MJ-12 documents being genuine:

  "Memorandum for General Twining, from Robert Cutler, Special Assistant to the President, Subject: "NSC/MJ-12 Special Studies Project" dated July 14, 1954. The memos advised Twining of a change of scheduling for a planned briefing following an already scheduled, unspecified "White House meeting" on July 16. Cutler was Eisenhower's National Security Adviser. The memorandum does not identify MJ-12 or the purpose of the briefing. However, arguments have been made against this document's authenticity.

- Regarding the Cutler memo, Jim Speiser writes, "The alleged maker of the memo, Robert Cutler, was out of the country when it was typed. Researchers counter that Cutler's assistants, James Lay and Patrick Coyne, routinely sent out memos under Cutler's name, and they point to the fact that the memo (extant now in carbon copy only) is unsigned." Stanton Friedman has argued that if the memo had the absent Cutler's signature on it, it would have proven that it was a hoax.

- Nuclear physicist and UFO researcher Stanton Friedman has offered other rebuttals of many arguments against the documents' authenticity. For example, Philip J. Klass suggested that the Cutler/Twining memo was fraudulent because it was typed on a typewriter set for pica spacing (10 characters per inch), while Klass insisted that genuine White House documents of that era were only typed using elite spacing (12 characters per inch). Klass offered $100 for every example of genuine pica type that could be presented. Friedman responded as Speiser wrote in the same article cited above, "Friedman provides no fewer than 20 such exemplars, more than enough to win the maximum prize." (Klass paid him $1000, though Speiser suggests the challenge might more accurately be called a draw: "Klass' letter specifically called for 'letters' and 'memoranda'; Friedman provides only headings and dates in his initial response.) Some other Friedman objections to Klass' arguments are provided further below.

- Citing work by Timothy Good, **C.D.B. Bryan** notes the existence of a secret memorandum written by Canadian radio engineer **Wilbert B. Smith,** who had long worked for the **Canadian Department of Transportation**. Dated November 21, 1950, the memo recommended that the Canadian government establish a formal investigation of UFOs (**Project Magnet** was this study). In part, Smith wrote that his own "discreet inquiries" through the Canadian embassy in Washington D.C. had uncovered the fact that "flying saucers exist", "the matter is the most highly classified subject in the United States Government, rating higher even than the H-bomb", and that "concentrated effort is being made by a small group headed by Doctor Vannevar Bush" into their "modus operandi" (Bryan, 186;) Smith's memo was authenticated by the Canadian government. Good concluded that this document is a major argument in favor of MJ-12's reality.

507

Additional documents from Smith and the Canadian embassy named Bush and the Research and Development Board (RDB) as being needed to clear a magazine article being written by **Donald Keyhoe** on Smith's flying saucer theories. Smith also made some public statements about being loaned UFO crash material for metallurgical analysis by some "highly classified group" which he would not name but indicated it was not the Air Force or CIA. In a 2002 interview, James Smith revealed that his father on his deathbed confessed to being shown actual craft and bodies by the U.S. government.

- In a letter from 1983, **Dr. Omond Solandt**, director of the Canadian **Defence Research Board (DRB)**, who had approved Smith's initial UFO study and lent support from the DRB (according to Smith's memo), confirmed meeting with Bush on a regular but "informal" basis to discuss flying saucers and Smith's UFO work.

- Smith's primary source in 1950 was **Dr. Robert Sarbacher**, a missile and electronics expert and a consultant for the RDB's guided missile committee. When contacted again in 1983 by **William Steinman**, Sarbacher in a letter confirmed having the 1950 meeting, reconfirmed that Bush and the RDB were definitely involved, added that mathematician **John von Neumann** was also definitely involved and he thought **Dr. Robert Oppenheimer** as well. He also reconfirmed that there had been flying saucer crashes and being told that the material recovered was extremely lightweight and strong. He was told about small alien bodies.

- In later interviews, Sarbacher would also implicate electrical engineer **Dr. Eric A. Walker**, the executive secretary of the RDB from 1950–1951 and later President of Penn State University. Walker was contacted by phone in 1987 by Steinman. He was asked first whether he had attended meetings at Wright-Patterson AFB concerning the military recovery of flying saucers and bodies of occupants. According to Steinman, he responded, "Yes, I attended meetings concerning that subject matter." When asked as to whether he knew about MJ-12, he responded, "Yes, I know of MJ-12. I have known of them for 40 years." In subsequent interviews and correspondence by other researchers, Walker became much more evasive. But in two interviews from 1990, **Walker**, while saying he thought the MJ-12 documents were not authentic, also ***admitted he had had nothing to do with MJ-12 "for a long time" but they still existed and were "a handful of elite", no longer military, and no longer all American. "We have learnt so much, and we are not working with them, only contact. The technology is far beyond what is known in ordinary terms of physics."*** [Bold italics added by author for emphasis].

- Another person to implicate Bush and Walker as likely being involved was **Dr. Fred Darwin,** who had been **Executive Director for the Guided Missile Committee** for the RDB from 1949 to 1954, to which both Sarbacher and Walker acted as consultants. Like Sarbacher, Darwin also suggested John von Neumann and added alleged MJ-12 member Lloyd Berkner and physicist Dr. Karl Compton.

- Following a famous close encounter with a 300-foot flying saucer while flying from Iceland to Newfoundland on February 10, 1951, Naval Reserve pilot **Commander Graham Bethune** relates that he and the entire crew were immediately debriefed by

USAF and Naval intelligence. In May 1951, Bethune was again questioned by a naval intelligence officer. Bethune says he then asked the officer where such reports ended up. He responded that they first went to *"a committee of twelve men"* screening them for *"national security impact"*. If deemed to have such impact, it would never be sent elsewhere. Otherwise, they would be sent to USAF or Naval offices handling ordinary UFO cases.

- Although he never used the name "MJ 12", **Air Force Brig. Gen. Arthur E. Exon** (Commanding Officer of Wright-Patterson Air Force Base from 1964–1966) reported that a secret group of mostly high-ranking Pentagon officers were somehow involved with UFO studies; he nicknamed this group the **"Unholy Thirteen"**. However, this does not necessarily mean Exon's **"Unholy 13"** and **"MJ 12"** were the same group. When Stanton Friedman sent Exon a copy of his 1990 "Final Report on Operation Majestic 12," he reported Exon "strongly approved" the contents and that the names of the "Unholy 13" group "were those of high-level personnel he thought would know about what was happening, not of people he knew to be involved in a control group."

- Author Whitley Strieber in his books *Breakthrough* (1995) and *Confirmation* (1998) claimed his uncle Colonel Edward Strieber, who had spent much of his career at Wright-Patterson AFB, knew of MJ-12: "My uncle informed me that he had knowledge of the Majestic project. He spoke of the delivery of alien materials, artifacts, and biological remains to Wright Field from the Roswell Army Air Base in the summer of 1947. He felt sure that the existence of these materials and what to do about them had been debated at the highest levels of the government. ...In 1991, after I had written *Majestic* [a partly fictionalized account of the Roswell incident], my uncle put me into contact with a general [Arthur Exon] – an old and trusted friend of his..." Strieber said Exon told him that everybody "from Truman on down" had known about the Roswell incident from the day it happened, and that it was known to be an alien spacecraft "almost as soon as we got on the scene."

- **Edward J. Ruppelt**, the director of the Air Force's public UFO investigation Project Blue Book, several times in his 1956 book *The Report on Unidentified Flying Objects* hinted that there was another highly secretive UFO government group (or groups) operating parallel UFO investigations outside the public eye. For example, in discussing the demoralization of Project Sign personnel following the rejection by Gen. Hoyt Vandenberg of their 1948 Estimate of the Situation that UFOs were extraterrestrial, Ruppelt wrote that Sign personnel hardly investigated UFO sightings anymore and instead "More and more work was being pushed off onto *the other investigative organization* that was helping **ATIC [Air Technical Intelligence Center** at Wright-Patterson AFB]." Regarding the 1950 investigation of the so-called Lubbock Lights, Ruppelt wrote, "The only other people outside Project Blue Book who have studied the complete case of the **Lubbock Lights** were a group who, due to their associations with the government, had complete access to our files. ...they were scientists—rocket experts, nuclear physicists, and intelligence experts. They had banded together to study our UFO reports because they were convinced that some of the UFO's that were being reported were interplanetary spaceships..."

- UFO researcher and MJ-12 skeptic Brad Sparks, however, says evidence points to the group described by Ruppelt investigating the "Lubbock Lights" as being the CIA's **Office of Scientific Investigation (OSI),** not "MJ-12". However, Sparks has also found evidence that the CIA OS/I division (today called the Directorate of Science and Technology) became the primary investigative group for the DOD's **Research and Development Board (RDB)** starting in January 1949. Researcher **David Rudiak** has pointed out that the 1950-51 Canadian documents mentioning an MJ-12-type group under Vannevar Bush's direction have them operating precisely out of the RDB, which would again directly link "MJ-12" to the secret group investigating the Lubbock Lights, as described by Ruppelt. Furthermore, MJ-12 was supposed to be the control group, and it would be very much in Bush's management style to assign investigative responsibility to others rather than MJ-12 conducting the detailed investigations themselves.

- UFO and paranormal researcher Ethan A. Blight has presented a refutation of many of the arguments put forth by critics of the documents, especially those of UFO debunker Philip J. Klass, which are used in the "Arguments against" section below. Stanton Friedman has likewise presented arguments that many of Klass' and other objections are either weak or completely bogus. Both Blight and Friedman argue that there exists no conclusive evidence against the authenticity of the documents, which, while not proving the documents' authenticity, removes much doubt. Both also argue that such false or misleading arguments are in fact characteristic of UFO debunkers in general.

Arguments Against Authenticity of the MJ 12 Documents

Below are a number of arguments against the authenticity of various MJ-12 documents:

- The FBI investigated the matter and quickly formed doubts as to the documents' authenticity. FBI personnel contacted the U.S. Air Force, asking if MJ-12 had ever existed. The Air Force reported that no such committee had ever been authorized, and had never been formed. The FBI presently declares that "The investigation was closed after it was learned that the document was completely bogus."

- Critics note that the documents are of suspicious provenance. Shandera and Good both claimed to have received documents from anonymous senders, and most subsequent MJ-12 documents have surfaced under equally questionable circumstances.

- Though Good initially thought the documents were genuine, he has since, according to Philip Klass, expressed "suspicions about the new ... documents" due to "some factual anomalies in their content."

- UFO researcher Jerome Clark discusses the MJ-12 documents in the "Hoaxes" section of his *The UFO Book* and strongly favors a hoax interpretation. He notes that as of 1998, a mere "handful" of ufologists support the documents' authenticity.

- Scientific forensic linguistic testing was applied to select Majestic Documents in 2007 by Dr. Carol Chaski and evidence was found to disprove attributed authorship. Dr. Chaski is

510

the founder of The Institute for Linguistic Evidence (ILE), a research organization that validates reliable document authentication techniques and provides assistance to investigators and attorneys in criminal and civil trials whenever the authorship of any document is questioned or suspicious."

- The format of the Majestic-12 Documents, with justification and different fonts and type sizes, generates some doubts: the first typewriter with IBM typeballs (selector compensator), and with it replaceable fonts, was the IBM 72, built from 1961 — and only the successors of this machine also had the memory necessary for justification.

- Page 11 of the **Special Operations Manual SOM1-01** refers to "Area 51 S4", which is the same nomenclature and location referenced by **Robert Lazar** 35 years later in 1989. Lazar's unproven background and claims shouldn't have any credibility in this discussion.

http://en.wikipedia.org/wiki/Majestic_12
Howard Blum, *Out There*, 1990, Pocket Books (Simon & Schuster), ISBN 0-671-66261-9
C.D.B. Bryan, *Close Encounters of the Fourth Kind: Alien Abduction, UFOs and the Conference at M.I.T.*, 1995, Alfred A. Knopf, ISBN 0-679-42975-1
Jerome Clark, The UFO Book: Encyclopedia of the Extraterrestrial; Visible Ink, 1998; ISBN 1-57859-029-9
Richard M. Dolan, *UFOs and the National Security State: Chronology of a Coverup, 1941-1973*, 2002, Hampton Roads Publishing Company, ISBN 1-57174-317-0
Stanton T. Friedman, *TOP SECRET/MAJIC*, 1997, Marlowe & Co., ISBN 1-56924-741-2
Timothy Good, *Above Top Secret: The Worldwide UFO Cover-up*, 1988, Quill (William Marlow), ISBN 0-688-09202-0
Timothy Good, *Beyond Top Secret: The Worldwide UFO Security Threat*, 1997, Pan Books (MacMillan Publishers), ISBN 0-330-34928-7
 Steven M. Greer, *Disclosure: Military and Government Witnesses Reveal the Greatest Secrets in Modern History*, 2001, ISBN 0-9673238-1-9
Michael Hesemann and Philip Mantle, *Beyond Roswell: The Alien Autopsy Film, Area 51, & the U.S. Government Coverup of UFOs*, 1997, Marlowe & Company, ISBN 1-56924-781-1
John Spencer, *The UFO Encyclopedia*, 1991, Avon Books, ISBN 0-380-76887-9, pp 199–200

TOP SECRET / MAJIC
EYES ONLY

NATIONAL SECURITY INFORMATION

* TOP SECRET *

EYES ONLY

COPY ONE OF ONE.

BRIEFING DOCUMENT: OPERATION MAJESTIC 12

PREPARED FOR PRESIDENT-ELECT DWIGHT D. EISENHOWER: (EYES ONLY)

18 NOVEMBER, 1952

WARNING: This is a TOP SECRET - EYES ONLY document containing compartmentalized information essential to the national security of the United States. EYES ONLY ACCESS to the material herein is strictly limited to those possessing Majestic-12 clearance level. Reproduction in any form or the taking of written or mechanically transcribed notes is strictly forbidden.

* TOP SECRET *

TOP SECRET / MAJIC
EYES ONLY

EYES ONLY

T52-EXEMPT (E)

002

COPY ONE OF ONE.

SUBJECT: OPERATION MAJESTIC-12 PRELIMINARY BRIEFING FOR
   PRESIDENT-ELECT EISENHOWER.

DOCUMENT PREPARED 18 NOVEMBER, 1952.

BRIEFING OFFICER: ADM. ROSCOE H. HILLENKOETTER (MJ-1)

NOTE: This document has been prepared as a preliminary briefing
only. It should be regarded as introductory to a full operations
briefing intended to follow.

* * * * * *

OPERATION MAJESTIC-12 is a TOP SECRET Research and Development/
Intelligence operation responsible directly and only to the
President of the United States. Operations of the project are
carried out under control of the Majestic-12 (Majic-12) Group
which was established by special classified executive order of
President Truman on 24 September, 1947, upon recommendation by
Dr. Vannevar Bush and Secretary James Forrestal. (See Attachment
"A".) Members of the Majestic-12 Group were designated as follows:

          Adm. Roscoe H. Hillenkoetter
          Dr. Vannevar Bush
          Secy. James V. Forrestal*
          Gen. Nathan F. Twining
          Gen. Hoyt S. Vandenberg
          Dr. Detlev Bronk
          Dr. Jerome Hunsaker
          Mr. Sidney W. Souers
          Mr. Gordon Gray
          Dr. Donald Menzel
          Gen. Robert M. Montague
          Dr. Lloyd V. Berkner

The death of Secretary Forrestal on 22 May, 1949, created
a vacancy which remained unfilled until 01 August, 1950, upon
which date Gen. Walter B. Smith was designated as permanent
replacement.

Majestic 12 Documents pg. 2
http://www.ufocasebook.com/documents.html

513

On 24 June, 1947, a civilian pilot flying over the Cascade Mountains in the State of Washington observed nine flying disc-shaped aircraft traveling in formation at a high rate of speed. Although this was not the first known sighting of such objects, it was the first to gain widespread attention in the public media. Hundreds of reports of sightings of similar objects followed. Many of these came from highly credible military and civilian sources. These reports resulted in independent efforts by several different elements of the military to ascertain the nature and purpose of these objects in the interests of national defense. A number of witnesses were interviewed and there were several unsuccessful attempts to utilize aircraft in efforts to pursue reported discs in flight. Public reaction bordered on near hysteria at times.

In spite of these efforts, little of substance was learned about the objects until a local rancher reported that one had crashed in a remote region of New Mexico located approximately seventy-five miles northwest of Roswell Army Air Base (now Walker Field).

On 07 July, 1947, a secret operation was begun to assure recovery of the wreckage of this object for scientific study. During the course of this operation, aerial reconnaissance discovered that four small human-like beings had apparently ejected from the craft at some point before it exploded. These had fallen to earth about two miles east of the wreckage site. All four were dead and badly decomposed due to action by predators and exposure to the elements during the approximately one week time period which had elapsed before their discovery. A special scientific team took charge of removing these bodies for study. (See Attachment "C".) The wreckage of the craft was also removed to several different locations. (See Attachment "B".) Civilian and military witnesses in the area were debriefed, and news reporters were given the effective cover story that the object had been a misguided weather research balloon.

****
* TOP SECRET *
****
<u>EYES ONLY</u> TOP SECRET / MAJIC
EYES ONLY

T52-EXEMPT (E)

00°

TOP SECRET / MAJIC
EYES ONLY
* TOP SECRET *

EYES ONLY

COPY ONE OF ONE.

A covert analytical effort organized by Gen. Twining and Dr. Bush acting on the direct orders of the President, resulted in a preliminary concensus (19 September, 1947) that the disc was most likely a short range reconnaissance craft. This conclusion was based for the most part on the craft's size and the apparent lack of any identifiable provisioning. (See Attachment "D".) A similar analysis of the four dead occupants was arranged by Dr. Bronk. It was the tentative conclusion of this group (30 November, 1947) that although these creatures are human-like in appearance, the biological and evolutionary processes responsible for their development has apparently been quite different from those observed or postulated in homo-sapiens. Dr. Bronk's team has suggested the term "Extra-terrestrial Biological Entities", or "EBEs", be adopted as the standard term of reference for these creatures until such time as a more definitive designation can be agreed upon.

Since it is virtually certain that these craft do not originate in any country on earth, considerable speculation has centered around what their point of origin might be and how they get here. Mars was and remains a possibility, although some scientists, most notably Dr. Menzel, consider it more likely that we are dealing with beings from another solar system entirely.

Numerous examples of what appear to be a form of writing were found in the wreckage. Efforts to decipher these have remained largely unsuccessful. (See Attachment "E".) Equally unsuccessful have been efforts to determine the method of propulsion or the nature or method of transmission of the power source involved. Research along these lines has been complicated by the complete absence of identifiable wings, propellers, jets, or other conventional methods of propulsion and guidance, as well as a total lack of metallic wiring, vacuum tubes, or similar recognizable electronic components. (See Attachment "F".) It is assumed that the propulsion unit was completely destroyed by the explosion which caused the crash.

* TOP SECRET *

```
****************
*  TOP SECRET  *
****************
```

EYES ONLY   COPY ONE OF ONE

A need for as much additional information as possible about
these craft, their performance characteristics and their
purpose led to the undertaking known as U.S. Air Force Project
SIGN in December, 1947. In order to preserve security, liason
between SIGN and Majestic-12 was limited to two individuals
within the Intelligence Division of Air Materiel Command whose
role was to pass along certain types of information through
channels. SIGN evolved into Project GRUDGE in December, 1948.
The operation is currently being conducted under the code name
BLUE BOOK, with liason maintained through the Air Force officer
who is head of the project.

On 06 December, 1950, a second object, probably of similar
origin, impacted the earth at high speed in the El Indio -
Guerrero area of the Texas - Mexican boder after following
a long trajectory through the atmosphere. By the time a
search team arrived, what remained of the object had been almost
totally incinerated. Such material as could be recovered was
transported to the A.E.C. facility at Sandia, New Mexico, for
study.

Implications for the National Security are of continuing importance in that the motives and ultimate intentions of these
visitors remain completely unknown. In addition, a significant
upsurge in the surveillance activity of these craft beginning
in May and continuing through the autumn of this year has caused
considerable concern that new developments may be imminent.
It is for these reasons, as well as the obvious international
and technological considerations and the ultimate need to
avoid a public panic at all costs, that the Majestic-12 Group
remains of the unanimous opinion that imposition of the
strictest security precautions should continue without interruption into the new administration. At the same time, con-
tingency plan MJ-1949-04P/78 (Top Secret - Eyes Only) should
be held in continued readiness should the need to make a
public announcement present itself. (See Attachment "G".)

516

TOP SECRET
EYES ONLY
THE WHITE HOUSE
WASHINGTON

September 24, 1947.

MEMORANDUM FOR THE SECRETARY OF DEFENSE

Dear Secretary Forrestal:

As per our recent conversation on this matter, you are hereby authorized to proceed with all due speed and caution upon your undertaking. Hereafter this matter shall be referred to only as Operation Majestic Twelve.

It continues to be my feeling that any future considerations relative to the ultimate disposition of this matter should rest solely with the Office of the President following appropriate discussions with yourself, Dr. Bush and the Director of Central Intelligence.

Harry Truman

TOP SECRET
EYES ONLY.

The Jason Society Theory

As the story goes ... allegedly former **President Eisenhower** commissioned a secret society known as the **Jason Society** under the leadership of the following; **Director of Central Intelligence, Allen Welsh Dulles, Dr. Zbigniew Brzezinski, President of the Trilateral Commission** from 1973 until 1976, and **Dr. Henry Kissenger**, leader of the scientific effort, to sift through all the facts, evidence, technology, lies and deceptions and find the truth of the Alien question. The society was made up of thirty-two (32) of the most prominent men in the USA.

MJ-12 allegedly operates inside the Jason Society. The top 12 members of the 32 members of the Jason Society were designated as MJ-12. MJ-12 has control of everything. They are designated by the code J-1, J-2, J-3, etc. all the way through the members of the Jason Society. The director of Central Intelligence was appointed J-1 and is the Director of the MJ-12 group. MJ-12 use to only be responsible to the President of the United States (not true anymore). The actual cost of funding the Alien connected projects is higher than anything you could imagine.

Some believe that MJ-12 runs most of the world's illegal drug trade to hide funding and thus keep the secret from congress and the people of the United States. It was justified in that it would identify and eliminate the weak and undesired elements of our society.

A secret meeting place was constructed for the MJ-12 group in Maryland and is only accessible by air. It contains full living, recreational, and other facilities for the MJ-12 group and the Jason Society. It is code named **"The Country Club"**. The land for The Country Club was donated by the Rockefeller family. Only those with Ultra Top Secret - MAJI clearances are allowed to attend.

MAJI - Majority Agency for Joint Intelligence: All information, disinformation, and intelligence is gathered and evaluated by this agency. This agency is responsible for all disinformation and operates in conjunction with the CIA, NSA, DIA, and the Office of Naval Intelligence. This is a very powerful organization and all Alien projects are under its control. MAJI is responsible only to MJ-12. MAJIC is the security classification and clearance of all Alien connected material, projects, and information.

MAJIC is also MAJI.

MJ-1 is the classification for the director of MAJI, who is the director of the CIA and reports only to the President. Other members of MAJI are designated MJ-2, MJ-3, etc. This is why there is some confusion about references of MJ-12, the group or MJ-12 the person.

Designation for MJ-12, the group are MAJI or MAJIC Designation in official documents about MJ-12 means the person only.

In 1947, PROJECT SIGN was created to acquire as much information as possible about UFO's, their performance characteristics and their purposes. In order to preserve security, liaison between Project Sign and MJ-12 was limited to two individuals within the intelligence division of the Air Material Command whose role was to pass along certain types of information through

518

channels. **Project Sign** evolved **Project Grudge** in December of 1948. Project Grudge had an overt civilian counterpart named **PROJECT BLUE BOOK**, with which we are all familiar. Only "Safe" reports were passed to Project Blue Book.

Project Grudge was a short-lived project by the U.S. Air Force to investigate unidentified flying objects (UFOs). Grudge succeeded Project Sign in February 1949 and was then followed by Project Blue Book. The project formally ended in December 1949, but actually continued on in a very minimal capacity until late 1951. Project Sign had been active from 1947 to 1949. Some of Sign's personnel including director **Robert Sneider** favored the extraterrestrial hypothesis as the best explanation for UFO reports. They prepared the **Estimate of the Situation** arguing their case. This theory was ultimately rejected by high-ranking officers, and Project Sign was dissolved and replaced by Project Grudge. http://www.crystalinks.com/mj12.html

Project Blue Book was one of a series of systematic studies of Unidentified flying objects (UFOs) conducted by the United States Air Force (U.S.A.F.). Started in 1952, it was the second revival of such a study. A termination order was given for the study in December 1969, and all activity under its auspices ceased in January 1970. Project Blue Book had two goals: to determine if UFOs were a threat to national security, and to scientifically analyze UFO-related data. Thousands of UFO reports were collected, analyzed and filed. As the result of the Condon Report, which concluded there was nothing anomalous about any UFOs, Project Blue Book was ordered shut down in December 1969. This project was the last publicly known UFO research project led by the USAF.

By the time Project Blue Book ended, it had collected 12,618 UFO reports and concluded that most of them were misidentifications of natural phenomena (clouds, stars, et cetera) or conventional aircraft. A few were considered hoaxes. 701 of the reports - about six percent - were classified as unknowns, defying detailed analysis. The UFO reports were archived and are available under the **Freedom of Information Act**, but names and other personal information of all witnesses have been redacted. Though many accepted Blue Book's final conclusions that there was nothing extraordinary about UFOs, critics - then and now - have charged that Blue Book, especially in its later years, was engaging in dubious research, or even perpetuating a cover-up of UFO evidence. Some evidence suggests that not only did some UFO reports bypass Blue Book entirely but, that the U.S. Air Force continued collecting and studying UFO reports after Blue Book had been discontinued, despite official claims to the contrary.
http://www.crystalinks.com/mj12.html

MJ-12 was originally organized by General George C. Marshall in July 1947 to study the Roswell-Magdalena UFO crash recovery and debris. Admiral Hillenkoetter, director of the CIA from May 1, 1947, until September 1950, decided to activate the "Robertson Panel," which was designed to monitor civilian UFO study groups that were appearing all over the country. He also joined NICAP in 1956 and was chosen as a member of its board of directors. It was from this position that he was able to act as the MJ-12 "Mole", along with his team of other covert experts. They were able to steer NICAP in any direction they wanted to go. With the "Flying Saucer Program" under complete control of MJ-12 and with the physical evidence hidden away, General Marshall felt more at ease with this very bizarre situation. These men and their successors have most successfully kept most of the public fooled up to the present, including much of the western world, by setting up false experts and throwing their influence behind them to make their plan work, with considerable success until now.

Within 6 months of the Roswell crash on July 2, 1947, and the finding of another crashed UFO at San Augustine Flats near Magdelena, New Mexico on July 3, 1947, a great deal of reorganization of agencies and shuffling of people took place. The main thrust behind the original "Security Lid", and the very reason for its construction, was the analysis and attempted duplication of the technologies of the discs. The activity was headed up by the following groups.

1. The Research and Development Board **(R&DB)**
2. Air Force Research and Development **(AFRD)**
3. The Office of Naval Research **(ONR)**
4. CIA Office of Scientific Intelligence **(CIA-OSI)**
5. NSA Office of Scientific Intelligence **(NSA-OSI)**

No single one of these groups knew the whole story, which is not uncommon in conspiratorial circles. Each group was to know only the parts that MJ-12 allowed them to know. MJ-12 also operates through the various civilian intelligence and investigative groups. The CIA and FBI are manipulated by MJ-12 to carry out their purposes. The NSA was created in the first place to protect the secret of the recovered flying discs, and eventually got complete control over all communication intelligence. This control allows the NSA to monitor any individual through mail, telephone, telexes, telegrams, and now through on-line computers, monitoring private and personal communications as they may desire. http://www.crystalinks.com/mj12.html

Alleged Majic Projects

The following projects are associated with the Majestic Twelve organization but, like everything about the UFO/ETI subject, much of the information has to be thoroughly corroborated from many sources in order to be accepted as authentic documented evidence, thus, some or all of the listed projects are considered as alleged until fully proven by other researchers.

SIGMA is the project which first established communications with the Aliens and is responsible for communications.

PLATO is the project responsible for **Diplomatic Relations with the Aliens.** This project secured a formal treaty with the aliens.

AQUARIUS is the project which compiled the history of the **Alien presence** and interaction on Earth and the homo sapiens.

GARNET is the project responsible for control of all information and documents regarding the Alien subjects and accountability of their information and documents.

PLUTO is a project responsible for evaluating all UFO and IAC information pertaining to space technology.

520

POUNCE project was formed to recover all downed and/or crashed craft and Aliens. This project provided cover stories and operations to mask the true endeavor, whenever necessary. Covers which have been used were crashed experimental aircraft, construction, Mining, etc. This project has been successful and is ongoing today.

NRO is the **National Reconnaissance Organization** based at Fort Carson, Colorado. It's responsible for security on all Alien or **Alien Spacecraft** connected to the projects.

DELTA is the designation for the specific arm of NRO which is specially trained and tasked with security of all **MAJI projects**. It's a security team and task force from NRO especially trained to provide Alien tasked projects and **LUNA security** (Also has the code name: "**Men In Black" (MIB)**. This project is ongoing.

BLUE TEAM is the first project responsible for reaction and/or recovery of downed and/or **crashed Alien craft and/or Aliens**. This was a U.S. Air Force Material Command project.

SIGN is the second project responsible for the collection of intelligence and determining whether Alien presence constitutes a threat to the U.S. National Security. SIGN absorbed the BLUE TEAM project. This was a U.S. Air Force and CIA project.

REDLIGHT was the project to test fly recovered Alien craft. This project was postponed after every attempt resulted in the destruction of the craft and death of the pilots. This project was carried out at **AREA 51, Groom Lake, (Dreamland)** in Nevada. Project REDLIGHT was resumed in 1972. This project has been partially successful. UFO sightings of craft accompanied by **Black Helicopters** are **Project REDLIGHT** assets. This project is now ongoing at AREA 51 in Nevada. (Believed to have moved to Mexico at this time for testing).

SNOWBIRD was established as a cover for Project Redlight. A "Flying Saucer" type aircraft was built using conventional technology. It was unveiled to the press and flown in public on several occasions. The purpose was to explain accidental sightings or disclosures of Redlight as having been Snowbird crafts. This was a very successful disinformation operation. This project is only activated when needed. This deception has not been used for many years. This project is currently in mothballs until it is needed again.

BLUE BOOK was a U.S. Air Force, UFO, and Alien Intelligence collection and disinformation project. This project was terminated and its collected information and duties were absorbed by project Aquarius. A classified report named "Grudge/Blue Book, Report Number 13" is the only significant information derived from the project and is unavailable to the public.
http://www.crystalinks.com/mj12.html

Does MJ-12 Have a New Identity - PI 40?

Majestic Twelve, as it has been understood by most Ufologists is chiefly an administrative body whose primary function is to oversee all matters related to UFOs and Extraterrestrials, the recovery of saucers, alien bodies (dead or alive), the research and development of alien technology, its implementation and disbursement into the various sectors of the military and private industry with its eventual filtering of selected technology down into the mainstream of society. Initially and on occasion, they would get their hands dirty with the actual investigation, recovery and research of a retrieved ET craft like the Roswell and Aztec saucer crashes. But, like most organizations that come into existence to meet the exigencies of the time, they evolve, undergo membership change, suffer setbacks and outcome to eventual dissolution or are replaced by something else more effective.

Therefore, at this point in time, certain questions must be asked, if the MJ 12 organization is real does it still exist? Would it still have the same basic mission statements? Would it follow the same agenda as originally formulated by its predecessors? Would there be a new core of members? Would it even be known by the same name or would the organization have been replaced by a new agency with some new code name?

Many theories suggest that MJ-12's efforts continue to the present. For example, UFO researcher **Bill Hamilton** says he has identified the present-day members of MJ-12. **Gordon Novel**, a shadowy figure associated with various CIA conspiracies, **Watergate**, and the **Jim Garrison** investigation of the Kennedy assassination, in a recent interview, further adds that most are Americans with a few foreigners. Allegedly they were involved with Kennedy's murder because Kennedy wanted to end the cover-up. They are major world power brokers and manipulate events behind the scenes in a bid for total world power. Supposedly a key motivation behind the cover-up is back-engineering captured alien technology in order to obtain such domination. Moreover, many criminal acts have been committed towards this end, including numerous murders to maintain security and control of international drug trafficking to pay for the huge research and security costs.

Another person to say that MJ-12 still existed was **Dr. Eric Walker**. When originally contacted, Walker said he had known of their existence since their creation in 1947. Similar to Novel, Walker in a later 1990 interview said the current membership was mostly American but had added some foreigners. They were a highly elite group of individuals and Walker repeatedly discouraged interviewers from trying to learn more, saying there was nothing the average person could do. http://www.crystalinks.com/mj12.html

Dr. Steven Greer has stated in his book, "Extraterrestrial Contact: The Evidence and the Implications" that the last code name of this covert organization was related to him by reliable sources was **PI 40**. Most recently in 2015 Dr. Greer has heard that it is now called **SIIG (Senior Interagency Intelligence Group).** Is this the new nom de guerre for Majestic 12, an organization whose very existence is argumentatively questionable or is it the real covert organization to which the top-secret **Wilbur Smith Document** referred to in 1950? A document in which it is stated that a secret small US group was working hard on the UFO matter trying to discover the technology behind the UFOs, which was the most secret undertaking in the US government, exceeding even the secrecy surrounding the development of the H-bomb.

Not much is known about this code name to date. It is not known what the **"PI"** stands for or whether the **"40"** indicates the number of its core membership. What is known about it comes from Dr. Greer's research work as of the spring of 1996. It characteristics are described as quasi-governmental, USAPS related, quasi-private operating international/transnational, hierarchal,

522

extremely powerful, without financial limitations, it has its own military force, it answers to no one but itself, and it is the **"Varsity Team"** of all black projects! It is the mystery within an enigma wrapped in a riddle.

When one speaks of this **PI 40** organization, in the same breathe one must speak of **Unacknowledged Special Access Projects (USAPs)** and **Plausible Deniability**.

A USAP is a top-secret, compartmentalized project requiring special access even for those with a top secret clearance, and is unacknowledged. Dr. Greer has been told by his inside sources that if someone, anyone including your superiors, including the Commander-in-Chief, the President, asks you about it, you reply that no such project exists. You lie!

According to Greer, "people in these **USAPs** are dead serious in keeping their project secret, and will do nearly anything to keep the story covered and to keep both other officials and the public disinformed". Even murder is not out of the question for those who get too close to the truth!
Greer, S. (1999). Extraterrestrial Contact: The Evidence and Implications. Crossing Point, Inc. Publications, Afton, VA. ISBN: 0-9673238-0-0

Whether it is the President of the United States, members of congress, court judges, or world leaders or the UN, they simply don't have a need to know when it comes to the UFO/ET matter. By their not knowing, by keeping them out of the loop on the subject, their ignorance is innocence of any knowledge of these secret projects. In a sense, they have plausible deniability by ignorance of these unacknowledged projects. Not being told the truth behind USAPs keeps these people in a relative state of security.

PI 40s function is purely administrative while decisions are carried out by military and intelligence operatives in compartmentalized units. "Related compartmentalized units, which are also USAPs, are involved in disinformation, public deception, active disinformation, so-called abductions and mutilations, reconnaissance and UFO tracking, space-based weapons systems, and specialized liaison groups (for example to media, political leaders, the scientific community, the corporate world etc). Think of this entity as a hybrid between government USAPs and private industry".

PI-40 consists primarily of mid-level USAPS-related military and intelligence operatives, USAPS or black units within certain high-tech corporate entities, and select liaisons within the international policy analysis community, certain religious groups, the scientific community, and the media, among others. The identities of some of these entities and individuals are known, though most remain unidentified.

Nearly half of the **PI 40** membership is said to be made up of younger members who are in favour of public disclosure because they are less complicit in past excesses while the rest of the core group oppose or are ambivalent to a near-term disclosure.

Actual policy and decision-making seem to rest predominantly at this time in the private, civilian sector, as opposed to USAP-related military and intelligence officials, though some information indicates that there is significant relative autonomy in certain areas of operations. Increasing

523

debate regarding certain covert operations and the advisability of a disclosure seems to be in free-fall within PI-40 at this time according to Greer's assessment based upon one of his Disclosure witnesses.

Many compartmentalized operations within 'black' or USAPS projects are structured so that those working on the task may be unaware that it is UFO/ET related. For example, some aspects of the so-called *Star Wars* *(SDI)* effort, re intended to target extraterrestrial spacecraft which come into close proximity to earth, but the vast majority of scientists and workers in the SDI program are unaware of this. Greer, S. (1999). Extraterrestrial Contact: The Evidence and Implications. Crossing Point, Inc. Publications, Afton, VA. ISBN: 0-9673238-0-0

Greer and his team are aware of three separate, corroborating sources that since the early 1990s, at least 2 extraterrestrial spacecraft have been targeted and destroyed by experimental space-based weapons systems.

The vast majority of political leaders, including White House officials, military leaders, congressional leaders, UN leaders, other world leaders are not routinely briefed on this matter. When and if inquiries are made, they are told nothing about the operations, nor is the existence of any operation confirmed to them. In general, the nature of this covert entity ensures that such leaders do not even know to whom such' inquiries should be addressed.

International cooperation exists to a wide extent, though some witnesses state that certain countries, particularly China, have aggressively pursued somewhat independent agendas.

Major bases of operations, apart from widely diversified private sites, include:

- Edwards Air Force Base in California
- Nellis Air Force Base in Nevada, particularly S4 and adjacent facilities
- Los Alamos New Mexico
- Fort Huachuca Arizona (Army Intelligence Headquarters)
- the Redstone Arsenal in Alabama
- a relatively new, expanding underground facility accessible only by air in a remote area of Utah, among others.

Additional facilities and operations centers exist in a number of other countries, including the United Kingdom, Australia, and Russia.

Numerous agencies have deep cover, black, USAPS related units involved with these operations, including:

- the National Reconnaissance Office **(NRO)**
- the National Security Agency **(NSA)**
- the CIA
- the Defense Intelligence Agency **(DIA)**
- the Air Force Office of Special Investigations **(AFOSI**
- Naval Intelligence

524

- Army Intelligence
- Air Force Intelligence
- the FBI
- and others...

An even more extensive list of private, civilian and corporate entities have significant involvement.

The majority of scientific, technical and advanced technology operations are centered in the civilian industrial and research firms. Significant - and lethal - security is provided by private contractors.

The majority of personnel as well as the leadership of most if not all of these agencies and private groups are uninvolved and unaware of these compartmentalized, unacknowledged operations. For this reason, sweeping accusations related to any particular agency or corporate entity are wholly unwarranted. **"Plausible deniability"** exists at many levels. Moreover, specialization and compartmentalization allow a number of operations to exist without those involved knowing that their task is related to the UFO/ET subject.

Greer, S. (1999). Extraterrestrial Contact: The Evidence and Implications. Crossing Point, Inc. Publications, Afton, VA. ISBN: 0-9673238-0-0

PI 40 is an organization that is extremely tight and rigid in its operation as to be impenetrable; its pecking order of responsibility is so secure that plausible deniability is its mantra, and its identity is so unfathomable that it can only be speculated upon with uncertainty. Such is the nature of the beast of covert darkness!

Yet, that mystery appears to be quickly dissipating as every year that goes by, this elite Harvard Team of covert intelligence appears to be stepping forward and out from their sanctum sanctorum. Their identity becomes more obvious to the world at large with every nefarious scheme perpetrated on the public which is always self-serving and benefitting only a minority. Such actions add yet, one more characteristic by which PI 40 may be known by, that of being "arrogantly secure in their knowledge of being untouchable"!

https://www.youtube.com/watch?v=MNQ8_q-Bxto

CHAPTER 51

WHEN IT COMES TO NATIONAL SECURITY
LIE, LIE, LIE, DENY, DENY, DENY!

During the turbulent time of the Kennedy administration, the CIA would routinely withhold sensitive information from senior White house officials including the President in order to protect them from repercussions that ensued from illegal or disreputable CIA activities which became public knowledge. In light of such unpopular activities, the CIA coined the term **plausible deniability** as a way to deflect or deny blame for any acts of wrongdoing thus, leaving little or no evidence of involvement or abuse.

The term most often refers to the denial of blame in (formal or informal) chains of command, where senior figures assign responsibility to the lower ranks, and records of instructions given do not exist or are inaccessible, meaning independent confirmation of responsibility for the action is nearly impossible. In the case, that illegal or otherwise disreputable and unpopular activities become public, high-ranking officials may deny any awareness of such act or any connection to the agents used to carry out such acts. It typically implies forethought, such as intentionally setting up the conditions to plausibly avoid responsibility for one's (future) actions or knowledge. http://en.wikipedia.org/wiki/Plausible_deniability

In politics and espionage, deniability refers to the ability of a "powerful player" or intelligence agency to avoid "blowback" by secretly arranging for an action to be taken on their behalf by a third party ostensibly unconnected with the major player. In political campaigns, plausible deniability enables candidates to stay "clean" and denounce third-party advertisements that use unethical approaches or potentially libelous innuendo.

Plausible deniability is a clever way to lie or deny the incontrovertible evidence. If you happen to be a lawyer who is defending a client whom you suspect certain evidence and the knowledge of that evidence would convict your client then, you would choose not to investigate it, as you would then have to disclosure such evidence that may find your client guilty. If your opponent lacks incontrovertible proof (evidence) of their allegation, you can "plausibly deny" the allegation even though it may be true. http://en.wikipedia.org/wiki/Plausible_deniability

Plausible deniability is a powerful tool for politicians, intelligence agencies, and their operatives, for lawyers and industrial espionage agents. In the field of UFO investigations, where chains of command within the military are essential to keeping advanced technical information top secret and secured away from public awareness, denial of the existence of such top secret information becomes critical to maintaining the military as a "powerful player".

Where there is a serious leak of UFO information and a possible "blowback" of culpability leading to a military coverup then, high-ranking officials may deny any awareness of the information and disavow any agents working on their behalf. All ties are severed and knowledge of their existence denied or lied to those trying to penetrate such cover-ups. It becomes a "Mission Impossible" for those seeking, for example, to look within the inner sanctum of the Military Industrial Complex.

526

Plausible deniability involving cover-ups usually centered on statements of **"National Security"** which soon became the "catch-all basket" statement for anything the US military was involved in but didn't want the public or even the government to know about. It would also apply to any inventor trying to receive a patent for an invention from the US Patent Office that may involve a new method of energy generation.

That there is a UFO coverup is uncontestable (Fawcett & Greenwood, 1984; Good, 1988; Greer, 1999, pp. 301-312; Sheffield, 1996). It has been maintained through the voluntary wishes of about half the population, particularly persons in positions of authority, as well as by the various governmental agencies that established the UFO ridicule factor. This factor has been especially effective in causing the UFO topic to be taboo for scientists.

Fawcett, L., and Greenwood, B. J. (1984). Clear Intent: The Government Coverup of the UFO Experience. Englewood Cliffs, NJ: Prentice-Hall.
Good, T. (1988). Above Top Secret: The Worldwide UFO Cover-up. New York: William Morrow.
Greer, S. (1999). Extraterrestrial Contact: The Evidence and Implications. Crossing Point, Inc. Publications. Afton, VA.
Sheffield, D. (1996). UFO a Deadly Concealment: The Official Cover-up? London, UK: Blandford

The Roswell Saucer Crash of 1947, the alleged existence of the Majestic 12 organization and Majestic documents, reverse engineering of alien technology, the development of manmade flying saucers or the existence of Area 51, all of these are prime examples of plausible deniability in action. These examples are typical of the coverup of the existence of the UFO and ETI phenomenon in America but, the cover-up extends beyond the borders of the United States.

The Belgium UFO Wave - November1989 to March 1990

As a "good neighbor to the North", Canada is certainly duplicitous in the coverup program as a partner in arms with Americans through the NORAD agreements, as is Britain and most of the countries of Europe through the NATO alliance agreements. In fact, most nations around the planet have an involvement of cooperation with the US whether politically, militarily or through trade agreements, this makes them potential partners with the US in any UFO coverup programs.

A classic example of this duplicitous coverup involvement is the now famous Belgium wave of UFO sightings that began in November1989 to March 1990 with the sudden appearances of large triangular shaped ET craft over the countries of Belgium, Germany and Holland.

Derek Sheffield, a British UFO researcher details his attempts to get to the bottom of a cover-up by the **British Ministry of Defence (MOD)** of radar tracking records during the time of the Belgium wave of UFO sightings in his book, **"UFO a Deadly Concealment: The Official Cover-up?"** This is his investigation into one of the greatest wave of UFO sightings in history that nearly went unnoticed in British and American news media due to those countries' on-going censorship policies regarding the UFO subject matter.

Sheffield sees the whole subject of UFO cover-up as an on-going program of confusion and deception not only in the USA but, to a lesser degree in his homeland of Great Britain. The commonality in both countries is that "all matters relating to this phenomenon involve national security and as such are held to be classified information". **Sheffield, D. (1996). UFO a Deadly Concealment: The Official Cover-up? London, UK: Blandford**

In Sheffield's mind to the extent that this campaign of disinformation, misinformation, and obfuscation that surrounds the UFO subject which has been operating since the early 1950s up to the present time, feels that the whole American UFO scene must be disregarded as completely unreliable. By this is meant the media manipulation and spin doctoring of UFO stories which are ultimately controlled by the US military and in particular by the CIA, as well as the deliberate hoaxing of events and misidentification of actual UFO sightings from possible manmade highly advanced military aircraft.

Requests for UFO information in the United States are channeled through the CIA who informs you, "…there has not been any attempt to collect information on unidentified flying objects since the 1950s. Any further requests are referred to the National Archives and Records Administration in Washington, DC."

In Britain, a similar policy for requests for information is channeled through the British Ministry of Defence. In the case of Sheffield's request for radar information on the Belgium wave of triangular UFOs over that country, he was told that "…any radar inquiries that occur beyond the British boundaries cannot be answered as they become the subject of national security and as such are classified information".

The reply from the **Mr. Nicolas (Nick) G. Pope,** Secretariat (Air Staff) to the British MOD to Derek Sheffield was the usual "clap-trap" response about "*ghost*" radar returns or false returns which are easily detected by experienced radar operators or that radar signals are like ultra short radio wavelengths that behave like light which only reflect a signal from an object that has density and opacity. If radar does not pick up radar signals, phantom or otherwise or cannot be positively identified as legitimate objects then, they do not exist and therefore, are not deemed to be a threat to national security by the MOD. **Derek Sheffield, UFO a Deadly Concealment: The Official Cover-up?" 1996, Blandford Publishing, UK**

To make sense of Sheffield's investigative request for radar information of the UFO wave that had occurred over Belguim, it is necessary to step back and recall the UFO events leading up to the many letters of enquiry sent by Sheffield to the CIA, the MOD, the Cabinet Office, and the letters of response from the Belgium police in Hainaut, the Belgium Air Force and from the Belgium Ministry of Defence.

The Belgian UFO wave peaked with the events of the night of 30/31 March 1990. On that night unknown objects were tracked on radar, chased by two Belgian Air Force F-16's, photographed, and were sighted by an estimated 13,500 people on the ground – 2,600 of whom filed written statements describing in detail what they had seen. Following the incident, the Belgian air force released a report detailing the events of that night. "'Sunday Express' article on Belgium UFO". *Sunday Express*. 17 September 1995. **Retrieved 21 March 2008** and http://www.ufoevidence.org/documents/doc418.htm

528

At around 23:00 on 30 March the supervisor for the **Control Reporting Center (CRC)** at Glons received reports that three unusual lights were seen moving towards Thorembais-Gembloux, which lies to the South-East of Brussels. The lights were reported to be brighter than stars, changing color between red, green and yellow, and appeared to be fixed at the vertices of an equilateral triangle. At this point, Glons CRC requested the Wavre gendarmerie send a patrol to confirm the sighting.

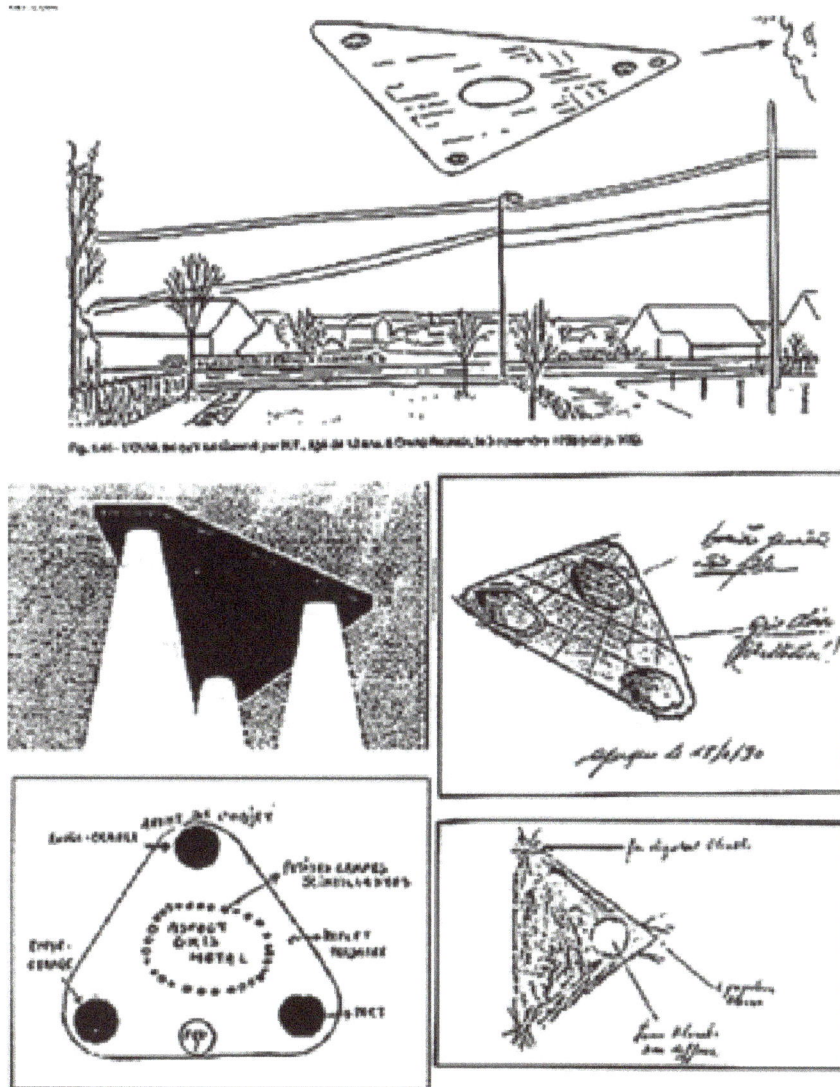

Eyewitness drawings and reconstructions of triangular craft seen in Belgium between 1989 and 1993. (Courtesy of SOBEPS) (Credit: Berliner)
http://www.ufoevidence.org/cases/case1125.htm

Approximately 10 minutes later, a second set of lights was sighted moving towards the first triangle. By around 23:30 the Wavre gendarmerie had confirmed the initial sightings and Glons CRC had been able to observe the phenomenon on radar. During this time the second set of lights, after some erratic maneuvers, had also formed themselves into a smaller triangle. After tracking the targets and after receiving a second radar confirmation from the Traffic Center

Control at Semmerzake, Glons CRC gave the order to scramble two F-16 fighters from Beauvechain Air Base shortly before midnight. Throughout this time the phenomenon was still clearly visible from the ground, with witnesses describing the whole formation as maintaining their relative positions while moving slowly across the sky. Witnesses also reported two dimmer lights towards the municipality of Eghezee displaying similar erratic movements to the second set of lights.

Over the next hour, the two scrambled F-16s attempted nine separate interceptions of the targets. On three occasions they managed to obtain a radar lock for a few seconds but each time the targets changed position and speed so rapidly that the lock was broken. During the first radar lock, the target accelerated from 240 km/h to over 1,770 km/h while changing altitude from 2,700 m to 1,500 m, then up to 3,350 m before descending to almost ground level – the first descent of more than 900 m taking less than two seconds. Similar maneuvers were observed during both subsequent radar locks. On no occasion were the F-16 pilots able to make visual contact with the targets and at no point, despite the speeds involved, was there any indication of a sonic boom. During this time, ground witnesses broadly corroborate the information obtained by radar. They described seeing the smaller triangle completely disappear from sight at one point, while the larger triangle moved upwards very rapidly as the F-16s flew past. After 00:30 radar contact became much more sporadic and the final confirmed lock took place at 00:40. This final lock was once again broken by an acceleration from around 160 km/h to 1,120 km/h after which the radar of the F-16s and those at Glons and Semmerzake all lost contact. Following several further unconfirmed contacts, the F-16s eventually returned to base shortly after 01:00.

The final details of the sighting were provided by the members of the Wavre gendarmerie who had been sent to confirm the original report. They describe four lights now being arranged in a square formation, all making short jerky movements, before gradually losing their luminosity and disappearing in four separate directions at around 01:30.
Report concerning the observation of UFOs in the night from March 30 to March 31, 1990 - ufoevidence.org". Retrieved 21 March 2008.
http://en.wikipedia.org/wiki/Belgian_UFO_wave

One good image (see below) was finally captured on videotape in April 1990. This image showed the underbelly of the craft with spotlights on the three corners. A still frame from this tape has been seen worldwide and is a classic UFO photograph. The Belgium wave has obtained classic status in UFO lore. With over 1,000 witnesses, confirmed radar sightings, plane radar lock-ins, and military confirmations, the fact that an unknown craft moved across the country of Belgium cannot be denied.

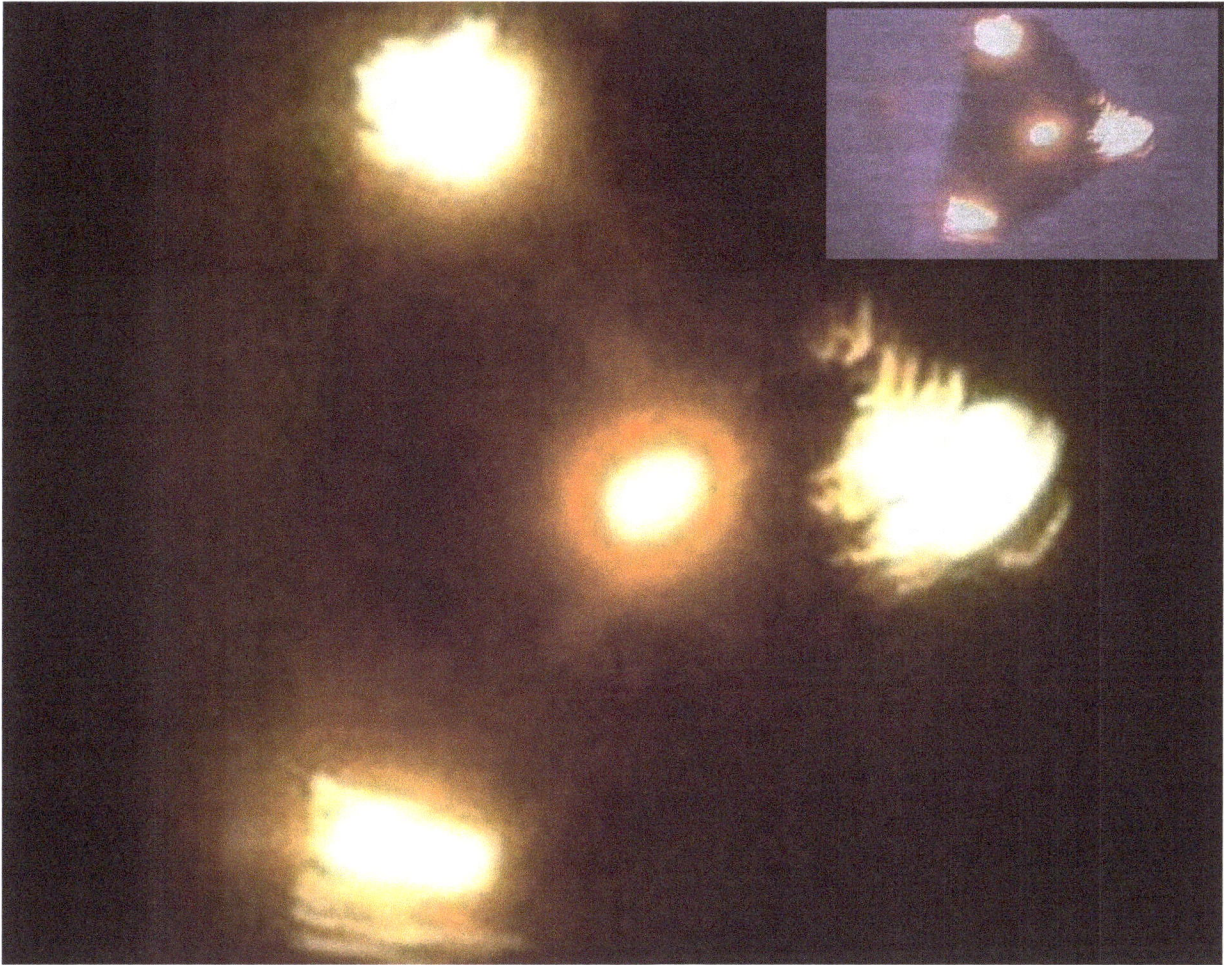

This famous triangular UFO photo was taken at Petit Rechain, Belgium, 1990. The photograph, when lightened, shows the triangle shape outline of the craft (see insert).

http://www.ufoevidence.org/cases/case1126.htm and http://badufos.blogspot.ca/2011/07/classic-ufo-photo-from-belgian-wave.html

The fact is that the Belgium Air Force took a unique position of open and honest public disclosure when **Major-General Wilfried de Brouwer** stated categorically in an interview that the many sightings that were reported by both the public and the military were not:

- Balloons
- Ultralight aircraft (ULM)
- Unmanned aerial vehicles (UAV)
- Aircraft (including Stealth)
- Laser projections or holograms
- Mirages or other meteorological phenomena

531

Major-General Wilfried de Brouwer
sporting a pair of Aviator glasses

Major-General de Brouwer went further to say that initial atmospheric phenomena and electromagnetic interference were ruled out almost right away and that the rapid accelerations and de-accelerations of these triangular craft were beyond human capabilities of technology or human tolerances to survive rapid 40G maneuvers. However, he did stop short in the interview from stating that the Belgium Air Force was dealing with Extraterrestrial aircraft. http://www.latest-ufo-sightings.net/2010/03/famous-ufo-cases-belgian-ufo-wave.html

This brings us back to Derek Sheffield's inquiries (remember him?) for information from his own government, who could not or would not provide confirmation that the British MOD had of radar tracking records of the Belgium UFO events. British radar stations have had the capability of long range radar penetration throughout most of continental Europe, since World War II. They are also a partner with the **NATO Air Defence Ground Environment (NADGE)** that covers all of Europe. Sheffield believed that there may be hostility from these large triangular craft (a false assumption by Sheffield) which were within six minutes of penetrating British airspace, there had to be radar tracking records held by British air defense radar stations.

Maj.-Gen. W. de Brouwer stated in a letter of reply to Sheffield that the UFO of March 30, 1990, was not headed to the UK but moving on an EASTERLY heading and it did not fulfill NATO's identification- criteria to be treated as a HOSTILE. If there were any determination of hostility, all surrounding radar stations including Neatshead, U.K. would automatically have been alerted. At least this response from the Belgium Air Force was positive and surprisingly forthcoming, but this was not to be the case with other letters sent to America or the European Parliament or some of its Members, and nor would it continue to be the case with Maj.-Gen. W. de Brouwer, who would lose patience with Sheffield persistent inquiries.

Sheffield felt that to bring the truth of the phenomenon into the open barring any type of censorship, it had to be done at the topmost level. He thought only then will governments and the media take serious notice of what was going on. Sheffield's approach was to contact the European Parliament; to obtain a serious question in the House of Commons, and to take the matter to the United Nations.

However, the response from many other letters of inquiries were either unanswered, went missing completely or were intercepted and but returned, opened. Sheffield had sent five letters to the United States of America; a letter sent to President Jimmy Carter had been opened and returned; three copies of his report and three reminders to the Belgium Prime Minister, the President of the Council of Ministers, the President of the European Parliament; letters to three European Members of Parliament; three letters to the **Societe Beige d'Etude des Phenomenes Spatiaux (SOBEP);** two letters to Patrick Ferryn (an agent to the owner of the now famous photograph that appears in this book section); four letters to the Belgium Royal Military Academy; and fourteen press releases to all press agencies in the United Kingdom and Western Europe – all went missing mysteriously!

It was becoming apparent to Sheffield that a cover-up was in play, that the evidence was irrefutable – UFOS had been seen and detected above Belgium skies by **NATO Air Defence Ground Environment**. The British MOD whether by ignorance or deliberate intention were obstructing the release of information to the public. Letters to the USA to gain inside information as to possible American military involvement ended up being channeled through the CIA stating non-involvement by the the US Military or the CIA.

It was possible that British MOD had instructions to obstruct the release of information, or **NADGE (NATO)** is withholding details of the phenomenon from them. The fact that the Ministry of Defence knew nothing about the event as they claimed because they were not informed by NATO meant that anything they stated on the subject could be truthful which would mean "plausible deniability" from the MOD's position. According to Sheffield, this must mean that the MOD or the CIA, or both *knew what it was…UFOs,* Extraterrestrial spacecraft intruding into the political heart of continental Europe!

Stealth aircraft had been ruled out because of the incredible acceleration rates performed by the UFO that produced G forces that would have crushed a human pilot in such violent maneuvers.

Sheffield was now more than ever, determined to seek answers to a now obvious deception involving the MOD by going directly to the top level of NATO. He contacted Admiral Lord Hill-Norton, a Lord of the realm, a former Chief of Defence Staff, and he had also been a senior NATO Commander. When questioned about the denial of MOD's knowledge of the Belgium detections, the fact that NADGE radars alerted the Belgium Air Force into action, therefore, the Ministry of Defence must have been advised of the situation, to which Admiral hill-Norton unhestitantly replied:

"Of course, they would have known. The UK Air Defence Ground Environment are part of the European complex, that's how the system works!"

In a letter from **Lord Hill-Norton** to Sheffield, he re-affirmed his position on the matter:

*"I must make my position clear. If NADGE had this object via the Belgium and German radar stations I consider it inconceivable that UK NADGE would not have been aware of it. I also consider it inconceivable that the Operations Division of **SHAPE** at Mons was not aware of it. I disregard your letter from Pope because he is a very junior civil servant... No admission will be made by the MOD until they are absolutely forced to it by one means or another."*

This acknowledgment by a former NATO Commander was damning to the contradictions that Derek Sheffield had received from the MOD. It didn't resolve the confusing retraction by Belgium authorities to the whole UFO event ever having taken place, a contradiction to what had been originally televised in a news press conference. It would appear as if denials from all quarters were being co-ordinated and controlled from one central source. The Belgium official statements and position were starting to look more like the British MOD statements and position. Sheffield was quickly concluding near the end of his investigation that there was a conspiracy of enormous proportions that was reaching into almost every sector of society, from the military, the British MOD, the defence ministry of other nations' governments, the news media, politicians in Britain and in continental Europe and even within the European Parliament.
The doors to open and honest UFO transparency as it related to the Belgium UFO Wave was slamming shut and the non-responsiveness from the British, Belgium, US and even Japanese governments, their various ministries, agencies and militaries, as well as the European Parliament and even the United Nations was to ignore the persistent enquiries of Sheffield (it seemed his correspondence had been tagged with a red flag) or to provide lame excuses or weak explanations. For the most part, their position on the whole UFO matter was a deafening roar of silence.
Although Sheffield's documented research into the validity of the Belgium UFO Wave did not elicit the outcome he sought for, it did leave the question of discussion in the Europen Parliament to still be resolved and it proved the lengths to which censorship and cover-up of a major UFO event that was well publicized in Europe was almost totally unknown in the UK and the USA until finally, Ufologist got wind of it.

It also established the international scale to which the suppression of UFO knowledge and the question of possible Extraterrestrial visitation to our planet is being kept by covert powers on this planet, even to the point that they can influence decisions by the UN and thus, making it difficult for the UN to deal impartially with the matter.

534

As Sheffield and other Ufologists have found out, that there was a symposium arranged to take place at the UN Headquarters in New York on 22 October 1993. The principal aim of the meeting was to press for the United Nations to implement decision GA33/426, which was made on 18 December 1978. This called for the establishment of an agency or a department of the United Nations for undertaking, coordinating and disseminating the results of research into unidentified flying objects and related phenomenon. An item on the agenda was that Mr. Johnson Takano, presenting a personal message from the Japanese Deputy Prime Minister, stated that "the Japanese Government has embarked upon a policy which will change the world forever. They will tell the world all that they know about the UFO." Mr. Johnson Takano never delivered that message. His contribution was cancelled and as far as is known his revelation have never been heard.

When one bears in mind the statement made by **Uthant, Secretary-General of the United Nations**, at a private meeting that *"Apart from the Vietnam War, unidentified flying objects are the most important problem confronting the United Nations,"* it is remarkable that after 15 years, this subject has still not been heard.

What is even more remarkable is that on 19 November 1993 a demonstration of some significance was held outside the United Nations building. News media was on hand to cover the event. The purpose of the demonstration was to get the United Nations to implement draft GA 33/426 on research into unidentified flying objects!

After World War II, moves towards European integration were seen by many as an escape from the extreme forms of nationalism which had devastated the continent. One such attempt to unite Europeans was the **European Coal and Steel Community,** which was declared to be "a first step in the federation of Europe", starting with the aim of eliminating the possibility of further wars between its member states by means of pooling the national heavy industries. The founding members of the Community were **Belgium, France, Italy, Luxembourg, the Netherlands,** and **West Germany**. In 1957, the six countries signed the **Treaty of Rome,** which extended the earlier cooperation within the **European Coal and Steel Community (ECSC)** and created the **European Economic Community, (EEC)** establishing a customs union. They also signed another treaty on the same day creating the **European Atomic Energy Community (Euratom)** for cooperation in developing nuclear energy. Both treaties came into force in 1958. http://en.wikipedia.org/wiki/European_Union

The EU operates through a system of supranational independent institutions and intergovernmental negotiated decisions by the member states. Important **institutions of the EU** include the **European Commission**, the **Council of the European Union**, the **European Council**, the **Court of Justice of the European Union**, and the **European Central Bank**. The **European Parliament** is elected every five years by EU citizens.

The **European Union** is composed of 27 sovereign member states: **Austria, Belgium, Bulgaria, Cyprus, the Czech Republic, Denmark, Estonia, Finland, France, Germany, Greece, Hungary, Ireland, Italy, Latvia, Lithuania, Luxembourg, Malta, the Netherlands, Poland, Portugal, Romania, Slovakia, Slovenia, Spain, Sweden,** and the **United Kingdom**. The

Union's membership has grown from the original six founding states—**Belgium, France, (then-West) Germany, Italy, Luxembourg and the Netherlands**—to the present day 27 by successive enlargements as countries acceded to the treaties and by doing so, pooled their sovereignty in exchange for representation in the institutions.
http://en.wikipedia.org/wiki/European_Union

This last fact is truly enlightening in that when countries pool or surrender their sovereignty in exchange for representation in the institutions they benefit in ways beyond bringing peace and stability to previously war-torn areas of the continent. They become a role model to the rest of the world where a commonwealth of countries demonstrate that the "whole becomes greater than its parts".

Now, this is not to detract from the prestige of the United Nations as an international body that represents most of the nations on this planet. However, the EU functions in a way that the UN does not, that being that there is only one vote for every country represented in the EU, whereas the UN has five nations (**Unites States of America, the United kingdom, France, Russia and China**, the founding members of the **UN Security Council** that established the UN from its **League of Nations** origins) which have a **Veto Power**. Any one of these five nations is able to thwart any motion in the UN General Assembly by a veto of the vote if they do not agree with it, regardless if all other the nations are in agreement.

This can hardly be considered as playing fair ball on the international stage of politics as it weakens the UN 's power and authority to bring about democratic change upon the planet that benefits all of humanity. If the United Nations is to have any serious unchallengeable power to do good then, these five powerful nations must show magnanimity and surrender this veto power and have only one vote that cannot trump the decisions of the majority of other nations. For there to be global peace, planetary stability with security and prosperity, it can only be one country, one vote with no veto power for any nation!

Now, it would appear then, that in order to keep a tight control on the information of the UFO phenomenon and its associated ETI presence, one or more of these founding members (probably lead by the US and possibly supported by either the U.K. and/or France) of the Security Council had to have placed coercive pressure upon the EU (of which both Britain and France are members… *"an insider job"*) perhaps, via economic sanctions or with threats of withdrawal of NATO military support for Europe against potential communist incursions into Europe in order to elicit EU cooperation. These five member nations of the Security Council have made the United Nations a lame duck, impotent of any real global power and thus, incapable of disclosing any information on the UFO/ETI presence.

The events that unfolded in central Europe did not stop the debunkers or hoaxers from coming out with their own explanations to explain away what so many thousands of people had reportedly seen for six months between 1989 and 1990. The helicopter onboard lights theory or **"Neglected Hypothesis"** was used to try and explain away a good percentage of the reports but, failed miserably to factor in other pertinent evidence such as lack of sound and rapid acceleration as well as close visual sightings of the craft within 500 feet of the witnesses. They certainly were not helicopters from the Belgium Air Force! It has even been suggested that the whole wave of

536

sightings was due to a highly advanced stealth blimp developed by the US and flown out of Britain and into Europe to gauge people reactions to it. The Americans must be pretty confident of their technical abilities and their potential risk assessment to carry out such an audacious aerial display for six consecutive months.

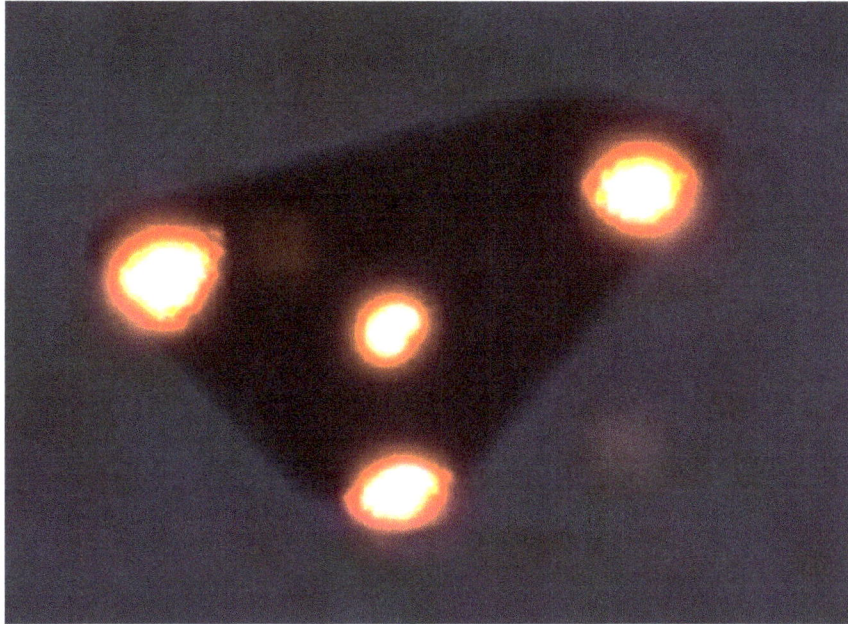

A black triangle, 15 June 1990, Wallonia, Belgium by J. S. Henrardi (now released thirteen years later into the public domain). Claimed to have been taken during the UFO wave
http://ancientufo.org/2015/10/the-belgian-triangular-ufos-wave/

Author's Rant: (Personally, I don't buy either explanation as it seems once again, debunkers were grasping at straws in the wind to dissuade people with misleading evidence thus, throwing doubt onto the whole UFO wave of Belgium.

There has even been an individual by the name of Patrick (last name unknown at the time of this writing) who came forward after nearly 21 years to confess that he had faked the now famous Petit-Rechain photo saying it was a polystyrene model on a string.

But some people are wondering, "Why now? Why wait 21 years to finally come clean?" Despite Patrick's confession, many still believed that this might be just another ploy from the government. A cover-up designed to sidetrack us from what really is going on. What if the **REAL** hoax is Patrick claiming to the public that the image is fake? What if the government persuaded the man to tell us a **DIFFERENT** story to conceal the shocking reality?
http://www.latest-ufo-sightings.net/2011/07/controversial-belgium-ufo-photo-proved.html

Certainly, this type of deception is within covert intelligence's bag of dirty tricks to create psychological confusion and ridicule whenever credible eyewitness and photographic evidence presents itself.

As Chilean-American journalist Antonio Huneeus and Ufologist has stated, "in a totally unprecedented move in the history of ufology, the Belgian Air Force and government has not only carefully documented the great UFO wave over Wallonia, but shared its results with civilian investigators and the public, in effect literally breaking down **"the Wall of UFO Silence"** that still stands in the western world" and thereby, setting a model for others to follow in their pursuit to uncover the mystery behind the UFO enigma.
http://www.ufoevidence.org/documents/doc406.htm

This particular UFO wave is also important in the annals of UFO research because it provides so many benchmarks to establishing a very high credibility of a genuine UFO phenomenon. It has ground radar and plane radar lock-ons of the same object and it has ground visual by over 13,000 witnesses from 30 different sectors of society. There is ground radar video of the triangular shape craft supported with plane radar video and private VHS video from the public of the same or similar objects seen at the time. If you wanted proof that Extraterrestrials are visiting the Earth this is the type of proof you would like to have as a scientist or a military official to support this fact!

NATO, The NSA, and UFOs

This wave of UFO sightings became unique for another reason when a large triangular ET craft was sighted over the international parliament buildings in Brussels which is the capital of Belgium and the *de facto* capital of the **European Union (EU).** Since the end of the Second World War, Brussels has been the main centre for international politics. Hosting principal EU institutions as well as the headquarters of the **North Atlantic Treaty Organization (NATO),** the city has become the polyglot home of numerous international organizations, politicians, diplomats and civil servants.

This fact was not lost on the Belgium military or the Belgium government or on NATO when a few later in 1993, the European Union was established in Brussels under the **Maastricht Treaty**. Could the ETs be trying to tell us something of our future? Or were ETI pointing out the obvious that the city of Brussels was more than the capital of the European Union that it had also become the nerve centre for NATO in Europe, an American CIA/NSA run military intelligence organization?

One more point of interest that came out of Sheffield's investigation was discovering the degree to which there was American intelligence and military involvement in Britain and in Europe. In an isolated moorland area known as Menwith Hill in the Yorkshire Dales, there is located a top secret facility surrounded by a five and a half mile impenetrable barbed wire fence. This facility is the home to electronic listening devices housed in immense multi-faceted spherical domes and its personnel is closed off from the outside world of Britain and British nationals are not permitted access into the area. The area is so well guarded and so top secret that any inquiries made by British Members of Parliament are unanswered and totally classified.

From this **National Security Agency (NSA)** run facility employing only American personnel, covert electronic surveillance is carried out in all sectors of Western Europe. This means that every form of electronic communication – mobile, private, business and continental telephone

538

calls, computer and satellite communication, radio and television communication – is subject to intense covert surveillance. It is a CIA operation and is directly involved with the British MOD GCHQ at Cheltenham.

This co-operative involvement by both countries intelligence agencies goes a long way to explaining the UK and US cover-up of the Belgium UFO sightings. Thus, it appears that the CIA is controlling surveillance operations via the NSA throughout Western Europe and in turn are, therefore, covertly controlling all defense and security systems through the **North Atlantic Treaty Organization (NATO)**. It is NATO then, that controls the release or suppression of all information on this phenomenon and it one arm of the multi-armed octopus of covert military intelligence that probably understands better than most the nature and reality of UFOs and ETI. Derek Sheffield, UFO a Deadly Concealment: The Official Cover-up?" 1996, Blandford Publishing, UK

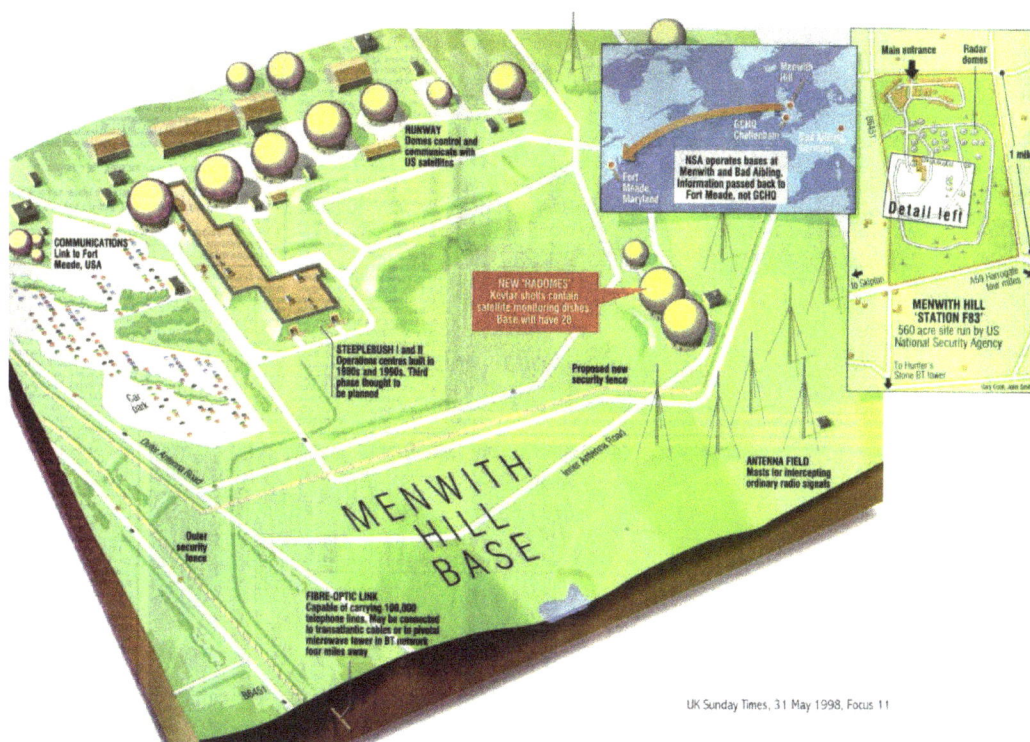

UK Sunday Times, 31 May 1998, Focus 11

**The top secret National Security Agency (NSA) run facility at
Menwith Hill in the Yorkshire Dales, U.K.**
http://www.c4i.org/erehwon/nsa-f83.html

It is often said that the (NATO) was founded in response to the threat posed by the Soviet Union. This is only partially true. In fact, the Alliance's creation was part of a broader effort to serve three purposes: deterring Soviet expansionism, forbidding the revival of nationalist militarism in Europe through a strong North American presence on the continent, and encouraging European political integration. http://www.nato.int/history/nato-history.html

To this mission statement should be added a fourth purpose: the monitoring, tracking, and targeting of UFOs or alien spacecraft that venture too close into Earth orbit or that overfly into

539

sensitive and restrictive military bases which essentially on this planet pretty much means any and all military bases.

The North Atlantic Treaty Organization also called the **(North) Atlantic Alliance** is an intergovernmental military alliance based on the North Atlantic Treaty which was signed on 4 April 1949. The organization constitutes a system of collective defense whereby its member states agree to mutual defense in response to an attack by any external party. **(This could also include potential Extraterrestrial threats from outer space)**. [Bracked bold statement added by author for emphasis] http://en.wikipedia.org/wiki/NATO

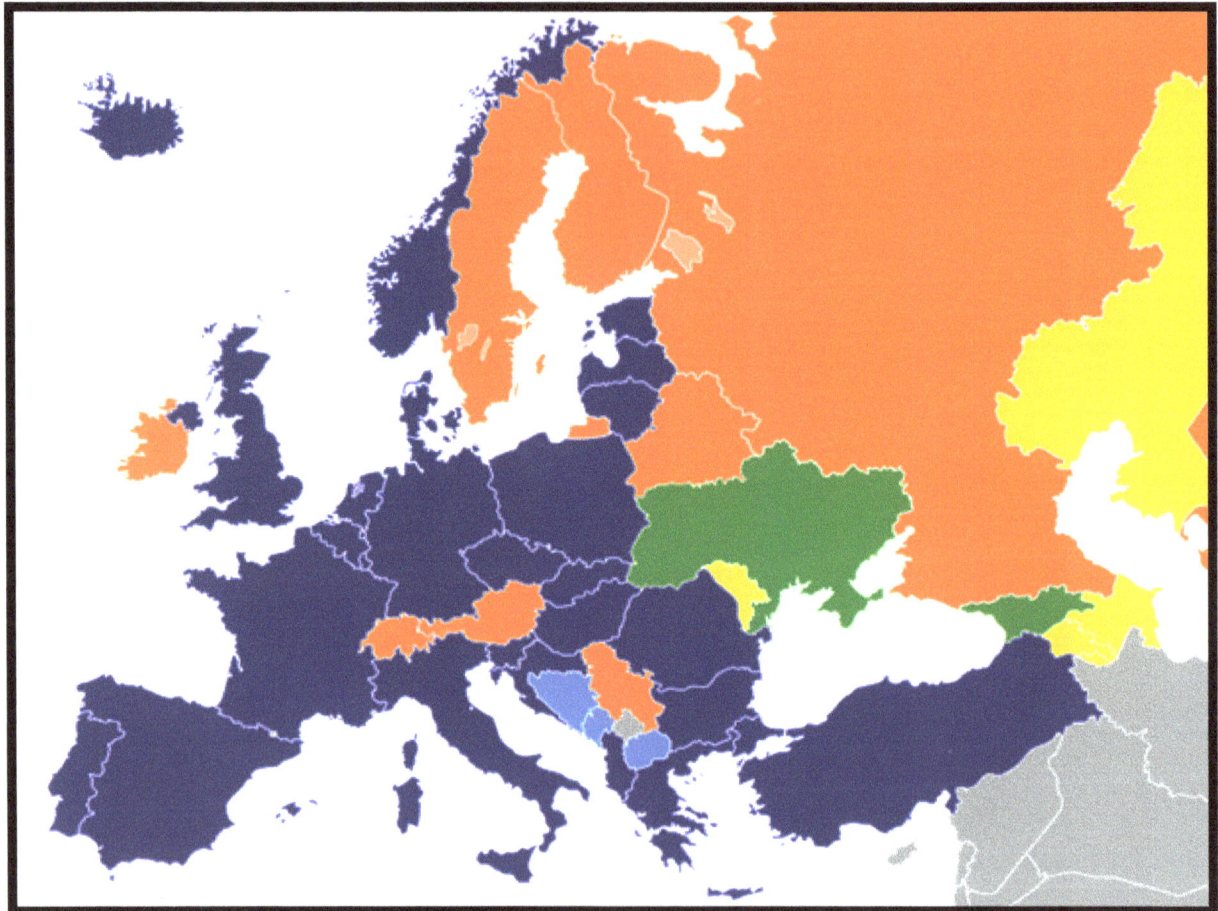

Map of NATO affiliations in Europe

▉ NATO member states	
▉ Membership Action Plan	
▉ Independent Partnership Action Plan	▉ Mediterranean Dialogue (MD)
▉ Partnership for Peace	▉ Instanbul Cooperation Initiative (ICI)
	▉ Contact countries (CC)

https://en.wikipedia.org/wiki/NATO

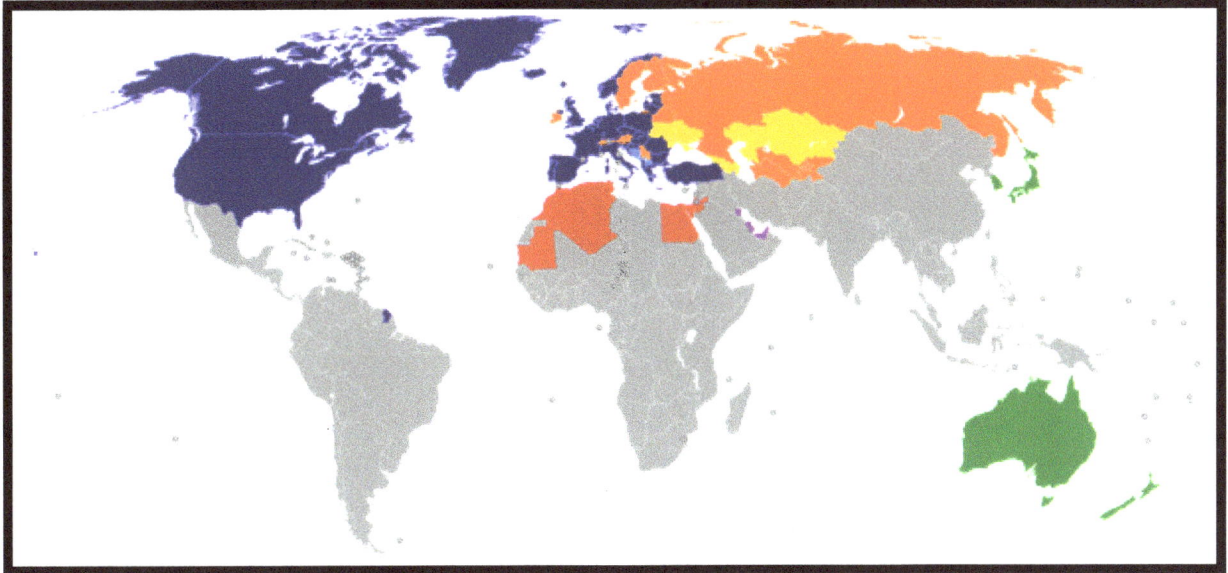

Map of NATO partnerships globally
https://en.wikipedia.org/wiki/NATO

The original 12 founding nations of the Alliance include Belgium, Canada, Denmark, France, Iceland, Italy, Luxembourg, the Netherland, Norway, Portugal, the United Kingdom and the United States. It has since grown to include Greece, Turkey, Germany, and Spain and with the fall of the Berlin Wall in 1989 and the collapse of Communism, it now also includes the Czech Republic, Hungary, Poland, Bulgaria, Estonia, Latvia, Lithuania, Romania, Slovakia, Slovenia, Albania, and Croatia. http://www.nato.int/history/nato-history.html

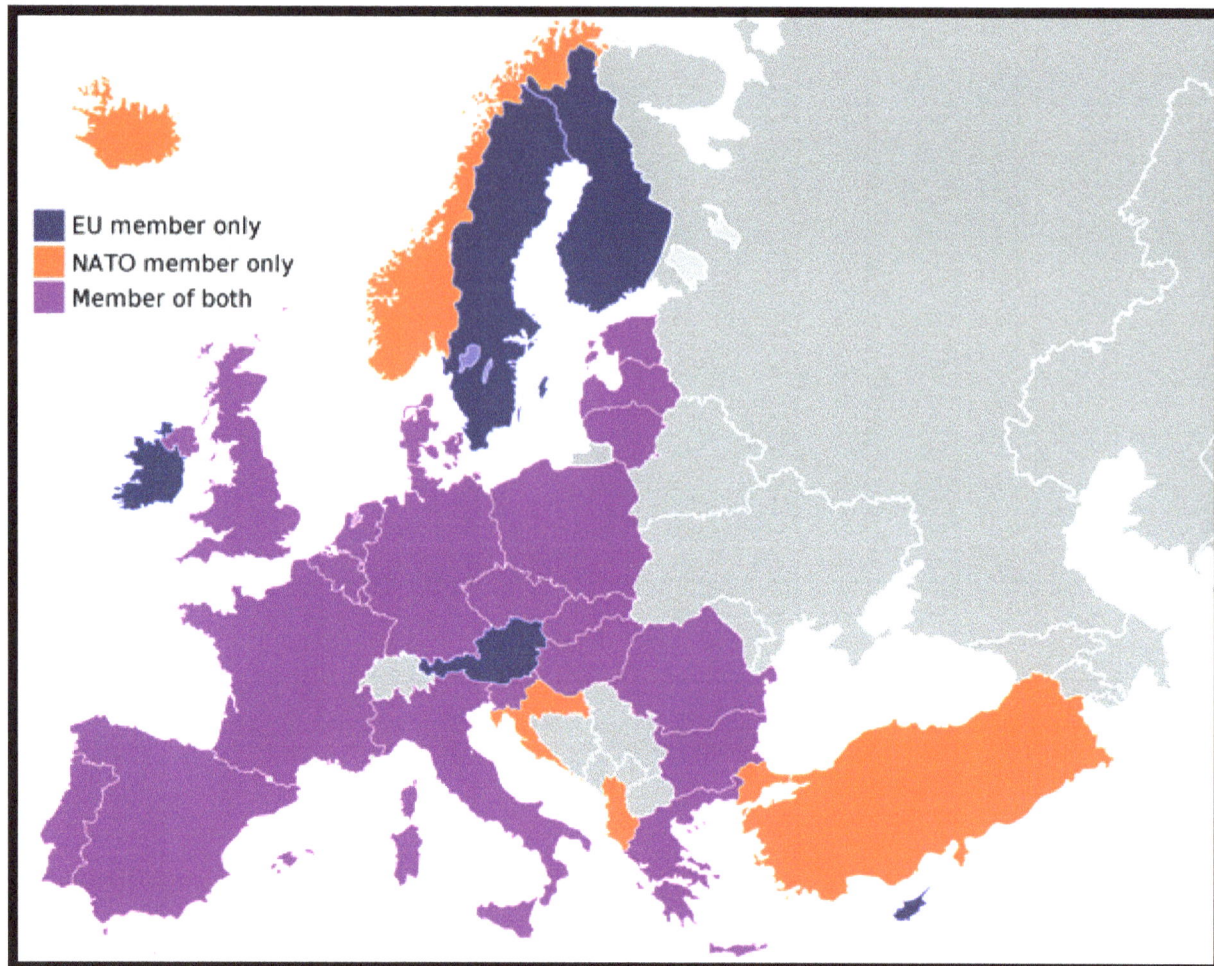

European membership of the European Union and NATO 2009

In each of these nations or in a group of nations situated close to each other and separated only by political boundaries, the US through NATO has established a listening post station to fulfill its multi-purpose function which is to listen, monitor, record and set into motion any defensive action necessary to deal with a potential hostile threat to the alliance. That means that no longer is any conversation between two people using some form of electronic communication device is considered private.

Now the logistics to monitor any and all communications would require literally tens of thousands maybe even hundreds of thousands of personnel working full time on individual computers around the clock. The US, Russia and no doubt China have this manpower to do this. However, in the US tying up tens of thousands of men in this type of tedious endeavour is a waste of manpower, probably mind-numbing to the extreme and prone to too many errors creeping into the database. Enter the supercomputer to solve this dilemma and bring back accuracy and time management to a state of manageable intelligence.

So sophisticated are these electronic listening posts in their ability to eavesdrop on your conversation that in order to monitor all global conversations, regardless of what language is

542

being spoken, a super fast central computer **(Cray Supercomputer)** was developed to receive, download and sort out individual conversations using keywords or phrases to trig an immediate monitoring and recording response that could pinpoint the location of the conversation and based on the sensitivity of the conversation, a priority level of response would determine the type action required to nullify the conversation or silence the conversationalists.
http://www.answers.com/topic/national-security-agency

Cray X-MP/24 (ser. no. 115) supercomputer
on display at the National Cryptologic Museum.
http://debarshibanerjee.blogspot.ca/

This is no different than the computers used by SETI astronomers to listen to and process the billions of signals that are received by radio telescopes every hour of every day. They just configured differently of course for their particular needs. SETI has even gone to the public to ask their help in receiving interstellar signals on their home computers, thus employing tens or hundreds of thousands of privately owned computers to crunch the numerical data that can then be sifted through later by SETI scientists.

This difference, of course, is that SETI just merely eavesdrops on potential alien signals from space, forever searching for that elusive **"WOW" signal**. The NSA through NATO, on the other hand, eavesdrops on you and on political leaders, on militaries and governments of enemy nations, potential terrorists, scientists and religious leaders.

The New York Times reported in an article dated, 7 September 1960: *"We know from working at NSA [that] the United States reads the secret communications of more than forty nations, including its own allies ... NSA keeps in operation more than 2000 manual intercept positions ... Both enciphered and plain text communications are monitored from almost every nation in the world, including the nations on whose soil the intercept bases are located."*

US Listening Posts Around the World

During the Cold War, there were hundreds of secret remote listening posts spread around the globe. From large stations in the moors of Scotland and mountains of Turkey that were complete with golf ball-like structures called **"radomes"** to singly operated stations in the barren wilderness of Saint Lawrence Island between Alaska and Siberia that had only a few antennae, these stations constituted the ground-based portion of the **United States Signals Intelligence (SIGINT) System or "USSS."**

Operated by the super-secret **National Security Agency (NSA)**, these stations were designed to intercept **Morse Code**, telephone, telex, radar, telemetry, and other signals emanating from behind the Iron Curtain. At one time, the NSA contemplated a worldwide, continuously operated array of 4120 intercept stations. While the agency never achieved that goal, it could still boast of several hundred intercept stations. These included its ground-based *"outstations"*, which were supplemented by other intercept units located on ships, submarines, aircraft (from U-2s to helicopters), unmanned drones, mobile vans, aerostats (balloons and dirigibles), and even large and cumbersome backpacks.

With the collapse of the Communist "bloc" and the advent of microwaves, fiber optics, and cellular phones, NSA's need for numerous ground-based intercept stations waned. It began to rely on a constellation of sophisticated **SIGINT satellites** with code names like **Vortex, Magnum, Jumpseat,** and **Trumpet** to sweep up the world's satellite, microwave, cellular, and high-frequency communications and signals. Numerous outstations met with one of three fates: they were shut down completely; remoted to larger facilities called **Regional SIGINT Operations Centers or "RSOCs"** or were turned over to host nation **SIGINT** agencies to be operated jointly with NSA.

However, NSA's jump to relying primarily on satellites proved premature. In 1993, Somali clan leader Mohammed Farah Aideed taught the agency an important lesson. Aideed's reliance on older and lower-powered walkie-talkies and radio transmitters made his communications virtually silent to the orbiting SIGINT "birds" of the NSA. Therefore, NSA technicians came to realize there was still a need to get in close in some situations to pick up signals of interest. In NSA's jargon, this is called improving **"hearability."**

544

As NSA outstations were closed or remoted, new and relatively smaller intercept facilities such as the "gateway" facility in Bahrain, reportedly used for retransmit signals intercepted in Baghdad last year to the U.S.A. sprang up around the world. In addition to providing NSA operators with fresh and exotic duty stations, the new stations reflected an enhanced mission for NSA economic intelligence gathering. Scrapping its old Cold War A and B Group SIGINT organization, NSA expanded the functions of its W Group to include SIGINT operations against a multitude of targets. Another unit, M Group, would handle intercepts from new technologies like the Internet. http://www.fas.org/irp/news/1999/02/radome.htm

A rethinking of satellite technology to eavesdrop on electronic communication eventually evolved to the same state that satellite photography can pick out minute detail of objects and people on Earth, to such a degree, it is reported that a newspaper can be read (even at night) from spy satellites several hundred miles up in orbit, through clouds! To such an extent has this technology advanced, it now means conversations between two people can be targeted and be audibly discerned to such a degree as to hear a whisper even, within a building! The NSA's **"hearability"** had now been greatly improved to the next level.

As incredible as this ability to see people and hear personal conversations from outer space seems to be appear, it is also rumoured that there are devices similar to those that can induce voices to be heard in one's head using an **RF Carrier wave** over a microwave beam (like those used on **Charles Hickson** and **Calvin Parkers** – the **Pasacougla UFO Abduction Incident**). These portable devices can not only send verbal-thought communications directly into your head, but the reverse ability of that technology can permit agents to hear your very intimate thoughts! This is a type of electronic assisted telepathy.

Big Brother can not only see you under any spectrum of light, day or night but now, can hear everything you say in normal conversation whether on the streets, in your home or in a highrise building. Gone are the days of assured privacy in conversation, verbal or electronic.

While a master list of SIGNIT and NSA secret listening posts operated by the agency and its partners around the world probably exists, somewhere in the impenetrable lair that is the NSA's Fort Meade, Maryland headquarters; it is assuredly stamped with one of the highest security classifications in the U.S. intelligence community.

If we are to believe the latest technical developments coming out of the **Black World of Science,** besides the military intelligence eavesdropping on other countries' governments and militaries as well as the general public from space, all cell phones have now become a listening device for the NSA and CIA to illegally tap into, anytime without you even being aware of it or without your acknowledgement.

As if that wasn't enough to get you "pissed off" with the intelligence community, they can make you see and believe things that don't really exist. Through the deployment of advanced laser holography projected from high flying aircraft or from space via orbiting satellites, 3D images can appear on the ground or in the air of anything you can image. It could easily be an image of Christ in the air or tanks on the ground as used initially in the Iraqi War against Saddam Husein's army or a fleet of military jets and bombers or even an armada of invading aliens. The image will

appear so real that people will easily be fooled or confused enough to not know what to make of it initially which is long enough for other offensive actions to be carried out against the enemy or the public. The world we knew my friends has changed forever. It is like a bad dream from which we seem unable to wake from and it is only becoming more nightmarish as each month and year of our lives tick by! It is like the Star Wars movie where the insidious "Dark side of the Force" has been steadily working away unnoticed initially in the background, carefully manipulating governments, militaries and people until it has all but taken over the world!
https://www.youtube.com/watch?v=RQ012kBoqiA

**A map showing some of the key strategic NSA listening sites
and orbiting satellites around the globe.**
https://www.bibliotecapleyades.net/ciencia/echelon04.htm

The National Security Agency headquartered at Fort Meade, Maryland, controls a global network of electronic interception stations.

The following list is the best-unclassified shot at describing the locations of the ground-based "ears" of the **Puzzle Palace**, the **United States SIGINT System (USSS)**

It is culled from press accounts, informed experts, and books written about the NSA and its intelligence partners. It does not include the numerous listening units on naval vessels and aircraft or those operating from U.S. and foreign embassies, consulates, and other diplomatic missions. http://www.fas.org/irp/news/1999/02/radome.htm

United States
- NSA Headquarters, Fort Meade, Maryland
- Buckley Air National Guard Ground Base, Colorado
- Fort Gordon, Georgia (RSOC)
- Imperial Beach, California
- Kunia, Hawaii (RSOC)
- Northwest, Virginia
- Sabana Seca, Puerto Rico
- San Antonio, Texas (RSOC)
- Shemya, Alaska [3]
- Sugar Grove, West Virginia

- Winter Harbor, Maine
- Yakima, Washington

Albania
- Durres [-6]
- Shkoder [-6]
- Tirana [-6]

Ascension Island
- Two Boats [-1]

Australia
- Bamaga [-6] [-7]
- Cabarlah [-7]
- Canberra (Defense Signals Directorate Headquarters) [-5]
- Harman [-7]
- Kojarena, Geraldton [-1]
- Nurunggar [-1]
- Pearce [-1]
- Pine Gap, Alice Springs [-1]
- Riverina [-7]
- Shoal Bay, Darwin [-1]
- Watsonia [-1]

Austria
- Konigswarte [-7]
- Neulengbach [-7]

Bahrain
- Al-Muharraq Airport [-3]

Bosnia and Herzegovina
- Tuzla

Botswana
- Mapharangwane Air Base

British Indian Ocean Territory
- Diego Garcia [-1]

Brunei
- Bandar Seri Begawan [-7]

Canada
- Alert [-7]
- Gander [-7]
- Leitrim [-1]
- Masset [-6] [-7]
- Ottawa [Communications Security Establishment (CSE) Headquarters] [-5]

China
- Korla [-1] [-6]
- Qitai [-1] [-6]

Croatia
- Brac Island, Croatia [-6]
- Zagreb-Lucko Airport [-7]

Cuba
- Guantanamo Bay

Cyprus
- Ayios Nikolaos [-1]

Denmark
- Aflandshage [-7]
- Almindingen, Bornholm [-7]
- Dueodde, Bornholm [-7]
- Gedser [-7]
- Hj rring [-7]
- Logumkloster [-7]

Eritrea
- Dahlak Island [-1] (NSA/Israel "8200" site)

Estonia
- Tallinn [-7]

Ethiopia
- Addis Ababa [-1]

Finland
- Santahamina [-7]

French Guiana
- Kourou [-7] (German Federal Intelligence Service station)

Germany

- Achern [-7]
- Ahrweiler [-7]
- Bad Aibling [-2]
- Bad Munstereifel [-7]
- Braunschweig [-7]
- Darmstadt [-7]
- Frankfurt [-7]
- Hof [-7]
- Husum [-7]
- Mainz [-7]
- Monschau [-7]
- Pullach (German Federal Intelligence Service Headquarters) [-5]
- Rheinhausen [-7]
- Stockdorf [-7]
- Strassburg [-7]
- Vogelweh, Germany

Gibraltar

- Gibraltar [-7]

Greece

- Iraklion, Crete

Guam

- Finegayan

Hong Kong

- British Consulate, Victoria ("The Alamo") [-7]

Iceland

- Keflavik [-3]

India

- Charbatia [-7]

Israel

- Herzliyya (Unit 8200 Headquarters) [-5]
- Mitzpah Ramon [-7]
- Mount Hermon, Golan Heights [-7]
- Mount Meiron, Golan Heights [-7]

Italy

- San Vito [-6]
- Sorico

550

Japan
- Futenma, Okinawa
- Hanza, Okinawa
- Higashi Chitose [7]
- Higashi Nemuro [7]
- Kofunato [7]
- Miho [7]
- Misawa
- Nemuro [7]
- Ohi [7]
- Rebunto [7]
- Shiraho [7]
- Tachiarai [7]
- Wakkanai

Korea (South)
- Kanghwa-do Island [7]
- Osan [1]
- Pyong-dong Island [7]
- P'yongt'aek [1]
- Taegu [1] [2] [6]
- Tongduchon [1]
- Uijongbu [1]
- Yongsan [1]

Kuwait
- Kuwait

Latvia
- Ventspils [7]

Lithuania
- Vilnius [7]

Netherlands
- Amsterdam (Technical Intelligence Analysis Center (TIVC) Headquarters) [5]
- Emnes [7]
- Terschelling [7]

New Zealand
- Tangimoana [7]
- Waihopai [1]
- Wellington (Government Communications Security Bureau Headquarters [5]

Norway
- Borhaug [-7]
- Fauske/Vetan [-7]
- Jessheim [-7]
- Kirkenes [-1]
- Randaberg [-7]
- Skage/Namdalen [-7]
- Vadso [-7]
- Vardo [-7]
- Viksjofellet [-7]

Oman
- Abut [-1]
- Goat Island, Musandam Peninsula [-3]
- Khasab, Musandam Peninsula [-3]
- Masirah Island [-3]

Pakistan
- Parachinar

Panama
- Galeta Island [-3]

Papua New Guinea
- Port Moresby [-7]

Portugal
- Terceira Island, Azores

Rwanda
- Kigali

Sao Tome and Principe
- Pinheiro

Saudi Arabia
- Araz [-7]
- Khafji [-7]

Singapore
- Kranji [-7]

Spain
- Pico de las Nieves, Grand Canary Island [-7]
- Manzanares [-7]

552

- Playa de Pals [-3]
- Rota

Solomon Islands
- Honiara [-7]

Sri Lanka
- Iranawilla

Sweden
- Karlskrona [-7]
- Loven (Swedish FRA Headquarters) [-7]
- Musko [-7]

Switzerland
- Merishausen [-7]
- Ruthi [-7]

Taiwan
- Quemoy [-7]
- Matsu [-7]
- Shu Lin Kuo [-5] (German Federal Intelligence Service/NSA/Taiwan J-3 SIGINT service site)

Turkey
- Adana
- Agri [-7]
- Antalya [-7]
- Diyarbakir
- Edirne [-7]
- Istanbul [-7]
- Izmir [-7]
- Kars
- Sinop [-7]

Thailand
- Aranyaprathet [-7]
- Khon Kaen [-1] [-3]
- Surin [-7]
- Trat [-7]

Uganda
- Kabale
- Galangala Island, Ssese Islands (Lake Victoria)

United Arab Emirates
- Az-Zarqa [3]
- Dalma [3]
- Ras al-Khaimah [3]
- Sir Abu Nuayr Island [3]

United Kingdom
- Belfast (Victoria Square) [7]
- Brora, Scotland [7]
- Cheltenham (Government Communications Headquarters) [5]
- Chicksands [7]
- Culm Head [7]
- Digby [7]
- Hawklaw, Scotland [7]
- Irton Moor [7]
- Menwith Hill, Harrogate [1] (RSOC)
- Molesworth [1]
- Morwenstow [1]
- Westminster, London [7]
- (Palmer Street)
- Yemen
- Socotra Island (planned)

KEY:
-1 Joint facility operated with a SIGINT partner.
-2 Joint facility partially operated with a SIGINT partner.
-3 Contractor-operated facility.
-4 Remoted facility.
-5 NSA liaison is present.
-6 Joint NSA-CIA site.
-7 Foreign-operated "accommodation site" that provides occasional SIGINT product to the USSS.

sweden = lov$nhttp://www.fas.org/irp/news/1999/02/radome.htm
The Radome Archipelago; February 24 - March 2, 1999, by Jason Vest and Wayne Madsen

From Sheffield's earlier account in this section, the NSA run Menwith Hill electronic surveillance facility is typical of most listening posts around the world. There is more than just the everyday covert monitoring of multi-media conversations being picked up from the British people, from enemies of the state, their governments, and military, or from the scheming machinations of terrorists plotting against the UK. UFO data is also being meticulously collected from reported sightings from sources all over Europe.

Electronic communications are also heavily monitored out of the Pine Gap facility near Alice Springs, Australia which is a highly sensitive electronic listening post strategically located in the

554

southern hemisphere. In fact, all listening post located around the world are strategically positioned to obtain the maximum surveillance coverage of the region.

All data received whether it is intel comes from governments, militaries, terrorist groups, the general public or covert operatives in the field is collected and channeled back to the NSA's Fort Meade, Maryland headquarters in the USA where it is processed quickly, sent to the appropriate agency or department of the military or government or covert private sector and the appropriate response is then, taken.

In the heart of Utah's desert, the National Security Agency is building the nation's largest, most expensive cyber-security spy center project. (Watch What You Say)
https://www.wired.com/2012/03/ff_nsadatacenter/

The collection and response to UFOs and/or ETI regardless, where in the world it originated from is given immediate priority over most other channeled inbound intelligence received. The collection and response is for the most part automatic in that keywords that are picked up from electronic surveillance such as in the case of our subject matter: UFO, ETI, ETs, alien spacecraft, flying saucers, triangular craft, abductions, extraterrestrials, or any such derivative of these words including code words, particular people in the UFO community or associated with ufology, etc., etc. triggers an immediate response to record or redirect the conversation to one of the listening posts or back directly to the NSA headquarters.

People like **Dr. Steven Greer** of **CSETI** and people associated with him have their conversations constantly monitored and their movements continuously tracked and under the

555

watchful eyes and ears of agents or high-flying surveillance drones, particularly when Dr. Greer and his team are trying to establish contact or communications with ETI.

Author's Rant: I have experienced this on a few occasions when in the presence of Dr. Greer in at least a couple of locations where this covert surveillance has taken place.

Big Brother Sees, Listens, Monitors, Tracks, Targets and Retrieves all Things UFO

When ET spacecraft venture into our atmosphere or are in near orbit of our planet, these listening posts are on high alert (which is usually all the time), key military, missile and high energy weapon bases in the US and around the world are always at a standby preparedness ready in launch mode deployment. The high regions of space around our planet are constantly monitored and tracked for US, Russian and Chinese satellites or spacecraft like the International Space Station (ISS), satellite space debris, near or inbound meteorites or asteroids and anything else that does not behave or response in the usual manner namely, Extraterrestrial spacecraft. From this type of monitoring, it is apparent the NSA and the **NRO** monitor more than electronic communications originating from Earth.

In the 1950s, **President Dwight D. Eisenhower** approved reconnaissance systems that included high-altitude balloons, airplanes, and satellites to gain strategic intelligence on the Soviet Union, China, and other potential threats to the United States. On August 31, 1960, **Secretary of the Air Force Dudley C. Sharp** established the **Office of Missile and Satellite Systems** to direct the Air Force satellite reconnaissance program. On September 6, 1961, Acting **Director of Central Intelligence General Charles P. Cabell** and **Deputy Secretary of Defense Roswell L. Gilpatric** officially established management arrangements for the **National Reconnaissance Program**. These arrangements consolidated many of America's national space and aerial reconnaissance projects under a covert, highly compartmented **National Reconnaissance Office**.

Headquartered in Chantilly, Virginia, the **National Reconnaissance Office (NRO)** develops and operates unique and innovative overhead reconnaissance systems **(Spy Satellites)** and conducts intelligence-related activities for U.S. national Security. The NRO works in close collaboration with the NSA at many of the NSA run sites around the world in electronic surveillance.

Formed in response to the Soviet launch of Sputnik, the NRO was secretly created on September 6, 1961, with the purpose of overseeing "all satellite and overflight reconnaissance projects whether overt or covert." Once considered a non-existent organization because of its covert activities, it is no longer a classified organization but is relentlessly working to foster "Innovative Overhead Intelligence Systems for National Security."

NRO Emblem
1973-1984

NRO Emblem
1984-1994

NRO Emblem
1994-Present

The NRO maintains ground stations at:

- Buckley Air Force Base, Colorado
- Fort Belvoir, Virginia,
- White Sands Missile Range, New Mexico,
- as well as a presence at the Joint Defense Facility Pine Gap, Australia and
- the Royal Air Force Base Menwith Hill Station, United Kingdom.
- NRO spacecraft launch offices reside at Cape Canaveral AFB, Florida and
- Vandenberg AFB, California.

The NRO, one of 16 Intelligence Community agencies, was officially established in September 1961 as a classified agency in the **Department of Defense (DoD).** The existence of the NRO and its mission were declassified in September 1992.

The NRO is a hybrid organization consisting of some 3000 personnel and is jointly staffed by members of the armed services, the **Central Intelligence Agency**, and DoD civilian personnel. http://www.nro.gov/about/nro/index.html

The sophistication of monitoring and tracking of ET spacecraft is beyond what most people are even remotely aware of; radar is not the primary means of detection of ET craft approaching the Earth. Some ET craft generate a magnetic disturbance or displacement in the energy field around the planet when they fly into the atmosphere or "pop" out of trans- dimensional flight near the Earth. This disturbance is picked up by advanced electromagnetic detection equipment that allows a ground-based high-energy scalar weapon to be fired at the UFO knocking it out of orbit and thus, sending it crashing towards earth for retrieval.

Such an incident did occur and was recorded on videotape from the space shuttle **STS 48A** when one or more ET craft orbiting the Earth was fired upon by some **Strategic Defence System (SDS)** or **Strategic Defence Initiative (SDI)** weapon. The ET craft is seen in the video moving above the Earth then, reversing its course by 135 degrees and moving away, just as a flash of light appears beneath it. A missile or energy weapon comes up from the surface of the planet and steaks by the area where the ET craft had just been literally moments before thus, avoiding a near-fatal hit. When NASA is questioned about this incident their usual standard reply has been "ice crystals", "space debris" or "satellites".

At the time the video was taken, the space shuttle was over Australia and it is estimated that the missile or beam weapon may have originated from an area close to Alice Springs. It is a well-established fact that Pine Gap, the largest US-run listening facility in the world is situated near Alice Springs and it may also be home to high-energy based weapons as well. The beam weapon fired upon the UFO may have come from this facility. Such aggressive action taken by the US military has on occasion ended in the fatality of a few alien visitors to our planet. Curiously, after this shuttle mission, NASA cut all live satellite feeds of its space missions to the public. No doubt this order in all probability came from the NRO. It is their mission statement after all to *"relentlessly work to foster **Innovative Overhead Intelligence Systems for National Security."***

This is certainly not the ***"Red Carpet"*** treatment you'd expect for potential interstellar dignitaries to our planet! Needless to say, every visiting Extraterrestrial Intelligence takes his life into his own hands coming to this planet. The outcome of visitation could result in the cost of ET life or possible capture by the US or Russian military. China has been known to be extremely aggressive in UFO shoot downs, although not necessarily as advanced in UFO space shoot downs as the Americans or the Russians. Laying down the "Red Carpet" has a whole different meaning to the military. It is the way of saying, ***"Welcome to the planet Earth! Let's see your invitation and your interstellar passport before you land!"*** [Bold italics added for emphasis.]

It's no wonder that Dr. Greer and Stanton Freidman have both stated repeatedly that such aggressive action from the militaries of this planet has resulted in a **"Cosmic Quarantine"** of the Earth which prevents us from leaving our Solar System for other regions of the universe! If this is indeed the situation facing us at this time then, it precludes any possibility that any humans are part of an exchange program with any neighbouring ET civilization given our current development on this planet.

If the reader accepts the premise that a Cosmic Quarantine has been placed upon this planet by a high order of Extraterrestrial civilizations working together in a commonwealth or federation of peaceful cooperation than, exchange programs like **Project Serpo,** the **Zeta Reticuli Exchange Program** is a probable disinformation campaign to confuse UFO researchers to the reality of possible ETI contact. It is also a clever subterfuge to ensnare unsuspecting UFO researchers with prejudicial inclinations toward certain aspects of UFO information in order to monitor the ebb and flow of disinformation, the connections, and interests within the UFO community. At any time the "rug of disinformation" can be "pulled" and be publically announced to illustrate the ineptitude and lack of professionalism within the UFO community. The ridicule factor would be extremely high and the result would be to further distance the public's interest away from the UFO/ETI subject matter thus, the phenomenon is effectively marginalized as nothing more than pseudo-science. Given the current status of the subject matter and the serious lack of interest by the science community to investigate the UFO phenomenon, it would be safe to say that the proponents of propaganda and disinformation are winning and are in control of the subject. Basically, Project Serpo refers to claims that a top secret exchange program in the 1960s and

558

'70s involved sending an American team of 12 military personnel to another planet, **Zeta Reticuli**. The similarity of the program is remarkable to the depictions in parts of the 1977 Steven Spielberg film ***Close Encounters of the Third kind.*** The account posted on the SERPO website reads very much like certain documents and confidential letters that were sent to Bennewitz from 1980 to '87. http://www.ufomystic.com/2007/02/22/serpo-was-disinformation/

All of the information on Project Serpo came to **Victor Martinez and Bill Ryan** by way of "Request Anonymous" who turned out to be none other than Richard C. Doty. Martinez may have suspected Doty's involvement, but Bill Ryan knew from the very start that he was getting the Serpo material directly from the former AFOSI security guard. Doty has continuously denied having any involvement in the story, despite the mountains of evidence to the contrary. (***Doty should have been suspected almost from the get go having been involved directly in the downfall of Paul Bennewitz.***) [Italicized sentence added by author for emphasis]. http://www.realityuncovered.net/ufology/articles/serpo/

Recall also, the famous **Betty and Barney Hill** case, and American bi-racial couple who were abducted on aboard an ET spacecraft on September 19–20, 1961 after leaving Quebec, Canada and crossing the border into New Hampshire, USA. Both Betty and Barney stated under many sessions of hypnosis that their abductors, who performed medical fertility procedures upon them, were small grayish coloured beings with somewhat large heads and eyes. Betty stated she was shown a star map of interstellar trade routes and the largest star on the map represented the star of the ETs home world. From this information, it was later deduced by Marjory Fish that the home world of the ETs must be in the star system of Zeta Reticuli.

This faulty deduction and leap of illogic, however, meritoriously investigated has led to all kinds of wild speculation and has contributed to an ever-growing body of false evidence and thus, a corrupt UFO database. Numerous similar stories and accounts have sprung up in epidemic proportions around the world ever since, most notably in the US and in some countries that have partnership agreements with the US.

The point here is that an exchange program between humans and ETs requires a very high level of trust and does not occur at the end of a gun barrel or from the launch of a salvo of missiles or whatever advanced scalar weapon is being targeted at you. It is more likely that from such action, you will incur the retaliatory wrath of your new found enemy and the only mutual exchange will be the volley after volley of death rays and God knows what!!! It would be an interstellar war that is not a fictional piece of celluloid entertainment but rather, an all too real, terrifying and horrific war that humanity never expected nor wanted.

Yet, this is precisely what is going on by most militaries with advanced weaponry. There is no real peace and stability on this planet when the most powerful nation continually vie with each other in a dangerous game of brinkmanship for superiority and ultimate control over the planet, it resources and its people. Monitoring, Tracking, targeting and shooting down ET craft from space in order to gain possession of alien technology to further bolster your military might over other nations is not an agenda that should be pursued when the potential outcome could be the demise of the Earth and her people.

CHAPTER 52

DEEP BLACK PROJECTS, SAPS, USAPS, MILITARY SUPER WEAPONS

If you had a really big secret, a top secret program or project that you want to keep *unacknowledged,* you would compartmentalize it in airtight, vacuum sealed (figuratively speaking) environment of absolute security requiring a special access even for those with a top secret clearance. It would become a **USAP (Unacknowledged Special Access Project)** where, if anyone including your superiors, whether in the military or government or even, no less a person the President of the United States, himself were to ask you about this top secret project, you would reply that no such project exists. You basically lie! End of discussion.

Any person in a USAP position will do anything to maintain the utmost security on a top secret project using disinformation and even, deadly force if necessary to keep the story concealed and away from any official or public inquiry. The crown jewel of all USAPs is, of course, the UFO/ETI subject

It doesn't matter what your classification type is in the U.S. whether Confidential, Secret, or Top Secret, if you don't have a **"need to know"** then you are not **"in the loop"** on the subject matter, regardless of your official position in the government or the military as explained above, access is denied. Almost every agency or department of the government, the Army, Navy and Air Force has a similar policy in place, eg. **(FBI, CIA, DIA, NSA, ONI, NRO, INSCOM,** and **AFOSI).** This subject is of the highest order of secrecy whether in Britain, Canada, Russia, China, France, Brazil and just about any nation you can think of with a military force.

USAPs become highly efficacious when combined with the proprietary power of the private contracting sectors of industry and private industry is known to keep secrets better than the military. It is an impregnable covert fortress and virtually unassailable. The gatehouse of the private sector of this bastion is guarded by a strong portcullis of **proprietary privilege** and the postturn of public sector-government in the back is slam shut and barred to entry by USAPs.

"…if you try to get at it through the private sector, it is protected by proprietary privilege. And if you try to get to it through the public sector-government it is hidden in USAPs…"
DISCLOSURE PROJECT BRIEFING DOCUMENT Prepared for Members of the Press, Members of United States Government, Members of the US Scientific Community; Written and Compiled by Steven M. Greer, M.D., Director and Theodore C. Loder III, Ph.D. April 2001

This means it's virtually impossible to get information about these projects, because private industry is protected by …**"proprietary privilege."** You normally can't get any information about a USAP by issuing a FOIA or by annoying a Congressman because of national security laws, but just in case anyone might be able to succeed, there's always the argument of *"proprietary privilege of the private industry."*

This group is a quasi-governmental, USAPS related, quasi-private entity operating internationally/transnationally. The majority of operations are centered in private industrial *"work for others"* contract projects related to the understanding and application of advanced extraterrestrial technologies. Related compartmentalized units, which are also USAPS, are involved in disinformation, public deception, active disinformation, so-called abductions and mutilations, reconnaissance and UFO tracking, space-based weapons systems and specialized liaison groups (for example to media, political leaders, the scientific community, the corporate world, etc.). Think of this entity as a hybrid between government, USAPS, and private industry.

The group consists primarily of mid-level USAPS-related military and intelligence operatives, USAPS or black units within certain high-tech corporate entities, and select liaisons within the international policy analysis community, certain religious groups, the scientific community, and the media, among others.

Actual policy and decision-making seems to rest predominantly at this time in the private, civilian sector, as opposed to USAP-related military and intelligence officials, though some information indicates that there is significant relative autonomy in certain areas of operations. It is our current assessment that a rising degree of debate exists regarding certain covert operations and the advisability of a disclosure. **DISCLOSURE PROJECT BRIEFING DOCUMENT by Steven M. Greer, M.D., Director and Theodore C. Loder III, Ph.D. April 2001**

This group is a quasi-governmental, USAPS related, quasi-private entity operating internationally/transnationally. The majority of operations are centered in private industrial "work for others" contract projects related to the understanding and application of advanced extraterrestrial technologies. Related compartmentalized units, which are also USAPS, are involved in disinformation, public deception, active disinformation, so-called abductions and mutilations, reconnaissance and UFO tracking, space-based weapons systems and specialized liaison groups (for example to media, political leaders, the scientific community, the corporate world, etc.). Think of this entity as a hybrid between government, USAPS, and private industry.

The group consists primarily of mid-level USAPS-related military and intelligence operatives, USAPS or black units within certain high-tech corporate entities, and select liaisons within the international policy analysis community, certain religious groups, the scientific community, and the media, among others. The identities of some of these entities and individuals are known to us, though most remain unidentified...

Actual policy and decision-making seems to rest predominantly at this time in the private, civilian sector, as opposed to USAP-related military and intelligence officials, though some information indicates that there is significant relative autonomy in certain areas of operations. It is our current assessment that a rising degree of debate exists regarding certain covert operations and the advisability of a disclosure. **DISCLOSURE PROJECT BRIEFING DOCUMENT by Steven M. Greer, M.D., Director and Theodore C. Loder III, Ph.D. April, 2001**

A three-tiered classification system is not enough to protect some of the more sensitive information. Therefore additional levels of compartmentalization have been created. After a very intensive background check, someone with a **Top Secret** clearance might obtain an additional **Sensitive Compartmented Information (SCI)** clearance, under which information is buried that needs to be restricted to even fewer individuals. This **TS-SCI** clearance had been introduced

561

mainly to stop some higher ranking officers from looking into Top Secret files they don't have any business with. http://www.bibliotecapleyades.net/sociopolitica/sociopol_USAP.htm

But even the **TS-SCI clearance** doesn't provide the secrecy needed for some of the most sensitive projects. This is the reason why **Special Access Programs (SAPs)** have been invented. In these cases only a predetermined group of authorized personnel have access to the project and additional security measures can be taken to keep outsiders away. Different congressional committees are informed about these SAPs, but very little time is reserved for questions.

Most SAPs start out as Unacknowledged Special Access Programs (USAPs), better known as **Black Projects**. The **F-117A Nighthawk** and the **B-2 Spirit** are examples of projects that started out as Unacknowledged SAPs. A DOD manual describes a USAP as follows:

"**Unacknowledged SAPs** require a significantly greater degree of protection than acknowledged SAPs. A SAP with protective controls that ensures the existence of the Program is not acknowledged, affirmed, or made known to any person not authorized for such information. All aspects (e.g., technical, operational, logistical, etc.) are handled in an unacknowledged manner."

If questioned about a particular USAP, the persons involved are under orders to deny such a program exists. It is not allowed to react with a "no comment" because that would immediately fuel suspicions that something is being hidden and is likely to cause further inquiries. Officers not accessed for a USAP, even superior ones, are to be given the same response. The more sensitive the program, the more protection the commanding officer can demand. He could even subject his personnel to regular lie detector tests to see whether or not a person has compromised the project. According to a 1997 Senate investigation:

"Additional security requirements to protect these special access programs can range from mere upgrades of the collateral system's requirements (such as rosters specifying who is to have access to the information) to entire facilities being equipped with added physical security measures or elaborate and expensive cover, concealment, deception, and operational security plans." http://www.bibliotecapleyades.net/sociopolitica/sociopol_USAP.htm

There are two versions of the Unacknowledged Special Access Programs. The first one is the regular USAP. These regular USAPs are reported in the same manner as their acknowledged versions. In closed sessions, the House National Security Committee, the Senate Armed Services Committee, and the defense subcommittees of the House and Senate Appropriations committees can get some basic information about them. The Secretary of Defense, however, can decide to 'waive' particularly sensitive USAPs. These are unofficially referred to as **Deep Black Programs**. According to **Jane's Defense Weekly** in January 2000:

"Among black programs, a further distinction is made for **"waived" programs**, considered to be so sensitive that they are exempt from standard reporting requirements to the Congress. The chairperson, ranking member, and, on occasion, other members and staff of relevant Congressional committees are notified only orally of the existence of these programs."

562

This leads to the conclusion that only very few people are aware of these waived USAPs - Unacknowledged Special Access Programs. Congress certainly doesn't get the information it needs to speak out against newly established waived USAPs nor it seems is their opinion is actually appreciated. You could also ask yourself if Congress is told the truth about many of the most sensitive Special Access Projects or if their successors are informed about previously activated (waived) USAPs. Even with regular SAPs Congress is ignored at times. Again in Jane's Defense Weekly:

"Last summer, the House Defense Appropriations Committee complained that "the air force acquisition community continues to ignore and violate a wide range of appropriations practices and acquisition rules". One of the alleged infractions was the launch of a SAP without Congressional notification."
http://www.bibliotecapleyades.net/sociopolitica/sociopol_USAP.htm

What makes **Unacknowledged Special Access Projects** even more impenetrable is the fact that a lot of these programs are located within private industry. The U.S. government generally doesn't develop a whole lot. If you look at the defense industry, you have companies like **Boeing, Lockheed, Northrop, McDonnell Douglas, TRW, Rockwell, Bechtel, SAIC, or Decision-Science Applications (DSA Inc.)**, who develop certain technologies for the U.S. government.

Investigating into any of these private companies is protected by **"proprietary privilege."** No amount of FOIA requests will reveal information on any particular USAP or by getting Congressional assistance to bypass national security laws, there is still the barrier of *"proprietary privilege of the private industry"*.

Category	Secrecy levels	
Additional levels of Compartmentalization	A USAP behind another SAP or USAP, combined with the protection the private industry enjoys.	
	'Waved' Unacknowledged Special Access Programs / 'Deep Black Programs' (details already completely invisible to congress and the president)	
	Unacknowledged Special Access Programs / 'Black Programs'	
	(acknowledged) Special Access Programs	
	Top Secret Sensitive Compartmented Information (TS-SCI)	
Basic secrecy levels	Top Secret	NATO Cosmic Top Secret
	Secret	NATO Secret
	Confidential	NATO Confidential
Public or semi-public	For Official Use Only	NATO Restricted
	Unclassified	NATO Unclassified

A case in point which was already mentioned earlier in this book is the investigations by the late New Mexico Republican, Congressman Steven Schiff. After hearing from frustrated constituents who wanted answers from the Pentagon about the "Roswell Incident," Schiff asked the Air Force to declassify and provide him with all its relevant material. When the Pentagon referred the congressman to the National Archives instead, Schiff smelled a cover-up and called in the General Accounting Office to look for documents and investigate whether the Air Force kept adequate records relating to the incident. The GAO concluded that many records of government activity in Roswell during the summer of 1947 were mysteriously missing. The apparent destruction of those records was unusual and unauthorized, the GAO said, but without them, the investigation came to a standstill.

No government official ever pushed harder or more publicly for disclosure of Roswell-related government records than Congressman Steve Schiff. And, although the news media claimed there was absolutely no evidence of a connection between his GAO investigation and his cancer, UFO researchers are compelled to wonder why this particular Congressman came down with an unusually vicious disease so soon after his Roswell investigation ended. When someone pushes this hard to get to the bottom of a coverup it frequently does end with that person losing their life. In the case of Congressman Schiff, his investigations did cost him his life, contrary to what was stated in the news media! Schiff was targeted as were Dr. Steven Greer and his CSETI partner Sheri Adamiak by sophisticated scalar field weaponry that induces diseases like cancer. Only Dr. Greer barely survived this targeted attack, Sheri Adamiak, unfortunately, died from breast cancer.

This is a classic example of the lengths in which USAPs are protected from investigation or the suppression of contact and communication with Extraterrestrials. Murder is not out of the question, even as a retaliatory means to send a clear message to all others who dare to cross the threshold of those who jealously guard the inner sanctum on the UFO/ETI matter.

And even in case the **National Security State** and proprietary privilege fails, there still seems to be at least one other (unverified) mechanism to protect the most sensitive projects from public exposure. This information comes from (many, very credible) **Disclosure Project** witnesses, all of them claiming to have some kind of experience with these type of projects.

Some of these people, with no one in the project disputing it, are saying that certain **Black Programs (USAPs)** act as covers for UFO / ET-related projects. This means that in an emergency situation a sensitive **Black or Deep Black Program** could be revealed to the public, while the program behind it remains undiscovered.

When a black project like the SR-71 Blackbird or the F-117A Nighthawk or the B-2 Spirit, all stealth aircraft that were once considered sensitive projects are deliberately revealed to the public, it is because other deep black projects, more advanced are being researched and developed. What is seen by the public is in reality, old technology. Most R & D programs are at least 50 years ahead of public perception and understanding.

In a 1997 speech, former astronaut **Edgar Mitchell** summarizes what the Disclosure Project is all about:

"I also think that the prevalence in the modern era of so many events - the sightings, the continual mutilation events, the so-called abduction events - that we are looking at likely reversed engineered technology in the hands of humans that are not under government control or any type of high-level control...

*So if there are **back engineered technologies** existing, they are probably in the hands of this group of individuals, formerly government, formerly perhaps intelligence, formerly, under private sector control with some sort of oversight by military or by government. But this (oversight) is likely no longer the case as a result of this access denied category that is now operating. I call it a **clandestine group.** The technology is not in our military arsenals anywhere in the world, but it does exist, and to me, that's quite disconcerting."*
http://www.bibliotecapleyades.net/sociopolitica/sociopol_USAP.htm

Major bases of operations, apart from widely diversified private sites, include Edwards Air Force Base in California, Nellis Air Force Base in Nevada, particularly S4 and adjacent facilities, Los Alamos New Mexico, Fort Huachuca Arizona (Army Intelligence Headquarters), the Redstone Arsenal in Alabama, and a relatively new, expanding underground facility accessible only by air in a remote area of Utah, among others. Additional facilities and operations centers exist in a number of other countries, including the United Kingdom, Australia, and Russia.

Numerous agencies have deep cover, black, USAPS related units involved with these operations, including the **National Reconnaissance Office (NRO),** the **National Security Agency (NSA), the CIA,** the **Defense Intelligence Agency (DIA), the Air Force Office of Special Investigations (AFOSI) , Naval Intelligence, Army Intelligence, Air Force Intelligence,** the **FBI,** and a group known as **MAJI** control. An even more extensive list of private, civilian and corporate entities have significant involvement. The majority of scientific, technical and advanced technology operations are centered in the civilian industrial and research firms. Significant — and lethal — security is provided by private contractors... **DISCLOSURE PROJECT BRIEFING DOCUMENT Written and Compiled by Steven M. Greer, M.D., Director and Theodore C. Loder III, Ph.D. April 2001** and https://www.youtube.com/watch?v=DJKvPipavs8 and https://www.youtube.com/watch?v=Aaf_61I7vZ0

What are these black and deep black projects that are 50 years ahead of current human technological development in the white world of science? Are they the stealth aircraft that were developed back in the 1960s? Are the reported sightings of these advanced aircraft the real UFOs that have confused so many people worldwide with their unusual aerodynamic designs? Were they developed from advanced WWII German captured technology or from reversed engineering of real alien technology?

With the outbreak of World War II in 1939, the outlook for flying wing development improved immeasurably. On both sides of the Atlantic, governments were more than willing to gamble funds and manpower in a search for the right combination of weapons and aircraft that could mean the difference between victory and defeat in the air war ahead. Most of the governments that were at war took a few tentative steps in the direction of tailless fighters, but only one aircraft of the type, the Messerschmitt Me 163, was used in combat. In the United States, only Jack Northrop worked vigorously to build a flying wing, but it was not until a year after the war that the first of his giant aircraft made its maiden flight. In Germany, despite the exigencies of war, the Hortens continued to create a series of imaginative flying wing designs that culminated in the world's first turbojet-powered flying wing. The Lippisch designed Me 163 became the world's first operational tailless fighter. http://www.century-of-flight.net/new%20site/frames/horten%20frame.htm

German Aircraft Development During WWII

Germany had many radical aircraft designs based on sound principles of aerodynamics and it was evident during the war that some of these new advanced aircraft saw action over Europe. Nazi Germany was not the only nation to develop advanced aircraft designs as many Allied nations were progressing along the same lines, however, the Germans were much further ahead in building such craft. It would be fair to say that the Germans continually thought outside the box when it came to wartime developments in aircraft and weaponry. It should come as no surprise then, that there was a strong impetus by British, American and Russian forces to enter German territory and capture their scientists and technology.

There is a revisionist debate over whether Northrop came up with the flying wing on its own or if they got the idea from captured German plans at the end of WW2. Clearly, **Jack Northrop** thought of the flying wing in the 1930s as evidenced by his contract with the USAAF to build the contra-rotating prop-driven B-35 in the 1940s--long before the US captured German scientists through Operation Paperclip. Yet, the greatest strides were made by the Americans after the war and with the cooperation of the captured Nazi scientists.

Focke-Wulf Fw 1000x1000x1000 Bomber Project B (model). Development cancelled
http://frank.bol.ucla.edu/jets.htm

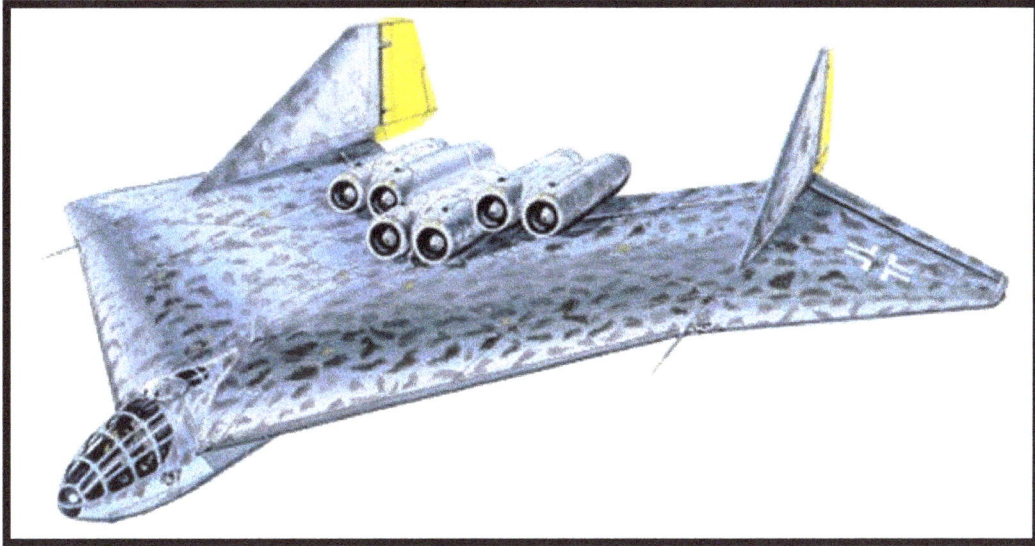

Arado Ar E.555-1
On December 28, 1944, Development Cancelled
http://www.luft46.com/arado/are555s.html

Arado Ar E.555-6
http://www.luft46.com/arado/arc555s.html

The world's first turbojet-powered flying wing, the Ho IX V2, is prepared for flight tests somewhere in Germany in January 1945.

Horten Ho 229

Horten Ho Vc

**Horten Ho XV111 A was a proposed design for a trans-Atlantic
six engine wing bomber but it was never developed**

Bachem Ba 349 Natter (Adder/Viper) Vertically-Launched Rocket-Powered Interceptor
http://www.luftarchiv.de/index.htm?/flugzeuge/sonstige/ba349.htm

The Messerschmitt Me 163 Komet. Shown here is one of the B-series prototypes, the Me 163BV2, which made its first rocket-powered flight on June 24, 1943
http://acepilots.com/german/me163.html

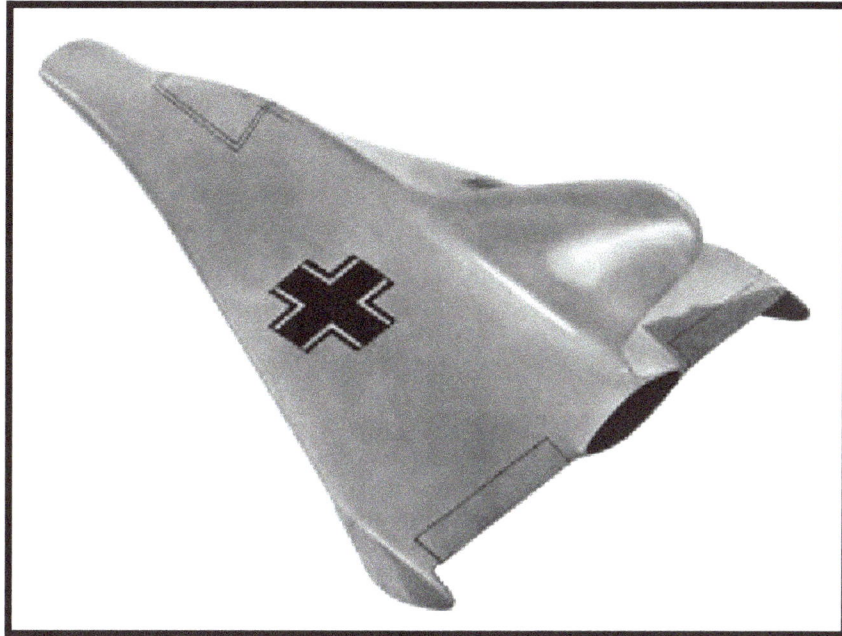

Project P12, (model) was to be an experimental aircraft equipped with a ramjet engine. Since ramjet engines do not produce thrust at zero speed, the aircraft would have to be accelerated to flying speed either by a "piggyback" arrangement or rocket assisted takeoff.

http://www.century-of-flight.net/new%20site/frames/horten%20frame.htm

Another German aircraft by Doenitz of early 1945 was the Gotha P.60A discovered by the Americans. Also with a triangular shape (actually, it's more of a flying wing).

http://www.kheichhorn.de/html/body_gotha_p_60_c.html

American Aircraft Development During WWII

In many ways American aircraft development paralleled the advanced WWII German aerodynamic designs, the difference being that American engineers improved upon the basic German concepts over the ensuing post-war years. Aircraft became stronger in construction, faster with newer and more improved engine design and thus, broke speed records and flew higher reaching the upper limits of Earth's atmosphere.

Basic aerodynamic designs indicated Americans favoured aircraft with swept- back wings tending toward delta or triangular shape and aircraft evolved toward being tailless unless intended toward operating within atmospheric parameters. Radar invisibility for aircraft as well as naval vessels was an ongoing program that began early in the Second World War and by the early '60s stealth capability became possible with radar absorbing or deflecting composite surface materials.

The fascination with flying higher and faster created many challenges and possibilities to develop space planes that could fly into space, return and land safely without the need of rocket launch assistance. The **X-1** and the **X-15** are examples of aircraft that flew higher and faster.

The **Bell X-1**, originally designated **XS-1**, was a joint NACA-U.S. Army/US Air Force supersonic research project built by Bell Aircraft. Conceived in 1944 and designed and built over 1945, it eventually reached nearly 1,000 mph (1,609 km/h) in 1948. A derivative of this same design, the Bell X-1A, having greater fuel capacity and thus longer engine burn time, exceeded 1,600 mph (2,574 km/h) in 1954.[1] The X-1 was the first aircraft to exceed the speed of sound in controlled, level flight, and was the first of the so-called X-planes, an American series of experimental aircraft designated for testing of new technologies and usually kept highly secret.
http://en.wikipedia.org/wiki/Bell_X-1

The **North American X-15** was a rocket-powered aircraft operated by the United States Air Force and the **National Aeronautics and Space Administration (NASA)** as part of the X-plane series of experimental aircraft. The X-15 was based on a concept study from Walter Dornberger for the NACA for a hypersonic research aircraft. The X-15 set speed and altitude records in the early 1960s, reaching the edge of outer space and returning with valuable data used in aircraft and spacecraft design. As of 2012, the X-15 holds the official world record for the fastest speed ever reached by a manned aircraft.

During the X-15 program, 13 different flights by eight pilots met the USAF spaceflight criterion by exceeding the altitude of 50 miles (80 km) thus qualifying the pilots for astronaut status. The USAF pilots qualified for USAF astronaut wings, while the civilian pilots were awarded NASA astronaut wings in 2005, 35 years after the last X-15 flight. Of all the X-15 missions, two flights (by the same pilot) qualified as space flights per the international (*Fédération Aéronautique Internationale*) definition of a spaceflight by exceeding 100 kilometers (62.1 mi, 328,084 ft) in altitude

Like many X- series aircraft, the X-15 was designed to be carried aloft, under the wing of a NASA B-52 mother ship, the *Balls 8*. Release took place at an altitude of about 8.5 miles

572

(13.7 km) and a speed of about 805 kilometers per hour (500 mph).
http://en.wikipedia.org/wiki/North_American_X-15

Bell X-1, (rocket plane) Nicknamed: "Glamorous Glennis"
Manufacturer Bell Aircraft First Flight: 19 January 1946
https://simple.wikipedia.org/wiki/Bell_X-1

Northrop N9MB Flying Wing Concept Demonstrator
https://www.pinterest.com/explore/flying-wing/

Northrop XB-35

Northrop YB-49 flying wing

A-12 Avenger II manufactured by McDonnell Douglas/General Dynamics
It was never flown and thus cancelled

B-1 Lancer manufactured by Rockwell International

B-2 Spirit "Stealth Bomber" manufactured by Northrop Grumman

Q-1 Predator, Warrior manufactured by General Atomics

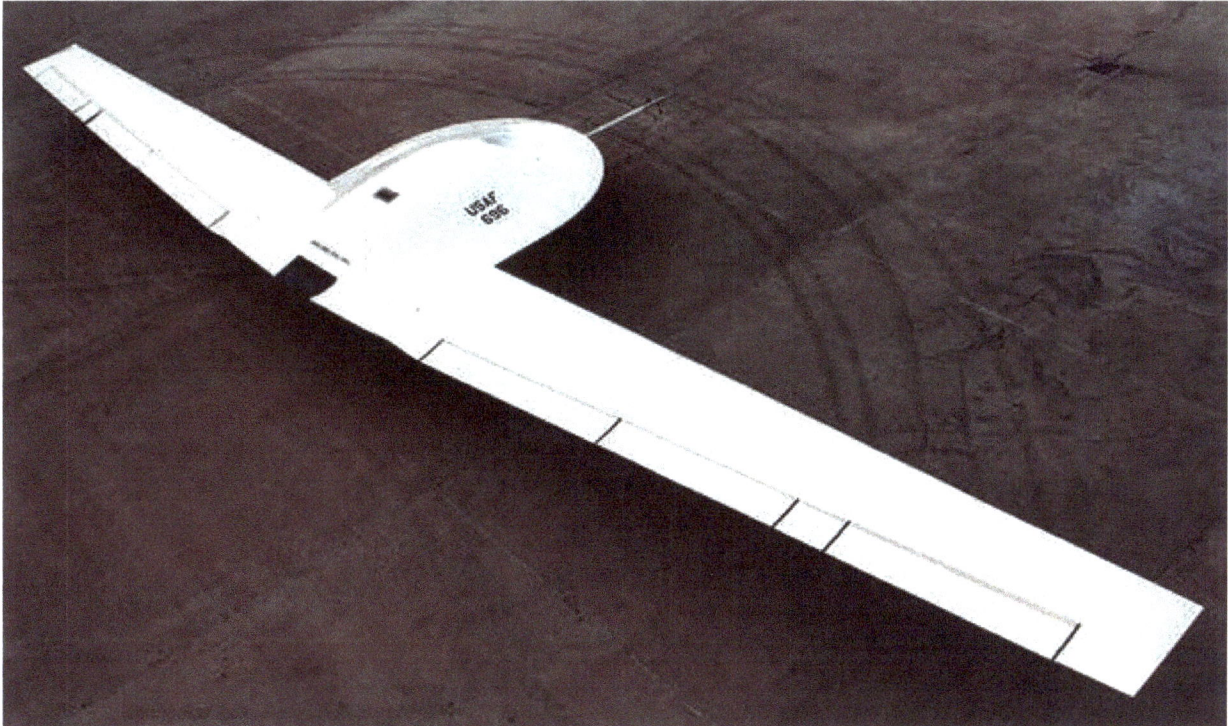

Q-3 Dark Star manufactured by Lockheed Martin and Boeing

Q-4 Global Hawk manufactured by Northrop Grumman

**Q-9 Reaper, Altair originally known as the" Predator B"
manufactured by General Atomics**

Q-15 Neptune manufactured by DRS Technologies

RQ-170 Sentinel manufactured by Lockheed Martin
http://avia.pro/blog/lockheed-martin-rq-170-sentinel

X-3 Stiletto manufactured by Douglas Aircraft USAF,
NACA first flown on October 27, 1952
https://www.nasa.gov/centers/armstrong/news/FactSheets/FS-077-DFRC.html

X-15 manufactured by North American Aviation USAF,
NASA first flown: June 8, 1959

X-24 manufactured by Martin Marietta USAF,
NASA first flown: April 17, 1969

X-24B manufactured by Martin Marietta USAF, NASA first flown: August 1, 1973

X-29 manufactured by Grumman DARPA, USAF, NASA flown: 1984

**X-36 manufactured by McDonnell Douglas/Boeing,
NASA first flown: May 17, 1997**

X-38 manufactured by Scaled Composites NASA first flown: 1999

X-43 Hyper-X manufactured by Microcraft NASA first flown: June 2, 2001
https://www.nasa.gov/centers/armstrong/news/FactSheets/FS-084-DFRC.html

X-44 MANTA manufactured by Lockheed Martin USAF, NASA
It was never flown (so we are told, see insert) and thus cancelled
http://www.f-16.net/forum/viewtopic.php?t=8840

583

X-45 manufactured by Boeing DARPA, USAF first flown: May 22, 2002
https://en.wikipedia.org/wiki/Boeing_X-45

**X-47A Pegasus manufactured by Northrop Grumman DARPA, USN
first flown: February 23, 2003**
https://en.wikipedia.org/wiki/Northrop_Grumman_X-47A_Pegasus

X-47B manufactured by Northrop Gruman DARPA, USN first flown: February 23, 2003

**Robot X-47B Stealth Bomber manufactured by Northrop Grumman
first flown: Feb.09, 2011?**

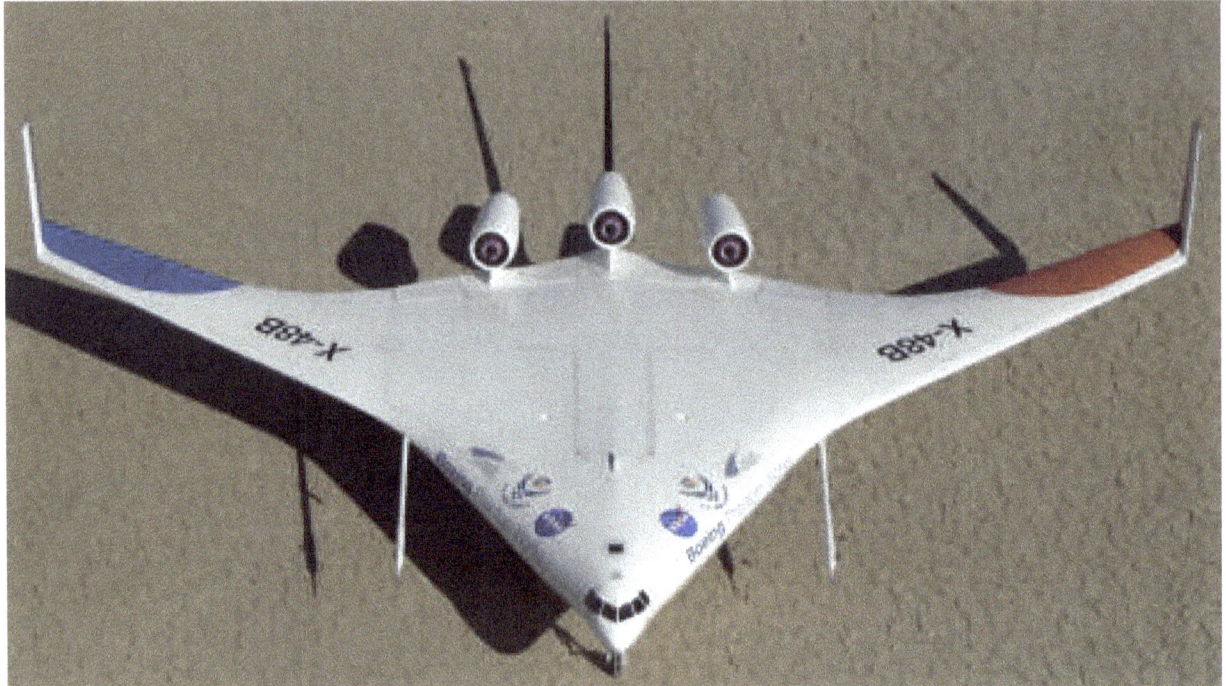

X-48 manufactured by Boeing NASA first flown: July 20, 2007

(F-12) SR-71 "Blackbird" manufactured by Lockheed Skunk Works

F-19A Specter manufactured by Northrup/Loral
http://www.keiththomsonbooks.com/blog/f-19

F-22 Raptor manufactured by Lockheed Martin
https://en.wikipedia.org/wiki/Lockheed_Martin_F-22_Raptor

F-117 Nighthawk manufactured by Lockheed
https://en.wikipedia.org/wiki/Lockheed_F-117_Nighthawk

Bird of Prey, Tacit Blue manufactured by Boeing DARPA
http://www.ufo-contact.com/ufo-inspired-black-projects-gallery-2/boeing-bird-of-prey-2

588

Clarence Leonard "Kelly" Johnson and Skunk Works

Clarence Leonard "Kelly" Johnson (February 27, 1910 – December 21, 1990) was an American system engineer and aeronautical innovator, renowned for his contributions to many noteworthy aircraft designs, especially the Lockheed U-2 and SR-71 Blackbird spy planes, but also including the P-38 Lightning, P-80 Shooting Star, and F-104 Starfighter among others. As a member and first team leader of the **Lockheed Skunk Works**, Johnson worked for more than four decades and is said to have been an "organizing genius". He played a leading role in the design of over forty aircraft, including several that were honored with the prestigious Collier Trophy, acquiring a reputation as one of the most talented and prolific aircraft design engineers in the history of aviation. In 2003, as part of its commemoration of the 100th anniversary of the Wright Brothers' flight, *Aviation Week & Space Technology* ranked Johnson 8th on its list of the top 100 "most important, most interesting, and most influential people" in the first century of aerospace. Hall Hibbard, Kelly's Lockheed boss, once remarked to Ben Rich: *"That damned Swede can actually see air".* http://en.wikipedia.org/wiki/North_American_X-15

**Clarence Leonard "Kelly" Johnson with the U2 and
the SR-71 Blackbird in the background**
https://www.pinterest.com/explore/kelly-johnson-engineer/

Johnson joined the **Lockheed Company** in 1933 and after several successful project designs for the company; he was given assignments as flight test engineer, stress analyst, aerodynamicist, and weight engineer, and in 1938he became chief research engineer. By 1952, he was appointed chief engineer of Lockheed's Burbank, California, plant, which later became the Lockheed-California Company. In 1956 he became Vice President of Research and Development.

He was the first team leader at the famous Lockheed Skunk Works and in 1955 personally scouted and picked the location for and initiated construction of the airbase at **Groom Lake, NV,** better known as **Area 51**, at the request of the Central Intelligence Agency. http://ufopartisan.blogspot.ca/2010/08/area-51-founders-ufo-sightings.html

Johnson became Vice President of **Advanced Development Projects (ADP**) in 1958. The first ADP offices were nearly uninhabitable; the stench from a nearby plastic factory was so vile that one of the engineers began answering the intra-Lockheed "house" phone "Skunk Works!" In Al Capp's comic strip *Li'l Abner,* Big Barnsmell's Skonk Works — spelled with an "o" — was where Kickapoo Joy Juice was brewed. When the name "leaked" out, Lockheed ordered it changed to **"Skunk Works"** to avoid potential legal trouble overuse of a copyrighted term. The term rapidly circulated throughout the aerospace community and became a common nickname for research and development offices; however, reference to "The Skunk Works" means the Lockheed ADP department. This project provided a secret location for flight testing the **U-2**.

The U-2 has a unique place in UFO history having become inextricably intertwined with fantastic rumors of captured UFOs, reverse-engineered alien technology, and endless other exotic ideas. It was first test-flown at **Groom Lake**, a secret airbase inside the restricted zone surrounding the **Atomic Energy Commission's** Nevada nuclear test range. While not commenting on such matters, the CIA has officially stated that the U-2 and its successor, the 2,400 mph **A-12 OXCART**, were themselves responsible for "over half of all UFO reports" in the 1950s and 60s.

At the very moment that the U-2's was being built for the CIA in December 1953, a group of Lockheed's top engineers and pilots were witnessed to a UFO sighting that is regarded with high scientific significance because of the professionalism of its unique witnesses. Yet, in the Project Blue Book files, it was labeled as "identified" for nearly half a century, then buried, and nearly forgotten. Recent study of the Blue Book case file yielded the names of seven eyewitnesses - a crew of Lockheed flight test engineers in the air, and Lockheed's Chief Engineer and his wife on the ground. The Chief Engineer was Clarence L ("Kelly") Johnson - the creator of the U-2.

Historically, Lockheed aircraft have played roles in many of the pivotal cases in UFO history. Its fighters have chased radar tracked UFOs over Washington, DC in the summer of 1952, other military planes were involved in significant events like the Great Falls UFO film case in 1950, the Ft Monmouth UFO chase of September 1951, the fatal Walesville incident of 1954, and numerous other UFO intercept events. In 1952, an Air Defense Command Colonel, harassed by continuing UFO sightings near his New Mexico base, proposed converting several of Lockheed's latest F-94C fighters into actual dedicated **"UFO interceptors." "Project POUNCE** was implemented and these fighters equipped with a battery of nose-mounted cameras were on 24-hour standby alert ready to go at a moments notice in an attempt to obtain clear, close-up photos of a saucer in flight. http://conspiracy101.com/ufos/skunkworks/index.html

Johnson also led the development of the **SR-71 Blackbird** family of aircraft. Through a number of significant innovations, Johnson's team was able to create an aircraft that flew so high and fast that it could not be intercepted nor shot down. No other jet airplane has matched the Blackbird's performance. http://en.wikipedia.org/wiki/Kelly_Johnson_%28engineer%29

One of Kelly Johnson more impressive aircraft designs, the M-21 with D-21 drone-mounted on the back of the aircraft.
https://en.wikipedia.org/wiki/Lockheed_D-21

He also witnessed UFOs with otherworldly capabilities on two occasions and wrote in his official report on his 1953 sighting, *"For at least five years I have definitely believed in the possibility that flying saucers exist — this in spite of a good deal of kidding from my technical associates. Having seen this particular object on December 16th, I am now more firmly convinced than ever that such devices exist, and I have some highly technical converts in this belief as of that date."* His credentials as a UFO witness are unimpeachable.

On December 16, 1953, Johnson was in his home three miles west of Agoura, CA and looking out the window, he describes seeing a solid object hovering over what he estimated to be Point Mugu, which was and is also the location of a Naval Air Station, not quite 20 miles west of his home. *"When it did not move or disintegrate, I asked my wife to get me our 8X binoculars, so I would not have to take my eyes off the object, which by now I recognized as a so-called 'saucer.'"*

The time was about 5 PM and the setting sun was below the horizon line, leaving the ellipsoid UFO silhouetted. By the time Johnson got the binoculars and went outside for a better view, the object was moving away from him and out over the Pacific Ocean.

591

F-117s Nighthawks in formation, another of Kelly Johnson's stealth aircraft designs.

"I gathered its' speed was very high because of the rate of foreshadowing of its' major axis. The object, even in the glasses, appeared black and distinct, but I could make out no detail, as I was looking toward the setting sun, which was, of course, below the horizon line," he wrote.

The UFO continued on its' path westward in a slight climb and disappeared in 90 seconds according to Johnson. He did not make an estimate of its' dimensions.

Scientifically, this case is significant not only because of Johnson expertise in aviation and aerodynamics but because there were four other witnesses and they were in flight several miles to the south! This allowed later investigation to triangulate the UFO's position, direction of flight and speed.

While testing out a Lockheed WV-2 Constellation over the Catalina Channel, four crew members, flight engineer R.L. Thoren, test pilot Roy Wimmer, aerodynamics engineer Philip Colman and flight test supervisor J.F. Hare, saw the same UFO Johnson did. They describe the exact same hovering for a couple minutes and then flying on a path westward and leaving their sight within a matter of seconds.

One curious discrepancy between the accounts is that the airborne witnesses were able to make out a crescent/flying wing shape from more than double the distance of Johnson and without the aid of binoculars, and Johnson was not. For Johnson to only see a saucer shape, I think the UFO would have had to remain at just the right angle to him the entire sighting. The official explanation was that they all saw a lenticular cloud, but the path westward the vehicle took and the speed it left the area in seem to rule that out. Whatever, Johnson and the flight crew really saw that day, filing an official report didn't seem to negatively effect his career. Johnson was already chief engineer at Lockheed's Burbank plant, would be selected to break ground on the mysterious Area 51 air base in 1955 and became Lockheed's vice president of Research Development Projects in 1958.

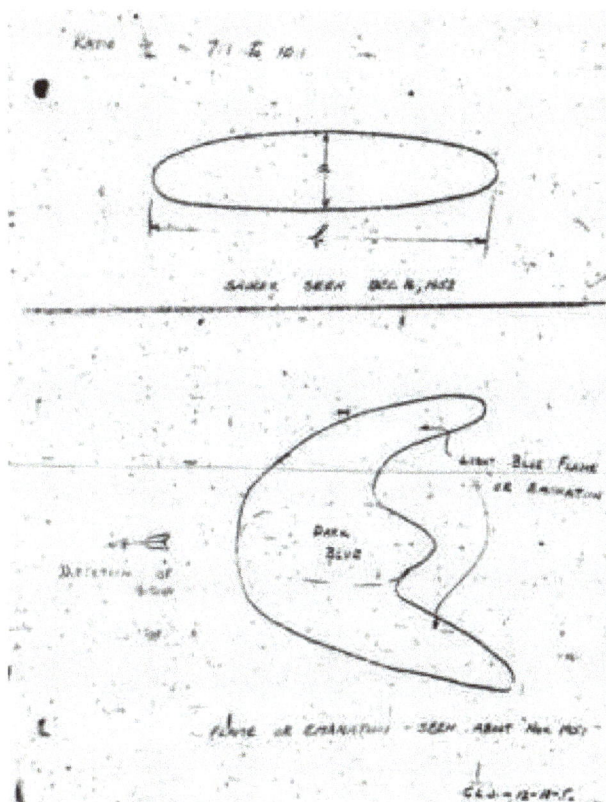

Possible inspiration of the B-2 Bomber from Kelly Johnson's UFO drawings made many years earlier. The drawing and the B-2 Bomber are reminiscent of Kenneth Arnold's "winged-shaped" UFO and the WWII German Horten Ho 229 and Ho Xv111 A
http://ufopartisan.blogspot.ca/2010/08/area-51-founders-ufo-sightings.html

The entire case becomes more interesting when Johnson's drawings (see above) from his two sightings are compared with the Northrop B-2. It's almost like a Johnson fan at Northrop read his UFO report, looked at the drawings and a light bulb appeared over his head. It's certainly a much more elegant design than Lockheed's boxy F-117. The bottom drawing was from the earlier incident in November 1951 and it is clearly described as an emanation from an object Johnson couldn't make out, not the object itself, but the gestalts are amazingly similar.

Clarence "Kelly" Johnson had his following, no question about that, and his influence reverberates through the aeronautics industry to this day. The question is what influence did alien spacecraft have on Johnson?

http://ufopartisan.blogspot.ca/2010/08/area-51-founders-ufo-sightings.html
Lockheed's official public pronouncements on UFOs had been negative. For example, on July 7, 1947, Hall Hibbard had made disparaging remarks to Los Angeles Times reporters who inquired about the saucers being reported nationwide:

They're either reflections from planes flying singly or in formation or mass hysteria and the desire of various persons to get their names in the paper. I know of no secret aviation projects which would have the slightest bearing on these so-called 'phenomena.'

Nevertheless, the company harbored at least two prominent engineers who were interested in UFOs - as well as two notorious UFO "contactees."
http://conspiracy101.com/ufos/skunkworks/index.html

Some "Skunks" in the Skunk Works

Beginning in January 1952, former Lockheed aircraft mechanic **George Van Tassel** experienced a series of mental communications with a retinue of extraterrestrials who, he said, were based on a spacecraft hovering 80,000 feet over Giant Rock, a remote desert airfield near Yucca Valley, California. Van Tassel produced a book later that year, and by early 1954 he was hosting public ET "channeling sessions" which evolved into the famous annual **Giant Rock** saucer conventions. In August 1953, one of the extraterrestrials finally landed and gave Van Tassel a tour of his saucer. **Clark, Jerome, *The UFOEncyclopedia*, Volume 2 - entries for Angelucci, Orfeo Matthew and Van Tassel, George W**

Orfeo Angelucci, a laborer who built fiberglass nose radar domes for the F-94, was another Lockheed employee who claimed contact with UFO occupants. One evening during May 1953, as he told it later, Angelucci was driving home from Burbank when he experienced an encounter with a red, oval object. The UFO deployed a screen-like construct that displayed images of majestic humanoid beings. A few weeks later he entered a landed UFO which took him a thousand miles into space, where he was shown a thousand-foot "mothership" and given a tutorial on the intentions of the aliens by a disembodied voice. In September, Angelucci went on a week-long "psychic journey" to the homeworld of the humanoid beings.
The alien leader's name was Orion, and he told Angelucci that his own true name was Neptune - both being names, oddly, of Lockheed airplanes. Angelucci's religiously-tinged story was so distinctive that legendary psychologist **Carl Jung** devoted considerable attention to it in his 1959 book "Flying Saucers: A Modern Myth Of Things Seen In The Skies."
Jung, Carl G, *Flying Saucers: A Modern Myth of Things Seen in the Skies*. Princeton University Press, 1978

Around the time Van Tassel and Angelucci were reporting their tales of encounters with astronauts from other worlds, Lockheed propulsion expert **Nathan Price** began to devote a considerable amount of effort to the creation of two fascinating concepts for disc-shaped aircraft of extreme performance. Price's place in the history of turbojet engine development in the US is

594

seldom highlighted, but given that he was working on an extremely sophisticated jet power plant as early as 1938; his visionary disc aircraft designs are noteworthy.

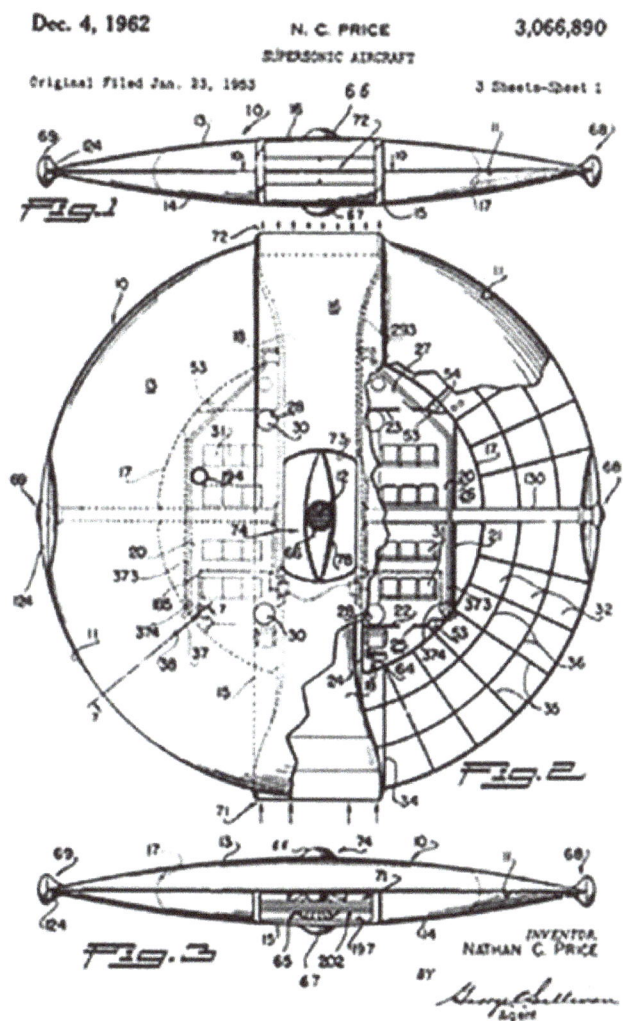

Lockheed propulsion expert Nathan Price's supersonic "saucer" aircraft design
http://conspiracy101.com/ufos/skunkworks/

As advanced as these ideas were, anticipating in some respects the engine systems of the famous "Blackbirds" that flew ten years later, perhaps the most striking aspect of Price's saucer is that it was intended to use propane, butane or liquid hydrogen fuels, and the patents cover some of the sophisticated engineering details and special tank technologies that cryogenic fuels require. Standard histories of Lockheed's work on hydrogen-powered aircraft indicate that the Skunk Works only attempted to actually build such components under the highly-classified Air Force-sponsored **SUNTAN reconnaissance aircraft project** circa 1956-7. Engine designer Price was thinking years ahead with his saucer, and given the amount of airframe detail that the patents show, it's tempting to speculate that the designs, like the L-133 of a decade before, were the result of more than his work alone.

While Kelly Johnson and other Lockheed engineers have long been rumored to have been interested in UFOs, the Blue Book sighting file on the **Lockheed UFO Incident**, as will be seen, provides definitive proof.

The following day, Kelly Johnson had returned to work and was discussing the WV-2 test flight with Thoren, who was still ruminating on the incident. A bit worried that Johnson would ridicule him; the pilot casually mentioned the sighting. Thoren was surprised when Johnson excitedly interrupted him and described his own sighting in detail. Both concluded that all the witnesses had been viewing the same object at the same time. Over the course of the next few weeks each of the pilots wrote a detailed personal account of the case, probably at Johnson's urging, and the Chief Engineer, in his typical meticulous style, assembled them into a file (Lockheed file LAC/149536) and drafted a personal cover letter addressed to the **"Air Force Investigation Group on Flying Saucers"** at Wright Field. Johnson hesitated to send the report hoping not to jeopardize his chances to work on the Air Force's new covert strategic reconnaissance aircraft competition and was very concerned that a UFO report might hurt his credibility. He may have sought the advice of his friend, **Lt General Donald Putt**.

ARDC commander Putt, who had the responsibility for reviewing and approving new Air Force reconnaissance projects like Johnson's CL-282, had also had a long, if intermittent, association with the UFO problem. A former test pilot with a solid Cal Tech engineering background, Putt was one of the stars of the Air Force, having risen to prominence as project officer for the B-29 bomber during WWII. Closely allied with Theodore von Karman's Scientific Advisory Board (he would be its military supervisor for several years), Putt had helped make science and advanced technology the backbone of the emerging air service. He had been commander of **T-2 Air Intelligence** during a crucial period following the end of hostilities in Europe and had been a strong backer of **Operation Paperclip**, the transfer of German aeronautical scientists to Wright Field. A widely-respected, politically savvy officer, Putt showed little evidence of the antipathy to UFOs that many other Air Force generals exhibited. In December 1948 **Project RAND** missile expert **James Lipp** had written a paper at Putt's request that examined the possibility that UFOs were extraterrestrial spacecraft that used propulsion principles comparable to known or foreseeable rocket technology. (The fascinating paper was published as an appendix to the final report of **Project SIGN**.)

Just months before the Lockheed UFO incident, Putt and twenty-five USAF scientists and officials had visited the plant of **Avro Aircraft Limited** in Toronto, Canada, where he had seen a full-scale engineering mockup of an advanced, supersonic saucer-like aircraft, **"Project Y,"** that had been conceived by a group of ex-British engineers.

He was impressed enough by the design that ARDC would begin to fund its development within the year. The Project Y vehicle was a virtual replica of Kenneth Arnold's initial June 1947 sketches of a "spade-shaped" UFO, probably because the Avro craft's chief designer was an avid collector of UFO data and considered Arnold's report to be reliable. The Project Y machine, which had been under development since the fall of 1951, was somewhat similar to Nathan Price's 1952-3 concept, and like the Lockheed saucer was intended to be able to take off vertically, hover in midair, and climb to extreme altitudes at incredible speeds. It seems likely that Lockheed, having originated one of the most sophisticated saucer-like aircraft concepts to

596

date, would have been aware of the Avro design prior to the UFO incident.
http://conspiracy101.com/ufos/skunkworks/index.html

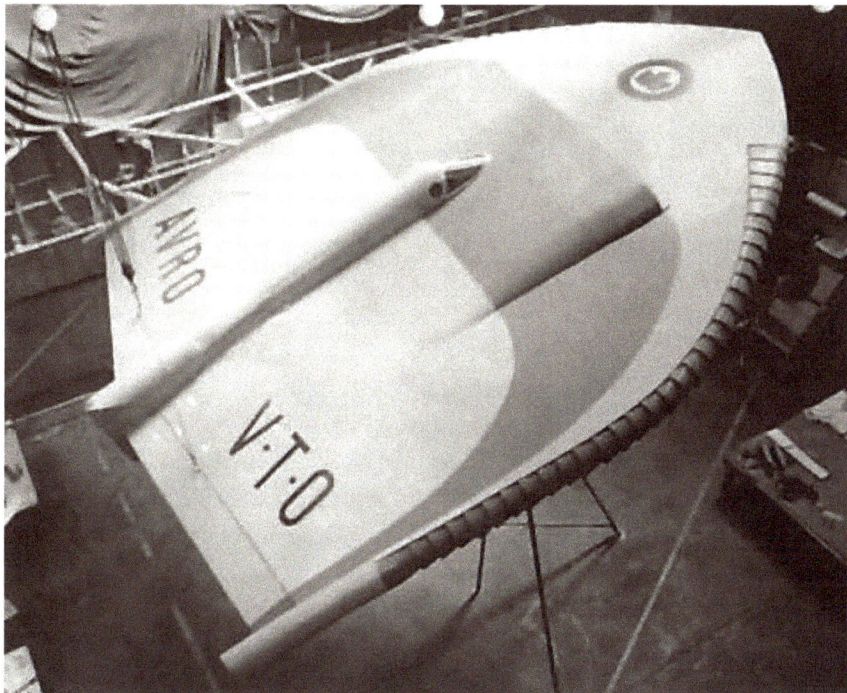

Avro Project Y mockup
http://conspiracy101.com/ufos/skunkworks/

In any case, Putt's visit to Avro was no secret. It was featured prominently in US newspapers, particularly in the *New York Times*. In two articles, one in September, at the time of the trip, and another in early October, the *Times* discussed Putt's interest in the **Avro VTOL disc**. Perhaps a more important indicator of the effect of Putt's Avro visit was **Project Blue Book Report 12**, which was issued about eleven weeks before the Lockheed UFO incident. Report 12 devoted several pages to a somewhat garbled history of the Avro project and its designer's attempts to interest the US in the concept. The report noted that the Avro engineers believed that the Soviets were responsible for many UFO reports and that Red saucers were being launched from submarines off the North American coasts on surreptitious overflights of the US. Avro, Report 12 said, wanted to install their saucer in submarines too. Zuk, Bill, *Canada's Flying Saucer: The Story of Avro Canada's Secret Projects.* **Erin, ON: Boston Mills Press, 2001**

Oddly enough, an article touting Putt's positive comments about the Avro saucer appeared in the December 16 edition of the magazine "People Today" -- the very day of the Lockheed sighting. http://conspiracy101.com/ufos/skunkworks/index.html

Putt clearly made his views on the Avro saucer known within the Air Force as well. On December 29, **Maj Gen John Samford, Air Force Director of Intelligence**, memoed **Col George L Wertenbaker, ATIC chief**, on the Avro program.

It is my understanding that you are continuing an active interest in the 'Flying Saucer' being developed by the Canadians. Also, you may have knowledge of General Putt's reaction to their program from his recent trip to that country. I would appreciate your analysis of this Canadian program. There is also an interest from both the possibility standpoint, and the time factor required by a foreign country to achieve results in this field. If you so desire, we might be able through our contacts with the Canadians here, to arrange ATIC representation during this development or phases thereof.

Kelly Johnson had close personal and professional ties with Putt and officers like him, and it's likely that sometime in early 1954 the two discussed the UFO sighting. Putt also received a copy of the formal Lockheed sighting report from the company via some unnamed intermediary. The general understood Johnson's concerns about the credibility problems a UFO report would cause and forwarded the file to Col George Wertenbaker at the **ATIC (Air Technical Intelligence Center)** at Wright-Patterson AFB in mid-February with personal cover letter.

"This report was handed to me by Lockheed personnel," Putt informed Wertenbaker, *with the explanation that Mr. Johnson was most reluctant to write the report in the first place and then refused to forward it on to you because of his belief that those who profess to have seen flying saucers are not usually considered to be logical and practical hard-headed engineers. However, I thought you should have the report for whatever value it may be in your overall studies.*

Putt's intervention would seem to have guaranteed that the incident would be taken seriously. Nevertheless, the eventual official Blue Book evaluation was that the object was just a lenticular cloud.

Although the sighting was not highly dramatic in comparison to many UFO reports, what makes it particularly unusual - and scientifically useful - is that it was a long-duration daylight sighting of an object seen by two independent groups of observers. Moreover, these groups - one stationary and one moving - were separated by very long baselines in both horizontal and vertical planes, facilitating triangulation of the object's position. One group even had the advantage of using optical instruments to view the UFO. Unlike many sensationalized UFO events, these observations were recorded in a calm, professional manner by the witnesses themselves while the details were still fresh in their minds. No media attention was sought or desired. And most importantly, the integrity and competence of this particular group of observers are almost unquestionable. A more qualified set of eyewitnesses would be hard to imagine.

Because of their precise observations and careful documentation of the incident, a great deal of information can be extracted from the account. Available weather data makes it clear that Blue Book's rubber-stamp identification of the object as a lenticular cloud is untenable...and therefore the object's identity is still a mystery. http://conspiracy101.com/ufos/skunkworks/index.html

EXCLUSIVE—AT LAST A REAL 'FLYING SAUCER'

The drawing above differs from all other "flying saucer" pictures published so far. It represents something that indisputably exists—and promises to revolutionize human flight before long. This as yet mysterious craft, actually seen so far by only a handful of leading Western scientists, generals and officials, is hidden behind tarpaulin screens in the experimental hangar of the Avro Canada plant at Marton, a suburb of Toronto. Security is so strict that when Britain's Field Marshal Montgomery, the Deputy Commander of NATO, came to look earlier this year, his Scotland Yard escorts had to wait outside. "Fantastic! I couldn't believe my eyes," Montgomery was quoted afterward.)

Lt. Gen. Donald L. Putt, head of USAF's Research and Development Command, at first denied and then admitted that he'd gone to see the top secret "saucer." Others who are looked: Canadian Prime Minister St. Laurent; Dr. O. M. Solandt, chairman of Canada's Defense Research Board ("I don't want to reveal details") and an important British Air Ministry official ("This craft is so revolutionary that if it flies everything else will be obsolete"). An executive of Avro Canada—which belongs to the Hawker Siddeley Group, largest aircraft combine in the British Commonwealth — says "our mouths are taped; we can't talk about the saucer, or even confirm that it exists." But one of the few who peeked behind the tarpaulin at Marton told PEOPLE TODAY: "I'm amazed how much the drawing looks like the real thing."

Here are as many facts about "the real thing" as can safely be revealed:

Project Y (also known as "Omega," because the craft is round with one side blunted, resembling the Greek letter) now is in the wood-metal-and-plastic full-scale mock-up stage. It measures about 40' in diameter. The design is by Englishman John Frost, formerly assistant to Sir Frank Whittle, pioneer of jet engine development. Plans call for the pilot to sit in a pressurized bubble in the center of the craft. Around him revolves the jet power plant, like a ring. It rotates several hundred times a minute, to set up a gyroscopic force stabilizing the craft at a top speed of 1,500 m.p.h. The horseshoe-shaped wing is stationary, with air intakes in front and exhausts along the sides.

Among the practical advantages for such drastic design is phenomenal maneuverability. "Omega" reportedly is planned to make U-turns without banking or losing altitude. "Omega" also is designed to take off straight up from a launching tripod, needing no runway—an important factor in the Alaskan wilds or African desert and difficult terrain generally. (The craft reportedly has no landing gear; how it lands must remain a matter of speculation.)

Most important: Rising vertically at high speed, thanks to additional rocket power, "Omega" might do away with today's biggest headache in interceptor strategy. That's how to get up high enough to engage fast bombers at very short notice, when literally every second counts.

"Omega" models reportedly will be tested soon in USAF wind tunnels. With maximum effort, a prototype could be flying within 2 years.

People Today
Dec. 16, 1953

'Omega'—The Interceptor of the Future?

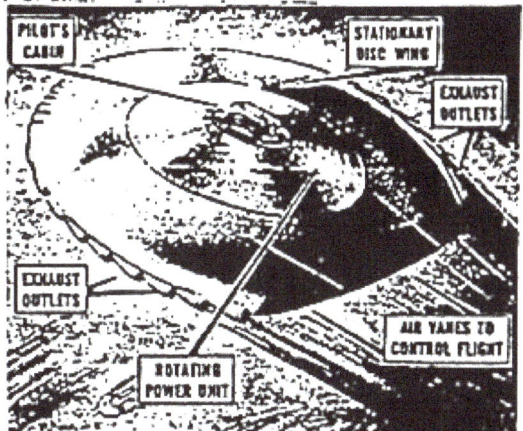

PILOT'S CABIN — STATIONARY DISC WING — EXHAUST OUTLETS — EXHAUST OUTLETS — ROTATING POWER UNIT — AIR VANES TO CONTROL FLIGHT

Interested in top secret "Project Y": USAF research chief Gen. Putt (l.), British Field Marshal Montgomery (r.) and leading Western men of science.

Avro's real "flying saucer" is designed to change direction abruptly at high speed, like the unverified saucers reported for years. Says Maj. Gen. Roger Ramey, USAF operations chief: "The Canadian project has 'mass.' What caused saucer reports so far was insubstantial, like electronic phenomena."

**News clipping of Canadian built Flying Saucer
from People Today magazine Dec. 16, 1953**
http://conspiracy101.com/ufos/skunkworks/

There was no question in about this in Kelly Johnson's view. He was absolutely certain that it was no cloud, aircraft or other mundane object.

The case file reveals a multitude of intriguing threads, including references to prior sightings by several of the engineers. Johnson furnished two signed sketches with his report. One showed the December 1953 object, a simple flattened ellipse, but the second, dated "about November 1951," shows a strange object resembling a rounded, swept-back flying wing aircraft. In his written account, Johnson elaborates:

I should also state that about two years ago Mrs. Johnson and I saw an object which I believed at the time, and still do, to be a saucer, flying west of Brents Junction, California, on a very dark night. I did not see the object itself but saw a clearly defined flame or emanation, as shown on the attached sketch. This object was travelling from east to west at a very high speed and with no noise. The flame or emanation was a beautiful light blue, having extremely well-defined edges. My first impression was that it was an afterburning airplane, but the lack of noise and the pure spread of the flame eliminated that possibility completely.
(Brents Junction is a small town just east of Johnson's Agoura ranch.)

Roy Wimmer mentioned in his report that he had seen mysterious lights over Santa Catalina Island sometime in 1951 or 1952 during a flight in Constellation 1961S, the original prototype of the Constellation line.

Joseph Ware reported that he had previously visited a group of UFO enthusiasts at Giant Rock. While he does not specify the group by name, it seems clear that this is a reference to contactee George Van Tassel and his followers. Van Tassel had been a mechanic for Howard Hughes, who was, via his airline TWA, the original customer for the Constellation airliner. Van Tassel had later worked directly for Lockheed on the Constellation program and it's almost a given that the Constellation test crew on board the WV-2 during the sighting would have known him.

For better or worse, Lockheed's contactees lurk in the background of the Johnson sighting. The likelihood is high that Johnson at least would have been aware of Van Tassel's beliefs and activities, and probably considered that ridicule aimed at the Giant Rock group could easily have spilled over onto his own sighting and onto the Skunk Works itself. This may explain the fact that his sighting report file was held back from ATIC and only reached Putt through a "back channel."

I should state that for at least five years I have definitely believed in the possibility that flying saucers exist - this in spite of a good deal of kidding from my technical associates. Having seen this particular object on December 16th, I am now more firmly convinced than ever that such devices exist, and I have some highly technical converts in this belief as of that date.

What happened circa 1948 to make Johnson a believer? Was he given intelligence briefings that convinced him? Had he spoken to reliable pilots who had had persuasive sightings? Had he heard of Project SIGN's legendary Estimate of the Situation that concluded that flying discs were extraterrestrial? Had CIA's Phil Strong used Johnson as an unofficial consultant or sounding board on UFO-related issues? [4] Since he felt that it was important to document the sighting and ultimately forward a report to the Air Force (in spite of any embarrassment it might cause himself or Lockheed), Johnson obviously was aware of the existence of an Air Force UFO investigation project, but he clearly was not intimately familiar with it, since his report is generically addressed to an "Air Force Investigating Group on Flying Saucers," rather than to Project Blue Book by name. http://conspiracy101.com/ufos/skunkworks/index.html

It is worth noting that Johnson does not specifically state that he believed in the extraterrestrial origin of these "devices." But it seems clear that he regarded them as vehicles of some type, and certainly their performance was beyond that of any known or foreseeable manmade craft - even Avro's or Lockheed's own super-saucer concept.

Johnson's formidable political and engineering skills overcame any negative effects the sighting may have had on his reputation, and even though the Air Force did reject his modified XF-104 reconnaissance aircraft concept, he was victorious in the end when, in late 1954, he was awarded a contract for a developed version of the CL-282 by the CIA. Code named AQUATONE, but known to Johnson's Skunk Works design group as the "Angel," the glider-like jet (later designated "U-2") would begin test flights at Nevada Test Site Area 51 in August 1955. Soon, according to the official CIA history of the AQUATONE program, the U-2's gleaming sunlit surfaces would begin generating dozens of high-altitude UFO sightings, particularly among airline pilots on the Chicago-Los Angeles routes. The CIA's history asserts that the U-2 itself was thus responsible for "over half of all UFO reports" of the 1950s. http://conspiracy101.com/ufos/skunkworks/index.html

Ben Rich - Shunk Works' Heir - Apparent to Kelly Johnson

Benjamin Robert Rich (June 18, 1925 – January 5, 1995) was the second director of Lockheed's Skunk Works from 1975 to 1991, succeeding its founder, Kelly Johnson. Regarded as the **"father of stealth,"** Ben Rich was responsible for leading the development of the F-117, the first production stealth aircraft. He also worked on the F-104, U-2, SR-71, A-12, and F-22 among others, many of which are still classified. http://en.wikipedia.org/wiki/Ben_Rich

Ben Rich **Lockheed Skunk Works** CEO had admitted in his Deathbed Confession that Extraterrestrial UFO visitors are real and the U.S. Military travels among stars.

According to an article published in May 2010 issue of the *Mufon UFO Journal* …Ben Rich, the "Father of the Stealth Fighter-Bomber" and former head of Lockheed Skunk Works, had once let out information about *Extraterrestrial UFO Visitors Are Real And U.S. Military Travel To Stars* http://www.realufos.net/2010/09/lockheed-skunk-works-chief-ben-rich.html

What he said might be new to many people today, but he revealed the information before his death in January 1995. His statements helped to give credence to reports that the U.S. military has been flying vehicles that mimic alien craft.

The article was written by **Tom Keller**, an aerospace engineer who has worked as a computer systems analyst for NASA's **Jet Propulsion Laboratory**.

601

1. *"Inside the Skunk Works (Lockheed's secret research and development entity), we were a small, intensely cohesive group consisting of about fifty veteran engineers and designers and a hundred or so expert machinists and shop workers. Our forte was building technologically advanced airplanes of small number and of high class for highly secret missions."*

2. *"We already have the means to travel among the stars, but these technologies are locked up in black projects, and it would take an act of God to ever get them out to benefit humanity. Anything you can imagine, we already know how to do."*

3. *"We now have the technology to take ET home. No, it won't take someone's lifetime to do it. There is an error in the equations. We know what it is. We now have the capability to travel to the stars. First, you have to understand that we will not get to the stars using chemical propulsion. Second, we have to devise a new propulsion technology. What we have to do is find out where Einstein went wrong."*

4. When Rich was asked at a conference for aerospace engineers by **Jan Harzan** how UFO propulsion worked, Ben Rich replied, *"Let me ask you. How does ESP work?"* Harzan responded off the top of his head with, ***"All points in time and space are connected?"*** Rich then said, *"That's how it works!"* https://www.youtube.com/watch?v=u9ZZekWMiUQ

Ben Rich Lockheed Former CEO knew of extraterrestrial UFO visitors
http://www.lockheedmartin.ca/us/100years/stories/rich.html

Author's Rant: Jan Harzan is one of Dr. Steven Greer's Disclosure Witnesses and is one of the main UFO investigators interviewed in the film documentary *"Sirius"*. Harzan has a

B.S. in Nuclear Engineering and also on the Board of Directors of MUFON. This author had the opportunity to meet and talk to Jan and his wife at the premiere opening of the movie *Sirius* in Los Angeles on April 22, 2013.

Lockheed "Skunk Works" former CEO knew the Roswell extraterrestrial UFO influenced designs of **Testor model kits** for Roswell UFO models, and U.S. top secret aircraft. According to a CNI News report by Colorado resident Michael Lindemann, the design information was derived from forensic illustrations and numerous witness testimonies about the Roswell UFO, provided by William L. "Bill" McDonald.

In an e-mail, dated July 29, 1999, apparently addressed to Lindemann, McDonald referenced an excerpt of a discussion with Harold Puthoff, founder of the highly classified U.S. "remote viewing" program. http://www.ufo-blogger.com/2010/08/ufo-are-real-ben-rich-lockheed-skunk.html

McDonald said: "Well Hal, you asked for it! Now that legendary Lockheed engineer and chief model kit designer for the **Testor Corporation**, **John Andrews**, is dead, I can announce that he personally confirmed the design connection between the Roswell Spacecraft and the Lockheed Martin **Unmanned Combat Air Vehicles (UCAVs),** spyplanes, Joint Strike Fighters, and Space Shuttles".

Andrews was a close personal friend of "Skunk Works" CEO Ben Rich — the hand-picked successor of Skunk Works founder Kelly Johnson and the man famous for the F-117 Nighthawk "Stealth" fighter, its "half-pint" prototype the "HAVE BLUE", and the top-secret F-19 Stealth Interceptor. Before Rich died of cancer, Andrews took my questions to him.

Dr. Ben R. Rich former Lockheed Skunk *Works* CEO confirmed:

1.   *There are 2 types of UFOs — the ones we build and ones 'they' build. We learned from both crash retrievals and actual "hand-me-downs." The Government knew and until 1969 took an active hand in the administration of that information. After a 1969 Nixon "purge", administration was handled by an international board of directors in the private sector...*

2.   *Nearly all "biomorphic" aerospace designs were inspired by the Roswell spacecraft — from Kelly's SR-71 Blackbird onward to today's drones, UCAVs, and aerospace craft...*

3.   *It was Ben Rich's opinion that the public should not be told [about UFOs and extraterrestrials]. He believed they could not handle the truth — ever. Only in the last months of his decline did he begin to feel that the "international corporate board of directors" dealing with the "Subject" could represent a bigger problem to citizens' personal freedoms under the United States Constitution than the presence of off-world visitors themselves."*

Lindemann added that "Bill McDonald received the above information from Andrews from 1994 until their last phone call near Christmas in 1998." Lindemann also noted "It should also be known that Dr. Ben R. Rich attended a public aerospace designers and engineers conference in 1993 before his illness overwhelmed him in which he stated — in the presence of MUFON

Orange County Section Director Jan Harzan and many others that – 'We' (i.e., the U.S. aerospace community/military industrial complex) had in its possession the technology to "take us to the stars". http://www.ufo-blogger.com/2010/08/ufo-are-real-ben-rich-lockheed-skunk.html

In the 1980s **John Lear**, son of **William Lear Sr.** an aviation Legend, who owned Lear Electronics Company which built the crucial autopilot for the AQUATONE (also known as the U2), started promoting stories concerning the existence of alien spacecraft being stored, or even flown, at Area 51. Some of the early versions of these tales claimed that these UFOs were being operated by Lockheed under the code name **REDLIGHT**.

And in fact, in the 1970s, when the Skunk Works, under its new head, Ben Rich, began studying methods of drastically reducing the radar cross-section of aircraft, its engineers had toyed with saucer-shaped designs. Stymied by the non-aerodynamic faceted surfaces dictated by the electrical engineers who were responsible for finding ways to scatter radar waves from the airplane's skin, some Skunk Works oldtimers pondered even more exotic forms.

"Several of our aerodynamics experts," Rich recalled, *"including Dick Cantrell, seriously thought that maybe we would do better trying to build an actual flying saucer. The shape itself was the ultimate in low observability. The problem was finding ways to make a saucer fly. Unlike our plates, it would have to be rotated and spun. But how? The Martians wouldn't tell us".*

Disclosure Project witness and Lockheed Skunkworks Engineer USAF and **CIA Contractor, Don Phillips** Admitted: *"UFO Are Real!"*

Don Phillips: *"These UFOs were huge and they would just come to a stop and do a 60 degree, 45 degree, 10 degree turn, and then immediately reverse this action".*

During the Apollo landing, **Neil Armstrong** says, ***"They're here. They are right over there and looking at the size of those ships, it is obvious they don't like us being here".*** *When I was working with the Skunkworks with Kelly Johnson, we signed an agreement with the government to keep very quiet about this.*

Anti-gravitational research was going on. We know that there were some captured craft from 1947 in Roswell, they were real. And, yes, we really did get some technology from them. And, yes, we really did put it to work. We knew each other from what we call an unseen industry. We can term it black, deep black, or hidden.

The knowledge I have of these technologies came from the craft that were captured here. I didn`t see the craft, nor did I see the bodies, but I certainly know some of the people that did. There was no question that there were beings from outside the planet.

Are these ET people hostile? Well, if they were hostile, with their weaponry they could have destroyed us a long time ago. We got these things that are handhold scanners that scan the body and determine what the condition is. We can also treat from the same scanner.

604

I can tell you personally that we've been working on them. And we have ones that can diagnose and cure cancer. One of the purposes I had for founding my technology corporation in 1998 was to bring forth these technologies that can clean the air and can help get rid of the toxins, and help reduce the need for so much fossil fuel. Yes, it is time. I can tell you personally that it has already started. pp. 375, 383. **Disclosure: Military and Government Witnesses Reveal the Greatest Secrets in Modern History by Steven M. Greer, MD; 2001; Crossing Point, Inc. Publications; Carden Jennings Publishing Co.; Charlottesville, VA.; ISBN0-9673238-1-9**

There are some interesting comments that were made by or at least attributed to the late Ben Rich that indicate that the US has technology that is both man-made and alien which can travel to the stars and "take ET back home". Some of these were first stated publically by Dr. Greer during his Disclosure Event in Washington, D.C. in 2001. Since then, other statements attributed to Ben Rich have surfaced on the internet and are that have been 100% verified and authenticated by aircraft historian, **Michael Schratt**.

"We did the F-104, C-130, U2, SR-71, F-117 and many other programs that I can't talk about. We are still working very hard, I just can't tell you what we are doing".
(Source: 1993 WPAFB slide presentation)

"The Air Force has just given us a contract to take ET back home".
(Source: 1993 WPAFB slide presentation)

"We already have the means to travel among the stars, but these technologies are locked up in black projects and it would take an act of GOD to ever get them out to benefit humanity".
(Source: statement made after UCLA presentation to three Disclosure Project witnesses)

"We now have the technology to take ET back home".
(Source: UCLA School of Engineering Alumni speech 3/23/93)

"We now know how to travel to the stars".
(Source: UCLA School of Engineering Alumni speech 3/23/93)

"There is an error in the equations, and we have figured it out, and now know how to travel to the stars, and it won't take a lifetime to do it".
(Source: UCLA School of Engineering Alumni speech 3/23/93)

"It is time to end all secrecy on this, as it no longer poses a national security threat, and make the technology available for use in the private sector".
(Source: UCLA School of Engineering Alumni speech 3/23/93)

"There are many in the intelligence community who would like to see this stay in the black, and not see the light of day".
(Source: UCLA School of Engineering Alumni speech 3/23/93)

"Jim, we have things out in the desert that are fifty (50) years beyond what you could possibly

comprehend. If you have seen it on Star Wars or Star Trek, we've been there done that, or decided it was not worth the effort".

(Source: direct comments by **Ben Rich** to **Jim Goodall** via telephone call at the USC medical center approximately one week before Ben passed away on January 5th, 1995).
"Dear John, Yes, I'm a believer in both categories. Feel anything is possible. Many of our man-made UFO's were Unfunded Opportunities. In both categories, there are a lot of books and charlatans, be careful. Best regards, Ben Rich".
(Source: 7/21/86 letter to John Andrews (Testors model Corporation) from Ben Rich who asked Ben if he was a believer in man-made UFOs and extraterrestrial UFOs)

"We have some new things. We are not stagnating. What we are doing is updating ourselves, without advertising. There are some new programs, and there are certain things some of them 20 or 30 years old that are still breakthroughs and appropriate to keep quiet about. Other people don't have them yet".
(Source: statement made by Ben Rich to Stuart F. Brown in an interview published in Popular Science October 1994)

"I wish I could tell you about the projects we are currently working on. They are both fascinating and fantastic. They call for technologies once only dreamed of by science fiction writers".
(Source: AIAA lecture Atlanta, Ga. September 7-9 1988)

Colonel Philip J. Corso

One of the most important milestones in the UFO disclosure process began with the publishing of the book "The Day After Roswell" by **Lt. Col. Philip J. Corso** (ret), with (co-author) UFO Magazine publisher - **William Birnes**. The allegations in this book rocked the foundations of the cover-up by revealing how Col. Corso, acting on behalf of the US government, directly oversaw the harvesting and reverse engineering of crashed spacecraft of ET origin, which led to the development of laser technology, fiber optics, polymers and anti-gravity technology. Lt. Col. Corso was a decorated officer who believed the public should know the truth about the extraterrestrial reality and not long before his death, he documented his experiences with ET and **Above Top Secret Technology,** for future generations.
http://www.youtube.com/watch?v=58xEbRI3UKI

Corso revealed the extent to which the US Military reverse engineered the artifacts of alien technology from the Roswell saucer crash. Besides the indirect developments of lasers and fibre optics, there was the R&D on night vision technology, the integrated circuit chip, Kevlar material, as well as accelerated particle beam devices.

Because of cold war tension between the UDA and Russia proposed the development of space-based weapons like the **Strategic Defense Initiative (SDI), or "Star Wars",** that was meant to achieve the destructive capacity of electronic guidance systems in incoming enemy warheads, as well as the disabling of enemy spacecraft, including targeting of extraterrestrial spacecraft.

Another proposal that Corso recommended to the Pentagon was establishing a permanent presence or military bases on the moon which would have nuclear capability.

It is strongly believed by some people in the UFO community that such bases both US Military and alien exist on the dark side of the Moon and perhaps on the Earth side near the lunar terminator.

Col. Philip J. Corso in the US Military and Corso in Rome, Italy in 1997.
https://en.wikipedia.org/wiki/Philip_J._Corso and http://www.openminds.tv/the-day-col-corso-died/4315
(image credit: Paola Harris)

CHAPTER 53

ARE THE LUCRATIVE BLACK PROJECTS AND PROGRAMS THE REAL CAUSE FOR THE COLLAPSE OF THE AMERICAN ECONOMY?

Unlike the Germans whose scientists stated on numerous occasions that they had help from Extraterrestrial beings during the Second World War, the Americans have never made such claims for its advanced aircraft development yet, it seems that they too have made some quantum leaps in aeronautics that seem beyond human capability. If we accept that the Roswell Saucer Incident did happen as it is reported in the annals of Ufology by people like the late Col. Phillip Corso then, US aeronautic developments did not happen as a matter of steady evolvement but had a major leap forward from Extraterrestrial influence, namely reversed engineering from captured ET technology.

If we look at the aeronautical development of Nazi Germany and at some of the American aeronautic and space developments, we could classify such advances in aeronautical engineering as Top Secret military programs in the **"White World of Science"** due to the exigencies of the time like a world war.

Robotic unmanned drones, delta or triangular aircraft, exotic aircraft like the X- 36, the ramjet Hyper X, the B2 stealth bomber, the F-119A "Specter", the F-117 "Nighthawk" stealth fighter, The Tacit Blue "Bird of Prey" and the granddaddy of the them all the SR-71 "Blackbird" are classic examples of **"Black World Science"**. The Lockheed SR-71 is a black project aircraft developed in the 1960s by the Lockheed **"Skunk Works"** one of many highly advanced aircraft to come out of this mysterious arm of the Lockheed company. The **SR-71 "Blackbird"** was an advanced, long-range, Mach 3+ strategic A-12 reconnaissance aircraft. It held the world records for being the fastest and highest flying jet for many decades.

Top Secret and Black program are always kept hidden and suppressed, away from public knowledge, until they become revealed usually during wartime or they simply become outdated in which case, they are downgraded into the public consciousness, because they have been replaced by more advanced aircraft. These black project aircraft may also have been observed over a period of years by the general public and when they are finally revealed by the military, it doesn't really come as a surprise to Ufologists or those who have an interest in exotic aircraft. In some ways, it's a vindication to the powers of observation by the public which in the opinion of this writer should never be underestimated nor ignored. Rarely does a well-trained professional observer ever makes an initial discovery of an unusual circumstance or event, it is historically made by an individual of common background or even by someone with low institutional education but, highly aware of his environment.

Black projects are never intended to remain secret or in covert programs forever, they have a "best before date" stamped on them which is usually determined by the existing political and military exigencies of the day. Of course, if the private industrial sector happens to gain possession of alien technology from the military and a quantum leap in knowledge results in advancing aeronautics or space based terrestrial technology then, these black project aircraft may see a longer service life than usual.

Funding top secret and black military projects comes annually from the national budget which increases significantly each year. Military budgets are a necessary financial expenditure in order to keep a nation secure from its enemies, both within and outside of its borders and to provide protection to its foreign interests and policies. It is a part of doing business, of maintaining a government and its military, paid wholly by the taxpayers of that country. Military budgets are controlled and regulated with no real unexpected cost overruns from mysterious projects or programs that cannot be accounted for.

Currently, the world's five largest military spenders (United States, China, Russia, the United Kingdom and France) are recognized to be world powers with each being a *veto-wielding permanent member* of the **United Nations Security Council**.

The first list is based on the **Stockholm International Peace Research Institute (SIPRI)** Yearbook 2012 which includes a list of the world's top 15 military spenders in 2011 at current market exchange rates. The second list is based on the SIPRI military expenditure database for the year 2011 (in constant 2009).
http://en.wikipedia.org/wiki/List_of_countries_by_military_expenditures

Rank	Country	Spending ($ Bn.)	% of GDP	World Share (%)	Spending ($ Bn. PPP)
—	**World Total**	**1,735**	**2.5**	**100**	1562.3
1	United States	711.0	4.7	41	711
2	China[y]	143.0	2.0	8.2	228
3	Russia	71.9	3.9	4.1	93.7
4	United Kingdom	62.7	2.6	3.6	57.5
5	France	62.5	2.3	3.6	50.1
6	Japan	59.3	1.0	3.4	44.7
7	Saudi Arabia	48.2	8.7	2.8	58.8
8	India	46.8	2.5	2.7	112
9	Germany	46.7	1.3	2.7	40.4
10	Brazil	35.4	1.5	2.0	33.8
11	Italy	34.5	1.6	2.0	28.5
12	South Korea	30.8	2.7	1.8	42.1
13	Australia	26.7	1.8	1.5	16.6
14	Canada	24.7	1.4	1.4	19.9
15	Turkey	17.9	2.3	1.0	25.2

SIPRI Yearbook 2012 - World's top 15 military spenders
https://en.wikipedia.org/wiki/List_of_countries_by_military_expenditures **(for the year 2012)**

The United States spends approximately $700-billion annually on its military, dwarfing the defense spending of any other country. It also currently faces a roughly $1-trillion budget deficit. Massive cuts to the defense budget will be a major blow to the job market; note that's to the job market, not to national security. That's because of two things. One is that the US military budget is as large as the next 17 national military budgets added together. No army, air force or navy could begin to contemplate an invasion of the United States. Second is that the past three wars have proven that *no army, no matter how well equipped or trained, can invade a nation with a hostile populace and establish American-style democracy*.
http://www.seacoastonline.com/articles/20120811-OPINION-208110312

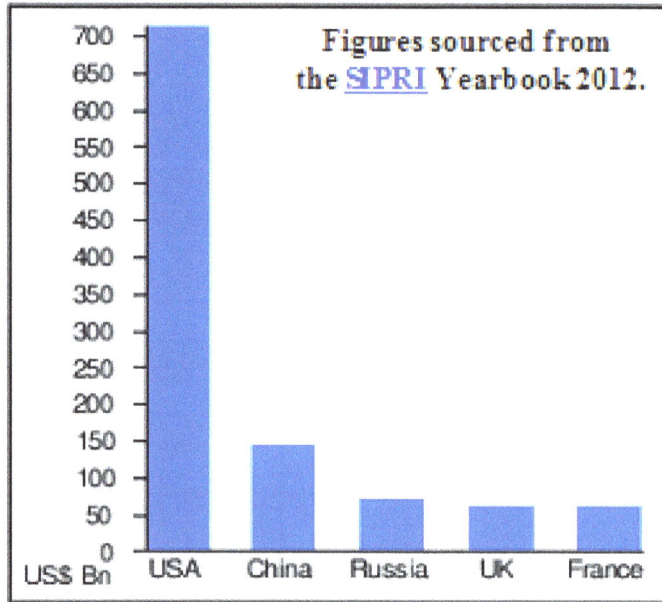

The world's top 5 military spenders in 2011
http://www.businessinsider.com/we-should-cut-military-spending-gradually-2013-2

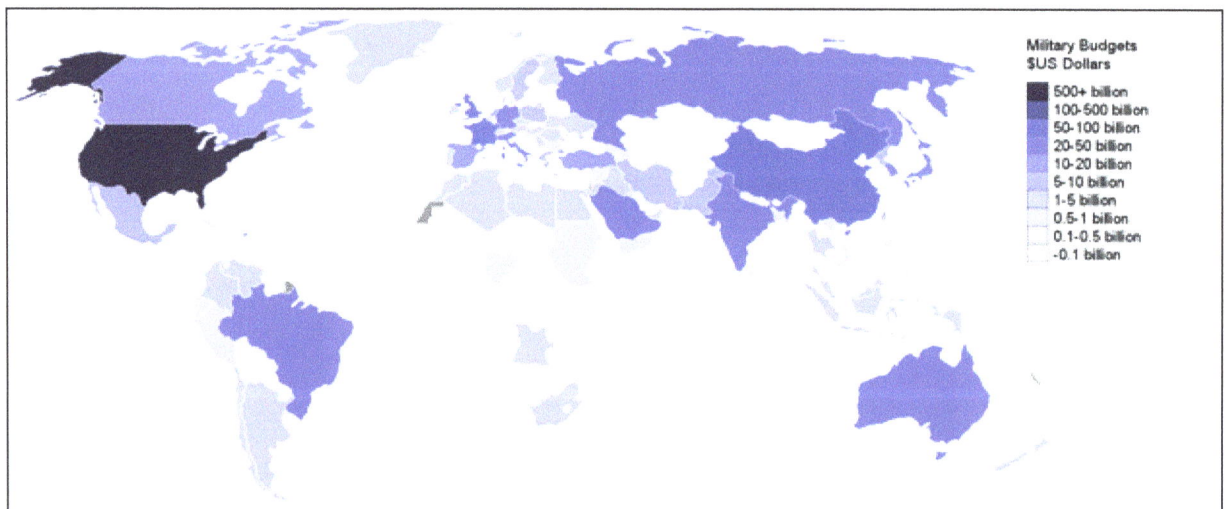

Military Spending
http://www.businessinsider.com/we-should-cut-military-spending-gradually-2013-2

610

Top 10 Defence Budgets 2011 US$bn

United States — 739.3
China — 89.8
United Kingdom — 62.7
France — 58.8
Japan — 58.4
Russia — 52.7
Saudi Arabia — 46.2
Germany — 44.2
India — 37.3
Brazil — 36.6

Other top 10 countries | United States

2011 Top 10 Defence Budgets as a % of GDP*

Saudi Arabia — 8.26%
Oman — 6.42%
Israel — 5.99%
Yemen — 5.50%
United States — 4.91%
Jordan — 4.82%
Algeria — 4.47%
Iraq — 4.46%
Myanmar — 4.42%
Armenia — 3.77%

*Analysis only includes countries for which sufficient comparable data is available. Notable exceptions include Cuba, Eritrea and North Korea.

Planned Global Defence Expenditure by Region 2011†

Latin America and the Caribbean 4.1%
Sub-Saharan Africa 1.0%
Middle East and North Africa 7.9%
North America 47.0%
Asia and Australasia 18.5%
Russia 3.3%
Europe 18.3%

†Figures may not sum due to rounding effects

Planned Defence Expenditure by Country 2011†

Other Latin America and the Caribbean 1.8%
Other Middle East and North Africa 5.0%
Brazil 2.3%
Sub-Saharan Africa 1.0%
Saudi Arabia 2.9%
Russia 3.3%
Other Asia and Australasia 7.0%
India 2.3%
Japan 3.6%
China 5.5%
Non-NATO Europe 1.6%
Other NATO 7.8%
Germany 2.7%
France 3.6%
United Kingdom 3.9%
United States 45.7%

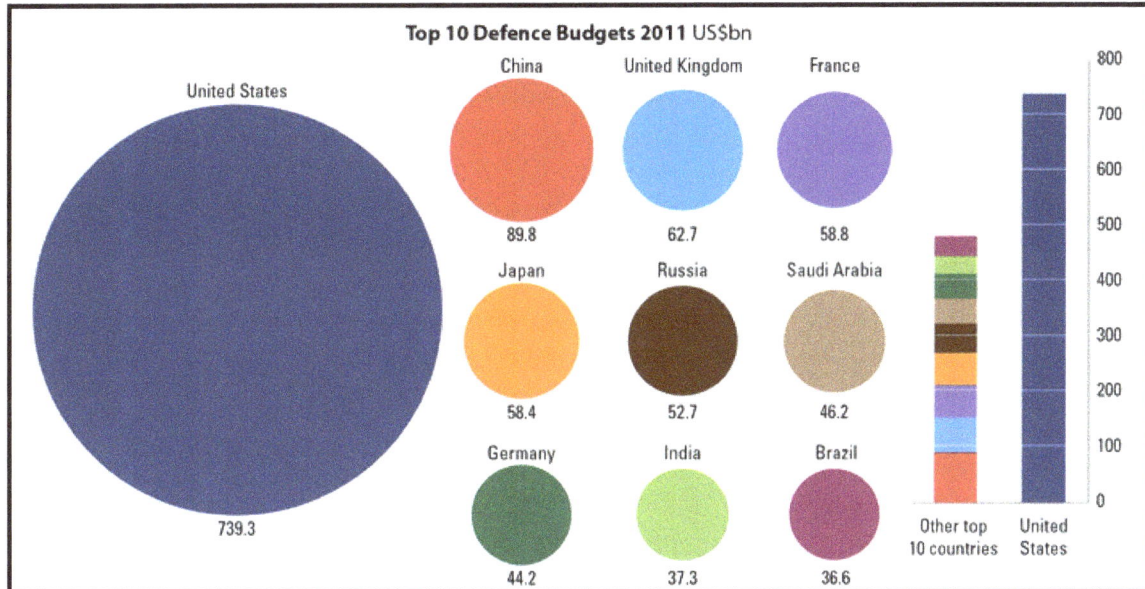

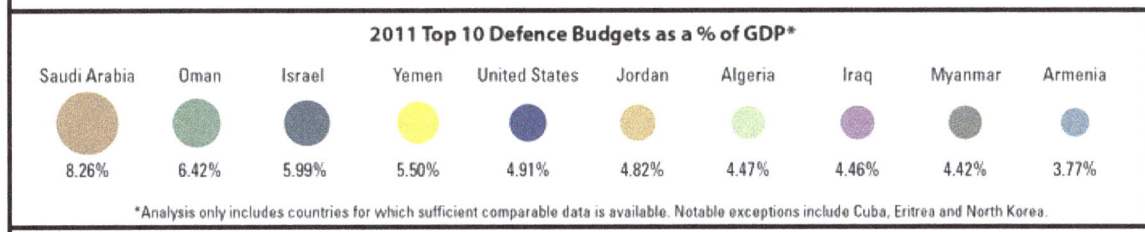

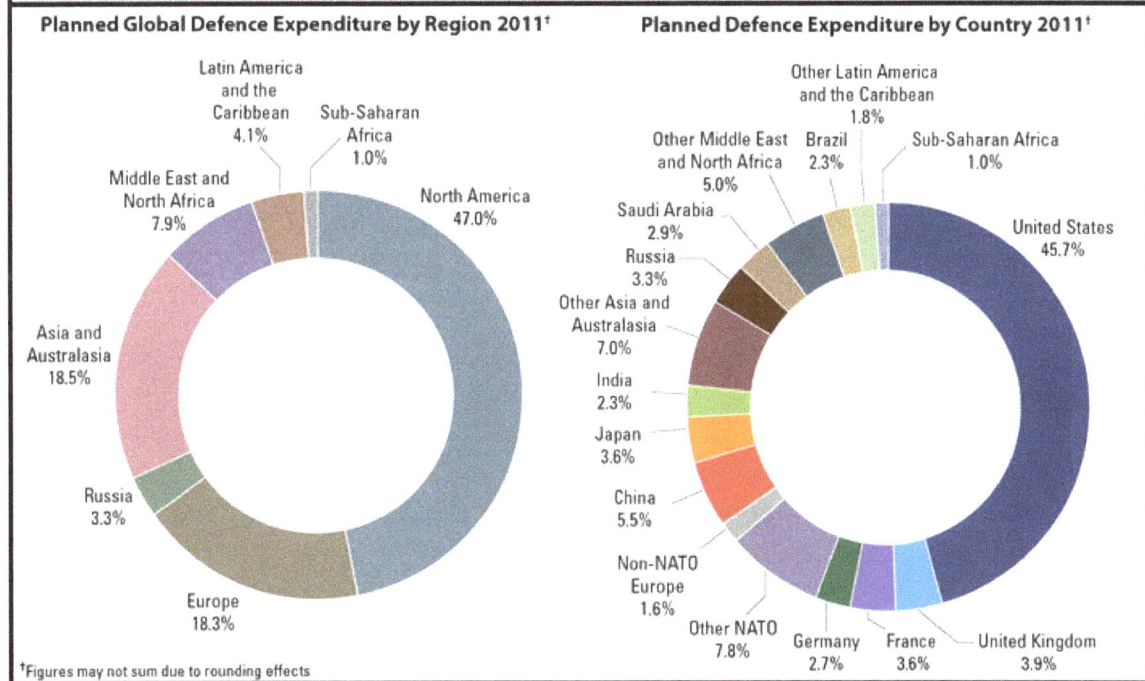

Comparative Defence Statistics - Defence Budgets and Expenditures

(Copyright ©2006 - 2012 The International Institute For Strategic Studies)

http://www.iiss.org/publications/military-balance/the-military-balance-2012/press-statement/figure-comparative-defence-statistics/

611

The trick then is how to sustain a healthy financial flow into the military coffers in order to keep the research and development programs viable and strong with finished products rolling continuously off the assembly lines of the military industrial sector, particularly those **deep black projects** which are so lucrative to many aerospace companies.

The plan is to always have an enemy more terrifying than the last one that was fought whether it is the Democracy of the USA vs. the Communists of the Soviet Union or the war on the drug cartels of Central and South American countries or the war on terrorism from some Middle Eastern nation. If those scenarios are not enough to keep the military expenditure high and continuously flowing then, create an alien threat which will raise the global alarm to financially support the military industrial sector with an even higher military budget.

If top secret and black programs in the US were considered expensive outspending the combined military budgets of the next 15 closest nations then, deep black programs are considered to be astronomically expensive by comparison, both in terms of financial costs and in terms of the out of this world type projects.

If Black Programs are kept under tight national security and secrecy from the public purview, even from competitors in the aeronautical industry then, Deep Black Programs are absolutely impenetrable as Unacknowledged Special Access Programs within Special Access Programs surrounded by walls of plausible deniability, lies, misinformation, disinformation, and obfuscation. Leaks of any kind from workers and insiders are dealt with extreme prejudice. Threats, imitations and even murder to one's own person or possibly to family and friends or associates are all means to maintain secrecy.

Work upon deep black projects is carried out in stifling conditions that allow for no communications with follower workers even if they may be working on some other aspect of the same project unless it is related to the particular task at hand. This is to prevent employees from putting pieces of the puzzle together in order to make sense of what it is that they are working on. Guards, who are close enough to breathe down your neck, are usually posted nearby to ensure that the work is carried out without the familiar chit chat.

Employees, who do undertake to work in such programs are paid exceedingly well. Million dollar plus paycheques per year is not unheard of for those who decide that working on highly advanced projects is worth the extreme secrecy and the high-stress levels needed to be endured.

The big question that remains, of course, is how all these projects are funded. The official **black budget of the DoD** would be the most likely explanation, but there are 'indications' that the U.S. economy is being plundered for at least $1 trillion every year (yes, about 10% of GNP). The days of hiding cost overruns in black military budgets with tools and equipment costing an arm and a leg are long over as the **General Accounting Office (GAO)** in the US scrupulously review all budget expenditures annually. Hammers costing $750.00 each, nails costing $2.00 to $5.00 a piece or toilet seats costing over $1000 each as portrayed in the Sci-Fi movie, *Independence Day*, although humorous in adding a light-hearted moment in the movie is nevertheless, in reality not far from the truth.

612

Following the money trail would often lead to the company or agency that were involved in the black programs or black projects. But it has become increasingly difficult to track down black budgets from their source, the government to their final user-destination. In fact, many black programs have evolved into deep black programs that are virtually untraceable at this current time are usually farmed out to large corporations that are sometimes owned outright by the military. Various aspects of a project may be further farmed out to smaller or dummy companies that appear to have nothing to do with the program. One aspect of a black project may require *"widgets"* and the parts contract is sent to a company that makes *"bloggets"* that are totally unrelated but, the tooling for the blogget is very similar to a widget and therefore, that company gets to build the black project widget, without really knowing what it is used for.

Essentially, every covert black program is funded by taxpayer's dollars which in essence make that project, the property of the taxpaying public or at least to the government that represents them. Now, take this one step further, where the deep black projects are developed in a similar fashion to the black programs but, the costs for the these program are astronomical in proportion to black world developments. Instead of billions being spent on Black Projects, trillions of dollars are now required for building and developing projects of Deep Black Programs. Billions of dollars in cost overruns is somewhat explainable when it comes to military budgets, even the expenditure of a trillion dollars on the Iraqi War, though totally unnecessary and an extreme financial burden on the American people can nonetheless, be explained. There's no mystery there, but how do you explain lost or missing trillions of dollars annually from the American Budget, since the last couple of decades that appear to be untraceable.

According to financial expert **Catherine Austin Fitts,** this has become possible due to the introduction of acts like the 1947 National Security Act and the 1949 CIA Act (6). Large New York banks like **J.P. Morgan Chase** and defense contractors like **Lockheed Martin**, who are running the systems of all the government departments, seem to be responsible for diverting and laundering billions of dollars every day from public and other undisclosed funds.

The question then, who is using who here? Is **J.P. Morgan Chase**, the core of the American part of the Anglo-American financial empire, being used as a milch cow to fund secret projects of the most unimaginable magnitude? Or, in line with the NWO conspiracy theories, are the bankers of the **Pilgrims Society** themselves really the ones in control? Or is there some kind of mutual interest here, whereby these bankers fund the Black Projects, while technology and services from these Black Projects keeps them on top of the world? Anything is possible at this moment.

USAPs don't always have to involve the development of new cutting edge technology. In the following case, a USAP is used as to a tool to circumvent national and international humanitarian laws. **Seymour Hersh**, **2004**:

Condoleezza Rice and **Donald Rumsfeld** know what many others involved in the prisoner discussions did not -- that sometime in late 2001 or early 2002, the President had signed a top-secret finding, as required by law, authorizing the Defense Department to set up a specially recruited clandestine team of **Special Forces** operatives and others who would defy diplomatic niceties and international law and snatch -- or assassinate, if necessary -- identified 'high-value' **Al Qaeda** operatives anywhere in the world.

Equally secret *interrogation centers* would be set up in allied countries where harsh treatments were meted out, unconstrained by legal limits of public disclosure. The program was hidden inside the Defense Department as an 'unacknowledged' **special-access program, or SAP**, whose operational details were known only to a few in the **Pentagon**, the **CIA** and the **White House**."
http://www.bibliotecapleyades.net/sociopolitica/sociopol_USAP.htm

Year	Missing	Sources (all cached at the bottom of this article)
1998	$3.4 trillion	Washington Times
1999	$2.3 trillion	Congressional meeting
2000	$1.1 trillion	Congressional meeting; Insight Magazine
2001	$2.3 trillion	CBS quoting Rumsfeld
2002	$1+ trillion	San Francisco Chronicle; CBS

http://www.bibliotecapleyades.net/sociopolitica/sociopol_USAP.htm

- April 1, 1999, Washington Times, '$3,400,000,000,000 Of Taxpayers' Money Is Missing'
- November 6, 2000, Insight Magazine, 'Why Is $59 Billion Missing From HUD?'
- June 25, 2001, Insight Magazine, 'THE CABINET - Inside HUD's Financial Fiasco'
- September 3, 2001, Insight Magazine, 'Rumsfeld Inherits Financial Mess'
- April 29, 2002, Insight Magazine, 'Government Fails Fiscal Fitness Test'
- May 18, 2003, San Francisco Chronicle, 'Military waste under fire $1 trillion missing -- Bush plan targets Pentagon accounting'
- May 19, 2003, CBS, 'Pentagon Fights For (Its) Freedom'
- May 22, 2003, The Guardian, 'So much for the peace dividend: Pentagon is winning the battle for a $400bn budget'
- June 28, 2003, NPR's Morning Edition, Congressman Dennis Kucinich mentions the missing trillions.
- April 6, 2004, USA Today, 'NASA costs can't be verified, GAO report says'
- March 2005, Senate Armed Services Committee, FY 2006 Defense Dept. Budget (congresswoman Cynthia McKinney asks some hard questions)

It has been estimated that $9 to $12 Trillion dollars has gone missing from the American economy since 1998 and even that particular estimate may be low as some think it could be as high as $28 to $40 Trillion dollars! It appears that some persons or some organization are blatantly stealing from the American public tout de suite to finance some hidden covert agenda which as yet, most people are totally unaware of. It has become financial terrorism with the US

614

economy under attack and middle-class America is the target. Much like the political assassination of **President John F. Kennedy** which was a coup d'état of the US government from within, with the removal of the Chief Executive Officer; this has all those same earmarkings except this time it is a financial hostage takeover of a country, a financial coup d'état of the US!

Catherine Austin Fitts

Catherine Austin Fitts is a former Assistant Secretary of Housing under **George H.W. Bush's** presidency, she is also a whistleblower "on how the financial terrorists have deliberately imploded the US economy and transferred gargantuan amounts of wealth offshore as a means of sacrificing the American middle class. Fitts documents how trillions of dollars went missing from government coffers in the 90's and how she was personally targeted for exposing the fraud.

Fitts explains how every dollar of debt is issued to service every war, building project, and government program since the **American Revolution** up to around 2 years ago - around $12 trillion - has been doubled again in just the last 18 months alone with the bank bailouts. "We're literally witnessing the leveraged buyout of a country and that's why I call it a financial coup d'état, and that's what the bailout is for," states Fitts.

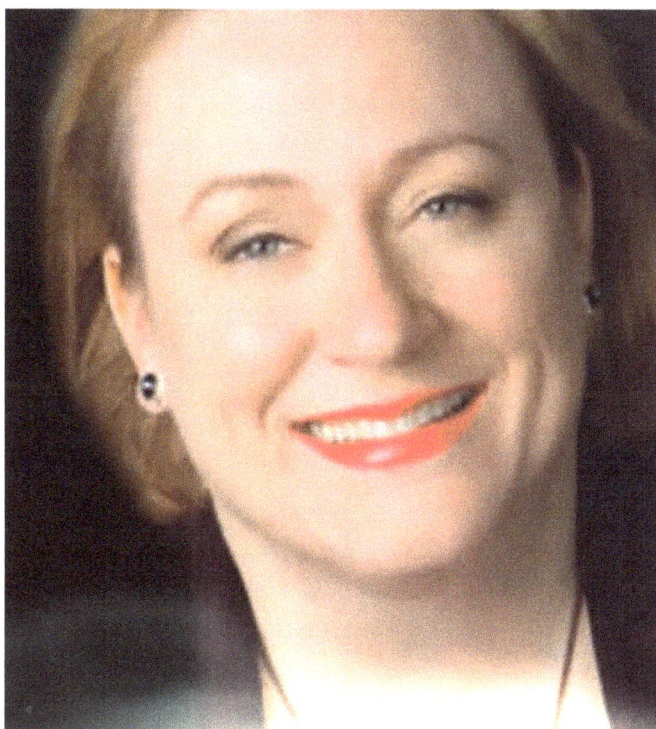

Catherine Austin Fitts Former Assistant Secretary of Housing under George H.W. Bush
http://solari.com/assets/images/press_kit/catherine_austin_fitts_700_pixels.jpg

Massive amounts of financial capital have been sucked out the United States and moved abroad, explains Fitts, ensuring that corporations have become more powerful than governments,

changing the very structure of governance on the planet and ensuring we are ruled by private corporations. Pension and social security funds have also been stolen and moved offshore, leading to the end of fiscal responsibility and sovereignty as we know it".
From Website: **TheAlexJonesChannel** on Nov 3, 2010.

Readers may recall the movie: ***Rollerball*** with James Cann as the lead actor, where large mega corporations have become the new governments of the world and the violent game of Rollerball has become the global pastime. Rollerball teams from around the world competed in international tournaments which act as distractions to prevent the public's attention being focused on financial or political issues. We seem to be on the verge of entering into that possible reality, while some people feel already that megacorporations rule the world!

On the ***Scoop Independent News*** website, **Catherine Fitts** relates how a think tank for the US Navy met with her in Arlington, Virgina and told her that there were aliens living among us and asked her to explain how the Navy could adjust its operations in light of this revelation with all political views aside.

Fitts's experience in dealing with black budget/slush fund operations include **FHA and Ginnie Mae at the Department of Housing and Urban Development (HUD), Justice, Treasury and the New York Federal Reserve Bank**. (See her article, "The Myth of the Rule of Law," **http://www.scoop.co.nz/mason/stories/HL0208/S00055.htm**)

The **Black Budget** is the 800 pound Gorilla sitting in the living room which appears to be growing in size and as impossible as it may be to ignore it and pretend that it doesn't exist, sooner or later, one must confront the 800 pound Gorilla!

For Catherine Austin Fitts it was time to confront the 800 lb. Gorilla or in her words, "the biggest crazy aunt in the living room" and ask the 64 dollar question.

The questions that need to be raised in any effort to find the truth about a significant event related to power and money in the US --- and answered if we are to ever find the truth of how our world works --- are:

1. *WHAT HAS BEEN THE SIZE OF THE U. S. BLACK BUDGET OVER THE LAST 50 YEARS?*

2. *WHAT ARE THE SOURCES AND USES OF FUNDS IN THE BLACK BUDGET?*

3. *WHO GOVERNS AND MANAGES THE BLACK BUDGET AND OPERATIONS?*

4. *WHY HAVE SO MANY DECENT PEOPLE SUPPORTED AN ILLEGAL BLACK BUDGET OPERATION OF SUCH MAGNITUDE FOR SO LONG?*

5. *IS THE WAR ON TERRORISM MOVING THE BLACK BUDGET ON BUDGET?*

Fitts doesn't know the answers to these questions but realizes that these questions raise many challenging uncertainties.

"When dealing with challenging uncertainty, I find that many researchers simply delete the existence of such uncertainty from their reality. That is a big mistake --- again pretending that we do not have a very big crazy aunt sitting in the living room. George Orwell once said that "Omission is the greatest lie." Hence, our crazy aunt has to be dealt with.

The way to deal with uncertainty is with scenario design and probability -- ie. to create a framework of the possible explanations with your best estimate of the plausible scenarios and the probability (all adding up to 100%) of each one being true based on intuition. Maintaining such a framework allows data collection on key variables and assumptions and an evolution to occur that can enlighten --- even determine --- concrete questions and answers over time.
http://www.scoop.co.nz/stories/HL0209/S00126.htm

In 1998, I was approached by **John Peterson,** head of the Arlington Institute, a small high quality military think tank in Washington, DC. I had gotten to know John through Global Business Network and had been impressed by his intelligence, effectiveness, and compassion. John asked me to help him with a high-level strategic plan Arlington was planning to undertake for the Undersecretary of the Navy.

At the time I was the target of an intense smear campaign that would lead the normal person to assume that I would be in jail shortly or worse. John explained that the Navy understood that it was all politics ---- they did not care.

I met with a group of high-level people in the military in the process --- including the Undersecretary. According to John, the purpose of the plan --- discussed in front of several military or retired military officers and former government officials--- was to help the Navy adjust their operations for a world in which it was commonly known that aliens exist and live among us.

When John explained this purpose to me, I explained that I did not know that aliens existed and lived among us. John asked me if I would like to meet some aliens. For the only time in my life, I declined an opportunity to learn about something important. I was concerned that my efforts with Arlington could boomerang and be connected with the smear campaign and the effects that I was managing. I regret that decision. At John's suggestion, I started to read books on the topic and read about 25 books over the next year on the alien question, the black budget, and alien technology.

I had to drop from the project due to the need to attend to litigation and the physical harassment and surveillance of me and some of the people helping me. This process ---which turned out to be incredibly time-consuming --- I now believe was connected with the black budget/slush fund activities connected with FHA and Ginnie Mae at HUD. (See, "The Myth of the Rule of Law") John then asked me if I would join the board of the Arlington Institute.

When I attended one of my first meetings, I joined in discussion with about 10 people which included James Woolsey, former head of the CIA in the Clinton Administration, Napier Collyns, founder of Global Business Network and former senior Shell executive, Joe Firmage, John, and other members of the Arlington board. The main topic of discussion was whether or not the major project for the coming year should be a white paper on how to help the American people adjust to aliens existing and living among us. I said nothing -- just listened. Not that long after, I dropped from the board due to the continued demands related to litigation with the Department of Justice and their informant. http://www.scoop.co.nz/stories/HL0209/S00126.htm

To cut a series of additional long stories short, when I talk with my few sources from the military and intelligence community, I hear the same themes:

1. Aliens exist and live among us;

2. In part for this reason as well the accumulated investment over the last 50 years, the technology we have access to through the black budget is far more advanced than is commonly understood;

3. The black budget/slush fund construct was created to deal with this issue, which is why reasonable people thought selling drugs to the children who were US citizens was the better of several options --- including the option of telling the American public the truth and funding the expenses on budget.

As a result of these experiences, here is my framework for dealing with this very large pile of uncertainty.

One Of The Three Following Possible Scenarios Must Be True

Scenario #1. Entre Nous: The alien question is the single largest and most expensive disinformation campaign in the history of our race. A portion of the human race has advanced technologically so far beyond the rest of us --- and is attributing various things to aliens as a way of managing their resulting risk---that we have become as aliens to each other. To fully understand this scenario, we need to try to understand the use of individual mind control such as criminal hypnosis to effect financial and government fraud and corruption and the truth, whatever it is, of wider subliminal programming and brainwashing. These are slippery subjects for most people to deal with unless they have the training and ability to do so.

% Probability Scenario #1 Is True _____ (You fill in)

Scenario #2 --- Holy Cow! Aliens exist and live among us. Planet earth is subject to the politics, economics and laws and/or lawlessness of a larger system or systems. The transfer of advanced technology into a society that has not evolved governance and legal systems to manage a world with the presence of such technology and the influence of such other system(s) helps to explain current events. To fully understand this scenario, we must also try to understand the use of individual mind control such as criminal hypnosis to effect financial and government fraud and corruption and the truth, whatever it is, of wide subliminal programming and brainwashing. Again, these are slippery subjects for most people to deal with unless they have the training and ability to do so.

% Probability Scenario #2 Is True _____ (you fill in)

618

3. Scenario #3 --- Muddled: Some combination of (1) and (2) above.

% Probability Scenario #3 Is True _____ (you fill in)

Total % Probability Of Scenarios #1, 2 & 3 **Being True 100%**

Could my experiences with Arlington and the Navy or any subsequent contacts fit with a disinformation scenario? Absolutely! I have no evidence to support any scenario. The only evidence I believe I have is my experience dealing with tremendous amounts of money siphoned off over the years, whether through Iran-Contra S&L, HUD and BCCI fraud, or the latest round of money missing from the federal government in the last five years. This is all documented in the articles below. This cash flow and the operations and syndicates it appears to fund are important to what drives the governance and allocation of resources in this country and around the globe. There is a reason that power and money are centralizing and the rich are getting richer. I want to know what it is. http://www.scoop.co.nz/stories/HL0209/S00126.htm

Whichever scenario is true ---and I do not know which one is ---- if we had the truth it will help us grapple with 9-11 and the War on Terrorism. It would also help us better understand all the efforts to press for centralization of economic and political power that have grown as the black budget has grown since WWII. The reality is the possibilities of why 9-11 occurred and how it was operationalized are impacted by the facts of the US black budget and the advanced technology it has developed. Why would sane and even decent people think it was the best option for the good of the whole?

We must sit in the shoes of the person or people who gave the order ---whoever they may be and wherever they may be ---- and understand how their power and money worked. Why would a group with the operational and financial capacity to effect a 9-11 operation give the green light to plan 9-11, finance 9-11, do 9-11, stand down so 9-11 could happen, take advantage of 9-11 or to make sure the truth was not illuminated?

My experience in Washington and Wall Street would indicate that there is a reasonable chance that all these various people and many in the food chain **Are Not Bad or Irrational People**. That is, we have a better chance of finding the truth if for purposes of our explorations we assume that the people involved in all these roles were people like you and me --- just with a different map of certain parts of the world and dealing with a variety of stress and pressures trying to build cooperation or achieve goals that we do not understand.

Watching the federal government contract announcements and without benefit of seeing the full budget and assuming The Disclosure Project allegations regarding the estimates of the black budget size, if I had to guess today my guess would be that 9-11 has been highly successful in permitting the black budget to be moved on budget, partly in the normal budget and partly in the budget that is only disclosed to the Congressional intelligence committees.

The solution to continued growth and funding of the black budget may have been to steal as much money and assets as possible from the federal budget and then to move the black budget on budget when control of media, Congress, Justice and Securities & Exchange Commission

619

ensured that the legal budget was non-transparent, controlled by private government contractors and could not be harmed or fiddled with by Congress. (See various stories on the disappearance of over $3.3 trillion from the federal government from fiscal 1998-2001). http://www.scoop.co.nz/mason/stories/HL0207/S00031.htm#a

Grossly oversimplified, by the end of the 2004 fiscal year, covert operations and the black budget may be fully legal, run and controlled by corporations and banks without interference from knowledgeable career civil service, taxpayer funded, and non-transparent.

Why do I say all of this? We do not have time to pay attention to people who use smears like "conspiracy theorist" to waste our time. Money, budgets, finance...these are all facts. If something does not work, and it continues to not work, there is a reason. Systems have a logic. They have a rationale. Money helps explain and illuminate that rationale. Anyone who would smear a serious researcher or writer for trying to understand where a great deal of money is disappearing to as a "conspiracy theorist" is someone who would never be taken seriously or associated within the circles in which I travel. The exception is if the person doing the smearing is a professional being paid to smear those who have the scent of "the real deal" --- **Cointelpro** or related public relations. The number of people paid to do such things are many and growing. http://www.scoop.co.nz/stories/HL0209/S00126.htm

9-11 was an operation that was implemented by a conspiracy. To get to the bottom of what happened and why one must develop and use conspiracy theories to research in what can only be described as a conspiracy reality. We know in regards to 9-11 one of two scenarios must be true - --- some portion of the US military and intelligence are (i) complicit and/or (ii) guilty of criminal gross negligence. We also know that the official story is not truthful. Either scenario and the reality of the official story beg the question, "why?" This course always leads us to consider the fact that the US military and intelligence have enjoyed significant expansion of their budgets and powers as a result of 9-11. We are rewarding failure. 9-11 was great for business. Lockheed Martin's stock is up over 50% in a falling stock market. The oil companies have their pipeline.

When a system does not work and continues to not work, there is something that prevents its learning. Ultimately what we must understand that the problem is not that the system does not work and that it is not learning. The problem is that the system is controlled and operated by the most powerful forces in that system and that the system is indeed working and learning in the general direction they want it to go. This happens in a manner that is highly organic and requires lots of planning, trial and error and factionalism that involves competition and cooperation over time in the pursuit of the system's real goals. Hence, the system is working and is learning. It is our construct that is wrong. We do not understand the system and the goals of the system. Once we are clear that the system is working and is learning, once we stop falling back on scapegoating as a technique to explain events, we can start to understand the real system that we are a part of. This is how power and money work in reality.

The amount of money and power spent to keep the general population from understanding this --- and focused on things like scapegoating which keep us from building a good map of "the real deal" or gaining the power to impact events by using our money and time as a vote in the marketplace in which the banks and corporations must thrive ----is immense. That is why I

620

strongly encourage reading Jon Rappoport's back interviews on his Info monster and becoming a subscriber for his ongoing interviews.

For those would like to know more, I would refer you to articles on all the money missing from the US Treasury and the manipulation of the gold markets as well as the **Disclosure Project** video of their event at the **National Press Club**. You can find them on the web. As for me, I do not have the answers. I wish I did. That is why I am so appreciative of efforts like **Unanswered Questions** (http://www.unansweredquestions.org) and the effort of researchers networking and collaborating globally to illuminate, research and answer the right questions.

Let me close with the unanswered questions with which I began and which I believe are critical to our ongoing collaboration:

1. What Has Been The Size Of The U. S. Black Budget Over The Last 50 Years?

2. What Are The Sources And Uses Of Funds In The Black Budget?

3. Who Governs And Manages The Black Budget And Operations?

4. Why Have So Many Decent People Supported An Illegal Black Budget Operation Of Such Magnitude For So Long?

5. Is The War On Terrorism Moving The Black Budget On Budget?
http://www.scoop.co.nz/stories/HL0209/S00126.htm **and**
https://www.youtube.com/watch?v=w0mimIp8mr8

The Monolithic Monster of Capitalism and the High Price of Corruption

According to a recent report by the **Tax Justice Network** (established by the British Houses of Parliament in 2003), the reason for the lack of money in the failing western economies, is that the elites are hoarding it, to the tune of $12.5 trillion to $20 trillion USD (that's according to data from the world's top 50 banks)!!! Blind Bat News; Post July 24, 2012
http://www.blindbatnews.com/2012/07/evil-elites-20-trillion-stolen-away-and-horded-by-global-elites-thats-why-the-economy-sucks-money-hording-causes-poverty-destroys-economies-is-a-threat-to-national-security-too-big-to-fai/14459

The world's richest individuals have placed $11.5 trillion of assets in offshore havens, mainly as a tax avoidance measure. Governments appear unable, or unwilling, to prevent the rich employing aggressive strategies to minimize their tax liabilities.

The **OECD** (The **Organisation for Economic Co-operation and Development**) … confirmed that international tax avoidance is a growing problem that troubles governments not just of rich countries, but middle-income ones as well.

'This is one of the defining crises of our times,' said **John Christensen**, coordinator of the **Tax Justice Network** and a former economic adviser to the Jersey government. 'One of the most

621

fundamental changes in our society in recent years is how money and the rich have become more mobile. This has resulted in the wealthy becoming less inclined to associate with normal society and feeling no obligation to pay taxes.'

Individuals such as **Rupert Murdoch, Philip Green, Lakshmi Mittal** and **Hans Rausing** - among the world's richest men - all make extensive use of tax havens.

There is nothing illegal about placing assets and cash offshore, but campaigners are promising to attack tax avoidance by the world's richest people in much the same way that they currently target environment and trade issues.

The $11.5trn does not include the vast amount of money stashed in tax havens by multinational corporations, which are using increasingly sophisticated techniques to run rings round the authorities. http://politics.guardian.co.uk/economics/story/0,11268,1446127,00.html

James Henry, former chief economist at consultancy **McKinsey** and an expert on tax havens, has compiled the most detailed estimates yet of the size of the offshore economy in a new report, "The Price of Offshore Revisited", released exclusively to the *Observer*.

He shows that at least £13tn – perhaps up to £20tn – has leaked out of scores of countries into secretive jurisdictions such as **Switzerland** and the **Cayman Islands** with the help of private banks, which vie to attract the assets of so-called high net-worth individuals. Their wealth is, as Henry puts it, "protected by a highly paid, industrious bevy of professional enablers in the private banking, legal, accounting and investment industries taking advantage of the increasingly borderless, frictionless global economy". **Heather Stewart, business editor;** guardian.co.uk, **Saturday 21 July 2012 21.00 BST** http://www.guardian.co.uk/business/2012/jul/21/global-elite-tax-offshore-economy

According to Henry's research, the top 10 private banks involved:

* **UBS**
* **Credit Suiss**
* **HSBC**
* **Deutche Bank**
* **BNP Paribas**
* **JP MorganChase**
* **Morgan Stanely/SSB**
* **Wells Fargo**
* **Goldman Sachs**
* **Pitchet**

http://www.blindbatnews.com/2012/07/evil-elites-20-trillion-stolen-away-and-horded-by-global-elites-thats-why-the-economy-sucks-money-hording-causes-poverty-destroys-economies-is-a-threat-to-national-security-too-big-to-fai/14459

Distribution of Wealth in the U.S., 2001

Top 1% own 33%
Next 4% own 26%
Next 5% own 12%

Next 10% own 13%

Next 20% own 11%

Middle 20% own 4%

Next 22% own 0.3%

Bottom 18% have zero or negative net worth

percentile

99th
95th
90th
80th
60th
40th
18th

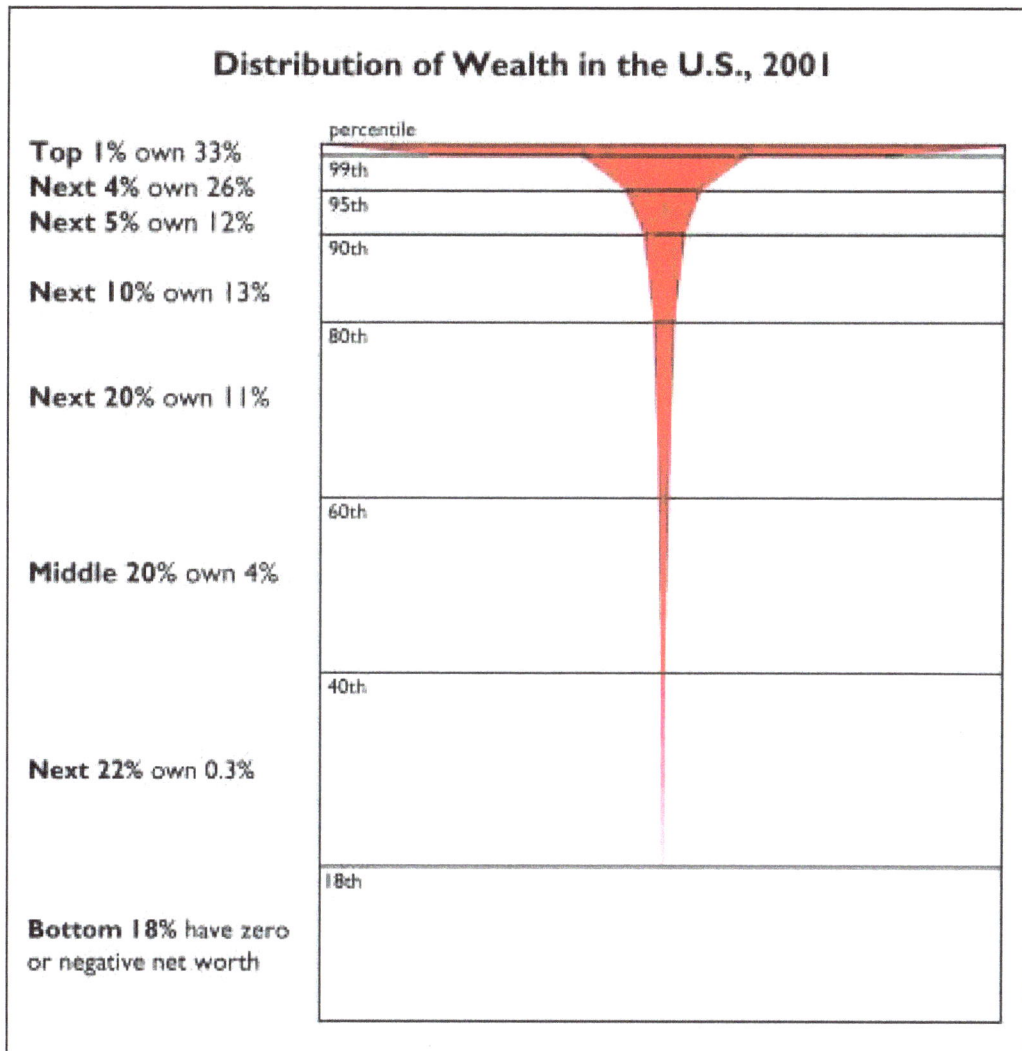

**About 75 percent of the U.S.'s wealth is owned by
the top 10 percent of the country's wealthy**
http://www.oilempire.us/trillions.html

But, the robbery doesn't stop there! It now appears that many other countries besides the US are also, missing enormous amounts of money annually, to the tune of hundreds of billions of dollars!

Oil-rich states with an internationally mobile elite have been especially prone to watching their wealth disappear into offshore bank accounts instead of being invested at home, the research suggests. Once the returns on investing the hidden assets are included, almost £500bn has left **Russia** since the early 1990s when its economy was opened up. **Saudi Arabia** has seen £197bn flood out since the mid-1970s, and **Nigeria** £196bn.

"The problem here is that the assets of these countries are held by a small number of wealthy individuals while the debts are shouldered by the ordinary people of these countries through their governments," the report says.

The sheer size of the cash pile sitting out of reach of tax authorities is so great that it suggests standard measures of inequality radically underestimate the true gap between rich and poor. According to Henry's calculations, £6.3tn of assets is owned by only 92,000 people, or 0.001% of the world's population – a tiny class of the mega-rich who have more in common with each other than those at the bottom of the income scale in their own societies.

Heather Stewart, business editor; guardian.co.uk, Saturday 21 July 2012 21.00 BST
http://www.guardian.co.uk/business/2012/jul/21/global-elite-tax-offshore-economy

India is one of the largest producers of coal in the world and has lost $33 billions of dollars by selling coalfields cheaply. One of the latest in a series of financial scandals has caused one of the opposition's MPs to call for the resignation of India's Prime Minister Manmohan Singh to quit over a recent auditor's report on the sale of coal. India's national auditor said the government lost huge sums of money by selling coal fields to private companies without competitive bidding and in a deal for Delhi's international airport, adding to massive losses from dubious auctions of other state assets. http://www.businessweek.com/ap/2012-08-17/india-auditor-says-billions-lost-in-coal-scandal

AP News; "India auditor says billions lost in coal scandal" by Nirmala George on August 17, 2012

Stolen Assets by Corrupt Political Leaders:

The wealth accumulated by Libya, Egypt, and Tunisia's leaders stands at an estimated $190 billion.
Source: TI paper based on news reports (2011)

General Sani Abacha of Nigeria is suspected to have looted between US $3 billion to US$ 5 billion of public money.
Source: Basel Institute of Governance/ICAR, "Managing Proceeds of Asset Recovery: The Case of Nigeria, Peru, The Philippines and Kazakhstan" (2009)

Ferdinand Marcos of the Philippines siphoned off between US $ 5 to 10 billion during his reign in the Philippines from 1965 to 1986.
Source: Basel Institute of Governance/ICAR, "Managing Proceeds of Asset Recovery: The Case of Nigeria, Peru, The Philippines and Kazakhstan" (2009)

The level of assets stolen by corrupt leaders and moved to offshore accounts is estimated at US $180 billion. Only US $5 billion has ever been returned.
Source: Comité Catholique contre la Faim et pour le Développement
The amount of money laundered through the UK each year is estimated to be £48 billion (2% of UK GDP).
Source: Money Laundering Bulletin, June 2011

The Cost of High Tax Evasion:

The Virgin Islands has over 40 companies registered for every citizen.
Source: **NORAD, 'Commission on Capital Flight From Developing Countries: Tax Havens and Development' (2009)**

The world's shadow economy, or the legal economic activities that go untaxed, now accounts for 17% of the world economy
Source: **University of Linz, 'Shadow Economies around the World: Novel Insights, Accepted Knowledge and New Estimates' (2011)**

Greece loses €15 billion a year to tax evasion and has a shadow economy twice the size of its budget deficit. Tax evasion has cost the USA US$3 trillion over the last decade.
Sources: **Economist, "Dues and Don'ts" (August 2010) and Demos website, 'Tax Evasion: The Real Costs'**

Assets placed by wealthy private individuals in tax havens represent an estimated annual loss of roughly USD 255 billion in tax revenues.
Source: **Tax Justice Network, 'Briefing Paper: The Price of Offshore'(2005)**
http://www.transparency.org.uk/corruption/statistics-and-quotes/stolen-assets-a-tax-evasion

These are the most infamous of the political leaders caught with their hands in their countries' cookie jars. The ones not listed are the ones who have not been caught, yet!

- **Sani Abacha**
- **Ferdinand Marcos**
- **Vladmiro Montesinos**
- **Jean-Claude "Baby Doc" Duvalier**
- **Pavlo Lazarenko**
- **Joshua Dariye**
- **Dieprieye Alamieyeseigha**
- **Hendra Rahardja**
- **Ao Man Long**

http://en.wikipedia.org/wiki/International_asset_recovery

The point being made here is that much of the planet's wealth is being placed into offshore banks like the Cayman Islands and in Switzerland by the world's wealthiest elite and has become untaxable and untouchable. Some questions need to be asked now, is there a purpose behind this money? Is there a hidden agenda being developed by the wealthy corporate elite? How is this money being used? What is being developed with this money?
The money can't be just simply being "saved for a rainy day"! It has to be for a special purpose. Tens to hundreds of millions of dollars would see to anyone's unique lifestyle needs. Hundreds of billions of dollars is enough to buy a small country or government. Tens of trillions of dollars become unimaginable in the efficacy of its usage but, one thing is certain, whatever its potential

usage, whether beneficial or for malevolence, it could have an incredible global impact on everyone!

This brings us back to the very **Deep Black Programs** which require incredible amounts of money to develop and build some kind of covert space program perhaps, to fulfill a hidden agenda in which a **False Flag alien invasion scenario** will be perpetrated against an unsuspecting humanity.

If we acknowledge the testimony of many of **Dr. Steven Greer's** witnesses from the **Disclosure Event** held at the **National Press Club** back in May 2001that there is Extraterrestrial visitations to our planet, that the US Military is targeting and shooting down ET vehicles, and are reverse engineering the captured alien technology, that the US Treasury and gold markets have been manipulated to fund that technology to develop a secret black space program, all the while hiding the evidence for the existence of Extraterrestrial visitors and their spacecraft from the public and if we accept **Catherine Austin Fitts'** insight that the US Navy acknowledges the existence of aliens living amongst us and the inescapable fact that the US military and intelligence have enjoyed significant expansion of their black budgets and powers as a result of the 9-11disaster, and if we also accept the testimony of **Dr. Carol Rosin**, one of the **Disclosure Project** witnesses who stated that **Wernher von Braun** warned her repeatedly that there is an ever-evolving missile and space development program with particular emphasis on space-based "star wars"-type weapon platforms then, by a process of logical deduction does the grossly exorbitant financial resources stolen from the global economy of many nations and particularly from the US, makes any sense. The financial resources of such deep black programs and projects with all its associated infrastructures and while kept under the most unimaginablly tightest security measures would have to be astronomical in its undertaking for an agenda that is literally out of this world!

Distillation of the evidence in this book so far points to the inescapable conclusion that many deep black programs centres around the development and construction of spacecraft and space-based weaponry utilizing reversed engineered alien technology gained from captured alien spacecraft during the 1940s and 50s. There were incredible technological breakthroughs in science prior to World War II, during and after the War in propulsion and in aircraft and spacecraft design. Most of the breakthroughs and developments occurred in Nazi Germany and then, in post-war US with captured Nazi scientists and their technology. This is indisputable fact and is now a part of history.

But, more questions are raised in order to understand and explain the thinking behind some of these deep black programs such as, where did this alien technology come from? How was it obtained? How was it utilized? What programs came out reverse engineering of the alien technology? What specifically were some of these projects? What about the programs of genetic engineering of ET biology? What associated laboratories and technology were needed to support the captured alien technology? What spin-offs in spacecraft developments were created? Is there any evidence for any of these projects? Is NASA a public front for the real **Black Space Program** being carried on behind closed doors under a lid of absolute tight national security? Is there a US military **Space Force**? Is there a planned **False Flag** scenario utilizing reversed engineered alien technology about to be unleashed upon the world? When discussing Deep Black

626

Programs and Black Projects inevitably the names of certain aerospace engineers, scientists, and military individuals come up in the discussion like **Kelly Johnson, Ben Rich, Col. Philip J. Corso, Nicola Tesla, Townsend T. Brown**, and many others who have advanced the aerospace industry beyond the public's current knowledge and understanding. Much of their work may never be acknowledged as it has either been suppressed or expunged from public records.

The Global Economy is a Reflection of the State of the Spiritual, Physical and Mental Health of Humanity

If we compare the world to the human body and the world's monetary resources to the life-blood of the human body and in turn, all the major organs to the major social, geopolitical institutes of the planet then, the human populace would, in turn, be represented as the red life-sustaining blood cells that do all the work and maintain the proper function of the major organs of that body, we would then have a very good idea of the flow of financial resources on the planet.

In a strong, vibrant, and healthy human body, blood flows constantly through the pumping action of heart to all parts of the body. Organs, limbs, muscles, bones, the nervous and respiratory systems, the brain as well as all the cells of the body receive their portion of blood flow which maintains and provides health and vigor to each of these bodily parts. Thus, the whole system is invigorated and sustained and enjoys long life! The human body takes on an outer positive countenance of health and vitality because the internal system is working as it was created to do so with marvelous efficiency!

All these inner and outer signs of health allows the human body do perform great and astonishing things to occur in movement and agility. With such health, even the mental faculties become keener, more astute and insightful into the mysteries of life thus, permitting the body of humanity to make great leaps in mental understanding that are reflected in an advanced civilization. One would say without contestation that humanity's spiritual health and higher consciousness is mirrored in his physical health!

This is the perfect state of health of the human body and of the collective body of humanity!!!

Now, consider the opposite condition to this perfect state of health. Metaphorically, in our example, the comparison of money is likened to that of the flow of blood within the human body. Initially, there may be no outward sign to indicate anything is wrong within the body. Such outward signs may be subtle at first and apparent only to the most observant but, over a period of time, the outer body starts to show a lack of physical energy and vitality, mental acuity is diminished, perhaps, due to the invasion of a spiritual malaise or disease whether from outside or within. As this disease takes hold upon the body, greater signs of ill health manifest outwardly and within indicating that something cancerously unhealthy has invaded the human body.

The body begins to fight internally with itself, each organ vying for attention and control of the life-sustaining flow of blood as they become increasingly diseased. The physical action of the limbs become weaker, mental deficiency and confusion escalate, the automatic response is for the heart to pump harder to force blood into all areas of the body, but the organs divert most of the blood flow to themselves to stay alive. The red blood cells are expending so much energy to the corrupt and diseased organs that the body begins to shut down on itself! The brain goes into a comatose state to function on a basic autonomic level but, in reality, it is dying and the diseased organs are gorged with blood but, at the expense of the rest of the body. The body has collapsed and now lays prone and lifeless. The forensic post mortem has shown that lividity has settled to the back of the body, this is the blood's final resting place! The human body is dead!!!

In our example, when the flow of financial/monetary resources fail to circulate properly through the collective commonwealth of humanity collecting instead into the pockets and offshore banks of the wealthy corporate elite, ultimately, the body of humanity dies and taking irresistibly with it, the corrupted sources of its demise, the diseased wealthy corporate elite!!! No one escapes the inevitable outcome... death. The death is the physical outcome but, it is a spiritual disease that is the cause of death!

For humanity to be in the perfect state of health, no one individual or persons or mega corporations or government can ever be in control of the planet's financial and natural resources. They are not the brains or the heart of a civilization whether national or global but, merely the organs performing one of many important functions of health for the body of humanity.

Like the flow of blood in the human body, when money does not flow through the hands everyone, through small and medium size businesses as well as large businesses within society then, ultimately, the body of humanity dies!!! Hoarding of wealth kills a society, this is ill-health and a spiritual disease. Perfect health requires neither extremes of poverty nor extremes of wealth to exist. Perfect health is a balance that must be maintained on a daily basis and never taken for granted!

Chapter 54

BEHIND THE HALL OF MIRRORS: WHAT'S REALLY GOING ON?

Groom Lake/Area 51

Groom Lake, also known as **"The Ranch," "Watertown Strip,"** **"Dreamland"**, S4 or more popularly as **"Area 51,"** *(although, no one in the know or who works there calls it by that name)* is located in Nevada about 150 miles north of Las Vegas and has the same land size area as Switzerland. Until recently, its very existence was officially denied for decades by U.S. government; unofficially, since the early 1950s, *(although the Russian military has known about it almost as long as its existence, which begs the question from whom were the Americans trying to keep it a secret from? Its own people, maybe?)* it has been the operational testing range for cutting edge technological developments of America's military. There are many test facilities, labs, radar sites and hangars with some of the longest runways in the world, all centered in and around the Groom Lake area. It has become the most secretive military testing base in the world requiring an extremely high level of security, to the point that Area 51 has developed a modern-day clandestined status of mythological proportions. (Bold italics added by author for emphasis).

S4 is separated from the more infamous flight test facility at Groom Lake, Nevada, known also as Area 51, by the Papoose Range of mountains. The site is said to be built into the western slopes of the mountains, east of and adjacent to Papoose Dry Lakebed. It is very remote and virtually inaccessible to the public. Even Congressmen attempting to visit the area are told they can't go because the area is "radioactive". So we are told…

Area 51 has become the touchstone location for all things alien, strange and exotic in the way of advanced aircraft. It is a place that could be viewed at one time in relative safety from a distance, on a small mountain peak known as Papoose Mountain which overlooked the whole Area 51 facility. Its very existence was denied for decades.

Groom Lake's non-existence has become a thing of the past when a Russian spy satellite photographed extensively the dry lake test range area and their pictures were later released into the public domain. The US military was hard pressed at that point to continue its official position of the site's non-existence.

Entry into the site is only accessible either by an authorized vehicle along a lonely dirt road that is monitored by camouflaged guards in white jeeps with binoculars as well as by electronic surveillance systems, motion sensors, military units and air support. There is also, a carefully worded warning sign indicating that any entry by trespassers beyond that point would be met with authorized deadly force. The only way into Area 51 is as an employee in a privately contracted airline owned by **EG &G** on the northwest side of McCarran Airport in Las Vegas, Nevada.

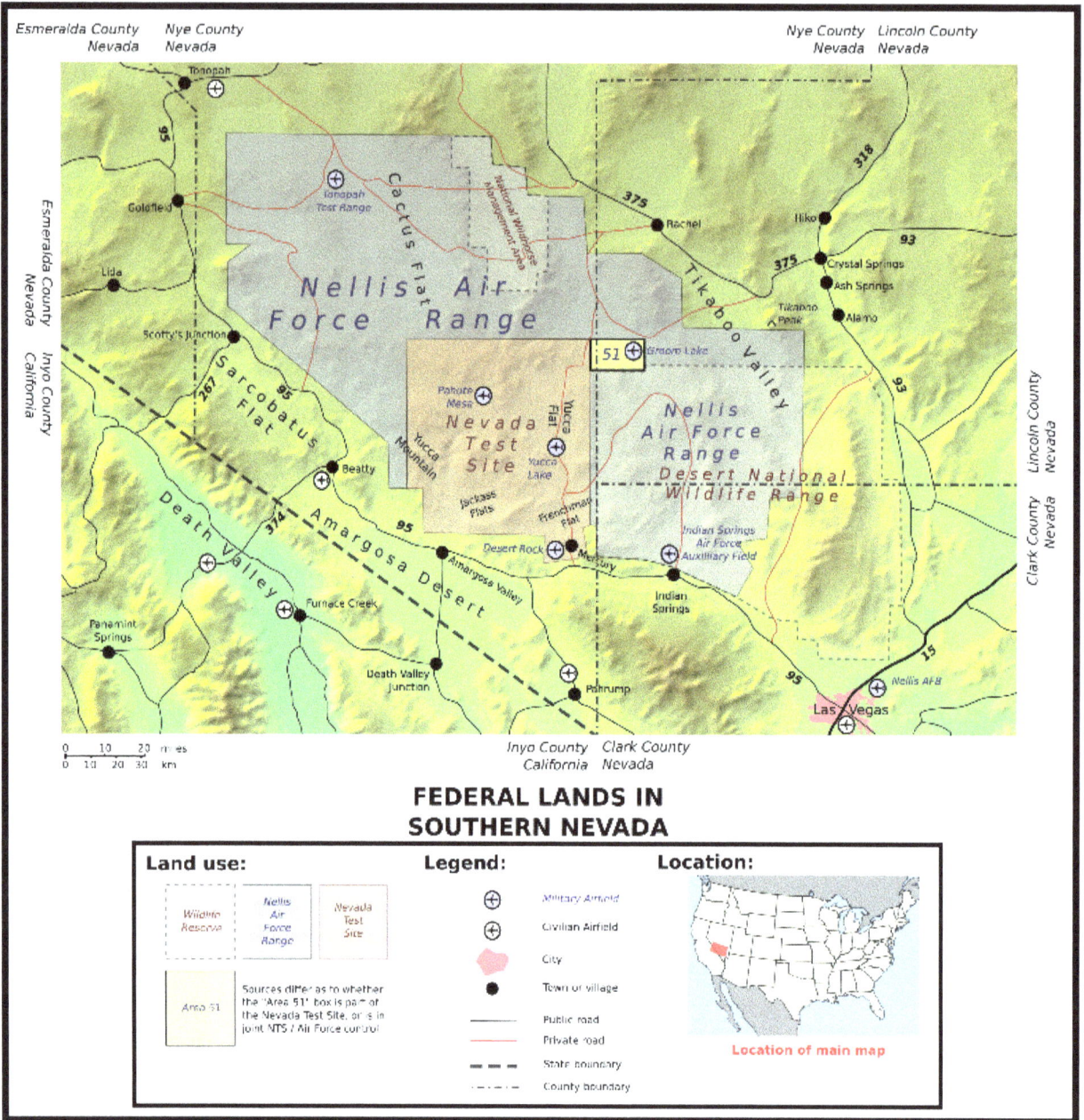

Map of Area 51
https://en.wikipedia.org/wiki/Area_51

Area 51 does exist as this "Bird's Eye View" captured by a high-flying Russian spy satellite clearly shows! Note the very long runway across the dry lakebed.

**A close up photo of the Area 51 complex and runways which are long enough
for any aircraft to land including the Space Shuttle**

https://www.pinterest.com/capper01/area-51/?lp=true

Are there new, highly advanced aircraft being flown out of Area 51 at Groom Lake as reported by John Lear and his friends? According to **Jane's Intelligence Briefing**, the world's leading impartial provider of military intelligence, the US has developed a variety of spacecraft and propulsion technology. Some of this technological development has come from human invention like ramjet and aerospike engines, pulse detonation thrust propulsion derived from small nuclear power plant engines and other propulsion systems are a combination of manmade and reversed engineered alien technology utilizing electrogravitics, electromagnetic and anti-gravity that mimic alien flight characteristics.

An exploration of some of these manmade flying saucers, discs, diamond and triangular shape aerospace vehicles will help to lay out concrete proof that America has leaped ahead of almost every nation in the development, design and engineering of exotic spacecraft. The development of these spacecraft has come from not only the smartest minds in the fields of aerospace and

632

rocket science in America but, also from reverse engineering captured alien technology which has been married with terrestrial technology to produce new hybrid space vehicles.

Jane's Intelligence Briefing

Your inside guide to the world of military developments

The aero-diamond's aerospike engine and diamond platform could propel the craft to speeds of Mach 14.

Americans Chase Hypersonic Flight

Aerospace giant McDonnell Douglas is developing aircraft that will look and fly like the UFOs in *THE X-FILES*.

Research into hypersonic craft that are capable of speeds of Mach 14 is now being undertaken by McDonnell Douglas Aerospace Advance Systems and Technology at its Phantom Works. There are several design studies under way.

One prototype that shows promise is Project Diamond. Researchers have produced a diamond-shaped, proof of concept model for a hypersonic craft.

Jerry Ennis, head of the Phantom Works prototype center, says the design is still at the early stage of research and development and remains highly secret. "A lot of people are looking at diamond shapes for aircraft, some of them we can't even discuss," Ennis said.

Aero-diamonds are one of several configurations proposed by aerodynamicists for conquering hypersonic flight. The aircraft would be powered by a type of rocket engine known as an aerospike. The entire rear half of the vehicle becomes the "plug nozzle" under a process called "external surface burning." Fuel sprayed from the center of the craft is ignited, launching the vehicle into hypersonic flight. Speeds of Mach 14 could be possible.

At the moment, these aircraft are represented by subscale prototypes of flying models, but if they work, full-size craft will follow. ➤

Inside This Issue

2 News Update

Chinese F-7MG Puzzle Solved • Joint Strike Fighter Contracts Awarded • Qatar Buys Hawks • American Airlines Orders Boeing Jets • Smiths Industries Supplies Royal Australian Air Force • Argentina Restructures Defense Ministry • U.S. Air Force Awards Laser Contract

3 International News

Iranian Arms Build-up • Red Cross Seeks Ban On "Tumbling" Bullet • SH-2G(A) Glass Cockpit For Australian Helicopter Bid

4 Armed Forces Briefing

Can the Israeli Air Force remain first class and keep its edge?

6 Weapons & Equipment

Major retrofits of avionics and weapons ensure the MiG-21 will fly into the 21st century

7 Focus

Ahmadshah Massoud, current Afghan military leader, continues the fight

8 Research & Development

Wearable Information Systems • Bio-engineered Spider Silk

Jane's article on hypersonic flight allegedly still considered hypothetical but may in fact be reality

http://ufosworld.blogspot.ca/2007_11_01_archive.html

**The F-19 - hypersonic prototype of a stealth diamond shape aircraft based
upon similar flight characteristics of the B-2 Stealth Bomber**
http://rense.com/general57/f19.htm **and** http://hamzalippisch.deviantart.com/art/A-storm-is-raising-through-the-desert-386454513

The US military has always been obsessed with aircraft that could soar into space since the first successful test flight of the X-15 rocket plane back in 1959. Most **TAVs (Trans-Atmospheric Vehicles)** to date have a triangular aerodynamic design while other craft are decidedly diamond or saucer shape. All are designed to leave the Earth's atmosphere, fly into orbital space then, return and land like conventional aircraft. These manmade spacecraft may also have the capability to go into deep space or even to the Moon!

Most of these craft are manmade, while others are actually alien spacecraft that are typically the basic flying saucer or disc-shaped craft. All these spacecraft are super deep black projects contracted by the military, chiefly the USAir Force and Navy from secret departments within the aerospace industry. There is often an overlap of military defense projects which would explain some of the similarities in development and appearances of these covert secret aircraft. It also explains some of the confusion in aerospace craft identification, classification, and designation.
http://www.sushi-x.com/gallery/nonseq/aurora/aurora.html

634

Project Diamond (aka. Project Pumpkinseed) is a proof of concept hypersonic space platform from Phantom Works that uses aerospike technology to travel at Mach 14 or better into space. This craft may currently be in operation
http://www.456fis.org/AREA-51_&_AR.htm

USAF Nuclear Powered Flying Saucer also known as a Lenticular Re-entry Vehicle (LRV) may actually have flown in the mid 1960s.
http://www.military.com/Content/MoreContent/?file=PMsaucer

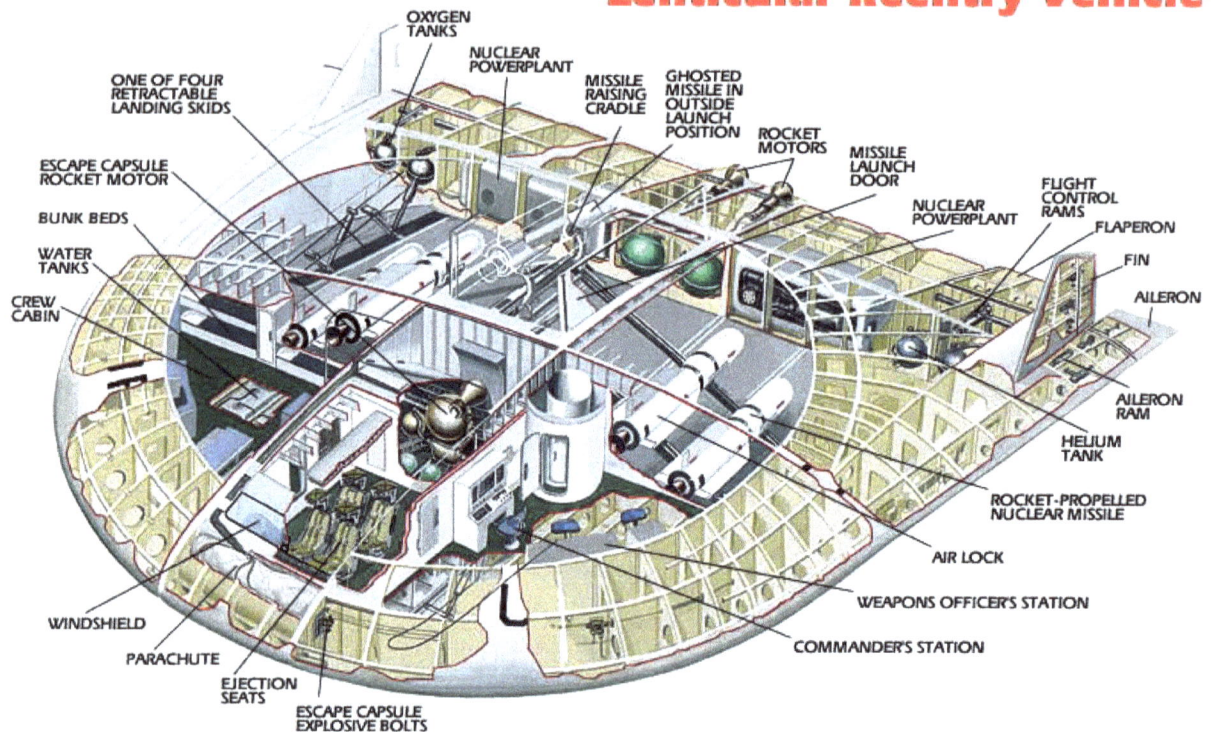

Labels (clockwise from top):
OXYGEN TANKS · NUCLEAR POWERPLANT · MISSILE RAISING CRADLE · GHOSTED MISSILE IN OUTSIDE LAUNCH POSITION · ROCKET MOTORS · MISSILE LAUNCH DOOR · NUCLEAR POWERPLANT · FLIGHT CONTROL RAMS · FLAPERON · FIN · AILERON · AILERON RAM · HELIUM TANK · ROCKET-PROPELLED NUCLEAR MISSILE · AIR LOCK · WEAPONS OFFICER'S STATION · COMMANDER'S STATION · ESCAPE CAPSULE EXPLOSIVE BOLTS · EJECTION SEATS · PARACHUTE · WINDSHIELD · CREW CABIN · WATER TANKS · BUNK BEDS · ESCAPE CAPSULE ROCKET MOTOR · ONE OF FOUR RETRACTABLE LANDING SKIDS

**The interior layout structure of the Lenticular Re-entry Vehicle (LRV)
from the November 2000 issue of Popular Mechanics**
http://www.military.com/Content/MoreContent/?file=PMsaucer

From 300 miles in space, the LRV would be able to rain nuclear destruction on the Soviet Union, Red China, and North Korea. At least that is the thinking in US military circles should any of these nations act with aggression against the USA. In other words, whoever holds the high ground in war controls the battle and space is about as high as you can get!

With the end of WWII and the rush by Allied Forces to raid Germany's most advanced weapons research laboratories, weapon prototypes, and Nazi scientists became the ultimate prize objectives. Soldiers and military intelligence from Communist Russia were in on the race to get to Germany and in those frantic hours of World War Two went from a low agrarian and industrial society to becoming a superpower in just four years, after the dropping of the atom bomb on Hiroshima.

America now realized that it had competition and the nuclear arms race was born on Aug. 29, 1949, with Soviet Russia detonating its own atomic bomb. American way of life was now potentially threatened by a clear and present danger coming from the Soviets who were looking to develop a missile delivery system and nuclear bombers. In order to counter this potential threat, America considered a new radical aircraft bomber design based on a Nazi swept or curved wing design but also, based on previous U.S. Navy concepts, ten years earlier of a circular wing aircraft. Such a design would allow for greater lift and greater bomb capacity providing an advantage to the US.

636

In **Popular Mechanics (P.M.)** of July 1997, a cover story entitled, "Roswell Plus 50," detailed how Air Force interest in duplicating Nazi technology led to two American flying disc projects. **Project Silver Bug** sought to build a vertical takeoff and landing aircraft. **Project Pye Wacket** was to create small discs for use as air-to-air missiles. Documents declassified since then point to a third secret project, a 40-ft. **"flying saucer"** designed to rain nuclear destruction on the Soviet Union from 300 miles in space.
http://www.timstouse.com/ScienceNews/nuclearflyingsaucer.htm

Project Pye Wacket was the inspiration to develop a larger version of the disc missile into a manned vehicle, as extensive wind tunnel experiments proved the stability of such a semi-disc shape craft.

The official designation for America's nuclear-flying saucer was the **Lenticular Reentry Vehicle (LRV).** It was designed by engineers at the Los Angeles Division of **North American Aviation**, under a contract with the U.S. Air Force. The project was managed out of Wright-Patterson Air Force Base, in Dayton, Ohio, where German engineers (from **Operation Paperclip**) who had worked on rocket plane and flying disc technology had been resettled.

The LRV escaped public scrutiny because it was hidden away as one of the Pentagon's so-called **"black budget"** items — that is, a secret project that is incorporated into some piece of non-classified work. On Dec. 12, 1962, security officers at Wright-Patterson classified the LRV as secret because: "It describes an offensive weapon system." The project remained classified until May 1999, when a congressionally- mandated review of old documents changed the project's status as a government secret, downgrading it to public information. The **Department of Defense** did, however, successfully seek to have the document's distribution restricted to defense contractors. **P.M.** obtained its copy as the result of a **Freedom of Information Act** request.
http://www.timstouse.com/ScienceNews/nuclearflyingsaucer.htm

A typical operational mission would be of six-week duration at an orbital altitude of 300 nautical miles, with a crew of four men. The weapons bay would hold "four winged weapons" that could be either launched or detached and parked on orbit. There are repeated references The LRV was capable of launching "weapons-carrying clusters", a type of multiple warhead delivery devices similar to **MIRVs (Multiple Independent Reentry Vehicles)**

A considerable part of the design study focuses on the details of building a 40 ft diameter airframe and strengthening it against the acceleration of 8 g's and wind shear it would experience during launch. However, no mention is made of the type of booster the disc would ride into space.

The four-man crew would ride a wedge-shaped capsule built inside the LRV. The capsule would divide the front portion of the disc into separate work and off-duty areas. The nuclear-tipped rockets would be stored in the rear segments.

In normal operations, the capsule would function as the LRV's flight control center. In an emergency, the crew could fire the capsule's independent 50,000-pound-thrust solid-fuel rocket motor and return to Earth. The capsule's final descent would be slowed by a parachute, much like the X-38 "lifeboat" planned for the international space station now under construction.

A textbook mission would conclude with the entire LRV returning to Earth. It would fire it's nuclear or liquid-fueled main rocket to brake, then travel edge-first into the atmosphere. Its disc form would dissipate the heat of re-entry then, act as a wing. Its flattened tail structure would provide directional stability and control. A minute or so before landing skids would extend and the LRV would settle onto a stretch of dry lakebed.
http://www.timstouse.com/ScienceNews/nuclearflyingsaucer.htm

In 1997, as part of its effort to debunk the Roswell alien landing myth, the Air Force revealed details of several heavy-lift balloon research projects. Among those were experiments in which 15,000-pound payloads were lifted to 170,000 ft. While not specifically acknowledging the LRV by name, an Air Force spokesman conceded that during the Cold War it routinely used high-altitude balloons to lift unusual airframes for aerodynamic tests. Airframe tests of secret planes were most likely the cause of still-unexplained UFO sightings. And a balloon-lifted LRV test flight would certainly match the classic UFO reports of a silvery disc hovering motionless in the sky then, silently shooting upward.

The engineering study obtained by **Popular Mechanics** contains language that describes a re-entry heating test that, at the time, could have been accomplished by only a high-altitude drop of a flying prototype. A further indication that the LRV flew comes from a retired Air Force contractor. He tells PM he personally saw a craft fitting the description of the LRV at a Florida base that he had been visiting on unrelated business in the late 1960s. However, what is by far the most compelling evidence that the LRV, or a flying prototype, was actually built comes from Australia. http://www.timstouse.com/ScienceNews/nuclearflyingsaucer.htm

It is believed in 1966 that a LRV had been test flown but, for unknown reasons, it had unfortunately crashed leaving honeycomb debris from the explosion over a secret Australian test range where the British and Americans conducted some of their most secret atomic experiments. Since the LRV was to carry a small nuclear reactor to provide electricity for flight systems, it is conceivable that tests would have been conducted at this isolated location.

Years later in 1975, the daughter of a rancher found an odd bit of this honeycomb-like debris on the family's ranch which happens to be near the test range site. Thinking it was debris from an extraterrestrial spacecraft it was analyzed but was found to be common aircraft-grade fiberglass panels. Could this be the proof that the US had tested a Lenticular Reentry Vehicle only to have it crash in the outbacks of Australia? Only time and further research will hold the answer.

All Things Flying Black and Triangular

The question that is often asked by the general public and particularly by Ufologist is "Just what the hell are Black Triangles?" The question may also be asked is "Have they replaced the original Flying Saucer craft and why do they appear all over the world, with little regard to who sees them?"

638

Usually seen at night, witnesses report sightings of huge, dark triangular shapes, moving across the sky, blotting out the stars. Sometimes they are entirely dark, sometimes they show lights. During the early 1990's there was a wave of over 4,000 sightings across the north of England, Belgium, Holland, and Germany, with fighters being sent to intercept the craft. Sightings have also been reported from the USA, where witnesses describe similar events.

As one might expect, every sort of theory has been tried to explain the phenomena, but no-one knows. The UFO brigade has typically jumped on board, but most of their theories are the 'content-free noise' variety. More credible is the theory that these are earth-based aircraft of some description. There have been many sightings of dark, triangular aircraft over the years, which may or may be attributable to projects such as Aurora, **TR-3B**, **Astra** and others. It's possible that the new sightings are of a **"stealth blimp"** - a lighter than air craft of huge proportions. Certainly, such a vehicle would be attractive to the military. It would have large cargo carrying capacity, could loiter for long periods at altitude for reconnaissance, and may employ exotic technologies for propulsion and camouflage. Many sightings have described a form of 'visual stealth' where it appeared that the craft was displaying an image of the sky above it in order to mask its shape and size.

An illustration of a Stealth Blimp
http://deepbluehorizon.blogspot.ca/2010/05/case-for-stealth-blimps.html

It could also be related to the mythical **"Senior Citizen"** program to develop a high-speed, stealthy, VTOL transport aircraft. As usual, the people who know are saying nothing, leaving an empty space inhabited by all kinds of theories, which serves to create confusion, misinformation and effectively conceal their true identity.
http://www.bisbos.com/rocketscience/aircraft/black/triang/triang.html

With the development of flying wing aircraft during and after the Second World War by both the Allied and Axis Forces and the current trend toward delta or wedge-shaped craft with stealth capabilities by the Americans, British, Russian and Chinese and by other countries to a lesser degree, it seem inevitable that this basic design find its greatest aerodynamic expression in triangular winged spacecraft. The **Aurora/TR-3B** is perhaps the most heavily emulated antigravity aerospace craft which uses a combination of advanced human aerodynamic design and reverse-engineered ET technology. Many top aerospace manufacturers have created similar aerodynamic triangular shapes and many pictures can be found on the internet with craft having the same name designation of TR-3B and Aurora.

This has lead to some confusion as to which manufacturer has the rightful claim to the TR-3B or the Aurora designations as there appears to be three or four aircraft that are distinctly different from each other, but with some similarities, yet all are called the Aurora/TR-3B. There is either a serious error being made by researchers as to what aircraft is the correct Aurora/TR-3B or it is a common nomenclature for this type of aerospace craft with an overlap in aerodynamic design as mentioned earlier. It is also conceivable that this is a deliberate disinformation campaign by US aircraft manufacturers to confuse the militaries of other countries and possibly UFO researchers as well, as to its true nature.

The **Aurora** and the **ASTRA (Advanced Stealth Technology Recon Aircraft)** have also been confused as their basic triangular shapes are very similar to each other, except for the canted vertical tail planes that are described as either canted "inwards" or "outwards" as are often described by civilian witnesses, particularly over the highlands of Scotland, where they are seen to be either test flown or are in fact in actual operational flight missions.

The **TR-3A "Black Manta"** is another aircraft that is often confused with the Northrop **AMC (Advanced Manned Concept)** because of similar shape and design.
http://www.dreamlandresort.com/black_projects/aircraft.htm

Below are black stealth aircraft which may or may not be the manufacturer's true aircraft designation as many aerospace craft that have been posted on the internet have been labelled **Astra, Manta, TR-3A, TR-3B** and **Aurora**, etc. yet, they all appear to have different physical aerodynamic structures either with small wings and ailerons or without them or they may simply be proposed computer designs for future aerospace craft.

640

Lockheed Martin P- 791 demonstrator LTAV (Lighter Than Air Vehicle)
(Source: Lockheed Martin)
http://www.compositesworld.com/news/lockheed-martins-hybrid-airship-targets-heavy-lift-market

Lockheed Martin's SR-91 Aurora
https://uk.pinterest.com/explore/aurora-aircraft/

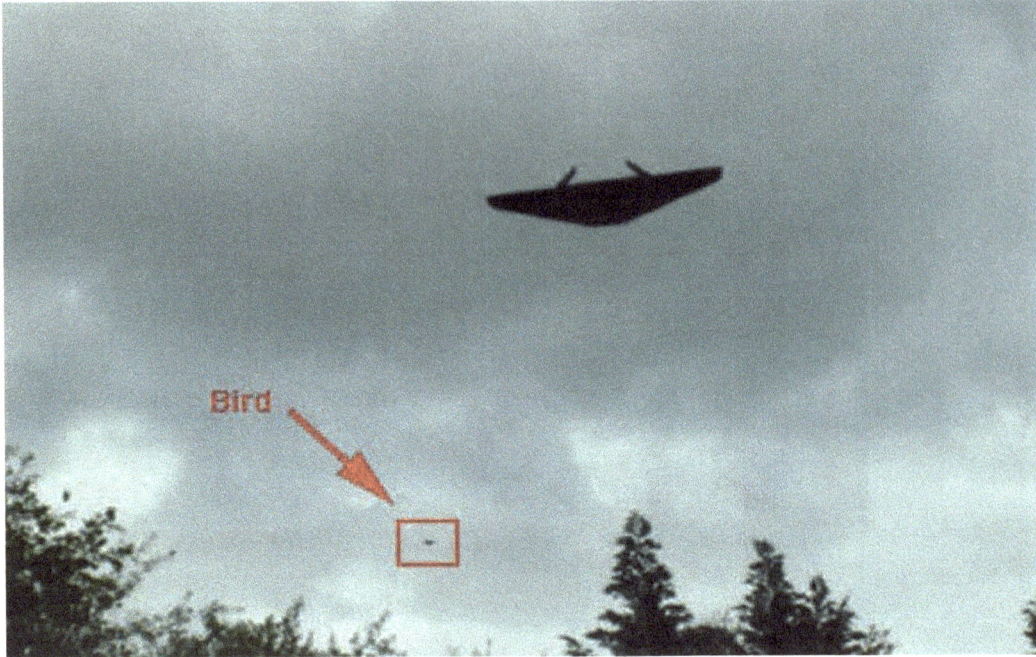

**A Triangle shape craft taken on April 19, 2008 at Greenville, South Carolina.
Is this a TR-3A Black Manta or the TR-3B or something else?**
https://www.rcgroups.com/forums/showthread.php?287752-Mystery-Aircraft/page645

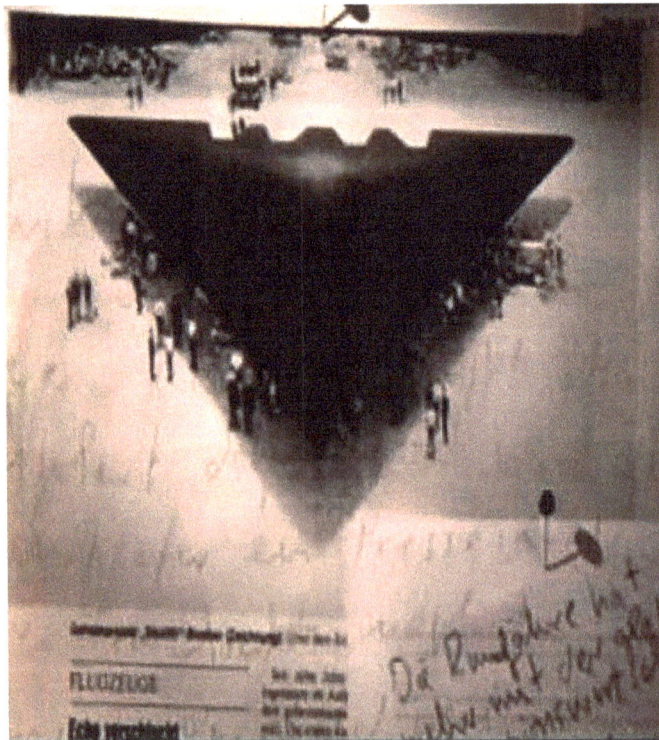

**Black Project Spacecraft Stealth Technology, the TR-3B Area S-4 Triangular Shaped
Astra or Avenger Black Manta Echo – McDonnell Douglas A-12 Avenger I?**
http://tr3aurora.blogspot.ca/2013/02/exposed-usafs-tr-3-aurora.html

Is this the Astra or Black Manta or is it merely a computer-generated aircraft?

https://codigooculto.com/2015/10/flota-operativa-antigravedad-actual-del-gobierno-de-estados-unidos/

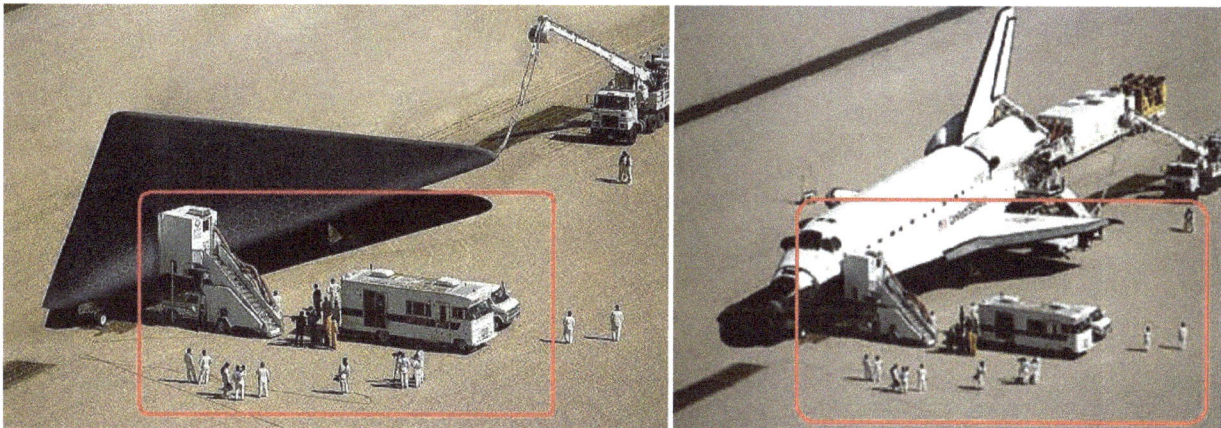

The TR-3B left being serviced after landing is a CGI hoax but, the space shuttle endeavour right is the real photograph. Note the service vehicles and people are exactly the same in each photo image including the portable stair ramp. This illustrates the extent of hoaxing to either convince or to deceive the reality of the TR-3B Manta

https://codigooculto.com/2015/10/flota-operativa-antigravedad-actual-del-gobierno-de-estados-unidos/ and
https://spaceflight.nasa.gov/shuttle/reference/basics/landing.html

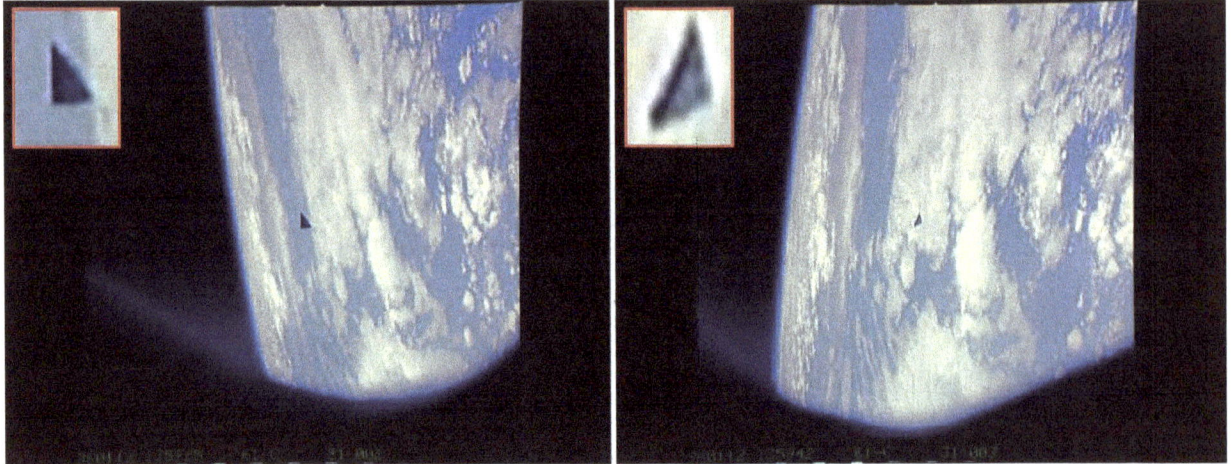

Is this a TR-3B or the SR-91 Aurora spyplane in orbit? Photographed by the Space Shuttle Endeavor during STS-61 and the International Space Station in 1986.

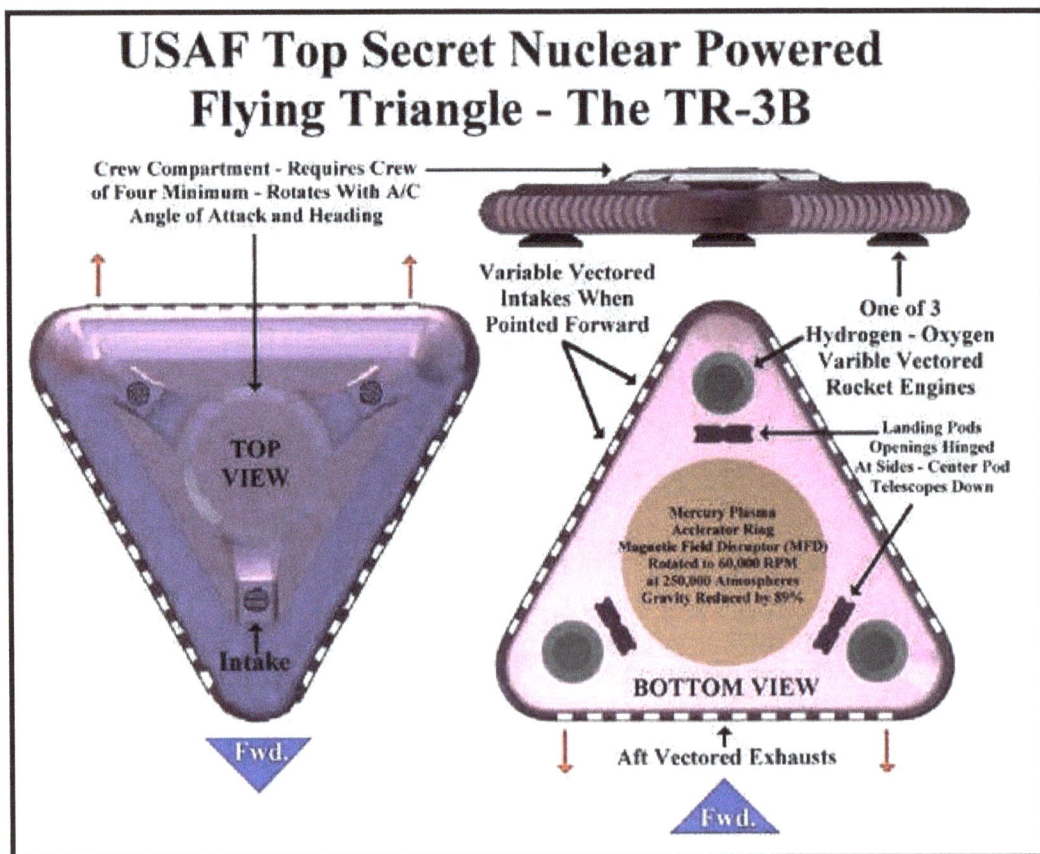

The nuclear-powered TR-3B or is it just a hypothetical spacecraft?

644

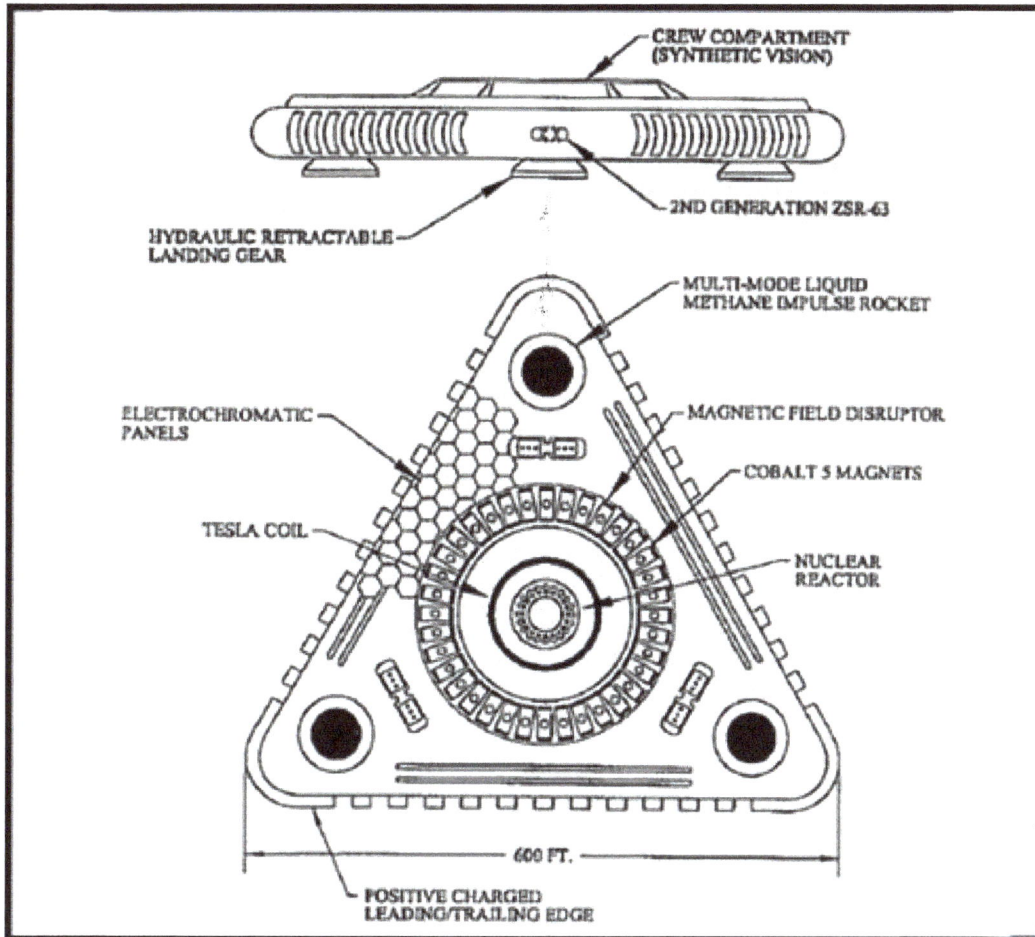

The 600-foot nuclear propulsion system of the TR-3B

Senior Citizen craft was developed by Northrop Grumman to meet this requirement, mainly due to its work on the X-21 laminar flow research aircraft. The stealthy shape was developed during testing at Northrop's microwave testing facility in southern California and at the USAF's facility at Holloman AFB in New Mexico. The clipped diamond shape that was tested also appeared in the **YF-23 ATF prototype. The Senior Citizen** features many advanced technologies:

Stealth - While Senior Citizen has a very low **RCS Radar Cross Section),** it is not in the same class as the B2 'Spirit' bomber. As it is required to operate at low levels, much attention has been paid to reducing engine noise, and infra-red emissions. Also, Senior Citizen employs visual stealth. At night, it has a pattern of three lights to disguise it's characteristic shape, while in daylight 'Yehudi' type lights are used to make it difficult to spot beyond 1.5 miles.

STOL Performance - The C41-SR employs the enhanced **Coanda effect** with an energized upper wing surface - also thought to have been used on the YF-23 - to give exceptional STOL performance.

Circulation Control - By using an arrangement of valves and ducts in the leading and trailing edges of the wing, air is accelerated over the wing, increasing lift and decreasing drag. This technique is also employed on the B2 and the YF-23.

Author's concept of Northrop Grumman's "Senior Citizen" in orbit
http://stargazer2006.online.fr/aircraft/pages/citizen.htm

Robert Scott Lazar

The name, Groom Lake or Area 51 or Dreamland, etc. when mentioned, conjures up images of hidden alien technology being reversed engineered, strange saucer-shaped craft being built and flown over the nearby test range under the cover of night. Ufologist, also associate the name of **Bob Lazar** with Groom Lake and his claim that he had worked there on secret flying saucer projects.

According to **Bob Lazar**, a free-lance physicist and engineer, Groom Lake activities include reverse-engineering extra-terrestrial craft! He first came forward with these claims in March 1989, when appearing on **George Knapp's** news program on Channel 8, based in Las Vegas. He described his brief time at a facility known as S-4, where he worked on back-engineering exotic craft built to accommodate small beings.
http://www.ufos-aliens.co.uk/cosmicarea51.html

When asked why he had decided to do the interview he said he wanted to share his work with the scientific community and to do that he felt the best way was to go public to insure himself

646

against any mysterious sudden demise for exposing classified information. http://www.where-is-area-51.com/bob-lazar.html

Robert Scott Lazar, whistleblower to the existence of Area 51
and reverse engineering of Alien spacecraft
http://www.thelivingmoon.com/41pegasus/12insiders/Bob_Lazar_001.html

Lazar said his tasks consisted in the scientific investigation of the propulsion system of one of nine disc-shaped aircraft, as a general part of the ongoing reverse engineering project taking place at S-4. In August 1990, Knapp investigated Lazar's and was provided with a W-2 tax slip showing payments from the Department of Naval Intelligence, and other pieces of evidence Knapp believes supports Lazar's claims.

In his interview with Knapp, Lazar said he first thought the saucers were secret, terrestrial aircraft, whose test flights must have been responsible for many UFO reports. Gradually, on closer examination and from having been shown multiple briefing documents, Lazar came to the conclusion that the discs must have been of Extraterrestrial origin. In his filmed testimony, Lazar explains how this impression first hit him after he boarded the craft under study and examined their interior. *(The navigational control room had small seats or chairs designed to accommodate not human beings but, small child-size ET beings).* [Italics added by author]. http://parascientifica.com/articles.php?a=155

The "Sport Model" ET Spacecraft

The craft Lazar claimed he worked on - dubbed the **"Sport Model"** *"...because it was just so damn sleek!"* - is approximately 52 feet in diameter. Power is said to be generated from an anti-matter reactor in near perfect thermodynamic balance. The propulsion force is gravity, created, amplified and vector-directed aboard the craft. He added that time, as we know it, is altered and, at full power, light is distorted around the disc.

The "Sports Model" Flying Saucer that Bob Lazar claimed to have worked upon to discover the physics of its propulsion system

This one rested in the hangar on its lower surface (no landing gear). Lazar said he was on the main floor of the craft, viewed the lower area with the **Gravity Amplifiers** - even moved them with his hands - but stated he was never in the upper floor of the craft. He thought the upper area might contain navigation and electronic equipment but he was not certain.

Lazar described the interior of the craft as being like a well-harmonized symphony. "There is a flow to all aspects". Lazar stated that he felt the internal design was an integrated blend of both structure and power transmission. He said that nothing was wasted; that perhaps it was a device with components and systems "tuned" or in harmony with itself and the universe. It is an

648

observation that **Bruce Cathie** has also detailed in his *"Harmonic"* series of books on universal harmonics and Earth Energy grids. http://projectavalon.net/forum4/showthread.php?33746-Bob-Lazar-Captured-UFOs-at-Top-Secret-S4-Base

The structure and surface of the craft was described as flowing smoothly, like one piece and every corner had a radius. The material was like a huge seamless piece of stainless steel. He could not fit in the seats because they were too small. The question that follows is: Then who are these seats designed for? Lazar is very firm in stating that he never saw a live "alien." He did say that he has read summary reports on aliens while at S4.

People working with him jokingly referred to the aliens as "The Kids." He could stand in the centre of the craft but not at the edges, and his hair tended to stand up while he was inside, indicating an electrostatic charge. Lazar also mentioned that the disc was showing signs of rough treatment; the "human intervention," as he termed it.

He claims that he saw posters on the lab walls that humorously said, "They're here!"

According to Lazar's testimony, he had worked on this UFO between 1988-1989 and was witness to a test flight of this alien ship along with other workers who saw the alien craft rise and hover over the desert floor.

"...It was evening... they called me outside to see something. It rose about 40 feet in the air. There was some distortion on the bottom - like a corona. It stood there. Stable. Silent. Just a little wobble, no rotation. It was awesome!"- Robert Lazar, 25 November 1990

In 1989, Physicist Robert Lazar claimed that the US Government had not one, but several captured alien UFOs at its top-secret **S4** base. Since then, several other "whistleblowers" from **S4 & Area 51** have stepped forward to corroborate Lazar's claims. *(**NOTE**: Although numerous secret stealth aircraft have operated out of Area 51 for decades, many mistakenly think the super-secret base is under Air Force control. Groom Lake is actually a **NAVY** facility).*

Rumours that captured UFOs exist in a secret U.S. military hangar have been around since the infamous Roswell crash of 1947. Lazar shocked the UFO community and the public in 1989 when he stated publicly that the U.S. government *did* indeed have alien spacecraft in its possession, and that he had personally witnessed and examined captured alien UFOs at a top secret base known as S4.

Lazar said nine UFOs were housed in hangars cut into the mountainside at S4. One of the craft appeared to have been damaged by a projectile passing through it.

Diagram labels (cross-section):
- collocated structure
- shell (15 cm thick)
- wave guide
- positronic emitter
- seat
- gravity amplifier head
- all forms inside are smoothed for electronic charge retention
- upper deck
- lower deck
- maximum angle of "lens" effect
- extent of hull structure as metallized waveguide
- gravity wave guides (focus ring stacks)
- rings
- millimeter electromagnetic wave is pulsed THROUGH the shell casing

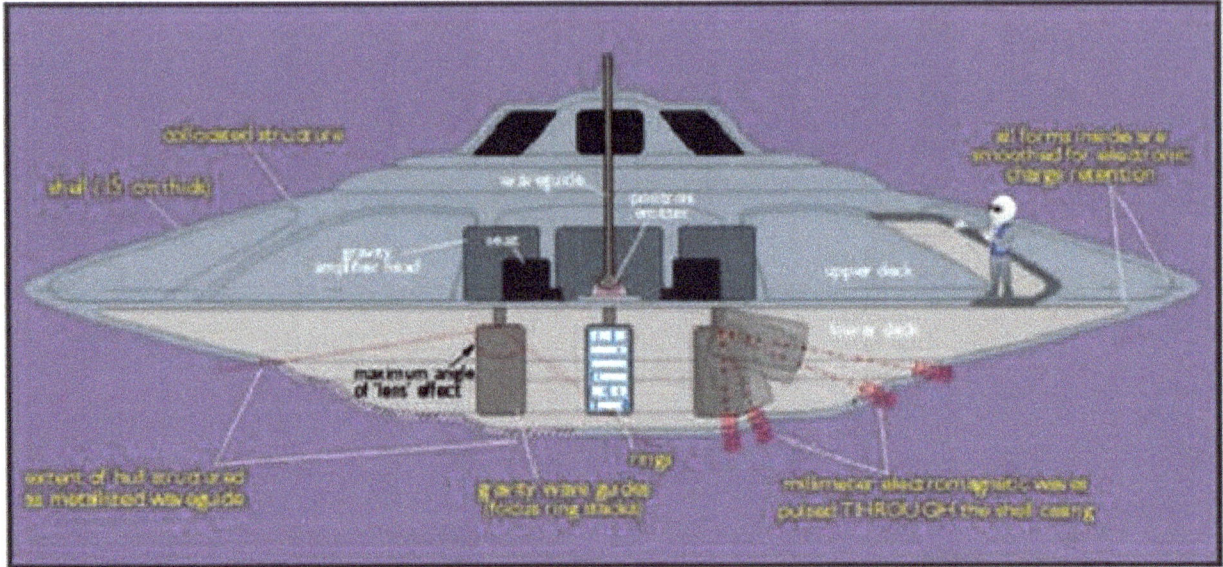

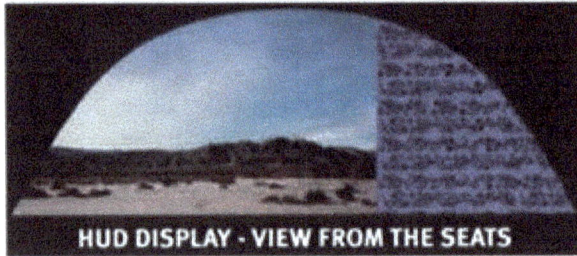

HUD DISPLAY - VIEW FROM THE SEATS

ABOVE: Lazar examined the upper and lower decks of the "Sport Model," - he thought the very upper area (with rectangular "windows") might contain navigation and electronic equipment but he was not certain.

Transparent Metal "Arches"

Lazar remembers metal in one section of the craft turn a clear blue colour and strange writing scrolled on its interior surface reminiscent of the HUD (Head Up Display) used in modern military aircraft.

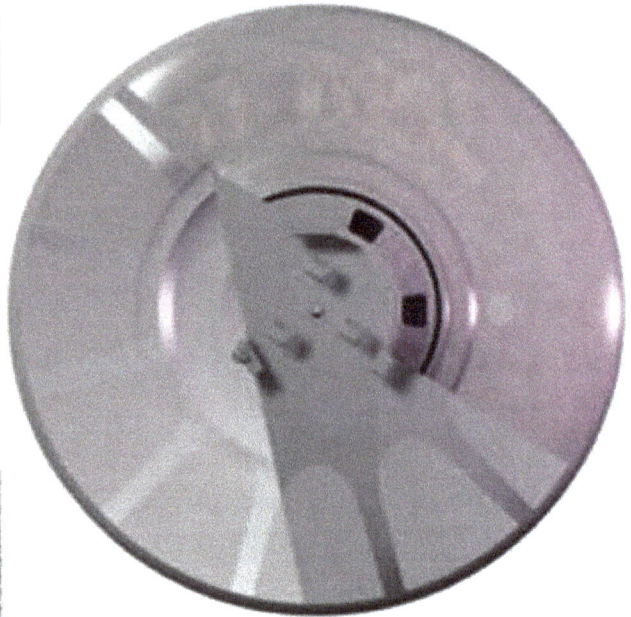

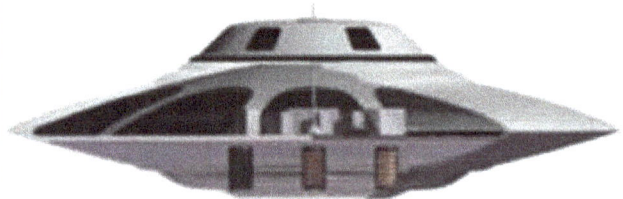

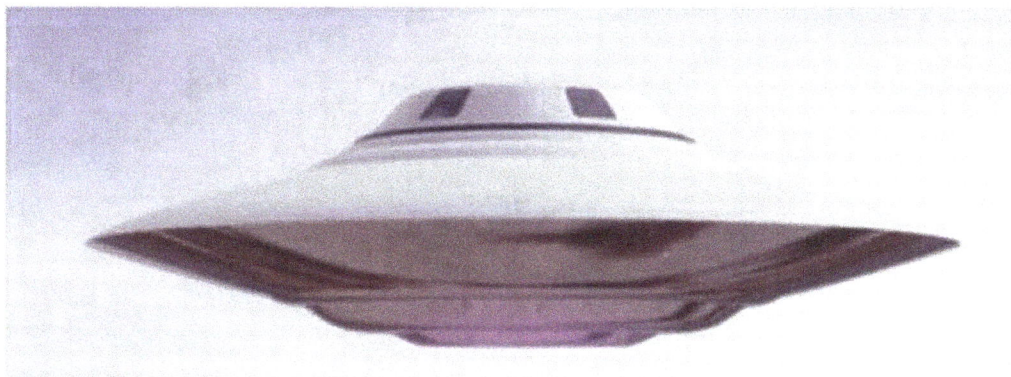

ABOVE: Disc #1 - The "Sports Model" Lazar worked on bears a striking resemblance to one of the Billy Meier UFOs and Disc #7 (lower right)

ABOVE: Disc #2

ABOVE: Disc #3

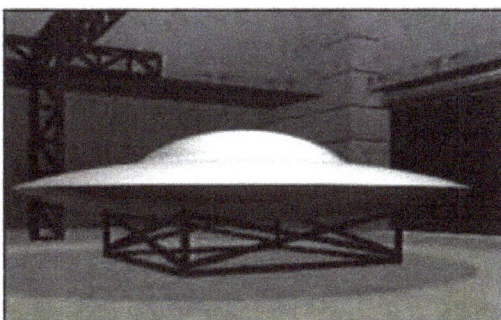

ABOVE: Disc #4

ABOVE: Craft #5 - fits the description of the craft recovered at the 2nd Roswell site, Socorro, NM

ABOVE: Disc #6 - "Top Hat" model with damaged Disc # 7 in background (at right)

ABOVE: Disc #7 - this UFO was damaged from a projectile passing through the upper section

http://projectavalon.net/forum4/showthread.php?33746-Bob-Lazar-Captured-UFOs-at-Top-Secret-S4-Base

651

S4, in the Papoose Dry Lakebed, is located just a few miles South of **Area 51/Groom Lake** (inset below), and is connected by high-speed "subterréne" tunnels linking other underground bases in the US.
The red arrow (above) shows a disc in flight, photographed by a Russian Spy satellite, (enlarged below).

The hangar doors match the outside of the mountain, roll up when opened, and are camouflaged to match the terrain and also avoid spy satellite technology. *(However, not quick enough to show*

652

*the remarkable enlargement of a disc-shaped craft in the air approaching S4, taken by a Russian Spy satellite after the Cold War (**See image above**)).* Security is extremely high at S4. The many guards are civilian, wear desert uniforms and are heavily armed.
http://projectavalon.net/forum4/showthread.php?33746-Bob-Lazar-Captured-UFOs-at-Top-Secret-S4-Base

This disc is capable of two distinct modes of travel (below): The "**Omicron**" **configuration** is used for short-range travel near a source of gravity such as a star or planet, (and within a planet's atmosphere). The "**Delta**" **configuration** is used for travelling larger, more vacuous areas of space/time as would be required for interstellar travel. The gravity amplifiers of the disc can be focused independently; they are pulsed and do not stay on continuously. When all three of these amplifiers are being used for travel they're in the "Delta" configuration. When only one is being used for travel it's in the "Omicron" configuration. When the disc is near another source of gravity like earth, the Gravity-A wave which propagates outward from the disc, is phase-shifted into the Gravity-B wave which propagates outward from the planet creating lift. The ball-shaped end of the reactor (above) houses a cone-shaped layered mass of heavy material designated **Element 115**. It is a low-level alpha emitter. This material is used in an almost perfect 100% matter annihilation reaction.

The **Waveguide Terminator** at the top of the craft is very similar to other UFOs described in the past, as are the three gravity amplifiers, observed in both the **Adamski Saucer** and Early German designs, notably the **HAUNEBU I, II & III** Series, housed in three spheres or balls under the craft. The "sport model" also bears a close resemblance to one of the **Billy Meier** discs.

The **Gravity Amplifiers** attach to the floor with what seems to be a "flexible metal." Lazar said that he could move the amplifiers easily, yet felt their weight. This also corroborates the description **Frank Scully** made in his book "Behind The Flying Saucers," and **George Adamski** in the 1950's. Scully described those ships as "having three ball or spherical devices on the underside" and "were easily able to be rotated by hand but had virtually no weight".

Lazar has been criticized and challenged by many suggesting that the discs are simply Earth-made advanced technology. His rebuttal to this is, *"well why would we be reverse-engineering an existing [Earth-made] craft with seats only a child could sit in?"* He also adds that **Element 115** doesn't even exist on Earth yet or any of our Periodic Tables.
http://projectavalon.net/forum4/showthread.php?33746-Bob-Lazar-Captured-UFOs-at-Top-Secret-S4-Base

Although his credibility has since been questioned by many, notably the nuclear physicist **Stanton Friedman**, supporting evidence indicates that Lazar did indeed work at S-4, as well as other scientific establishments who initially denied all knowledge of him. It seems that somebody within the Government wanted to remove all evidence of Lazar working at the base, but his name can be found in one of the on-site phone books which date back to the time that Lazar said he was working there. His academic credentials remain in doubt, although his scientific knowledge is undeniable. It seems that his free-lance working pattern makes him a lower security risk to places like Area 51, simply because he is so difficult to authenticate as a scientist. http://www.ufos-aliens.co.uk/cosmicarea51.html

"DELTA" CONFIGURATION

"OMICRON" CONFIGURATION

GRAVITY AMPLIFIER HEAD (3)
SEATS (3)
WAVEGUIDE
REACTOR
GRAVITY AMPLIFIER OUTPUTS (3)
CORE PLATES

WAVEGUIDE
REACTOR

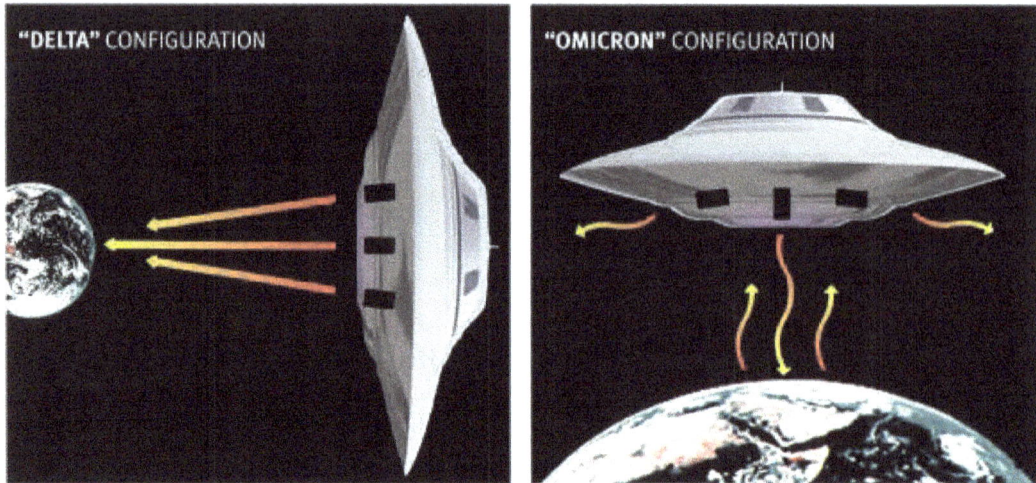

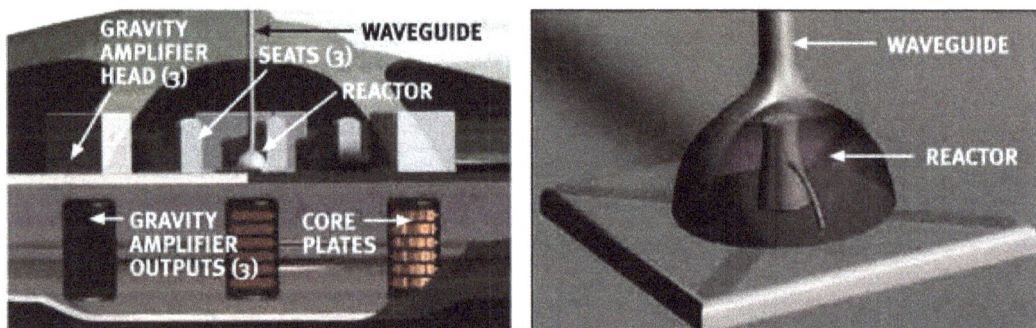

Beamships used by the Pleiadian races. Image created using the bitmap program from line drawings in the Billy Meier case in 'UFO Contact From The Pleiades' by Wendelle Stevens.

Pleiadian Type 1	Pleiadian Type 2	Pleiadian Type 3
7 Meter Diameter	7 Meter Diameter	7 Meter Diameter
3 Person Crew	3 Person Crew	
Reddish bottom Emits humming sound	White top dome Reddish/orange ports	Gold/silver finish Dimensional travel

ABOVE: Billy Meier UFOs - similar to the "Sport Model"

BELOW LEFT: George Adamski's **"Venusian Scout Craft"** showing similar Waveguide & Gravity Amplifiers, is almost identical in design to the German **HAUNEBU II** series of the 1940's (BELOW RIGHT)

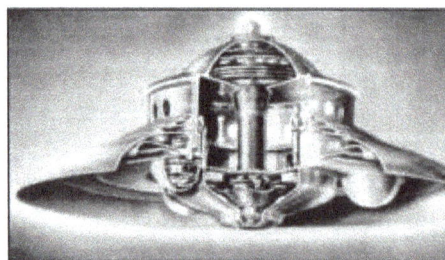

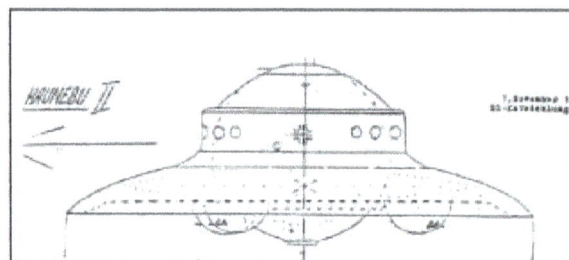

http://projectavalon.net/forum4/showthread.php?33746-Bob-Lazar-Captured-UFOs-at-Top-Secret-S4-Base

Other individuals have since come forward to corroborate a lot of what Lazar claims, although, most of them have chosen to remain anonymous. As a result, Area 51 is now an intrinsic part of

654

UFO folklore and has done much to support the extra-terrestrial hypothesis. One thing is for sure; many bizarre, seemingly exotic craft have been witnessed, photographed and filmed in the immediate area. The question is, are they the result of extraterrestrial technology or simply our own independent development? Also, in recent years, 2009 to be exact, element 115 has been created in the laboratory via atom smashing to produce about 4 atoms worth of Elemnet115! *(Although, not enough raw material to make even a common fruit fly, fly without wings!!)* (Italics added by author).

*(As any "Trekker" will tell you, the legendary "USS Enterprise" in **STAR TREK** also uses "anti-matter" engines and literally "warps" space. But this UFO (the Lazar "Sports Model") does it for real! (Which also validates just how visionary STAR TREK's creator - the late Gene Roddenberry - really was).* http://projectavalon.net/forum4/showthread.php?33746-Bob-Lazar-Captured-UFOs-at-Top-Secret-S4-Base *and* https://www.youtube.com/watch?v=K0jdCER8ypA

Star Trek's Starship USS Enterprise NCC -1701

A Slight Detour of Thought

*(Actually, **Irving Block** and **Allen Adler** originated most of the sci-fi concepts ten years earlier in the 1956 science fiction cult classic, **"The Forbidden Planet"** which were later adapted to the TV series, **Star Trek** in 1966 by **Gene Roddenberry**. You could say that Block and Adler were the true visionaries and Roddenberry was ten years too late or merely a copycat!* Even many years

later, in the biography of Gene Roddenberry, *Star Trek Creator*, notes that *Forbidden Planet* was one of the inspirations for the series *Star Trek*. http://en.wikipedia.org/wiki/Forbidden_Planet

Still, don't believe it is true? Well, here are the original science fiction concepts found in The ***Forbidden Planet*** *and later copied in* ***Star Trek****:*

- *Both movie and TV series took place in the **23rd Century** time period*
- *In Forbidden Planet, it was considered **Shakespeare ("The Tempest") in space** and in Star Trek it was considered a **Western in space**;*
- *Forbidden Planet had a **flying saucer type spaceship**, Star Trek had a **saucer-shaped section** in the forward part of the ship;*
- *Each ship could travel **faster than the speed of light**; in the case of the F.P. **Spacecruiser C-57D**, their ship used **"hyperdrive"** and in S.T. the **Starship USS Enterprise NCC-1701** used **"warp drive"**;*
- *In F.P. the ship had a room containing **"de-accelerator pods"** to condition the crew safely out of hyperdrive; in S.T. the ship had a room containing **"teleportation pods"** to transport people to a planet's surface;*
- *In F.P. the C-57D had a **viewing screen** to see what was out in space and in S.T. the Enterprise also, had a **viewing screen** to see what was out in space.*
- *In F.P. the space cruiser belongs to the **United Planets** and in S.T. the starship belongs to the **United Federation of Planets**;*
- ***Canadian actor Leslie Nelson** played **Commander John J. Adams** in F.P. and **Canadian actor William Shatner** played **Captain James T. Kirk** of the USS Enterprise in S.T. Both men's names seem to flow off the tongue easily!*
- *In F.P **Second in command Lt. Jerry Farman** was played by an **American, Jack Kelly** and in S. T. Second in command and the Science Officer was American **Leonard Nimoy** and in Star Trek's Next Generation (STNG,) the **Second in command (No. 2) Commander William T, Riker** was played by **American actor Johnathan Frakes**;*
- *In F.P. they had an emotionless, very logical and intelligent robot called **"Robbie, the Robot"** and in S.T. they had an emotionless, very logical and intelligent Vulcan named **"Mr' Spock"** played by **Leonard Nimoy**;*
- *In F. P. they had a ship's doctor or **Medical Officer** named **Lt. "Doc" Ostrow** played by **Warren Stevens** and in S. T. the ship's **Medical Officer** was **"Bones" McCoy** played by **DeForest Kelley**;*
- *In F. P. the **Chief** and **Communication Officer, Chief Quinn** was played by **Richard Anderson** (father of Dean Anderson of "McGiver" and "Stargate 1" fame), in S.T. the ship's **Communication Officer** was **Uhura** played by **Nichelle Nicols**;*
- *In F.P. they had a **Bosun** or the equivalent of an **Engineering Officer** played by **George Wallace** and in S.T. their **Engineering Officer** was **Montgomery Scott ("Scotty")** played by **Canadian actor, James Doohan**;*
- *In F.P. they used a small **handheld computer device** (which became the forerunner to today's Palm Pilots, Smartphones, and iPhones) and in S.T. they had the **Tri-Quarter** device.*
- *The **highly intelligent antagonist** in F.P. was **Dr. Edward Morbius** played by **Canadian actor, Walter Pigeon** and in S.T. "The Wrath of Khan" the **highly intelligent antagonist** was **Khan Noonien Singh** played by **Mexican, actor Ricardo Montalban**;*

656

- *Altair IV was a planet visited by both starships in F.P. and in S.T.*
- *Many incredibly large "alien sets" were used with effectiveness in both F.P. and in the many S.T. movies. Forbidden Planet was the first movie to use animation with live actors!*
- *No doubt, there are probably other similarities.*

Still, think Gene Roddenberry was the inspiration and visionary to the sci-fi TV series Star Trek? And remember there was a ten year time span difference between the movie and the TV series when they aired for the first time). http://en.wikipedia.org/wiki/Forbidden_Planet

From *Forbidden Planet*: United Planets Cruiser C-57D landing on Altair IV
http://recentlyviewedmovies.blogspot.ca/2011_07_01_archive.html

ARVs – Alien Reproduction Vehicles

While Lazar may have worked on actual ET space vehicles at Area 51 during his short work career there, **Mark McCandlish**, another one of Dr. Greer's **Disclosure Project** witnesses has come forward with testimony revealing a drawing of a reverse engineered disc shape craft derived from captured alien technology.

Mark McCandlish is an accomplished aerospace illustrator and has worked for many of the top aerospace corporations in the United States. His colleague, **Brad Sorenson**, an aerospace inventor with whom he studied, has been inside a facility at **Norton Air Force Base**, where he witnessed **Alien Reproduction Vehicles ARVs)**, that were fully operational and hovering.

It was Brad Sorenson along with a gentleman known a **Frank Carlucci** who, on November 12, 1988, actually saw the **ARVs** or "**Flux Liners**" as they were nicknamed in a large partitioned

657

hangar at Norton AFB during a public airshow. The large hangar was not open to the public for viewing but only to a select invited group of officials.

Sorenson describes that he saw three manmade saucers ranging in size from 24 feet in diameter to 60 feet in diameter at the base up to 120 or 103 feet in diameter. All of the saucers were hovering off the ground with no attachments from the ceiling and no supports underneath from the floor. The smallest one had open sections exposed to allow people to see the inside construction and mechanism of it. All of the components in the system were off-the-shelf components - things that you could find right in the inventory.

Sorenson was told that the US not only has operational antigravity propulsion devices, but has had them for many, many years, and they have been developed through the study, in part, of extraterrestrial vehicles over the past fifty years. **Disclosure: Military and Government Witnesses Reveal the Greatest Secrets in Modern History by Steven M. Greer, MD; 2001; Crossing Point, Inc. Publications; Carden Jennings Publishing Co.; Charlottesville, VA.; ISBN0-9673238-1-9**

Mark McCandlish presents testimony at the National Press Club along with Dr. Steven Greer holding a schematic of man-made flying saucer
https://www.youtube.com/watch?v=VrRwTAEvkX0

The drawing and schematic below is one of these alien reproduction vehicles in which Brad Sorenson described in detail to Mark McCandlish. Together, they were able to illustrate in some remarkable detail what Sorenson had witnessed.

Also, included are photos taken by a US pilot over Utah on July 1966 which I have found on the internet which supports the claim by McCandlish and Sorenson and gives a base timeline as to when the US actually had built and flown these disc shape craft. No hoaxing or disputing its existence! This is a real manmade Flying Saucer!!!

All these **Flux Liners** operate on the principle of **Zero Point Energy (ZPE)** that is by utilizing the quantum flux of space that surrounds all atoms and focusing it through a Tesla-like coil around the craft and into the capacitor plates in the base of the craft which provide the omnidirectional vector of movement. This tapping of the ZPE is kick-started by two marine type batteries sending voltage through the system where it can be potentiated into millions of volts of energy. The result is levitation to either hover or move omnidirectionally in small increments, to "hop" or to make sharp right-hand turns or to perform 180 changes of direction without slowing down. Larger increments of energy into the capacitor plates would enable one to zip about the atmosphere with impunity or to leave the Earth and fly out into space and if need be to travel faster than the speed of light!

All this is possible because of the principle of Zero Point Energy which is absorbed into the craft's propulsion system *(I'll leave the actual technical explanation to the physicists to explain)* creating a diminishment in mass and inertia of the atoms of the spacecraft. It prevents it from interacting with the atomic structure of the vehicle without destabilizing the actual atoms, merely de-energizing or slowing the atoms down. At the same time, it's providing additional power to the capacitor section. It means as that system and all the electrons flowing through that big Tesla coil become mass-canceled, it also becomes the perfect superconductor, which means the efficiency of the systems goes right through the ceiling. The whole electrical system that is going on in the vehicle is in a closed loop, self-contained, ever-increasing, self-perpetuating or infinite energy generation system. In effect, the faster you go, the lighter the craft gets and the easier it becomes to go up to and exceed the speed of light.

It was **James Clerk Maxwell** who speculated that there's enough of this flux, this electrical charge, in the nothingness of space, that if you could capture all the energy that was embedded in just a cubic yard of space, you'd have enough energy to boil the oceans of the entire world. That's how much energy is sitting there waiting to be tapped. Disclosure: by Steven M. Greer, MD; 2001; Crossing Point, Inc. Publications; Carden Jennings Publishing Co.; Charlottesville, VA.; ISBN0-9673238-1-9

The **ARV** even has a mechanical arm that can extend out from these little trap doors that opened up on the side of the ARV, much like the Canadarm on the Space Shuttle. This particular craft has been operational since the early 60s and maybe even longer.

Author's Rant: As a teenager growing up in the Space Age of competitive rocket development and space exploration between America and Russia, I have always wondered, particularly after Neil Armstrong first stepped foot upon the Moon, if there wasn't perhaps, a secret space program going on behind closed doors away from public scrutiny. It always seems strange that after the Apollo 17's lunar excursions that America nor any other nation has ever gone back to the Moon with the intention of building a permanent lunar outpost for further exploration.

It appears that **Tom Bearden** in a conversation with **Mark McCandlish** had also expressed the same sentiments. One of the things that he said, just off the top of his head, was, *"Have you ever wondered why the NASA budget has been cut back so severely? It's because they've got all this other technology that is so much better, so much faster. They are so much better than rocket-propelled spaceships that take months, sometimes years, to get to the outer reaches of the solar system. Why would you put millions and millions of dollars [into] what [amounts to] a public works program for scientists? Why invest all that money when you have this classified system that's used exclusively by the National Security Agency, the CIA, or Air Force Intelligence? It will go anywhere in the solar system in hours, compared to months or years. Why spend all that money on NASA when you've got something that will go there right now?"* http://www.markmccandlish.com/Default.aspx?tabid=147

When people speculate that there might be manned bases on the back side of the moon or there might be bases on Mars, given the current technology in the black world of science, the probability that this is true is almost 100% positive.

According to **McCandlish**, a patent filed by a **James King Jr.** indicated an **ARV** with a similar system except that instead of having a dome for a crew compartment, it has a cylinder in the center. It had the same shape, the flat bottom, and the sloping sides. It has the coils around the circumference, and it has the capacitor plates that are all radially-oriented. This patent was filed initially in 1960 and was secured in 1967 - the same year that a photo was taken near Provo, Utah that looks just like this craft (See photos below).

James King had worked with **Townsend T. Brown** and he, in turn, had worked at a laboratory near Princeton, New Jersey with a scientist by the name of **Agnew Bahnson** in the **Bahnson Laboratories**. These scientists performed experiments in electrogravitic propulsion and on one occasion a video titled "Daddy's Laboratory", that was shot by Agnew Bahnson's daughter on 16 mm film (later converted to VHS) shows all these experiments by Bahnson, Thomas Townsend Brown and their assistant James King. That film shows little discs levitating and shooting off sparks and stuff.

For McCandlish as a **Disclosure Witness** the technology not only exists, they've got the technology in deployment. Not only does it fly but it looks just like the patents that were filed back in the 1960s - the same year the photos were taken near Area 51 - between Area 51 and Provo, Utah, by a military pilot. It shows all the same features; it shows all the same shapes and people that have seen it. I have seen these things myself, so to me, it's just really a matter of time before they bring this technology out of the black and begin to let us use it for other things like pollution-free production of energy. You could probably take a couple of little things that look like those flying saucers and put them around a crankshaft and use them to drive an engine, pollution-free - no use of fuel. http://www.markmccandlish.com/Default.aspx?tabid=147

660

Schematic drawings of ARV (Alien Reproduction Vehicle) (above and below)

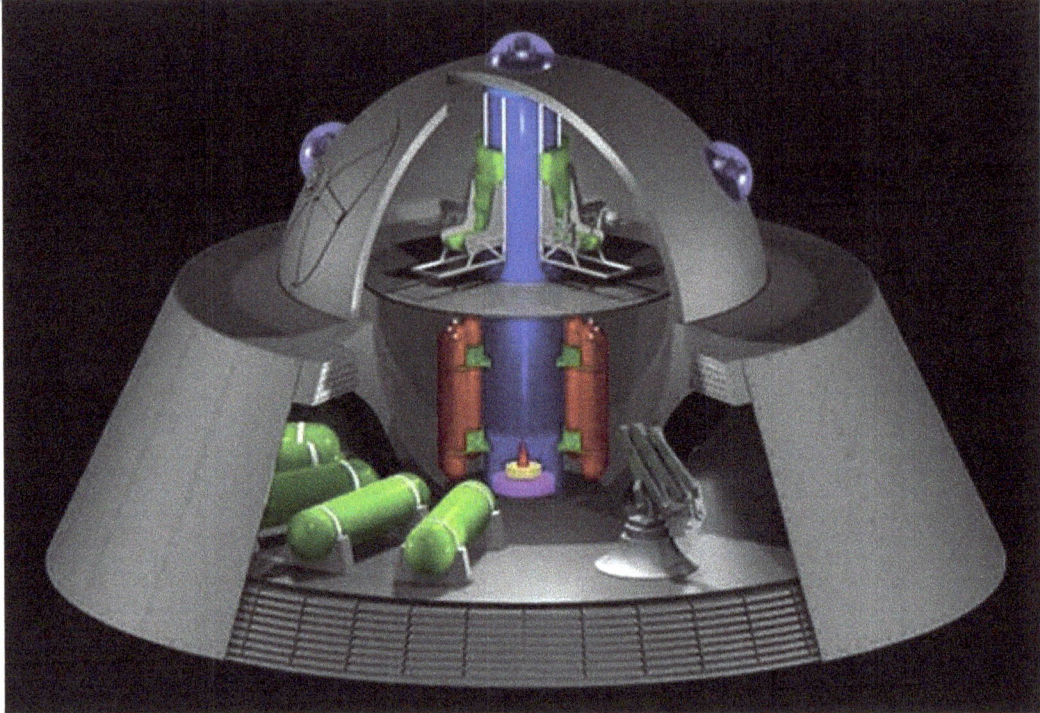

Digital layout of some of the interior components of the ARV
http://www.siriusdisclosure.com/evidence/best-evidence-new-energy/

This photo was taken in July 1966 by an American military pilot when he saw a dark red UFO at 15,000 feet over Provo, Utah. Although, no official explanation was offered, comparison with the drawings of the ARV above, indicate that it is the same craft in flight!
http://nationalufocenter.com/2014/08/filers-files-33-2014-faster-space-travel/

662

The coloured photo of the ARV (Alien Reproduction Vehicle), the "Flux Liner"
http://ufologie.patrickgross.org/htm/usafprovo66.htm

The emergence of **stealth aircraft** as claimed by ufologists is one possible explanation as to why Area 51 exists. Although stealth aircraft first saw action during the Gulf War, Lockheed Martin had been secretly developing the technology for many years at **"Skunkworks"**. The assumption, then, is that the more advanced, exotic vehicles will emerge in the near future when required by the next military enterprise. The problem with this stance is that these vehicles have been witnessed for many, many years and have never been seen to be used. Their advanced capabilities would certainly be of immeasurable benefit to Allied armed forces, but they remain closeted away.

This fuels speculation that these craft are indeed recovered UFOs, or at least our best efforts at replicating the alien technology. After all, the billions of dollars clearly spent on these black projects must have resulted in breakthrough technology, so their obvious absence from the theatres of war can only be attributed to their sensitive, indeed paradigm-shifting, nature. To use these craft is to admit to their existence and to the **Big Secret**. In essence, the whole covert agenda for this technology is at stake. To make matters worse, the American public would then realize that billions of their tax dollars have been covertly spent on reverse-engineering UFOs, with no material benefit to themselves. http://www.ufosaliens.co.uk/cosmicarea51.html

This also begs the question, that if the Big Secret is to be kept from the public, why has all this time, energy and financial resources gone into so much research and development of advanced military aircraft and in reverse engineering of captured alien spacecraft? If it is not going to see combat action, particularly in the Middle East or in some other military war arena, as one would

663

expect then, what other reason would advanced military technology be used for unless there is a more sinister agenda in the near future in which this technology will come into play? This is a serious question that needs a serious answer as it lies directly at the heart of the covert agenda of the Military Industrial Complex

The late **Colonel Corso** would then have claimed that the use of alien technology was indeed seeded into American industry over several decades, but surely this can only be the tip of the iceberg. Consider the potential uses of advanced propulsion, anti-gravity and alternative energy sources inevitably involved in alien technology. How would the voters react if they were aware that the U.S. Government and military has had access to this potential technology for decades, especially considering the damage to our planet inflicted by the misuse of our current energy resources? The potential ramifications of this discovery are enormous. What President in their right mind would let the cat out of this bag? http://www.ufos-aliens.co.uk/cosmicarea51.html

Author's Rant: What has been demonstrated in the construction of an Alien Reproduction Vehicle is a basic understanding of a fundamental principle of energy science that needs to be taught in every high school in the country and around the world. Zero Point Energy is a God-given right of every living being and the ARV is the practical utilization of this infinite energy source which is all around us and is in every living thing. Zero Point Energy is the "universal cement or glue" that bonds the atoms of life together in the universe!

In essence, this energy source is the energy of the future and the future is now! We have been deliberately lied to about its existence and its very nature and technological application into everyday society has been suppressed for over a hundred years or more from its initial discovery. This fact alone should have everyone on this planet pissed off with those who have deliberately kept it off the public radar for as long as they have. The usurpers of this energy generation system are none other than the wealthy corporate elite, the oligarchs of the world, who run most of the global Oil companies. They are the real terrorists of our generation and their outrageous self-centered behavior toward the environment is nothing short of planeticide to the Earth and its people!

Their day in the light is OVER! They have hijacked this planet of its good hopeful future and brought nothing but black inky darkness from this fuel source. Oil companies have polluted this planet and destroyed many lifeforms in the search and distribution of this thick black viscous goop! It has brought nothing but misery to billions of people while benefiting merely a small handful of its inhabitants in corrupt wealth.

Zero Point Energy must and will inevitably replace all other energy generation forms upon this planet, particularly those sources of energy derived from fossil fuel namely, OIL and COAL, but also, other forms of energy generation derived from Thermal Nuclear power, Wind, Solar, Hydro, Geothermal and any combination of hybridization power generation. It should now become obvious to the reader if it hasn't already, why the existence of UFOs and Extraterrestrial visitors has been denied from official public disclosure and why any associated knowledge with this phenomenon has been suppressed, hidden and denied credibility in higher institutions of learning and academia. More will be stated regarding this critical aspect of the UFO/ETI phenomenon in a later section.

Before leaving this sub-section, we need to examine the evidence that there are even stranger, more exotic looking craft, capable of shape-shifting, able to expand in size or fold in upon themselves like some child's articulating mechano-construction set or something out of a **"Star Wars"** movie (see picture below).

Some of these deep black craft are enormous in size reaching one mile in length and half a mile in width, they are probably **LTAVs (Lighter Than Air Vehicles)** (see picture below).

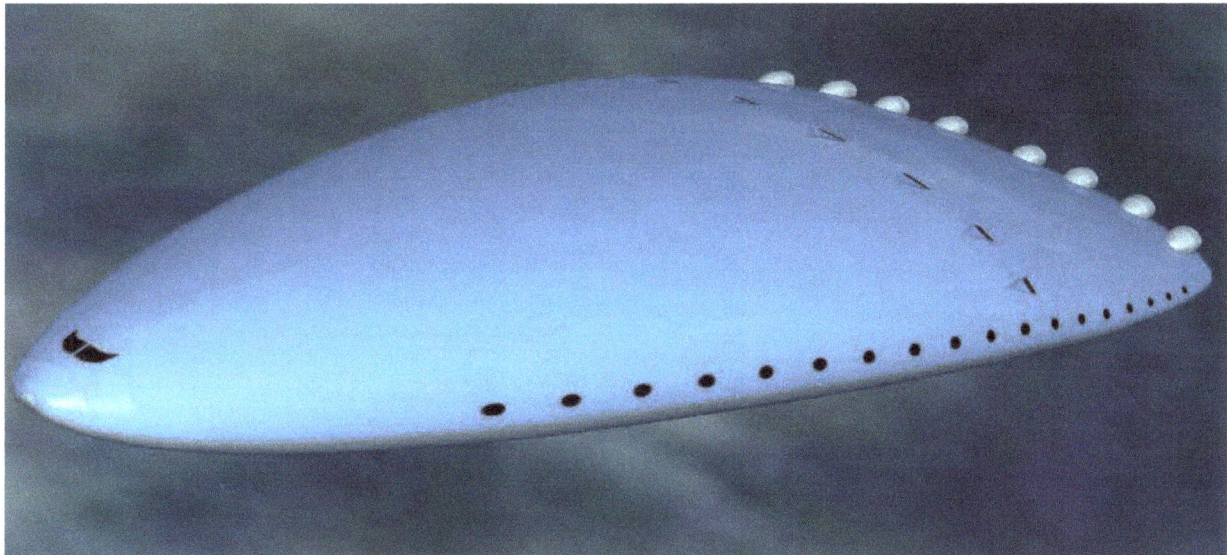

STOL-340 Lockheed Stealth Blimp an enormous deltoid LTAV (Lighter Than Air Vehicle)
by Paul AZ http://www.stratacafe.com/image.asp?imageID=23573

Some craft have straight edges and tubes combined with curves and domes that move independently of each other thus, altering the spaceship's physical configuration. Of course, this could be a description of almost any type of craft, battleship, submarine or even a building, except that there may very well be photographic evidence to support such claims of extraordinary top-secret military spacecraft.

An example of an articulating craft is this detailed sketch based on a Polaroid photo taken of this vehicle hovering near a roadside just north of Cedarville, CA in 1982, in broad daylight. There could be a dozen other witnesses in view in the foreground of the photo, *(coming soon, maybe)*. http://www.markmccandlish.com/Default.aspx?tabid=147

The craft was immense in size, estimated at somewhere between 300 and 600 feet in length along one side. It appeared to be in trouble, with some kind of flight control problem. It was positioned about 250 feet in the air, nose down at a 30 degree angle to the horizontal, banked about 40 degrees to the starboard. It emitted a low frequency, pulsating sound that could be felt and heard even from a quarter of a mile away, in the estimation of the witness.
When the vehicle finally sorted out its problem, it pitched end over end, with the front rising and the trailing "wings" dropping while at the same time rolling on its longitudinal axis clockwise. It ended this manoeuvre pointing up and away from the witnesses. It began to move slowly away,

665

ascending, and once it reached an altitude of perhaps 12,000 feet, it suddenly shot away at a tremendous velocity, and out of sight.

The witness who took the Polaroid returned to the location several days to a week later hoping to find some evidence of the vehicle's presence and was flagged down almost immediately and asked to leave the area by a local county sheriff's deputy.
http://www.markmccandlish.com/Default.aspx?tabid=147

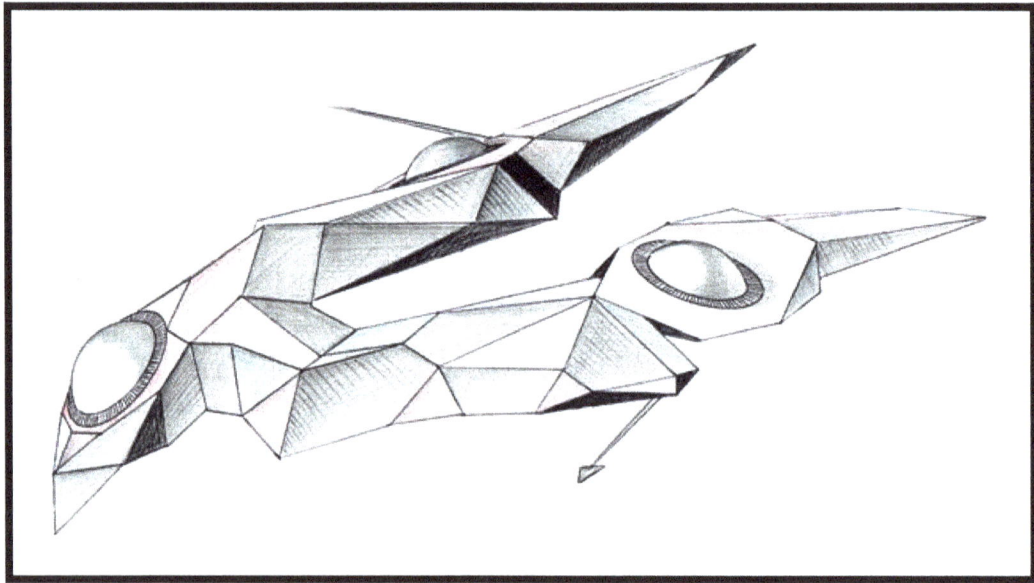

A drawing of a UFO from a photo taken in Cedarville, CA. in 1982. The dome sections on this exotic craft could be the propulsion system or the crew navigation cockpits. The angled sections appear to articulate and lend themselves to stealth capability
http://markmccandlish.com/Unusual-Craft/1982-Sighting-Cedarville

Is there highly advanced US military spacecraft flying routinely into space, outside the traditional publicly aware NASA space program?

To understand the incredible technology that exists in orbit around the Earth we need to look at the strange case of **Gary McKinnon** that has been all the buzz in the UFO world, the major newspapers, and television media in both the US and Britain in the last couple of year.

We will also look at and consider the photographic evidence of **John Lenard Walson** who took alleged photographs of bizarre and unusual objects in space through a special camera and telescope he invented.

Finally, we need to look at the **US Space Command** that is either a new military department or a combined joint venture of the USAF, the US Navy, and the US Marines. The US Space Command has been operating for over thirty years under the auspices of the USAF tracking satellites, space vehicles both domestic and foreign (that includes Unidentified Flying Object) and astronomical bodies like meteors, meteorites, asteroids, and comets.

666

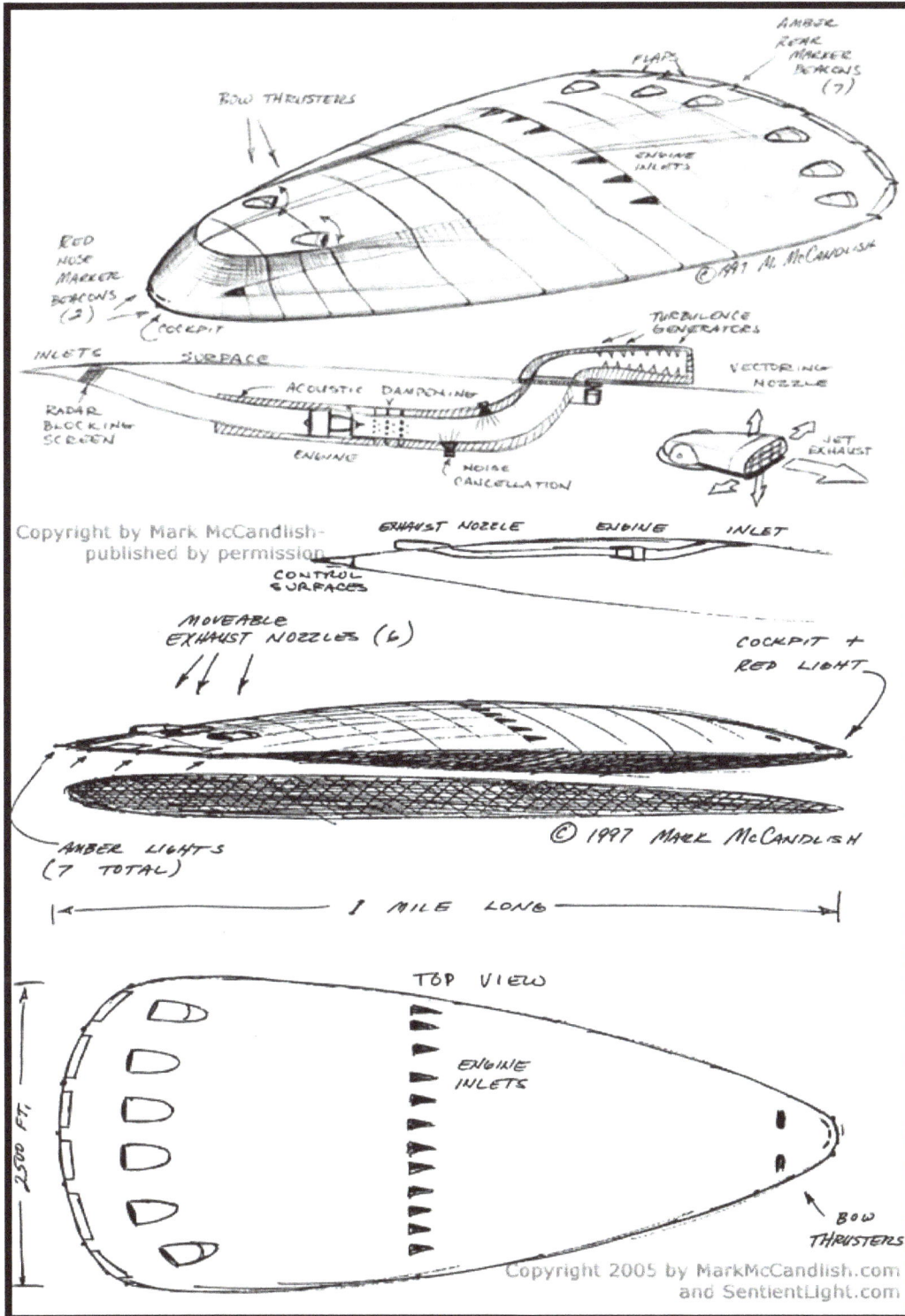

An illustration of an enormous stealth airship or LTAV (Lighter Than Air Vehicle) one mile long as described by Sean David Morton to Mark McCandlish seen near Groom Lake, Nevada overlooking the airfield at Area-51around 1992.

CHAPTER 55

UNITED STATES SPACE COMMAND

Gary McKinnon Hacks into America's Secret Deep Black Space Program

In what US prosecutors have called the biggest military hack of all time, Scottish hacker, **Gary McKinnon** says it was all done in an effort to end secrecy regarding UFOs and Free Energy technology. McKinnon has been accused of hacking into computer systems belonging to NASA, the US Army, US Navy, Department of Defense, and the US Air Force. He is fighting extradition to the United States to be held on trial, and if extradited faces spending the rest of his life in prison, but were his efforts in vain, or did he really find something?

In all of his interviews, McKinnon talks about two UFO-related finds. He told the *Guardian* newspaper that he thought what he found was so important that he tried to barter with the government. When first caught he was offered the chance to take a plea bargain and get a three to four year sentence. He turned the offer down to get a lesser sentence, threatening to release everything he found if they didn't give him a better deal. Unfortunately for Gary, the US government wasn't too worried about his revelations. Now he faces spending a 70 year sentence in a US prison, where they don't serve tea and crumpets.

Gary McKinnon

McKinnon was inspired by physician **Dr. Steven Greer's Disclosure Project**. Greer had brought together a number of very credible witnesses to testify in front of the **Washington National Press Club** that they had knowledge of the existence of Extraterrestrial visitation and that it was being hidden from the public. These people were all saying there is UFO technology, there's anti-gravity, there's free energy, and it's extraterrestrial in origin and they've captured spacecraft and reverse engineered it. He said, he investigated a NASA photographic expert's claim that at the **Johnson Space Center's Building 8**, images from space were regularly "cleaned" or altered; evidence of UFO craft were routinely taken out from the pictures. McKinnon confirmed this, comparing the raw originals with the "processed" images.

McKinnon "stumbled" into Johnson's systems and said he found a high definition picture of "something not man-made", a cigar shaped object floating above the northern hemisphere. He said that he was so shocked by the picture that he didn't think to immediately save it. He also said that the file size was so large that is was difficult to view it on his computer. McKinnon stated the image was approximately 256 megabytes in size, yet that the craft's details were still distinct in 4-bit color and low resolution on his mere 56k modem connection. Eventually, his connection was lost, and so was the picture.
http://www.nationalufocenter.com/artman/publish/article_465.php

Could this digital representation be a reasonable facsimile of one of the US Space Command's secret Black Spacecraft as described by Gary McKinnon
http://www.nationalufocenter.com/artman/publish/article_465.php

The most shocking find to McKinnon, the one he thought would be his ace in the hole in negotiating with the US government, was what found in the systems of the **US Space Command**. McKinnon says he found a log that listed **non-terrestrial officers**. He doesn't believe that these were aliens, but he believes this to be evidence that the US military has a secret battalion in space. Some of these logs were ***ship to ship transfers.***

If Mr. McKinnon's data is correct, it validates the assumption the US Navy may well be operating off-planet via back-engineered ET technology. His data indicates the Navy and probably the Air Force operate a fleet of spacecraft and officers to either man them or otherwise control them. Gary McKinnon states, "***I found a list of fleet-to-fleet transfers and a list of ship names. I looked them up. They weren't U.S. Navy ships. What I saw made me believe they have some kind of spaceship, off-planet***" The US Navy's **Clementine** mission was cited as a project that mapped the entire Moon. Here we see clear evidence of the US Navy's interest in space exploration and mapping of planetary bodies. Below is probably one of the best interviews of NASA/ Pentagon hacker Gary McKinnon, very revealing!!
http://www.disclose.tv/action/viewvideo/198559/UFO_Hacker_Gary_McKinnon_talks_about_NASA_Hack/?utm_medium=email

There are rumors that he has talked about the names of two of the ships he saw on the transfer logs, the names of the ships being the **USSS** *LeMay* and the **USSS** *Hillenkoetter*. Typically Navy ship names just have two S' as in the designation **USS**, an acronym for **United States Ship**, however, there are three S' here as in **USSS**, presumably standing for **United States Space Ship**. The names of the ships are also significant.
http://www.nationalufocenter.com/artman/publish/article_465.php

General Curtis LeMay

General Curtis LeMay was friends with retired Air Force Reserve Major General and former

U.S. Senator from Arizona, **Barry Goldwater**. Goldwater believed there was a UFO cover-up deep within the government, and suspected that his friend LeMay knew about it. There were rumors that there was UFO evidence being held in a secret room at Wright-Patterson Air Force Base called the **"Blue Room"**. Goldwater told the media several times that when he asked LeMay about this room, LeMay got upset and told him,

"Not only can't you get into it but don't you ever mention it to me again."

The second ship's namesake, **Admiral Roscoe Hillenkoetter**, was the first director of the CIA and was also a member of a UFO research organization, the **National Investigations Committee on Aerial Phenomena (NICAP)**. In 1960 the *New York Times* reported that Hillenkoetter had sent a letter to Congress that included this statement:

"Behind the scenes, high-ranking Air Force officers are soberly concerned about UFOs. But through official secrecy and ridicule, many citizens are led to believe the unknown flying objects are nonsense."

Although Hillenkoetter fought for the end of UFO secrecy, he eventually stopped commenting on the matter. Alleged secret documents that were leaked to UFO researchers, list Hillenkoetter as a member of the infamous **Majestic 12** group, an organization rumored to have been made up of high-ranking military officers and civilians that was supposedly created by President Truman to initially manage the UFO issue.

http://www.nationalufocenter.com/artman/publish/article_465.php

Admiral Hillenkoetter
https://fr.wikipedia.org/wiki/Roscoe_Henry_Hillenkoetter

So that is it, the UFO picture that McKinnon saw and the ship rosters were all he had, and unfortunately for him, they were not enough to scare the government into going easy on him. Instead, he has been fighting a long multi-year battle to keep from being extradited to the US, a fight that he is losing. So far every British court he has appealed to has denied his stay.

The US government is really throwing the book at him, alleging that he took down military computers making the US vulnerable soon after 911. McKinnon denies those claims and says that he was able to observe many hackers from around the world accessing the networks he was on at the same time he was on them. Many believe that McKinnon may just be a scapegoat.
http://www.nationalufocenter.com/artman/publish/article_465.php

The inevitable extradition of Gary McKinnon has taken another bizarre turn as it has come to light that he suffers from Asperger's syndrome and may also be suicidal with the thought of knowing that he may end up in a US prison. His mother has spoken to the news media and with British authorities regarding his Aspergers but, the expert chosen by the British Home Office had no experience with Asperger's syndrome.

Computer hacker McKinnon "has no choice" but to refuse a medical test to see if he is fit to be extradited to the US, his mother has said. At least this action would delay the extradition for a while longer.

The home secretary, Theresa May, has delayed until October any announcement on whether to extradite the computer hacker Gary McKinnon to the US.

His family says the Home Office expert, **Professor Thomas Fahy**, has no experience in uncovering suicidal tendencies in Asperger's syndrome patients. Mr. McKinnon, 46, had three medical examinations in April by three leading experts in Aspergers and suicidal risk. They concluded Mr. McKinnon was at extreme risk of suicide if extradited and he was currently unfit for trial.

Prime Minister David Cameron has stated, *"I do recognize the seriousness of this case, and the Deputy Prime Minister and I actually raised it with **President Obama** when he visited. I think the point is that it is not so much about the alleged offence, which everyone knows is a very serious offence, and we can understand why the Americans feel so strongly about it."*

The search for truth on the UFO matter is a difficult one, and some may argue that the secrets being kept are illegal. However, taking illegal steps to get to the truth is ill-advised, and unfortunately, McKinnon is learning this the hard way.
http://www.nationalufocenter.com/artman/publish/article_465.php

There is a lot more going on here that is not being revealed by US authorities and I feel that Gary McKinnon is indeed a scapegoat for this type of government hacking into sensitive websites. Of course, the US government and their military are within their rights to maintain security from such hacking practices by individuals and other government intelligence agencies. The problem, however, is that ***China and Russia are the two largest hackers*** into the US government, its agencies and the various US military departments but also, they routinely computer hack into

672

many of the largest corporations in the States as well as those of many western nations. *Yet, we do not see the same kind of threats and intimidations used on these nations* as we see being played out with McKinnon case.

If McKinnon hacked into sensitive files and websites that revealed a secret military space force and this author strongly feels he has, but, was unable to download the files or photos to back up his claim then, rest assure that Russians and Chinese and God knows whoever else, also has accessed this same information and knows what the US military is up to. Their weak-kneed excuse that he was trying to take down their military computers to make the US vulnerable so soon after 911is bogus and merely a deflection away from the real truth of the matter.

The truth is that Gary McKinnon had discovered something very few people in a lifetime of UFO research come across and that is a highly significant and important piece of the UFO puzzle. He discovered that the US has indeed, a secret Black Space Program outside the public awareness of NASA's space program. He found that some of their spaceships have names of prominent people high up in military circles. These are men who have distinguished themselves throughout their military careers as go-getters, hard-nosed, no-nonsense soldiers who get the job done; men who have reached the top of their professional careers and are considered as heroes of their country. What better way to honour and recognize these men, than by naming a highly advanced spacecraft after them! This is what Gary McKinnon discovered and for his poor lack of judgement, the US government wants his "head on a platter" as an object lesson to all would-be hackers. **Todate, the US has failed to extradite Mckinnon. He walks as a free man!**

Author's Rant: The best thing for Gary to do, if he is going down for a breach of international espionage is to go down fighting and broadcast load and wide to every world news media what he found. Put the spotlight of truth on the whole UFO/ETI subject and the involvement of the US in the cover-up with the exploitation of alien technology in covert deep black programs and projects.

To many, this sounds absolutely ludicrous and just plain conspiratorial in thinking, right? The reader at this point may feel that the author must have some personal vendetta to roast the US military over the coals, that they represent everything wrong and evil in the world, today. Wrong! I state categorically that it is not the US military per say, but rather those rogue elements within it that have created confusion within the military mindset and have used the various departments of the military as puppets to do its bidding and to carry out its hidden agendas.

Gary McKinnon is just another computer hacker looking for hidden UFO and ETI information within the government, the Military, NASA and the Pentagon and then, ineptly got caught by leaving a traceable tag leading ultimately back to himself. However, it has been shown through photographs taken from the space shuttle that the US military does indeed have secret and covert aircraft and spacecraft that are both triangular and disc shape able to travel into space and beyond as easy as a jet flies from New York to London and back. These spacecraft have been identified as the **TR-3B**, the **Aurora**, the **ASTRA**, the **Black Manta**, and various **ARVs**, nuclear-powered saucers, some have designations as the **X-37B** (a spacecraft similar to the Space Shuttle operated by remote control), **Tin Can** and **NOTE**, and there are even actual ET space vehicles.

Is there other photographic evidence of America's secret hidden Black Space Program? According to many UFO websites that believe that there is evidence, one, in particular, is the Jeff Rense's "Sightings" section from his website: **Rense.com** which first brought public awareness to the photographic work of John Lenard Walson.

Photographic Evidence of America's Secret Black Space Program

In the time period of 2007 – 2008, a young man by the name of John Lenard Walson suddenly became known to the UFO community with his release of some remarkable photographs over the Jeff Rense website. He had discovered a new way to take a digital video camera and a small telescope and extend its capabilities to enhance the optical resolutions enabling him to photograph some unusual machine-like objects in space. Over the next four or five years, Walson has videotaped these strange objects both night and day, many of which have never been seen before in Earth orbit. The video footage resulting from Walson's astrophotography reveals a multitude of unusually large machines, possible satellites, spacecraft like the **ISS (Internation Space Station)** and possibly space ships or orbiting military weapons platforms which otherwise would to the casual observer appear as "stars".

UFO researchers and astronomy buffs have stated on various websites that John Walson's video and photos do leave a lot to be desired in terms of "clear" photographic evidence. That they are too fuzzy and indistinct, that it could be photographs of almost anything videotaped under dark lighting.

A good pair of binoculars like 10 x50"s or higher will give good optic resolution to low orbital satellites or even the ISS, or whatever it is that is flying overhead in orbit. The resolution won't be crystal clear to make out a lot of detail but, good enough to distinguish some of the solar panels on the space station.

Author's Rant: I have a pair of binoculars (10 x 50) with a phenomenal 7.1 degree field of view and I have been able to see Jupiter and her four largest moons with them. In summer of 1996, I used these same binoculars to observe very large white saucer-shaped craft with a dome top in the Crestone, Colorado area which had appeared high up above the Earth to take up a geosynchronous orbit with the other stars. To any casual observer without a telescope or binoculars it would have been just another "star", but was, in fact, miles in diameter!!!

Walson stated that he used an 8 inch Meade telescope which should have been sufficient optical technology to have photographed with reasonable clarity the large orbiting machine structures. In addition to discovering and refining his optical telescope videotaping technique, John has also discovered how to actually hear and record the sounds in real time coming from the particular craft he is videotaping. By carefully aligning a satellite dish receiver with his telescope, he has been able to record some very unusual and intriguing sound from the different spacecraft. http://rense.com/general79/wdx.htm

The point here is that filming with good clarity through a small reflector telescope such as a 4.5 or 6 inch mirror wouldn't be a problem unless you are photographing through a higher powered

674

secondary objective lens which will cut down on the amount of light received if you are attempting to magnify the image larger. This may have been the case with most of Walson's videotapes which have been discredited among some Ufologist and earned him a status as a possible hoaxer. It has been stated by some UFO researchers hat his video is nothing more than tin-foil spaceship-designed constructions filmed through mirrors and glass with low or poor lighting to give the perception of distant orbiting spacecraft.

Needless to say, the skeptics and debunkers have had a field day at Walson's expense but then, that is nothing new in this quasi-science of Ufology.

What can be said without argument, is that John Lenard Walson has been harassed by numerous unmarked large helicopters near or above his home from either government or military agencies routinely at different times of the day or night. He has obviously irked the wrath of someone or some agency who feel that he needs an object lesson by intimidation to get him to back off from his telescopic investigation and videotaping of these large, mysterious orbiting machines.

If Walson was, as his accusers were claiming that he was deliberately hoaxing videotape of high orbiting spacecraft then, why the harassment from the military or intelligence agency? Would not a deliberate hoax by Walson be playing right into the hands of the covert agenda of the military intelligence complex? Would it not be an added bonus from a civilian that they could use to exploit or discredit the UFO community?

Surely, the helicopter harassments which are typical intelligence ploys only go to further John Walson claims that he was indeed videotaping actual military spacecraft in high orbit! Why attract that kind of unwanted government or military attention? And then, there is the testimony made by Gary McKinnon that he hacked into the Pentagon database years ago and found a secret US Navy file titled **"Non-Terrestrial Officers"**. Could John **Lenard Walson** like **Gary McKinnon**, have also found an important piece of the UFO puzzle?

It does give one time to pause and consider the information that has come to light in the last few years with regard to a possible US Military **Black Space Program**.

There has been more creditability given to the dubious information coming from alien abductions that lack far less evidence based upon the circumstantial or subjective information derived from hypnotic regression testimony. Such abduction testimony runs the gambit from alien implants, anal probes, sexual fertility and medical procedures, biopsies, downloading of new scientific information to having hybrid babies floating around in space somewhere and the American populace seem to be in the forefront of this bizarre phenomenon. Yet, it is given the highest credibility rating in the UFO subculture.

Should we not also, take the testimony of McKinnon and the photographic evidence of Walson respectively, and added it to the photographs of strange triangular objects taken from the various Space Shuttles then, process it thoroughly to distil any useful information from it? This then is tangible hard-core evidence that is not subjective, imagined or hoaxed!
What did John Walson photograph from his newly enhanced telescopic system that he invented?

Many of the stills from the videotapes show objects that are long in length similar to the "Death Star Cruisers" from the **Star Wars** movies or as one of the battle cruisers from the TV show, **Battlestar Galactica,** with their large command and navigational sections in the aft section of the spaceship. Some of the machines are box shape with interlacing support structures that can expand in size or collapse in upon itself in a matter of mere moments all the while travelling at great velocity. Some ships appear as bug or crustacean-like, others look like flying assault rifles with high powered scopes (you get the idea).

Author's Rant: (I have personally remote viewed this type of spacecraft, back in 1996 which resembled, what I refer to as a flying "Skeleton Key" in appearance.

The jury of the UFO community is still out as to whether these spacecraft are alien in nature parking themselves in near-Earth proximity or whether they are manmade secret orbiting weapons platforms from the US or Russia or both countries. The odds on favourite is that these are manmade nut and bolts craft tinkered together using a hybrid technology derived from alien recovered craft from such crash saucer incidents as Roswell, Aztec, Kingman, Kecksburg, etc. as well as from highly advanced terrestrial technology developed out of Nazi Germany and from good old fashion American and Russian aeronautical and naval ingenuity.

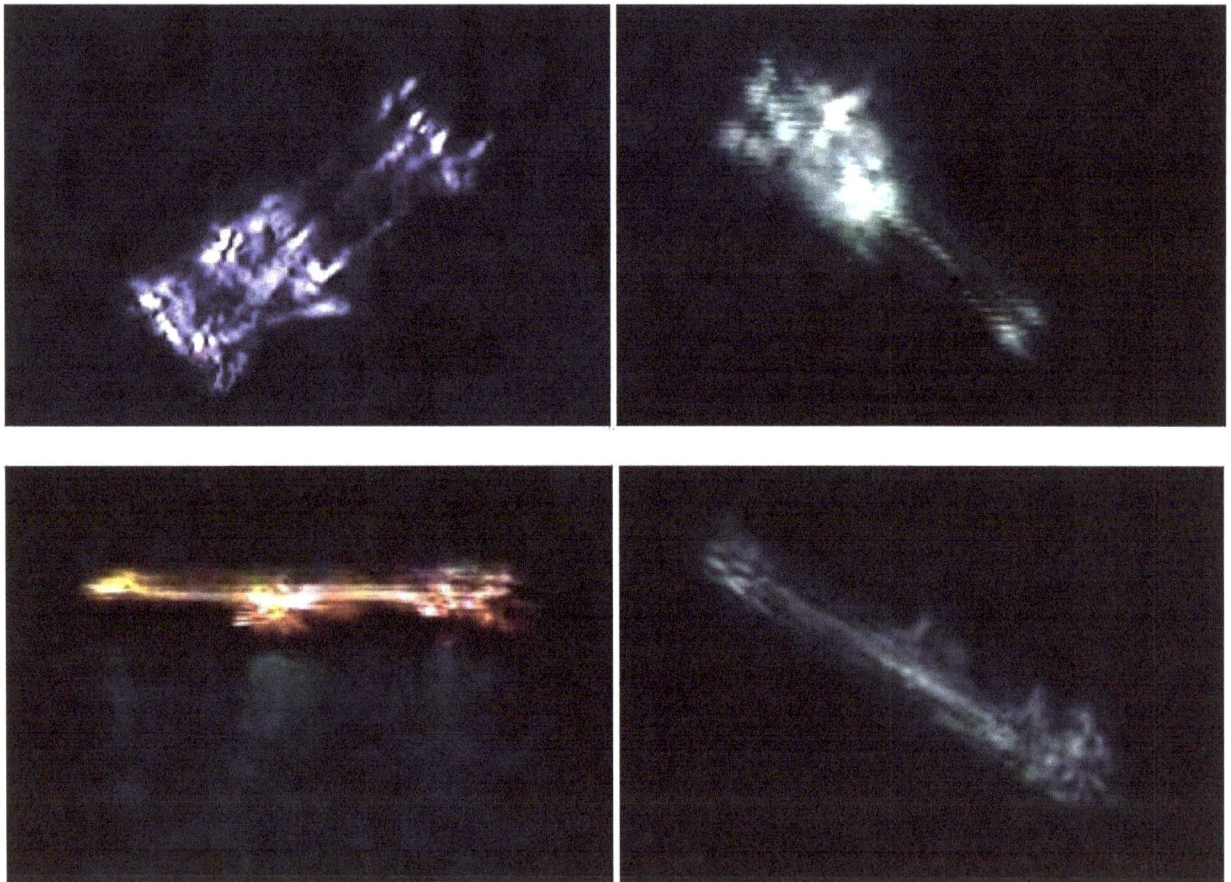

John Walson's photographs of strange battle ship-like spacecraft in orbit around Earth (© John Lenard Walson)
http://www.tarrdaniel.com/documents/Ufology/spacemachines.html

676

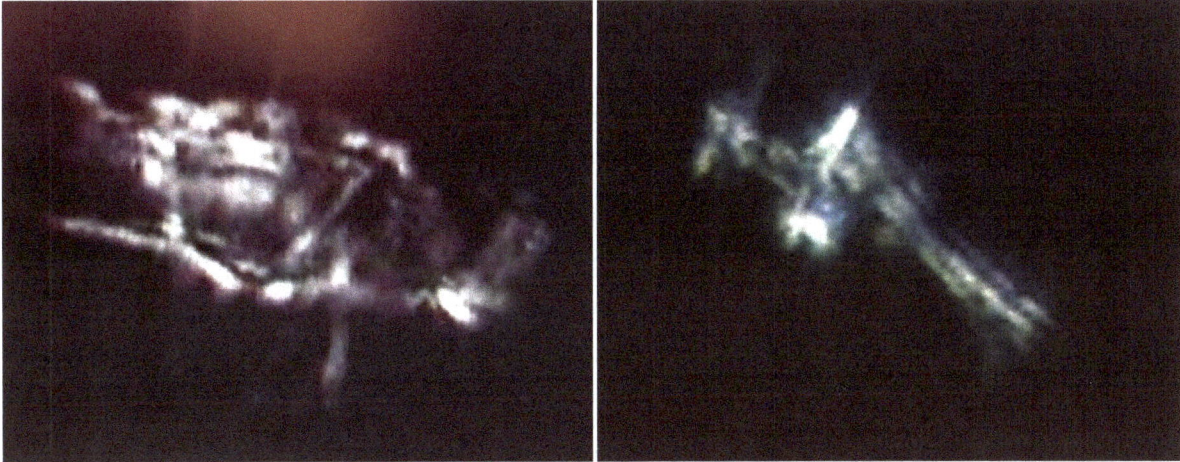

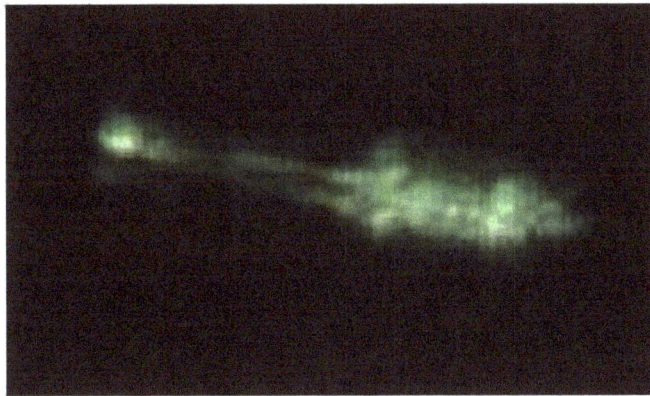

Sample photos of unusual spacecraft orbiting the Earth (© John Lenard Walson)
http://www.tarrdaniel.com/documents/Ufology/spacemachines.html

We need to carefully consider that there have been major scientific breakthroughs in reverse engineered alien technology married with highly advanced terrestrial technology to produce spacecraft that are decades, if not centuries beyond conventional white world science. Sciences as far advanced beyond our current understanding as taught in our universities and institutes of higher learning that **Ben Rich**, the former head of **"Skunkworks"** of Lougheed Martin has said, that, *"Whatever you can imagine, we have already done it!"*

There are many departments of the US Military as well as top US Intelligence agencies that have a space program building satellites and spacecraft to enable monitoring , surveillance and first strike response to any situation whether foreign or domestic. These departments and agencies have many branches within them that utilize conventional technology like rockets, computers, communication and navigational technology that are all within the frame of current knowledge of science. These operate at times on a Black Program level with a need to know basis. For the most part, these agencies and departments, the people and officers within them are out of the loop of information when it comes to **Deep Black Projects and Programs** that involve anything to do with Extraterrestrial research and development.

677

There are, however, within these organizations, other branches that are privy to things that are Extraterrestrial in nature with their associated programs and projects that have been implemented to exploit these alien technologies. These branches have a **Cosmic Top Secret** level requirement that is given to very few people who have a need to know with respect to the truth about the **"ETI Question"**.

There are rogue elements within these top secret branches of government or military intelligence that do the operational work, the crash retrievals of ETI craft, the security enforcement, the actual dirty work of threats, intimidations and murder when required. These rogue elements seem to operate from different levels of covert activity behind hidden agendas that appear to be outside of the control of the Deep Black Programs. They seem to operate independently of military and government intelligence and take orders from a cabal of transnational power and ultimate authority.

British UFO author **Timothy Good,** retired **Major Robert O. Dean, Dr. Steven Greer of the Disclosure Project, Astronauts Dr. Edgar Mitchell** and **Gordon Cooper, Dr. Carol Rosin, Army Sgt. Clifford Stone, Col. Philip Corso,** and **Mark McCandlish** are just a few of the many UFO researchers, Disclosure witnesses, astronauts, and military officers and government officials that have talked and written about the levels of secrecy behind the UFO/ETI phenomenon.

Below are a few of the USA's top secret military and intelligence agencies that have multi-billion dollar budgets to carry out almost anything they desire including their own space programs and who have many branch levels that operate in **Black and Deep Black Programs** Some of these agencies have rogue groups who operate above these agencies control and oversight but, who use these agencies as a cover for their own covert hidden agenda.

These agencies below indicate the known parameters from which they operate and no doubt there is a gray line or area outside of these parameters which the public knows very little or anything about.

United States Space Command

The **United States Space Command (USSPACECOM)** was a **Unified Combatant Command** of the **United States Department of Defense**, created in 1985 to help institutionalize the use of outer space by the United States Armed Forces. The **Commander in Chief of U.S. Space Command (CINCUSSPACECOM),** with headquarters at Peterson Air Force Base, Colorado was also the **Commander in Chief of the binational U.S.-Canadian North American Aerospace Defense Command (CINCNORAD),** and for the majority of time during USSPACECOM's existence, also the **Commander of the U.S. Air Force major command Air Force Space Command**. Military space operations coordinated by USSPACECOM proved to be very valuable for the U.S.-led coalition in the 1991 Persian Gulf War.

The U.S. military has relied on satellite communications, intelligence, navigation, missile warning and weather systems in areas of conflict since at least the early 1990s, including the

Balkans, Southwest Asia, and Afghanistan. Space systems have since then been considered as indispensable providers of tactical information to U.S. forces.

U.S. Space Command emblem
https://en.wikipedia.org/wiki/United_States_Space_Command

As part of the ongoing initiative to transform the U.S. military, on June 26, 2002, **Secretary of Defense Donald Rumsfeld** announced that **U.S. Space Command** would merge with **USSTRATCOM (U.S. Strategic Command)** . The Unified Command Plan directed that Unified Combatant Commands be capped at ten, and with the formation of the new United States Northern Command, one would have to be deactivated in order to maintain that level. Thus the USSPACECOM merger into an expanded USSTRATCOM, which would retain the **U.S. Strategic Command** name and would be headquartered at Offutt Air Force Base. The merger was intended to improve combat effectiveness and speeds up information collection and assessment needed for strategic decision-making.

Within STRATCOM, responsibilities for space were first held by the **Joint Functional Component Command for Space and Global Strike** until July 2006 when the command was divided. Space operations are now overseen by the **Joint Functional Component Command for Space.** http://en.wikipedia.org/wiki/United_States_Space_Command

US Space Command was disbanded on October 1, 2002, and its responsibilities were handed over to **US STRATCOM.**

As one of the nation's nine unified (multi-service) commands, **U.S. Space Command** coordinated the use of *Army, Naval, and Air Force space forces.* These space forces provide the information needed by the US military to out-maneuver the enemy, attack with precision and protect themselves from attack.

The men and women of U.S. Space Command put the satellites that provide these capabilities in orbit, operated them, protected them, and ensured that the information they provide is exactly what America's warfighters need to protect national security interests today and tomorrow.

USSPACECOM coordinated the use of the Department of Defense's military space forces in providing:

- **Missile Warning** - Defense Support Program satellites and ground-based radars provide both strategic and theater ballistic missile warning to our nation's leadership and to deployed troops worldwide.
- **Communications** - Communication satellites provide constant global connectivity with deployed forces.
- **Navigation** - The Air Force Space Command's Global Positioning System (GPS) constellation of 28 satellites provides precise navigation and timing support to coordinate the positioning and maneuver of U.S. and allied aircrews, naval forces, and ground forces.
- **Weather** - Defense Meteorological Satellite Program collects and distributes global weather data.
- **Imagery & Signals Intelligence** - U.S. military space operators coordinate space-based imagery between intelligence agencies and planners within Unified Commands. http://www.globalsecurity.org/space/agency/usspacecom.htm

USSPACECOM's major functions were to support the **North American Aerospace Defense Command (NORAD)** warning and assessment, conduct space operations and, computer network operations:

- **Space Support** - Launching and operating satellites, including satellite operations and telemetry, tracking and commanding and, spare activation. All launches occur at Cape Canaveral Air Force Station, FL, or Vandenberg Air Force Base, CA
- **Force Enhancement** - Satellite communications, navigation, weather, missile warning and intelligence.
- **Space Control** - Assuring U.S. access to and freedom of operation in space, and denying enemies the same.
- **Force Application** - Researching and developing space-based capabilities that have the potential to engage adversaries from space. Requires policy change before implementation.
- **Computer Network Defense/Computer Network Attack (CND/CNA)** - Computer Network Defense includes protecting and defending information, computers, and networks from disruption, denial, degradation, or destruction. Computer Network Attack includes developing the capabilities to disrupt, deny, degrade or destroy information resident in computers, computer networks or, computers and networks themselves. The USSPACECOM component command for these missions is **Joint Task Force - Computer Network Operations.**

680

- **Information Operations (IO)** - Although not a specific mission, USSPACECOM is the sponsor for the **Joint Information Operations Center (JIOC)** at Lackland AFB in San Antonio, Texas. The JIOC maintains specialized expertise in IO systems engineering, operational applications, capabilities, and vulnerabilities. The JIOC also assists with the development of IN doctrine, tactics and procedures.
  http://www.globalsecurity.org/space/agency/usspacecom.htm

Air Force Space Command

In 1982, the Air Force established **Air Force Space Command**, with space operations as its primary mission. During the Cold War, space operations focused on missile warning, launch operations, satellite control, space surveillance and command and control for national leadership. In 1991, **Operation Desert Storm** validated the command's continuing focus on support to the warfighter. The **Space Warfare Center**, now named the **Space Innovation and Development Center**, was created to ensure space capabilities reached the warfighters who needed it. ICBM forces joined AFSPC in July 1993.

AFSPC's mission is to provide resilient and cost-effective Space and Cyberspace capabilities for the Joint Force and the Nation.

Air Force Space Command Shield
http://www.globalsecurity.org/space/agency/afspc.htm

Space Capabilities: Spacelift operations at the East and West Coast launch bases provide services, facilities and range safety control for the conduct of DOD, NASA, and commercial launches. Through the command and control of all DOD satellites, satellite operators provide force-multiplying effects -- continuous global coverage, low vulnerability and autonomous

operations. Satellites provide essential in-theater secure communications, weather and navigational data for ground, air and fleet operations and threat warning.
Ground-based radar, Space-Based Infrared System and Defense Support Program satellites monitor ballistic missile launches around the world to guard against a surprise missile attack on North America. Space surveillance radars provide vital information on the location of satellites and space debris for the nation and the world. Maintaining space superiority is an emerging capability required to protect U.S. space assets.

Cyberspace Capabilities: The Air Force's overall goal in cyberspace operations is to assure the mission - finding and using the best tools, skills, and capabilities to ensure the ability to fly, fight, and win in air, space, and cyberspace. Cyberspace is critical to joint and Air Force operations. AFSPC conducts cyberspace operations through its subordinate units within 24th Air Force, including the 67th Network Warfare Wing, the 688th Information Operations Wing, both headquartered at Lackland AFB, Texas, as well as the 689th Combat Communications Wing headquartered at Robins AFB, Ga.

Collectively, these units are the warfighting organizations that establish, operate, maintain and defend Air Force networks and conduct full-spectrum operations. These organizations, made up of cyberspace professionals, a diverse blend of career fields including cyber operators, intelligence professionals, acquisitions personnel, aviators and many more, ensure the Air Force and joint force ability to conduct operations in, through and from cyberspace.
http://www.globalsecurity.org/space/agency/usspacecom.htm

Naval Space Command

Not wanting to be left out from operating from the "high ground", the US Navy as one of the major arms of the American military decided that it made military sense to have its own space command which could communicate with naval ships and submarines on any ocean or seas anywhere in the world

Naval Space Command is a military command of the United States Navy. It was headquartered at Dahlgren, Va., and began operations Oct. 1, 1983. Naval Space Command used the medium of space and its potential to provide essential information and capabilities to shore and afloat naval forces by a variety of means:

- Operating surveillance, navigation, communication, environmental, and information systems;
- Advocating naval warfighting requirements in the joint arena; and
- Advising, supporting, and assisting the naval services through training, and by developing space plans, programs, policies, concepts, and doctrine.

The command was merged into **Naval Network and Space Operations Command**, itself part of **Naval Network Warfare Command**, about July 2002.

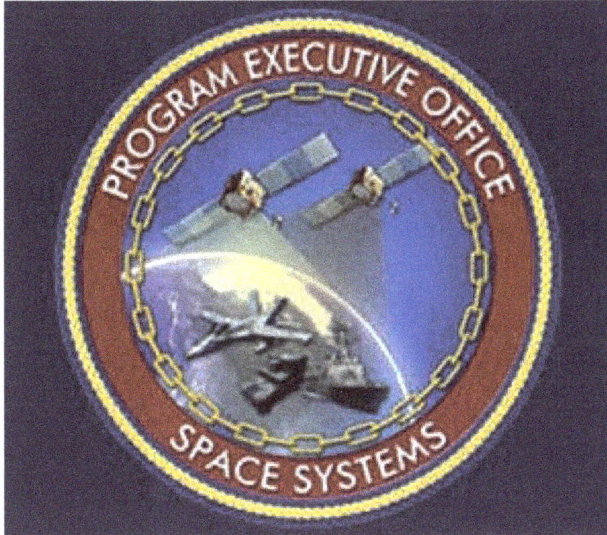

Naval Space Command Emblems

Naval Space Command's headquarters staff and operational element numbers approximately 350 Navy military and civilian personnel. Their component commands include the **Naval Satellite Operations Center** and the **Fleet Surveillance Support Command**.

- Naval Space Command, a component of **USSPACECOM**, operates assigned space systems to provide surveillance and warning and provides spacecraft telemetry and on-orbit engineering support. In addition, Naval Space Command serves as the **Alternate Space Control Center [AASC]** for USSPACECOM's primary centers located at **Cheyenne Mountain** AS.

- ASCC missions include operational direction of the entire global space surveillance network (SSN) for commander in chief, USSPACECOM (USCINCSPACE). The ASCC also detects, tracks, identifies, and catalogs all man-made objects in space and provides position information on these objects to about 1,000 customers. In addition, ASCC is charged with monitoring the space environment and informing owners and operators of U.S. and allied space systems of potential threats to their assets by continuous liaison with the systems' operations centers.

- The heartbeat of **Naval Space Command** revolves around providing space support to day-to-day operations of the **Fleet and Fleet Marine Forces** worldwide, whether for routine deployments, exercises, or actions in response to a crisis situation. This space support to terrestrial and naval forces can be categorized across a broad spectrum of activities that encompass communications, surveillance, and indication, and warning, intelligence, navigation, and remote sensing.
http://en.wikipedia.org/wiki/Naval_Space_Command

In a recent news article of the Washington Post that has probably slip by mainstream media, and which is quite revealing with this segment of a secretive space program, the budget-constrained

NASA having left those in the public interested in the space program wondering what will replace the space shuttle, will we ever return to the Moon again, whether the **James Webb Telescopes** will ever get into space, etc., became the unexpected beneficiaries of two highly advanced spy telescopes from the **NRO (National Reconnaissance Office).** The news article reads:

One of the emblems of the NRO
http://www.collectspace.com/news/news-083100a.html

NASA Gets Two Military Spy Telescopes for Astronomy

(**Washington Post**) The secretive government agency that flies spy satellites has made a stunning gift to NASA: two exquisite telescopes as big and powerful as the Hubble Space Telescope. They've never left the ground and are in storage in Rochester, N.Y.

It's an unusual technology transfer from the military-intelligence space program to the better-known civilian space agency. It could be a boost for NASA's troubled science program, which is groaning under the budgetary weight of the James Webb Space Telescope, still at least six years from launch.

Or it could be a gift that becomes a burden. NASA isn't sure it can afford to put even one of the two new telescopes into orbit.

The telescopes were built by private contractors for the National Reconnaissance Office, one of 16 U.S. intelligence agencies. The telescopes have 2.4-meter (7.9-foot) mirrors, just like the Hubble, but they have 100 times the field of view. Their structure is shorter and squatter.

684

They're "space qualified," as NASA puts it, but they're a long way from being functioning space telescopes. They have no instruments — there are no cameras, for example. More than that, they lack a funded mission and all that entails, such as a scientific program, support staff, data analysis and office space. They will remain in storage while NASA mulls its options.

"It's great news," said NASA astrophysics director **Paul Hertz**. "It's real hardware, and it's got really impressive capabilities."

The announcement Monday raised the obvious question of why the intelligence agency would no longer want, or need, two Hubble-class telescopes. A spokeswoman, Loretta DeSio, provided information sparingly.

"They no longer possessed intelligence-collection uses," she said of the telescopes.

She confirmed that the hardware represents an upgrade of Hubble's optical technology.

Author's Rant: The reader should be questioning the generosity of the NRO with their gift of two mothball telescopes that have never left the ground for orbital space, being many times better than Hubble. Also, the question arises, what does the NRO have in their position to replace these unused telescopes since they are not limited by budgetary constraints like NASA?

Bear in mind that we never really get the full story with anything regarding military or technological advancements.

There are insiders in the infamous Area 51 who have come out in secret to reveal that there is technology that has been developed since the '60s that still has not seen the light of day. The rumours that are bantered about the UFO community regarding the probable existence of bases on the Moon and Mars with regulars missions to these solar bodies speaks volumes as to the possible technology that exists in the black world.

It's thought that black-op advances are decades if not centuries ahead of "known" technologies.
http://en.wikipedia.org/wiki/Naval_Space_Command

On September 1, 2009, Jeff Rense and a guest with the blog name of Trunkneck discussed the topic of Space planes on his Radio Show. **Aviation Week and Space Technology magazine** investigated myriad sightings of a two-stage-to-orbit system that could place a small military spaceplane in orbit during the 1990s. After the shuttle **Challenger** disaster in January 1986, and a subsequent string of expendable-booster failures, the military needed a quick method to get satellites into space to keep tabs on its Cold War adversaries. Considerable evidence supports the existence of a highly classified system, and top Pentagon officials have hinted that it's "out there". It appears the **"Blackstar" system** may have replaced the shelved U.S. **SR-71 reconnaissance aircraft**. It was composed of a large **"Mothership",** closely resembling the Air Force's historic **XB-70 supersonic bomber** with the orbital component under its fuselage, accelerating to supersonic speeds at high altitude before dropping the space plane. The orbiter's engines fire and boost the vehicle into space. The shuttle could also carry a small space plane into orbit that could then be launched.

The **National Aero-Space Plane (NASP)** was under development during the **Carter Administration** that could reach low Earth orbit but was allegedly cancelled.
http://ufoweek.com/tag/reconnaissance-aircraft/

Satellite orbits are predictable and activities having intelligence value can be scheduled to avoid overflights. The Navy and **Air Force Space Command (AFSPC)** have wanted an operational space plane for years. Small maneuverable spacecraft could be carried into orbit by the Shuttle or launched from a Mothership and conduct key reconnaissance. The space plane is capable of carrying an advanced imaging suite that could be configured to deliver specialized microsatellites to low Earth orbit or, perhaps, be fitted with no warhead hypervelocity weapons– what military visionaries have called "rods from god." Launched from the fringes of space, these high-Mach weapons could destroy deeply buried bunkers and weapons facilities.

The "Blackstar" Sr-3 and the XOV in flight together
http://www.spyflight.co.uk/blackstar.htm

The "Blackstar" Sr-3 and the XOV in the underbelly of the "Mothership

The XOV in orbit

"In 1986, Boeing filed a U.S. patent application for an advanced two-stage space transportation system. Patent No. 4,802,639, was awarded on Feb. 7, 1989, details how a small orbiter could be air-dropped from the belly of a large delta-winged carrier at Mach 3.3 and 103,800-ft. altitude. Tons of material–including long-lead structural items–for a third XB-70 Valkyrie had been stored in California warehouses years before, and a wealth of data from the **X-20 DynaSoar** military space plane program was readily available for application to a modern orbiter. The **DynaSoar program** in November 1959 was the first effort to use a manned boost-glider to fly in near-orbital space and return. The B-70 was to carry the 10,000-lb. DynaSoar glider and a 40,000-lb. liquid rocket booster to 70,000 ft. and release them while traveling at Mach 3. With this lofty start, the booster could then push the glider into its final 300-mile orbit.

The Dynasoar glider
http://www.astronautix.com/d/dynasoar.html

On October 4, 1998, an XB-70 carrier-like aircraft was spotted flying over Salt Lake City at about 2:35 PM, by **James Petty**, the president of JP Rocket Engine Co. He saw a small highly swept-winged vehicle nestled under the belly of the XB-70-like aircraft climbing slowly on a west-southwest heading. The sky was clear enough to see both vehicles' leading edges, which Petty described as a dark gray or black. http://ufoweek.com/tag/reconnaissance-aircraft/

The XB-70 was believed to be cancelled and never saw actual flight operation being replaced by the SR-71, Although, there is nothing wrong with the design and concept of piggy-backing a space shuttle to high altitude and then, releasing the shuttle into sub-orbital space around the Earth. In fact, proof of concept and design were use in **Spaceship One** and the **White Knight,** the aerospace craft owned by **Sir Richard Brandson's** company **Virgin Galactic** to take civilians into space.

688

SpaceShip One connected to its mother ship White Knight in flight

SpaceShip One, note this aerodynamic design with the USAF Dynasoar above.

When one thinks about secret military bases that are working on deep black aerospace projects or reverse engineering UFOs and alien technology, Area 51 is the first secret military base that most people think of, due in part from many TV documentary shows on UFOs or sci-fi alien movies like "Independence Day". There are however, many more military bases where reverse engineering is also being carried on and where biological engineering on alien DNA and the hybridization of DNA from humans, animals and extraterrestrial beings is taking place beyond

the current knowledge of white world science. Below are some of the military bases suspected of having highly advanced reversed engineering and biological black programs that are not the usual secret aircraft development and the typical research into chemical and biological weapons.
https://www.youtube.com/watch?v=RaghqW1QWQQ

690

CHAPTER 56

SECRET DEEP UNDERGROUND MILITARY BASES (DUMBS

If you are going to build super secret aircraft or spacecraft and its associated weaponry you can simply build it out in the open where your enemy can easily find it even, if it heavily guarded with military and high powered weapons. Given the ever-evolving satellite technology in use by many countries today, militaries have resulted to using the ancient tactical advantage of hiding things and themselves underground.

Almost every culture has had stories, traditions, and myths of using caves, digging tunnels and excavating large caverns for underground cities or emergency shelters. These stories date back to ancient times telling of a vast subterranean world or hollow earth civilizations as told by Socrates while some people like the Hopis claimed to have emerged from deep within the earth through a tunnel at the base of the San Francisco peaks near Flagstaff, Arizona.
http://subterraneanbases.com/plugins/content/content.php?content.166

The 8th–7th centuries B.C **Derinkuyu Underground City** is one such ancient multi-level underground city in the Derinkuyu district in Nevşehir Province, Turkey. With its five floors extending to a depth of approximately 60 m, it was large enough to shelter approximately 20,000 people together with their livestock and food stores. The city was connected with other underground cities through miles of tunnels. It is the largest excavated underground city in Turkey and is one of several underground complexes found across Cappadocia.
http://en.wikipedia.org/wiki/Derinkuyu_Underground_City

Are underground facilities real? Of course, they are real and the concept of mining, tunneling and carving out temples and homes or storage sites is as old as mankind itself. The difference in this day and age is the purpose in which tunneling is used for and some will argue that their purpose has never really changed but, the method of creating tunnels on a massive scale has. The old salt mines of Europe and the Middle East and the copper, gold and silver mines of the Americas and in the continent of Africa were all tunneled out by hand implements and with the sweat and blood of tens of thousands. All these ancient mines still leave historians, archeologists and modern day mining engineers in awe given their size and complexity. In this modern era, however, there are nuclear-powered tunnel boring machines that can cut through bedrock like a hot knife through butter using laser drilling machine heads. These amazing tunnel borers can carve out seven miles of tunnel in a single day leaving little rock debris to be removed as the tunnels are instantly vitrified with a glass shell-like surface.

In this modern age, what you can't see or photograph from satellites makes it harder for your enemies to know what you possess even if they know your exact location. It is a fact that many governments have built underground tunnels and facilities for a variety of reasons but, most are for military purposes. The Russians, Chinese, Koreans, and Vietnamese have all built subterranean tunnels and bases while on the opposite side of the planet, America, Britain and Canada have also, been actively tunneling out its own underground cities and facilities.
The US military has built at least 140 **Deep Underground Military Bases** (known as **D.U.M.B.s** because the general public is too dumb to realize). Worldwide there are nearly 15,000.

Reportedly under construction since the 1940s, these bases are on average the size of a small city, 10 - 30 miles across with an average depth of 4 miles. They are carved out by massive nuclear-powered laser drilling machines and connected by underground mag-lev train lines. The bases are stocked with food/supplies and have the ability to grow crops with artificial lighting. Why do you think governments around the world are building huge networks of underground cities? Why are they secret? Where do you think the money comes from to build these?

There are huge boring machines with 30-40 feet diameters and larger that are used in constructing the many miles of tunnels which connect one underground military facility to another. One has only to surf the internet for proof of the existence of this equipment and its usage and Richard Sauder, PhD. has also published several books for anyone interested in learning more about this topic. His first book "Underground Bases and Tunnels (What is the Government trying to Hide)", his second book is "Underwater and Underground Bases" and most recently his third book on the subject is "The Sauder Report: Notes from the Underground". All of his books are true eye-openers as to what is taking place below your feet with many photos, diagrams, and documents from the Federal archives across the country.
http://subterraneanbases.com/plugins/content/content.php?content.166

NORAD in the **Cheyenne Mountain** of Colorado is a prime example of an underground complex seen frequently in movies and on television and there is the NORAD Canadian equivalent, the **Sage Site** in the mountain next to **Trout Lake** in North Bay, Ontario. In recent years most people have heard of **Area 51**, the existence of which the government has denied for some 40 years and of course the many missile silos built during the 60's along the US – Canada border. All of these facilities have living quarters for hundreds to thousands of people underground for long periods of time.

There are also, the underground facilities of **FEMA (Federal Emergency Management Agency)** one of which is **Mount Weather**, Virginia, 46 miles from Washington, DC that can house over 200 people underground for a month or longer. Like Area 51, Mount Weather's existence had for many years been officially denied. Awareness of these facilities is virtually of the public radar scope and their purpose raises serious questions about who holds the reins of power in this country. There is a misconception in the common belief that these facilities are owned by taxpayers through their tax dollars. Taxpayer dollars certainly funded these deep black projects but ownership belongs to the corporate elite via the military industrial complex.
http://subterraneanbases.com/plugins/content/content.php?content.166

In **Richard Sauder**'s book "Underwater and Underground Bases", he lists 200 underground and submarine tunnels and bases, all of which can be found from the **U.S. Bureau of Reclamation**, on **http://www.usbr.gov/wcg/tunnels/tdata.html** (*this website may no longer be active*).

Underground tunnels and bases are common knowledge amongst the UFO community but little is known about secret submarine bases that are literally under the sea floor, off the shores of most of the world's superpowers. Yet, these too are not unimaginable given the fact that many countries have been doing undersea mining for over a hundred years and more.

A US Air Force tunnel boring machine at Little Skull Mountain, Nevada, December 1982. Question: why would the US Air Force have need for a tunnel boring machine?

One of the tunnel boring machines used in excavating the Chunnel, the tunnel that connects England and France deep beneath the English Channel.

A huge boring machine about 40 feet in diameter
http://www.sheepletv.com/nuclear-tunnel-boring-machines-switzerland-has-nothing-on-us-2/

In Britain, coal mining under the North Sea and the Firth of Clyde has long been developed and worked for over a century with mine shafts as deep as 1800 feet under the sea floor. In Canada, there have been many undersea mines. Off the coast of Nova Scotia for over sixty-five years coal has been mined three miles out to the sea and 1600 feet beneath the sea floor. Other undersea mines are the **Wabana Iron Mine** on Bell Island, Newfoundland and the coal mine off of Cape Breton Island, Nova Scotia as well as the submarine coal mine off of Vancouver Island (near Nanaimo - Departure Bay?). **Ripple Rock** was at one time a grave hazard to shipping in the **Seymour Narrows** of British Columbia, Canada. Ripple Rock was actually twin peaks that lurked just below the surface of the water of Seymour Narrows which has been the cause of more than 100 ships to sink in these treacherous waters. The twin peaks were situated between Maud and Vancouver Islands in swift, turbulent and unpredictable waters. **"Underwater and Underground Bases", 2001 by Richard Sauder, Ph.D., published by Adventures Unlimited Press, Kempton, Illinois, USA, ISBN 0-932813-88-7**

Drilling down on to the peaks in 1943 and 1945 proved to be disastrous on two attempts with the loss of nine men on the second attempt by drowning. It wasn't until the 1950s that conventional mining techniques were employed by drilling a 570 foot shaft down through Maud Island and then, tunneling 2500 feet under the channel into the twin peaks with two parallel, 300 foot elevator shaft upwards. A series of tunnels were created off the two vertical shafts which were then pacted with 1375 tons of explosives. On April 5, 1958, the explosives were donated in what

694

has been described as the biggest peacetime, non-nuclear explosion on record for that time with over 370,000 tons of rock removed from the top of the peaks.

Entrance to undersea mine in Canada's Maritime Provinces. The mine shaft continues straight ahead, out under the Atlantic Ocean, beyond the horizon.

(Source: Carl Austin) . "Underwater and Underground Bases", 2001 by Richard Sauder, Ph.D.,

The twin peaks were sheared off from their original 9 feet below the surface water to 47 feet below the surface at low tide. The Seymour Narrows were now safe for any large ship to navigate through without harm.

This project had obvious implications for the construction of clandestine underwater activities demonstrating the feasibility of tunneling under a shipping channel and then excavating a network of tunnels inside a submerged sea mount. If it could be done off the coast of Vancouver Island, it could easily be done anywhere, wherever, there are seamounts in the world and in all probability, no doubt there have been some clandestine undersea bases tunneled out in this fashion. **"Underwater and Underground Bases", 2001 by Richard Sauder, Ph.D., published by Adventures Unlimited Press, Kempton, Illinois, USA, ISBN 0-932813-88-7**

Many of the underground bases are connected by tens or even hundreds of miles of tunnels for the express purpose of utilizing a high speed (300mph) **Magnetic Levitation (Maglev)** transportation system - trains that levitate above a magnetic field - to shuttle people and freight back and forward between military bases.

There are reports that these Maglev tube trains are capable of supersonic speed and engineers and scientists have stated that speeds of 2000 mph can be achieved and even, speeds of 4000 to 5000 mph are obtainable by evacuating the air out of the tube tunnels to form a vacuum environment. Air acts as drag against the surface of the train much like air drag against an aeroplane's wings and fuselage so, having tunnels with little or no air pressure would greatly increase the velocity of the Maglev tube train. Many countries currently have high speed above ground Maglev trains like Germany and Japan and China is seriously contemplating a Maglev transportation system as is Sweden, France, Russia, Australia and others.
http://pinterest.com/pin/218776494367987573/

Maglev train motors are embedded in the track which creates a traveling magnetic field beneath the train, lifting the cars and propelling them at 300-plus mph. The train's onboard systems are powered by induction from the track and only the section of track under the train is energized. Each section of the track switches on and off as the train moves over it causing the train to be propelled forward (much like an electromagnetic rail gun weapon).

This is real technology in existence now, with incredible potential to move massive numbers of people across countries in a matter of a few hours or even, under the oceans from continent to continent. An underground supersonic Maglev train could cross the USA from Los Angeles to New York in 45 minutes or from Washington, DC to Beijing, China in 2 hours! Whether travel is inter-continental or trans-continental, the potential of this type of transportation is almost unlimited and truly astonishing.

But, like most good things developed in science, the technology usually comes under military control, is often weaponized or is outright sequestered away from the public domain for reasons of national security! Keep in mind that the US Military-Industrial Complex is functioning 50 years ahead of anything that is currently being developed or operating in the known white world of science. It is conceivable then, that Maglev tube trains in the black world could be already travelling at speeds between 2000 to 5000 miles per hour intercontinentally.

The use of supersonic Maglev tube trains means that men, equipment, weapons and recovered Extraterrestrial spacecraft can now be moved around the country from facility to facility and from military base to military base, all out of the sight of the public, who would be none the wiser as to what was really going on below their feet.

Maglev trains travel upon a magnetic field beneath the train which is embedded in the track, lifting the cars and propelling them at 300-plus mph.
http://www.maglev.net/news/maglev-trains-for-australia

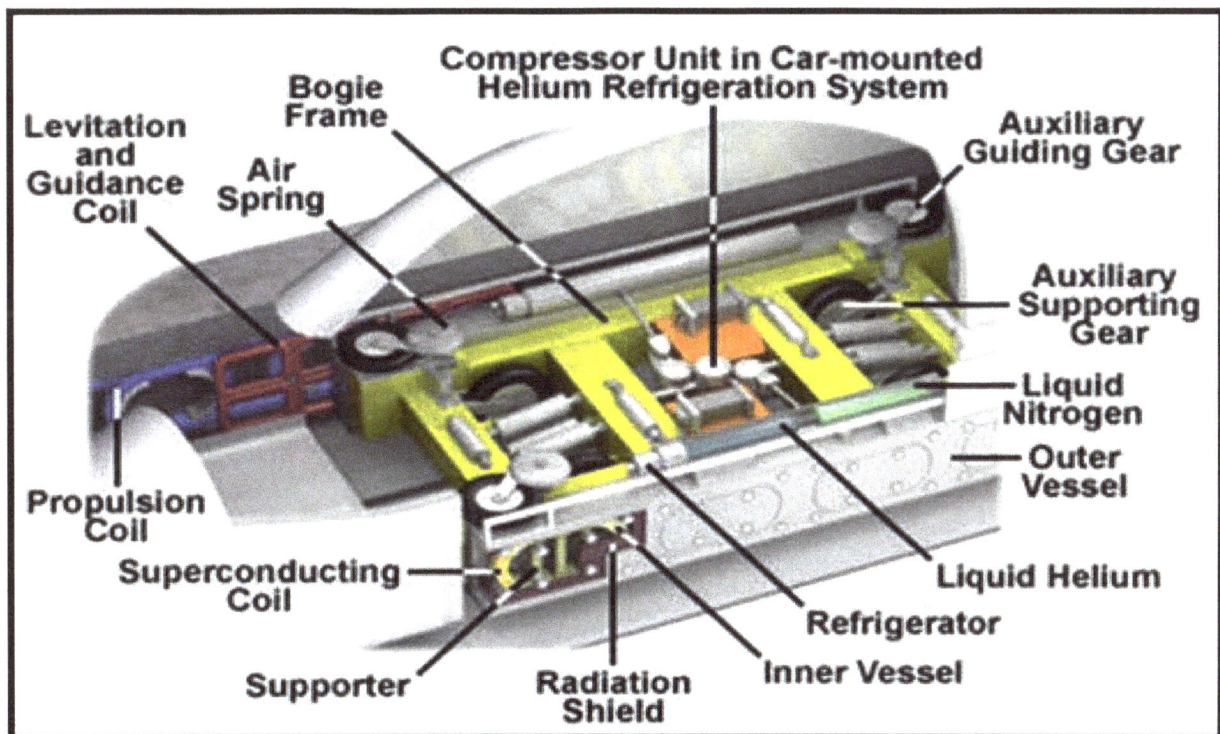

The Internal Workings of the Maglev Train. This is one configuration of a Maglev train riding upon a magnetic field track
http://future.wikia.com/wiki/RyansWorld:_Maglev_train

Author's Rant: During Expo '84 in Vancouver, B.C. Canada, I had the opportunity along with my wife to ride a Maglev train from Japan as a demonstration of the technology. The track was relatively short so, we were unable to develop high-speed however; the ride itself was very smooth and quite. Acceleration and braking were seamless without the customary passenger jolting backwards or forwards as one would experience in a bus, car or conventional diesel train.

This photo is allegedly of an unusual tunnel boring machine the size of a house that was unearthed near a military base. It is claimed to have strange hieroglyphs covering it.
http://www.darkstar1.co.uk/gregjenner11.html

The above photo if authentic as claimed represents tangible proof that an ancient culture or unknown race may have inhabited the planet long before modern Homo sapiens and had highly advanced technology capable of at least, boring underground tunnels. It would be like unearthing a modern 747 passenger jet built over 2000 years ago before mankind ever knew how to fly. It shouldn't exist but, it does, however, the construction and mechanism look too modern and familiar. It is hard to imagine how exactly it would work or by what means it is powered, given that this is the only photograph available of the object dated from the sixties. As some people suggest, it may be one of many buckets taken off the giant excavating wheel of the world's largest earth moving machinery for repair *(see photo below for comparison purposes).* This does

698

not appear to be the case and todate, the photo still remains a picture of an unusual mysterious object!

The largest earth moving machinery in the world being used in Germany. Each bucket is about the size of a large bulldozer or dump truck and the overall size of the machinery is about the length of a football or soccer field.

Some Deep Underground Military Bases in America.

Below are the locations of some **Deep Underground Military Bases (DUMBs)** in America. It is not a complete list as new DUMBs continue to come online every year and many more underground sites can be found in Appendix B of Richard Sauder's book," Underwater and Underground Bases". Each DUMB serves a purpose militarily and corporately and is connected to other underground bases by subterranean tunnels that run for hundreds of miles. It is also conceivable that low pitch hums or high pitch screeches and "ethereal "horn blowing" sounds may originate from these subterrene tunnel borers that carve out miles of tunnels in a day!

ALASKA

1. **Brooks Range, Alaska**

2. **Delta Junction, Alaska**

2a. **Fort Greeley, Alaska.** In the same Delta Junction area.

ARIZONA

1. **Arizona (Mountains)** (not on map)
   Function: Genetic work.
   Levels: Multiple

2. **Fort Huachuca, Arizona** (also reported detainment camp)
   Function: NSA Facility

2a. **Luke Air Force Base**

3. **Page, Arizona**
   Tunnels to: Area 51, Nevada Dulce base, New Mexico

4. **Sedona, Arizona** (also reported detainment camp)
   Notes: Located under the Enchantment Resort in Boynton Canyon. There have been many reports by people in recent years of "increased military presence and activity" in the area.

5. **Wikieup, Arizona**
   Tunnels to: Area 51 6. Yucca (Mtns.), Arizona

CALIFORNIA

1. **29 Palms, California**
   Tunnels to: Chocolate Mts., Fort Irwin, California (possibly one more site due west a few miles)

2. Benicia, California

3. **Catalina Island, California**
   Tunnels to: it is a 'common rumor' that there is a tunnel from the base to this Island, and also to Edwards Air Force Base, possibly utilizing old mines.

4. **China Lake Naval Weapons Testing Center**

5. **Chocolate Mountains, California**
   Tunnels to: Fort Irwin, California

6. **Death Valley, California**
   Function: The entrance to the Death Valley
   Tunnel: is in the Panamint Mountains down on the lower edge of the range near Wingate Pass, in the bottom of an abandoned mine shaft. The bottom of the shaft opens into an extensive tunnel system

7. **Deep Springs, California**
   Tunnels to: Death Valley, Mercury, NV, Salt Lake City

8. **Edwards AFB, California**
   Function: Aircraft Development - antigravity research and vehicle development
   Levels: Multiple
   Tunnels to: Catalina Island Fort Irwin, California Vandenburg AFB, California
   Notes: Delta Hanger - North Base, Edwards AFB, Ca. Haystack Butte - Edwards, AFB, Ca.

9. **Fort Irwin, California** (also reported detainment camp)
   Tunnels to: 29 Palms, California Area 51, Nevada Edwards AFB. California Mt. Shasta, California

10. **Helendale, California**
    Function: Special Aircraft Facility Helendale has an extensive railway/shipping system through it from the Union Pacific days which runs in from Salt Lake City, Denver, Omaha, Los Angeles and Chicago

11. **Lancaster, California**
    Function: New Aircraft design, anti-gravity engineering, Stealth craft and testing
    Levels: 42
    Tunnels To: Edwards A.F.B., Palmdale

12. **Lawrence-Livermore International Labs, California** The lab has a Human Genome Mapping project on chromosome #19 and a newly built $1.2 billion laser facility

13. **Moreno Valley, California Function unknown**

14. **Mt. Lassen, California**
    Tunnels to: Probably connects to the Mt. Shasta main tunnel.

15. **Mt. Shasta**
    Function: Genetic experiments, magnetic advance, space and beam weaponry.
    Levels: 5
    Tunnels to: Ft. Irwin, California North

16. **Napa, California**
    Functions: Direct Satellite Communications, Laser Communications. Continuation of Government site.

Levels: Multi-level
Tunnels to: Unknown
Notes: Located on Oakville Grade, Napa County, Ca. 87 Acres

17. **Needles, California**
    Function: unknown

18. **Palmdale, California**
    Function: New Aircraft Design, anti-gravity research

19. **Tehachapi Facility (Northrop, California - Tejon Ranch**
    Function: Unknown
    Levels: 42
    Tunnels to: Edwards, Llona and other local areas
    Notes: 25 miles NW of Lancaster California, in the Tehachapi mountains.

20. **Ukiah, California**
    Function: Unknown

COLORADO

1. **Near Boulder, Co. in the mountains Function unknown**

2. **Cheyenne Mountain -NORAD -Colorado Springs, Colorado**
   Function: Early Warning systems - missile defense systems - Space tracking
   Levels: Multiple
   Tunnels to: Colorado Springs
   Function: Early warning systems, military strategy, satellite operations
   Levels: Multiple NORAD is a massive self-sustaining 'city' built inside the mountain
   Tunnels to: Creede, Denver, Dulce Base, Kinsley.

3. **Creede,** Colorado
   Function: Unknown
   Tunnels to: Colorado Springs, Colorado - Delta, Colorado - Dulce Base, New Mexico

4. **Delta, Colorado**
   Function: Unknown
   Tunnels to: Creede Salt Lake, Utah

5. **Denver International Airport** (also a detainment camp)
   Function: Military research, construction, detainment camp facilities
   Levels: 7 reported
   Tunnels to: Denver proper, Colorado and Rocky Mountain "safe housing", Colorado Springs, Colorado (Cheyenne Mtn.)

6. **Falcon Air Force Base, Falcon, Colorado**
   Function: SDI, Satellite Control
   Levels: Multiple
   Tunnels to: Colorado Springs, possibly more.

7. **Fort Collins, Colorado**
   Function: Suspect high precision equipment manufacturing for space.

8. **Grand Mesa, Colorado**
   Function: Unknown

9. **Gore Range Near Lake, west of Denver, Co.**
   Function: Library and Central Data Bank

10. **San Juan Valley, Colorado** Hidden beneath and in an operating Buffalo Ranch
    Function: Unknown

11. **Telluride, Colorado**
    Function: Unknown

12. **University of Denver, Co (Boulder area)**
    Function: Genetics, geology/mining as related to tunneling and underground construction.

13. **Warden Valley West of Fort Collins, CO**
    Function Unknown
    Tunnels to: Montana

GEORGIA

Marrietta GA. Dobbins Air Force Base
Function: test site for plasma and antigravity air craft, experimental crafts, and weapons

INDIANA

Kokomo, Indiana
Function: Unknown
Notes: for years people in that area have reported a "hum" that has been so constant that some have been forced to move and it has made many others sick. It seems to come from underground, and "research" has turned up nothing although it was suggested by someone that massive underground tunneling and excavating is going on, using naturally occurring caverns, to make an underground containment and storage facility.

KANSAS

1. **Hutchinson, Kansas**

Function: Unknown
Tunnels to: Kinsley, Nebraska.
Notes: The entrance to the tunnel is underneath Hutchinson Hospital and is huge. It is thought that inhuman and malevolent experiments are being conducted deep below in this facility.

2. **Kansas City, Kansas**
Function Unknown
Notes: Entrance near Worlds of Fun

3. **Kinsley, Kansas**
Function: Unknown
Tunnels to: Colorado Springs, Colorado; Hutchinson, Kansas; Tulsa Kokoweef Peak, SW California
Notes: Gold stored in huge cavern, blasted shut. Known as the "midway city" because it's located halfway between New York and San Francisco.

MARYLAND

Edgewood Arsenal, Maryland. Martins AFB, Aberdeen Proving Ground, Maryland

MASSACHUSETTS

Maynard MA, FEMA regional center. Wackenhut is here too.

MONTANA

Bozeman, Mont.
Function: Genetics **NEVADA**
Area 51 - Groom Lake - Dreamland - Nellis Air Force Base Area 51 was said to exist only in our imaginations until Russian satellite photos were leaked to US sources.
Function: Stealth and cloaking Aircraft research & development. **'Dreamland (Data Repository Establishment and Maintenance Land), Elmint (Electromagnetic Intelligence),** Biological weapons research and genetic manipulation/warfare storage, **Cold Empire, EVA, Program HIS (Hybrid Intelligence System), BW/CW; IRIS (Infrared Intruder Systems),**
Security: Above ground cameras, underground pressure sensors, ground and air patrol

2. **Blue Diamond, Nevada**
Function unknown

3. **Fallon Air Force Base area** (the flats, near Reno) "American City" restricted military sites southwest of Fallon

4. **Mercury, Nevada Function unknown**

5. **Tonopah, Nevada**
   Function: Unknown

6. **San Gabriel (mountains)** On Western side of Mojave Desert
   Function: Unknown
   Notes: Heavy vibrations coming from under the forest floor which sounds like geared machinery. These vibrations and sounds are the same as heard in Kokomo, Indiana and are suspected underground building/tunneling operations.

NEW MEXICO

1. **Albuquerque, New Mexico (AFB)**
   Function: Unknown
   Levels: Multiple
   Tunnels to: Carlsbad, New Mexico Los Alamos, New Mexico Possible connections to Datil, and other points.

2. **Carlsbad, New Mexico**
   Functions: Underground Nuclear Testing
   Tunnels to: Fort Stockton, Texas. Roswell

3. **Cordova, New Mexico**
   Function: Unknown

4. **Datil, New Mexico**
   Function: Unknown
   Tunnels to: Dulce Base

5. **Dulce Base, New Mexico.**
   Tunnels to: Colorado Springs, Colorado Creed, Colorado Datil, N.M. Los Alamos. Page, Arizona Sandia Base Taos, NM

6. **Los Alamos, New Mexico**
   Functions: Psychotronic Research, Psychotronic Weapons
   Levels: Multiple
   Tunnels to: ALB AFB, New Mexico Dulce, New Mexico Connections to Datil, Taos

7. **Sandia Base, New Mexico**
   Functions: Research in Electrical/magnetic Phenomena
   Levels: Multiple
   Tunnels to: Dulce Base
   Notes: Related Projects are studied at Sandia Base by **'The Jason Group'** (of 55 Scientists). They have secretly harnessed the **'Dark Side of Technology'** and hidden the beneficial technology from the public.

8. **Sunspot, NM**
   Function: Unknown

9. **Taos, New Mexico**
   Function: Unknown
   Tunnels to: Dulce, New Mexico; Cog, Colorado
   Notes: Several other sidelines to area where Uranium is mined or processed.

10. **White Sands, NM**
    Function: Missile testing/design
    Levels: Seven known

NEW HAMPSHIRE

There may be as many as three underground installations in New Hampshire's hills, according to reports.

NEW YORK

New York, New York
Function: Unknown
Tunnels to: Capitol Building, D.C.

OHIO

Wright-Patterson Air Force Base - Dayton, Ohio
Function: Air Force Repository. Rumored to house stealth technology and prototype craft

OREGON

1. **Cave Junction, Oregon**
   Function: Suspected Underground UFO Base
   Levels: At least one
   Notes: Suspected location is in or near Hope Mountain. Near Applegate Lake, Oregon, just over into California. Multiple shafts, access areas to over 1500 feet depth. Built using abandoned mine with over 36 known miles of tunnels, shafts.

2. **Crater Lake, Oregon**
   Tunnels: possible to Cave Junction

3. **Klamath Falls, Oregon**

4. **Wimer, Oregon (Ashland Mt. area)**
   Function: Underground Chemical Storage
   Levels: At least one

PENNSYLVANIA

Raven Rock, Pa (near Ligonier)
Function: working back up underground Pentagon - sister site of Mt. Weather
Notes: 650' below summit, 4 entrances.

TEXAS

1. **Calvert, Texas**
Function: Unknown

2. **Fort Hood, Texas** (also reported detainment camp)
Levels: Multiple

3. **Fort Stockton, Texas**
Function: Unknown
Tunnels to: Carlsbad, New Mexico

UTAH

1. **Dugway, Utah**
Function: Chemical Storage, Radiation storage.

2. **Salt Lake City Mormon Caverns**
Function: Religions archives storage.
Levels: Multiple
Tunnels to: Delta, Colorado & Riverton, Wyoming

VIRGINIA

Mount Poney - Near Culpepper, Virginia
Function: Unknown

WASHINGTON

Mt. Rainier, Washington
Function: Unknown.
Levels: Multiple
Tunnels to: Unknown Yakima Indian Reservation
Function: Unknown
Notes: Southeast of Tacoma Washington, on the Reservation, in an area 40 by 70 miles. Unusual sounds from underground (Toppenish Ridge). Low flying Silver Cigar shaped craft seen to disappear into the Middle fork area of Toppenish creek.

WASHINGTON, D.C.

Function: Part of a massive underground relocation system to house select government and military personnel in the event of a cataclysmic event.
Tunnels to: New York City; Mt. Weather.

WEST VIRGINIA

White Sulfer Springs, West Virginia. Greenbrier Facility under the Greenbriar Resort.

WYOMING

Riverton, Wyoming
Function: Unknown
Tunnels to: Salt Lake, Utah Denver, Colorado.
http://nstarzone.com/CODERED.html

Follow the Money - U.S. Department of Defense Fiscal Year 2012

The following is a listing of the Top-100 U.S. **Department of Defense (DoD) Prime Contractors** in FY 2012 ranked by the total amount of money awarded. In fiscal 2012, as of August 15, 2012, the DoD has awarded a total of $202.85 billion in defense contracts. In FY 2012 so far, Lockheed Martin has received $19.12 billion in contract awards (prime contracts) or 9.4% of total contract funds awarded by the DoD. Runner-up is Boeing with $17.24 billion (8.5%) followed by **General Dynamics** in third place with $10.01 billion (4.9%). Raytheon has received $8.44 billion (4.2%) followed by Northrop **Grumman, L-3 Communications**, **BAE Systems, and United Technologies**. BAE Systems, #7 on the list, is the largest foreign contractor in FY 2012 (to date).

Top 100 Defence Contractors by Ranking for the Year 2011

2011	Company	Contracts	2010
1	Lockheed Martin Corp.	$17,344,113,000	1
2	Northrop Grumman	$10,800,453,000	2
3	Boeing Co.	$8,400,115,000	3
4	Raytheon Co.	$6,206,515,000	4
5	General Dynamics Corp.	$5,493,414,000	6
6	Science Applications International Corp.	$5,159,739,000	5
7	Hewlett-Packard Co.	$3,831,520,000	11
8	L-3 Communications Corp.	$3,815,873,000	8
9	Booz Allen Hamilton	$3,718,644,000	9
10	KBR Inc.	$3,546,605,000	7
11	Computer Sciences Corp.	$3,532,784,000	10
12	DynCorp International Inc.	$3,047,513,000	23
13	Harris Corp.	$2,893,847,000	13
14	CACI International Inc.	$2,517,616,000	16
15	Dell Inc.	$2,180,421,000	12

708

16	ITT Corp.	$2,061,343,000	14
17	BAE Systems Inc.	$1,986,983,000	15
18	Verizon Communications Inc.	$1,844,453,000	19
19	Fluor Corp.	$1,804,871,000	47
20	Jacobs Engineering Group Inc.	$1,703,308,000	20
21	IBM Corp.	$1,594,165,000	17
22	ManTech International Corp.	$1,467,181,000	31
23	United Technologies Corp.	$1,465,147,000	21
24	Battelle Memorial Institute	$1,330,333,000	22
25	Deloitte	$1,190,904,000	18
26	Honeywell International Inc.	$1,143,030,000	28
27	URS Corp.	$1,133,733,000	24
28	Serco Inc.	$992,211,000	29
29	AECOM Technology Corp.	$983,154,000	43
30	SRA International Inc.	$936,812,000	30
31	Rockwell Collins Inc.	$917,973,000	26
32	The Aerospace Corp.	$842,638,000	34
33	Accenture	$837,671,000	27
34	Apptis Inc.	$826,280,000	35
35	Sprint Corp.	$820,557,000	32
36	General Electric Co.	$813,760,000	33
37	Bechtel Marine Propulsion Corp.	$793,298,000	36
38	Wyle	$772,933,000	45
39	DRS Technologies Inc.	$767,390,000	25
40	Unisys Corp.	$749,850,000	38
41	CGI Group Inc.	$735,933,000	82
42	Mission Essential Personnel LLC	$681,304,000	62
43	Qinetiq Group PLC	$678,017,000	40
44	AT&T Corp.	$664,057,000	46
45	General Atomics International	$634,573,000	39
46	World Wide Technology Inc.	$557,693,000	53
47	Alliant Techsystems Inc.	$532,641,000	48
48	Vangent Inc.	$529,059,000	57
49	CDW-Government	$506,229,000	52
50	NCI Inc.	$492,477,000	71
51	Alion Science and Technology Corp.	$478,236,000	41
52	Development Alternatives Inc.	$455,721,000	61
53	Chemonics International Inc.	$454,521,000	51
54	VSE Corp.	$443,919,000	37
55	Qwest Communications International	$435,395,000	56
56	Xerox Corp.	$427,916,000	64
57	Arctic Slope Regional Corp.	$405,373,000	54
58	RTI International	$402,469,000	67
59	Nana Regional Corporation Inc.	$401,781,000	55
60	ARINC Inc.	$394,402,000	49
61	Westat Inc.	$361,987,000	72

62	U.S. Investigation Services Inc.	$356,150	
63	SGT Inc.	$338,665,000	42
64	ICF International Inc.	$331,879,000	66
65	Chenega Corp.	$309,457,000	68
66	immixGroup Inc.	$306,840,000	58
67	Alutiiq LLC	$299,640,000	84
68	GTSI Corp.	$287,668,000	59
69	Artel Inc.	$284,749,000	73
70	Cubic Corp.	$277,846,000	75
71	Microtech LLC	$273,857,000	95
72	Telos Corp.	$273,575,000	83
73	Creative Associates International	$267,518,000	92
74	Carahsoft Technology Corp.	$246,375,000	99
75	Eyak Technology	$239,986,000	65
76	Scientific Research Corp.	$235,710,000	77
77	Navmar Applied Sciences Corp.	$233,034,000	--
78	Energy Enterprise Solutions LTD	$230,353,000	85
79	Babcock International Group PLC	$229,421,000	--
80	Oracle Corp.	$229,159,000	88
81	ABT Associates Inc.	$226,404,000	--
82	Non-Intrusive Inspection Technology Inc.	$214,889,000	--
83	Concurrent Technologies Corp.	$211,824,000	97
84	InDyne Inc.	$205,005,000	89
85	Mythics Inc.	$197,639,000	
86	Tetra Tech Inc.	$196,537,000	60
87	Camber Corp.	$205,747,000	90
88	STG Inc.	$193,890,000	96
89	CH2M Hill	$191,615,000	69
90	Trax International	$190,331,000	74
91	DLT Solutions Inc.	$190,121,000	--
92	Ball Corp.	$188,280,000	86
93	Dynamics Research Corp.	$185,208,000	--
94	Orbital Sciences Corp.	$183,322,000	93
95	Louis Berger Group	$181,740,000	76
96	Jorge Scientific Corp.	$181,239,000	--
97	SRI International	$181,032,000	--
98	Chickasaw Nation Industries Inc.	$180,603,000	
99	Sierra Nevada Corp.	$173,991,000	78
100	Teledyne Technologies Inc.	$169,170,000	94

Washington Technology, Federal Procurement Data System and Houlihan
Lokeyhttp://washingtontechnology.com/toplists/top-100-lists/2011.aspx

These defense contract awards are easily tracked and are routinely budgeted annually but, there are probably other monies that are awarded but which are never tracked and these contract awards run in the hundreds of billions or trillions of dollars. These untraceable financial awards are the black budgets that go to super-secret, deep black programs and projects such as reverse engineered alien spacecraft and the associated technology as well as the infrastructure necessary to support these back programs.

It is probable that the missing and unaccountable loss of trillions of dollars annually from the American fiscal budget has been deliberately siphoned off into these deep black projects. Such cutting edge technology in the black world that is at least 50 years ahead of the white world of science would indeed require not billions but, trillions of dollars to maintain and sustain.

Consider the infrastructure required to build and hide such advanced technology away from the nation's enemies as well as keeping it hidden from the public. Besides the usual airbase structures of hangars, office buildings, laboratories, various storage facilities, mess halls, officers and junior rank quarters, there would also be underground complexes to conceal the real work and development of these exotic alien craft. Entrances to the underground complex would have camouflage surfacing to make it harder to see from the air and tunneling out of bedrock of multiple levels to house these underground facilities and then, interconnecting these and other underground facilities hundreds of miles away with a supersonic **"subterrene"** transportation system are just some of the massive expenditures needed to maintain this covert and hidden infrastructure. https://www.youtube.com/watch?v=uEDAE_9v4h0

CHAPTER 57

WHAT GOES ON DEEP DOWN IN A FEW D.U.M.B.s

Dugway Proving Ground

Dugway Proving Ground (DPG) is a US Army facility located approximately 85 miles (140 km) southwest of Salt Lake City, Utah in southern Tooele County and just north of Juab County. It encompasses 801,505 acres (3,243 km², or 1,252 sq mi) of the Great Salt Lake Desert, an area the size of the state of Rhode Island, and is surrounded on three sides by mountain ranges. It had a resident population of 795 persons as of the 2010 United States Census, all of whom lived in the community of Dugway, Utah, at its extreme eastern end. The name "Dugway" comes from a technique of digging a trench into a hillside to create a flat surface along which a wagon can travel. Dugway Proving Ground is located 13 miles south of the 2,624 sq mi Utah Test and Training Range. Combined, they form the largest military space in the United States.

Following the public attention drawn to Area 51 in the early 1990s, Ufologists and concerned citizens have suggested that whatever covert operations, if any, may have been underway at that location were subsequently transferred to **DPG.**

The Deseret News reported that **Dave Rosenfeld**, president of **Utah UFO Hunters**, stated:

"Numerous UFOs have been seen and reported in the area in and around Dugway...[military aircraft can't account for] all the unknowns seen in the area. It might be that our star visitors are keeping an eye on Dugway too...[Dugway is] the new area 51. And probably the new military spaceport." http://en.wikipedia.org/wiki/Dugway_Proving_Ground

Tonopah Test Range

Tonopah Test Range (TTR), also known as **Area 52** is a restricted military installation located about 30 miles (48 km) southeast of Tonopah, Nevada. It is part of the northern fringe of the Nellis Range, measuring 625 sq mi (1,620 km²). Tonopah Test Range is located about 70 miles (110 km) northwest of Groom Dry Lake, home of the Area 51 facility. Like the Groom Lake facility, Tonopah is a site of interest to conspiracy theorists, mostly for its use of experimental and classified aircraft. As such, it is not the focus of alien enthusiasts, unlike its neighbor. It is currently used for nuclear weapons stockpile reliability testing, research and development of fusing and firing systems, and testing nuclear weapon delivery systems. The airspace comprises restricted area R-4809 of the Nevada Test and Training Range and is often used for military training.

F-117s of the 4450th Tactical Group operated from Tonopah in secret from 1982 through 1989 while the program was still classified. During this period Main camp was connected to the airfield by shuttle bus service, while the airfield, in turn, was connected to Nellis Air Force Base by between five to twenty Key Air Boeing 727 and/or Boeing 737 flights per day from Nellis to Tonopah. The airfield was also serviced by one or two JANET Boeing 737 flights daily, which were presumably from McCarran International Airport and served Sandia National Laboratories

712

employees. In early 1991, Key Air lost the contract and the service was taken over by American Trans Air Boeing 727 aircraft."

Area 52" has been featured in television newscasts, such as presented by Las Vegas Channel 8 News. Area 52 has also been featured on **The History Channel's UFO Hunters**. In both of these presentations, Area 52 is speculated to facilitate the use of highly advanced aircraft technology including craft such as flying saucers or UFOs.
http://en.wikipedia.org/wiki/Tonopah_Test_Range

Dulce, New Mexico

Dulce Base is an alleged secret underground joint 'U.S. military-alien" biogenetic laboratory designed to carry out bizarre experiments on humans and animals. The facility is located under Archuleta Mesa on the Colorado-New Mexico border near the town of Dulce, New Mexico in the United States. Claims of alien activity there first arose from Albuquerque businessman **Paul Bennewitz**. http://rense.com/general75/dulcde.htm

In 1979, Bennewitz became convinced he was intercepting electronic communications from alien spacecraft and installations outside of Albuquerque. By the 1980s he believed he'd discovered an underground base near Dulce. The story spread rapidly within the UFO community and by 1990, Ufologist **John Lear** claimed he had independent confirmations of the base's existence.
http://en.wikipedia.org/wiki/Dulce_Base

According to the book, "Entrances to Subterranean Tunnels - Underground Alien Bases" an underground military base/laboratory in **Dulce, New Mexico** connects with the underground network of tunnels which honeycombs our planet, and the lower levels of this base are allegedly under the control of **Inner Earth** beings or Aliens. This base is connected to Los Alamos research facilities via an underground "tube-shuttle." (It can be assumed that such a shuttle way would be a straight-line construction).

New Mexico State Police Officer **Gabe Valdez** was drawn into the mysteries of Dulce when called to investigate a mutilated cow in a pasture 13 miles east of Dulce on the Manual Gomez ranch. Gomez had lost four cattle to mutilations between 1976 and June 1978 when a team of investigators which included Tom Adams arrived from Paris, Texas to examine the site of the carcasses.

Curious as to how cattle were being selected by the mysterious mutilators, an interesting experiment was conducted on July 5, 1978, by Valdez, Gomez, and retired scientist Howard Burgess. They pinned up about 120 of the Gomez beef cattle and moved them through a squeeze chute under an ultraviolet light. They found a "glittery substance on the right side of the neck, the right ear, and the right leg." Samples of the affected hides were removed as well as control samples from the same animals. Schoenfeld Clinical Laboratories in Albuquerque analyzed the samples and found significant deposits of potassium and magnesium. The potassium content was 70 times above normal. http://hyperboreanvibrations.blogspot.ca/2011/04/secret-dulce-underground-base-and-grey.html

Some investigators attribute the mutilations to aliens from UFOs. UFOs have been seen frequently around Dulce. Sightings of strange lights and other aerial phenomena have been reported in many areas where the cows have been found at the time of the reported mutilation.

Investigation of the Gomez Ranch, the road by the Navajo River, and the imposing Archuleta Mesa revealed landing tracks and crawler marks near the site of the mutes, lending credence to scientist Paul Bennewitz of Thunder Scientific Labs in Albuquerque, who was definitely on the right track in his attempts to locate the underground alien facility in the vicinity of Dulce. No one knows for sure where the facility was located or how humans or aliens gained secret entry to the facility. Paul Bennewitz had not only photographed UFOs but had established a communication link with their underground base at Dulce. Bennewitz had first come to prominence during the August 1980, sightings over the **Manzano Weapons Storage Area at Kirtland AFB**. A Kirtland AFB incident report dated October 28, 1980, mentions that Bennewitz had taken film of the UFOs over Kirtland. Paul, who was president of the Thunder Scientific Labs which was adjacent to Kirtland, gave a briefing in Albuquerque detailing how he had seen the aliens on a video screen. At the time, the aliens were transmitting signals to him from a base underneath Archuleta Mesa.

Researcher **William Moore** claims that the government agents became interested in Bennewitz's activities and were trying to defuse him by pumping as much disinformation through him as he could absorb. Whether Paul's communication with supposed aliens at the Dulce Base was part of this disinformation campaign is unclear. If we believe that Paul is the single source of reports on the Dulce Facility, then discounting Paul's story and discrediting him could be a tactical maneuver. The actual disinformation maneuver would result in making the public believe there was nothing to the Dulce story."

UFO hunter **Bill Birnes** says that his popular TV series was canceled because of what the show uncovered. In a recent interview on **Paranormal Podcast** with Jim Harold, Birnes recounts the shocking details about what his UFO team uncovered in Dulce, New Mexico on the Jicarilla Apache Reservation.

Birnes says that one episode on **UFO Hunters** had uncovered evidence that seemed to back up the strange **hybrid/chimera technology** reportedly going on at Dulce.A retired New Mexico State Trooper, Gabe Valdez, showed the TV **UFO Hunter Team** what looked like a human head on a cow.

After the broadcast of the "human-headed cow" episode, Birnes says they "evidently stepped on the big one. We were told what did you do? You guys are off the air. You guys are bad guys." Birnes also says he was told by somebody that claimed to be from an intelligence agency that the show went too far when they aired the human-headed cow.

Interestingly, Dulce is right in the middle of an area known for cattle mutilations. As I have previously reported, Northern New Mexico and Southern Colorado have been a hot spot for not only UFO activity but for animal mutilations.
https://indianinthemachine.wordpress.com/2011/04/16/the-cow-with-a-human-head-a-horrific-story-of-hybrids-ufo-aliens-and-the-dulce-underground-base/#more-1299

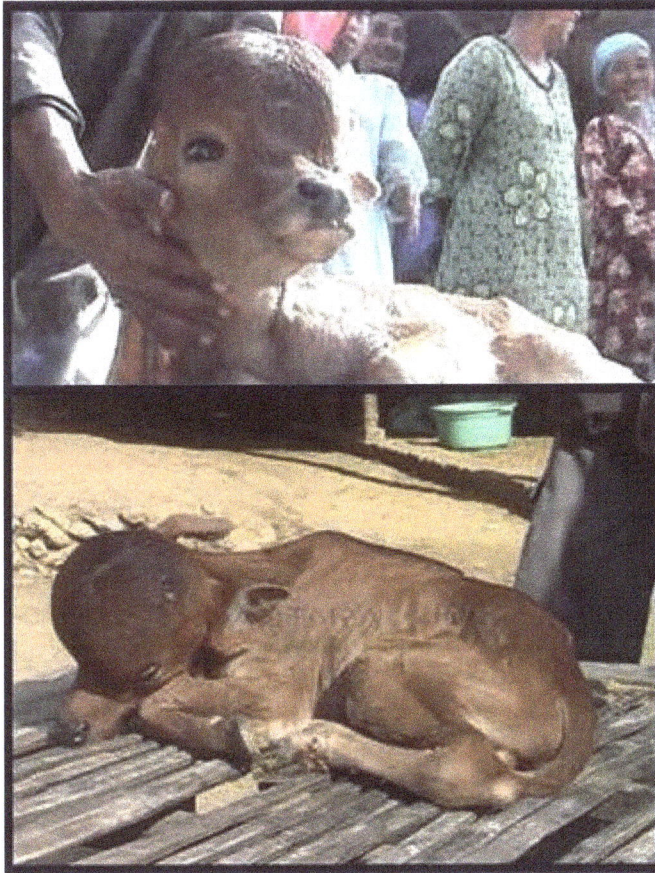

A human-headed cow or a birth defect?

Locals of Dulce have also stated they had frequently observed strange lights in the sky, especially flying over and around, and even appearing to disappear into the sides of the Mesa. Witnesses have testified that they also frequently observed military helicopters over Dulce, along with reports of cattle mutilations in the ranches nearby.
http://rense.com/general75/dulcde.htm

There have been sightings of huge, silent delta-shaped dark objects emitting extremely bright lights, as well as a daytime sighting of a silver, saucer-shaped object over the Archuleta Mesa. There was even a sighting of a huge, flying "triangle" estimated to be about half-a-mile in length with some type of a "cloaking device" that almost appeared to have a transparent body!!

The biggest and most impressive sighting took place in May of 2004. It involved hundreds of brilliant objects in the night sky that literally filled up the entire sky, according to the testimony of the former Dulce police dispatcher. There were close to 100 witnesses to this incredible incident. Some even said that there were probably close to several hundred objects in the night sky. The "armada" of UFOs moved en masse slowly from one end of the sky to the other. It was reminiscent of the well-documented 1950 mass sightings of UFOs over Farmington, near the Four Corners area of New Mexico.

What was particularly fascinating about this sighting was that everyone also saw a small fleet of military helicopters which seemed to follow the objects. A now very familiar side effect with UFO sightings was that car radios also went dead throughout the sighting.

Last but not the least of the impressive Dulce sightings involved a Jicarilla Apache Forest Service ranger who witnessed a 'craft' of some kind enter the east side of the Archuleta Mesa through several large rocks that appeared to open (almost like a door) and in went the craft into the side of the mesa. He excitedly reported this sighting live on his microphone while he was communicating on his radio with the Forest Service station across the south side of Dulce. The ranger was stationed at the top of the Archuleta Mesa in the look-out building next to the radio communications tower. This took place a few years after a big fire destroyed many of the trees on and around the mesa. (What is still strange about the aftermath of the fire, which they say happened about 10 years ago, is the fact that all attempts for the re-forestation have so far failed on and around the Archuleta Mesa. The trees just don't seem to grow for some strange reason or other.

Lastly, Dulce is known for a rather 'strange' 1967 government project called **Project Gasbuggy** which involved a large underground nuclear explosion (29 kilotons of TNT) deep inside the high plateau area 25 miles south of Dulce, allegedly to release natural gas from deep under the ground. It was a joint project with **El Paso Natural Gas Company**. What is not frequently mentioned in association with this curious project was that the huge nuclear explosion had created, deep, huge underground extensive caverns all over the area along with extensive natural "tunnels". http://rense.com/general75/dulcde.htm

Inside Dulce Underground Base

The following material cannot be verified. It is third-party information which has been collected from supposed. Dulce personnel, construction workers who were employed there, inner-Earth researcher and former abductees.

Located almost two miles beneath Archuleta Mesa on the Jicarilla Apache Indian Reservation near Dulce, New Mexico is an installation classified so secret, its existence is one of the least known in the world. Here is Earth's first and main joint United States Government/alien biogenetics laboratory. Others exist in Colorado, Nevada, and Arizona.

The multi-level facility at Dulce goes down for at least seven known levels and is reported to have a central HUB which is controlled by base security. The level of security required to access different sections rises as one goes further down the facility. There are over 3000 real-time video cameras throughout the complex at high-security locations (entrances and exits). There are over 100 secret exits near and around Dulce with many around Archuleta Mesa, others to the south around Dulce Lake and even as far east as Lindrith. Deep sections of the complex connect into natural cavern systems. http://www.abovetopsecret.com/forum/thread61316/pg1

The Grey and Reptillian Species

There are unsubstantiated rumours and tales based on second and third-hand person accounts that this underground military base/ laboratory houses an alien race known by the prejudiced monika as the **Greys**, who have taken over most of the complex. These are the beings which supposedly entered into an illicit agreement with the US government during the Eisenhower presidential years of the Fifties. **"The Agreement"** alleges that the Greys could abduct/take cows and some humans for experiments while keeping the US government informed of how many humans were taken and their names.

As the story goes, the relationship broke down due to deception and deviousness by the Greys who were abducting too many humans and cattle, but a **"Special Group" of the Government** still kept covering up for them. By the Eighties, the Government realized there was no defense against the Greys. So, programs were enacted to prepare the public for open contact with non-human ET beings.

Depending on who tells the story, there is another ET race, perhaps an ally called the Reptoids, a kind of highly evolved 6 foot tall humanoid-dinosaur, who are considered an enemy species of the Greys. They may intervene on behalf of humans to meet out some form of galactic justice and end the lop-sided relationship. The Greys only known enemy is the Reptillian Race, and they are on their way to Earth.

An advance guard Reptoids seem to have shown a great interest in research maps of New Mexico and Colorado indicating sites of animal mutilations, caverns, locations of high UFO activity, repeated flight paths, abduction sites, ancient ruins and suspected alien underground bases. Were the Reptoids merely replacing one enemy for another with themselves?

Some accounts state that the Greys and the Reptoids are working together and that there are over 125 other interstellar civilizations looking to get a piece of the human **"MetaGene"** action as they don't seem to be interested in anything else on this planet.
http://nar.oxfordjournals.org/content/34/19/5623.short and
http://www.pnas.org/content/101/12/4164.long

 Some forces in the Government want the public to be aware of what is happening. Other forces "The Collaborators" want to continue making "whatever deals are necessary" for an Elite few to survive the conflicts. http://www.abovetopsecret.com/forum/thread61316/pg1

Then, there is the story of **Philip Schneider's** controversial life and death. In a lecture videotaped in May 1996, Philip Schneider claimed that his father, Oscar Schneider, a Captain in the United States Navy, worked in nuclear medicine and helped design the first nuclear submarines. Captain Schneider was also part of **Operation Crossroads**, which was responsible for the testing of nuclear weapons in the Pacific at Bikini Island. Phil also claimed that his father was also involved with the infamous "**Philadelphia Experiment**."

In addition, Schneider claimed to be an ex-government structural engineer who was involved in building **deep underground military bases (DUMB)** around the country, and to be one of only three people to survive the 1979 incident between tall alien Grays and U.S. military forces at the Dulce underground base. Philip was critically injured losing some fingers and suffered a serious abdominal wound, but fortunately was able to survive the shootout. He recovered from his injuries and went on lecture tours to reveal his involvement in building secret military underground bases like Dulce in New Mexico. Schneider's ex-wife, **Cynthia Drayer** believes that Philip was murdered because he publicly revealed the truth about the U.S. government's involvement with UFOs. http://www.ufodigest.com/mystery.html

Such stories keep the mystery of UFOs and the ET phenomenon an unsolvable conundrum that fulfills **Winston Churchill**'s 1939 proverbial hyperbole *"A riddle wrapped in a mystery inside an enigma."* https://www.youtube.com/watch?v=xn2XT09M6xA

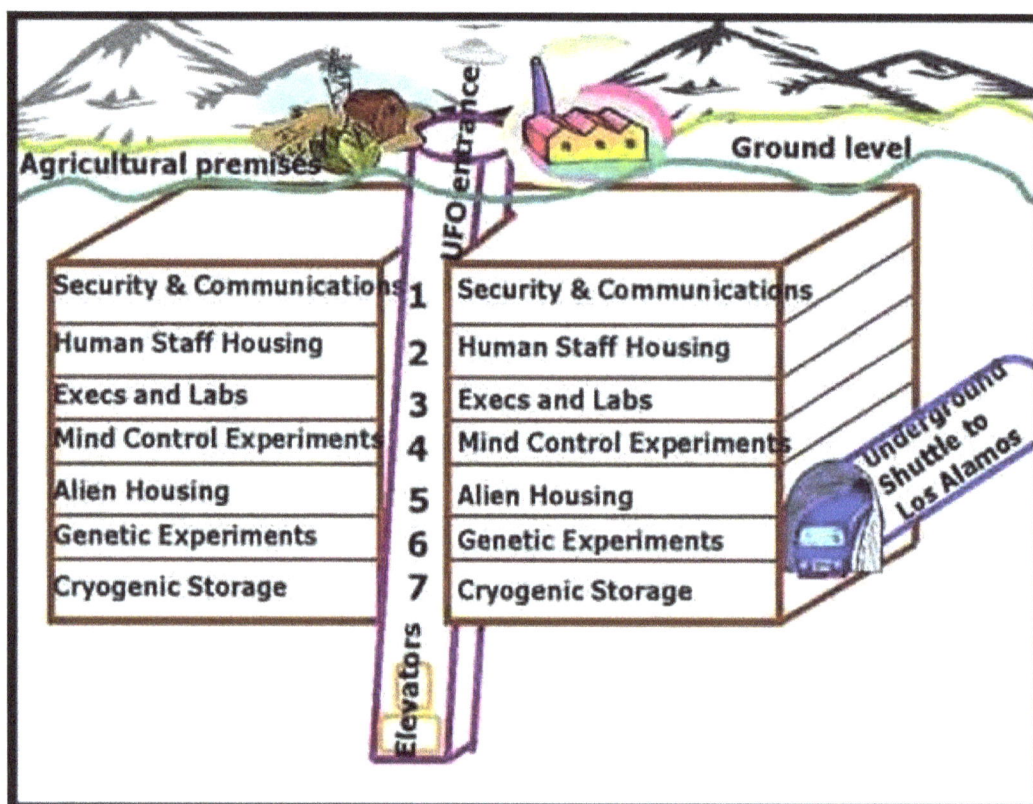

Schematic of the underground Dulce Base
http://esoteric-research.tumblr.com/post/110364394766/dumbs-deep-underground-military-bases-leaks

Author's rant: I will go on record to say that these heretical stories are a carefully orchestrated agenda of disinformation and misinformation to constantly keep the public and primarily UFO researchers confused and off-balanced in their investigations to seek out the truth of this phenomenon. Such tales are just another propaganda campaign to keep UFO researchers pursuing a false trail of disinformation down a blind rat hole while unwittingly promoting such nonsense and prejudicial hatred toward one or two alien species.

718

What this clearly shows is our blind hatred and mistrust for anything different than ourselves. We buy into the loose, sketchy and tantalizing tidbits of information such as human abductions, cattle mutilations and small gray little aliens with big heads and large black wrap-around eyes and within a veil of anthropocentric illogic connect the dots of disinformation.

Note the carefully planned psychological brainwashing and subliminal conditioning of the public perception and consciousness toward Extraterrestrials in an evolution of scary alien science fiction movies and so-called factual TV documentaries that feed upon our fear of xenophobia and alien invasions upon this planet. The cornerstone of that fear has become the Greys and Reptoids and the campaign of fear is being ramped up toward a crescendo, a false flag alien invasion that will see us all giving up even more of our power, our privacy, our civil rights and our financial resources.

The problem with most UFO researchers is that they do not seem willing to consider that humanity is its own worst enemy in which old shibbolethic standards and prejudices of religious beliefs, class, race, culture and economic status have held mankind back from achieving true peace and unity within itself. Men of corruption with spiritually bankrupted and moral values have taken hold of the reins of our destiny and hijacked our good hopeful future. Worst of all, we blindly have accepted that they have our best interests at heart and have allowed them to lead us down a doomed path of chaos and ruin.

Few UFO researchers seem to not want to ask the obvious question: *"Could there be at least in part, a human answer to the UFO/ETI phenomenon"?* The answer is …YES! The problem is to sort out what is ET and what is human about the phenomenon. We need to ask ourselves when investigating the UFO/ETI phenomenon is there a terrestrial explanation or a terrestrial technology or a terrestrial knowledge base that could explain what is going on around us, above us or beneath us!

Genetic Engineering and Hybridization of Humans and Extraterrestrial Biological Lifeforms

The Secret Government cloned humans by a process perfected in the world's largest and most advanced bio-genetic research facility, Los Alamos. The elite humans now have their own disposable slave-race. Like the alien Greys, the US Government secretly impregnated females, and then removed the hybrid fetus after a three month time period, before accelerating their growth in laboratories. **Biogenetic (DNA Manipulation)** programming is then installed - they are implanted and controlled at a distance through **RF (Radio Frequency)** transmissions. **[Recall the section in this book on cloning and genetic engineering and manipulation also, Martin Cannon's mind control paper on *"The Controllers"*].**

Many Humans are also being implanted with brain transceivers. These act as telepathic communication "channels" and telemetric brain manipulation devices. This network was developed and initiated by **DARPA.** Two of the procedures were **RHIC (Radio-Hypnotic Intercerebral Control)** and **EDOM (Electronic Dissolution of Memory).** [Recall the section in this book on the methodology of mind control technology].

They also developed **ELF** and **EM wave** propagation equipment which affect the nerves and can cause nausea, fatigue, irritability, even death. This research into biodynamic relationships within organisms has produced a technology that can change the genetic structure and heal.
http://www.abovetopsecret.com/forum/thread61316/pg1

Overt and Covert Research

U.S. Energy Secretary John Herrington named the Lawrence Berkeley Laboratory and New Mexico's Los Alamos National Laboratory to house new advanced genetic research centers as part of a project to decipher the human genome. The genome holds the genetically coded instructions that guide the transformation of a single cell, a fertilized egg, into a biological organism.

"The **Human Genome Project** may well have the greatest direct impact on humanity of any scientific initiative before us today", said David Shirley, Director of the Berkeley Laboratory.

Covertly, this research has been going on for years at the Dulce bio-genetics labs. Level 6 is hauntingly known by employees as **"Nightmare Hall".** It holds the genetic labs at Dulce. Reports from workers, who have seen bizarre experimentation, are as follows:

"I have seen multi-legged 'humans' that look like half-human/half-octopus. Also reptilian-humans, and furry creatures that have hands like humans and cries like a baby, it mimics human words... also huge mixture of lizard-humans in cages. There are fish, seals, birds and mice that can barely be considered those species. There are several cages (and vats) of winged humanoids, grotesque bat-like creatures 3 1/2 to 7 feet tall; **Gargoyle**-like beings and **Draco-Reptoids**."

"**Level 7** is worse, row after row of thousands of humans and human mixtures in cold storage. Here too are embryo storage vats of humanoids in various stages of development. I frequently encountered humans in cages, usually dazed or drugged, but sometimes they cried and begged for help. We were told they were hopelessly insane, and involved in high-risk drug tests to cure insanity. We were told to never try to speak to them at all. At the beginning, we believed that story. Finally, in 1978 a small group of workers discovered the truth. It began the Dulce Wars".
http://www.abovetopsecret.com/forum/thread61316/pg1

When the truth was evident that humans were being produced from abducted females, impregnated against their will, a secret resistance group formed. This did little, though, over time they were assassinated or "died under mysterious circumstances".

As previously stated, there are over 18,000 "aliens" at the Dulce complex. In late 1979, there was a confrontation, primarily over weaponry and the majority of human scientists and military personnel were killed. The facility was closed for awhile but is currently active.

Human and animal abductions slowed in the mid-1980s when the Livermore Berkeley Labs began production of artificial blood for Dulce. **William Cooper** states: "A clash occurred wherein 66 people, of our people, from the **National Recon Group**, the **DELTA Group**, which is responsible for security of all alien connected projects, were killed."

The DELTA Group (within Intelligence Support Activity) has been seen with badges which have a black Triangle on a red background. DELTA is the fourth letter of the Greek alphabet. It has the form of a triangle and figures prominently in certain Masonic Signs. Each base has its own symbol. The Dulce Base symbol is a triangle with the Greek letter "Tau" (T) within it and then the symbol is inverted, so the triangle points down.

- The Insignia of "a triangle and 3 lateral lines" has been seen on "Saucer (transport) Craft", The Tri-Lateral Symbol.
- Other symbols mark landing sights and alien craft.
  http://www.abovetopsecret.com/forum/thread61316/pg1

Again, remember to keep in mind that this third-party information cannot yet be verified as factual, so the reader must draw his own conclusions. The caveat to this precaution is that circumstantial evidence is steadily growing and the reader should peruse once again, the previous sections in this book dealing with cloning and hybridization engineering and mind control. The reader should also recall the video testimony of former UK Security Guard, Barry King earlier in this book, one of Dr. Greer's Disclosure witnesses, who stated that there is a secret underground rogue program in Peasemore, UK and throughout the US and Canada, etc. of cloning and manufacturing manmade "Greys" and "Reptoids"!!!

Inside the Dulce Complex

Security Officers wear jumpsuits, with the Dulce Symbol on the front upper left side. The standard hand weapon at Dulce is a **"Flash Gun"**, which is good against humans and aliens. The ID card (used in card slots, for the doors and elevators) has the Dulce Symbol above the ID photo. **"Government Honchos"** use cards with the **Great Seal of the U.S.** on it, stating the words **New World Order** in Latin.

After the **2nd Level**, everyone is weighed in the nude then, given a uniform. Visitors are given an 'off white' uniform. In front of ALL sensitive areas are scales built under the doorway, by the door control. The person's card must match with the weight and code or the door won't open. Any discrepancy in weight (any change over three pounds) will summon security. No one is allowed to carry anything into or out of sensitive areas. All supplies are put through a security conveyor system. The **Alien Symbol language** appears a lot at the Facility.

During the construction of the facility (which was done in stages, over many years) the aliens assisted in the design and construction materials. Many of the things assembled by the workers were of a technology they could not understand, yet it would function when fully put together. Example: The elevators have no cables. They are controlled magnetically. The magnetic system is inside the walls. *(Elevator maglevs!)* [Author's italics added].

There are no conventional electrical controls. All is controlled by advanced magnetics. That includes a magnetically induced (phosphorescent) illumination system. There are no regular light bulbs. All exits are magnetically controlled. It has been reported that "If you place a large magnet on an entrance, it will affect an immediate interruption. They will have to come out and reset the system." http://www.abovetopsecret.com/forum/thread61316/pg1

Mind Manipulation Experiments

Dulce has studied mind control implants, Bio-Psi Units, ELF devices capable of mood, sleep and heartbeat control.

DARPA is using these technologies to manipulate people. They establish 'The Projects', set priorities, coordinate efforts and guide the many participants in these undertakings. Related projects are studied at Sandia Base by **"The Jason Group"** (of 55 scientists). They have secretly harnessed the dark side of technology and hidden the beneficial technology from the public.

Other projects take place at the Groom Lake installation in Nevada, also known as Area 51. **ELMINT (Electro-Magnetic Intelligence), Code Empire, Code Eva, Program His (Hybrid Intelligence System), BW/CW, IRIS (Infrared Intruder System)**, **BI-PASS, REP-TILES.**

The studies on **Level 4** at Dulce include Human-Aura research, as well as all aspects of dreams, hypnosis, and telepathy. They know how to manipulate the "**bioplasmic body**" *(aka. the Astral Body)* of humans. They can lower your heartbeat, with deep sleep-inducing delta waves, induce a static shock, and then re-program via a neurological-computer link. They can introduce data and programmed reactions into your mind (information impregnation - the **"Dream Library"**).

We are entering an era of the technologicalization of psychic powers. The development of techniques to enhance man/machine communications, nanotechnology, bio-technological micro-machines, PSI-War, **E.D.O.M. (Electronic Dissolution of Memory), R.H.I.C. (Radio-Hypnotic Intra-Cerebral Control)** and various forms of behavior control (by chemical agents, ultrasonics, optical and other forms of EM radiation). This is the physics of "Consciousness."
http://www.abovetopsecret.com/forum/thread61316/pg1

Dr. Steven Greer has referred to this type of technology which is employed extensively by Extraterrestrial Intelligences in the operation of their spacecraft and throughout their civilizations as **C.A.T. (Consciousness Assisted Technology)** or **T.A.C. (Technology Assisted Consciousness)**. In other words, we are talking about **Transdimensional** or **Hyperdimensional Physics,** the metaphysics of the mind or of consciousness.

Below is a list of other military bases suspected of reverse engineering alien spacecraft, developing hybridization of alien-human technology and advanced genetic research and development and bio-engineering of alien and human DNA. Some of the bases may also be storage facilities and underground hangars for **ARVs**, advanced terrestrial spacecraft and **PLF (Programable Life Forms)** and other genetically enhanced bio-lifeforms.

Military Bases Suspected of UFO Activity

UNITED STATES

Andrews Air Force Base:

Location: Southeast of Suitland, Maryland. Very close to Washington, D.C.

722

Description: UFOs have been spotted over this base by military personnel and public.

Barksdale Air Force Base:

Location: West of Shreveport, Louisiana.
Description: This base was put on alert during the 1975 Loring AFB intrusions.

Bolling Air Force Base:

Location: Washington D.C.
Description: UFOs have been spotted at this major Air Force Base.

Brunswick Naval Air Station:

Location: West of Brunswick, in southern Maine.
Description: A UFO was sighted at this base just before the Loring AFB intrusions in 1975. A strange abduction occurred less than 40 miles from this base.

Cannon Air Force Base:

Location: East of Clovis, on the eastern edge New Mexico.
Description: UFOs have been sighted over this secretive air base.

Carswell Air Force Base:

Location: Fort Worth, Texas.
Description: It is believed that some of the UFO wreckage from the 1947 Roswell, New Mexico, crash was sent here.

Dugway Proving Ground (DPG):

Location: A US Army facility located approximately 85 miles (140 km) southwest of Salt Lake City, Utah in southern Tooele County and just north of Juab County.
Description: It encompasses 801,505 acres (3,243 km², or 1,252 sq mi) of the Great Salt Lake Desert, an area the size of the state of Rhode Island, and is surrounded on three sides by mountain ranges. Dugway Proving Ground is located 13 miles south of the 2,624 sq mi Utah Test and Training Range. Combined, they form the largest military space in the United States. "Numerous UFOs have been seen and reported in the area in and around Dugway. Military aircraft can't account for] all the unknowns seen in the area. Dugway is the new area 51and probably the new military spaceport."

Edwards Air Force Base:

Location: A large section of land east of Rosamond, California, that includes Rogers Lake (Dry) and Rosamond Lake (Dry).

Description: A well-known experimental aircraft testing range, this area has tested many saucer-shaped aircraft. A UFO landed here over a 2-day period in April of 1954. The occupants of the UFO referred to as the **"Etherians"**, gave amazing demonstrations to a select group of individuals (including **President Eisenhower**). The Etherians allowed the military to inspect the UFO and even helped them.

Eglin Air Force Base:

Location: Located on the Western edge of Choctawhatchee Bay, Florida. Just south of Valparaiso, Florida.
Description: In 1976, UFOs were spotted over the base's Armament Development and Test Center. This base is just east of Gulf Breeze, Florida, a major UFO hotspot.

Ellington Air Force Base (a.k.a. NASA Ellington Field):

Location: Near NASA's Johnson Space Center in Houston, Texas.
Description: Bob Oechsler (a former NASA Engineer and now a UFOlogist) was flown (by helicopter), in 1990, to a NASA facility 20 miles southwest of Ellington AFB. The facility had a microgravity chamber that, Oechsler is convinced, was made using alien technology.

Ellsworth Air Force Base:

Location: South Dakota.
Description: UFOs have been spotted over this air base.

Fairchild Air Force Base:

Location: Washington.
Description: This base was put on alert during the 1975 Loring AFB intrusions.

Fort Ritchie:

Location: Maryland.
Description: During 1976, UFOs were spotted over Weapons Storage Facilities on the base.

Grand Forks Air Force Base:

Location: This base is located 16 miles west of Grand Forks, North Dakota.
Description: This base was put on alert during the 1975 Loring AFB intrusions. It experienced its own intrusions also during 1975.

Groom Lake Test Facility:

Location: About 120 Miles northwest of Las Vegas, Nevada, in the Nellis Air Force Range. The base is constructed on the edge of the Groom Lake salt bed in the dry Emigrant Valley between two jagged mountain ranges.

724

Description: Officially, the base that created the U-2 and SR-71 Blackbird spy planes, doesn't exist. The base cannot be seen on any public map. Old Government maps will list it as Area 51. It is a common hotspot for UFOs and it is believed that a new Mach-8 spyplane, nicknamed Aurora, has been developed there. It is also believed that an aircraft made from UFO technology, the **Human- Piloted Alien Craft (H-PAC),** was developed there. Bob Lazar, a nuclear physicist who claims he was hired by the government to study an alien propulsion system, said that another facility, called **S-4 (Section-4),** is located just south of Groom Lake at Papoose Lake (also dry). There have been many UFO sightings and cattle mutilations in and around the area. Base personnel have disappeared during night time UFO activity. Area 51 has a formidable security system (the area is protected by a group of Green berets). Recently, the USAF gained control over a key ridge near Groom Lake, from which you could see the installation (read USAF Claims Freedom Ridge for more information). A 1988 Soviet Satellite photo shows a high-security compound at one end of the base, purpose unknown.

Holloman Air Force Base:

Location: Approximately 15 km. southwest of Alamogordo, New Mexico.
Description: This air base is a well-known testing site for experimental aircraft. It is also near where the first A-bomb was tested. On April 25, 1964 (12 hours after the famous Socorro, New Mexico, UFO landing), a UFO (escorted by 2 others that remained in the air) landed at this base. Three aliens got out of the craft and talked with base officials. This story was confirmed by Richard C. Doty (possible pseudonym), a retired counterintelligence officer with the Air Force's **Office of Special Investigations (OSI).**

Homestead Air Force Base:

Location: Florida.
Description: This base is suspected of keeping alien bodies in a Top-secret, underground repository.

Hunter Army Air Field:

Location: Georgia.
Description: In 1976, 2 base security policemen spotted a UFO near this base.

Kinchloe Air Force Base:

Location: Michigan.
Description: This base was put on alert during the 1975 Loring AFB intrusions.

Kinross Air Force Base:

Location: Michigan.
Description: UFOs have been sighted over this base and jets have also been scrambled from this base to chase UFOs.

Kirtland Air Force Base:

Location: Albuquerque, New Mexico.
Description: This Airbase is a well-known landing site for space shuttles and is suspected to be harboring alien bodies. During 1980, UFOs were spotted over Kirtland AFB by Base Security Personnel. The UFOs would fly solo and would hover over an area known as Coyote Canyon (part of a restricted testing range on the base). This testing range is used by the **Air Force Weapons Laboratories, Sandia Laboratories, Defense Nuclear Agency**, and the **Department of Energy**.

Langley Air Force Base:

Location: Just Northeast of Hampton, Virginia.
Description: Also known as Langley Field, this highly-secretive base is suspected to be harboring alien bodies. UFOs have been spotted near this base. Jets have been scrambled from this base to intercept UFOs.

Loring Air Force Base:

Location: Just Northeast of Limestone, Maine.
Description: During 1975, a rash of UFO sightings occurred over this base. The base, along with others, was put on a "Security Option 3" and was told to prepare to "Defend against a helicopter assault". However, the UFOs were never identified as helicopters, and many witnesses said they looked like bright "stretched-football" crafts about the size of a car. Jets were scrambled to intercept the UFOs, but their attempts proved unsuccessful. This base has nuclear weapons installations.

Los Alamos Research Facility:

Location: Los Alamos (north of Santa Fe), New Mexico.
Description: This facility is owned and operated by the **Nuclear Regulatory Commission (NRC)**. It was the site of the famous **"Manhattan Project"** that developed the Atom Bomb. UFOs have been sighted over this facility. This facility is suspected to have analyzed UFO wreckage and EBEs from the 1947 Roswell Crash. Between 1948 and 1952, UFOs were spotted over the facility by personnel.

Luke Air Force Base:

Location: Arizona.
Description: UFOs have been sighted over this air base.

Malstrom Air Force Base:

Location: Montana.
Description: This is a base that was put on alert in 1975 after the Loring AFB intrusions. In 1975, UFOs were spotted (both visually and by radar) hovering over the base's Nuclear Missile

Installations. Closer examination of the missiles revealed that the UFOs had somehow tampered with the missiles. Jets were scrambled to intercept these UFOs, but the UFOs were far superior to any military aircraft and therefore could not be shot down.

March Air Force Base:

Location: California.
Description: This installation is a **Strategic Air Command (SAC)** base. UFOs have been spotted over this base.

Maxwell Air Force Base:

Location: Near Prattville, Alabama.
Description: Two base security policemen spotted a UFO over this base.

Minot Air Force Base:

Location: The base is located 13 miles north of Minot, North Dakota.
Description: This was another base put on alert after the Loring AFB intrusions of 1975. This base is a **Strategic Air Command (SAC)** base and has a large amount of land reserved for missile sites. In 1975 it experienced UFO intrusions.

NORAD Headquarters:

Location: Inside Cheyenne Mountain, near Colorado Springs, Colorado.
Description: This highly-secretive base is headquarters to the **North American Air Defense (NORAD) System**. It is built inside a mountain, and can withstand a nuclear blast. Its **Deep Space Surveillance Center** is believed to have radar tracking of UFOs.

Norton Air Force Base:

Location: California.
Description: Fighter jets have been dispatched from this base to intercept UFOs.

Oakdale Armory:

Location: Near Greater Pittsburgh International Airport, Pennsylvania.
Description: This armory is home to the 662nd Radar Squadron. This squadron was the military group involved in the cover-up of the Kecksburg, Pennsylvania UFO crash of 1965.

Pease Air Force Base:

Location: Near Portsmouth, New Hampshire.
Description: This base was put on alert during the 1975 Loring AFB intrusions. UFOs have been spotted over this installation.

The Pentagon:

Location: Arlington, Virginia.
Description: The Pentagon is The United States Military Headquarters. It is home to the **National Military Command Center (NMCC).** The NMCC is believed to have radar tracking of UFOs.

Peterson Air Force Base:

Location: Colorado Springs, CO
Description: This Air Base is the home of the USAF **Space Command Center (SCC).** The SCC is in charge of the **Defense Support Program (DSP) Satellites**. These satellites are highly classified. They were designed to warn the U.S. Military of **Inter-continental Ballistic Missile (ICBM)** launches. They are believed to have recordings of UFOs.

Plattsburgh Air Force Base:

Location: New York.
Description: This installation was another base that was put on alert during the 1975 Loring AFB intrusions. UFOs have been sighted at this base (especially over the base's Weapons Storage Area).

Sawyer Air Force Base:

Location: Michigan.
Description: This base was put on alert during the 1975 Loring AFB intrusions.

Truax Air Force Base:

Location: Michigan.
Description: UFOs have been spotted over this base. Fighter jets have been scrambled from this base to chase UFOs.

Wright-Patterson Air Force Base:

Location: About 5 km. west of Fairborn, Ohio.
Description: An extremely secretive Air Base, security officials once refused entrance to the defense secretary. Wright-Patterson is home to the infamous **Hangar 18** which is suspected of harbouring UFO wreckage and dead **Extraterrestrial Biological Entities (EBEs)**. The alien bodies and some of the UFO wreckage retrieved from the 1947 Roswell crash was sent to this base. UFO wreckage from around the world (Australia, etc.) has been sent here. W-PAFB is headquarters to **Project Moondust**. Moondust is a "foreign space debris" analysis program of the Air Force Systems Command's **Foreign Technology Division (FTD.** Hangar 18 is known to have huge Cryogenic freezing chambers, purpose unknown.

728

Wurtsmith Air Force Base:

Location: Michigan.
Description: This was another base put on UFO alert during the 1975 Loring AFB intrusions. Many UFOs were spotted in and around this base during the 1975 UFO-flap (especially over the base's Weapons Storage Area). Fighter jets were scrambled to intercept the UFOs, but their attempts were useless.

OTHER COUNTRIES

AUSTRALIA

Pine Gap Research Facility:

Location: Located 12 miles from Alice Springs in Australia's Northern Territory. **Description:** This base is described as a **"Defense Space Research Facility"** sponsored by both the American and the Australian defense departments. It is run by the American **National Security Agency (NSA)**. This secret communications base serves primarily as a downlink for geosynchronous **SIGINT (Signals Intelligence) Satellites.** It is suspected to be monitoring UFO communications. UFOs have been sighted in and around the base. It is believed that there is an underground alien base on the facility's grounds.

CANADA

Falconbridge Air Force Station:

Location: Near Northbay, Ontario, Canada.
Description: In 1975, this Radar Station experienced a rash of UFO sightings. American jets were scrambled to intercept the UFOs after the Canadians requested assistance.

ENGLAND

Bentwaters Air Force Base (a.k.a. RAF Bentwaters):

Location: Near Bentwaters, England.
Description: In 1980, UFOs were sighted (one hovering above the ground), at very close range, by base personnel in Rendlesham forest (a few miles from the base). The personnel were threatened into silence and the public was misinformed about the incident.

IRAN

Shahrokhi Air Force Base:

Location: In Hamadan, Iran.
Description: In 1976, jets were dispatched from this base to pursue UFOs. The UFOs, which were spotted by the public, and the pilots (and subsequently confirmed by radar), outmaneuvered these jets. Every time the jets approached the crafts, their systems failed. The smaller of the two

UFOs engaged one jet, causing it to break off the chase. The smaller UFO landed behind a hill while the larger **"Mother Ship"** disappeared into the sky. Military Helicopters dispatched later to find the UFO that apparently landed reported no evidence of a landing.

RUSSIA

Plesetsk Military Cosmodrome:

Location: Plesetsk, in northwest Russia.
Description: This super-secret Russian military space center has been responsible for many UFO sightings in Russia. The base's satellite and missile launches along with experimental plane testing have often been mistaken for UFOs throughout Russia.
http://www.theforbiddenknowledge.com/hardtruth/military_bases_ufos.htm

National Security and Murder Inc. (Incorporated)

How do you enforce secrecy and cover up on the greatest event phenomenon in human history and how do you ensure its legitimacy behind the wall of **National Security**, when it steps outside the boundaries of government oversight and control?

To answer this question, it is necessary to have a brief but, basic understanding of the concept of what National Security means, how it functions within the USA as well as its relationship to Murder Incorporated.

The term "National Security" is a rather ambiguous concept at best with no single universally accepted definition and was developed in the U.S.A. after World War II to maintain the survival of the state through the use of economic, diplomacy, power projection and political power. Initially focusing on military might, it emphasized the freedom from military threat and political coercion but has increased in sophistication to include other forms of non-military security as suited the circumstances of the time.

A security threat would normally involve conventional foes such as other nation-states but now, it also includes violent non-state actors (terrorists), narcotic cartels, multinational corporations and non-governmental organizations; some authorities include natural disasters and events causing severe environmental damage in this category.

Measures taken to ensure national security include:

- using diplomacy to rally allies and isolate threats
- marshalling economic power to facilitate or compel cooperation
- maintaining effective armed forces
- implementing civil defense and emergency preparedness measures (including anti-terrorism legislation)
- ensuring the resilience and redundancy of critical infrastructure
- using intelligence services to detect and defeat or avoid threats and espionage, and to protect classified information

730

- using counterintelligence services or secret police to protect the nation from internal threats. http://en.wikipedia.org/wiki/National_security

The United States Armed Forces defines national security (of the United States) in the following manner:

National Security — A collective term encompassing both national defense and foreign relations of the United States. Specifically, the condition provided by:

a. a military or defense advantage over any foreign nation or group of nations;
b. a favorable foreign relations position; or
c. a defense posture capable of successfully resisting hostile or destructive action fromwithin or without, overt or covert.

In 2010, **Barack Obama** included an all-encompassing world-view in his definition of America's national security interests as:

- The security of the United States, its citizens, and U.S. allies and partners;
- A strong, innovative, and growing U.S. economy in an open international economic system that promotes opportunity and prosperity;
- Respect for universal values at home and around the world; and
- An international order advanced by U.S. leadership that promotes peace, security, and opportunity through stronger cooperation to meet global challenges.

In the same National Security Strategy, Obama mentioned that polarization between the Republicans and Democrats could pose a threat to national security. In the President's estimation, polarization had the potential to affect the country's policies and posture around the world. **(This is an interesting prophetic usage of national security, given the recent 2012 presidential elections with both political parties polarized over the US economy which has brought the country to the "fiscal precipice" of another recession).** (Author's bold italics added for emphasis).

Traditionally, the earliest recognized form of national security has been "military security" implying the capability of a nation to defend itself, and/or deter military aggression. Alternatively, military security implies the capability of a nation to enforce its policy choices by use of military force. The term "military security" is considered synonymous with "security" and therefore with National Security. Http

From the above understanding of the definition of National Security: *"Military Security is National Security"!* This, in essence, gives the US military incredible sweeping powers with a virtual impenetrable wall to hide behind any matter of scientific knowledge, recent discoveries of science and technology. Primary examples of this would be the "crown jewels of secrecy" like the existence of UFOs, Extraterrestrial life forms, alien technology, exotic off-planet power generation and propulsion systems, terrestrial inventions and their patents, all hidden and suppressed in the name of National Security! And it's all legal, backed by the Executive Order of the US President!!

What is **Murder Incorporated (Murder Inc.)? Murder, Inc.** was a name used back in the 1930s through the 1940s given by the press to organized crime groups like the **Brownsville Boys**; known in syndicate circles as **The Combination.** Murder Inc. acted as the "enforcement arm" of the **American Mafia** and **Jewish Mafia**, the early organized crime groups in New York and elsewhere. Murder, Inc. was responsible for between 400 and 1,000 contract killings.

Today, it considered as the "enforcement arm" for any covert black program that requires a final solution arising from a troublesome situation, event or person that cannot be resolved in a peaceful manner of threats and intimidation. It is a hit squad of paramilitary people who are ordered to act with extreme prejudice and ruthlessness while ensuring that no incriminating evidence can be traced back to the operatives or the rogue group that they represent. These rogue groups are a part of the wealthy corporate elite, the private industrial sector, the CIA, the NSA, the NRO, DARPA or any other intelligence agency of the US Government.

These industrialists, these military departments, this alphabet soup of intelligence agencies are the **Military Industrial Complex (MIC)** in which President Eisenhower warned us of in his last public speech. Murder Inc. is the covert strong arm group that acts when needed on behalf of the Military Industrial Complex!

Murder is sanctioned when it becomes a matter of national security and matters of UFO/ETI reality and related alien technology ranks as the nation's most super secretive black program requiring the absolute strictest level of national security imposed upon the subject matter.

Matters that potentially threaten the national security (by this is meant "military security" or the security of a powerful mega corporation or person) or the interests of the nation may result in the murder of an individual or the death of many people. It matters not that the individual is a blue collar worker or whether he was a former CIA Director, a Senator of New Mexico or even a US President. If you are perceived as a potential or a real threat, who happens to have gotten too close to exposing the hidden truth behind the MIC agenda then, you are eliminated or your days are shortly numbered as a matter of National Security!

National Security has become a catchall phrase for all things covert that seeks legitimacy and yet, is beyond the nation's oversight and control.

The Military Industrial Complex makes its "living" killing!

The business of defense contractors even those only peripherally connected with the production of weapons themselves is death! Yet another way of putting it: The Military Industrial Complex is Murder Inc.

Our country has become Murder Incite has become its euphemism: The National Security State. The origin of Murder, Inc. is the National Security Act of 1947 a blueprint for fascism. It granted to the Pentagon powers of unlimited defense spending. **How the Military/Industrial complex turned America into Murder, Inc.; *By Len Hart;* OpEdNews Op Eds 8/18/2006 at 13:12:48** http://www.opednews.com/articles/opedne_len_hart_060818_how_the_military_2find.htm

It would not have been possible to justify such largesse without a bogeyman to be dragged out whenever the population is roused by truth to suspicion. In 1947, the bogeyman was the specter of communism a tactic urged upon President Harry Truman by GOP Senator Arthur Vandenburg. The climate of fear is maintained by both parties but with less embarrassment by the GOP. In what Gore Vidal called "a fit of conscience" never witnessed among modern Republicans, President Eisenhower warned us that such a Military Industrial Complex might establish permanent control over the state itself.

Quoting **Jean-Paul Sartre**: *"Fascism is not defined by the number of its victims, but by the way it kills them"*.

Sartre was, of course, correct. Killing someone for revenge, an act of passion, even a petty theft is one thing. But a society is morally lost when mass murder becomes its number one export. In such a society, everyone is guilty. That is Fascism. That is absurdity. The rationale du jour is terrorism.

Terrorism can only be exploited in an atmosphere of irrational and endemic hatred. Hatred is nurtured by a party that could not possibly exist without the various strawmen upon which it directs the bile and hatred of the American right wing hatred of science, hatred of the humanities, especially art, literature, and most of all: philosophy. Right-wing hatred of philosophy is most certainly based on its inherent distrust of critical thought itself. The **GOP ("Grand Old Party")** aka. the **Republican Party** finds the process of critical thinking subversive of its own inflexible, unquestioned dogma. This is dogma that they would, of course, impose upon you in various and subtle ways. It is a brave and rare person who dares think outside the box.

It would appear that America's *raison d'etre* has become the mass murder of people from Viet Nam to Iraq. The collective will of America is doing precisely what it really wants to do. There are no rationales. There are no excuses. And the world will not be safe until America faces that ugly fact about itself. **How the Military/Industrial complex turned America into Murder, Inc.;** *By Len Hart;* **OpEdNews Op Eds 8/18/2006 at 13:12:48**
http://www.opednews.com/articles/opedne_len_hart_060818_how_the_military_2find.htm

The Reason for Secrecy and Cover-up

What could possibly be the reason for all the secrecy, the suppression, and cover-up that surrounds the existence of Extraterrestrial Civilizations visiting the Earth? This is a world-wide phenomenon witnessed by millions of people so unofficially; the public has accepted the evidence that we are not alone in the universe! Humanity in truth has never been alone, although globally, we do appear to suffer from some long-term collective amnesia and are now, just awakening to this reality.

Open contact with an interstellar civilization would certainly rank as the single most important event in human history. It would have such profound implications, not only on the global human psyche and our place in the universe but, it would also be a major factor in determining the direction in which humanity will evolve towards in the next 500,000 years!

The real reason for the secrecy of Extraterrestrial Intelligences visiting the Earth, once we get past the ridicule that plagues this subject and acknowledge its reality, is to ask, "How are they able to get here?" They are able to travel between star systems faster than the speed of light without using rocket power or any known terrestrial science. How is that scientifically possible? Everything that has been orchestrated up to this current time by the **Military Industrial Complex** has been to suppress the existence, the knowledge and the recovered alien technology to gain the upper hand over other nations in understanding and utilizing the propulsion system of Extraterrestrial Intelligences.

To be able to travel among the stars has always been a dream of mankind's that until now, remained only in the realm of science fiction. Knowing the secret to the infinite energy generation system of flying saucers or any other ET spacecraft would put America in the driver's seat so to speak and give it a technological and military advantage over other nations by hundreds of years or more.

An infinite source of energy generation has been the real reason for the decades of UFO secrecy and cover-up. It would explain the real reason behind many of the coverup tactics used by the US Air Force and by various intelligence agencies.

It explains why certain pre and post-World War II scientific knowledge, particularly from Nazi Germany's R&D work on flying saucers and other disc-shaped craft was seized through **Operation Paperclip** and has been continuously suppressed out of the history books and from the white world of science. It explains the coverup story of the 1947 **Roswell Saucer Crash** and all subsequent crash saucer retrievals. It would also explain why military and commercial pilots are prevented by military regulations **JANAP 147** and **AFR 200-2** from talking publicly about UFO sightings. It would explain why the news media and the film industry have been seized and controlled by inside operatives of the **MIC** to spin-doctor news about UFO sightings and ET contact with a "Giggle Factor" or to provide a particular negative slant to alien type science fiction movies. It explains the false analysis and conclusions reached by such pseudo-scientific UFO investigation programs like **Project Blue Book** and the **Condon Committee Report on UFOs**, etc. It also explains the need to build underground and undersea bases with an interconnecting supersonic subterrene train system, so that alien spacecraft and its associated technology could be studied, reverse engineered, built, test flown and put into production.

CHAPTER 58

THE REAL REASON FOR THE UFO COVER-UP AND SECRECY - CLEAN, CHEAP, POLLUTION-FREE, INFINITE ENERGY!!!

In a Nutshell: incredible as this may be, there is still a lot of mystery that surrounds this whole phenomenon. Most major military nations already know this fact and have poured vast financial resources into unlocking the science behind an infinite energy source, commonly referred to as **Zero Point Energy (ZPE)** aka. **quantum vacuum energy** is an infinite energy form from the quantum vacuum of space, which is clean, dirt cheap, pollution-free, environmentally friendly and would make all other forms of energy generation that are currently known obsolete, almost overnight!

Most people have never heard of such an energy other than those who research unidentified flying objects or new alternative energy generation systems. The reason for this public ignorance is that ZPE is a jealously guarded US military secret!

This energy form has been known with demonstrated application for *over a hundred years* by the likes of such inventors as **Nikola Tesla** and **T. Townsend Brown**, however, this technology has been publicly suppressed and covered up, but is secretly being utilized by the US Military in deep black weaponization programs and projects!

This imaginary photo is what Zero Point Energy may look like on the quantum level if we had a microscope strong enough to view
https://www.bbc.com/news/business-27071303

America and some of her allies already have a major head start over other nations in the development of this new energy science. Thus, secrecy continues to be a part of military brinkmanship among countries but as already stated, most nations know about ZPE which also explains the aggressive shoot down orders employed by the world militaries in playing catch up with the USA.

So, what exactly is **zero-point energy**?

In quantum field theory, the vacuum state is the quantum state with the lowest possible energy; it contains no physical particles and is the energy of the ground state. This is also called the zero point energy; the energy of a system at a temperature of zero. But quantum mechanics says that, even in their ground state, all systems still maintain fluctuations and have an associated zero-point energy as a consequence of their wave-like nature. Thus, even a particle cooled down to absolute zero will still exhibit some vibrations.

 Liquid helium-4 is a great example: Under atmospheric pressure, even at absolute zero, it does not freeze solid and will remain a liquid. This is because its zero-point energy is great enough to allow it to remain as a liquid, even if a very cold one. Everything everywhere has a zero-point energy, from particles to electromagnetic fields, and any other type of field. Combine them all together and you have the vacuum energy, or the energy of all fields in space.
https://www.huffpost.com/entry/zero-point-field_b_913831

Zero-point energy, also known as ground state energy and it is a by-product of the fact that subatomic particles don't really behave like single particles, but like waves constantly flitting between different energy states. This means even ***the seemingly empty vacuum of space is actually a roiling sea of virtual particles fluctuating in and out of existence, and all those fluctuations require energy.***

However, we can only guess how much energy is actually contained in the vacuum, with legendary physicists in fierce disagreement on this point. **Richard Feynman** and **John Wheeler** calculated the zero-point radiation of the vacuum was so powerful that even a small cup of it would be enough to set all of Earth's oceans to a boil. But **Albert Einstein's theory of general relativity** suggests zero-point radiation would "gravitate" — spreading out throughout the universe and be mitigated to a weak power, ***however, because this is an infinite energy source it is not weaken.*** (Bold italic added by author for emphasis)
https://www.inverse.com/article/35077-wtf-is-zero-point-energy

This would seem to imply that a vacuum state -- or simply vacuum -- is not empty at all, but the ground state energy of all fields in space and may collectively be called the zero-point field. The vacuum state contains, according to **quantum mechanics**, fleeting **electromagnetic waves** and **virtual particles** that pop into and out of existence at a whim. So, we must then ask, can this energy be measured? Or even calculated?

In physics, there is something called the . In this experiment two conducting plates are brought parallel to each other with an electromagnetic field held between them. The cavity between the plates cannot sustain all frequency modes of the electromagnetic field, particularly wavelengths comparable to the plate separation. This creates a zero-point pressure on the outside of the plates, trying to push the plates together, much like how radiation pressure from the sun pushes a comet's tail away from its nucleus. The resulting effect is called the Casmir force, and it increases in strength the closer the plates get but vanishes once the plates make actual physical contact -- or when the plates are so close together that zero-point wavelengths no longer see a perfectly conducting surface.

This Casmir effect is often sited as evidence of a sea of zero-point energy throughout the universe. Another possible manifestation of the ZPF might be the cosmological constant so well used in cosmology; some say it might be a measure of this zero-point energy. One calculation even puts the energy of a cubic centimeter of empty space at around a trillionth of an erg; not much but collect that over all of space and you still get infinity. In 1913, **Albert Einstein** and **Otto Stern** performed an analysis of the specific heat of hydrogen at low temperatures, and discovered the available data was best fit if the vibrational energy was represented by the equation:

Even with the temperature, T, at absolute zero, you can see how the first term drops out to zero as well, but we are still left with the second term as the minimum energy retained. This is the zero-point energy for hydrogen -- and space being filled with the stuff -- that alone would fill the vacuum with zero-point electromagnetic radiation.

Another derivation of the **ZPF** comes, as mentioned before, from the uncertainty principle. For a given particle, one cannot know both its position and momentum at the same time -- with the least possible uncertainty being proportional to **Planck's Constan**t. This uncertainty relates to the inherent quantum fuzziness of energy and matter due to their wave-like nature. Thus, one cannot have a particle lying motionless at the bottom of its potential well, for then you would know both its position and energy with absolute certainty.

So, the lowest possible energy of a given system must be greater than the minimum potential of the well -- its zero-point energy. This leads us to postulate the collective potential of all particles everywhere with their individual zero-point energies merging into one universal . The theories and scientific research in this particular area of quantum physics lay the groundwork for attempting to explain how the mind/brain/brain waves initiate transactions in the natural; how our thoughts commingle with everything else, and cause matter to manifest in our lives. The more we look at this area, the clearer the God/science connection becomes. If thoughts equal energy and energy equals matter, then thoughts become matter. Observe your thoughts as they will manifest themselves in your life in the natural via the ZPF.
https://www.huffpost.com/entry/zero-point-field_b_913831

If there's as much energy in those fluctuations as some physicists believe, and if we could ever learn how to tap into this phenomenon, we would gain access to an unparalleled source of energy. Zero-point energy could power the planet with the strength of multiple suns, making it easy for us to solve Earth's energy problems forever. It could be miniaturized into small electronic devices and large power plants or used to power all types of industries, transportation systems from automobiles to buses, trains, aircraft, or to travel beyond the solar system and take our place among the stars.

Perhaps the clearest application would be super-fast spaceflight, the kind that could take you across the solar system in mere hours or minutes. NASA scientists have looked into developing batteries and engines which could theoretically produce a gargantuan amount of energy by harnessing a zero-point energy system based on a notion in quantum mechanics called the Casimir effect. This effect is small, but if there's a way to observe and intervene with these very small-scale forces, they could work as a potential source of energy for allowing spacecraft to move through space.

It's a bit strange to think that centuries after the idea of an "**ether**" permeating through the world was debunked, physics has come around to say that perhaps there is a universal energy stuck in the empty space all over.

NASA's Eagleworks Laboratories, which claims to have successfully tested a **Quantum Vacuum Plasma Thruster**. This "**Q-thruste**r" takes advantage of the Casimir effect to create propulsion. https://www.inverse.com/article/35077-wtf-is-zero-point-energy

As we have stated earlier in this book and other textbooks in this series that the US military through covert black projects and programs, USAPs, and alien saucer crash retrievals have developed via reverse engineering of alien technology advance aero to space craft that have the ability for interplanetary travel! This was stated by the late **Ben Rich** of **Lockheed Skunk Works**:

"We already have the means to travel among the stars, but these technologies are locked up in black projects, and it would take an act of God to ever get them out to benefit humanity. Anything you can imagine, we already know how to do."

"We now have the technology to take ET home. No, it won't take someone's lifetime to do it. There is an error in the equations. We know what it is. We now have the capability to travel to the stars. First, you have to understand that we will not get to the stars using chemical propulsion. Second, we have to devise a new propulsion technology. What we have to do is find out where Einstein went wrong."

When **Jan Harzan** asked **Ben Rich** how do UFOs work, he replied *"Let me ask you. How does ESP work?"* Harzan responded off the top of his head with, **"All points in time and space are connected?"** Rich then said, *"That's how it works!*

The late **Mark McCandlish**, an aerospace illustrator who diagrammed the **ARV (Alien Reproduction Vehicle** aka. **Advanced Research Vehicle**, aka. the "**Fluxliner**") based upon the description from eyewitnesses and a Congressman who saw three similar craft in various sizes, one of which had part of its outer shell exposed to see the interior of the craft. The Fluxliner is powered by **pulsed electokinetics** as the distributor cap rotates and fire three equidistant capacitor banks at a time.

Mark McCandlish claimed that the ARV Fluxliner was capable of travelling at a staggering 300 times the speed of light which would mean interstellar travel between stars was not only possible

738

but the travel time was reduced to minutes, hours or days depending on which planet or star was being visited!

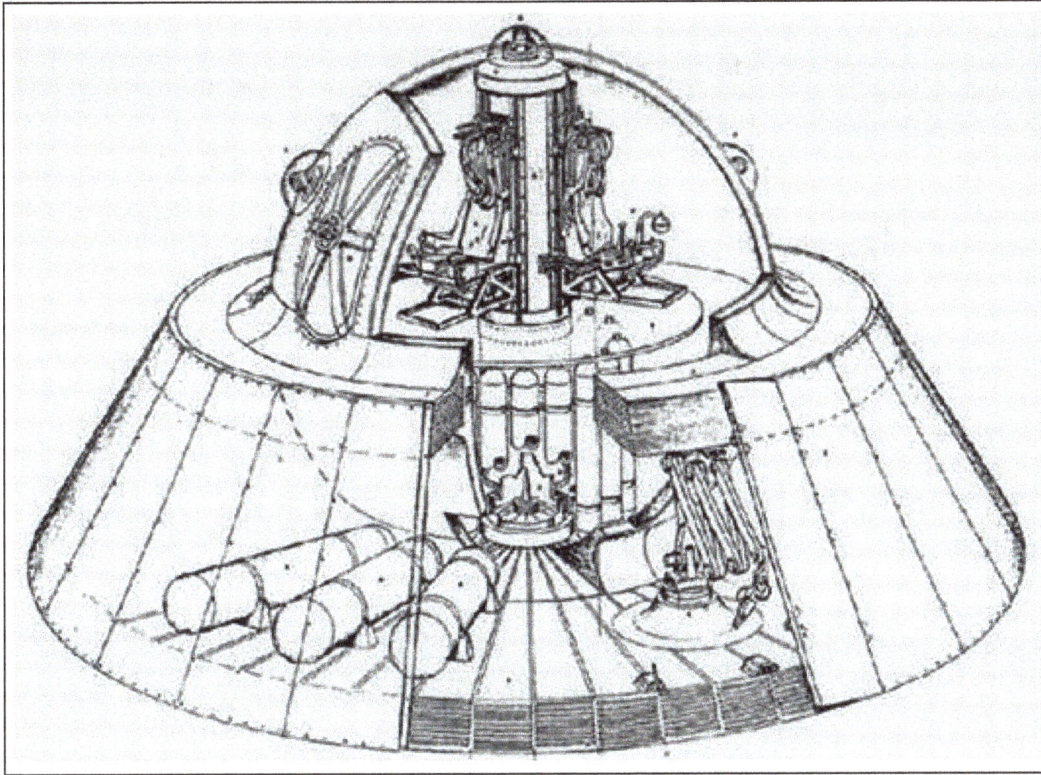

The impressive Advanced Research Vehicle (ARV) hovercraft diagrammed by aerospace illustrator, Mark McCandlish, which he calls the "Fluxliner"

This ARV "Fluxliner" opens up a myriad of questions as to what is the physics, math and science required to build it. Did it come from alien technology, how long has the military had this technology, where in space did they travel with this craft, what planet and star systems were explored and what did the astronauts discover in their journeys into space???

So, what's the real question of secrecy in keeping a new infinite energy source hidden? If it was keeping this knowledge away from other nations and their military then, from who is the secrecy being kept from and why?

The answer to the first part of the question is easy, it is **YOU!**

The answer to the second part of the question of why, strongly implies several probabilities that are related to each other.

We have reached a point in our evolution on this planet where we may be witnessing the beginning of a **Breakaway Society** or a **Breakaway Civilization** moving independently in a new direction away from the rest of humanity. It may be the beginning of a worldwide indentured servitude campaign, a kind of global economic slavery overshadowed by a possible massive depopulation of the planet or possibly, it may also, be some combination of these two potential outcomes! This revolutionary and world-shaking event according to new age rumour mongers is said to be tied to a timeline corresponding with the end of an epoch period in human civilization referred to in the **Mayan Calendar.** An event said to end on **December 21, 2012!**

That time has come and gone with no global upheavals to befall humanity, however, that doesn't mean a worse case red flag scenario may yet be implemented at some time in the near future. Either way, it does not look good for the majority of us on this planet and Extraterrestrial visitations would seem to be the least of our worries.

How is this possible? All the clues and the evidence provided so far in this textbook points in this direction, as incredible as this may appear, unless there is yet, another scenario that still remains veiled and elusive to us. How would this be accomplished?
https://www.youtube.com/watch?v=oHxGQjirV-c

CHAPTER 59

ORCHESTRATION OF A FALSE FLAG SCENARIO:
A'LA "INDEPENDENCE DAY' ALIEN INVASION

There is speculation among Ufologists as well as by aviation experts like those in **Jane's Defence Week,** who strongly suggest that highly advanced aerospace craft may have been used in the Iraq-Afghanistan war and in former wars. This would preclude the most highly advanced black project aircraft except those that disclosure of their existence would not comprise future programs or campaigns in a wartime scenario. These super advanced aerospacecraft have had a steady growth and development over the decades yet, they have never seen wartime service, why is that?

The only logical explanation for a massive build up in space-based weaponry and spaceplanes is an implementation of a **false flag scenario.** Recall, the admonitions and warnings by **Wernher Von Braun** through his protégé, **Dr. Carol Rosin**, who stated that since the late sixties and early seventies space-based weapons had been continuously placed into orbital space. They are aimed not only at targets on Earth but, the majority of weapons are aimed outwards into space away from the Earth! Why? Because repeatedly Von Braun warned Dr. Rosin that the "final card that would be played was the alien invasion card!"

The orchestration of a **False Flag Alien Invasion** could explain the development of the **USAF Space Command**, the missing trillions of US dollars annually to support and maintain an ambitious **Deep Black Space Program** along with all its infrastructure, its many military bases and as well as it growing network of underground and undersea bases all connected by a subterrene supersonic tube train system.

One of **Dr. Steven Greer's Disclosure Project** witnesses, an insider of the **Majestic group** revealed to Dr. Greer and his team that such a false flag alien invasion upon the Unites States and other countries was in the final planning stages. It primary purpose is to get the people of the planet to give up more of their rights and freedoms by globally supporting a military space-based defense program against any and all future alien invasions of our planet. The financial outpouring globally into the coffers of the armament dealers and the private industrial sector would be in the hundreds of trillions of dollars making all past wartime expenditures pale by comparison.

This witness was to testify at the **National Press Club** Disclosure Project event in Washington, D.C. in May, 2001but at the last moment, he never showed up to the event. The following day a note was found on one of the team member's hotel door explaining his sudden disappearance.

He had been called back to his handlers in the Majestic group who debriefed him and placed a gag order on him not to disclose anything of what he knew about a false flag operation at the National Press Club.

The note revealed that many of the younger members within Majestic wanted the whole UFO/ETI subject to come out to the public but that the "old guard" in the organization wanted to proceed as planned. The witness said that the whole subject was in freefall because of **Dr. Greer's Disclosure** event.

Disclosure: Military and Government Witnesses Reveal the Greatest Secrets in Modern History by Steven M. Greer; 2001; Crossing Point, Inc. Publications; Crozet, VA. USA; ISBN0-9673238-1-9

How would a false flag alien invasion be staged and made believable in its impact upon an unsuspecting global populace?

Alien Reproduction Vehicles (ARVs) sometimes referred as **Stagecraft** have been built since the late fifties and early sixties piloted by genetically engineered **Programmable Lifeforms (PLFs)** (the Greys and Reptoid aliens) created by the **Military Industrial Complex** in military underground bases such as Dulce, New Mexico and in Britain, Canada, and Australia would be used to in an *"Independence Day"* type alien invasion. The USA and her partners along with other militaries from around the world would come to our rescue to defend us against the alien invasion of our planet. Some "Doomsday" people have speculated that the date for this false flag alien invasion has been set for **December 21, 2012** to tie into the end of the **Mayan Calendar**! But that date has come and gone, eventfully; it was just another year like every other year.

Every hidden technology developed and built over the decades from reversed and hybridization of alien technology would be thrown into the orchestrated invasion. This would include 3D holographic beam projection of stagecraft and ETs, scalar and particle beam weapons, mind control weapons to create panic, astral energy extraction, and negative bioform energy generation weapons, etc. etc. Literally, everything in the covert black ops and space command arsenal will be thrown at us to scare the living hell and bejesus of us, that we will more than gladly give control over of our governments and institutions to the Military Industrial Complex to save us.

The outcome will be predictable with much loss of human life but, with an ultimate victory for the newly united forces of planet Earth. Global cooperation, if not willingly given will become mandatory under severe penalty from new sweeping laws to control human daily livelihood. Financial support on an unprecedented global scale will be required for the newly formed *"World Police Force"* to control people in every country and a *"World Military Space Force"* will take the lion's share of financial resources against any further future alien aggression and invasions.

World leaders and no doubt military and wealthy industrialists will give speeches stating something to the effect that they had been aware of an alien presence for decades based upon military observation of alien spacecraft in near-Earth orbit and flying in our skies with the reports of hundreds of thousands of human abductions by little Grey aliens. They will claim that it became necessary to build a secret space force with the cooperation of other powerful nations to defend the Earth while not creating undue fear and panic among the planet's populace. We poor ignorant people will hopefully buy into this **BIG LIE,** as we have already bought into other false government lies and rally around our military heroes and leaders giving them whatever support they ask for or need to defend us against the "Bad Aliens!"

742

Fear and lies will control and rule the masses. Truth will be denied to us while being kept in the darkness of ignorance as to what is really going on around us!

Where have we heard this before? It has been with us since man fought against man and nation against nation but, in more recent times, it was used with complete effectiveness by the Nazis against Germans to enlist their undying support for their campaigns of terror and war throughout Europe! If the lessons of pre-war Nazi Germany are to be learned, we should be aware that events don't just happened but are carefully planned, unfolding like a skillful game of chess. One insidious event followed another and another, until the populace falls in line with their begrudging compliance and support.

In a world awakening from its heedlessness of apathy, complacency, and timidity they may still find themselves with global crises and problems from the negative forces of creed, corruption and power control that have plagued humanity in its advancement towards a greater stage of civilization.

There is however, an ever-growing bright light at the end of this tunnel of global chaos, gloom and doom!

A more hopeful future is destined for humanity, it will require great effort from everyone on the planet by discarding the old, tire and worn-out values of the past for the new positive values of the future that will benefit, not just the few, but for all of mankind. A positive good future awaits us and through the trials and tribulations of evolving, a quantum leap forward may propel us toward a transformation of global consciousness and spiritual evolvement on this planet!

Accordingly, it will be a time when the foundations of a global civilization on this planet will be layed that is destined to last for 500,000 years as foretold in most of the world's religious texts, in partnership with open Extraterrestrial contact events.

The world is in the pangs of birthing a new divine civilization! In the end, the planet will finally be united in all areas of the human experience which will redound in a world commonwealth that is at peace with itself. With an ever-advancing civilization that is destined to travel among the stars, we will become the Extraterrestrials to other interstellar civilizations!

This is the good future that awaits a maturing humanity if we can but take it!
https://www.youtube.com/watch?v=6psv6f3jF3g and
https://www.youtube.com/watch?v=qf5SvJpGuv0 and
https://www.youtube.com/watch?v=naaBXzG89UY and
https://www.youtube.com/watch?v=sza0WuB6l_U and
https://www.youtube.com/watch?v=8xMfQX0SmTQ

BIBLIOGRAPHY, WEBLIOGRAPHY AND VIDEOGRAPHY

The following list includes all books and major journals, newspapers and web based material, including other reference ebooks and materials such as web links and video links found on the internet in researching this book. Not all chapters use reference material and therefore, these chapter are not listed.

Bibliography listed refers to books marked in RED,
Webliography refers to websites marked in BLUE,
Videography refer to video websites marked in GREEN,
News Service websites are marked in LIGHT BLUE, and
Newspaper websites, magazines and professional papers are marked in PURPLE.

It does not include specific government documents, archival repositories, or various journals or other web based material.

VOLUME THREE (BOOK THREE)
MEDIA CONTROL, POWER PLAYERS, THE MILITARY INDUSTRIAL COMPLEX AND SPOOK CENTRAL

CHAPTER 38
THE NEWS MEDIA: THE PEOPLE'S "WATCH DOG OR A PUPPET OF THE INTELLIGENCE COMMUNITY?

"Operation Mockingbird: CIA Media Manipulation" Mary Louise, 03 August, 2007, The National Expositor
http://www.bibliotecapleyades.net/sociopolitica/sociopol_mediacontrol05.htm

http://tmh.floonet.net/articles/cia_press.html

https://www.youtube.com/watch?v=nWQHAOUKgJk

http://www.mediafreedominternational.org/2009/12/21/inside-the-military-media-industrial-complex-impacts-on-movements-for-peace-and-social-justice/

"MEDIA PLAY" By Steven M. Greer MD; 29 April 2004

Video News Release (VNR). http://en.wikipedia.org/wiki/News_propaganda

http://www.mediafreedominternational.org/2009/12/21/inside-the-military-media-industrial-complex-impacts-on-movements-for-peace-and-social-justice/

http://www.nowfoundation.org/issues/communications/tv/mediacontrol.html

https://www.youtube.com/watch?v=9ona0jYWa6s

"The Missing Times, News Media Complicity in the UFO Cover-up" by Terry Hansen; 2000; USA; published by Xlibris Corporation; ISBN 0-7388-3611-7 (Hard) and ISBN 0-7388-3612-5 (soft)

https://www.youtube.com/watch?v=VD7ewtykM98

https://www.youtube.com/watch?v=1qAYxmFsxOM Terry Hansen

http://en.wikipedia.org/wiki/US_intelligence_community

http://socioecohistory.wordpress.com/2010/07/29/coast-to-coast-am-norio-hayak

http://www.examiner.com/exopolitics-in-honolulu/hollywood-ufos-and-extraterrestrial-disclosure

http://www.earthfiles.com/news.php?ID=1760&category=Environment

http://www.sightings.com/1.reports2010/spielsaucersecrets.html

UFO/ET Headlines Spielberg's Saucer Secrets By Robbie Graham & Matthew Alford; ©2010 Robbie Graham & Matthew Alford Reprinted By Permission

http://www.sightings.com/1.reports2010/spielsaucersecrets.html

http://en.wikipedia.org/wiki/History_of_the_Internet

http://www.wordiq.com/definition/CFB_North_Bay

http://en.wikipedia.org/wiki/ARPANET

http://www.imdb.com/title/tt0163521

 Beyond Roswell: The Alien Autopsy Film, Area 51, & the U.S. Government Coverup of UFOs by Michael Hesemann (Author), Philip Mantle (Author), Bob Shell (Author) Paperback: 303 pages publisher: Marlowe & Company (August 1998) ISBN 1-56924-709-9 ISBN 978-1-56924-709-9

CHAPTER 39
DISCLOSURE VS. THE MILITARY INDUSTRIAL COMPLEX

http://blogs.myspace.com/index.cfm?fuseaction=blog.view&friendId=23406090&blogId=535986038

http://www.computerweekly.com/Articles/2009/04/30/235865/Who-controls-the-Internet.htm

http://www.slate.com/id/2131182/

"Who Controls the Internet?" "Why it doesn't matter if the United States is in charge."By Adam L. Penenberg Updated Tuesday, Nov. 29, 2005, at 3:31 PM ET

http://www.foreignaffairs.com/articles/61192/kenneth-neil-cukier/who-will-control-the-internet

"Who Will Control the Internet?" By Kenneth Neil Cukier November/December 2005 Published by the Council on Foreign Affairs

http://www.militaryindustrialcomplex.com/what-is-the-military-industrial-complex.asp

http://en.wikipedia.org/wiki/Military_industrial_complex

http://www.h-net.org/~hst306/documents/indust.html

https://www.youtube.com/watch?v=KyrlFPD1fLE

http://www.telegraph.co.uk/news/newstopics/howaboutthat/ufo/7926037/UFO-files-Winston-Churchill-feared-panic-over-Second-World-War-RAF-incident.html

UFO files: Winston Churchill 'feared panic' over Second World War RAF incident; By Andrew Hough, and Peter Hutchison; Published: 12:01AM BST 05 Aug 2010

http://www.ideamarketers.com/library/printarticle.cfm?articleid=331462

Above Top Secret" by Timothy Good

http://www.ufologie.net/htm/m.htm

CHAPTER 40
COUNTRIES THAT HAVE HAD FLYING SAUCER PROGRAMS

https://www.youtube.com/watch?v=MDvZu50HvLs

https://www.youtube.com/watch?v=HLLWH56uULY

http://www.globalfirepower.com/

http://en.wikipedia.org/wiki/Coand%C4%83_effect

The Arrow Scrapbook – Rebuilding a Dream and a Nation by Peter Zuuring; Arrow Alliance Press; Dalkeith, Ontario; 1999 ISBN1-55056-690-3)

http://en.wikipedia.org/wiki/John_Carver_Meadows_Frost

Campagna, Palmiro. The UFO Files: The Canadian Connection Exposed. Toronto: Stoddart Publishing, 1998. ISBN 0-7737-5973-5

https://www.youtube.com/watch?v=UI0Z6qZkFYo

https://www.youtube.com/watch?v=0EtFOEkf75E

http://www.naziufos.com/#join

Henry Stevens, Hitler's Flying Saucers: A Guide to German Flying Discs of the Second World War, 2003, Adventures Unlimited Press, Kempton, Illinois, 60946   ISBN 1-931862-13-4

http://www.bibliotecapleyades.net/ufo_aleman/rfz/index.htm#menu

http://www.bibliotecapleyades.net/ufo_aleman/rfz/chapter1.htm

http://www.stevequayle.com/High.Jump/Vril.and.Andromeda.html

Secret Societies and their Power in The 20th Century by Jan Van Helsing (pseudonym); Translation and typesetting: Urs Thoenen, Zurich. Original Title: Geheimgesellschaften und ihre Macht im 20. Jahrhundert

http://www.bibliotecapleyades.net/sociopolitica/secretsoc_20century/secretsoc_20century06.htm#CHAPTER%2033

http://www.bibliotecapleyades.net/ufo_aleman/rfz/chapter3b.htm

http://www.bibliotecapleyades.net/ufo_aleman/rfz/foofighter.htm

Renato Vesco; Intercept UFO; Zebra Publications, Inc. 275 Madison Ave, New York, NY. 10016; Grove Press, Inc.; 1974. ISBN: 0-8468-0010-1 (Recently reissued as Man-Made UFOs 1944-1994 by Adventures Unlimited Press)

Corso, Phillip J., Col., 1997, page 161, The Day After Roswell. Pocket Books, a division of Simon & Schuster Inc., 1230 Avenue of the Americans, New York, NY. 10020

http://www.bibliotecapleyades.net/ufo_aleman/rfz/chapter3d.htm

http://www.bibliotecapleyades.net/ufo_aleman/rfz/schauberger.htm

http://www.bibliotecapleyades.net/ufo_aleman/rfz/rocket.htm

http://www.bibliotecapleyades.net/ufo_aleman/rfz/schappellerchapter4b.htm

http://www.rexresearch.com/schapp/schapp.htm

https://www.youtube.com/watch?v=KyrlFPD1fLE

 "Secret Societies and their Power in the 20th Century" by Jan Van Helsing (Pseudonym); copyrighted 1995; ISBN 3-89478-654 –X; Translated by EWERTVERLAG S.L.; P.O. Box 35290, Playa del Ingles, Gran Canaria, Spain

http://www.bibliotecapleyades.net/sociopolitica/secretsoc_20century/secretsoc_20century06.htm#CHAPTER%2033

http://discaircraft.greyfalcon.us/HAUNEBU.htm

 "Secret Societies and their Power in the 20th Century" by Jan Van Helsing

http://greyfalcon.us/restored/Secret%20flying%20discs%20of%20the%20Third%20Reich.htm

https://www.youtube.com/watch?v=IClVotk8ff8

https://www.youtube.com/watch?v=YoX_3kPCQAA

http://www.bibliotecapleyades.net/tierra_hueca/esp_tierra_hueca_6c.htm

The Antarctic Enigma from Violations.org Website; Book IV - Antarctica Enigma

http://www.rense.com/general35/op.htm

"How High Can You Jump? Operation "Highjump & The UFO Connection by Erich J. Choron

www.wintersteel.com

kommissar@mtu-net.ru hmerik16@yahoo.com

http://www.bibliotecapleyades.net/tierra_hueca/esp_tierra_hueca_6c.htm

http://www.bibliotecapleyades.net/antarctica/antartica11.htm.

http://en.wikipedia.org/wiki/Master_race#Nazi_beliefs_about_the_Aryans

https://www.youtube.com/watch?v=DkH4TcVD_CM

CHAPTER 41
POSTWAR SCIENTISTS, SECRET PROJECTS AND PROGRAMS – THE BEGINNING OF THE MILITARY INDUSTRIAL COMPLEX IN AMERICA

http://en.wikipedia.org/wiki/Operation_paperclip

http://en.wikipedia.org/wiki/T-Force

http://www.youtube.com/watch?v=bh37r3alzzs

http://www.hnet.org/~hst306/documents/indust.html

http://en.wikipedia.org/wiki/Hermann_Oberth

http://en.wikipedia.org/wiki/Wernher_von_Braun

http://en.wikipedia.org/wiki/Dr._Carol_Rosin

Carol Rosin. (2001-05-09), National Press Club Conference, [Recording], Washington DC: Disclosure Project.

748

http://www.amazon.com/review/R1C5F8WI3FLI5M

http://en.wikipedia.org/wiki/J._Robert_Oppenheimer

http://www.amazon.com/review/R1C5F8WI3FLI5M

Oppenheimer and UFO Crash Retrievals: A Possibility, November 3, 2008 By Robert B. Lelieuvre (Book Review)

http://www.cosmicparadigm.com/ufonews/edward-h-bomb-teller-consulted-re-ufos/

Excerpts from The Presidents UFO Web Site - Star Wars, UFOs, and Dr. Edward Teller
http://www.bibliotecapleyades.net/exopolitica/esp_exopolitics_F_h.htm

CHAPTER 42
THE FORMATIVE AGE OF ETI CONTACT AND
THE MODERN AGE OF UFO SIGHTINGS

http://www.gilderlehrman.org/history-by-era/jackson-lincoln/essays/technology-1800s

God Passes By, Shoghi Effendi, © 1944 by the National Spiritual Assembly of the United States of America, Sixth Printing 1970, SBN: 0-87743-020-9

https://www.youtube.com/watch?v=RStc-wncVQ4

http://en.wikipedia.org/wiki/Balloon_(aircraft)

http://en.wikipedia.org/wiki/Montgolfier_brothers

http://www.centennialofflight.gov/essay/Lighter_than_air/Civil_War_balloons/LTA5.htm

http://www.unmuseum.org/airship.htm

http://en.wikipedia.org/wiki/Mystery_airship

The source for this file is the June 1970 issue of Flying Saucers magazine, published and edited by Ray Palmer. Reprinted from BUFORA Journal, British U.F.O. Research Association

http://www.unexplainable.net/artman/publish/article_782.shtml

http://ufomania.proboards.com/index.cgi?board=history&action=display&thread=715

Passport To Magonia On UFOS, Folklore, and Parallel Worlds; 1969 by Jacques Vallee; published by Contemporary Books Inc.; ISBN: 0-8092-3796-2

http://en.wikipedia.org/wiki/Early_flying_machines

CHAPTER 43
BIG BROTHER SEES, LISTENS, MONITORS, TRACKS, TARGETS AND RETRIEVES ALL THINGS UFO

http://users.erols.com/mwhite28/war-1900.htm

http://necrometrics.com/20c5m.htm

http://users.erols.com/mwhite28/warstat8.htm

http://en.wikipedia.org/wiki/Trinity_(nuclear_test)#First_deployment

http://ufocasebook.com/missouricrash.html

http://www.where-is-area-51.com/ufo-sightings-west-coast-air-raid.html

http://www.rense.com/general27/battle.htm

Long Beach Independent EXTRA February 25, 1942

http://www.majesticdocuments.com/documents/pre1948.php

UFOs and the National Security State' By Richard M. Dolan Keyhole Publishing, 2000

http://www.where-is-area-51.com/ufo-sightings-west-coast-air-raid.html

http://www.slideshare.net/scalarenergyproducts

The Truth About the UFO Crash at Roswell by Kevin D. Randle and Donald R. Schmitt; 1994; Published by M. Evans and Company, New York, New York; ISBN 0-87131-761-3

Crash At Corona by Stanton T. Freidman; 1992; Paragon House Publishers, New York, NY; ISBN 1-55778-449-3

http://en.wikipedia.org/wiki/Air_Force_reports_on_the_Roswell_UFO_incident

http://www.subversiveelement.com/roswell_crash_dummies.html

 Poll: U.S. hiding knowledge of aliens - UFO Evidence

http://www.abductee.ca/seven/roswell_article4.html

http://www.fas.org/sgp/othergov/roswell.html.

http://www.anomalies.net/archive/cni-news/CNI.0999.html

http://www.v-j-enterprises.com/gao.html

UFO Crash at Aztec: A Well Kept Secret" By William S. Steinman, Contribution by Wendelle C. Stevens; Copyright 1986 by UFO Photo Archives; Privately Published by Wendelle C. Stevens; Distributed by America West Distributors; ISBN 0-934269-05-X

http://www.aztecufo.com/crash.htm

http://arizonaufosightings.com/kingman-arizona-1953-ufo-crash.html

http://thechurchofufology.blogspot.com/2010/03/ufo-crash-near-kingman-az

Crash: When UFOs Fall From the Sky by Kevin Randle; (c) 2010; The Career Press; ISBN 978-1-60163-100-8 and ISBN 978-1-60163-736-9

http://www.ufodigest.com/news/0508/theotherroswell.html

www.ufodigest.com/news/0310/images/del-rio.jpg

https://www.youtube.com/watch?v=9u7yd9j1qR0

http://ufocasebook.com/Kecksburg.html

http://www.cseti.org/crashes/crash.htm

http://www.cseti.org/crashes/crash.htm

https://www.youtube.com/watch?v=HYv6TFuB0A4

http://ufocasebook.conforums.com/index.cgi?board=general&action=display&num=1

http://www.angelfire.com/journal/alienseek/ufocrash.html
https://www.youtube.com/watch?v=WgMY8wTNsdU

CHAPTER 44
THE US MILITARY'S PROACTIVE APPROACH TO UFOs

Flying Saucers: Top Secret; copyright © 1960 by Donald E. Keyhoe, U.S. Marine Corps, Ret.; Published by G. P. Putnam's Sons; New York

Need To Know - UFOS, the Military and Intelligence" by Timothy Good; 2007; Published by Pegasus Books LLC; New York; ISBN 978-1-933648-38-5

Edward W. Condon, Director, and Daniel S Gillmor, Editor; Final Report of the Scientific Study of Unidentified Flying Objects; Bantam Books, 1968

CHAPTER 45
U.S. MILITARY'S UFO INTERCEPTION AND SHOOT DOWN PROGRAM AND AIRCRAFT CASUALITIES AND LOSSES

http://www.coasttocoastam.com/

http://www.coasttocoastam.com/show/2007/12/06

http://www.paranormal-encyclopedia.com/u/ufo/sightings/1948/mantell.html

Flying Saucers: Top Secret by Major Donald E. Keyhoe; © 1960; Published by G.P. Putnam's Sons; New York

http://ufos.about.com/od/bestufocasefiles/a/gorman.htm

http://ufos.about.com/od/classicufocases/a/1951newfoundland.htm

Aliens From Space – The Real Story of Unidentified Flying Objects By Major Donald E. Keyhoe; © 1973; Published by Doubleday & Company, Inc; New York

http://science.howstuffworks.com/space/aliens-ufos/ufo-government4.htm

By Daily Telegraph Reporter; 8:14AM GMT 26 Jan 2009

http://www.nationalmuseum.af.mil/factsheets/factsheet.asp?id=362

http://www.airforcetimes.com/news/2008/10/airforce_ufo_shootdown_102008/

http://en.wikipedia.org/wiki/1976_Tehran_UFO_incident

http://en.wikipedia.org/wiki/Milton_Torres_1957_UFO_Encounter

By Raphael G. Satter of AP Associated Press; Updated 10/20/2008 3:50:07 PM ET

http://www.theufochronicles.com/2007/11/transcipt-of-witness-declarations-from.html

http://ufos.about.com/od/bestufocasefiles/p/iran1976.htm

http://en.wikipedia.org/wiki/1976_Tehran_UFO_incident

Jerome Clark, *The UFO Book: Encyclopedia of the Extraterrestrial*, 1998, Visible Ink Press, ISBN 1-57859-029-9

Lawrence Fawcett and Barry J. Greenwood, *The UFO Cover-up* (formerly titled *Clear Intent*), Fireside Book/Simon & Schuster, ISBN 0-671-76555-8

Timothy Good, *Above Top Secret*, 1988, William Morrow and Co., ISBN 978-0-688-09202-3 (contains copy of full DIA report of incident)

Timothy Good, *Beyond Top Secret*, 1996, Pan Books, ISBN 0-330-34928-7 (contains copy of MIJI Quarterly report of incident, also below)

http://ufos.about.com/od/ufohistory/a/decade_2.htm

http://en.wikipedia.org/wiki/Michael_Faraday

Thanks to Timothy Good new book, *"Need to Know"* P.172

http://www.amazon.com/Shoot-Them-Down-Flying-Saucer/dp/0615155537

http://www.flatwoodsmonster.com/

www.amazon.co.uk/Flying-Saucers-Science-Scientist

http://www.nationalufocenter.com/artman/publish/article_280.php

http://www.nicap.org/moondust.htm

http://www.majesticdocuments.com/official.investigations.projectmoondust.bluefly.php

http://www.bibliotecapleyades.net/sociopolitica/esp_sociopol_mj12_3k.htm

https://www.youtube.com/watch?v=pBSTT7qZGig

CHAPTER 46
UFO INTRUSIONS AND OVER-FLIGHTS OF MISSILE BASES AND WEAPONS STORAGE SITES

Creighton G.W.; 1962; *Amazing News from Russia.* Flying Saucer Review 8, no. 6 (Nov. – Dec.): 27-28

Richard F. Haines Ph.D.; CE-5 - Close Encounters of the Fifth Kind, *Case No. 112;* 1999; Sourcebooks, Inc.; ISBN 1-570571-427-4

http://www.bibliotecapleyades.net/ciencia/ufo_briefingdocument/1989.htm

www.iraap.org...

http://www.iraap.org/Martin/PR.htm

http://www.alien-ufos.com/ufo-alien-discussions/20220-unidentified-aerial-phenomena-2.html

v.soodin@the-sun.co.uk

https://www.flyingmag.com/five-reasons-to-be-skeptical-about-that-new-york-times-ufo-story/

CHAPTER 47
IS THE NAVY REWRITING THE UFO NARRATIVE

https://www.flyingmag.com/five-reasons-to-be-skeptical-about-that-new-york-times-ufo-story/

 https://en.wikipedia.org/wiki/To_the_Stars_(company)

https://en.wikipedia.org/wiki/Advanced_Aerospace_Threat_Identification_Program

https://en.wikipedia.org/wiki/Tom_DeLonge

https://www.theguardian.com/tv-and-radio/2020/sep/15/star-of-bethlehem-spaceship-tom-delonges-new-career-ufo-expert-blink-182

"Contact: Countdown to Transformation - The CSETI Experience 1992 - 2009" by Steven M. Greer; September 2009; ISBN 9780967323831; Printed in USA; Produced by 123Printfinder, Inc.

https://news.sky.com/story/tom-delonge-on-ufo-research-i-wouldnt-have-left-blink-182-for-something-pie-in-the-sky-12061013

https://www.youtube.com/watch?v=Jmhf_TQ_w2M&feature=youtu.be

https://en.wikipedia.org/wiki/Pentagon_UFO_videos

https://www.washingtonpost.com/news/checkpoint/wp/2017/12/18/former-navy-pilot-describes-encounter-with-ufo-studied-by-secret-pentagon-program/

FLIR1_Official_UAP_Footage_from_the_USG_for_Public_Release.webm

Gimbal_The_First_Official_UAP_Footage_from_the_USG_for_Public_Release.webm

Go_Fast_Official_USG_Footage_of_UAP_for_Public_Release.webm

https://www.nytimes.com/2017/12/16/us/politics/pentagon-program-ufo-harry-reid.html

https://theintercept.com/2019/06/01/ufo-unidentified-history-channel-luis-elizondo-pentagon/

https://www.auforg.ca/recent-posts/disclosure/

"UFOs in Wartime What They didn't Want You to Know"; 2011; Published by Berkley Publishing Group a division of the Penguin Group, New York, New York; ISBN: 978-0-425-24011-3
https://www.history.com/news/uk-ufos-mainbrace-nato

CHAPTER 48
AMERICAN, SOVIET AND CHILEAN NAVAL ENGAGEMENTS WITH UFOS

http://en.wikipedia.org/wiki/Exercise_Mainbrace

http://www.nicap.dabsol.co.uk/mainbrace.htm source: Richard Hall

http://www.youtube.com/watch?v=tBXOBAigIIc

http://www.metacafe.com/watch/576930/operation_mainbrace_ufos/

http://www.realufos.net/2012/01/operation-mainbrace-ufo-encounters-1952.html

http://www.aliensthetruth.com/UFO_sightings_famous.php?view=1&ID=33

Flying Saucer Review, Volume 49/2, Summer 2004, pp. 21-23 From the NICAP records, by Major Donald E. Keyhoe

http://www.ufocasebook.com/navy1956.html

Conspiracy Shack website by Unknown - Reformatted by Kidd 11/2000
http://j_kidd.tripod.com/b/140.html

http://www.alieneight.com/alien-encounters-the-russian-navy-and-aquatic-ufos.htm

http://www.nationalufocenter.com/artman/publish/article_291.php

http://www.ufoevidence.org/cases/case1016.htm

http://www.zuko.com/Inexplicable/US_Navy_USO_Encounters.asp

https://www.youtube.com/watch?v=u3FTpJJkcbE

https://www.youtube.com/watch?v=k92OKqporaE

CHAPTER 49
ETI SEND A CLEAR MESSAGE TO THE MILITARIES OF THE WORLD – "WE WILL NOT LET YOU DESTROY THIS PLANET!"

http://www.nicap.org/610110capecanav_dir.htm

The Hynek UFO Report p. 44; 1977; Dell Publishing Co., Inc.; ISBN 0-440-19201-3

Disclosure: Military and Government Witnesses Reveal the Greatest Secrets in Modern History; 2001; Publisher's Cataloging-in-Publication by Quality Books, Inc.; ISBN 0-9673238-1-9

http://www.guardian.co.uk/news/datablog/2009/sep/06/nuclear-weapons-world-us-north-korea-russia-iran

Condon Report, Case 2; p.387;

http://en.wikipedia.org/wiki/Lakenheath-Bentwaters_incident

Thayer, Condon Report, Optical and Radar Analysis, p.246;
http://en.wikipedia.org/wiki/Lakenheath-Bentwaters_incident

http://en.wikipedia.org/wiki/Nuclear_weapons_and_the_United_Kingdom

http://en.wikipedia.org/wiki/RAF_Lakenheath

http://en.wikipedia.org/wiki/Project_E

"Minister warned over 'UK Roswell'". BBC News. 2009-08-17. Retrieved 2012-05-21.

http://en.wikipedia.org/wiki/Rendlesham_Forest_incident

http://www.theblackvault.com/wiki/index.php/Rendlesham_Forest_Incident

http://www.abovetopsecret.com/forum/thread546470/pg1

Lawrence Fawcett and Barry J. Greenwood; The UFO Cover-up (Originally published as Clear Intent); 1984; Published by Simon and Schuster Inc.; ISBN 0-671-76555-8

Greenwood, Barry and Fawcett, Lawrence. Clear Intent: The Government Cover-up of the UFO Experience, Prentice-Hall Inc., 1984. pp. 41-45

24th NORAD Region Senior Director's log excerpt reproduced on page 29, and National Military Command Center memorandum reproduced on pp. 30-31.

Raymond Fowler, Casebook of a UFO Investigator, Englewood Cliffs, N.J., Prentice-Hall, 1981, pp. 190-191.

Fawcett and Greenwood; The UFO Cover-up (Originally published as Clear Intent); 1984, pp. 16-19, 46-47.

Copies of Government documents in Fund for UFO Research files.

http://www.ufocasebook.com/sacbaseweapons1975.html

http://noufors.com/falconbridge_canadian_forces_radar_station.html

Clark, Jerome. The UFO Book, Visible Ink Press, 1998, p. 259

http://www.wired.co.uk/news/archive/2010-10/06/mirage-men

http://www.sott.net/articles/show/245414-UFOs-Filmed-Hovering-Over-U-S-Air-Force-Nuclear-Weapons-Storage-Area

http://www.wired.co.uk/news/archive/2010-10/06/mirage-men

CHAPTER 50
THE AGENCY OF UFO SECRECY - MAJESTIC TWELVE
(AKA. MAJESTIC 12, MAJIC 12, MJ 12)

http://en.wikipedia.org/wiki/Majestic_12

http://en.wikipedia.org/wiki/Majestic_12#Arguments_for

http://majesticdocuments.com

http://majesticdocuments.com/documents/intro.php

https://www.earthfiles.com/news.php?ID=2276&category=Environment

http://www.rotten.com/library/conspiracy/majestic-12

Howard Blum, *Out There*, 1990, Pocket Books (Simon & Schuster), ISBN 0-671-66261-9

C.D.B. Bryan, *Close Encounters of the Fourth Kind: Alien Abduction, UFOs and the Conference at M.I.T.*, 1995, Alfred A. Knopf, ISBN 0-679-42975-1

Jerome Clark, The UFO Book: Encyclopedia of the Extraterrestrial; Visible Ink, 1998; ISBN 1-57859-029-9

Richard M. Dolan, *UFOs and the National Security State: Chronology of a Coverup, 1941-1973*, 2002, Hampton Roads Publishing Company, ISBN 1-57174-317-0

Stanton T. Friedman, *TOP SECRET/MAJIC*, 1997, Marlowe & Co., ISBN 1-56924-741-2

Timothy Good, *Above Top Secret: The Worldwide UFO Cover-up*, 1988, Quill (William Marlow), ISBN 0-688-09202-0

Timothy Good, *Beyond Top Secret: The Worldwide UFO Security Threat*, 1997, Pan Books (MacMillan Publishers), ISBN 0-330-34928-7

Steven M. Greer, *Disclosure: Military and Government Witnesses Reveal the Greatest Secrets in Modern History*, 2001, ISBN 0-9673238-1-9

Michael Hesemann and Philip Mantle, *Beyond Roswell: The Alien Autopsy Film, Area 51, & the U.S. Government Coverup of UFOs*, 1997, Marlowe & Company, ISBN 1-56924-781-1

John Spencer, *The UFO Encyclopedia*, 1991, Avon Books, ISBN 0-380-76887-9, pp 199–200

http://www.crystalinks.com/mj12.html

Greer, S. (1999). Extraterrestrial Contact: The Evidence and Implications. Crossing Point, Inc. Publications, Afton, VA. ISBN: 0-9673238-0-0

https://www.youtube.com/watch?v=MNQ8_q-Bxto

CHAPTER 51
WHEN IT COMES TO NATIONAL SECURITY
LIE, LIE. LIE. DENY, DENY, DENY!

http://en.wikipedia.org/wiki/Plausible_deniability

Fawcett, L., and Greenwood, B. J. (1984). Clear Intent: The Government Coverup of the UFO Experience. Englewood Cliffs, NJ: Prentice-Hall.

Good, T. (1988). Above Top Secret: The Worldwide UFO Cover-up. New York: William Morrow.

Greer, S. (1999). Extraterrestrial Contact: The Evidence and Implications. Crossing Point, Inc. Publications. Afton, VA.

Sheffield, D. (1996). UFO a Deadly Concealment: The Official Cover-up? London, UK: Blandford

Derek Sheffield, UFO a Deadly Concealment: The Official Cover-up?" 1996, Blandford Publishing, UK

"'Sunday Express' article on Belgium UFO", Sunday Express. 17 September 1995. Retrieved 21 March 2008.

http://www.ufoevidence.org/documents/doc418.htm

Report concerning the observation of UFOs in the night from March 30 to March 31, 1990 - ufoevidence.org". Retrieved 21 March 2008.

http://en.wikipedia.org/wiki/Belgian_UFO_wave

http://www.latest-ufo-sightings.net/2010/03/famous-ufo-cases-belgian-ufo-wave.html

http://en.wikipedia.org/wiki/European_Union

http://www.latest-ufo-sightings.net/2011/07/controversial-belgium-ufo-photo-proved.html

http://www.ufoevidence.org/documents/doc406.htm

http://www.nato.int/history/nato-history.html

http://en.wikipedia.org/wiki/NATO

http://www.answers.com/topic/national-security-agency

http://www.fas.org/irp/news/1999/02/radome.htm

https://www.youtube.com/watch?v=RQ012kBoqiA

sweden = lov$nhttp://www.fas.org/irp/news/1999/02/radome.htm

The Radome Archipelago; February 24 - March 2, 1999 by Jason Vest and Wayne Madsen

http://www.nro.gov/about/nro/index.html

http://www.ufomystic.com/2007/02/22/serpo-was-disinformation/

http://www.realityuncovered.net/ufology/articles/serpo/

CHAPTER 52
DEEP BLACK PROJECTS, SAPS, USAPS, MILITARY SUPER WEAPONS

DISCLOSURE PROJECT BRIEFING DOCUMENT Prepared for: Members of the Press, Members of United States Government, Members of the US Scientific Community; Written and Compiled by Steven M. Greer, M.D., Director and Theodore C. Loder III, Ph.D. April, 2001

DISCLOSURE PROJECT BRIEFING DOCUMENT by Steven M. Greer, M.D., Director and Theodore C. Loder III, Ph.D. April, 2001

http://www.bibliotecapleyades.net/sociopolitica/sociopol_USAP.htm

https://www.youtube.com/watch?v=DJKyPipays8

https://www.youtube.com/watch?v=Aaf_61I7yZ0

http://www.century-of-flight.net/new%20site/frames/horten%20frame.htm

http://en.wikipedia.org/wiki/Bell_X-1

http://en.wikipedia.org/wiki/North_American_X-15

http://ufopartisan.blogspot.ca/2010/08/area-51-founders-ufo-sightings.html

http://conspiracy101.com/ufos/skunkworks/index.html

http://en.wikipedia.org/wiki/Kelly_Johnson_%28engineer%29

http://conspiracy101.com/ufos/skunkworks/index.html

Clark, Jerome, *The UFO Encyclopedia*, Volume 2 - entries for Angelucci, Orfeo Matthew and Van Tassel, George W

Jung, Carl G, *Flying Saucers: A Modern Myth of Things Seen in the Skies.* Princeton University Press, 1978

Zuk, Bill, *Canada's Flying Saucer: The Story of Avro Canada's Secret Projects.* Erin, ON: Boston Mills Press, 2001

http://en.wikipedia.org/wiki/Ben_Rich

http://www.realufos.net/2010/09/lockheed-skunk-works-chief-ben-rich.html

https://www.youtube.com/watch?v=u9ZZekWMiUQ

http://www.ufo-blogger.com/2010/08/ufo-are-real-ben-rich-lockheed-skunk.html

Disclosure: Military and Government Witnesses Reveal the Greatest Secrets in Modern History by Steven M. Greer, MD; 2001; Crossing Point, Inc. Publications; Carden Jennings Publishing Co.; Charlottesville, VA.; ISBN0-9673238-1-9

http://www.youtube.com/watch?v=58xEbRI3UKI

CHAPTER 53
ARE THE LUCRATIVE BLACK PROJECTS AND PROGRAMS THE REAL CAUSE FOR THE COLLAPSE OF THE AMERICAN ECONOMY?

http://en.wikipedia.org/wiki/List_of_countries_by_military_expenditures
http://www.seacoastonline.com/articles/20120811-OPINION-208110312

http://www.bibliotecapleyades.net/sociopolitica/sociopol_USAP.htm

TheAlexJonesChannel

http://www.scoop.co.nz/mason/stories/HL0208/S00055.htm)

http://www.scoop.co.nz/stories/HL0209/S00126.htm

http://www.scoop.co.nz/mason/stories/HL0207/S00031.htm#a

https://www.youtube.com/watch?v=w0mimIp8mr8

http://www.blindbatnews.com/2012/07/evil-elites-20-trillion-stolen-away-and-horded-by-global-elites-thats-why-the-economy-sucks-money-hording-causes-poverty-destroys-economies-is-a-threat-to-national-security-too-big-to-fai/14459

http://politics.guardian.co.uk/economics/story/0,11268,1446127,00.html

Heather Stewart, business editor; guardian.co.uk, **Saturday 21 July 2012 21.00 BST**
http://www.guardian.co.uk/business/2012/jul/21/global-elite-tax-offshore-economy

http://www.businessweek.com/ap/2012-08-17/india-auditor-says-billions-lost-in-coal-scandal

http://www.transparency.org.uk/corruption/statistics-and-quotes/stolen-assets-a-tax-evasion

http://en.wikipedia.org/wiki/International_asset_recovery

CHAPTER 54
BEHIND THE HALL OF MIRRORS: WHAT'S REALLY GOING ON?

http://www.sushi-x.com/gallery/nonseq/aurora/aurora.html

http://www.timstouse.com/ScienceNews/nuclearflyingsaucer.htm

http://www.bisbos.com/rocketscience/aircraft/black/triang/triang.html

http://www.dreamlandresort.com/black_projects/aircraft.htm

http://www.ufos-aliens.co.uk/cosmicarea51.html

http://www.where-is-area-51.com/bob-lazar.html

http://parascientifica.com/articles.php?a=155

http://projectavalon.net/forum4/showthread.php?33746-Bob-Lazar-Captured-UFOs-at-Top-Secret-S4-Base

http://www.ufos-aliens.co.uk/cosmicarea51.html

https://www.youtube.com/watch?v=K0jdCER8ypA

http://en.wikipedia.org/wiki/Forbidden_Planet

Disclosure: by Steven M. Greer, MD; 2001; Crossing Point, Inc. Publications; Carden Jennings Publishing Co.; Charlottesville, VA.; ISBN0-9673238-1-9

761

http://www.markmccandlish.com/Default.aspx?tabid=147

http://www.ufosaliens.co.uk/cosmicarea51.html

CHAPTER 55
UNITED STATES SPACE COMMAND

http://www.nationalufocenter.com/artman/publish/article_465.php

http://www.disclose.tv/action/viewvideo/198559/UFO_Hacker_Gary_McKinnon_talks_about_NASA_Hack/?utm_medium=email

http://rense.com/general79/wdx.htm

http://en.wikipedia.org/wiki/United_States_Space_Command

http://www.globalsecurity.org/space/agency/usspacecom.htm

http://ufoweek.com/tag/reconnaissance-aircraft/

https://www.youtube.com/watch?v=RaghqW1QWQQ

CHAPTER 56
IN SEARCH OF SECRET (DUMBS) DEEP UNDERGROUND
MILITARY BASES

http://subterraneanbases.com/plugins/content/content.php?content.166

http://en.wikipedia.org/wiki/Derinkuyu_Underground_City

"Underwater and Underground Bases", 2001 by Richard Sauder, Ph.D., published by Adventures Unlimited Press, Kempton, Illinois, USA, ISBN 0-932813-88-7

http://pinterest.com/pin/218776494367987573/

http://nstarzone.com/CODERED.html

Lokeyhttp://washingtontechnology.com/toplists/top-100-lists/2011.aspx

https://www.youtube.com/watch?v=uEDAE_9v4h0

CHAPTER 57
WHAT GOES ON DEEP DOWN IN A FEW D.U.M.B.s

http://en.wikipedia.org/wiki/Dugway_Proving_Ground
http://en.wikipedia.org/wiki/Tonopah_Test_Range

http://rense.com/general75/dulcde.htm

http://en.wikipedia.org/wiki/Dulce_Base

http://hyperboreanvibrations.blogspot.ca/2011/04/secret-dulce-underground-base-and-grey.html

https://indianinthemachine.wordpress.com/2011/04/16/the-cow-with-a-human-head-a-horrific-story-of-hybrids-ufo-aliens-and-the-dulce-underground-base/#more-12990

http://rense.com/general75/dulcde.htm

http://www.abovetopsecret.com/forum/thread61316/pg1

http://nar.oxfordjournals.org/content/34/19/5623.short

http://www.pnas.org/content/101/12/4164.long

https://www.youtube.com/watch?v=xn2XT09M6xA

http://www.theforbiddenknowledge.com/hardtruth/military_bases_ufos.htm

How the Military/Industrial complex turned America into Murder, Inc.; *By Len Hart;* OpEdNews Op Eds 8/18/2006 at 13:12:48

http://www.opednews.com/articles/opedne_len_hart_060818_how_the_military_2find.htm

How the Military/Industrial complex turned America into Murder, Inc.; *By Len Hart;* OpEdNews Op Eds 8/18/2006 at 13:12:48

http://www.opednews.com/articles/opedne_len_hart_060818_how_the_military_2find.htm

CHAPTER 58
THE REAL REASON FOR THE UFO COVER-UP AND SECRECY - CLEAN, CHEAP, POLLUTION-FREE, INFINITE ENERGY!!!

https://www.youtube.com/watch?v=oHxGQjirV-c

CHAPTER 59
ORCHESTRATION OF A FALSE FLAG SCENARIO:
A'LA INDEPENDENCE DAY ALIEN INVASION

Disclosure: Military and Government Witnesses Reveal the Greatest Secrets in Modern History by Steven M. Greer; 2001; Crossing Point, Inc. Publications; Crozet, VA. USA; ISBN0-9673238-1-9

https://www.youtube.com/watch?v=6psv6f3jF3g

https://www.youtube.com/watch?v=qf5SvJpGuv0

https://www.youtube.com/watch?v=naaBXzG89UY

https://www.youtube.com/watch?v=sza0WuB6l_U

https://www.youtube.com/watch?v=8xMfQX0SmTQ

INDEX (VOLUME THREE

B

C

770

J

K

O

P_____

T_____

U

About the Author

As a new author, Terry Tibando's background experience and understanding of this phenomenon spans 65 years of personal UFO sightings and ET contact that began at the age of five years. This childhood experience initiated a lifetime of many other-worldly sightings and encounters into a mysterious universe of Unidentified Flying Objects, Extraterrestrial Intelligence and the paranormal. As an experiencer, researcher, and investigator in Ufology he brings a unique and refreshing perspective on this subject based on a world view.

While attending Victoria High School in the mid sixties, Terry began attending UFO lectures meeting such people as Dr. Edward Edwards, a linguist from the University of Victoria and a fellow member of APRO (Aerial Phenomenon Research Organization and also Daniel Fry from New Mexico, USA, well known contactee and UFO author.

Terry attended the University of Victoria majoring in astronomy, physics, math and other sciences. During those university years other alien craft were sighted near his family's home in Victoria leading Terry to theorized that a possible undersea ET base existed off the coast of Vancouver Island which may account for the numerous UFO sightings seen over the Island.

He was a former member of APRO and its Canadian sister organization CAPRO during the sixties. His investigative research culminated back in the summer of 1996 when he met with Dr. Steven M. Greer during a one week "Ambassadors to the Universe" training seminar. They soon discovered that they shared similar UFO/ETI experiences during their early life.

793

Terry was a speaker at the Bellingham UFO Group (BUFOG) UFO seminar in 1996, and as a panel speaker along with Peter Davenport from NUFORC and Sharon Filip, alien abduction researcher.

He has talked on the Grimerica blog talk radio and been interviewed on the Discovery Channel during their "Alien Week" series in 1997 which had two ET spacecraft show up during the TV interview; he has been interviewed on BCTV News, and appeared briefly in Dr. Greer's successful documentary movie "Sirius" and was a major financial contributor to the current documentary "Unacknowledged"!

He was instrumental in coordinating, hosting and emceeing the first Disclosure Project event on UFOs and ETS in Canada as a part of Dr. Greer's Disclosure Witness Tour held at Simon Fraser University in Vancouver on September 9, 2001, which included guest speakers Dr. Steven Greer, Dr. Carol Rosin and Dr, Alfred Webre.

 For the last 25 years, Terry has been the field coordinator of CSETI Vancouver leading teams of people on field expeditions to successfully establish contact and communications with extraterrestrial intelligences visiting the Earth.

www.ingramcontent.com/pod-product-compliance
Lightning Source LLC
Chambersburg PA
CBHW080134240326
41458CB00128B/6462

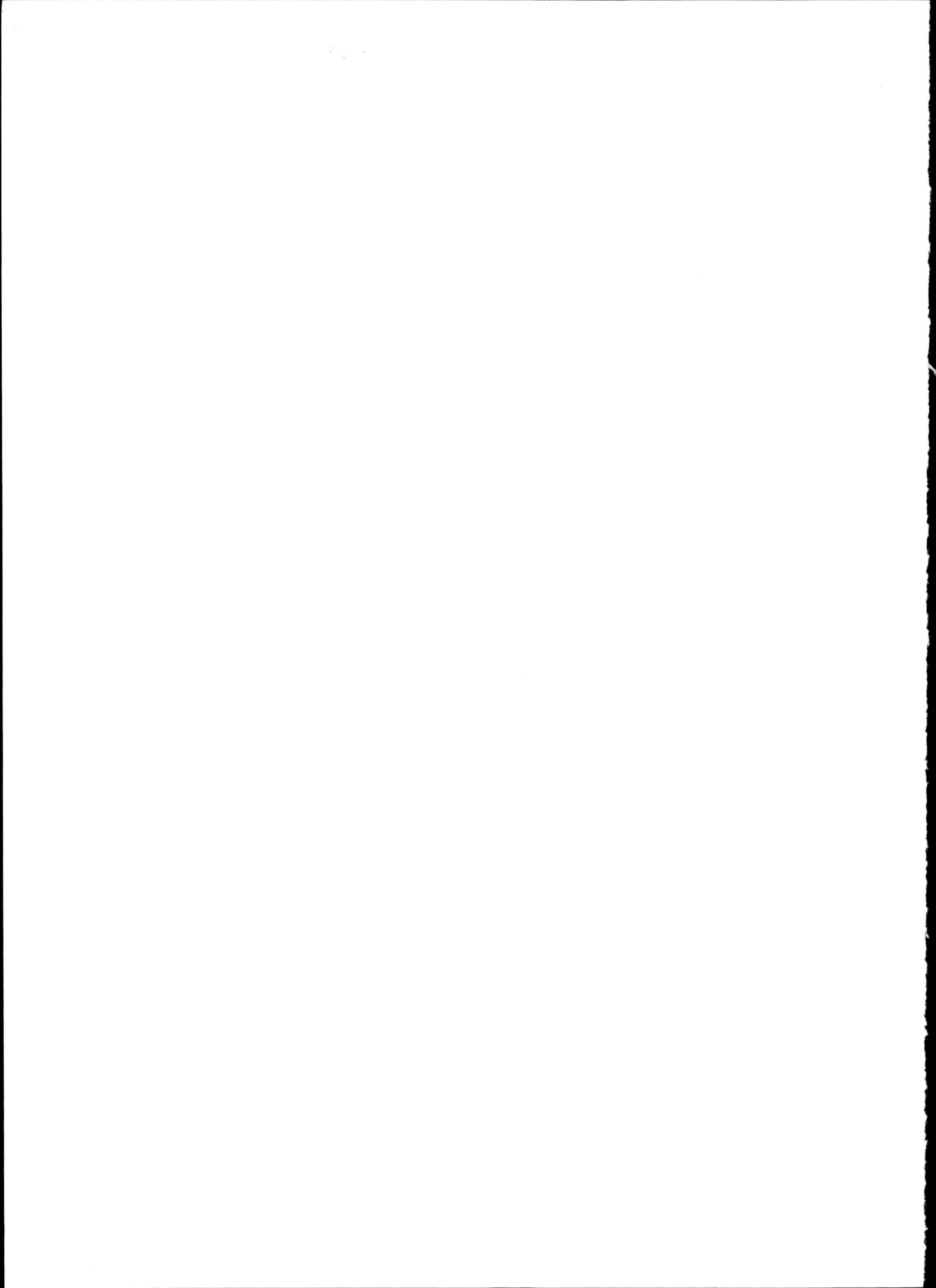